Elementary Differential Equations and Boundary Value Problems

EIGHTH EDITION

Elementary Differential Equations and Boundary Value Problems

William E. Boyce
Edward P. Hamilton Professor Emeritus

Richard C. DiPrima
formerly Eliza Ricketts Foundation Professor

Department of Mathematical Sciences
Rensselaer Polytechnic Institute

WILEY John Wiley & Sons, Inc.

ASSOCIATE PUBLISHER	Laurie Rosatone
PROJECT EDITOR	Jenny Welter
EDITORIAL ASSISTANT	Kelly Boyle
EXECUTIVE MARKETING MANAGER	Julie Z. Lindstrom
SENIOR PRODUCTION EDITOR	Ken Santor
COVER DESIGN	Carol C. Grobe
SENIOR DESIGNER	Karin Kincheloe
ILLUSTRATION EDITOR	Sigmund Malinowski

This book was set in Times Ten by Techsetters, Inc. and printed and bound by Von Hoffmann, Inc. The cover was printed by Von Hoffmann, Inc.

This book is printed on acid free paper. ∞

To order books or for customer service please, call 1(800)-CALL-WILEY (225-5945).

Library of Congress Cataloging-in-Publication Data

Boyce, William E.
 Elementary differential equations and boundary value problems / William E. Boyce, Richard C. DiPrima.
 p. cm.
 Includes index.
 ISBN 0-471-43338-1 (cloth)
 1. Differential equations. 2. Boundary value problems. I. DiPrima, Richard C. II. Title.

QA371.B773 2004
515′.35–dc22

2003070545

Printed in the United States of America

10 9 8 7 6 5 4 3 2 1

To Elsa and Maureen
To Siobhan, James, Richard, Jr., Carolyn, and Ann
And to the next generation:
Charles, Aidan, Stephanie, Veronica, and Deirdre

The Authors

William E. Boyce received his B.A. degree in Mathematics from Rhodes College, and his M.S. and Ph.D. degrees in Mathematics from Carnegie-Mellon University. He is a member of the American Mathematical Society, the Mathematical Association of America, and the Society for Industrial and Applied Mathematics. He is currently the Edward P. Hamilton Distinguished Professor Emeritus of Science Education (Department of Mathematical Sciences) at Rensselaer. He is the author of numerous technical papers in boundary value problems and random differential equations and their applications. He is the author of several textbooks including two differential equations texts, and is the coauthor (with M.H. Holmes, J.G. Ecker, and W.L. Siegmann) of a text on using Maple to explore Calculus. He is also coauthor (with R.L. Borrelli and C.S. Coleman) of *Differential Equations Laboratory Workbook* (Wiley 1992), which received the EDUCOM Best Mathematics Curricular Innovation Award in 1993. Professor Boyce was a member of the NSF-sponsored CODEE (Consortium for Ordinary Differential Equations Experiments) that led to the widely-acclaimed *ODE Architect*. He has also been active in curriculum innovation and reform. Among other things, he was the initiator of the "Computers in Calculus" project at Rensselaer, partially supported by the NSF. In 1991 he received the William H. Wiley Distinguished Faculty Award given by Rensselaer.

Richard C. DiPrima (deceased) received his B.S., M.S., and Ph.D. degrees in Mathematics from Carnegie-Mellon University. He joined the faculty of Rensselaer Polytechnic Institute after holding research positions at MIT, Harvard, and Hughes Aircraft. He held the Eliza Ricketts Foundation Professorship of Mathematics at Rensselaer, was a fellow of the American Society of Mechanical Engineers, the American Academy of Mechanics, and the American Physical Society. He was also a member of the American Mathematical Society, the Mathematical Association of America, and the Society for Industrial and Applied Mathematics. He served as the Chairman of the Department of Mathematical Sciences at Rensselaer, as President of the Society for Industrial and Applied Mathematics, and as Chairman of the Executive Committee of the Applied Mechanics Division of ASME. In 1980, he was the recipient of the William H. Wiley Distinguished Faculty Award given by Rensselaer. He received Fulbright fellowships in 1964–65 and 1983 and a Guggenheim fellowship in 1982–83. He was the author of numerous technical papers in hydrodynamic stability and lubrication theory and two texts on differential equations and boundary value problems. Professor DiPrima died on September 10, 1984.

PREFACE

Audience and Prerequisites

This edition, like its predecessors, is written from the viewpoint of the applied mathematician, whose interest in differential equations may sometimes be quite theoretical, sometimes intensely practical, and often somewhere in between. We have sought to combine a sound and accurate (but not abstract) exposition of the elementary theory of differential equations with considerable material on methods of solution, analysis, and approximation that have proved useful in a wide variety of applications.

The book is written primarily for undergraduate students of mathematics, science, or engineering, who typically take a course on differential equations during their first or second year of study. The main prerequisite for reading the book is a working knowledge of calculus, gained from a normal two- or three-semester course sequence or its equivalent. Some familiarity with matrices will also be helpful in the chapters on systems of differential equations

A Changing Learning Environment

The environment in which instructors teach, and students learn, differential equations has changed enormously in the past several years and continues to evolve at a rapid pace. Computing equipment of some kind, whether a graphing calculator, a notebook computer, or a desktop workstation is available to most students of differential equations. This equipment makes it relatively easy to execute extended numerical calculations, to generate graphical displays of a very high quality, and, in many cases, to carry out complex symbolic manipulations. A high-speed internet connection offers an enormous range of further possibilities.

The fact that so many students now have these capabilities enables instructors, if they wish, to modify very substantially their presentation of the subject and their

expectations of student performance. Not surprisingly, instructors have widely vary-ing opinions as to how a course on differential equations should be taught under these circumstances. Nevertheless, at many colleges and universities courses on differen-tial equations are becoming more visual, more quantitative, more project-oriented, and less formula-centered than in the past.

A Flexible Approach

To be widely useful a textbook must be adaptable to a variety of instructional strate-gies. This implies at least two things. First, instructors should have maximum flexi-bility to choose both the particular topics that they wish to cover and also the order in which they want to cover them. Second, the book should be useful to students having access to a wide range of technological capability.

Modular Chapters

With respect to content, we provide this flexibility by making sure that, so far as possible, individual chapters are independent of each other. Thus, after the basic parts of the first three chapters are completed (roughly Sections 1.1 through 1.3, 2.1 through 2.5, and 3.1 through 3.6) the selection of additional topics, and the order and depth in which they are covered, is at the discretion of the instructor. For example, an instructor who wishes to emphasize a systems approach to differential equations can take up Chapter 7 (Linear Systems) and perhaps even Chapter 9 (Nonlinear Autonomous Systems) immediately after Chapter 2. Or, while we present the basic theory of linear equations first in the context of a single second order equation (Chapter 3), many instructors have combined this material with the corresponding treatment of higher order equations (Chapter 4) or of linear systems (Chapter 7). Or, while the main discussion of numerical methods is in Chapter 8, an instructor who wishes to emphasize this approach can introduce some of this material in conjunction with Chapter 2. Many other choices and combinations are also possible and have been used effectively with earlier editions of this book.

Technology

With respect to technology, we note repeatedly in the text that computers are ex-tremely useful for investigating differential equations and their solutions, and many of the problems are best approached with computational assistance. Nevertheless, the book is adaptable to courses having various levels of computer involvement, ranging from little or none to intensive. The text is independent of any particular hardware platform or software package. More than 450 problems are marked with the symbol 🖉 to indicate that we consider them to be technologically intensive. These problems may call for a plot, or for substantial numerical computation, or for extensive symbolic manipulation, or for some combination of these requirements. Naturally, the designation of a problem as technologically intensive is a somewhat subjective judgment, and the 🖉 is intended only as a guide. Many of the marked problems can be solved, at least in part, without computational help, and a computer can be used effectively on many of the unmarked problems.

Homework Problems

From a student's point of view, the problems that are assigned as homework and that appear on examinations drive the course. We believe that the most outstanding feature of this book is the number, and above all the variety and range, of the problems that it contains. Many problems are entirely straightforward, but many others are more challenging, and some are fairly open-ended, and can serve as the basis for independent student projects. There are far more problems than any instructor can use in any given course, and this provides instructors with a multitude of possible choices in tailoring their course to meet their own goals and the needs of their students.

One of the choices that an instructor now has to make concerns the role of computing in the course. For instance, many more or less routine problems, such as those requesting the solution of a first or second order initial value problem, are now easy to solve by a computer algebra system. This edition includes quite a few such problems, just as its predecessors did. We do not state in these problems how they should be solved, because we believe that it is up to each instructor to specify whether their students should solve such problems by hand, with computer assistance, or perhaps both ways. Also, there are many problems that call for a graph of the solution. Instructors have the option of specifying whether they want an accurate computer-generated plot or a hand-drawn sketch, or perhaps both.

Mathematical Modeling

Building from Basic Models

The main reason for solving many differential equations is to try to learn something about an underlying physical process that the equation is believed to model. It is basic to the importance of differential equations that even the simplest equations correspond to useful physical models, such as exponential growth and decay, spring-mass systems, or electrical circuits. Gaining an understanding of a complex natural process is usually accomplished by combining or building upon simpler and more basic models. Thus a thorough knowledge of these basic models, the equations that describe them, and their solutions, is the first and indispensable step toward the solution of more complex and realistic problems. We describe the modeling process in detail in Sections 1.1, 1.2, and 2.3. Careful constructions of models appear also in Sections 2.5, 3.8, and in the appendices to Chapter 10. Differential equations resulting from the modeling process appear frequently throughout the book, especially in the problem sets.

A Combination of Tools—Analytical and Numerical

Nonroutine problems often require the use of a variety of tools, both analytical and numerical. Paper and pencil methods must often be combined with effective use of a computer. Quantitative results and graphs, often produced by a computer, serve to illustrate and clarify conclusions that may be obscured by complicated analytical expressions. On the other hand, the implementation of an efficient numerical procedure typically rests on a good deal of preliminary analysis—to determine the

qualitative features of the solution as a guide to computation, to investigate limit-ing or special cases, or to discover which ranges of the variables or parameters may require or merit special attention. Thus, a student should come to realize that investi-gating a difficult problem may well require both analysis and computation; that good judgment may be required to determine which tool is best-suited for a particular task; and that results can often be presented in a variety of forms.

Gaining Insight into the Behavior of a Process

We believe that it is important for students to understand that (except perhaps in courses on differential equations) the goal of solving a differential equation is seldom simply to obtain the solution. Rather, one is interested in the solution in order to obtain insight into the behavior of the process that the equation purports to model. In other words, the solution is not an end in itself. Thus, we have included a great many problems, as well as some examples in the text, that call for conclusions to be drawn about the solution. Sometimes this takes the form of asking for the value of the independent variable at which the solution has a certain property, or to determine the long term behavior of the solution. Other problems ask for the effect of variations in a parameter, or for the determination of a critical value of a parameter at which the solution experiences a substantial change. Such problems are typical of those that arise in the applications of differential equations, and, depending on the goals of the course, an instructor has the option of assigning few or many of these problems.

Notable Changes in the Eighth Edition

Readers familiar with the preceding edition will observe that the general structure of the book is unchanged. The revisions that we have made in this edition have several goals: to enlarge the range of applications that are considered, to make the presen-tation more visual by adding some new figures, and to improve the exposition by including several new or improved examples. More specifically, the most important changes are the following:

1. There are approximately 65 new problems scattered throughout the book. There are also about 15 new figures and 8 new or modified examples.
2. Section 2.1, "Linear Equations; Method of Integrating Factors," has been substantially rewritten, with two new examples, to reduce repetition.
3. Section 2.5, "Autonomous Equations and Population Dynamics," has been modified to give more prominence to the phase line as an aid to sketching solutions.
4. In Section 3.9 the general case of damped vibrations is considered before the special case of undamped vibrations, reversing the order of previous editions. The presentation is more detailed and there are three new figures.
5. The proof of the convolution theorem in Section 6.6 has been rewritten and six new problems on integral and integro-differential equations have been added.
6. To illustrate the occurrence of systems of higher than second order a new example on coupled oscillators has been added to Section 7.6, with three accompanying figures and several related problems.
7. An example has been added to Section 7.9 to demonstrate the use of Laplace transforms for nonhomogeneous linear systems.

8. There are several new problems in Sections 2.5, 9.4, and 9.7 to illustrate the occurrence of bifurcations in one- and two-dimensional nonlinear systems.

9. There are new problems in Section 10.6 on heat conduction in the presence of external heat sources, in Section 10.7 on dispersive waves, and in Section 10.8 on the flow through an aquifer.

As the subject matter of differential equations continues to grow, as new technologies become commonplace, as old areas of application are expanded, and as new ones appear on the horizon, the content and viewpoint of courses and their textbooks must also evolve. This is the spirit we have sought to express in this book.

William E. Boyce
Grafton, New York
February 23, 2004

Supplemental Resources for Instructors and Students

The *ODE Architect* CD is included with every copy of the text. *ODE Architect* is a prize-winning, state-of-the-art NSF-sponsored learning software package, which is Windows-compatible. A solver tool allows you to build your own models with ODEs and study them in a truly interactive point-and-click environment. The *Architect* includes an interactive library of more than one hundred model differential equation systems with graphs of solutions. The *Architect* also has 14 interactive multimedia modules, which provide a range of models and phenomena, from a golf game to chaos.

An Instructor's Solutions Manual, ISBN 0-471-67972-0, includes solutions for all problems in the text.

A Student Solutions Manual, ISBN 0-471-43340-3, includes solutions for selected problems in the text.

A Companion Web site, www.wiley.com/college/boyce, provides a wealth of resources for students and instructors, including:

- PowerPoint slides of important ideas and graphics for study and note taking.
- Review and Study Outlines to help students prepare for quizzes and exams.
- Online Review Quizzes to enable students to test their knowledge of key concepts. For further review diagnostic feedback is provided that refers to pertinent sections in the text.
- Getting Started with *ODE Architect*. This guide introduces students and professors to *ODE Architect*'s simulations and multimedia.
- Additional problems for use with Mathematica, Maple, and MATLAB, allowing opportunities for further exploration of important ideas in the course utilizing these computer algebra and numerical analysis packages.

eGrade Plus

eGrade Plus is a powerful online tool that provides instructors with an integrated suite of teaching and learning resources in one easy-to-use Web site.

eGrade Plus is organized around the essential activities you perform in class:

- **Prepare & Present:** Create class presentations using a wealth of Wiley-provided resources–such as an online version of the textbook, PowerPoint slides, and interactive simulations–making your class preparation more efficient. You may easily adapt, customize, and add to this content to meet the needs of your course.
- **Create Assignments:** Automate the assigning and grading of homework or quizzes by using Wiley-provided question banks, or by writing your own. Student responses will be graded automatically and the results recorded in your gradebook. eGrade Plus can link homework problems to the relevant section of the online text, providing students with context-sensitive help.
- **Track Student Progress:** Keep track of your students' progress via an instructor's gradebook, which allows you to analyze individual and overall class results to determine their progress and level of understanding.

- **Administer Your Course:** eGrade Plus can easily be integrated with another course management system, gradebook, or other resources you are using in your class, providing you with the flexibility to build your course, your way.

For more information about the features and benefits of eGrade Plus, please view our online demo at www.wiley.com/college/egradeplus.

ACKNOWLEDGMENTS

It is a pleasure to offer my grateful appreciation to the many people who have generously assisted in various ways in the preparation of this book.

To the individuals listed below who reviewed the manuscript and provided numerous valuable suggestions for its improvement:

Deborah Brandon, Carnegie Mellon University
James R. Brannan, Clemson University
Philip Crooke, Vanderbilt University
Dante DeBlassie, Texas A&M University
Juan B. Gil, Penn State Altoona
Moses Glasner, Pennsylvania State University
Murli M. Gupta, The George Washington University
Donald Hartig, California Polytechnic State University, San Luis Obispo
Thomas Hill, Lafayette College
Richard Hitt, University of South Alabama
Melvin D. Lax, California State University, Long Beach
Gary M. Lieberman, Iowa State University
Rafe Mazzeo, Stanford University
Diego A. Murio, University of Cincinnati
Martin Nakashima, California Polytechnic State University, Pomona
David Nicholls, University of Notre Dame
Bent E. Petersen, Oregon State University
Chris Schneider, University of Missouri – Rolla
Avy Soffer, Rutgers University
Steve Zelditch, Johns Hopkins University

To my colleagues and students at Rensselaer whose suggestions and reactions through the years have done much to sharpen my knowledge of differential equations as well as my ideas on how to present the subject.

To James Brannan (Clemson University), Bent Petersen (Oregon State University), and Josef Torok (Rochester Institute of Technology), who contributed many suggestions for new problems, mainly oriented toward applications.

To those readers of the preceding edition who called errors or omissions to my attention and especially to George Bergman (University of California at Berkeley) for his detailed list of comments and corrections.

To Lawrence Shampine (Southern Methodist University) for his consultation and to William Siegmann (Rensselaer) who made time for several lengthy conversations about the subject matter of this book from a pedagogical viewpoint.

To Charles Haines (Rochester Institute of Technology) who once again revised the Student Solutions Manual and in the process checked the solutions to many problems.

To Josef Torok (Rochester Institute of Technology) who updated the Instructor's Solutions Manual.

To David Ryeburn (Simon Fraser University) who carefully read the page proofs and was responsible for a number of corrections and clarifications.

To the editorial and production staff of John Wiley and Sons, and of Techsetters, Inc., who have always been ready to offer assistance and have displayed the highest standards of professionalism.

Finally, and most important, to my wife Elsa for many hours spent proofreading and checking details, for raising and discussing questions both mathematical and stylistic, and above all for her unfailing support and encouragement during the revision process. In a very real sense this book is a joint product.

William E. Boyce

CONTENTS

Introduction

In this chapter we try in several different ways to give perspective to your study of differential equations. First, we use two problems to illustrate some of the basic ideas that we will return to and elaborate upon frequently throughout the remainder of the book. Later, we indicate several ways of classifying equations, in order to provide organizational structure for the book. Finally, we outline some of the major trends in the historical development of the subject and mention a few of the outstanding mathematicians who have contributed to it. The study of differential equations has attracted the attention of many of the world's greatest mathematicians during the past three centuries. Nevertheless, it remains a dynamic field of inquiry today, with many interesting open questions.

1.1 Some Basic Mathematical Models; Direction Fields

Before embarking on a serious study of differential equations (for example, by reading this book or major portions of it), you should have some idea of the possible benefits to be gained by doing so. For some students the intrinsic interest of the subject itself is enough motivation, but for most it is the likelihood of important applications to other fields that makes the undertaking worthwhile.

Many of the principles, or laws, underlying the behavior of the natural world are statements or relations involving rates at which things happen. When expressed in mathematical terms, the relations are equations and the rates are derivatives. Equations containing derivatives are **differential equations**. Therefore, to understand and to investigate problems involving the motion of fluids, the flow of current in electric circuits, the dissipation of heat in solid objects, the propagation and detection

1

of seismic waves, or the increase or decrease of populations, among many others, it is necessary to know something about differential equations.

A differential equation that describes some physical process is often called a **mathematical model** of the process, and many such models are discussed throughout this book. In this section we begin with two models leading to equations that are easy to solve. It is noteworthy that even the simplest differential equations provide useful models of important physical processes.

EXAMPLE

1

A Falling
Object

Suppose that an object is falling in the atmosphere near sea level. Formulate a differential equation that describes the motion.

We begin by introducing letters to represent various quantities that may be of interest in this problem. The motion takes place during a certain time interval, so let us use t to denote time. Also, let us use v to represent the velocity of the falling object. The velocity will presumably change with time, so we think of v as a function of t; in other words, t is the independent variable and v is the dependent variable. The choice of units of measurement is somewhat arbitrary, and there is nothing in the statement of the problem to suggest appropriate units, so we are free to make any choice that seems reasonable. To be specific, let us measure time t in seconds and velocity v in meters/second. Further, we will assume that v is positive in the downward direction—that is, when the object is falling.

The physical law that governs the motion of objects is Newton's second law, which states that the mass of the object times its acceleration is equal to the net force on the object. In mathematical terms this law is expressed by the equation

$$F = ma, \tag{1}$$

where m is the mass of the object, a is its acceleration, and F is the net force exerted on the object. To keep our units consistent, we will measure m in kilograms, a in meters/second2, and F in newtons. Of course, a is related to v by $a = dv/dt$, so we can rewrite Eq. (1) in the form

$$F = m(dv/dt). \tag{2}$$

Next, consider the forces that act on the object as it falls. Gravity exerts a force equal to the weight of the object, or mg, where g is the acceleration due to gravity. In the units we have chosen, g has been determined experimentally to be approximately equal to 9.8 m/sec^2 near the earth's surface. There is also a force due to air resistance, or drag, that is more difficult to model. This is not the place for an extended discussion of the drag force; suffice it to say that it is often assumed that the drag is proportional to the velocity, and we will make that assumption here. Thus the drag force has the magnitude γv, where γ is a constant called the drag coefficient. The numerical value of the drag coefficient varies widely from one object to another; smooth streamlined objects have much smaller drag coefficients than rough blunt ones.

In writing an expression for the net force F, we need to remember that gravity always acts in the downward (positive) direction, whereas drag acts in the upward (negative) direction, as shown in Figure 1.1.1. Thus

$$F = mg - \gamma v \tag{3}$$

and Eq. (2) then becomes

$$m\frac{dv}{dt} = mg - \gamma v. \tag{4}$$

Equation (4) is a mathematical model of an object falling in the atmosphere near sea level. Note that the model contains the three constants m, g, and γ. The constants m and γ depend very much on the particular object that is falling, and they are usually different for different

objects. It is common to refer to them as parameters, since they may take on a range of values during the course of an experiment. On the other hand, the value of g is the same for all objects.

FIGURE 1.1.1 Free-body diagram of the forces on a falling object.

To solve Eq. (4) we need to find a function $v = v(t)$ that satisfies the equation. It is not hard to do this, and we will show you how in the next section. For the present, however, let us see what we can learn about solutions without actually finding any of them. Our task is simplified slightly if we assign numerical values to m and γ, but the procedure is the same regardless of which values we choose. So, let us suppose that $m = 10$ kg and $\gamma = 2$ kg/sec. If the units for γ seem peculiar, remember that γv must have the units of force, namely, kg-m/sec^2. Then Eq. (4) can be rewritten as

$$\frac{dv}{dt} = 9.8 - \frac{v}{5}. \tag{5}$$

EXAMPLE

2

─────────

A Falling
Object
(continued)

Investigate the behavior of solutions of Eq. (5) without solving the differential equation.

We will proceed by looking at Eq. (5) from a geometrical viewpoint. Suppose that v has a certain value. Then, by evaluating the right side of Eq. (5), we can find the corresponding value of dv/dt. For instance, if $v = 40$, then $dv/dt = 1.8$. This means that the slope of a solution $v = v(t)$ has the value 1.8 at any point where $v = 40$. We can display this information graphically in the tv-plane by drawing short line segments with slope 1.8 at several points on the line $v = 40$. Similarly, if $v = 50$, then $dv/dt = -0.2$, so we draw line segments with slope -0.2 at several points on the line $v = 50$. We obtain Figure 1.1.2 by proceeding in the same way with other values of v. Figure 1.1.2 is an example of what is called a **direction field** or sometimes a **slope field**.

The importance of Figure 1.1.2 is that each line segment is a tangent line to the graph of a solution of Eq. (5). Thus, even though we have not found any solutions, and no graphs of solutions appear in the figure, we can nonetheless draw some qualitative conclusions about the behavior of solutions. For instance, if v is less than a certain critical value, then all the line segments have positive slopes, and the speed of the falling object increases as it falls. On the other hand, if v is greater than the critical value, then the line segments have negative slopes, and the falling object slows down as it falls. What is this critical value of v that separates objects whose speed is increasing from those whose speed is decreasing? Referring again to Eq. (5), we ask what value of v will cause dv/dt to be zero. The answer is $v = (5)(9.8) = 49$ m/sec.

In fact, the constant function $v(t) = 49$ is a solution of Eq. (5). To verify this statement, substitute $v(t) = 49$ into Eq. (5) and observe that each side of the equation is zero. Because it does not change with time, the solution $v(t) = 49$ is called an **equilibrium solution**. It is the solution that corresponds to a balance between gravity and drag. In Figure 1.1.3 we show the equilibrium solution $v(t) = 49$ superimposed on the direction field. From this figure we

can draw another conclusion, namely, that all other solutions seem to be converging to the equilibrium solution as t increases.

FIGURE 1.1.2 A direction field for Eq. (5).

FIGURE 1.1.3 Direction field and equilibrium solution for Eq. (5).

The approach illustrated in Example 2 can be applied equally well to the more general Eq. (4), where the parameters m and γ are unspecified positive numbers. The results are essentially identical to those of Example 2. The equilibrium solution of Eq. (4) is $v(t) = mg/\gamma$. Solutions below the equilibrium solution increase with time, those above it decrease with time, and all other solutions approach the equilibrium solution as t becomes large.

Direction Fields. Direction fields are valuable tools in studying the solutions of differential equations of the form

$$\frac{dy}{dt} = f(t, y), \tag{6}$$

where f is a given function of the two variables t and y, sometimes referred to as the **rate function**. A useful direction field for equations of the form (6) can be constructed by evaluating f at each point of a rectangular grid consisting of at least a few hundred points. Then, at each point of the grid, a short line segment is drawn whose slope is the value of f at that point. Thus each line segment is tangent to the graph of the solution passing through that point. A direction field drawn on a fairly fine grid gives a good picture of the overall behavior of solutions of a differential equation. The construction of a direction field is often a useful first step in the investigation of a differential equation.

Two observations are worth particular mention. First, in constructing a direction field, we do not have to solve Eq. (6), but merely to evaluate the given function $f(t, y)$ many times. Thus direction fields can be readily constructed even for equations that may be quite difficult to solve. Second, repeated evaluation of a given function is a task for which a computer is well suited, and you should usually use a computer to draw a direction field. All the direction fields shown in this book, such as the one in Figure 1.1.2, were computer-generated.

Field Mice and Owls. Now let us look at another, quite different example. Consider a population of field mice who inhabit a certain rural area. In the absence of predators we assume that the mouse population increases at a rate proportional to the current population. This assumption is not a well-established physical law (as Newton's law of motion is in Example 1), but it is a common initial hypothesis[1] in a study of population growth. If we denote time by t and the mouse population by $p(t)$, then the assumption about population growth can be expressed by the equation

$$\frac{dp}{dt} = rp, \tag{7}$$

where the proportionality factor r is called the **rate constant** or **growth rate**. To be specific, suppose that time is measured in months and that the rate constant r has the value 0.5/month. Then each term in Eq. (7) has the units of mice/month.

Now let us add to the problem by supposing that several owls live in the same neighborhood and that they kill 15 field mice per day. To incorporate this information into the model, we must add another term to the differential equation (7), so that it becomes

$$\frac{dp}{dt} = 0.5p - 450. \tag{8}$$

Observe that the predation term is -450 rather than -15 because time is measured in months and the monthly predation rate is needed.

[1] A somewhat better model of population growth is discussed in Section 2.5.

EXAMPLE

3

Investigate the solutions of Eq. (8) graphically.

A direction field for Eq. (8) is shown in Figure 1.1.4. For sufficiently large values of p it can be seen from the figure, or directly from Eq. (8) itself, that dp/dt is positive, so that solutions increase. On the other hand, for small values of p the opposite is the case. Again, the critical value of p that separates solutions that increase from those that decrease is the value of p for which dp/dt is zero. By setting dp/dt equal to zero in Eq. (8) and then solving for p, we find the equilibrium solution $p(t) = 900$ for which the growth term and the predation term in Eq. (8) are exactly balanced. The equilibrium solution is also shown in Figure 1.1.4.

FIGURE 1.1.4 Direction field and equilibrium solution for Eq. (8).

Comparing Examples 2 and 3, we note that in both cases the equilibrium solution separates increasing from decreasing solutions. In Example 2 other solutions converge to, or are attracted by, the equilibrium solution, so that after the object falls far enough, an observer will see it moving at very nearly the equilibrium velocity. On the other hand, in Example 3 other solutions diverge from, or are repelled by, the equilibrium solution. Solutions behave very differently depending on whether they start above or below the equilibrium solution. As time passes, an observer might see populations either much larger or much smaller than the equilibrium population, but the equilibrium solution itself will not, in practice, be observed. In both problems, however, the equilibrium solution is very important in understanding how solutions of the given differential equation behave.

A more general version of Eq. (8) is

$$\frac{dp}{dt} = rp - k, \tag{9}$$

where the growth rate r and the predation rate k are unspecified. Solutions of this more general equation behave very much like those of Eq. (8). The equilibrium solution of Eq. (9) is $p(t) = k/r$. Solutions above the equilibrium solution increase, while those below it decrease.

You should keep in mind that both of the models discussed in this section have their limitations. The model (5) of the falling object is valid only as long as the

object is falling freely, without encountering any obstacles. The population model (8) eventually predicts negative numbers of mice (if $p < 900$) or enormously large numbers (if $p > 900$). Both these predictions are unrealistic, so this model becomes unacceptable after a fairly short time interval.

Constructing Mathematical Models. In applying differential equations to any of the numerous fields in which they are useful, it is necessary first to formulate the appropriate differential equation that describes, or models, the problem being investigated. In this section we have looked at two examples of this modeling process, one drawn from physics and the other from ecology. In constructing future mathematical models yourself, you should recognize that each problem is different, and that successful modeling is not a skill that can be reduced to the observance of a set of prescribed rules. Indeed, constructing a satisfactory model is sometimes the most difficult part of the problem. Nevertheless, it may be helpful to list some steps that are often part of the process:

1. Identify the independent and dependent variables and assign letters to represent them. Often the independent variable is time.

2. Choose the units of measurement for each variable. In a sense the choice of units is arbitrary, but some choices may be much more convenient than others. For example, we chose to measure time in seconds in the falling-object problem and in months in the population problem.

3. Articulate the basic principle that underlies or governs the problem you are investigating. This may be a widely recognized physical law, such as Newton's law of motion, or it may be a more speculative assumption that may be based on your own experience or observations. In any case, this step is likely not to be a purely mathematical one, but will require you to be familiar with the field in which the problem originates.

4. Express the principle or law in step 3 in terms of the variables you chose in step 1. This may be easier said than done. It may require the introduction of physical constants or parameters (such as the drag coefficient in Example 1) and the determination of appropriate values for them. Or it may involve the use of auxiliary or intermediate variables that must then be related to the primary variables.

5. Make sure that each term in your equation has the same physical units. If this is not the case, then your equation is wrong and you should seek to repair it. If the units agree, then your equation at least is dimensionally consistent, although it may have other shortcomings that this test does not reveal.

6. In the problems considered here, the result of step 4 is a single differential equation, which constitutes the desired mathematical model. Keep in mind, though, that in more complex problems the resulting mathematical model may be much more complicated, perhaps involving a system of several differential equations, for example.

PROBLEMS

In each of Problems 1 through 6 draw a direction field for the given differential equation. Based on the direction field, determine the behavior of y as $t \to \infty$. If this behavior depends on the initial value of y at $t = 0$, describe the dependency.

 1. $y' = 3 - 2y$ 2. $y' = 2y - 3$

 3. $y' = 3 + 2y$ 4. $y' = -1 - 2y$

 5. $y' = 1 + 2y$ 6. $y' = y + 2$

In each of Problems 7 through 10 write down a differential equation of the form $dy/dt = ay + b$ whose solutions have the required behavior as $t \to \infty$.

7. All solutions approach $y = 3$. 8. All solutions approach $y = 2/3$.

9. All other solutions diverge from $y = 2$. 10. All other solutions diverge from $y = 1/3$.

In each of Problems 11 through 14 draw a direction field for the given differential equation. Based on the direction field, determine the behavior of y as $t \to \infty$. If this behavior depends on the initial value of y at $t = 0$, describe this dependency. Note that in these problems the equations are not of the form $y' = ay + b$, and the behavior of their solutions is somewhat more complicated than for the equations in the text.

11. $y' = y(4 - y)$ 12. $y' = -y(5 - y)$

13. $y' = y^2$ 14. $y' = y(y - 2)^2$

Consider the following list of differential equations, some of which produced the direction fields shown in Figures 1.1.5 through 1.1.10. In each of Problems 15 through 20 identify the differential equation that corresponds to the given direction field.

(a) $y' = 2y - 1$ (b) $y' = 2 + y$

(c) $y' = y - 2$ (d) $y' = y(y + 3)$

(e) $y' = y(y - 3)$ (f) $y' = 1 + 2y$

(g) $y' = -2 - y$ (h) $y' = y(3 - y)$

(i) $y' = 1 - 2y$ (j) $y' = 2 - y$

15. The direction field of Figure 1.1.5.

16. The direction field of Figure 1.1.6.

17. The direction field of Figure 1.1.7.

18. The direction field of Figure 1.1.8.

19. The direction field of Figure 1.1.9.

20. The direction field of Figure 1.1.10.

21. A pond initially contains 1,000,000 gal of water and an unknown amount of an undesirable chemical. Water containing 0.01 gram of this chemical per gallon flows into the pond at a rate of 300 gal/hr. The mixture flows out at the same rate, so the amount of water in the pond remains constant. Assume that the chemical is uniformly distributed throughout the pond.

 (a) Write a differential equation for the amount of chemical in the pond at any time.

 (b) How much of the chemical will be in the pond after a very long time? Does this limiting amount depend on the amount that was present initially?

22. A spherical raindrop evaporates at a rate proportional to its surface area. Write a differential equation for the volume of the raindrop as a function of time.

23. Newton's law of cooling states that the temperature of an object changes at a rate proportional to the difference between the temperature of the object itself and the temperature of its surroundings (the ambient air temperature in most cases). Suppose that the ambient temperature is 70°F and that the rate constant is 0.05 (min)$^{-1}$. Write a differential equation for the temperature of the object at any time.

24. A certain drug is being administered intravenously to a hospital patient. Fluid containing 5 mg/cm^3 of the drug enters the patient's bloodstream at a rate of 100 cm^3/hr. The drug is absorbed by body tissues or otherwise leaves the bloodstream at a rate proportional to the amount present, with a rate constant of 0.4 (hr)$^{-1}$.

 (a) Assuming that the drug is always uniformly distributed throughout the bloodstream, write a differential equation for the amount of the drug that is present in the bloodstream at any time.

 (b) How much of the drug is present in the bloodstream after a long time?

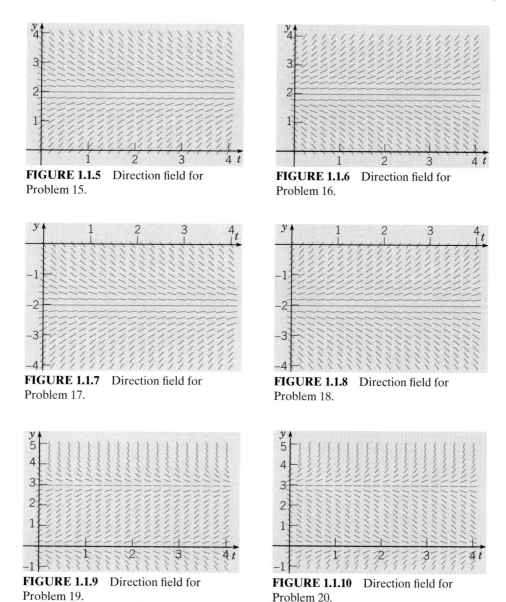

FIGURE 1.1.5 Direction field for Problem 15.

FIGURE 1.1.6 Direction field for Problem 16.

FIGURE 1.1.7 Direction field for Problem 17.

FIGURE 1.1.8 Direction field for Problem 18.

FIGURE 1.1.9 Direction field for Problem 19.

FIGURE 1.1.10 Direction field for Problem 20.

 25. For small, slowly falling objects, the assumption made in the text that the drag force is proportional to the velocity is a good one. For larger, more rapidly falling objects, it is more accurate to assume that the drag force is proportional to the square of the velocity.[2]

(a) Write a differential equation for the velocity of a falling object of mass m if the drag force is proportional to the square of the velocity.

(b) Determine the limiting velocity after a long time.

[2]See Lyle N. Long and Howard Weiss, "The Velocity Dependence of Aerodynamic Drag: A Primer for Mathematicians," *American Mathematical Monthly* 106 (1999), 2, pp. 127–135.

(c) If $m = 10$ kg, find the drag coefficient so that the limiting velocity is 49 m/sec.

(d) Using the data in part (c), draw a direction field and compare it with Figure 1.1.3.

In each of Problems 26 through 33 draw a direction field for the given differential equation. Based on the direction field, determine the behavior of y as $t \to \infty$. If this behavior depends on the initial value of y at $t = 0$, describe this dependency. Note that the right sides of these equations depend on t as well as y; therefore their solutions can exhibit more complicated behavior than those in the text.

26. $y' = -2 + t - y$

27. $y' = te^{-2t} - 2y$

28. $y' = e^{-t} + y$

29. $y' = t + 2y$

30. $y' = 3\sin t + 1 + y$

31. $y' = 2t - 1 - y^2$

32. $y' = -(2t + y)/2y$

33. $y' = y^3/6 - y - t^2/3$

1.2 Solutions of Some Differential Equations

In the preceding section we derived the differential equations

$$m\frac{dv}{dt} = mg - \gamma v \tag{1}$$

and

$$\frac{dp}{dt} = rp - k. \tag{2}$$

Equation (1) models a falling object and Eq. (2) a population of field mice preyed on by owls. Both these equations are of the general form

$$\frac{dy}{dt} = ay - b, \tag{3}$$

where a and b are given constants. We were able to draw some important qualitative conclusions about the behavior of solutions of Eqs. (1) and (2) by considering the associated direction fields. To answer questions of a quantitative nature, however, we need to find the solutions themselves, and we now investigate how to do that.

EXAMPLE 1

Field Mice and Owls (continued)

Consider the equation

$$\frac{dp}{dt} = 0.5p - 450, \tag{4}$$

which describes the interaction of certain populations of field mice and owls [see Eq. (8) of Section 1.1]. Find solutions of this equation.

To solve Eq. (4) we need to find functions $p(t)$ that, when substituted into the equation, reduce it to an obvious identity. Here is one way to proceed. First, rewrite Eq. (4) in the form

$$\frac{dp}{dt} = \frac{p - 900}{2}, \tag{5}$$

or, if $p \neq 900$,

$$\frac{dp/dt}{p - 900} = \frac{1}{2}. \tag{6}$$

By the chain rule the left side of Eq. (6) is the derivative of $\ln|p - 900|$ with respect to t, so we have

$$\frac{d}{dt} \ln|p - 900| = \frac{1}{2}. \tag{7}$$

Then, by integrating both sides of Eq. (7), we obtain

$$\ln|p - 900| = \frac{t}{2} + C, \tag{8}$$

where C is an arbitrary constant of integration. Therefore, by taking the exponential of both sides of Eq. (8), we find that

$$|p - 900| = e^{(t/2)+C} = e^C e^{t/2}, \tag{9}$$

or

$$p - 900 = \pm e^C e^{t/2}, \tag{10}$$

and finally

$$p = 900 + c e^{t/2}, \tag{11}$$

where $c = \pm e^C$ is also an arbitrary (nonzero) constant. Note that the constant function $p = 900$ is also a solution of Eq. (5) and that it is contained in the expression (11) if we allow c to take the value zero. Graphs of Eq. (11) for several values of c are shown in Figure 1.2.1.

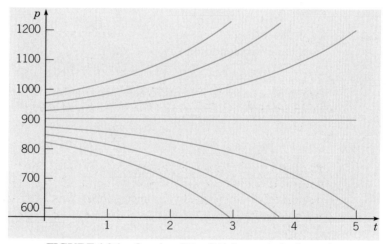

FIGURE 1.2.1 Graphs of Eq. (11) for several values of c.

Note that they have the character inferred from the direction field in Figure 1.1.4. For instance, solutions lying on either side of the equilibrium solution $p = 900$ tend to diverge from that solution.

In Example 1 we found infinitely many solutions of the differential equation (4), corresponding to the infinitely many values that the arbitrary constant c in Eq. (11) might have. This is typical of what happens when you solve a differential equation. The solution process involves an integration, which brings with it an arbitrary constant, whose possible values generate an infinite family of solutions.

Frequently, we want to focus our attention on a single member of the infinite family of solutions by specifying the value of the arbitrary constant. Most often, we do this indirectly by specifying instead a point that must lie on the graph of the solution. For example, to determine the constant c in Eq. (11), we could require that the population have a given value at a certain time, such as the value 850 at time $t = 0$. In other words, the graph of the solution must pass through the point $(0, 850)$. Symbolically, we can express this condition as

$$p(0) = 850. \tag{12}$$

Then, substituting $t = 0$ and $p = 850$ into Eq. (11), we obtain

$$850 = 900 + c.$$

Hence $c = -50$, and by inserting this value in Eq. (11), we obtain the desired solution, namely,

$$p = 900 - 50e^{t/2}. \tag{13}$$

The additional condition (12) that we used to determine c is an example of an **initial condition**. The differential equation (4) together with the initial condition (12) form an **initial value problem**.

Now consider the more general problem consisting of the differential equation (3)

$$\frac{dy}{dt} = ay - b$$

and the initial condition

$$y(0) = y_0, \tag{14}$$

where y_0 is an arbitrary initial value. We can solve this problem by the same method as in Example 1. If $a \neq 0$ and $y \neq b/a$, then we can rewrite Eq. (3) as

$$\frac{dy/dt}{y - (b/a)} = a. \tag{15}$$

By integrating both sides, we find that

$$\ln |y - (b/a)| = at + C, \tag{16}$$

where C is arbitrary. Then, taking the exponential of both sides of Eq. (16) and solving for y, we obtain

$$y = (b/a) + ce^{at}, \tag{17}$$

where $c = \pm e^C$ is also arbitrary. Observe that $c = 0$ corresponds to the equilibrium solution $y = b/a$. Finally, the initial condition (14) requires that $c = y_0 - (b/a)$, so the solution of the initial value problem (3), (14) is

$$y = (b/a) + [y_0 - (b/a)]e^{at}. \tag{18}$$

The expression (17) contains all possible solutions of Eq. (3) and is called the **general solution**. The geometrical representation of the general solution (17) is an infinite family of curves called **integral curves**. Each integral curve is associated with a particular value of c and is the graph of the solution corresponding to that value of c. Satisfying an initial condition amounts to identifying the integral curve that passes through the given initial point.

To relate the solution (18) to Eq. (2), which models the field mouse population, we need only replace a by the growth rate r and b by the predation rate k. Then the solution (18) becomes

$$p = (k/r) + [p_0 - (k/r)]e^{rt}, \tag{19}$$

where p_0 is the initial population of field mice. The solution (19) confirms the conclusions reached on the basis of the direction field and Example 1. If $p_0 = k/r$, then from Eq. (19) it follows that $p = k/r$ for all t; this is the constant, or equilibrium, solution. If $p_0 \neq k/r$, then the behavior of the solution depends on the sign of the coefficient $p_0 - (k/r)$ of the exponential term in Eq. (19). If $p_0 > k/r$, then p grows exponentially with time t; if $p_0 < k/r$, then p decreases and eventually becomes zero, corresponding to extinction of the field mouse population. Negative values of p, while possible for the expression (19), make no sense in the context of this particular problem.

To put the falling-object equation (1) in the form (3), we must identify a with $-\gamma/m$ and b with $-g$. Making these substitutions in the solution (18), we obtain

$$v = (mg/\gamma) + [v_0 - (mg/\gamma)]e^{-\gamma t/m}, \tag{20}$$

where v_0 is the initial velocity. Again, this solution confirms the conclusions reached in Section 1.1 on the basis of a direction field. There is an equilibrium, or constant, solution $v = mg/\gamma$, and all other solutions tend to approach this equilibrium solution. The speed of convergence to the equilibrium solution is determined by the exponent $-\gamma/m$. Thus, for a given mass m, the velocity approaches the equilibrium value more rapidly as the drag coefficient γ increases.

EXAMPLE

2

A Falling Object (continued)

Suppose that, as in Example 2 of Section 1.1, we consider a falling object of mass $m = 10$ kg and drag coefficient $\gamma = 2$ kg/sec. Then the equation of motion (1) becomes

$$\frac{dv}{dt} = 9.8 - \frac{v}{5}. \tag{21}$$

Suppose this object is dropped from a height of 300 m. Find its velocity at any time t. How long will it take to fall to the ground, and how fast will it be moving at the time of impact?

The first step is to state an appropriate initial condition for Eq. (21). The word "dropped" in the statement of the problem suggests that the initial velocity is zero, so we will use the initial condition

$$v(0) = 0. \tag{22}$$

The solution of Eq. (21) can be found by substituting the values of the coefficients into the solution (20), but we will proceed instead to solve Eq. (21) directly. First, rewrite the equation as

$$\frac{dv/dt}{v - 49} = -\frac{1}{5}. \tag{23}$$

By integrating both sides we obtain

$$\ln|v - 49| = -\frac{t}{5} + C, \tag{24}$$

and then the general solution of Eq. (21) is

$$v = 49 + ce^{-t/5}, \tag{25}$$

where c is arbitrary. To determine c, we substitute $t = 0$ and $v = 0$ from the initial condition (22) into Eq. (25), with the result that $c = -49$. Then the solution of the initial value problem (21), (22) is

$$v = 49(1 - e^{-t/5}). \tag{26}$$

Equation (26) gives the velocity of the falling object at any positive time (before it hits the ground, of course).

Graphs of the solution (25) for several values of c are shown in Figure 1.2.2, with the solution (26) shown by the heavy curve. It is evident that all solutions tend to approach the equilibrium solution $v = 49$. This confirms the conclusions we reached in Section 1.1 on the basis of the direction fields in Figures 1.1.2 and 1.1.3.

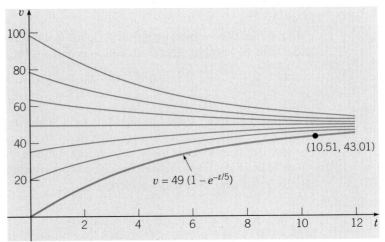

FIGURE 1.2.2 Graphs of the solution (25) for several values of c.

To find the velocity of the object when it hits the ground, we need to know the time at which impact occurs. In other words, we need to determine how long it takes the object to fall 300 m. To do this, we note that the distance x the object has fallen is related to its velocity v by the equation $v = dx/dt$, or

$$\frac{dx}{dt} = 49(1 - e^{-t/5}). \tag{27}$$

Consequently, by integrating both sides of Eq. (27), we have

$$x = 49t + 245e^{-t/5} + c, \tag{28}$$

where c is an arbitrary constant of integration. The object starts to fall when $t = 0$, so we know that $x = 0$ when $t = 0$. From Eq. (28) it follows that $c = -245$, so the distance the object has fallen at time t is given by

$$x = 49t + 245e^{-t/5} - 245. \tag{29}$$

Let T be the time at which the object hits the ground; then $x = 300$ when $t = T$. By substituting these values in Eq. (29), we obtain the equation

$$49T + 245e^{-T/5} - 545 = 0. \tag{30}$$

The value of T satisfying Eq. (30) can be readily approximated by a numerical process using a scientific calculator or computer, with the result that $T \cong 10.51$ sec. At this time, the corresponding velocity v_T is found from Eq. (26) to be $v_T \cong 43.01$ m/sec. The point $(10.51, 43.01)$ is also shown in Figure 1.2.2.

Further Remarks on Mathematical Modeling. Up to this point we have related our discussion of differential equations to mathematical models of a falling object and of a hypothetical relation between field mice and owls. The derivation of these models may have been plausible, and possibly even convincing, but you should remember that the ultimate test of any mathematical model is whether its predictions agree with observations or experimental results. We have no actual observations or experimental results to use for comparison purposes here, but there are several sources of possible discrepancies.

In the case of the falling object, the underlying physical principle (Newton's law of motion) is well established and widely applicable. However, the assumption that the drag force is proportional to the velocity is less certain. Even if this assumption is correct, the determination of the drag coefficient γ by direct measurement presents difficulties. Indeed, sometimes one finds the drag coefficient indirectly—for example, by measuring the time of fall from a given height and then calculating the value of γ that predicts this observed time.

The model of the field mouse population is subject to various uncertainties. The determination of the growth rate r and the predation rate k depends on observations of actual populations, which may be subject to considerable variation. The assumption that r and k are constants may also be questionable. For example, a constant predation rate becomes harder to sustain as the field mouse population becomes smaller. Further, the model predicts that a population above the equilibrium value will grow exponentially larger and larger. This seems at variance with the behavior of actual populations; see the further discussion of population dynamics in Section 2.5.

If the differences between actual observations and a mathematical model's predictions are too great, then you need to consider refining the model, making more careful observations, or perhaps both. There is almost always a tradeoff between accuracy and simplicity. Both are desirable, but a gain in one usually involves a loss in the other. However, even if a mathematical model is incomplete or somewhat inaccurate, it may nevertheless be useful in explaining qualitative features of the problem under investigation. It may also give satisfactory results under some circumstances but not others. Thus you should always use good judgment and common sense in constructing mathematical models and in using their predictions.

PROBLEMS

1. Solve each of the following initial value problems and plot the solutions for several values of y_0. Then describe in a few words how the solutions resemble, and differ from, each other.

 (a) $dy/dt = -y + 5$, $y(0) = y_0$
 (b) $dy/dt = -2y + 5$, $y(0) = y_0$
 (c) $dy/dt = -2y + 10$, $y(0) = y_0$

 2. Follow the instructions for Problem 1 for the following initial value problems:

 (a) $dy/dt = y - 5,$ $\qquad y(0) = y_0$
 (b) $dy/dt = 2y - 5,$ $\qquad y(0) = y_0$
 (c) $dy/dt = 2y - 10,$ $\qquad y(0) = y_0$

3. Consider the differential equation

$$dy/dt = -ay + b,$$

 where both a and b are positive numbers.

 (a) Solve the differential equation.
 (b) Sketch the solution for several different initial conditions.
 (c) Describe how the solutions change under each of the following conditions:
 i. a increases.
 ii. b increases.
 iii. Both a and b increase, but the ratio b/a remains the same.

4. Consider the differential equation $dy/dt = ay - b$.

 (a) Find the equilibrium solution y_e.
 (b) Let $Y(t) = y - y_e$; thus $Y(t)$ is the deviation from the equilibrium solution. Find the differential equation satisfied by $Y(t)$.

5. **Undetermined Coefficients.** Here is an alternative way to solve the equation

$$dy/dt = ay - b. \qquad\qquad (i)$$

 (a) Solve the simpler equation

$$dy/dt = ay. \qquad\qquad (ii)$$

 Call the solution $y_1(t)$.
 (b) Observe that the only difference between Eqs. (i) and (ii) is the constant $-b$ in Eq. (i). Therefore it may seem reasonable to assume that the solutions of these two equations also differ only by a constant. Test this assumption by trying to find a constant k such that $y = y_1(t) + k$ is a solution of Eq. (i).
 (c) Compare your solution from part (b) with the solution given in the text in Eq. (17).

 Note: This method can also be used in some cases in which the constant b is replaced by a function $g(t)$. It depends on whether you can guess the general form that the solution is likely to take. This method is described in detail in Section 3.6 in connection with second order equations.

6. Use the method of Problem 5 to solve the equation

$$dy/dt = -ay + b.$$

7. The field mouse population in Example 1 satisfies the differential equation

$$dp/dt = 0.5p - 450.$$

 (a) Find the time at which the population becomes extinct if $p(0) = 850$.
 (b) Find the time of extinction if $p(0) = p_0$, where $0 < p_0 < 900$.
 (c) Find the initial population p_0 if the population is to become extinct in 1 year.

8. Consider a population p of field mice that grows at a rate proportional to the current population, so that $dp/dt = rp$.

 (a) Find the rate constant r if the population doubles in 30 days.
 (b) Find r if the population doubles in N days.

9. The falling object in Example 2 satisfies the initial value problem

$$dv/dt = 9.8 - (v/5), \qquad v(0) = 0.$$

(a) Find the time that must elapse for the object to reach 98% of its limiting velocity.

(b) How far does the object fall in the time found in part (a)?

10. Modify Example 2 so that the falling object experiences no air resistance.

(a) Write down the modified initial value problem.

(b) Determine how long it takes the object to reach the ground.

(c) Determine its velocity at the time of impact.

 11. Consider the falling object of mass 10 kg in Example 2, but assume now that the drag force is proportional to the square of the velocity.

(a) If the limiting velocity is 49 m/sec (the same as in Example 2), show that the equation of motion can be written as

$$dv/dt = [(49)^2 - v^2]/245.$$

Also see Problem 25 of Section 1.1.

(b) If $v(0) = 0$, find an expression for $v(t)$ at any time.

(c) Plot your solution from part (b) and the solution (26) from Example 2 on the same axes.

(d) Based on your plots in part (c), compare the effect of a quadratic drag force with that of a linear drag force.

(e) Find the distance $x(t)$ that the object falls in time t.

(f) Find the time T it takes the object to fall 300 meters.

12. A radioactive material, such as the isotope thorium-234, disintegrates at a rate proportional to the amount currently present. If $Q(t)$ is the amount present at time t, then $dQ/dt = -rQ$, where $r > 0$ is the decay rate.

(a) If 100 mg of thorium-234 decays to 82.04 mg in 1 week, determine the decay rate r.

(b) Find an expression for the amount of thorium-234 present at any time t.

(c) Find the time required for the thorium-234 to decay to one-half its original amount.

13. The **half-life** of a radioactive material is the time required for an amount of this material to decay to one-half its original value. Show that for any radioactive material that decays according to the equation $Q' = -rQ$, the half-life τ and the decay rate r satisfy the equation $r\tau = \ln 2$.

14. Radium-226 has a half-life of 1620 years. Find the time period during which a given amount of this material is reduced by one-quarter.

15. According to Newton's law of cooling (see Problem 23 of Section 1.1), the temperature $u(t)$ of an object satisfies the differential equation

$$du/dt = -k(u - T),$$

where T is the constant ambient temperature and k is a positive constant. Suppose that the initial temperature of the object is $u(0) = u_0$.

(a) Find the temperature of the object at any time.

(b) Let τ be the time at which the initial temperature difference $u_0 - T$ has been reduced by half. Find the relation between k and τ.

16. Suppose that a building loses heat in accordance with Newton's law of cooling (see Problem 15) and that the rate constant k has the value 0.15 hr^{-1}. Assume that the interior

temperature is 70°F when the heating system fails. If the external temperature is 10°F, how long will it take for the interior temperature to fall to 32°F?

17. Consider an electric circuit containing a capacitor, resistor, and battery; see Figure 1.2.3. The charge $Q(t)$ on the capacitor satisfies the equation[3]

$$R\frac{dQ}{dt} + \frac{Q}{C} = V,$$

where R is the resistance, C is the capacitance, and V is the constant voltage supplied by the battery.

(a) If $Q(0) = 0$, find $Q(t)$ at any time t, and sketch the graph of Q versus t.

(b) Find the limiting value Q_L that $Q(t)$ approaches after a long time.

(c) Suppose that $Q(t_1) = Q_L$ and that at time $t = t_1$ the battery is removed and the circuit closed again. Find $Q(t)$ for $t > t_1$ and sketch its graph.

FIGURE 1.2.3 The electric circuit of Problem 16.

 18. A pond containing 1,000,000 gal of water is initially free of a certain undesirable chemical (see Problem 21 of Section 1.1). Water containing 0.01 g/gal of the chemical flows into the pond at a rate of 300 gal/hr, and water also flows out of the pond at the same rate. Assume that the chemical is uniformly distributed throughout the pond.

(a) Let $Q(t)$ be the amount of the chemical in the pond at time t. Write down an initial value problem for $Q(t)$.

(b) Solve the problem in part (a) for $Q(t)$. How much chemical is in the pond after 1 year?

(c) At the end of 1 year the source of the chemical in the pond is removed; thereafter pure water flows into the pond, and the mixture flows out at the same rate as before. Write down the initial value problem that describes this new situation.

(d) Solve the initial value problem in part (c). How much chemical remains in the pond after 1 additional year (2 years from the beginning of the problem)?

(e) How long does it take for $Q(t)$ to be reduced to 10 g?

(f) Plot $Q(t)$ versus t for 3 years.

19. Your swimming pool containing 60,000 gal of water has been contaminated by 5 kg of a nontoxic dye that leaves a swimmer's skin an unattractive green. The pool's filtering system can take water from the pool, remove the dye, and return the water to the pool at a flow rate of 200 gal/min.

(a) Write down the initial value problem for the filtering process; let $q(t)$ be the amount of dye in the pool at any time t.

(b) Solve the problem in part (a).

[3]This equation results from Kirchhoff's laws, which are discussed in Section 3.8.

(c) You have invited several dozen friends to a pool party that is scheduled to begin in 4 hr. You have also determined that the effect of the dye is imperceptible if its concentration is less than 0.02 g/gal. Is your filtering system capable of reducing the dye concentration to this level within 4 hr?

(d) Find the time T at which the concentration of dye first reaches the value 0.02 g/gal.

(e) Find the flow rate that is sufficient to achieve the concentration 0.02 g/gal within 4 hr.

1.3 Classification of Differential Equations

The main purpose of this book is to discuss some of the properties of solutions of differential equations, and to present some of the methods that have proved effective in finding solutions or, in some cases, approximating them. To provide a framework for our presentation, we describe here several useful ways of classifying differential equations.

Ordinary and Partial Differential Equations. One of the more obvious classifications is based on whether the unknown function depends on a single independent variable or on several independent variables. In the first case, only ordinary derivatives appear in the differential equation, and it is said to be an **ordinary differential equation**. In the second case, the derivatives are partial derivatives, and the equation is called a **partial differential equation**.

All the differential equations discussed in the preceding two sections are ordinary differential equations. Another example of an ordinary differential equation is

$$L\frac{d^2Q(t)}{dt^2} + R\frac{dQ(t)}{dt} + \frac{1}{C}Q(t) = E(t), \tag{1}$$

for the charge $Q(t)$ on a capacitor in a circuit with capacitance C, resistance R, and inductance L; this equation is derived in Section 3.8. Typical examples of partial differential equations are the heat conduction equation

$$\alpha^2 \frac{\partial^2 u(x,t)}{\partial x^2} = \frac{\partial u(x,t)}{\partial t} \tag{2}$$

and the wave equation

$$a^2 \frac{\partial^2 u(x,t)}{\partial x^2} = \frac{\partial^2 u(x,t)}{\partial t^2}. \tag{3}$$

Here, α^2 and a^2 are certain physical constants. The heat conduction equation describes the conduction of heat in a solid body, and the wave equation arises in a variety of problems involving wave motion in solids or fluids. Note that in both Eqs. (2) and (3) the dependent variable u depends on the two independent variables x and t.

Systems of Differential Equations. Another classification of differential equations depends on the number of unknown functions that are involved. If there is a single function to be determined, then one equation is sufficient. However, if there are two or more unknown functions, then a system of equations is required. For example, the

Lotka–Volterra, or predator–prey, equations are important in ecological modeling. They have the form

$$dx/dt = ax - \alpha xy$$
$$dy/dt = -cy + \gamma xy, \tag{4}$$

where $x(t)$ and $y(t)$ are the respective populations of the prey and predator species. The constants a, α, c, and γ are based on empirical observations and depend on the particular species being studied. Systems of equations are discussed in Chapters 7 and 9; in particular, the Lotka–Volterra equations are examined in Section 9.5. In some areas of application it is not unusual to encounter very large systems containing hundreds, or even many thousands, of equations.

Order. The **order** of a differential equation is the order of the highest derivative that appears in the equation. The equations in the preceding sections are all first order equations, whereas Eq. (1) is a second order equation. Equations (2) and (3) are second order partial differential equations. More generally, the equation

$$F[t, u(t), u'(t), \ldots, u^{(n)}(t)] = 0 \tag{5}$$

is an ordinary differential equation of the nth order. Equation (5) expresses a relation between the independent variable t and the values of the function u and its first n derivatives $u', u'', \ldots, u^{(n)}$. It is convenient and customary in differential equations to write y for $u(t)$, with $y', y'', \ldots, y^{(n)}$ standing for $u'(t), u''(t), \ldots, u^{(n)}(t)$. Thus Eq. (5) is written as

$$F(t, y, y', \ldots, y^{(n)}) = 0. \tag{6}$$

For example,

$$y''' + 2e^t y'' + yy' = t^4 \tag{7}$$

is a third order differential equation for $y = u(t)$. Occasionally, other letters will be used instead of t and y for the independent and dependent variables; the meaning should be clear from the context.

We assume that it is always possible to solve a given ordinary differential equation for the highest derivative, obtaining

$$y^{(n)} = f(t, y, y', y'', \ldots, y^{(n-1)}). \tag{8}$$

We study only equations of the form (8). This is mainly to avoid the ambiguity that may arise because a single equation of the form (6) may correspond to several equations of the form (8). For example, the equation

$$y'^2 + ty' + 4y = 0 \tag{9}$$

leads to the two equations

$$y' = \frac{-t + \sqrt{t^2 - 16y}}{2} \quad \text{or} \quad y' = \frac{-t - \sqrt{t^2 - 16y}}{2}. \tag{10}$$

Linear and Nonlinear Equations. A crucial classification of differential equations is whether they are linear or nonlinear. The ordinary differential equation

$$F(t, y, y', \ldots, y^{(n)}) = 0$$

is said to be **linear** if F is a linear function of the variables $y, y', \ldots, y^{(n)}$; a similar definition applies to partial differential equations. Thus the general linear ordinary differential equation of order n is

$$a_0(t)y^{(n)} + a_1(t)y^{(n-1)} + \cdots + a_n(t)y = g(t). \tag{11}$$

Most of the equations you have seen thus far in this book are linear; examples are the equations in Sections 1.1 and 1.2 describing the falling object and the field mouse population. Similarly, in this section, Eq. (1) is a linear ordinary differential equation and Eqs. (2) and (3) are linear partial differential equations. An equation that is not of the form (11) is a **nonlinear** equation. Equation (7) is nonlinear because of the term yy'. Similarly, each equation in the system (4) is nonlinear because of the terms that involve the product xy.

A simple physical problem that leads to a nonlinear differential equation is the oscillating pendulum. The angle θ that an oscillating pendulum of length L makes with the vertical direction (see Figure 1.3.1) satisfies the equation

$$\frac{d^2\theta}{dt^2} + \frac{g}{L}\sin\theta = 0, \tag{12}$$

whose derivation is outlined in Problems 29 through 31. The presence of the term involving $\sin\theta$ makes Eq. (12) nonlinear.

FIGURE 1.3.1 An oscillating pendulum.

The mathematical theory and methods for solving linear equations are highly developed. In contrast, for nonlinear equations the theory is more complicated, and methods of solution are less satisfactory. In view of this, it is fortunate that many significant problems lead to linear ordinary differential equations or can be approximated by linear equations. For example, for the pendulum, if the angle θ is small, then $\sin\theta \cong \theta$ and Eq. (12) can be approximated by the linear equation

$$\frac{d^2\theta}{dt^2} + \frac{g}{L}\theta = 0. \tag{13}$$

This process of approximating a nonlinear equation by a linear one is called **linearization**; it is an extremely valuable way to deal with nonlinear equations. Nevertheless, there are many physical phenomena that simply cannot be represented adequately by linear equations. To study these phenomena it is essential to deal with nonlinear equations.

In an elementary text it is natural to emphasize the simpler and more straight-forward parts of the subject. Therefore the greater part of this book is devoted to linear equations and various methods for solving them. However, Chapters 8 and 9, as well as parts of Chapter 2, are concerned with nonlinear equations. Whenever it is appropriate, we point out why nonlinear equations are, in general, more difficult and why many of the techniques that are useful in solving linear equations cannot be applied to nonlinear equations.

Solutions. A **solution** of the ordinary differential equation (8) on the interval $\alpha < t < \beta$ is a function ϕ such that $\phi', \phi'', \ldots, \phi^{(n)}$ exist and satisfy

$$\phi^{(n)}(t) = f[t, \phi(t), \phi'(t), \ldots, \phi^{(n-1)}(t)] \tag{14}$$

for every t in $\alpha < t < \beta$. Unless stated otherwise, we assume that the function f of Eq. (8) is a real-valued function, and we are interested in obtaining real-valued solutions $y = \phi(t)$.

Recall that in Section 1.2 we found solutions of certain equations by a process of direct integration. For instance, we found that the equation

$$\frac{dp}{dt} = 0.5p - 450 \tag{15}$$

has the solution

$$p = 900 + ce^{t/2}, \tag{16}$$

where c is an arbitrary constant. It is often not so easy to find solutions of differential equations. However, if you find a function that you think may be a solution of a given equation, it is usually relatively easy to determine whether the function is actually a solution simply by substituting the function into the equation. For example, in this way it is easy to show that the function $y_1(t) = \cos t$ is a solution of

$$y'' + y = 0 \tag{17}$$

for all t. To confirm this, observe that $y_1'(t) = -\sin t$ and $y_1''(t) = -\cos t$; then it follows that $y_1''(t) + y_1(t) = 0$. In the same way you can easily show that $y_2(t) = \sin t$ is also a solution of Eq. (17). Of course, this does not constitute a satisfactory way to solve most differential equations, because there are far too many possible functions for you to have a good chance of finding the correct one by a random choice. Nevertheless, you should realize that you can verify whether any proposed solution is correct by substituting it into the differential equation. For any problem that is important to you, this can be a very useful check. It is one that you should make a habit of considering.

Some Important Questions. Although for the equations (15) and (17) we are able to verify that certain simple functions are solutions, in general we do not have such solutions readily available. Thus a fundamental question is the following: Does an equation of the form (8) always have a solution? The answer is "No." Merely writing down an equation of the form (8) does not necessarily mean that there is a function $y = \phi(t)$ that satisfies it. So, how can we tell whether some particular equation has a solution? This is the question of *existence* of a solution, and it is answered by theorems stating that under certain restrictions on the function f in Eq. (8), the equation always has solutions. However, this is not a purely mathematical concern,

for at least two reasons. If a problem has no solution, we would prefer to know that fact before investing time and effort in a vain attempt to solve the problem. Further, if a sensible physical problem is modeled mathematically as a differential equation, then the equation should have a solution. If it does not, then presumably there is something wrong with the formulation. In this sense an engineer or scientist has some check on the validity of the mathematical model.

If we assume that a given differential equation has at least one solution, the question arises as to how many solutions it has, and what additional conditions must be specified to single out a particular solution. This is the question of *uniqueness*. In general, solutions of differential equations contain one or more arbitrary constants of integration, as does the solution (16) of Eq. (15). Equation (16) represents an infinity of functions corresponding to the infinity of possible choices of the constant c. As we saw in Section 1.2, if p is specified at some time t, this condition will determine a value for c; even so, we have not yet ruled out the possibility that there may be other solutions of Eq. (15) that also have the prescribed value of p at the prescribed time t. The issue of uniqueness also has practical implications. If we are fortunate enough to find a solution of a given problem, and if we know that the problem has a unique solution, then we can be sure that we have completely solved the problem. If there may be other solutions, then perhaps we should continue to search for them.

A third important question is: Given a differential equation of the form (8), can we actually determine a solution, and if so, how? Note that if we find a solution of the given equation, we have at the same time answered the question of the existence of a solution. However, without knowledge of existence theory we might, for example, use a computer to find a numerical approximation to a "solution" that does not exist. On the other hand, even though we may know that a solution exists, it may be that the solution is not expressible in terms of the usual elementary functions—polynomial, trigonometric, exponential, logarithmic, and hyperbolic functions. Unfortunately, this is the situation for most differential equations. Thus, we discuss both elementary methods that can be used to obtain exact solutions of certain relatively simple problems, and also methods of a more general nature that can be used to find approximations to solutions of more difficult problems.

Computer Use in Differential Equations. A computer can be an extremely valuable tool in the study of differential equations. For many years computers have been used to execute numerical algorithms, such as those described in Chapter 8, to construct numerical approximations to solutions of differential equations. These algorithms have been refined to an extremely high level of generality and efficiency. A few lines of computer code, written in a high-level programming language and executed (often within a few seconds) on a relatively inexpensive computer, suffice to approximate to a high degree of accuracy the solutions of a wide range of differential equations. More sophisticated routines are also readily available. These routines combine the ability to handle very large and complicated systems with numerous diagnostic features that alert the user to possible problems as they are encountered.

The usual output from a numerical algorithm is a table of numbers, listing selected values of the independent variable and the corresponding values of the dependent variable. With appropriate software it is easy to display the solution of a differential equation graphically, whether the solution has been obtained numerically or as the result of an analytical procedure of some kind. Such a graphical display is often

much more illuminating and helpful in understanding and interpreting the solution of a differential equation than a table of numbers or a complicated analytical formula. There are on the market several well-crafted and relatively inexpensive special-purpose software packages for the graphical investigation of differential equations. The widespread availability of personal computers has brought powerful computational and graphical capability within the reach of individual students. You should consider, in the light of your own circumstances, how best to take advantage of the available computing resources. You will surely find it enlightening to do so.

Another aspect of computer use that is very relevant to the study of differential equations is the availability of extremely powerful and general software packages that can perform a wide variety of mathematical operations. Among these are *Maple*, *Mathematica*, and MATLAB, each of which can be used on various kinds of personal computers or workstations. All three of these packages can execute extensive numerical computations and have versatile graphical facilities. *Maple* and *Mathematica* also have very extensive analytical capabilities. For example, they can perform the analytical steps involved in solving many differential equations, often in response to a single command. Anyone who expects to deal with differential equations in more than a superficial way should become familiar with at least one of these products and explore the ways in which it can be used.

For you, the student, these computing resources have an effect on how you should study differential equations. To become confident in using differential equations, it is essential to understand how the solution methods work, and this understanding is achieved, in part, by working out a sufficient number of examples in detail. However, eventually you should plan to delegate as many as possible of the routine (often repetitive) details to a computer, while you focus on the proper formulation of the problem and on the interpretation of the solution. Our viewpoint is that you should always try to use the best methods and tools available for each task. In particular, you should strive to combine numerical, graphical, and analytical methods so as to attain maximum understanding of the behavior of the solution and of the underlying process that the problem models. You should also remember that some tasks can best be done with pencil and paper, while others require a calculator or computer. Good judgment is often needed in selecting a judicious combination.

PROBLEMS

In each of Problems 1 through 6 determine the order of the given differential equation; also state whether the equation is linear or nonlinear.

1. $t^2 \dfrac{d^2y}{dt^2} + t\dfrac{dy}{dt} + 2y = \sin t$

2. $(1+y^2)\dfrac{d^2y}{dt^2} + t\dfrac{dy}{dt} + y = e^t$

3. $\dfrac{d^4y}{dt^4} + \dfrac{d^3y}{dt^3} + \dfrac{d^2y}{dt^2} + \dfrac{dy}{dt} + y = 1$

4. $\dfrac{dy}{dt} + ty^2 = 0$

5. $\dfrac{d^2y}{dt^2} + \sin(t+y) = \sin t$

6. $\dfrac{d^3y}{dt^3} + t\dfrac{dy}{dt} + (\cos^2 t)y = t^3$

In each of Problems 7 through 14 verify that each given function is a solution of the differential equation.

7. $y'' - y = 0;$ $y_1(t) = e^t,$ $y_2(t) = \cosh t$

8. $y'' + 2y' - 3y = 0;$ $y_1(t) = e^{-3t},$ $y_2(t) = e^t$

9. $ty' - y = t^2; \qquad y = 3t + t^2$

10. $y'''' + 4y''' + 3y = t; \qquad y_1(t) = t/3, \quad y_2(t) = e^{-t} + t/3$

11. $2t^2 y'' + 3ty' - y = 0, \quad t > 0; \qquad y_1(t) = t^{1/2}, \quad y_2(t) = t^{-1}$

12. $t^2 y'' + 5ty' + 4y = 0, \quad t > 0; \qquad y_1(t) = t^{-2}, \quad y_2(t) = t^{-2} \ln t$

13. $y'' + y = \sec t, \quad 0 < t < \pi/2; \qquad y = (\cos t) \ln \cos t + t \sin t$

14. $y' - 2ty = 1; \qquad y = e^{t^2} \int_0^t e^{-s^2} \, ds + e^{t^2}$

In each of Problems 15 through 18 determine the values of r for which the given differential equation has solutions of the form $y = e^{rt}$.

15. $y' + 2y = 0$

16. $y'' - y = 0$

17. $y'' + y' - 6y = 0$

18. $y''' - 3y'' + 2y' = 0$

In each of Problems 19 and 20 determine the values of r for which the given differential equation has solutions of the form $y = t^r$ for $t > 0$.

19. $t^2 y'' + 4ty' + 2y = 0$

20. $t^2 y'' - 4ty' + 4y = 0$

In each of Problems 21 through 24 determine the order of the given partial differential equation; also state whether the equation is linear or nonlinear. Partial derivatives are denoted by subscripts.

21. $u_{xx} + u_{yy} + u_{zz} = 0$

22. $u_{xx} + u_{yy} + uu_x + uu_y + u = 0$

23. $u_{xxxx} + 2u_{xxyy} + u_{yyyy} = 0$

24. $u_t + uu_x = 1 + u_{xx}$

In each of Problems 25 through 28 verify that each given function is a solution of the given partial differential equation.

25. $u_{xx} + u_{yy} = 0; \qquad u_1(x, y) = \cos x \cosh y, \quad u_2(x, y) = \ln(x^2 + y^2)$

26. $\alpha^2 u_{xx} = u_t; \qquad u_1(x, t) = e^{-\alpha^2 t} \sin x, \quad u_2(x, t) = e^{-\alpha^2 \lambda^2 t} \sin \lambda x, \quad \lambda$ a real constant

27. $a^2 u_{xx} = u_{tt}; \qquad u_1(x, t) = \sin \lambda x \sin \lambda at, \quad u_2(x, t) = \sin(x - at), \quad \lambda$ a real constant

28. $\alpha^2 u_{xx} = u_t; \qquad u = (\pi/t)^{1/2} e^{-x^2/4\alpha^2 t}, \quad t > 0$

29. Follow the steps indicated here to derive the equation of motion of a pendulum, Eq. (12) in the text. Assume that the rod is rigid and weightless, that the mass is a point mass, and that there is no friction or drag anywhere in the system.

(a) Assume that the mass is in an arbitrary displaced position, indicated by the angle θ. Draw a free-body diagram showing the forces acting on the mass.

(b) Apply Newton's law of motion in the direction tangential to the circular arc on which the mass moves. Then the tensile force in the rod does not enter the equation. Observe that you need to find the component of the gravitational force in the tangential direction. Observe also that the linear acceleration, as opposed to the angular acceleration, is $L d^2\theta/dt^2$, where L is the length of the rod.

(c) Simplify the result from part (b) to obtain Eq. (12) in the text.

30. Another way to derive the pendulum equation (12) is based on the principle of conservation of energy.

(a) Show that the kinetic energy T of the pendulum in motion is

$$T = \frac{1}{2} mL^2 \left(\frac{d\theta}{dt} \right)^2.$$

(b) Show that the potential energy V of the pendulum, relative to its rest position, is

$$V = mgL(1 - \cos\theta).$$

(c) By the principle of conservation of energy, the total energy $E = T + V$ is constant. Calculate dE/dt, set it equal to zero, and show that the resulting equation reduces to Eq. (12).

31. A third derivation of the pendulum equation depends on the principle of angular momentum: the rate of change of angular momentum about any point is equal to the net external moment about the same point.

(a) Show that the angular momentum M, or moment of momentum, about the point of support is given by $M = mL^2 d\theta/dt$.

(b) Set dM/dt equal to the moment of the gravitational force, and show that the resulting equation reduces to Eq. (12). Note that positive moments are counterclockwise.

1.4 Historical Remarks

Without knowing something about differential equations and methods of solving them, it is difficult to appreciate the history of this important branch of mathematics. Further, the development of differential equations is intimately interwoven with the general development of mathematics and cannot be separated from it. Nevertheless, to provide some historical perspective, we indicate here some of the major trends in the history of the subject and identify the most prominent early contributors. Other historical information is contained in footnotes scattered throughout the book and in the references listed at the end of the chapter.

The subject of differential equations originated in the study of calculus by Isaac Newton (1642–1727) and Gottfried Wilhelm Leibniz (1646–1716) in the seventeenth century. Newton grew up in the English countryside, was educated at Trinity College, Cambridge, and became Lucasian Professor of Mathematics there in 1669. His epochal discoveries of calculus and of the fundamental laws of mechanics date from 1665. They were circulated privately among his friends, but Newton was extremely sensitive to criticism and did not begin to publish his results until 1687 with the appearance of his most famous book, *Philosophiae Naturalis Principia Mathematica*. Although Newton did relatively little work in differential equations as such, his development of the calculus and elucidation of the basic principles of mechanics provided a basis for their applications in the eighteenth century, most notably by Euler. Newton classified first order differential equations according to the forms $dy/dx = f(x)$, $dy/dx = f(y)$, and $dy/dx = f(x,y)$. For the latter equation he developed a method of solution using infinite series when $f(x,y)$ is a polynomial in x and y. Newton's active research in mathematics ended in the early 1690s, except for the solution of occasional "challenge problems" and the revision and publication of results obtained much earlier. He was appointed Warden of the British Mint in 1696 and resigned his professorship a few years later. He was knighted in 1705 and, upon his death, was buried in Westminster Abbey.

Leibniz was born in Leipzig and completed his doctorate in philosophy at the age of 20 at the University of Altdorf. Throughout his life he engaged in scholarly work in several different fields. He was mainly self-taught in mathematics, since his interest in this subject developed when he was in his twenties. Leibniz arrived at the fundamental results of calculus independently, although a little later than Newton, but was the first to publish them, in 1684. Leibniz was very conscious of the power

of good mathematical notation, and our notation for the derivative, dy/dx, and the integral sign are due to him. He discovered the method of separation of variables (Section 2.2) in 1691, the reduction of homogeneous equations to separable ones (Section 2.2, Problem 30) in 1691, and the procedure for solving first order linear equations (Section 2.1) in 1694. He spent his life as ambassador and adviser to several German royal families, which permitted him to travel widely and to carry on an extensive correspondence with other mathematicians, especially the Bernoulli brothers. In the course of this correspondence many problems in differential equations were solved during the latter part of the seventeenth century.

The brothers Jakob (1654–1705) and Johann (1667–1748) Bernoulli of Basel did much to develop methods of solving differential equations and to extend the range of their applications. Jakob became professor of mathematics at Basel in 1687, and Johann was appointed to the same position upon his brother's death in 1705. Both men were quarrelsome, jealous, and frequently embroiled in disputes, especially with each other. Nevertheless, both also made significant contributions to several areas of mathematics. With the aid of calculus, they solved a number of problems in mechanics by formulating them as differential equations. For example, Jakob Bernoulli solved the differential equation $y' = [a^3/(b^2 y - a^3)]^{1/2}$ in 1690 and in the same paper first used the term "integral" in the modern sense. In 1694 Johann Bernoulli was able to solve the equation $dy/dx = y/ax$. One problem which both brothers solved, and which led to much friction between them, was the *brachistochrone* problem (see Problem 32 of Section 2.3). The brachistochrone problem was also solved by Leibniz, Newton and the Marquis de L'Hôpital. It is said, perhaps apocryphally, that Newton learned of the problem late in the afternoon of a tiring day at the Mint and solved it that evening after dinner. He published the solution anonymously, but upon seeing it, Johann Bernoulli exclaimed, "Ah, I know the lion by his paw."

Daniel Bernoulli (1700–1782), son of Johann, migrated to St. Petersburg as a young man to join the newly established St. Petersburg Academy but returned to Basel in 1733 as professor of botany and, later, of physics. His interests were primarily in partial differential equations and their applications. For instance, it is his name that is associated with the Bernoulli equation in fluid mechanics. He was also the first to encounter the functions that a century later became known as Bessel functions (Section 5.8).

The greatest mathematician of the eighteenth century, Leonhard Euler (1707–1783), grew up near Basel and was a student of Johann Bernoulli. He followed his friend Daniel Bernoulli to St. Petersburg in 1727. For the remainder of his life he was associated with the St. Petersburg Academy (1727–1741 and 1766–1783) and the Berlin Academy (1741–1766). Euler was the most prolific mathematician of all time; his collected works fill more than 70 large volumes. His interests ranged over all areas of mathematics and many fields of application. Even though he was blind during the last 17 years of his life, his work continued undiminished until the very day of his death. Of particular interest here is his formulation of problems in mechanics in mathematical language and his development of methods of solving these mathematical problems. Lagrange said of Euler's work in mechanics, "The first great work in which analysis is applied to the science of movement." Among other things, Euler identified the condition for exactness of first order differential equations (Section 2.6) in 1734–35, developed the theory of integrating factors (Section 2.6) in the same paper, and gave the general solution of homogeneous linear equations with

First Order Differential Equations

This chapter deals with differential equations of first order,

$$\frac{dy}{dt} = f(t, y), \tag{1}$$

where f is a given function of two variables. Any differentiable function $y = \phi(t)$ that satisfies this equation for all t in some interval is called a solution, and our object is to determine whether such functions exist and, if so, to develop methods for finding them. Unfortunately, for an arbitrary function f, there is no general method for solving the equation in terms of elementary functions. Instead, we will describe several methods, each of which is applicable to a certain subclass of first order equations. The most important of these are linear equations (Section 2.1), separable equations (Section 2.2), and exact equations (Section 2.6). Other sections of this chapter describe some of the important applications of first order differential equations, introduce the idea of approximating a solution by numerical computation, and discuss some theoretical questions related to the existence and uniqueness of solutions. The final section includes an example of chaotic solutions in the context of first order difference equations, which have some important points of similarity with differential equations and are simpler to investigate.

2.1 Linear Equations; Method of Integrating Factors

If the function f in Eq. (1) depends linearly on the dependent variable y, then Eq. (1) is called a first order linear equation. In Sections 1.1 and 1.2 we discussed a restricted type of first order linear equation in which the coefficients are constants. A typical

example is

$$\frac{dy}{dt} = -ay + b, \tag{2}$$

where a and b are given constants. Recall that an equation of this form describes the motion of an object falling in the atmosphere. Now we want to consider the most general first order linear equation, which is obtained by replacing the coefficients a and b in Eq. (2) by arbitrary functions of t. We will usually write the general **first order linear equation** in the form

$$\frac{dy}{dt} + p(t)y = g(t), \tag{3}$$

where p and g are given functions of the independent variable t.

Equation (2) can be solved by the straightforward integration method introduced in Section 1.2. That is, provided that $a \neq 0$ and $y \neq b/a$, we rewrite the equation as

$$\frac{dy/dt}{y - (b/a)} = -a. \tag{4}$$

Then, by integration we obtain

$$\ln|y - (b/a)| = -at + C,$$

from which it follows that the general solution of Eq. (2) is

$$y = (b/a) + ce^{-at}, \tag{5}$$

where c is an arbitrary constant.

Unfortunately, this direct method of solution cannot be used to solve the general equation (3), so we need to use a different method of solution for it. We owe this method to Leibniz; it involves multiplying the differential equation (3) by a certain function $\mu(t)$, chosen so that the resulting equation is readily integrable. The function $\mu(t)$ is called an **integrating factor**, and the main difficulty is to determine how to find it. We will introduce this method in a simple example, later showing how to extend it to other first order linear equations, including the general equation (3).

EXAMPLE
1

Solve the differential equation

$$\frac{dy}{dt} + \tfrac{1}{2}y = \tfrac{1}{2}e^{t/3}. \tag{6}$$

Plot several solutions, and find the particular solution whose graph contains the point $(0, 1)$.

The first step is to multiply Eq. (6) by a function $\mu(t)$, as yet undetermined; thus

$$\mu(t)\frac{dy}{dt} + \tfrac{1}{2}\mu(t)y = \tfrac{1}{2}\mu(t)e^{t/3}. \tag{7}$$

The question now is whether we can choose $\mu(t)$ so that the left side of Eq. (7) is recognizable as the derivative of some particular expression. If so, then we can integrate Eq. (7), even though we do not know the function y. To guide our choice of the integrating factor $\mu(t)$, observe that the left side of Eq. (7) contains two terms and that the first term is part of the result of differentiating the product $\mu(t)y$. Thus, let us try to determine $\mu(t)$ so that the left

side of Eq. (7) becomes the derivative of the expression $\mu(t)y$. If we compare the left side of Eq. (7) with the differentiation formula

$$\frac{d}{dt}[\mu(t)y] = \mu(t)\frac{dy}{dt} + \frac{d\mu(t)}{dt}y, \tag{8}$$

we note that the first terms are identical and that the second terms also agree, provided we choose $\mu(t)$ to satisfy

$$\frac{d\mu(t)}{dt} = \tfrac{1}{2}\mu(t). \tag{9}$$

Therefore our search for an integrating factor will be successful if we can find a solution of Eq. (9). Perhaps you can readily identify a function that satisfies Eq. (9): What well-known function from calculus has a derivative that is equal to one-half times the original function? More systematically, rewrite Eq. (9) as

$$\frac{d\mu(t)/dt}{\mu(t)} = \tfrac{1}{2}, \tag{10}$$

which is equivalent to

$$\frac{d}{dt}\ln|\mu(t)| = \tfrac{1}{2}. \tag{11}$$

Then it follows that

$$\ln|\mu(t)| = \tfrac{1}{2}t + C, \tag{12}$$

or

$$\mu(t) = ce^{t/2}. \tag{13}$$

The function $\mu(t)$ given by Eq. (13) is an integrating factor for Eq. (6). Since we do not need the most general integrating factor, we will choose c to be one in Eq. (13) and use $\mu(t) = e^{t/2}$.

Now we return to Eq. (6), multiply it by the integrating factor $e^{t/2}$, and obtain

$$e^{t/2}\frac{dy}{dt} + \tfrac{1}{2}e^{t/2}y = \tfrac{1}{2}e^{5t/6}. \tag{14}$$

By the choice we have made of the integrating factor, the left side of Eq. (14) is the derivative of $e^{t/2}y$, so that Eq. (14) becomes

$$\frac{d}{dt}(e^{t/2}y) = \tfrac{1}{2}e^{5t/6}. \tag{15}$$

By integrating both sides of Eq. (15) we obtain

$$e^{t/2}y = \tfrac{3}{5}e^{5t/6} + c, \tag{16}$$

where c is an arbitrary constant. Finally, on solving Eq. (16) for y, we have the general solution of Eq. (6), namely,

$$y = \tfrac{3}{5}e^{t/3} + ce^{-t/2}. \tag{17}$$

To find the solution passing through the point $(0, 1)$, we set $t = 0$ and $y = 1$ in Eq. (17), obtaining $1 = (3/5) + c$. Thus $c = 2/5$, and the desired solution is

$$y = \tfrac{3}{5}e^{t/3} + \tfrac{2}{5}e^{-t/2}. \tag{18}$$

Figure 2.1.1 includes the graphs of Eq. (17) for several values of c with a direction field in the background. The solution passing through $(0, 1)$ is shown by the heavy curve.

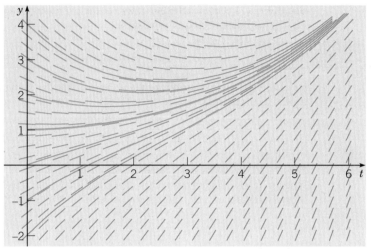

FIGURE 2.1.1 Integral curves of $y' + \frac{1}{2}y = \frac{1}{2}e^{t/3}$.

Let us now extend the method of integrating factors to equations of the form

$$\frac{dy}{dt} + ay = g(t), \tag{19}$$

where a is a given constant, and $g(t)$ is a given function. Proceeding as in Example 1, we find that the integrating factor $\mu(t)$ must satisfy

$$\frac{d\mu}{dt} = a\mu, \tag{20}$$

rather than Eq. (9). Thus the integrating factor is $\mu(t) = e^{at}$. Multiplying Eq. (19) by $\mu(t)$, we obtain

$$e^{at}\frac{dy}{dt} + ae^{at}y = e^{at}g(t),$$

or

$$\frac{d}{dt}(e^{at}y) = e^{at}g(t). \tag{21}$$

By integrating both sides of Eq. (21) we find that

$$e^{at}y = \int e^{at}g(t)\,dt + c, \tag{22}$$

where c is an arbitrary constant. For many simple functions $g(t)$ we can evaluate the integral in Eq. (22) and express the solution y in terms of elementary functions, as in Example 1. However, for more complicated functions $g(t)$, it is necessary to leave the solution in integral form. In this case

$$y = e^{-at}\int_{t_0}^{t} e^{as}g(s)\,ds + ce^{-at}. \tag{23}$$

Note that in Eq. (23) we have used s to denote the integration variable to distinguish it from the independent variable t, and we have chosen some convenient value t_0 as the lower limit of integration.

EXAMPLE
2

Solve the differential equation

$$\frac{dy}{dt} - 2y = 4 - t \tag{24}$$

and plot the graphs of several solutions. Discuss the behavior of solutions as $t \to \infty$.

Equation (24) is of the form (19) with $a = -2$; therefore the integrating factor is $\mu(t) = e^{-2t}$. Multiplying the differential equation (24) by $\mu(t)$, we obtain

$$e^{-2t}\frac{dy}{dt} - 2e^{-2t}y = 4e^{-2t} - te^{-2t}, \tag{25}$$

or

$$\frac{d}{dt}(e^{-2t}y) = 4e^{-2t} - te^{-2t}. \tag{26}$$

Then, by integrating both sides of this equation, we have

$$e^{-2t}y = -2e^{-2t} + \tfrac{1}{2}te^{-2t} + \tfrac{1}{4}e^{-2t} + c,$$

where we have used integration by parts on the last term in Eq. (26). Thus the general solution of Eq. (24) is

$$y = -\tfrac{7}{4} + \tfrac{1}{2}t + ce^{2t}. \tag{27}$$

A direction field and graphs of the solution (27) for several values of c are shown in Figure 2.1.2. The behavior of the solution for large values of t is determined by the term ce^{2t}. If $c \neq 0$, then the solution grows exponentially large in magnitude, with the same sign as c itself. Thus the solutions diverge as t becomes large. The boundary between solutions that ultimately grow positively from those that ultimately grow negatively occurs when $c = 0$. If we substitute $c = 0$ into Eq. (27) and then set $t = 0$, we find that $y = -7/4$ is the separation point on the y-axis. Note that, for this initial value, the solution is $y = -\tfrac{7}{4} + \tfrac{1}{2}t$; it grows positively, but linearly rather than exponentially.

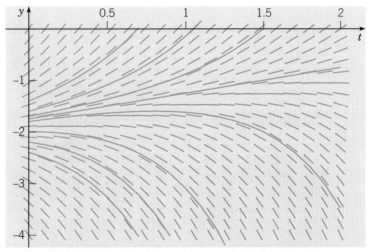

FIGURE 2.1.2 Integral curves of $y' - 2y = 4 - t$.

EXAMPLE 3

Solve the initial value problem

$$ty' + 2y = 4t^2,\tag{34}$$

$$y(1) = 2.\tag{35}$$

In order to determine $p(t)$ and $g(t)$ correctly, we must first rewrite Eq. (34) in the standard form (3). Thus we have

$$y' + (2/t)y = 4t,\tag{36}$$

so $p(t) = 2/t$ and $g(t) = 4t$. To solve Eq. (36) we first compute the integrating factor $\mu(t)$:

$$\mu(t) = \exp\int\frac{2}{t}\,dt = e^{2\ln|t|} = t^2.$$

On multiplying Eq. (36) by $\mu(t) = t^2$, we obtain

$$t^2 y' + 2ty = (t^2 y)' = 4t^3,$$

and therefore

$$t^2 y = t^4 + c,$$

where c is an arbitrary constant. It follows that

$$y = t^2 + \frac{c}{t^2}\tag{37}$$

is the general solution of Eq. (34). Integral curves of Eq. (34) for several values of c are shown in Figure 2.1.3. To satisfy the initial condition (35) it is necessary to choose $c = 1$; thus

$$y = t^2 + \frac{1}{t^2},\qquad t > 0\tag{38}$$

is the solution of the initial value problem (34), (35). This solution is shown by the heavy curve in Figure 2.1.3. Note that it becomes unbounded and is asymptotic to the positive y-axis as $t \to 0$ from the right. This is the effect of the infinite discontinuity in the coefficient $p(t)$ at the origin. The function $y = t^2 + (1/t^2)$ for $t < 0$ is not part of the solution of this initial value problem.

This is the first example in which the solution fails to exist for some values of t. Again, this is due to the infinite discontinuity in $p(t)$ at $t = 0$, which restricts the solution to the interval $0 < t < \infty$.

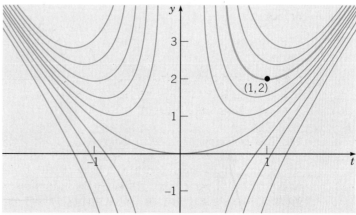

FIGURE 2.1.3 Integral curves of $ty' + 2y = 4t^2$.

Looking again at Figure 2.1.3, we see that some solutions (those for which $c > 0$) are asymptotic to the positive y-axis as $t \to 0$ from the right, while other solutions (for which $c < 0$) are asymptotic to the negative y-axis. The solution for which $c = 0$, namely, $y = t^2$, remains bounded and differentiable even at $t = 0$. If we generalize the initial condition (35) to

$$y(1) = y_0, \tag{39}$$

then $c = y_0 - 1$ and the solution (38) becomes

$$y = t^2 + \frac{y_0 - 1}{t^2}, \qquad t > 0. \tag{40}$$

As in Example 2, this is another instance where there is a critical initial value, namely, $y_0 = 1$, that separates solutions that behave in two quite different ways.

EXAMPLE 4

Solve the initial value problem

$$2y' + ty = 2, \tag{41}$$

$$y(0) = 1. \tag{42}$$

First divide the differential equation (41) by two, obtaining

$$y' + (t/2)y = 1. \tag{43}$$

Thus $p(t) = t/2$, and the integrating factor is $\mu(t) = \exp(t^2/4)$. Then multiply Eq. (43) by $\mu(t)$, so that

$$e^{t^2/4}y' + \frac{t}{2}e^{t^2/4}y = e^{t^2/4}. \tag{44}$$

The left side of Eq. (44) is the derivative of $e^{t^2/4}y$, so by integrating both sides of Eq. (44) we obtain

$$e^{t^2/4}y = \int e^{t^2/4}\, dt + c. \tag{45}$$

The integral on the right side of Eq. (45) cannot be evaluated in terms of the usual elementary functions, so we leave the integral unevaluated. However, by choosing the lower limit of integration as the initial point $t = 0$, we can replace Eq. (45) by

$$e^{t^2/4}y = \int_0^t e^{s^2/4}\, ds + c, \tag{46}$$

where c is an arbitrary constant. It then follows that the general solution y of Eq. (41) is given by

$$y = e^{-t^2/4}\int_0^t e^{s^2/4}\, ds + ce^{-t^2/4}. \tag{47}$$

The initial condition (42) requires that $c = 1$.

The main purpose of this example is to illustrate that sometimes the solution must be left in terms of an integral. This is usually at most a slight inconvenience, rather than a serious obstacle. For a given value of t the integral in Eq. (47) is a definite integral and can be approximated to any desired degree of accuracy by using readily available numerical integrators. By repeating this process for many values of t and plotting the results, you can obtain a graph of a solution. Alternatively, you can use a numerical approximation method,

such as those discussed in Chapter 8, that proceed directly from the differential equation and need no expression for the solution. Software packages such as Maple and Mathematica readily execute such procedures and produce graphs of solutions of differential equations.

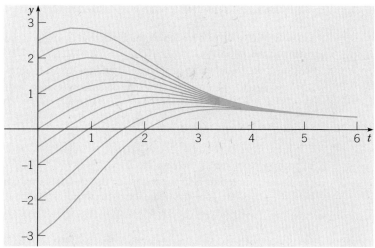

FIGURE 2.1.4 Integral curves of $2y' + ty = 2$.

Figure 2.1.4 displays graphs of the solution (47) for several values of c. From the figure it may be plausible to conjecture that all solutions approach a limit as $t \to \infty$. The limit can be found analytically (see Problem 32).

PROBLEMS

In each of Problems 1 through 12:

(a) Draw a direction field for the given differential equation.

(b) Based on an inspection of the direction field, describe how solutions behave for large t.

(c) Find the general solution of the given differential equation, and use it to determine how solutions behave as $t \to \infty$.

1. $y' + 3y = t + e^{-2t}$

2. $y' - 2y = t^2 e^{2t}$

3. $y' + y = te^{-t} + 1$

4. $y' + (1/t)y = 3\cos 2t, \qquad t > 0$

5. $y' - 2y = 3e^t$

6. $ty' + 2y = \sin t, \qquad t > 0$

7. $y' + 2ty = 2te^{-t^2}$

8. $(1 + t^2)y' + 4ty = (1 + t^2)^{-2}$

9. $2y' + y = 3t$

10. $ty' - y = t^2 e^{-t}, \qquad t > 0$

11. $y' + y = 5\sin 2t$

12. $2y' + y = 3t^2$

In each of Problems 13 through 20 find the solution of the given initial value problem.

13. $y' - y = 2te^{2t}, \qquad y(0) = 1$

14. $y' + 2y = te^{-2t}, \qquad y(1) = 0$

15. $ty' + 2y = t^2 - t + 1, \qquad y(1) = \frac{1}{2}, \quad t > 0$

16. $y' + (2/t)y = (\cos t)/t^2, \qquad y(\pi) = 0, \quad t > 0$

17. $y' - 2y = e^{2t}, \qquad y(0) = 2$

18. $ty' + 2y = \sin t$, $y(\pi/2) = 1$, $t > 0$

19. $t^3 y' + 4t^2 y = e^{-t}$, $y(-1) = 0$, $t < 0$

20. $ty' + (t+1)y = t$, $y(\ln 2) = 1$, $t > 0$

In each of Problems 21 through 23:

(a) Draw a direction field for the given differential equation. How do solutions appear to behave as t becomes large? Does the behavior depend on the choice of the initial value a? Let a_0 be the value of a for which the transition from one type of behavior to another occurs. Estimate the value of a_0.

(b) Solve the initial value problem and find the critical value a_0 exactly.

(c) Describe the behavior of the solution corresponding to the initial value a_0.

21. $y' - \frac{1}{2}y = 2\cos t$, $y(0) = a$

22. $2y' - y = e^{t/3}$, $y(0) = a$

23. $3y' - 2y = e^{-\pi t/2}$, $y(0) = a$

In each of Problems 24 through 26:

(a) Draw a direction field for the given differential equation. How do solutions appear to behave as $t \to 0$? Does the behavior depend on the choice of the initial value a? Let a_0 be the value of a for which the transition from one type of behavior to another occurs. Estimate the value of a_0.

(b) Solve the initial value problem and find the critical value a_0 exactly.

(c) Describe the behavior of the solution corresponding to the initial value a_0.

24. $ty' + (t+1)y = 2te^{-t}$, $y(1) = a$, $t > 0$

25. $ty' + 2y = (\sin t)/t$, $y(-\pi/2) = a$, $t < 0$

26. $(\sin t)y' + (\cos t)y = e^t$, $y(1) = a$, $0 < t < \pi$

27. Consider the initial value problem

$$y' + \tfrac{1}{2}y = 2\cos t, \qquad y(0) = -1.$$

Find the coordinates of the first local maximum point of the solution for $t > 0$.

28. Consider the initial value problem

$$y' + \tfrac{2}{3}y = 1 - \tfrac{1}{2}t, \qquad y(0) = y_0.$$

Find the value of y_0 for which the solution touches, but does not cross, the t-axis.

29. Consider the initial value problem

$$y' + \tfrac{1}{4}y = 3 + 2\cos 2t, \qquad y(0) = 0.$$

(a) Find the solution of this initial value problem and describe its behavior for large t.

(b) Determine the value of t for which the solution first intersects the line $y = 12$.

30. Find the value of y_0 for which the solution of the initial value problem

$$y' - y = 1 + 3\sin t, \qquad y(0) = y_0$$

remains finite as $t \to \infty$.

31. Consider the initial value problem

$$y' - \tfrac{3}{2}y = 3t + 2e^t, \qquad y(0) = y_0.$$

Find the value of y_0 that separates solutions that grow positively as $t \to \infty$ from those that grow negatively. How does the solution that corresponds to this critical value of y_0 behave as $t \to \infty$?

32. Show that all solutions of $2y' + ty = 2$ [Eq. (41) of the text] approach a limit as $t \to \infty$, and find the limiting value.
 Hint: Consider the general solution, Eq. (47), and use L'Hospital's rule on the first term.

33. Show that if a and λ are positive constants, and b is any real number, then every solution of the equation

$$y' + ay = be^{-\lambda t}$$

 has the property that $y \to 0$ as $t \to \infty$.
 Hint: Consider the cases $a = \lambda$ and $a \neq \lambda$ separately.

In each of Problems 34 through 37 construct a first order linear differential equation whose solutions have the required behavior as $t \to \infty$. Then solve your equation and confirm that the solutions do indeed have the specified property.

34. All solutions have the limit 3 as $t \to \infty$.
35. All solutions are asymptotic to the line $y = 3 - t$ as $t \to \infty$.
36. All solutions are asymptotic to the line $y = 2t - 5$ as $t \to \infty$.
37. All solutions approach the curve $y = 4 - t^2$ as $t \to \infty$.
38. **Variation of Parameters.** Consider the following method of solving the general linear equation of first order:

$$y' + p(t)y = g(t). \qquad \text{(i)}$$

 (a) If $g(t) = 0$ for all t, show that the solution is

$$y = A \exp\left[-\int p(t)\,dt\right], \qquad \text{(ii)}$$

 where A is a constant.
 (b) If $g(t)$ is not everywhere zero, assume that the solution of Eq. (i) is of the form

$$y = A(t) \exp\left[-\int p(t)\,dt\right], \qquad \text{(iii)}$$

 where A is now a function of t. By substituting for y in the given differential equation, show that $A(t)$ must satisfy the condition

$$A'(t) = g(t) \exp\left[\int p(t)\,dt\right]. \qquad \text{(iv)}$$

 (c) Find $A(t)$ from Eq. (iv). Then substitute for $A(t)$ in Eq. (iii) and determine y. Verify that the solution obtained in this manner agrees with that of Eq. (33) in the text. This technique is known as the method of **variation of parameters**; it is discussed in detail in Section 3.7 in connection with second order linear equations.

In each of Problems 39 through 42 use the method of Problem 38 to solve the given differential equation.

39. $y' - 2y = t^2 e^{2t}$ 40. $y' + (1/t)y = 3\cos 2t, \quad t > 0$
41. $ty' + 2y = \sin t, \qquad t > 0$ 42. $2y' + y = 3t^2$

EXAMPLE

3

Solve the equation

$$\frac{dy}{dx} = \frac{4x - x^3}{4 + y^3} \tag{22}$$

and draw graphs of several integral curves. Also find the solution passing through the point $(0, 1)$ and determine its interval of validity.

Rewriting Eq. (22) as

$$(4 + y^3)\, dy = (4x - x^3)\, dx,$$

integrating each side, multiplying by 4, and rearranging the terms, we obtain

$$y^4 + 16y + x^4 - 8x^2 = c, \tag{23}$$

where c is an arbitrary constant. Any differentiable function $y = \phi(x)$ that satisfies Eq. (23) is a solution of the differential equation (22). Graphs of Eq. (23) for several values of c are shown in Figure 2.2.3.

To find the particular solution passing through $(0, 1)$, we set $x = 0$ and $y = 1$ in Eq. (23) with the result that $c = 17$. Thus the solution in question is given implicitly by

$$y^4 + 16y + x^4 - 8x^2 = 17. \tag{24}$$

It is shown by the heavy curve in Figure 2.2.3. The interval of validity of this solution extends on either side of the initial point as long as the function remains differentiable. From the figure we see that the interval ends when we reach points where the tangent line is vertical. It follows from the differential equation (22) that these are points where $4 + y^3 = 0$, or $y = (-4)^{1/3} \cong -1.5874$. From Eq. (24) the corresponding values of x are $x \cong \pm 3.3488$. These points are marked on the graph in Figure 2.2.3.

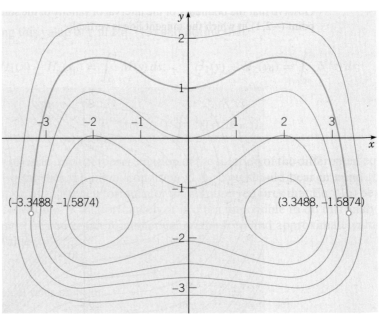

FIGURE 2.2.3 Integral curves of $y' = (4x - x^3)/(4y + y^3)$. The solution passing through $(0, 1)$ is shown by the heavy curve.

Sometimes an equation of the form (2),

$$\frac{dy}{dx} = f(x, y),$$

has a constant solution $y = y_0$. Such a solution is usually easy to find because if $f(x, y_0) = 0$ for some value y_0 and for all x, then the constant function $y = y_0$ is a solution of the differential equation (2). For example, the equation

$$\frac{dy}{dx} = \frac{(y - 3) \cos x}{1 + 2y^2} \tag{25}$$

has the constant solution $y = 3$. Other solutions of this equation can be found by separating the variables and integrating.

The investigation of a first order nonlinear equation can sometimes be facilitated by regarding both x and y as functions of a third variable t. Then

$$\frac{dy}{dx} = \frac{dy/dt}{dx/dt}. \tag{26}$$

If the differential equation is

$$\frac{dy}{dx} = \frac{F(x, y)}{G(x, y)}, \tag{27}$$

then, by comparing numerators and denominators in Eqs. (26) and (27), we obtain the system

$$dx/dt = G(x, y), \qquad dy/dt = F(x, y). \tag{28}$$

At first sight it may seem unlikely that a problem will be simplified by replacing a single equation by a pair of equations, but, in fact, the system (28) may well be more amenable to investigation than the single equation (27). Chapter 9 is devoted to nonlinear systems of the form (28).

Note: In Example 2 it was not difficult to solve explicitly for y as a function of x. However, this situation is exceptional, and often it will be better to leave the solution in implicit form, as in Examples 1 and 3. Thus, in the problems below and in other sections where nonlinear equations appear, the words "solve the following differential equation" mean to find the solution explicitly if it is convenient to do so, but otherwise to find an equation defining the solution implicitly.

PROBLEMS

In each of Problems 1 through 8 solve the given differential equation.

1. $y' = x^2/y$
2. $y' = x^2/y(1 + x^3)$
3. $y' + y^2 \sin x = 0$
4. $y' = (3x^2 - 1)/(3 + 2y)$
5. $y' = (\cos^2 x)(\cos^2 2y)$
6. $xy' = (1 - y^2)^{1/2}$
7. $\dfrac{dy}{dx} = \dfrac{x - e^{-x}}{y + e^y}$
8. $\dfrac{dy}{dx} = \dfrac{x^2}{1 + y^2}$

In each of Problems 9 through 20:
(a) Find the solution of the given initial value problem in explicit form.
(b) Plot the graph of the solution.
(c) Determine (at least approximately) the interval in which the solution is defined.

9. $y' = (1 - 2x)y^2$, $y(0) = -1/6$ 10. $y' = (1 - 2x)/y$, $y(1) = -2$

11. $x\,dx + ye^{-x}dy = 0$, $y(0) = 1$ 12. $dr/d\theta = r^2/\theta$, $r(1) = 2$

13. $y' = 2x/(y + x^2 y)$, $y(0) = -2$ 14. $y' = xy^3(1 + x^2)^{-1/2}$, $y(0) = 1$

15. $y' = 2x/(1 + 2y)$, $y(2) = 0$ 16. $y' = x(x^2 + 1)/4y^3$, $y(0) = -1/\sqrt{2}$

17. $y' = (3x^2 - e^x)/(2y - 5)$, $y(0) = 1$

18. $y' = (e^{-x} - e^x)/(3 + 4y)$, $y(0) = 1$

19. $\sin 2x\,dx + \cos 3y\,dy = 0$, $y(\pi/2) = \pi/3$

20. $y^2(1 - x^2)^{1/2}dy = \arcsin x\,dx$, $y(0) = 1$

Some of the results requested in Problems 21 through 28 can be obtained either by solving the given equations analytically or by plotting numerically generated approximations to the solutions. Try to form an opinion as to the advantages and disadvantages of each approach.

21. Solve the initial value problem

$$y' = (1 + 3x^2)/(3y^2 - 6y), \qquad y(0) = 1$$

and determine the interval in which the solution is valid.
Hint: To find the interval of definition, look for points where the integral curve has a vertical tangent.

22. Solve the initial value problem

$$y' = 3x^2/(3y^2 - 4), \qquad y(1) = 0$$

and determine the interval in which the solution is valid.
Hint: To find the interval of definition, look for points where the integral curve has a vertical tangent.

23. Solve the initial value problem

$$y' = 2y^2 + xy^2, \qquad y(0) = 1$$

and determine where the solution attains its minimum value.

24. Solve the initial value problem

$$y' = (2 - e^x)/(3 + 2y), \qquad y(0) = 0$$

and determine where the solution attains its maximum value.

25. Solve the initial value problem

$$y' = 2\cos 2x/(3 + 2y), \qquad y(0) = -1$$

and determine where the solution attains its maximum value.

26. Solve the initial value problem

$$y' = 2(1 + x)(1 + y^2), \qquad y(0) = 0$$

and determine where the solution attains its minimum value.

27. Consider the initial value problem

$$y' = ty(4 - y)/3, \qquad y(0) = y_0.$$

(a) Determine how the behavior of the solution as t increases depends on the initial value y_0.

(b) Suppose that $y_0 = 0.5$. Find the time T at which the solution first reaches the value 3.98.

 28. Consider the initial value problem

$$y' = ty(4 - y)/(1 + t), \qquad y(0) = y_0 > 0.$$

(a) Determine how the solution behaves as $t \to \infty$.

(b) If $y_0 = 2$, find the time T at which the solution first reaches the value 3.99.

(c) Find the range of initial values for which the solution lies in the interval $3.99 < y < 4.01$ by the time $t = 2$.

29. Solve the equation

$$\frac{dy}{dx} = \frac{ay + b}{cy + d},$$

where a, b, c, and d are constants.

Homogeneous Equations. If the right side of the equation $dy/dx = f(x, y)$ can be expressed as a function of the ratio y/x only, then the equation is said to be homogeneous.[1] Such equations can always be transformed into separable equations by a change of the dependent variable. Problem 30 illustrates how to solve first order homogeneous equations.

 30. Consider the equation

$$\frac{dy}{dx} = \frac{y - 4x}{x - y}. \tag{i}$$

(a) Show that Eq. (i) can be rewritten as

$$\frac{dy}{dx} = \frac{(y/x) - 4}{1 - (y/x)}; \tag{ii}$$

thus Eq. (i) is homogeneous.

(b) Introduce a new dependent variable v so that $v = y/x$, or $y = xv(x)$. Express dy/dx in terms of x, v, and dv/dx.

(c) Replace y and dy/dx in Eq. (ii) by the expressions from part (b) that involve v and dv/dx. Show that the resulting differential equation is

$$v + x\frac{dv}{dx} = \frac{v - 4}{1 - v},$$

or

$$x\frac{dv}{dx} = \frac{v^2 - 4}{1 - v}. \tag{iii}$$

Observe that Eq. (iii) is separable.

(d) Solve Eq. (iii), obtaining v implicitly in terms of x.

(e) Find the solution of Eq. (i) by replacing v by y/x in the solution in part (d).

(f) Draw a direction field and some integral curves for Eq. (i). Recall that the right side of Eq. (i) actually depends only on the ratio y/x. This means that integral curves have the same slope at all points on any given straight line through the origin, although the slope changes from one line to another. Therefore the direction field and the integral curves are symmetric with respect to the origin. Is this symmetry property evident from your plot?

[1]The word "homogeneous" has different meanings in different mathematical contexts. The homogeneous equations considered here have nothing to do with the homogeneous equations that will occur in Chapter 3 and elsewhere.

The method outlined in Problem 30 can be used for any homogeneous equation. That is, the substitution $y = xv(x)$ transforms a homogeneous equation into a separable equation. The latter equation can be solved by direct integration, and then replacing v by y/x gives the solution to the original equation. In each of Problems 31 through 38:

(a) Show that the given equation is homogeneous.

(b) Solve the differential equation.

(c) Draw a direction field and some integral curves. Are they symmetric with respect to the origin?

31. $\dfrac{dy}{dx} = \dfrac{x^2 + xy + y^2}{x^2}$

32. $\dfrac{dy}{dx} = \dfrac{x^2 + 3y^2}{2xy}$

33. $\dfrac{dy}{dx} = \dfrac{4y - 3x}{2x - y}$

34. $\dfrac{dy}{dx} = -\dfrac{4x + 3y}{2x + y}$

35. $\dfrac{dy}{dx} = \dfrac{x + 3y}{x - y}$

36. $(x^2 + 3xy + y^2)\, dx - x^2\, dy = 0$

37. $\dfrac{dy}{dx} = \dfrac{x^2 - 3y^2}{2xy}$

38. $\dfrac{dy}{dx} = \dfrac{3y^2 - x^2}{2xy}$

2.3 Modeling with First Order Equations

Differential equations are of interest to nonmathematicians primarily because of the possibility of using them to investigate a wide variety of problems in the physical, biological, and social sciences. One reason for this is that mathematical models and their solutions lead to equations relating the variables and parameters in the problem. These equations often enable you to make predictions about how the natural process will behave in various circumstances. It is often easy to vary parameters in the mathematical model over wide ranges, whereas this may be very time-consuming or expensive, if not impossible, in an experimental setting. Nevertheless, mathematical modeling and experiment or observation are both critically important and have somewhat complementary roles in scientific investigations. Mathematical models are validated by comparison of their predictions with experimental results. On the other hand, mathematical analyses may suggest the most promising directions to explore experimentally, and they may indicate fairly precisely what experimental data will be most helpful.

In Sections 1.1 and 1.2 we formulated and investigated a few simple mathematical models. We begin by recapitulating and expanding on some of the conclusions reached in those sections. Regardless of the specific field of application, there are three identifiable steps that are always present in the process of mathematical modeling.

Construction of the Model. In this step you translate the physical situation into mathematical terms, often using the steps listed at the end of Section 1.1. Perhaps most critical at this stage is to state clearly the physical principle(s) that are believed to govern the process. For example, it has been observed that in some circumstances heat passes from a warmer to a cooler body at a rate proportional to the temperature difference, that objects move about in accordance with Newton's laws of motion, and

that isolated insect populations grow at a rate proportional to the current population. Each of these statements involves a rate of change (derivative) and consequently, when expressed mathematically, leads to a differential equation. The differential equation is a mathematical model of the process.

It is important to realize that the mathematical equations are almost always only an approximate description of the actual process. For example, bodies moving at speeds comparable to the speed of light are not governed by Newton's laws, insect populations do not grow indefinitely as stated because of eventual limitations on their food supply, and heat transfer is affected by factors other than the temperature difference. Alternatively, one can adopt the point of view that the mathematical equations exactly describe the operation of a simplified physical model, which has been constructed (or conceived of) so as to embody the most important features of the actual process. Sometimes, the process of mathematical modeling involves the conceptual replacement of a discrete process by a continuous one. For instance, the number of members in an insect population changes by discrete amounts; however, if the population is large, it seems reasonable to consider it as a continuous variable and even to speak of its derivative.

Analysis of the Model. Once the problem has been formulated mathematically, you are often faced with the problem of solving one or more differential equations or, failing that, of finding out as much as possible about the properties of the solution. It may happen that this mathematical problem is quite difficult, and if so, further approximations may be indicated at this stage to make the problem mathematically tractable. For example, a nonlinear equation may be approximated by a linear one, or a slowly varying coefficient may be replaced by a constant. Naturally, any such approximations must also be examined from the physical point of view to make sure that the simplified mathematical problem still reflects the essential features of the physical process under investigation. At the same time, an intimate knowledge of the physics of the problem may suggest reasonable mathematical approximations that will make the mathematical problem more amenable to analysis. This interplay of understanding of physical phenomena and knowledge of mathematical techniques and their limitations is characteristic of applied mathematics at its best, and it is indispensable in successfully constructing useful mathematical models of intricate physical processes.

Comparison with Experiment or Observation. Finally, having obtained the solution (or at least some information about it), you must interpret this information in the context in which the problem arose. In particular, you should always check that the mathematical solution appears physically reasonable. If possible, calculate the values of the solution at selected points and compare them with experimentally observed values. Or ask whether the behavior of the solution after a long time is consistent with observations. Or examine the solutions corresponding to certain special values of parameters in the problem. Of course, the fact that the mathematical solution appears to be reasonable does not guarantee that it is correct. However, if the predictions of the mathematical model are seriously inconsistent with observations of the physical system it purports to describe, this suggests that errors have been made in solving the mathematical problem, that the mathematical model itself needs refinement, or that observations must be made with greater care.

Now we return to the general first order linear equation (3),

$$\frac{dy}{dt} + p(t)y = g(t),$$

where p and g are given functions. To determine an appropriate integrating factor, we multiply Eq. (3) by an as yet undetermined function $\mu(t)$, obtaining

$$\mu(t)\frac{dy}{dt} + p(t)\mu(t)y = \mu(t)g(t). \tag{28}$$

Following the same line of development as in Example 1, we see that the left side of Eq. (28) is the derivative of the product $\mu(t)y$, provided that $\mu(t)$ satisfies the equation

$$\frac{d\mu(t)}{dt} = p(t)\mu(t). \tag{29}$$

If we assume temporarily that $\mu(t)$ is positive, then we have

$$\frac{d\mu(t)/dt}{\mu(t)} = p(t),$$

and consequently

$$\ln \mu(t) = \int p(t)\, dt + k.$$

By choosing the arbitrary constant k to be zero, we obtain the simplest possible function for μ, namely,

$$\mu(t) = \exp \int p(t)\, dt. \tag{30}$$

Note that $\mu(t)$ is positive for all t, as we assumed. Returning to Eq. (28), we have

$$\frac{d}{dt}[\mu(t)y] = \mu(t)g(t). \tag{31}$$

Hence

$$\mu(t)y = \int \mu(t)g(t)\, dt + c, \tag{32}$$

where c is an arbitrary constant. Sometimes the integral in Eq. (32) can be evaluated in terms of elementary functions. However, in general this is not possible, so the general solution of Eq. (3) is

$$y = \frac{1}{\mu(t)}\left[\int_{t_0}^{t} \mu(s)g(s)\, ds + c\right], \tag{33}$$

where again t_0 is some convenient lower limit of integration. Observe that Eq. (33) involves two integrations, one to obtain $\mu(t)$ from Eq. (30) and the other to determine y from Eq. (33).

The examples in this section are typical of applications in which first order differential equations arise.

EXAMPLE

1

Mixing

At time $t = 0$ a tank contains Q_0 lb of salt dissolved in 100 gal of water; see Figure 2.3.1. Assume that water containing $\frac{1}{4}$ lb of salt/gal is entering the tank at a rate of r gal/min and that the well-stirred mixture is draining from the tank at the same rate. Set up the initial value problem that describes this flow process. Find the amount of salt $Q(t)$ in the tank at any time, and also find the limiting amount Q_L that is present after a very long time. If $r = 3$ and $Q_0 = 2Q_L$, find the time T after which the salt level is within 2% of Q_L. Also find the flow rate that is required if the value of T is not to exceed 45 min.

rate in = $\frac{1}{4}$ *lb/gal ·* r *gal/min*

= $\frac{r}{4}$ *lb/min*

rate out = $\frac{Q}{100}$ *lb/gal ·* r *gal/min*

or $\frac{Qr}{100}$ *lb/min*

r gal/min, $\frac{1}{4}$ lb/gal

r gal/min

FIGURE 2.3.1 The water tank in Example 1.

We assume that salt is neither created nor destroyed in the tank. Therefore variations in the amount of salt are due solely to the flows in and out of the tank. More precisely, the rate of change of salt in the tank, dQ/dt, is equal to the rate at which salt is flowing in minus the rate at which it is flowing out. In symbols,

$$\frac{dQ}{dt} = \text{rate in} - \text{rate out}. \tag{1}$$

The rate at which salt enters the tank is the concentration $\frac{1}{4}$ lb/gal times the flow rate r gal/min, or $(r/4)$ lb/min. To find the rate at which salt leaves the tank, we need to multiply the concentration of salt in the tank by the rate of outflow, r gal/min. Since the rates of flow in and out are equal, the volume of water in the tank remains constant at 100 gal, and since the mixture is "well-stirred," the concentration throughout the tank is the same, namely, $[Q(t)/100]$ lb/gal. Therefore the rate at which salt leaves the tank is $[rQ(t)/100]$ lb/min. Thus the differential equation governing this process is

$$\frac{dQ}{dt} = \frac{r}{4} - \frac{rQ}{100}. \tag{2}$$

The initial condition is

$$Q(0) = Q_0. \tag{3}$$

Upon thinking about the problem physically, we might anticipate that eventually the mixture originally in the tank will be essentially replaced by the mixture flowing in, whose concentration is $\frac{1}{4}$ lb/gal. Consequently, we might expect that ultimately the amount of salt in the tank would be very close to 25 lb. We can also find the limiting amount $Q_L = 25$ by setting dQ/dt equal to zero in Eq. (2) and solving the resulting algebraic equation for Q.

To solve the problem analytically, note that Eq. (2) is both linear and separable. Rewriting it in the usual form for a linear equation, we have

$$\frac{dQ}{dt} + \frac{rQ}{100} = \frac{r}{4}. \tag{4}$$

Thus the integrating factor is $e^{rt/100}$ and the general solution is

$$Q(t) = 25 + ce^{-rt/100}, \tag{5}$$

where c is an arbitrary constant. To satisfy the initial condition (3) we must choose $c = Q_0 - 25$. Therefore the solution of the initial value problem (2), (3) is

$$Q(t) = 25 + (Q_0 - 25)e^{-rt/100}, \tag{6}$$

or

$$Q(t) = 25(1 - e^{-rt/100}) + Q_0 e^{-rt/100}. \tag{7}$$

From Eq. (6) or (7), you can see that $Q(t) \to 25$ (lb) as $t \to \infty$, so the limiting value Q_L is 25, confirming our physical intuition. Further, $Q(t)$ approaches the limit more rapidly as r increases. In interpreting the solution (7), note that the second term on the right side is the portion of the original salt that remains at time t, while the first term gives the amount of salt in the tank due to the action of the flow processes. Plots of the solution for $r = 3$ and for several values of Q_0 are shown in Figure 2.3.2.

Now suppose that $r = 3$ and $Q_0 = 2Q_L = 50$; then Eq. (6) becomes

$$Q(t) = 25 + 25e^{-0.03t}. \tag{8}$$

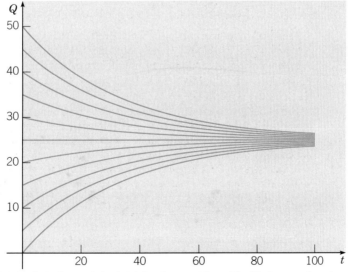

FIGURE 2.3.2 Solutions of the initial value problem (2), (3) for $r = 3$ and several values of Q_0.

Since 2% of 25 is 0.5, we wish to find the time T at which $Q(t)$ has the value 25.5. Substituting $t = T$ and $Q = 25.5$ in Eq. (8) and solving for T, we obtain

$$T = (\ln 50)/0.03 \cong 130.4 \text{ (min)}. \tag{9}$$

To determine r so that $T = 45$, return to Eq. (6), set $t = 45$, $Q_0 = 50$, $Q(t) = 25.5$, and solve for r. The result is

$$r = (100/45) \ln 50 \cong 8.69 \text{ gal/min}. \tag{10}$$

Since this example is hypothetical, the validity of the model is not in question. If the flow rates are as stated, and if the concentration of salt in the tank is uniform, then the differential equation (1) is an accurate description of the flow process. Although this particular example has no special significance, models of this kind are often used in problems involving a pollutant in a lake, or a drug in an organ of the body, for example, rather than a tank of salt water. In such cases the flow rates may not be easy to determine or may vary with time. Similarly, the concentration may be far from uniform in some cases. Finally, the rates of inflow and outflow may be different, which means that the variation of the amount of liquid in the problem must also be taken into account.

EXAMPLE

2

Compound
Interest

Suppose that a sum of money is deposited in a bank or money fund that pays interest at an annual rate r. The value $S(t)$ of the investment at any time t depends on the frequency with which interest is compounded as well as on the interest rate. Financial institutions have various policies concerning compounding: some compound monthly, some weekly, some even daily. If we assume that compounding takes place *continuously*, then we can set up a simple initial value problem that describes the growth of the investment.

The rate of change of the value of the investment is dS/dt, and this quantity is equal to the rate at which interest accrues, which is the interest rate r times the current value of the investment $S(t)$. Thus

$$dS/dt = rS \tag{11}$$

is the differential equation that governs the process. Suppose that we also know the value of the investment at some particular time, say,

$$S(0) = S_0. \tag{12}$$

Then the solution of the initial value problem (11), (12) gives the balance $S(t)$ in the account at any time t. This initial value problem is readily solved, since the differential equation (11) is both linear and separable. Consequently, by solving Eqs. (11) and (12), we find that

$$S(t) = S_0 e^{rt}. \tag{13}$$

Thus a bank account with continuously compounding interest grows exponentially.

Let us now compare the results from this continuous model with the situation in which compounding occurs at finite time intervals. If interest is compounded once a year, then after t years

$$S(t) = S_0(1 + r)^t.$$

If interest is compounded twice a year, then at the end of 6 months the value of the investment is $S_0[1 + (r/2)]$, and at the end of 1 year it is $S_0[1 + (r/2)]^2$. Thus, after t years we have

$$S(t) = S_0 \left(1 + \frac{r}{2}\right)^{2t}.$$

In general, if interest is compounded m times per year, then

$$S(t) = S_0 \left(1 + \frac{r}{m}\right)^{mt}. \tag{14}$$

The relation between formulas (13) and (14) is clarified if we recall from calculus that

$$\lim_{m\to\infty} S_0 \left(1 + \frac{r}{m}\right)^{mt} = S_0 e^{rt}.$$

The same model applies equally well to more general investments in which dividends and perhaps capital gains can also accumulate, as well as interest. In recognition of this fact, we will from now on refer to r as the rate of return.

Table 2.3.1 shows the effect of changing the frequency of compounding for a return rate r of 8%. The second and third columns are calculated from Eq. (14) for quarterly and daily compounding, respectively, and the fourth column is calculated from Eq. (13) for continuous compounding. The results show that the frequency of compounding is not particularly important in most cases. For example, during a 10-year period the difference between quarterly and continuous compounding is $17.50 per $1000 invested, or less than $2/year. The difference would be somewhat greater for higher rates of return and less for lower rates. From the first row in the table, we see that for the return rate $r = 8\%$, the annual yield for quarterly compounding is 8.24% and for daily or continuous compounding it is 8.33%.

TABLE 2.3.1 Growth of Capital at a Return Rate $r = 8\%$ for Several Modes of Compounding

	$S(t)/S(t_0)$ from Eq. (14)		$S(t)/S(t_0)$
Years	$m = 4$	$m = 365$	from Eq. (13)
1	1.0824	1.0833	1.0833
2	1.1717	1.1735	1.1735
5	1.4859	1.4918	1.4918
10	2.2080	2.2253	2.2255
20	4.8754	4.9522	4.9530
30	10.7652	11.0203	11.0232
40	23.7699	24.5239	24.5325

Returning now to the case of continuous compounding, let us suppose that there may be deposits or withdrawals in addition to the accrual of interest, dividends, or capital gains. If we assume that the deposits or withdrawals take place at a constant rate k, then Eq. (11) is replaced by

$$dS/dt = rS + k,$$

or, in standard form,

$$dS/dt - rS = k, \tag{15}$$

where k is positive for deposits and negative for withdrawals.

Equation (15) is linear with the integrating factor e^{-rt}, so its general solution is

$$S(t) = ce^{rt} - (k/r),$$

where c is an arbitrary constant. To satisfy the initial condition (12) we must choose $c = S_0 + (k/r)$. Thus the solution of the initial value problem (15), (12) is

$$S(t) = S_0 e^{rt} + (k/r)(e^{rt} - 1). \tag{16}$$

The first term in expression (16) is the part of $S(t)$ that is due to the return accumulated on the initial amount S_0, and the second term is the part that is due to the deposit or withdrawal rate k.

The advantage of stating the problem in this general way without specific values for S_0, r, or k lies in the generality of the resulting formula (16) for $S(t)$. With this formula we can readily compare the results of different investment programs or different rates of return.

For instance, suppose that one opens an individual retirement account (IRA) at age 25 and makes annual investments of \$2000 thereafter in a continuous manner. Assuming a rate of return of 8%, what will be the balance in the IRA at age 65? We have $S_0 = 0$, $r = 0.08$, and $k = \$2000$, and we wish to determine $S(40)$. From Eq. (16) we have

$$S(40) = (25{,}000)(e^{3.2} - 1) = \$588{,}313. \tag{17}$$

It is interesting to note that the total amount invested is \$80,000, so the remaining amount of \$508,313 results from the accumulated return on the investment. The balance after 40 years is also fairly sensitive to the assumed rate. For instance, $S(40) = \$508{,}948$ if $r = 0.075$ and $S(40) = \$681{,}508$ if $r = 0.085$.

Let us now examine the assumptions that have gone into the model. First, we have assumed that the return is compounded continuously and that additional capital is invested continuously. Neither of these is true in an actual financial situation. We have also assumed that the return rate r is constant for the entire period involved, whereas in fact it is likely to fluctuate considerably. Although we cannot reliably predict future rates, we can use expression (16) to determine the approximate effect of different rate projections. It is also possible to consider r and k in Eq. (15) to be functions of t rather than constants; in that case, of course, the solution may be much more complicated than Eq. (16).

The initial value problem (15), (12) and the solution (16) can also be used to analyze a number of other financial situations, including annuities, mortgages, and automobile loans.

EXAMPLE

3

Chemicals in
a Pond

Consider a pond that initially contains 10 million gal of fresh water. Water containing an undesirable chemical flows into the pond at the rate of 5 million gal/yr, and the mixture in the pond flows out at the same rate. The concentration $\gamma(t)$ of chemical in the incoming water varies periodically with time according to the expression $\gamma(t) = 2 + \sin 2t$ g/gal. Construct a mathematical model of this flow process and determine the amount of chemical in the pond at any time. Plot the solution and describe in words the effect of the variation in the incoming concentration.

Since the incoming and outgoing flows of water are the same, the amount of water in the pond remains constant at 10^7 gal. Let us denote time by t, measured in years, and the chemical by $Q(t)$, measured in grams. This example is similar to Example 1, and the same inflow/outflow principle applies. Thus

$$\frac{dQ}{dt} = \text{rate in} - \text{rate out},$$

where "rate in" and "rate out" refer to the rates at which the chemical flows into and out of the pond, respectively. The rate at which the chemical flows in is given by

$$\text{rate in} = (5 \times 10^6) \text{ gal/yr } (2 + \sin 2t) \text{ g/gal}. \tag{18}$$

The concentration of chemical in the pond is $Q(t)/10^7$ g/gal, so the rate of flow out is

$$\text{rate out} = (5 \times 10^6) \text{ gal/yr } [Q(t)/10^7] \text{ g/gal} = Q(t)/2 \text{ g/yr}. \tag{19}$$

Thus we obtain the differential equation

$$\frac{dQ}{dt} = (5 \times 10^6)(2 + \sin 2t) - \frac{Q(t)}{2}, \tag{20}$$

where each term has the units of g/yr.

To make the coefficients more manageable, it is convenient to introduce a new dependent variable defined by $q(t) = Q(t)/10^6$ or $Q(t) = 10^6 q(t)$. This means that $q(t)$ is measured in millions of grams, or megagrams. If we make this substitution in Eq. (20), then each term contains the factor 10^6, which can be canceled. If we also transpose the term involving $q(t)$ to the left side of the equation, we finally have

$$\frac{dq}{dt} + \tfrac{1}{2}q = 10 + 5\sin 2t. \tag{21}$$

Originally, there is no chemical in the pond, so the initial condition is

$$q(0) = 0. \tag{22}$$

Equation (21) is linear, and although the right side is a function of time, the coefficient of q is a constant. Thus the integrating factor is $e^{t/2}$. Multiplying Eq. (21) by this factor and integrating the resulting equation, we obtain the general solution

$$q(t) = 20 - \tfrac{40}{17}\cos 2t + \tfrac{10}{17}\sin 2t + ce^{-t/2}. \tag{23}$$

The initial condition (22) requires that $c = -300/17$, so the solution of the initial value problem (21), (22) is

$$q(t) = 20 - \tfrac{40}{17}\cos 2t + \tfrac{10}{17}\sin 2t - \tfrac{300}{17}e^{-t/2}. \tag{24}$$

A plot of the solution (24) is shown in Figure 2.3.3, along with the line $q = 20$. The exponential term in the solution is important for small t, but it diminishes rapidly as t increases. Later, the solution consists of an oscillation, due to the $\sin 2t$ and $\cos 2t$ terms, about the constant level $q = 20$. Note that if the $\sin 2t$ term were not present in Eq. (21), then $q = 20$ would be the equilibrium solution of that equation.

FIGURE 2.3.3 Solution of the initial value problem (21), (22).

Let us now consider the adequacy of the mathematical model itself for this problem. The model rests on several assumptions that have not yet been stated explicitly. In the first place, the amount of water in the pond is controlled entirely by the rates of flow in and out—none is lost by evaporation or by seepage into the ground, and none is gained by rainfall. The same is true of the chemical; it flows into and out of the pond, but none is absorbed by fish or other organisms living in the pond. In addition, we assume that the concentration of chemical in the pond is uniform throughout the pond. Whether the results obtained from the model are accurate depends strongly on the validity of these simplifying assumptions.

EXAMPLE
4

Escape
Velocity

A body of constant mass m is projected away from the earth in a direction perpendicular to the earth's surface with an initial velocity v_0. Assuming that there is no air resistance, but taking into account the variation of the earth's gravitational field with distance, find an expression for the velocity during the ensuing motion. Also find the initial velocity that is required to lift the body to a given maximum altitude ξ above the surface of the earth, and find the least initial velocity for which the body will not return to the earth; the latter is the **escape velocity**.

Let the positive x-axis point away from the center of the earth along the line of motion with $x = 0$ lying on the earth's surface; see Figure 2.3.4. The figure is drawn horizontally to remind you that gravity is directed toward the center of the earth, which is not necessarily downward from a perspective away from the earth's surface. The gravitational force acting on the body (that is, its weight) is inversely proportional to the square of the distance from the center of the earth and is given by $w(x) = -k/(x + R)^2$, where k is a constant, R is the radius of the earth, and the minus sign signifies that $w(x)$ is directed in the negative x direction. We know that on the earth's surface $w(0)$ is given by $-mg$, where g is the acceleration due to gravity at sea level. Therefore $k = mgR^2$ and

$$w(x) = -\frac{mgR^2}{(R + x)^2}. \tag{25}$$

Since there are no other forces acting on the body, the equation of motion is

$$m\frac{dv}{dt} = -\frac{mgR^2}{(R + x)^2}, \tag{26}$$

and the initial condition is

$$v(0) = v_0. \tag{27}$$

FIGURE 2.3.4 A body in the earth's gravitational field.

Unfortunately, Eq. (26) involves too many variables since it depends on t, x, and v. To remedy this situation, we can eliminate t from Eq. (26) by thinking of x, rather than t, as the independent variable. Then we must express dv/dt in terms of dv/dx by the chain rule; hence

$$\frac{dv}{dt} = \frac{dv}{dx}\frac{dx}{dt} = v\frac{dv}{dx},$$

and Eq. (26) is replaced by

$$v\frac{dv}{dx} = -\frac{gR^2}{(R + x)^2}. \tag{28}$$

Equation (28) is separable but not linear, so by separating the variables and integrating, we obtain

$$\frac{v^2}{2} = \frac{gR^2}{R + x} + c. \tag{29}$$

Since $x = 0$ when $t = 0$, the initial condition (27) at $t = 0$ can be replaced by the condition that $v = v_0$ when $x = 0$. Hence $c = (v_0^2/2) - gR$ and

$$v = \pm\sqrt{v_0^2 - 2gR + \frac{2gR^2}{R + x}}. \tag{30}$$

Note that Eq. (30) gives the velocity as a function of altitude rather than as a function of time. The plus sign must be chosen if the body is rising, and the minus sign if it is falling back to earth.

To determine the maximum altitude ξ that the body reaches, we set $v = 0$ and $x = \xi$ in Eq. (30) and then solve for ξ, obtaining

$$\xi = \frac{v_0^2 R}{2gR - v_0^2}. \tag{31}$$

Solving Eq. (31) for v_0, we find the initial velocity required to lift the body to the altitude ξ, namely,

$$v_0 = \sqrt{2gR\frac{\xi}{R + \xi}}. \tag{32}$$

The escape velocity v_e is then found by letting $\xi \to \infty$. Consequently,

$$v_e = \sqrt{2gR}. \tag{33}$$

The numerical value of v_e is approximately 6.9 miles/sec, or 11.1 km/sec.

The preceding calculation of the escape velocity neglects the effect of air resistance, so the actual escape velocity (including the effect of air resistance) is somewhat higher. On the other hand, the effective escape velocity can be significantly reduced if the body is transported a considerable distance above sea level before being launched. Both gravitational and frictional forces are thereby reduced; air resistance, in particular, diminishes quite rapidly with increasing altitude. You should keep in mind also that it may well be impractical to impart too large an initial velocity instantaneously; space vehicles, for instance, receive their initial acceleration during a period of a few minutes.

PROBLEMS

1. Consider a tank used in certain hydrodynamic experiments. After one experiment the tank contains 200 liters of a dye solution with a concentration of 1 g/liter. To prepare for the next experiment, the tank is to be rinsed with fresh water flowing in at a rate of 2 liters/min, the well-stirred solution flowing out at the same rate. Find the time that will elapse before the concentration of dye in the tank reaches 1% of its original value.

2. A tank initially contains 120 liters of pure water. A mixture containing a concentration of γ g/liter of salt enters the tank at a rate of 2 liters/min, and the well-stirred mixture leaves the tank at the same rate. Find an expression in terms of γ for the amount of salt in the tank at any time t. Also find the limiting amount of salt in the tank as $t \to \infty$.

3. A tank originally contains 100 gal of fresh water. Then water containing $\frac{1}{2}$ lb of salt per gallon is poured into the tank at a rate of 2 gal/min, and the mixture is allowed to leave at the same rate. After 10 min the process is stopped, and fresh water is poured into the tank at a rate of 2 gal/min, with the mixture again leaving at the same rate. Find the amount of salt in the tank at the end of an additional 10 min.

4. A tank with a capacity of 500 gal originally contains 200 gal of water with 100 lb of salt in solution. Water containing 1 lb of salt per gallon is entering at a rate of 3 gal/min, and the mixture is allowed to flow out of the tank at a rate of 2 gal/min. Find the amount of salt in the tank at any time prior to the instant when the solution begins to overflow. Find the concentration (in pounds per gallon) of salt in the tank when it is on the point of overflowing. Compare this concentration with the theoretical limiting concentration if the tank had infinite capacity.

5. A tank contains 100 gallons of water and 50 oz of salt. Water containing a salt concentration of $\frac{1}{4}(1 + \frac{1}{2}\sin t)$ oz/gal flows into the tank at a rate of 2 gal/min, and the mixture in the tank flows out at the same rate.

(a) Find the amount of salt in the tank at any time.

(b) Plot the solution for a time period long enough so that you see the ultimate behavior of the graph.

(c) The long-time behavior of the solution is an oscillation about a certain constant level. What is this level? What is the amplitude of the oscillation?

6. Suppose that a tank containing a certain liquid has an outlet near the bottom. Let $h(t)$ be the height of the liquid surface above the outlet at time t. Torricelli's[2] principle states that the outflow velocity v at the outlet is equal to the velocity of a particle falling freely (with no drag) from the height h.

(a) Show that $v = \sqrt{2gh}$, where g is the acceleration due to gravity.

(b) By equating the rate of outflow to the rate of change of liquid in the tank, show that $h(t)$ satisfies the equation

$$A(h)\frac{dh}{dt} = -\alpha a\sqrt{2gh}, \qquad\qquad (i)$$

where $A(h)$ is the area of the cross section of the tank at height h and a is the area of the outlet. The constant α is a contraction coefficient that accounts for the observed fact that the cross section of the (smooth) outflow stream is smaller than a. The value of α for water is about 0.6.

(c) Consider a water tank in the form of a right circular cylinder that is 3 m high above the outlet. The radius of the tank is 1 m and the radius of the circular outlet is 0.1 m. If the tank is initially full of water, determine how long it takes to drain the tank down to the level of the outlet.

7. Suppose that a sum S_0 is invested at an annual rate of return r compounded continuously.

(a) Find the time T required for the original sum to double in value as a function of r.

(b) Determine T if $r = 7\%$.

(c) Find the return rate that must be achieved if the initial investment is to double in 8 years.

8. A young person with no initial capital invests k dollars per year at an annual rate of return r. Assume that investments are made continuously and that the return is compounded continuously.

(a) Determine the sum $S(t)$ accumulated at any time t.

(b) If $r = 7.5\%$, determine k so that $1 million will be available for retirement in 40 years.

[2]Evangelista Torricelli (1608–1647), successor to Galileo as court mathematician in Florence, published this result in 1644. He is also known for constructing the first mercury barometer and for making important contributions to geometry.

(c) If $k = \$2000$/year, determine the return rate r that must be obtained to have $1 million available in 40 years.

9. A certain college graduate borrows $8000 to buy a car. The lender charges interest at an annual rate of 10%. Assuming that interest is compounded continuously and that the borrower makes payments continuously at a constant annual rate k, determine the payment rate k that is required to pay off the loan in 3 years. Also determine how much interest is paid during the 3-year period.

10. A home buyer can afford to spend no more than $800/month on mortgage payments. Suppose that the interest rate is 9% and that the term of the mortgage is 20 years. Assume that interest is compounded continuously and that payments are also made continuously.

(a) Determine the maximum amount that this buyer can afford to borrow.

(b) Determine the total interest paid during the term of the mortgage.

11. A recent college graduate borrows $100,000 at an interest rate of 9% to purchase a condominium. Anticipating steady salary increases, the buyer expects to make payments at a monthly rate of $800(1 + t/120)$, where t is the number of months since the loan was made.

(a) Assuming that this payment schedule can be maintained, when will the loan be fully paid?

(b) Assuming the same payment schedule, how large a loan could be paid off in exactly 20 years?

12. An important tool in archeological research is radiocarbon dating, developed by the American chemist Willard F. Libby.[3] This is a means of determining the age of certain wood and plant remains, hence of animal or human bones or artifacts found buried at the same levels. Radiocarbon dating is based on the fact that some wood or plant remains contain residual amounts of carbon-14, a radioactive isotope of carbon. This isotope is accumulated during the lifetime of the plant and begins to decay at its death. Since the half-life of carbon-14 is long (approximately 5730 years[4]), measurable amounts of carbon-14 remain after many thousands of years. If even a tiny fraction of the original amount of carbon-14 is still present, then by appropriate laboratory measurements the *proportion* of the original amount of carbon-14 that remains can be accurately determined. In other words, if $Q(t)$ is the amount of carbon-14 at time t and Q_0 is the original amount, then the ratio $Q(t)/Q_0$ can be determined, at least if this quantity is not too small. Present measurement techniques permit the use of this method for time periods of 50,000 years or more.

(a) Assuming that Q satisfies the differential equation $Q' = -rQ$, determine the decay constant r for carbon-14.

(b) Find an expression for $Q(t)$ at any time t, if $Q(0) = Q_0$.

(c) Suppose that certain remains are discovered in which the current residual amount of carbon-14 is 20% of the original amount. Determine the age of these remains.

13. The population of mosquitoes in a certain area increases at a rate proportional to the current population, and in the absence of other factors, the population doubles each week. There are 200,000 mosquitoes in the area initially, and predators (birds, bats, and

[3]Willard F. Libby (1908–1980) was born in rural Colorado and received his education at the University of California at Berkeley. He developed the method of radiocarbon dating beginning in 1947 while he was at the University of Chicago. For this work he was awarded the Nobel Prize in chemistry in 1960.

[4]*McGraw-Hill Encyclopedia of Science and Technology* (8th ed.) (New York: McGraw-Hill, 1997), Vol. 5, p. 48.

so forth) eat 20,000 mosquitoes/day. Determine the population of mosquitoes in the area at any time.

 14. Suppose that a certain population has a growth rate that varies with time and that this population satisfies the differential equation

$$dy/dt = (0.5 + \sin t)y/5.$$

(a) If $y(0) = 1$, find (or estimate) the time τ at which the population has doubled. Choose other initial conditions and determine whether the doubling time τ depends on the initial population.

(b) Suppose that the growth rate is replaced by its average value $1/10$. Determine the doubling time τ in this case.

(c) Suppose that the term $\sin t$ in the differential equation is replaced by $\sin 2\pi t$; that is, the variation in the growth rate has a substantially higher frequency. What effect does this have on the doubling time τ?

(d) Plot the solutions obtained in parts (a), (b), and (c) on a single set of axes.

 15. Suppose that a certain population satisfies the initial value problem

$$dy/dt = r(t)y - k, \qquad y(0) = y_0,$$

where the growth rate $r(t)$ is given by $r(t) = (1 + \sin t)/5$, and k represents the rate of predation.

(a) Suppose that $k = 1/5$. Plot y versus t for several values of y_0 between $1/2$ and 1.

(b) Estimate the critical initial population y_c below which the population will become extinct.

(c) Choose other values of k and find the corresponding y_c for each one.

(d) Use the data you have found in parts (b) and (c) to plot y_c versus k.

16. Newton's law of cooling states that the temperature of an object changes at a rate proportional to the difference between its temperature and that of its surroundings. Suppose that the temperature of a cup of coffee obeys Newton's law of cooling. If the coffee has a temperature of $200°$F when freshly poured, and 1 min later has cooled to $190°$F in a room at $70°$F, determine when the coffee reaches a temperature of $150°$F.

 17. Heat transfer from a body to its surroundings by radiation, based on the Stefan-Boltzmann[5] law, is described by the differential equation

$$\frac{du}{dt} = -\alpha(u^4 - T^4), \tag{i}$$

where $u(t)$ is the absolute temperature of the body at time t, T is the absolute temperature of the surroundings, and α is a constant depending on the physical parameters of the body. However, if u is much larger than T, then solutions of Eq. (i) are well approximated by solutions of the simpler equation

$$\frac{du}{dt} = -\alpha u^4. \tag{ii}$$

[5] Jozef Stefan (1835–1893), professor of physics at Vienna, stated the radiation law on empirical grounds in 1879. His student Ludwig Boltzmann (1844–1906) derived it theoretically from the principles of thermodynamics in 1884. Boltzmann is best known for his pioneering work in statistical mechanics.

Suppose that a body with initial temperature $2000°$K is surrounded by a medium with temperature $300°$K and that $\alpha = 2.0 \times 10^{-12}$ $°$K^{-3}/sec.

(a) Determine the temperature of the body at any time by solving Eq. (ii).

(b) Plot the graph of u versus t.

(c) Find the time τ at which $u(\tau) = 600$, that is, twice the ambient temperature. Up to this time the error in using Eq. (ii) to approximate the solutions of Eq. (i) is no more than 1%.

18. Consider an insulated box (a building, perhaps) with internal temperature $u(t)$. According to Newton's law of cooling, u satisfies the differential equation

$$\frac{du}{dt} = -k[u - T(t)]. \tag{i}$$

where $T(t)$ is the ambient (external) temperature. Suppose that $T(t)$ varies sinusoidally; for example, assume that $T(t) = T_0 + T_1 \cos \omega t$.

(a) Solve Eq. (i) and express $u(t)$ in terms of t, k, T_0, T_1, and ω. Observe that part of your solution approaches zero as t becomes large; this is called the transient part. The remainder of the solution is called the steady state; denote it by $S(t)$.

(b) Suppose that t is measured in hours and that $\omega = \pi/12$, corresponding a period of 24 hours for $T(t)$. Further, let $T_0 = 60°$F, $T_1 = 15°$F, and $k = 0.2$/hr. Draw graphs of $S(t)$ and $T(t)$ versus t on the same axes. From your graph estimate the amplitude R of the oscillatory part of $S(t)$. Also estimate the time lag τ between corresponding maxima of $T(t)$ and $S(t)$.

(c) Let k, T_0, T_1, and ω now be unspecified. Write the oscillatory part of $S(t)$ in the form $R \cos[\omega(t - \tau)]$. Use trigonometric identities to find expressions for R and τ. Let T_1 and ω have the values given in part (b), and plot graphs of R and τ versus k.

19. Consider a lake of constant volume V containing at time t an amount $Q(t)$ of pollutant, evenly distributed throughout the lake with a concentration $c(t)$, where $c(t) = Q(t)/V$. Assume that water containing a concentration k of pollutant enters the lake at a rate r, and that water leaves the lake at the same rate. Suppose that pollutants are also added directly to the lake at a constant rate P. Note that the given assumptions neglect a number of factors that may, in some cases, be important—for example, the water added or lost by precipitation, absorption, and evaporation; the stratifying effect of temperature differences in a deep lake; the tendency of irregularities in the coastline to produce sheltered bays; and the fact that pollutants are not deposited evenly throughout the lake but (usually) at isolated points around its periphery. The results below must be interpreted in the light of the neglect of such factors as these.

(a) If at time $t = 0$ the concentration of pollutant is c_0, find an expression for the concentration $c(t)$ at any time. What is the limiting concentration as $t \to \infty$?

(b) If the addition of pollutants to the lake is terminated ($k = 0$ and $P = 0$ for $t > 0$), determine the time interval T that must elapse before the concentration of pollutants is reduced to 50% of its original value; to 10% of its original value.

(c) Table 2.3.2 on page 64 contains data[6] for several of the Great Lakes. Using these data, determine from part (b) the time T necessary to reduce the contamination of each of these lakes to 10% of the original value.

[6]This problem is based on R. H. Rainey, "Natural Displacement of Pollution from the Great Lakes," *Science 155* (1967), pp. 1242–1243; the information in the table was taken from that source.

TABLE 2.3.2 Volume and Flow Data for the Great Lakes

Lake	V (km$^3 \times 10^3$)	r (km^3/year)
Superior	12.2	65.2
Michigan	4.9	158
Erie	0.46	175
Ontario	1.6	209

 20. A ball with mass 0.15 kg is thrown upward with initial velocity 20 m/sec from the roof of a building 30 m high. Neglect air resistance.

(a) Find the maximum height above the ground that the ball reaches.

(b) Assuming that the ball misses the building on the way down, find the time that it hits the ground.

(c) Plot the graphs of velocity and position versus time.

 21. Assume that the conditions are as in Problem 20 except that there is a force due to air resistance of $|v|/30$, where the velocity v is measured in m/sec.

(a) Find the maximum height above the ground that the ball reaches.

(b) Find the time that the ball hits the ground.

(c) Plot the graphs of velocity and position versus time. Compare these graphs with the corresponding ones in Problem 20.

 22. Assume that the conditions are as in Problem 20 except that there is a force due to air resistance of $v^2/1325$, where the velocity v is measured in m/sec.

(a) Find the maximum height above the ground that the ball reaches.

(b) Find the time that the ball hits the ground.

(c) Plot the graphs of velocity and position versus time. Compare these graphs with the corresponding ones in Problems 20 and 21.

23. A sky diver weighing 180 lb (including equipment) falls vertically downward from an altitude of 5000 ft and opens the parachute after 10 sec of free fall. Assume that the force of air resistance is $0.75|v|$ when the parachute is closed and $12|v|$ when the parachute is open, where the velocity v is measured in ft/sec.

(a) Find the speed of the sky diver when the parachute opens.

(b) Find the distance fallen before the parachute opens.

(c) What is the limiting velocity v_L after the parachute opens?

(d) Determine how long the sky diver is in the air after the parachute opens.

(e) Plot the graph of velocity versus time from the beginning of the fall until the skydiver reaches the ground.

24. A rocket sled having an initial speed of 150 mi/hr is slowed by a channel of water. Assume that, during the braking process, the acceleration a is given by $a(v) = -\mu v^2$, where v is the velocity and μ is a constant.

(a) As in Example 4 in the text, use the relation $dv/dt = v(dv/dx)$ to write the equation of motion in terms of v and x.

(b) If it requires a distance of 2000 ft to slow the sled to 15 mi/hr, determine the value of μ.

(c) Find the time τ required to slow the sled to 15 mi/hr.

25. A body of constant mass m is projected vertically upward with an initial velocity v_0 in a medium offering a resistance $k|v|$, where k is a constant. Neglect changes in the gravitational force.

 (a) Find the maximum height x_m attained by the body and the time t_m at which this maximum height is reached.

 (b) Show that if $kv_0/mg < 1$, then t_m and x_m can be expressed as

$$t_m = \frac{v_0}{g}\left[1 - \frac{1}{2}\frac{kv_0}{mg} + \frac{1}{3}\left(\frac{kv_0}{mg}\right)^2 - \cdots\right],$$

$$x_m = \frac{v_0^2}{2g}\left[1 - \frac{2}{3}\frac{kv_0}{mg} + \frac{1}{2}\left(\frac{kv_0}{mg}\right)^2 - \cdots\right].$$

 (c) Show that the quantity kv_0/mg is dimensionless.

26. A body of mass m is projected vertically upward with an initial velocity v_0 in a medium offering a resistance $k|v|$, where k is a constant. Assume that the gravitational attraction of the earth is constant.

 (a) Find the velocity $v(t)$ of the body at any time.

 (b) Use the result of part (a) to calculate the limit of $v(t)$ as $k \to 0$, that is, as the resistance approaches zero. Does this result agree with the velocity of a mass m projected upward with an initial velocity v_0 in a vacuum?

 (c) Use the result of part (a) to calculate the limit of $v(t)$ as $m \to 0$, that is, as the mass approaches zero.

27. A body falling in a relatively dense fluid, oil for example, is acted on by three forces (see Figure 2.3.5): a resistive force R, a buoyant force B, and its weight w due to gravity. The buoyant force is equal to the weight of the fluid displaced by the object. For a slowly moving spherical body of radius a, the resistive force is given by Stokes'[7] law, $R = 6\pi\mu a|v|$, where v is the velocity of the body, and μ is the coefficient of viscosity of the surrounding fluid.

FIGURE 2.3.5 A body falling in a dense fluid.

[7]George Gabriel Stokes (1819–1903), professor at Cambridge, was one of the foremost applied mathematicians of the nineteenth century. The basic equations of fluid mechanics (the Navier–Stokes equations) are named partly in his honor, and one of the fundamental theorems of vector calculus bears his name. He was also one of the pioneers in the use of divergent (asymptotic) series, a subject of great interest and importance today.

(a) Find the limiting velocity of a solid sphere of radius a and density ρ falling freely in a medium of density ρ' and coefficient of viscosity μ.

(b) In 1910 R. A. Millikan[8] studied the motion of tiny droplets of oil falling in an electric field. A field of strength E exerts a force Ee on a droplet with charge e. Assume that E has been adjusted so the droplet is held stationary ($v = 0$) and that w and B are as given above. Find an expression for e. Millikan repeated this experiment many times, and from the data that he gathered he was able to deduce the charge on an electron.

28. A mass of 0.25 kg is dropped from rest in a medium offering a resistance of $0.2|v|$, where v is measured in m/sec.

(a) If the mass is dropped from a height of 30 m, find its velocity when it hits the ground.

(b) If the mass is to attain a velocity of no more than 10 m/sec, find the maximum height from which it can be dropped.

(c) Suppose that the resistive force is $k|v|$, where v is measured in m/sec and k is a constant. If the mass is dropped from a height of 30 m and must hit the ground with a velocity of no more than 10 m/sec, determine the coefficient of resistance k that is required.

29. Suppose that a rocket is launched straight up from the surface of the earth with initial velocity $v_0 = \sqrt{2gR}$, where R is the radius of the earth. Neglect air resistance.

(a) Find an expression for the velocity v in terms of the distance x from the surface of the earth.

(b) Find the time required for the rocket to go 240,000 miles (the approximate distance from the earth to the moon). Assume that $R = 4000$ miles.

30. Let $v(t)$ and $w(t)$, respectively, be the horizontal and vertical components of the velocity of a batted (or thrown) baseball. In the absence of air resistance, v and w satisfy the equations

$$dv/dt = 0, \qquad dw/dt = -g.$$

(a) Show that

$$v = u\cos A, \qquad w = -gt + u\sin A,$$

where u is the initial speed of the ball and A is its initial angle of elevation.

(b) Let $x(t)$ and $y(t)$, respectively, be the horizontal and vertical coordinates of the ball at time t. If $x(0) = 0$ and $y(0) = h$, find $x(t)$ and $y(t)$ at any time t.

(c) Let $g = 32$ ft/sec^2, $u = 125$ ft/sec, and $h = 3$ ft. Plot the trajectory of the ball for several values of the angle A; that is, plot $x(t)$ and $y(t)$ parametrically.

(d) Suppose the outfield wall is at a distance L and has height H. Find a relation between u and A that must be satisfied if the ball is to clear the wall.

(e) Suppose that $L = 350$ ft and $H = 10$ ft. Using the relation in part (d), find (or estimate from a plot) the range of values of A that correspond to an initial velocity of $u = 110$ ft/sec.

(f) For $L = 350$ and $H = 10$, find the minimum initial velocity u and the corresponding optimal angle A for which the ball will clear the wall.

31. A more realistic model (than that in Problem 30) of a baseball in flight includes the effect of air resistance. In this case the equations of motion are

$$dv/dt = -rv, \qquad dw/dt = -g - rw,$$

where r is the coefficient of resistance.

[8]Robert A. Millikan (1868–1953) was educated at Oberlin College and Columbia University. Later he was a professor at the University of Chicago and California Institute of Technology. His determination of the charge on an electron was published in 1910. For this work, and for other studies of the photoelectric effect, he was awarded the Nobel Prize in 1923.

(a) Determine $v(t)$ and $w(t)$ in terms of initial speed u and initial angle of elevation A.

(b) Find $x(t)$ and $y(t)$ if $x(0) = 0$ and $y(0) = h$.

(c) Plot the trajectory of the ball for $r = 1/5, u = 125, h = 3$, and for several values of A. How do the trajectories differ from those in Problem 31 with $r = 0$?

(d) Assuming that $r = 1/5$ and $h = 3$, find the minimum initial velocity u and the optimal angle A for which the ball will clear a wall that is 350 ft distant and 10 ft high. Compare this result with that in Problem 30(f).

32. **Brachistochrone Problem.** One of the famous problems in the history of mathematics is the brachistochrone[9] problem: to find the curve along which a particle will slide without friction in the minimum time from one given point P to another Q, the second point being lower than the first but not directly beneath it (see Figure 2.3.6). This problem was posed by Johann Bernoulli in 1696 as a challenge problem to the mathematicians of his day. Correct solutions were found by Johann Bernoulli and his brother Jakob Bernoulli and by Isaac Newton, Gottfried Leibniz, and the Marquis de L'Hospital. The brachistochrone problem is important in the development of mathematics as one of the forerunners of the calculus of variations.

In solving this problem it is convenient to take the origin as the upper point P and to orient the axes as shown in Figure 2.3.6. The lower point Q has coordinates (x_0, y_0). It is then possible to show that the curve of minimum time is given by a function $y = \phi(x)$ that satisfies the differential equation

$$(1 + y'^2)y = k^2, \tag{i}$$

where k^2 is a certain positive constant to be determined later.

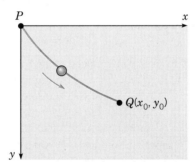

FIGURE 2.3.6 The brachistochrone.

(a) Solve Eq. (i) for y'. Why is it necessary to choose the positive square root?

(b) Introduce the new variable t by the relation

$$y = k^2 \sin^2 t. \tag{ii}$$

Show that the equation found in part (a) then takes the form

$$2k^2 \sin^2 t \, dt = dx. \tag{iii}$$

[9]The word "brachistochrone" comes from the Greek words *brachistos*, meaning shortest, and *chronos*, meaning time.

(c) Letting $\theta = 2t$, show that the solution of Eq. (iii) for which $x = 0$ when $y = 0$ is given by

$$x = k^2(\theta - \sin\theta)/2, \qquad y = k^2(1 - \cos\theta)/2. \tag{iv}$$

Equations (iv) are parametric equations of the solution of Eq. (i) that passes through $(0,0)$. The graph of Eqs. (iv) is called a **cycloid**.

(d) If we make a proper choice of the constant k, then the cycloid also passes through the point (x_0, y_0) and is the solution of the brachistochrone problem. Find k if $x_0 = 1$ and $y_0 = 2$.

2.4 Differences Between Linear and Nonlinear Equations

Up to now, we have been primarily concerned with showing that first order differential equations can be used to investigate many different kinds of problems in the natural sciences, and with presenting methods of solving such equations if they are either linear or separable. Now it is time to turn our attention to some more general questions about differential equations and to explore in more detail some important ways in which nonlinear equations differ from linear ones.

Existence and Uniqueness of Solutions. So far, we have discussed a number of initial value problems, each of which had a solution and apparently only one solution. This raises the question of whether this is true of all initial value problems for first order equations. In other words, does every initial value problem have exactly one solution? This may be an important question even for nonmathematicians. If you encounter an initial value problem in the course of investigating some physical problem, you might want to know that it has a solution before spending very much time and effort in trying to find it. Further, if you are successful in finding one solution, you might be interested in knowing whether you should continue a search for other possible solutions or whether you can be sure that there are no other solutions. For linear equations the answers to these questions are given by the following fundamental theorem.

Theorem 2.4.1 If the functions p and g are continuous on an open interval $I: \alpha < t < \beta$ containing the point $t = t_0$, then there exists a unique function $y = \phi(t)$ that satisfies the differential equation

$$y' + p(t)y = g(t) \tag{1}$$

for each t in I, and that also satisfies the initial condition

$$y(t_0) = y_0, \tag{2}$$

where y_0 is an arbitrary prescribed initial value.

Observe that Theorem 2.4.1 states that the given initial value problem *has* a solution and also that the problem has *only one* solution. In other words, the theorem asserts both the *existence* and *uniqueness* of the solution of the initial value problem (1),(2).

In addition, it states that the solution exists throughout any interval I containing the initial point t_0 in which the coefficients p and g are continuous. That is, the solution can be discontinuous or fail to exist only at points where at least one of p and g is discontinuous. Such points can often be identified at a glance.

The proof of this theorem is partly contained in the discussion in Section 2.1 leading to the formula [Eq. (32) in Section 2.1]

$$\mu(t)y = \int \mu(t)g(t)\,dt + c, \tag{3}$$

where [Eq. (30) in Section 2.1]

$$\mu(t) = \exp \int p(t)\,dt. \tag{4}$$

The derivation in Section 2.1 shows that if Eq. (1) has a solution, then it must be given by Eq. (3). By looking slightly more closely at that derivation, we can also conclude that the differential equation (1) must indeed have a solution. Since p is continuous for $\alpha < t < \beta$, it follows that μ is defined in this interval and is a nonzero differentiable function. Upon multiplying Eq. (1) by $\mu(t)$ we obtain

$$[\mu(t)y]' = \mu(t)g(t). \tag{5}$$

Since both μ and g are continuous, the function μg is integrable, and Eq. (3) follows from Eq. (5). Further, the integral of μg is differentiable, so y as given by Eq. (3) exists and is differentiable throughout the interval $\alpha < t < \beta$. By substituting the expression for y from Eq. (3) into either Eq. (1) or Eq. (5), you can easily verify that this expression satisfies the differential equation throughout the interval $\alpha < t < \beta$. Finally, the initial condition (2) determines the constant c uniquely, so there is only one solution of the initial value problem, thus completing the proof.

Equation (4) determines the integrating factor $\mu(t)$ only up to a multiplicative factor that depends on the lower limit of integration. If we choose this lower limit to be t_0, then

$$\mu(t) = \exp \int_{t_0}^{t} p(s)\,ds, \tag{6}$$

and it follows that $\mu(t_0) = 1$. Using the integrating factor given by Eq. (6), and choosing the lower limit of integration in Eq. (3) also to be t_0, we obtain the general solution of Eq. (1) in the form

$$y = \frac{1}{\mu(t)}\left[\int_{t_0}^{t} \mu(s)g(s)\,ds + c\right]. \tag{7}$$

To satisfy the initial condition (2) we must choose $c = y_0$. Thus the solution of the initial value problem (1), (2) is

$$y = \frac{1}{\mu(t)}\left[\int_{t_0}^{t} \mu(s)g(s)\,ds + y_0\right], \tag{8}$$

where $\mu(t)$ is given by Eq. (6).

Turning now to nonlinear differential equations, we must replace Theorem 2.4.1 by a more general theorem, such as the following.

Theorem 2.4.2 Let the functions f and $\partial f/\partial y$ be continuous in some rectangle $\alpha < t < \beta, \gamma < y < \delta$ containing the point (t_0, y_0). Then, in some interval $t_0 - h < t < t_0 + h$ contained in $\alpha < t < \beta$, there is a unique solution $y = \phi(t)$ of the initial value problem

$$y' = f(t, y), \qquad y(t_0) = y_0. \tag{9}$$

Observe that the hypotheses in Theorem 2.4.2 reduce to those in Theorem 2.4.1 if the differential equation is linear. For then $f(t, y) = -p(t)y + g(t)$ and $\partial f(t, y)/\partial y = -p(t)$, so the continuity of f and $\partial f/\partial y$ is equivalent to the continuity of p and g in this case. The proof of Theorem 2.4.1 was comparatively simple because it could be based on the expression (3) that gives the solution of an arbitrary linear equation. There is no corresponding expression for the solution of the differential equation (9), so the proof of Theorem 2.4.2 is much more difficult. It is discussed to some extent in Section 2.8 and in greater depth in more advanced books on differential equations.

Here we note that the conditions stated in Theorem 2.4.2 are sufficient to guarantee the existence of a unique solution of the initial value problem (9) in some interval $t_0 - h < t < t_0 + h$, but they are not necessary. That is, the conclusion remains true under slightly weaker hypotheses about the function f. In fact, the existence of a solution (but not its uniqueness) can be established on the basis of the continuity of f alone.

An important geometrical consequence of the uniqueness parts of Theorems 2.4.1 and 2.4.2 is that the graphs of two solutions cannot intersect each other. Otherwise, there would be two solutions that satisfy the initial condition corresponding to the point of intersection, in violation of Theorem 2.4.1 or 2.4.2.

We now consider some examples.

EXAMPLE 1

Use Theorem 2.4.1 to find an interval in which the initial value problem

$$ty' + 2y = 4t^2, \tag{10}$$

$$y(1) = 2 \tag{11}$$

has a unique solution.

Rewriting Eq. (10) in the standard form (1), we have

$$y' + (2/t)y = 4t,$$

so $p(t) = 2/t$ and $g(t) = 4t$. Thus, for this equation, g is continuous for all t, while p is continuous only for $t < 0$ or for $t > 0$. The interval $t > 0$ contains the initial point; consequently, Theorem 2.4.1 guarantees that the problem (10), (11) has a unique solution on the interval $0 < t < \infty$. In Example 3 of Section 2.1 we found the solution of this initial value problem to be

$$y = t^2 + \frac{1}{t^2}, \qquad t > 0. \tag{12}$$

Now suppose that the initial condition (11) is changed to $y(-1) = 2$. Then Theorem 2.4.1 asserts the existence of a unique solution for $t < 0$. As you can readily verify, the solution is again given by Eq. (12), but now on the interval $-\infty < t < 0$.

**EXAMPLE
2**

Apply Theorem 2.4.2 to the initial value problem

$$\frac{dy}{dx} = \frac{3x^2 + 4x + 2}{2(y-1)}, \qquad y(0) = -1. \tag{13}$$

Note that Theorem 2.4.1 is not applicable to this problem since the differential equation is nonlinear. To apply Theorem 2.4.2, observe that

$$f(x,y) = \frac{3x^2 + 4x + 2}{2(y-1)}, \qquad \frac{\partial f}{\partial y}(x,y) = -\frac{3x^2 + 4x + 2}{2(y-1)^2}.$$

Thus each of these functions is continuous everywhere except on the line $y = 1$. Consequently, a rectangle can be drawn about the initial point $(0, -1)$ in which both f and $\partial f/\partial y$ are continuous. Therefore Theorem 2.4.2 guarantees that the initial value problem has a unique solution in some interval about $x = 0$. However, even though the rectangle can be stretched infinitely far in both the positive and negative x directions, this does not necessarily mean that the solution exists for all x. Indeed, the initial value problem (13) was solved in Example 2 of Section 2.2 and the solution exists only for $x > -2$.

Now suppose we change the initial condition to $y(0) = 1$. The initial point now lies on the line $y = 1$ so no rectangle can be drawn about it within which f and $\partial f/\partial y$ are continuous. Consequently, Theorem 2.4.2 says nothing about possible solutions of this modified problem. However, if we separate the variables and integrate, as in Section 2.2, we find that

$$y^2 - 2y = x^3 + 2x^2 + 2x + c.$$

Further, if $x = 0$ and $y = 1$, then $c = -1$. Finally, by solving for y, we obtain

$$y = 1 \pm \sqrt{x^3 + 2x^2 + 2x}. \tag{14}$$

Equation (14) provides two functions that satisfy the given differential equation for $x > 0$ and also satisfy the initial condition $y(0) = 1$.

**EXAMPLE
3**

Consider the initial value problem

$$y' = y^{1/3}, \qquad y(0) = 0 \tag{15}$$

for $t \geq 0$. Apply Theorem 2.4.2 to this initial value problem and then solve the problem.

The function $f(t,y) = y^{1/3}$ is continuous everywhere, but $\partial f/\partial y$ does not exist when $y = 0$, and hence is not continuous there. Thus Theorem 2.4.2 does not apply to this problem and no conclusion can be drawn from it. However, by the remark following Theorem 2.4.2 the continuity of f does assure the existence of solutions, but not their uniqueness.

To understand the situation more clearly, we must actually solve the problem, which is easy to do since the differential equation is separable. Thus we have

$$y^{-1/3}dy = dt,$$

so

$$\tfrac{3}{2}y^{2/3} = t + c$$

and

$$y = \left[\tfrac{2}{3}(t+c)\right]^{3/2}.$$

The initial condition is satisfied if $c = 0$, so

$$y = \phi_1(t) = \left(\tfrac{2}{3}t\right)^{3/2}, \qquad t \geq 0 \tag{16}$$

satisfies both of Eqs. (15). On the other hand, the function

$$y = \phi_2(t) = -\left(\tfrac{2}{3}t\right)^{3/2}, \qquad t \geq 0 \tag{17}$$

is also a solution of the initial value problem. Moreover, the function

$$y = \psi(t) = 0, \qquad t \geq 0 \tag{18}$$

is yet another solution. Indeed, it is not hard to show that, for an arbitrary positive t_0, the functions

$$y = \chi(t) = \begin{cases} 0, & \text{if } 0 \leq t < t_0, \\ \pm\left[\tfrac{2}{3}(t - t_0)\right]^{3/2}, & \text{if } t \geq t_0 \end{cases} \tag{19}$$

are continuous, differentiable (in particular at $t = t_0$), and are solutions of the initial value problem (15). Hence this problem has an infinite family of solutions; see Figure 2.4.1, where a few of these solutions are shown.

As already noted, the nonuniqueness of the solutions of the problem (15) does not contradict the existence and uniqueness theorem, since the theorem is not applicable if the initial point lies on the t-axis. If (t_0, y_0) is any point not on the t-axis, however, then the theorem guarantees that there is a unique solution of the differential equation $y' = y^{1/3}$ passing through (t_0, y_0).

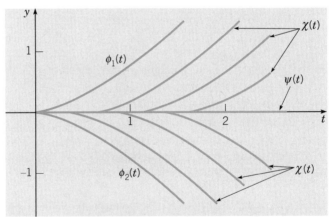

FIGURE 2.4.1 Several solutions of the initial value problem $y' = y^{1/3}$, $y(0) = 0$.

Interval of Definition. According to Theorem 2.4.1, the solution of a linear equation (1),

$$y' + p(t)y = g(t),$$

subject to the initial condition $y(t_0) = y_0$, exists throughout any interval about $t = t_0$ in which the functions p and g are continuous. Thus, vertical asymptotes or other discontinuities in the solution can occur only at points of discontinuity of p or g. For instance, the solutions in Example 1 (with one exception) are asymptotic to the y-axis, corresponding to the discontinuity at $t = 0$ in the coefficient $p(t) = 2/t$, but none of the solutions has any other point where it fails to exist and to be differentiable. The one exceptional solution shows that solutions may sometimes remain continuous even at points of discontinuity of the coefficients.

On the other hand, for a nonlinear initial value problem satisfying the hypotheses of Theorem 2.4.2, the interval in which a solution exists may be difficult to determine. The solution $y = \phi(t)$ is certain to exist as long as the point $[t, \phi(t)]$ remains within a region in which the hypotheses of Theorem 2.4.2 are satisfied. This is what determines the value of h in that theorem. However, since $\phi(t)$ is usually not known, it may be impossible to locate the point $[t, \phi(t)]$ with respect to this region. In any case, the interval in which a solution exists may have no simple relationship to the function f in the differential equation $y' = f(t, y)$. This is illustrated by the following example.

EXAMPLE 4

Solve the initial value problem

$$y' = y^2, \qquad y(0) = 1, \tag{20}$$

and determine the interval in which the solution exists.

Theorem 2.4.2 guarantees that this problem has a unique solution since $f(t, y) = y^2$ and $\partial f / \partial y = 2y$ are continuous everywhere. To find the solution we separate the variables and integrate with the result that

$$y^{-2}\, dy = dt \tag{21}$$

and

$$-y^{-1} = t + c.$$

Then, solving for y, we have

$$y = -\frac{1}{t + c}. \tag{22}$$

To satisfy the initial condition we must choose $c = -1$, so

$$y = \frac{1}{1 - t} \tag{23}$$

is the solution of the given initial value problem. Clearly, the solution becomes unbounded as $t \to 1$; therefore, the solution exists only in the interval $-\infty < t < 1$. There is no indication from the differential equation itself, however, that the point $t = 1$ is in any way remarkable. Moreover, if the initial condition is replaced by

$$y(0) = y_0, \tag{24}$$

then the constant c in Eq. (22) must be chosen to be $c = -1/y_0$, and it follows that

$$y = \frac{y_0}{1 - y_0 t} \tag{25}$$

is the solution of the initial value problem with the initial condition (24). Observe that the solution (25) becomes unbounded as $t \to 1/y_0$, so the interval of existence of the solution is $-\infty < t < 1/y_0$ if $y_0 > 0$, and is $1/y_0 < t < \infty$ if $y_0 < 0$. This example illustrates another feature of initial value problems for nonlinear equations; namely, the singularities of the solution may depend in an essential way on the initial conditions as well as on the differential equation.

General Solution. Another way in which linear and nonlinear equations differ concerns the concept of a general solution. For a first order linear equation it is possible to obtain a solution containing one arbitrary constant, from which all possible solutions follow by specifying values for this constant. For nonlinear equations this may not be the case; even though a solution containing an arbitrary constant may be found, there

may be other solutions that cannot be obtained by giving values to this constant. For instance, for the differential equation $y' = y^2$ in Example 4, the expression in Eq. (22) contains an arbitrary constant, but does not include all solutions of the differential equation. To show this, observe that the function $y = 0$ for all t is certainly a solution of the differential equation, but it cannot be obtained from Eq. (22) by assigning a value to c. In this example we might anticipate that something of this sort might happen because to rewrite the original differential equation in the form (21), we must require that y is not zero. However, the existence of "additional" solutions is not uncommon for nonlinear equations; a less obvious example is given in Problem 22. Thus we will use the term "general solution" only when discussing linear equations.

Implicit Solutions. Recall again that, for an initial value problem for a first order linear equation, Eq. (8) provides an explicit formula for the solution $y = \phi(t)$. As long as the necessary antiderivatives can be found, the value of the solution at any point can be determined merely by substituting the appropriate value of t into the equation. The situation for nonlinear equations is much less satisfactory. Usually, the best that we can hope for is to find an equation

$$F(t, y) = 0 \tag{26}$$

involving t and y that is satisfied by the solution $y = \phi(t)$. Even this can be done only for differential equations of certain particular types, of which separable equations are the most important. The equation (26) is called an integral, or first integral, of the differential equation, and (as we have already noted) its graph is an integral curve, or perhaps a family of integral curves. Equation (26), assuming it can be found, defines the solution implicitly; that is, for each value of t we must solve Eq. (26) to find the corresponding value of y. If Eq. (26) is simple enough, it may be possible to solve it for y by analytical means and thereby obtain an explicit formula for the solution. However, more frequently this will not be possible, and you will have to resort to a numerical calculation to determine the value of y for a given value of t. Once several pairs of values of t and y have been calculated, it is often helpful to plot them and then to sketch the integral curve that passes through them. You should arrange for a computer to do this for you, if possible.

Examples 2, 3, and 4 are nonlinear problems in which it is easy to solve for an explicit formula for the solution $y = \phi(t)$. On the other hand, Examples 1 and 3 in Section 2.2 are cases in which it is better to leave the solution in implicit form, and to use numerical means to evaluate it for particular values of the independent variable. The latter situation is more typical; unless the implicit relation is quadratic in y, or has some other particularly simple form, it is unlikely that it can be solved exactly by analytical methods. Indeed, more often than not, it is impossible even to find an implicit expression for the solution of a first order nonlinear equation.

Graphical or Numerical Construction of Integral Curves. Because of the difficulty in obtaining exact analytical solutions of nonlinear differential equations, methods that yield approximate solutions or other qualitative information about solutions are of correspondingly greater importance. We have already described, in Section 1.1, how the direction field of a differential equation can be constructed. The direction field can often show the qualitative form of solutions and can also be helpful in identi-

fying regions of the *ty*-plane where solutions exhibit interesting features that merit more detailed analytical or numerical investigation. Graphical methods for first order equations are discussed further in Section 2.5. An introduction to numerical methods for first order equations is given in Section 2.7, and a systematic discussion of numerical methods appears in Chapter 8. However, it is not necessary to study the numerical algorithms themselves in order to use effectively one of the many software packages that generate and plot numerical approximations to solutions of initial value problems.

Summary. The linear equation $y' + p(t)y = g(t)$ has several nice properties that can be summarized in the following statements:

1. Assuming that the coefficients are continuous, there is a general solution, containing an arbitrary constant, that includes all solutions of the differential equation. A particular solution that satisfies a given initial condition can be picked out by choosing the proper value for the arbitrary constant.

2. There is an expression for the solution, namely, Eq. (7) or Eq. (8). Moreover, although it involves two integrations, the expression is an explicit one for the solution $y = \phi(t)$ rather than an equation that defines ϕ implicitly.

3. The possible points of discontinuity, or singularities, of the solution can be identified (without solving the problem) merely by finding the points of discontinuity of the coefficients. Thus, if the coefficients are continuous for all t, then the solution also exists and is continuous for all t.

None of these statements is true, in general, of nonlinear equations. Although a nonlinear equation may well have a solution involving an arbitrary constant, there may also be other solutions. There is no general formula for solutions of nonlinear equations. If you are able to integrate a nonlinear equation, you are likely to obtain an equation defining solutions implicitly rather than explicitly. Finally, the singularities of solutions of nonlinear equations can usually be found only by solving the equation and examining the solution. It is likely that the singularities will depend on the initial condition as well as the differential equation.

PROBLEMS

In each of Problems 1 through 6 determine (without solving the problem) an interval in which the solution of the given initial value problem is certain to exist.

1. $(t - 3)y' + (\ln t)y = 2t, \qquad y(1) = 2$
2. $t(t - 4)y' + y = 0, \qquad y(2) = 1$
3. $y' + (\tan t)y = \sin t, \qquad y(\pi) = 0$
4. $(4 - t^2)y' + 2ty = 3t^2, \qquad y(-3) = 1$
5. $(4 - t^2)y' + 2ty = 3t^2, \qquad y(1) = -3$
6. $(\ln t)y' + y = \cot t, \qquad y(2) = 3$

In each of Problems 7 through 12 state where in the *ty*-plane the hypotheses of Theorem 2.4.2 are satisfied.

7. $y' = \dfrac{t - y}{2t + 5y}$

8. $y' = (1 - t^2 - y^2)^{1/2}$

9. $y' = \dfrac{\ln |ty|}{1 - t^2 + y^2}$

10. $y' = (t^2 + y^2)^{3/2}$

11. $\dfrac{dy}{dt} = \dfrac{1+t^2}{3y-y^2}$ 12. $\dfrac{dy}{dt} = \dfrac{(\cot t)y}{1+y}$

In each of Problems 13 through 16 solve the given initial value problem and determine how the interval in which the solution exists depends on the initial value y_0.

13. $y' = -4t/y$, $y(0) = y_0$ 14. $y' = 2ty^2$, $y(0) = y_0$

15. $y' + y^3 = 0$, $y(0) = y_0$ 16. $y' = t^2/y(1+t^3)$, $y(0) = y_0$

In each of Problems 17 through 20 draw a direction field and plot (or sketch) several solutions of the given differential equation. Describe how solutions appear to behave as t increases and how their behavior depends on the initial value y_0 when $t = 0$.

17. $y' = ty(3-y)$ 18. $y' = y(3-ty)$

19. $y' = -y(3-ty)$ 20. $y' = t - 1 - y^2$

21. Consider the initial value problem $y' = y^{1/3}, y(0) = 0$ from Example 3 in the text.

 (a) Is there a solution that passes through the point $(1,1)$? If so, find it.

 (b) Is there a solution that passes through the point $(2,1)$? If so, find it.

 (c) Consider all possible solutions of the given initial value problem. Determine the set of values that these solutions have at $t = 2$.

22. (a) Verify that both $y_1(t) = 1 - t$ and $y_2(t) = -t^2/4$ are solutions of the initial value problem

$$y' = \frac{-t + (t^2 + 4y)^{1/2}}{2}, \qquad y(2) = -1.$$

Where are these solutions valid?

 (b) Explain why the existence of two solutions of the given problem does not contradict the uniqueness part of Theorem 2.4.2.

 (c) Show that $y = ct + c^2$, where c is an arbitrary constant, satisfies the differential equation in part (a) for $t \geq -2c$. If $c = -1$, the initial condition is also satisfied, and the solution $y = y_1(t)$ is obtained. Show that there is no choice of c that gives the second solution $y = y_2(t)$.

23. (a) Show that $\phi(t) = e^{2t}$ is a solution of $y' - 2y = 0$ and that $y = c\phi(t)$ is also a solution of this equation for any value of the constant c.

 (b) Show that $\phi(t) = 1/t$ is a solution of $y' + y^2 = 0$ for $t > 0$ but that $y = c\phi(t)$ is not a solution of this equation unless $c = 0$ or $c = 1$. Note that the equation of part (b) is nonlinear, while that of part (a) is linear.

24. Show that if $y = \phi(t)$ is a solution of $y' + p(t)y = 0$, then $y = c\phi(t)$ is also a solution for any value of the constant c.

25. Let $y = y_1(t)$ be a solution of

$$y' + p(t)y = 0, \tag{i}$$

and let $y = y_2(t)$ be a solution of

$$y' + p(t)y = g(t). \tag{ii}$$

Show that $y = y_1(t) + y_2(t)$ is also a solution of Eq. (ii).

26. (a) Show that the solution (7) of the general linear equation (1) can be written in the form

$$y = cy_1(t) + y_2(t), \tag{i}$$

where c is an arbitrary constant. Identify the functions y_1 and y_2.

(b) Show that y_1 is a solution of the differential equation

$$y' + p(t)y = 0, \tag{ii}$$

corresponding to $g(t) = 0$.

(c) Show that y_2 is a solution of the full linear equation (1). We see later (for example, in Section 3.6) that solutions of higher order linear equations have a pattern similar to Eq. (i).

Bernoulli Equations. Sometimes it is possible to solve a nonlinear equation by making a change of the dependent variable that converts it into a linear equation. The most important such equation has the form

$$y' + p(t)y = q(t)y^n,$$

and is called a Bernoulli equation after Jakob Bernoulli. Problems 27 through 31 deal with equations of this type.

27. (a) Solve Bernoulli's equation when $n = 0$; when $n = 1$.

(b) Show that if $n \neq 0, 1$, then the substitution $v = y^{1-n}$ reduces Bernoulli's equation to a linear equation. This method of solution was found by Leibniz in 1696.

In each of Problems 28 through 31 the given equation is a Bernoulli equation. In each case solve it by using the substitution mentioned in Problem 27(b).

28. $t^2 y' + 2ty - y^3 = 0, \qquad t > 0$

29. $y' = ry - ky^2$, $r > 0$ and $k > 0$. This equation is important in population dynamics and is discussed in detail in Section 2.5.

30. $y' = \epsilon y - \sigma y^3$, $\epsilon > 0$ and $\sigma > 0$. This equation occurs in the study of the stability of fluid flow.

31. $dy/dt = (\Gamma \cos t + T)y - y^3$, where Γ and T are constants. This equation also occurs in the study of the stability of fluid flow.

Discontinuous Coefficients. Linear differential equations sometimes occur in which one or both of the functions p and g have jump discontinuities. If t_0 is such a point of discontinuity, then it is necessary to solve the equation separately for $t < t_0$ and $t > t_0$. Afterward, the two solutions are matched so that y is continuous at t_0; this is accomplished by a proper choice of the arbitrary constants. The following two problems illustrate this situation. Note in each case that it is impossible also to make y' continuous at t_0.

32. Solve the initial value problem

$$y' + 2y = g(t), \qquad y(0) = 0,$$

where

$$g(t) = \begin{cases} 1, & 0 \leq t \leq 1, \\ 0, & t > 1. \end{cases}$$

33. Solve the initial value problem

$$y' + p(t)y = 0, \qquad y(0) = 1,$$

where

$$p(t) = \begin{cases} 2, & 0 \leq t \leq 1, \\ 1, & t > 1. \end{cases}$$

2.5 Autonomous Equations and Population Dynamics

An important class of first order equations consists of those in which the independent variable does not appear explicitly. Such equations are called **autonomous** and have the form

$$dy/dt = f(y). \tag{1}$$

We will discuss these equations in the context of the growth or decline of the population of a given species, an important issue in fields ranging from medicine to ecology to global economics. A number of other applications are mentioned in some of the problems. Recall that in Sections 1.1 and 1.2 we considered the special case of Eq. (1) in which $f(y) = ay + b$.

Equation (1) is separable, so the discussion in Section 2.2 is applicable to it, but the main purpose of this section is to show how geometrical methods can be used to obtain important qualitative information directly from the differential equation, without solving the equation. Of fundamental importance in this effort are the concepts of stability and instability of solutions of differential equations. These ideas were introduced informally in Chapter 1, but without using this terminology. They are discussed further here and will be examined in greater depth and in a more general setting in Chapter 9.

Exponential Growth. Let $y = \phi(t)$ be the population of the given species at time t. The simplest hypothesis concerning the variation of population is that the rate of change of y is proportional[10] to the current value of y; that is,

$$dy/dt = ry, \tag{2}$$

where the constant of proportionality r is called the **rate of growth or decline**, depending on whether it is positive or negative. Here, we assume that $r > 0$, so the population is growing.

Solving Eq. (2) subject to the initial condition

$$y(0) = y_0, \tag{3}$$

we obtain

$$y = y_0 e^{rt}. \tag{4}$$

Thus the mathematical model consisting of the initial value problem (2), (3) with $r > 0$ predicts that the population will grow exponentially for all time, as shown in Figure 2.5.1 for several values of y_0. Under ideal conditions, Eq. (4) has been observed to be reasonably accurate for many populations, at least for limited periods of time. However, it is clear that such ideal conditions cannot continue indefinitely; eventually, limitations on space, food supply, or other resources will reduce the growth rate and bring an end to uninhibited exponential growth.

[10] It was apparently the British economist Thomas Malthus (1766–1834) who first observed that many biological populations increase at a rate proportional to the population. His first paper on populations appeared in 1798.

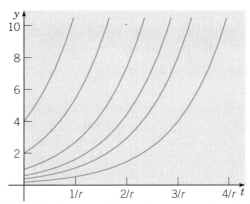

FIGURE 2.5.1 Exponential growth: y versus t for $dy/dt = ry$.

Logistic Growth. To take account of the fact that the growth rate actually depends on the population, we replace the constant r in Eq. (2) by a function $h(y)$ and thereby obtain the modified equation

$$dy/dt = h(y)y. \tag{5}$$

We now want to choose $h(y)$ so that $h(y) \cong r > 0$ when y is small, $h(y)$ decreases as y grows larger, and $h(y) < 0$ when y is sufficiently large. The simplest function that has these properties is $h(y) = r - ay$, where a is also a positive constant. Using this function in Eq. (5), we obtain

$$dy/dt = (r - ay)y. \tag{6}$$

Equation (6) is known as the Verhulst[11] equation or the **logistic equation**. It is often convenient to write the logistic equation in the equivalent form

$$\frac{dy}{dt} = r\left(1 - \frac{y}{K}\right)y, \tag{7}$$

where $K = r/a$. The constant r is called the **intrinsic growth rate**, that is, the growth rate in the absence of any limiting factors. The interpretation of K will become clear shortly.

We will investigate the solutions of Eq. (7) in some detail later in this section. Before doing that, however, we will show how you can easily draw a *qualitatively correct* sketch of the solutions. The same methods also apply to the more general Eq. (1).

[11]P. F. Verhulst (1804–1849) was a Belgian mathematician who introduced Eq. (6) as a model for human population growth in 1838. He referred to it as logistic growth; hence Eq. (6) is often called the logistic equation. He was unable to test the accuracy of his model because of inadequate census data, and it did not receive much attention until many years later. Reasonable agreement with experimental data was demonstrated by R. Pearl (1930) for *Drosophila melanogaster* (fruit fly) populations and by G. F. Gause (1935) for *Paramecium* and *Tribolium* (flour beetle) populations.

We first seek solutions of Eq. (7) of the simplest possible type, that is, constant functions. For such a solution $dy/dt = 0$ for all t, so any constant solution of Eq. (7) must satisfy the algebraic equation

$$r(1 - y/K)y = 0.$$

Thus the constant solutions are $y = \phi_1(t) = 0$ and $y = \phi_2(t) = K$. These solutions are called **equilibrium solutions** of Eq. (7) because they correspond to no change or variation in the value of y as t increases. In the same way, any equilibrium solutions of the more general Eq. (1) can be found by locating the roots of $f(y) = 0$. The zeros of $f(y)$ are also called **critical points**.

To visualize other solutions of Eq. (7) and to sketch their graphs quickly, we start by drawing the graph of $f(y)$ versus y. In the case of Eq. (7), $f(y) = r(1 - y/K)y$, so the graph is the parabola shown in Figure 2.5.2. The intercepts are $(0, 0)$ and $(K, 0)$, corresponding to the critical points of Eq. (7), and the vertex of the parabola is $(K/2, rK/4)$. Observe that $dy/dt > 0$ for $0 < y < K$; therefore, y is an increasing function of t when y is in this interval; this is indicated by the rightward-pointing arrows near the y-axis in Figure 2.5.2. Similarly, if $y > K$, then $dy/dt < 0$; hence y is decreasing, as indicated by the leftward-pointing arrow in Figure 2.5.2.

FIGURE 2.5.2 $f(y)$ versus y for $dy/dt = r(1 - y/K)y$.

In this context the y-axis is often called the **phase line**, and it is reproduced in its more customary vertical orientation in Figure 2.5.3a. The dots at $y = 0$ and $y = K$ are the critical points, or equilibrium solutions. The arrows again indicate that y is increasing whenever $0 < y < K$ and that y is decreasing whenever $y > K$.

Further, from Figure 2.5.2, note that if y is near zero or K, then the slope $f(y)$ is near zero, so the solution curves are relatively flat. They become steeper as the value of y leaves the neighborhood of zero or K.

To sketch the graphs of solutions of Eq. (7) in the ty-plane, we start with the equilibrium solutions $y = 0$ and $y = K$; then we draw other curves that are increasing when $0 < y < K$, decreasing when $y > K$, and flatten out as y approaches either of the values 0 or K. Thus the graphs of solutions of Eq. (7) must have the general shape shown in Figure 2.5.3b, regardless of the values of r and K.

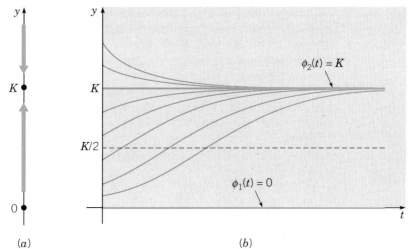

(a) (b)

FIGURE 2.5.3 Logistic growth: $dy/dt = r(1 - y/K)y$. (a) The phase line. (b) Plots of y versus t.

Figure 2.5.3*b* may seem to show that other solutions intersect the equilibrium solution $y = K$, but is this really possible? No, the uniqueness part of Theorem 2.4.2, the fundamental existence and uniqueness theorem, states that only one solution can pass through a given point in the ty-plane. Thus, although other solutions may be asymptotic to the equilibrium solution as $t \to \infty$, they cannot intersect it at any finite time.

To carry the investigation one step further, we can determine the concavity of the solution curves and the location of inflection points by finding d^2y/dt^2. From the differential equation (1) we obtain (using the chain rule)

$$\frac{d^2y}{dt^2} = \frac{d}{dt}\frac{dy}{dt} = \frac{d}{dt}f(y) = f'(y)\frac{dy}{dt} = f'(y)f(y). \tag{8}$$

The graph of y versus t is concave up when $y'' > 0$, that is, when f and f' have the same sign. Similarly, it is concave down when $y'' < 0$, which occurs when f and f' have opposite signs. The signs of f and f' can be easily identified from the graph of $f(y)$ versus y. Inflection points may occur when $f'(y) = 0$.

In the case of Eq. (7), solutions are concave up for $0 < y < K/2$ where f is positive and increasing (see Figure 2.5.2), so that both f and f' are positive. Solutions are also concave up for $y > K$ where f is negative and decreasing (both f and f' are negative). For $K/2 < y < K$, solutions are concave down since here f is positive and decreasing, so f is positive but f' is negative. There is an inflection point whenever the graph of y versus t crosses the line $y = K/2$. The graphs in Figure 2.5.3*b* exhibit these properties.

Finally, observe that K is the upper bound that is approached, but not exceeded, by growing populations starting below this value. Thus it is natural to refer to K as the **saturation level**, or the **environmental carrying capacity**, for the given species.

A comparison of Figures 2.5.1 and 2.5.3*b* reveals that solutions of the nonlinear equation (7) are strikingly different from those of the linear equation (1), at least

for large values of t. Regardless of the value of K, that is, no matter how small the nonlinear term in Eq. (7), solutions of that equation approach a finite value as $t \to \infty$, whereas solutions of Eq. (1) grow (exponentially) without bound as $t \to \infty$. Thus even a tiny nonlinear term in the differential equation has a decisive effect on the solution for large t.

In many situations it is sufficient to have the qualitative information about a solution $y = \phi(t)$ of Eq. (7) that is shown in Figure 2.5.3b. This information was obtained entirely from the graph of $f(y)$ versus y, and without solving the differential equation (7). However, if we wish to have a more detailed description of logistic growth—for example, if we wish to know the value of the population at some particular time— then we must solve Eq. (7) subject to the initial condition (3). Provided that $y \neq 0$ and $y \neq K$, we can write Eq. (7) in the form

$$\frac{dy}{(1 - y/K)y} = r \, dt.$$

Using a partial fraction expansion on the left side, we have

$$\left(\frac{1}{y} + \frac{1/K}{1 - y/K} \right) dy = r \, dt.$$

Then, by integrating both sides, we obtain

$$\ln|y| - \ln\left|1 - \frac{y}{K}\right| = rt + c, \tag{9}$$

where c is an arbitrary constant of integration to be determined from the initial condition $y(0) = y_0$. We have already noted that if $0 < y_0 < K$, then y remains in this interval for all time. Thus in this case we can remove the absolute value bars in Eq. (9), and by taking the exponential of both sides, we find that

$$\frac{y}{1 - (y/K)} = Ce^{rt}, \tag{10}$$

where $C = e^c$. In order to satisfy the initial condition $y(0) = y_0$, we must choose $C = y_0/[1 - (y_0/K)]$. Using this value for C in Eq. (10) and solving for y, we obtain

$$y = \frac{y_0 K}{y_0 + (K - y_0)e^{-rt}}. \tag{11}$$

We have derived the solution (11) under the assumption that $0 < y_0 < K$. If $y_0 > K$, then the details of dealing with Eq. (9) are only slightly different, and we leave it to you to show that Eq. (11) is also valid in this case. Finally, note that Eq. (11) also contains the equilibrium solutions $y = \phi_1(t) = 0$ and $y = \phi_2(t) = K$ corresponding to the initial conditions $y_0 = 0$ and $y_0 = K$, respectively.

All the qualitative conclusions that we reached earlier by geometrical reasoning can be confirmed by examining the solution (11). In particular, if $y_0 = 0$, then Eq. (11) requires that $y(t) = 0$ for all t. If $y_0 > 0$, and if we let $t \to \infty$ in Eq. (11), then we obtain

$$\lim_{t \to \infty} y(t) = y_0 K / y_0 = K.$$

Thus, for each $y_0 > 0$, the solution approaches the equilibrium solution $y = \phi_2(t) = K$ asymptotically as $t \to \infty$. Therefore we say that the constant solution $\phi_2(t) = K$ is

an **asymptotically stable solution** of Eq. (7) or that the point $y = K$ is an asymptotically stable equilibrium or critical point. After a long time, the population is close to the saturation level K regardless of the initial population size, as long as it is positive. Other solutions approach the equilibrium solution more rapidly as r increases.

On the other hand, the situation for the equilibrium solution $y = \phi_1(t) = 0$ is quite different. Even solutions that start very near zero grow as t increases and, as we have seen, approach K as $t \to \infty$. We say that $\phi_1(t) = 0$ is an **unstable equilibrium solution** or that $y = 0$ is an unstable equilibrium or critical point. This means that the only way to guarantee that the solution remains near zero is to make sure its initial value is *exactly* equal to zero.

EXAMPLE

1

The logistic model has been applied to the natural growth of the halibut population in certain areas of the Pacific Ocean.[12] Let y, measured in kilograms, be the total mass, or biomass, of the halibut population at time t. The parameters in the logistic equation are estimated to have the values $r = 0.71$/year and $K = 80.5 \times 10^6$ kg. If the initial biomass is $y_0 = 0.25K$, find the biomass 2 years later. Also find the time τ for which $y(\tau) = 0.75K$.

It is convenient to scale the solution (11) to the carrying capacity K; thus we write Eq. (11) in the form

$$\frac{y}{K} = \frac{y_0/K}{(y_0/K) + [1 - (y_0/K)]e^{-rt}} . \tag{12}$$

Using the data given in the problem, we find that

$$\frac{y(2)}{K} = \frac{0.25}{0.25 + 0.75e^{-1.42}} \cong 0.5797.$$

Consequently, $y(2) \cong 46.7 \times 10^6$ kg.

To find τ we can first solve Eq. (12) for t. We obtain

$$e^{-rt} = \frac{(y_0/K)[1 - (y/K)]}{(y/K)[1 - (y_0/K)]};$$

hence

$$t = -\frac{1}{r} \ln \frac{(y_0/K)[1 - (y/K)]}{(y/K)[1 - (y_0/K)]} . \tag{13}$$

Using the given values of r and y_0/K and setting $y/K = 0.75$, we find that

$$\tau = -\frac{1}{0.71} \ln \frac{(0.25)(0.25)}{(0.75)(0.75)} = \frac{1}{0.71} \ln 9 \cong 3.095 \text{ years.}$$

The graphs of y/K versus t for the given parameter values and for several initial conditions are shown in Figure 2.5.4.

[12]A good source of information on the population dynamics and economics involved in making efficient use of a renewable resource, with particular emphasis on fisheries, is the book by Clark listed in the references at the end of this chapter. The parameter values used here are given on page 53 of this book and were obtained from a study by H. S. Mohring.

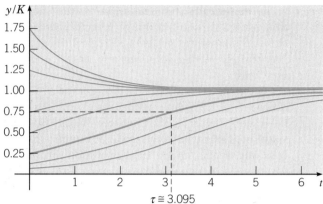

FIGURE 2.5.4 y/K versus t for population model of halibut in the Pacific Ocean.

A Critical Threshold. We now turn to a consideration of the equation

$$\frac{dy}{dt} = -r\left(1 - \frac{y}{T}\right)y, \tag{14}$$

where r and T are given positive constants. Observe that (except for replacing the parameter K by T) this equation differs from the logistic equation (7) only in the presence of the minus sign on the right side. However, as we will see, the solutions of Eq. (14) behave very differently from those of Eq. (7).

For Eq. (14) the graph of $f(y)$ versus y is the parabola shown in Figure 2.5.5. The intercepts on the y-axis are the critical points $y = 0$ and $y = T$, corresponding to the equilibrium solutions $\phi_1(t) = 0$ and $\phi_2(t) = T$. If $0 < y < T$, then $dy/dt < 0$, and y decreases as t increases. On the other hand, if $y > T$, then $dy/dt > 0$, and y grows as t increases. Thus $\phi_1(t) = 0$ is an asymptotically stable equilibrium solution and $\phi_2(t) = T$ is an unstable one. Further, $f'(y)$ is negative for $0 < y < T/2$ and positive for $T/2 < y < T$, so the graph of y versus t is concave up and concave down, respectively, in these intervals. Also, $f'(y)$ is positive for $y > T$, so the graph of y versus t is also concave up there.

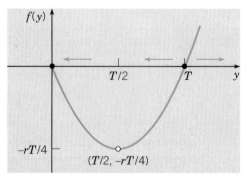

FIGURE 2.5.5 $f(y)$ versus y for $dy/dt = -r(1 - y/T)y$.

Figure 2.5.6a shows the phase line (the y-axis) for Eq. (14). The dots at $y = 0$ and $y = T$ are the critical points, or equilibrium solutions, and the arrows indicate where solutions are either increasing or decreasing.

Solution curves of Eq. (14) can now be sketched quickly. First draw the equilibrium solutions $y = 0$ and $y = T$. Then sketch curves in the strip $0 < y < T$ that are decreasing as t increases and change concavity as they cross the line $y = T/2$. Next draw some curves above $y = T$ that increase more and more steeply as t and y increase. Make sure that all curves become flatter as t approaches either zero or T. The result is Figure 2.5.6b, which is a qualitatively accurate sketch of solutions of Eq. (14) for any r and T. From this figure it appears that as time increases, y either approaches zero or grows without bound, depending on whether the initial value y_0 is less than or greater than T. Thus T is a **threshold level**, below which growth does not occur.

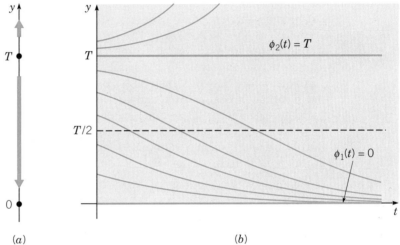

(a) (b)

FIGURE 2.5.6 Growth with a threshold: $dy/dt = -r(1 - y/T)y$. (a) The phase line. (b) Plots of y versus t.

We can confirm the conclusions that we have reached through geometrical reasoning by solving the differential equation (14). This can be done by separating the variables and integrating, just as we did for Eq. (7). However, if we note that Eq. (14) can be obtained from Eq. (7) by replacing K by T and r by $-r$, then we can make the same substitutions in the solution (11) and thereby obtain

$$y = \frac{y_0 T}{y_0 + (T - y_0)e^{rt}}, \tag{15}$$

which is the solution of Eq. (14) subject to the initial condition $y(0) = y_0$.

If $0 < y_0 < T$, then it follows from Eq. (15) that $y \to 0$ as $t \to \infty$. This agrees with our qualitative geometric analysis. If $y_0 > T$, then the denominator on the right side of Eq. (15) is zero for a certain finite value of t. We denote this value by t^* and

calculate it from

$$y_0 - (y_0 - T)e^{rt^*} = 0,$$

which gives

$$t^* = \frac{1}{r} \ln \frac{y_0}{y_0 - T} . \tag{16}$$

Thus, if the initial population y_0 is above the threshold T, the threshold model predicts that the graph of y versus t has a vertical asymptote at $t = t^*$; in other words, the population becomes unbounded in a finite time, whose value depends on y_0, T, and r. The existence and location of this asymptote were not apparent from the geometric analysis, so in this case the explicit solution yields additional important qualitative, as well as quantitative, information.

The populations of some species exhibit the threshold phenomenon. If too few are present, then the species cannot propagate itself successfully and the population becomes extinct. However, if a population larger than the threshold level can be brought together, then further growth occurs. Of course, the population cannot become unbounded, so eventually Eq. (14) must be modified to take this into account.

Critical thresholds also occur in other circumstances. For example, in fluid mechanics, equations of the form (7) or (14) often govern the evolution of a small disturbance y in a *laminar* (or smooth) fluid flow. For instance, if Eq. (14) holds and $y < T$, then the disturbance is damped out and the laminar flow persists. However, if $y > T$, then the disturbance grows larger and the laminar flow breaks up into a turbulent one. In this case T is referred to as the *critical amplitude*. Experimenters speak of keeping the disturbance level in a wind tunnel sufficiently low so that they can study laminar flow over an airfoil, for example.

Logistic Growth with a Threshold. As we mentioned in the last subsection, the threshold model (14) may need to be modified so that unbounded growth does not occur when y is above the threshold T. The simplest way to do this is to introduce another factor that will have the effect of making dy/dt negative when y is large. Thus we consider

$$\frac{dy}{dt} = -r \left(1 - \frac{y}{T}\right) \left(1 - \frac{y}{K}\right) y, \tag{17}$$

where $r > 0$ and $0 < T < K$.

The graph of $f(y)$ versus y is shown in Figure 2.5.7. In this problem there are three critical points, $y = 0$, $y = T$, and $y = K$, corresponding to the equilibrium solutions $\phi_1(t) = 0$, $\phi_2(t) = T$, and $\phi_3(t) = K$, respectively. From Figure 2.5.7 it is clear that $dy/dt > 0$ for $T < y < K$, and consequently y is increasing there. The reverse is true for $y < T$ and for $y > K$. Consequently, the equilibrium solutions $\phi_1(t)$ and $\phi_3(t)$ are asymptotically stable, and the solution $\phi_2(t)$ is unstable.

The phase line for Eq. (17) is shown in Figure 2.5.8a, and the graphs of some solutions are sketched in Figure 2.5.8b. You should make sure that you understand the relation between these two figures, as well as the relation between Figures 2.5.7 and 2.5.8a. From Figure 2.5.8b we see that if y starts below the threshold T, then y declines to ultimate extinction. On the other hand, if y starts above T, then y eventually approaches the carrying capacity K. The inflection points on the graphs of y versus t in Figure 2.5.8b correspond to the maximum and minimum points, y_1

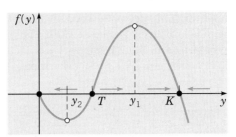

FIGURE 2.5.7 $f(y)$ versus y for $dy/dt = -r(1 - y/T)(1 - y/K)y$.

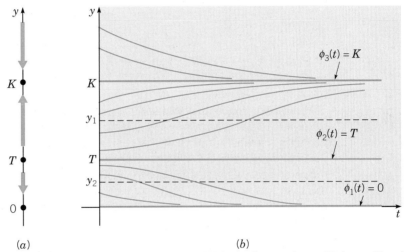

(a) (b)

FIGURE 2.5.8 Logistic growth with a threshold: $dy/dt = -r(1 - y/T)(1 - y/K)y$. (a) The phase line. (b) Plots of y versus t.

and y_2, respectively, on the graph of $f(y)$ versus y in Figure 2.5.7. These values can be obtained by differentiating the right side of Eq. (17) with respect to y, setting the result equal to zero, and solving for y. We obtain

$$y_{1,2} = (K + T \pm \sqrt{K^2 - KT + T^2})/3, \tag{18}$$

where the plus sign yields y_1 and the minus sign y_2.

A model of this general sort apparently describes the population of the passenger pigeon,[13] which was present in the United States in vast numbers until late in the nineteenth century. It was heavily hunted for food and for sport, and consequently its numbers were drastically reduced by the 1880s. Unfortunately, the passenger pigeon could apparently breed successfully only when present in a large concentration, corresponding to a relatively high threshold T. Although a reasonably large number of individual birds remained alive in the late 1880s, there were not enough in any one

[13]See, for example, Oliver L. Austin, Jr., *Birds of the World* (New York: Golden Press, 1983), pp. 143–145.

place to permit successful breeding, and the population rapidly declined to extinction. The last survivor died in 1914. The precipitous decline in the passenger pigeon population from huge numbers to extinction in a few decades was one of the early factors contributing to a concern for conservation in this country.

PROBLEMS

Problems 1 through 6 involve equations of the form $dy/dt = f(y)$. In each problem sketch the graph of $f(y)$ versus y, determine the critical (equilibrium) points, and classify each one as asymptotically stable or unstable. Draw the phase line, and sketch several graphs of solutions in the ty-plane.

1. $dy/dt = ay + by^2$, $a > 0$, $b > 0$, $y_0 \geq 0$
2. $dy/dt = ay + by^2$, $a > 0$, $b > 0$, $-\infty < y_0 < \infty$
3. $dy/dt = y(y - 1)(y - 2)$, $y_0 \geq 0$
4. $dy/dt = e^y - 1$, $-\infty < y_0 < \infty$
5. $dy/dt = e^{-y} - 1$, $-\infty < y_0 < \infty$
6. $dy/dt = -2(\arctan y)/(1 + y^2)$, $-\infty < y_0 < \infty$
7. **Semistable Equilibrium Solutions.** Sometimes a constant equilibrium solution has the property that solutions lying on one side of the equilibrium solution tend to approach it, whereas solutions lying on the other side depart from it (see Figure 2.5.9). In this case the equilibrium solution is said to be **semistable**.

FIGURE 2.5.9 In both cases the equilibrium solution $\phi(t) = k$ is semistable. (*a*) $dy/dt \leq 0$; (*b*) $dy/dt \geq 0$.

(a) Consider the equation

$$dy/dt = k(1 - y)^2, \tag{i}$$

where k is a positive constant. Show that $y = 1$ is the only critical point, with the corresponding equilibrium solution $\phi(t) = 1$.

(b) Sketch $f(y)$ versus y. Show that y is increasing as a function of t for $y < 1$ and also for $y > 1$. The phase line has upward-pointing arrows both below and above $y = 1$. Thus solutions below the equilibrium solution approach it, and those above it grow farther away. Therefore $\phi(t) = 1$ is semistable.

(c) Solve Eq. (i) subject to the initial condition $y(0) = y_0$ and confirm the conclusions reached in part (b).

Problems 8 through 13 involve equations of the form $dy/dt = f(y)$. In each problem sketch the graph of $f(y)$ versus y, determine the critical (equilibrium) points, and classify each one

as asymptotically stable, unstable, or semistable (see Problem 7). Draw the phase line, and sketch several graphs of solutions in the ty-plane.

8. $dy/dt = -k(y - 1)^2$, $k > 0$, $-\infty < y_0 < \infty$

9. $dy/dt = y^2(y^2 - 1)$, $-\infty < y_0 < \infty$

10. $dy/dt = y(1 - y^2)$, $-\infty < y_0 < \infty$

11. $dy/dt = ay - b\sqrt{y}$, $a > 0$, $b > 0$, $y_0 \geq 0$

12. $dy/dt = y^2(4 - y^2)$, $-\infty < y_0 < \infty$

13. $dy/dt = y^2(1 - y)^2$, $-\infty < y_0 < \infty$

14. Consider the equation $dy/dt = f(y)$ and suppose that y_1 is a critical point, that is, $f(y_1) = 0$. Show that the constant equilibrium solution $\phi(t) = y_1$ is asymptotically stable if $f'(y_1) < 0$ and unstable if $f'(y_1) > 0$.

15. Suppose that a certain population obeys the logistic equation $dy/dt = ry[1 - (y/K)]$.

(a) If $y_0 = K/3$, find the time τ at which the initial population has doubled. Find the value of τ corresponding to $r = 0.025$ per year.

(b) If $y_0/K = \alpha$, find the time T at which $y(T)/K = \beta$, where $0 < \alpha, \beta < 1$. Observe that $T \to \infty$ as $\alpha \to 0$ or as $\beta \to 1$. Find the value of T for $r = 0.025$ per year, $\alpha = 0.1$, and $\beta = 0.9$.

16. Another equation that has been used to model population growth is the Gompertz[14] equation:

$$dy/dt = ry \ln(K/y),$$

where r and K are positive constants.

(a) Sketch the graph of $f(y)$ versus y, find the critical points, and determine whether each is asymptotically stable or unstable.

(b) For $0 \leq y \leq K$, determine where the graph of y versus t is concave up and where it is concave down.

(c) For each y in $0 < y \leq K$, show that dy/dt as given by the Gompertz equation is never less than dy/dt as given by the logistic equation.

17. (a) Solve the Gompertz equation

$$dy/dt = ry \ln(K/y),$$

subject to the initial condition $y(0) = y_0$.
Hint: You may wish to let $u = \ln(y/K)$.

(b) For the data given in Example 1 in the text ($r = 0.71$ per year, $K = 80.5 \times 10^6$ kg, $y_0/K = 0.25$), use the Gompertz model to find the predicted value of $y(2)$.

(c) For the same data as in part (b), use the Gompertz model to find the time τ at which $y(\tau) = 0.75K$.

18. A pond forms as water collects in a conical depression of radius a and depth h. Suppose that water flows in at a constant rate k and is lost through evaporation at a rate proportional to the surface area.

(a) Show that the volume $V(t)$ of water in the pond at time t satisfies the differential equation

$$dV/dt = k - \alpha\pi(3a/\pi h)^{2/3}V^{2/3},$$

where α is the coefficient of evaporation.

[14]Benjamin Gompertz (1779–1865) was an English actuary. He developed his model for population growth, published in 1825, in the course of constructing mortality tables for his insurance company.

Integrating the first of Eqs. (20) gives

$$\psi(x, y) = \tfrac{3}{2}x^2y + xy^2 + h(y),\qquad(21)$$

where h is an arbitrary function of y only. To try to satisfy the second of Eqs. (20), we compute ψ_y from Eq. (21) and set it equal to N, obtaining

$$\tfrac{3}{2}x^2 + 2xy + h'(y) = x^2 + xy$$

or

$$h'(y) = -\tfrac{1}{2}x^2 - xy.\qquad(22)$$

Since the right side of Eq. (22) depends on x as well as y, it is impossible to solve Eq. (22) for $h(y)$. Thus there is no $\psi(x, y)$ satisfying both of Eqs. (20).

Integrating Factors. It is sometimes possible to convert a differential equation that is not exact into an exact equation by multiplying the equation by a suitable integrating factor. Recall that this is the procedure that we used in solving linear equations in Section 2.1. To investigate the possibility of implementing this idea more generally, let us multiply the equation

$$M(x, y)\,dx + N(x, y)\,dy = 0\qquad(23)$$

by a function μ and then try to choose μ so that the resulting equation

$$\mu(x, y)M(x, y)\,dx + \mu(x, y)N(x, y)\,dy = 0\qquad(24)$$

is exact. By Theorem 2.6.1, Eq. (24) is exact if and only if

$$(\mu M)_y = (\mu N)_x.\qquad(25)$$

Since M and N are given functions, Eq. (25) states that the integrating factor μ must satisfy the first order partial differential equation

$$M\mu_y - N\mu_x + (M_y - N_x)\mu = 0.\qquad(26)$$

If a function μ satisfying Eq. (26) can be found, then Eq. (24) will be exact. The solution of Eq. (24) can then be obtained by the method described in the first part of this section. The solution found in this way also satisfies Eq. (23), since the integrating factor μ can be canceled out of Eq. (24).

A partial differential equation of the form (26) may have more than one solution; if this is the case, any such solution may be used as an integrating factor of Eq. (23). This possible nonuniqueness of the integrating factor is illustrated in Example 4.

Unfortunately, Eq. (26), which determines the integrating factor μ, is ordinarily at least as hard to solve as the original equation (23). Therefore, although in principle integrating factors are powerful tools for solving differential equations, in practice they can be found only in special cases. The most important situations in which simple integrating factors can be found occur when μ is a function of only one of the variables x or y, instead of both. Let us determine necessary conditions on M and N so that Eq. (23) has an integrating factor μ that depends on x only. Assuming that μ is a function of x only, we have

$$(\mu M)_y = \mu M_y,\qquad (\mu N)_x = \mu N_x + N\,\frac{d\mu}{dx}.$$

Thus, if $(\mu M)_y$ is to equal $(\mu N)_x$, it is necessary that

$$\frac{d\mu}{dx} = \frac{M_y - N_x}{N}\mu. \tag{27}$$

If $(M_y - N_x)/N$ is a function of x only, then there is an integrating factor μ that also depends only on x; further, $\mu(x)$ can be found by solving Eq. (27), which is both linear and separable.

A similar procedure can be used to determine a condition under which Eq. (23) has an integrating factor depending only on y; see Problem 23.

EXAMPLE

4

Find an integrating factor for the equation

$$(3xy + y^2) + (x^2 + xy)y' = 0 \tag{19}$$

and then solve the equation.

In Example 3 we showed that this equation is not exact. Let us determine whether it has an integrating factor that depends on x only. On computing the quantity $(M_y - N_x)/N$, we find that

$$\frac{M_y(x, y) - N_x(x, y)}{N(x, y)} = \frac{3x + 2y - (2x + y)}{x^2 + xy} = \frac{1}{x}. \tag{28}$$

Thus there is an integrating factor μ that is a function of x only, and it satisfies the differential equation

$$\frac{d\mu}{dx} = \frac{\mu}{x}. \tag{29}$$

Hence

$$\mu(x) = x. \tag{30}$$

Multiplying Eq. (19) by this integrating factor, we obtain

$$(3x^2y + xy^2) + (x^3 + x^2y)y' = 0. \tag{31}$$

The latter equation is exact, and it is easy to show that its solutions are given implicitly by

$$x^3y + \tfrac{1}{2}x^2y^2 = c. \tag{32}$$

Solutions may also be readily found in explicit form since Eq. (32) is quadratic in y.

You may also verify that a second integrating factor of Eq. (19) is

$$\mu(x, y) = \frac{1}{xy(2x + y)},$$

and that the same solution is obtained, though with much greater difficulty, if this integrating factor is used (see Problem 32).

PROBLEMS Determine whether each of the equations in Problems 1 through 12 is exact. If it is exact, find the solution.

1. $(2x + 3) + (2y - 2)y' = 0$ 2. $(2x + 4y) + (2x - 2y)y' = 0$

3. $(3x^2 - 2xy + 2)\,dx + (6y^2 - x^2 + 3)\,dy = 0$

4. $(2xy^2 + 2y) + (2x^2y + 2x)y' = 0$

5. $\dfrac{dy}{dx} = -\dfrac{ax + by}{bx + cy}$ $\qquad\qquad$ 6. $\dfrac{dy}{dx} = -\dfrac{ax - by}{bx - cy}$

7. $(e^x \sin y - 2y \sin x)\, dx + (e^x \cos y + 2 \cos x)\, dy = 0$

8. $(e^x \sin y + 3y)\, dx - (3x - e^x \sin y)\, dy = 0$

9. $(ye^{xy} \cos 2x - 2e^{xy} \sin 2x + 2x)\, dx + (xe^{xy} \cos 2x - 3)\, dy = 0$

10. $(y/x + 6x)\, dx + (\ln x - 2)\, dy = 0, \qquad x > 0$

11. $(x \ln y + xy)\, dx + (y \ln x + xy)\, dy = 0; \qquad x > 0, \quad y > 0$

12. $\dfrac{x\, dx}{(x^2 + y^2)^{3/2}} + \dfrac{y\, dy}{(x^2 + y^2)^{3/2}} = 0$

In each of Problems 13 and 14 solve the given initial value problem and determine at least approximately where the solution is valid.

13. $(2x - y)\, dx + (2y - x)\, dy = 0, \qquad y(1) = 3$

14. $(9x^2 + y - 1)\, dx - (4y - x)\, dy = 0, \qquad y(1) = 0$

In each of Problems 15 and 16 find the value of b for which the given equation is exact, and then solve it using that value of b.

15. $(xy^2 + bx^2y)\, dx + (x + y)x^2\, dy = 0$

16. $(ye^{2xy} + x)\, dx + bxe^{2xy}\, dy = 0$

17. Assume that Eq. (6) meets the requirements of Theorem 2.6.1 in a rectangle R and is therefore exact. Show that a possible function $\psi(x, y)$ is

$$\psi(x, y) = \int_{x_0}^{x} M(s, y_0)\, ds + \int_{y_0}^{y} N(x, t)\, dt,$$

where (x_0, y_0) is a point in R.

18. Show that any separable equation

$$M(x) + N(y)y' = 0$$

is also exact.

Show that the equations in Problems 19 through 22 are not exact but become exact when multiplied by the given integrating factor. Then solve the equations.

19. $x^2y^3 + x(1 + y^2)y' = 0, \qquad \mu(x, y) = 1/xy^3$

20. $\left(\dfrac{\sin y}{y} - 2e^{-x} \sin x\right) dx + \left(\dfrac{\cos y + 2e^{-x} \cos x}{y}\right) dy = 0, \qquad \mu(x, y) = ye^x$

21. $y\, dx + (2x - ye^y)\, dy = 0, \qquad \mu(x, y) = y$

22. $(x + 2) \sin y\, dx + x \cos y\, dy = 0, \qquad \mu(x, y) = xe^x$

23. Show that if $(N_x - M_y)/M = Q$, where Q is a function of y only, then the differential equation

$$M + Ny' = 0$$

has an integrating factor of the form

$$\mu(y) = \exp \int Q(y)\, dy.$$

24. Show that if $(N_x - M_y)/(xM - yN) = R$, where R depends on the quantity xy only, then the differential equation

$$M + Ny' = 0$$

has an integrating factor of the form $\mu(xy)$. Find a general formula for this integrating factor.

In each of Problems 25 through 31 find an integrating factor and solve the given equation.

25. $(3x^2y + 2xy + y^3)\,dx + (x^2 + y^2)\,dy = 0$ 26. $y' = e^{2x} + y - 1$

27. $dx + (x/y - \sin y)\,dy = 0$ 28. $y\,dx + (2xy - e^{-2y})\,dy = 0$

29. $e^x\,dx + (e^x \cot y + 2y \csc y)\,dy = 0$

30. $[4(x^3/y^2) + (3/y)]\,dx + [3(x/y^2) + 4y]\,dy = 0$

31. $\left(3x + \dfrac{6}{y}\right) + \left(\dfrac{x^2}{y} + 3\dfrac{y}{x}\right)\dfrac{dy}{dx} = 0$

 Hint: See Problem 24.

32. Solve the differential equation

$$(3xy + y^2) + (x^2 + xy)y' = 0$$

using the integrating factor $\mu(x, y) = [xy(2x + y)]^{-1}$. Verify that the solution is the same as that obtained in Example 4 with a different integrating factor.

2.7 Numerical Approximations: Euler's Method

Recall two important facts about the first order initial value problem

$$\frac{dy}{dt} = f(t, y), \qquad y(t_0) = y_0. \tag{1}$$

First, if f and $\partial f/\partial y$ are continuous, then the initial value problem (1) has a unique solution $y = \phi(t)$ in some interval surrounding the initial point $t = t_0$. Second, it is usually not possible to find the solution ϕ by symbolic manipulations of the differential equation. Up to now we have considered the main exceptions to this statement: differential equations that are linear, separable, or exact or that can be transformed into one of these types. Nevertheless, it remains true that solutions of the vast majority of first order initial value problems cannot be found by analytical means such as those considered in the first part of this chapter.

Therefore it is important to be able to approach the problem in other ways. As we have already seen, one of these ways is to draw a direction field for the differential equation (which does not involve solving the equation) and then to visualize the behavior of solutions from the direction field. This has the advantage of being a relatively simple process, even for complicated differential equations. However, it does not lend itself to quantitative computations or comparisons, and this is often a critical shortcoming.

Another alternative is to compute approximate values of the solution $y = \phi(t)$ of the initial value problem (1) at a selected set of t-values. Ideally, the approximate solution values will be accompanied by error bounds that ensure the level of accuracy of the approximations. Today there are numerous methods that produce numerical approximations to solutions of differential equations, and Chapter 8 is devoted to a fuller discussion of some of them. Here, we introduce the oldest and simplest such

method, originated by Euler about 1768. It is called the **tangent line method** or the **Euler method**.

Let us consider how we might approximate the solution $y = \phi(t)$ of Eqs. (1) near $t = t_0$. We know that the solution passes through the initial point (t_0, y_0), and from the differential equation, we also know that its slope at this point is $f(t_0, y_0)$. Thus we can write down an equation for the line tangent to the solution curve at (t_0, y_0), namely,

$$y = y_0 + f(t_0, y_0)(t - t_0). \tag{2}$$

The tangent line is a good approximation to the actual solution curve on an interval short enough so that the slope of the solution does not change appreciably from its value at the initial point; see Figure 2.7.1. Thus, if t_1 is close enough to t_0, we can approximate $\phi(t_1)$ by the value y_1 determined by substituting $t = t_1$ into the tangent line approximation at $t = t_0$; thus

$$y_1 = y_0 + f(t_0, y_0)(t_1 - t_0). \tag{3}$$

FIGURE 2.7.1 A tangent line approximation.

To proceed further, we can try to repeat the process. Unfortunately, we do not know the value $\phi(t_1)$ of the solution at t_1. The best we can do is to use the approximate value y_1 instead. Thus we construct the line through (t_1, y_1) with the slope $f(t_1, y_1)$,

$$y = y_1 + f(t_1, y_1)(t - t_1). \tag{4}$$

To approximate the value of $\phi(t)$ at a nearby point t_2, we use Eq. (4) instead, obtaining

$$y_2 = y_1 + f(t_1, y_1)(t_2 - t_1). \tag{5}$$

Continuing in this manner, we use the value of y calculated at each step to determine the slope of the approximation for the next step. The general expression for y_{n+1} in terms of t_n, t_{n+1}, and y_n is

$$y_{n+1} = y_n + f(t_n, y_n)(t_{n+1} - t_n), \qquad n = 0, 1, 2, \ldots. \tag{6}$$

If we introduce the notation $f_n = f(t_n, y_n)$, then we can rewrite Eq. (6) as

$$y_{n+1} = y_n + f_n \cdot (t_{n+1} - t_n), \qquad n = 0, 1, 2, \ldots. \tag{7}$$

Finally, if we assume that there is a uniform step size h between the points $t_0, t_1, t_2, \ldots,$ then $t_{n+1} = t_n + h$ for each n, and we obtain Euler's formula in the form

$$y_{n+1} = y_n + f_n h, \qquad n = 0, 1, 2, \ldots. \tag{8}$$

To use Euler's method you simply evaluate Eq. (7) or Eq. (8) repeatedly, depending on whether or not the step size is constant, using the result of each step to execute the next step. In this way you generate a sequence of values y_1, y_2, y_3, \ldots that approximate the values of the solution $\phi(t)$ at the points t_1, t_2, t_3, \ldots. If, instead of a sequence of points, you need an actual function to approximate the solution $\phi(t)$, then you can use the piecewise linear function constructed from the collection of tangent line segments. That is, let y be given by Eq. (2) in $[t_0, t_1]$, by Eq. (4) in $[t_1, t_2]$, and in general by

$$y = y_n + f(t_n, y_n)(t - t_n) \tag{9}$$

in $[t_n, t_{n+1}]$.

EXAMPLE 1

Consider the initial value problem

$$\frac{dy}{dt} = 3 + e^{-t} - \tfrac{1}{2}y, \qquad y(0) = 1. \tag{10}$$

Use Euler's method with step size $h = 0.1$ to find approximate values of the solution of Eqs. (10) at $t = 0.1, 0.2, 0.3,$ and 0.4. Compare them with the corresponding values of the actual solution of the initial value problem.

Proceeding as in Section 2.1, we find the solution of Eqs. (10) to be

$$y = \phi(t) = 6 - 2e^{-t} - 3e^{-t/2}. \tag{11}$$

To use Euler's method, we note that in this case $f(t, y) = 3 + e^{-t} - y/2$. Using the initial values $t_0 = 0$ and $y_0 = 1$, we find that

$$f_0 = f(t_0, y_0) = f(0, 1) = 3 + e^0 - 0.5 = 3 + 1 - 0.5 = 3.5$$

and then, from Eq. (8) with $n = 0$,

$$y_1 = y_0 + f_0 h = 1 + (3.5)(0.1) = 1.35.$$

At the next step we have

$$f_1 = f(0.1, 1.35) = 3 + e^{-0.1} - (0.5)(1.35) \cong 3 + 0.904837 - 0.675 \cong 3.229837$$

and then

$$y_2 = y_1 + f_1 h \cong 1.35 + (3.229837)(0.1) \cong 1.672984.$$

Repeating the computation two more times, we obtain

$$f_2 \cong 2.982239, \qquad y_3 \cong 1.971208$$

and

$$f_3 \cong 2.755214, \qquad y_4 \cong 2.246729.$$

Table 2.7.1 shows these computed values, the corresponding values of the solution (11), and the differences between the two, which is the error in the numerical approximation.

TABLE 2.7.1 A Comparison of Exact Solution with Euler's Method for $h = 0.1$ for $y' = 3 + e^{-t} - \frac{1}{2}y$, $y(0) = 1$

t	Exact	Euler with $h = 0.1$	Error
0.0	1.0000	1.0000	0.0000
0.1	1.3366	1.3500	0.0134
0.2	1.6480	1.6730	0.0250
0.3	1.9362	1.9712	0.0350
0.4	2.2032	2.2467	0.0435

The purpose of Example 1 is to show you the details of implementing a few steps of Euler's method so that it will be clear exactly what computations are being executed. Of course, computations such as those in Example 1 are usually done on a computer. Some software packages include code for the Euler method, while others do not. In any case, it is easy to write a computer program to carry out the calculations required to produce results such as those in Table 2.7.1. The outline of such a program is given below; the specific instructions can be written in any high-level programming language.

The Euler Method

Step 1.	**define** $f(t, y)$
Step 2.	**input** initial values $t0$ and $y0$
Step 3.	**input** step size h and number of steps n
Step 4.	**output** $t0$ and $y0$
Step 5.	**for** j from 1 to n **do**
Step 6.	$k1 = f(t, y)$
	$y = y + h * k1$
	$t = t + h$
Step 7.	**output** t and y
Step 8.	**end**

The output of this algorithm can be numbers listed on the screen or printed on a printer, as in the third column of Table 2.7.1. Alternatively, the calculated results can be displayed in graphical form.

EXAMPLE 2

Consider again the initial value problem (10),

$$\frac{dy}{dt} = 3 + e^{-t} - \frac{1}{2}y, \qquad y(0) = 1.$$

Use Euler's method with various step sizes to calculate approximate values of the solution for $0 \le t \le 5$. Compare the calculated results with the corresponding values of the exact solution (11),

$$y = \phi(t) = 6 - 2e^{-t} - 3e^{-t/2}.$$

We used step sizes $h = 0.1, 0.05, 0.025$, and 0.01, corresponding respectively to 50, 100, 200, and 500 steps, to go from $t = 0$ to $t = 5$. The results of these calculations, along with the values

of the exact solution, are presented in Table 2.7.2. All computed entries are rounded to four decimal places, although more digits were retained in the intermediate calculations.

TABLE 2.7.2 A Comparison of Exact Solution with Euler's Method for Several Step Sizes h for $y' = 3 + e^{-t} - \frac{1}{2}y$, $y(0) = 1$

t	Exact	$h = 0.1$	$h = 0.05$	$h = 0.025$	$h = 0.01$
0.0	1.0000	1.0000	1.0000	1.0000	1.0000
1.0	3.4446	3.5175	3.4805	3.4624	3.4517
2.0	4.6257	4.7017	4.6632	4.6443	4.6331
3.0	5.2310	5.2918	5.2612	5.2460	5.2370
4.0	5.5574	5.6014	5.5793	5.5683	5.5617
5.0	5.7403	5.7707	5.7555	5.7479	5.7433

What conclusions can we draw from the data in Table 2.7.2? In the first place, for a fixed value of t, the computed approximate values become more accurate as the step size h decreases. This is what we would expect, of course, but it is encouraging that the data confirm our expectations. For example, for $t = 1$ the approximate value with $h = 0.1$ is too large by about 2%, whereas the value with $h = 0.01$ is too large by only 0.2%. In this case, reducing the step size by a factor of 10 (and performing 10 times as many computations) also reduces the error by a factor of about 10. A second observation from Table 2.7.2 is that, for a fixed step size h, the approximations become more accurate as t increases. For instance, for $h = 0.1$ the error for $t = 5$ is only about 0.5%, compared with 2% for $t = 1$. An examination of data at intermediate points not recorded in Table 2.7.2 would reveal where the maximum error occurs for a given step size and how large it is.

All in all, Euler's method seems to work rather well for this problem. Reasonably good results are obtained even for a moderately large step size $h = 0.1$, and the approximation can be improved by decreasing h.

Let us now look at another example.

EXAMPLE
3

Consider the initial value problem

$$\frac{dy}{dt} = 4 - t + 2y, \qquad y(0) = 1. \tag{12}$$

The general solution of this differential equation was found in Example 2 of Section 2.1, and the solution of the initial value problem (12) is

$$y = -\tfrac{7}{4} + \tfrac{1}{2}t + \tfrac{11}{4}e^{2t}. \tag{13}$$

Use Euler's method with several step sizes to find approximate values of the solution on the interval $0 \le t \le 5$. Compare the results with the corresponding values of the solution (13).

Using the same range of step sizes as in Example 2, we obtain the results presented in Table 2.7.3.

The data in Table 2.7.3 again confirm our expectation that, for a given value of t, accuracy improves as the step size h is reduced. For example, for $t = 1$ the percentage error diminishes from 17.3% when $h = 0.1$ to 2.1% when $h = 0.01$. However, the error increases fairly rapidly

TABLE 2.7.3 A Comparison of Exact Solution with Euler's Method for Several Step Sizes h for $y' = 4 - t + 2y$, $y(0) = 1$

t	Exact	$h = 0.1$	$h = 0.05$	$h = 0.025$	$h = 0.01$
0.0	1.000000	1.000000	1.000000	1.000000	1.000000
1.0	19.06990	15.77728	17.25062	18.10997	18.67278
2.0	149.3949	104.6784	123.7130	135.5440	143.5835
3.0	1109.179	652.5349	837.0745	959.2580	1045.395
4.0	8197.884	4042.122	5633.351	6755.175	7575.577
5.0	60573.53	25026.95	37897.43	47555.35	54881.32

as t increases for a fixed h. Even for $h = 0.01$, the error at $t = 5$ is 9.4%, and it is much greater for larger step sizes. Of course, the accuracy that is needed depends on the purpose for which the results are intended, but the errors in Table 2.7.3 are too large for most scientific or engineering applications. To improve the situation, one might either try even smaller step sizes or else restrict the computations to a rather short interval away from the initial point. Nevertheless, it is clear that Euler's method is much less effective in this example than in Example 2.

To understand better what is happening in these examples, let us look again at Euler's method for the general initial value problem (1)

$$\frac{dy}{dt} = f(t, y), \qquad y(t_0) = y_0,$$

whose solution we denote by $\phi(t)$. Recall that a first order differential equation has an infinite family of solutions, indexed by an arbitrary constant c, and that the initial condition picks out one member of this infinite family by determining the value of c. Thus $\phi(t)$ is the member of the infinite family of solutions that satisfies the initial condition $\phi(t_0) = y_0$.

At the first step Euler's method uses the tangent line approximation to the graph of $y = \phi(t)$ passing through the initial point (t_0, y_0), and this produces the approximate value y_1 at t_1. Usually $y_1 \neq \phi(t_1)$, so at the second step Euler's method uses the tangent line approximation not to $y = \phi(t)$, but to a nearby solution $y = \phi_1(t)$ that passes through the point (t_1, y_1). So it is at each following step. Euler's method uses a succession of tangent line approximations to a sequence of different solutions $\phi(t), \phi_1(t), \phi_2(t), \ldots$ of the differential equation. At each step the tangent line is constructed to the solution passing through the point determined by the result of the preceding step, as shown in Figure 2.7.2. The quality of the approximation after many steps depends strongly on the behavior of the set of solutions that pass through the points (t_n, y_n) for $n = 1, 2, 3, \ldots$.

In Example 2 the general solution of the differential equation is

$$y = 6 - 2e^{-t} + ce^{-t/2} \tag{14}$$

and the solution of the initial value problem (10) corresponds to $c = -3$. This family of solutions is a converging family since the term involving the arbitrary constant c approaches zero as $t \to \infty$. It does not matter very much which solutions we are approximating by tangent lines in the implementation of Euler's method, since all the solutions are getting closer and closer to each other as t increases.

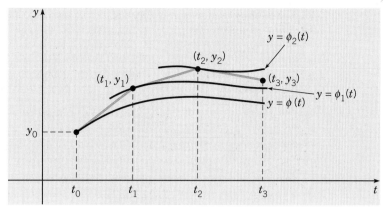

FIGURE 2.7.2 The Euler method.

On the other hand, in Example 3 the general solution of the differential equation is

$$y = -\tfrac{7}{4} + \tfrac{1}{2}t + ce^{2t}, \tag{15}$$

and this is a diverging family. Note that solutions corresponding to two nearby values of c separate arbitrarily far as t increases. In Example 3 we are trying to follow the solution for $c = 11/4$, but in the use of Euler's method we are actually at each step following another solution that separates from the desired one faster and faster as t increases. This explains why the errors in Example 3 are so much larger than those in Example 2.

In using a numerical procedure such as the Euler method, one must always keep in mind the question of whether the results are accurate enough to be useful. In the preceding examples, the accuracy of the numerical results could be ascertained directly by a comparison with the solution obtained analytically. Of course, usually the analytical solution is not available if a numerical procedure is to be employed, so what is needed are bounds for, or at least estimates of, the error that do not require a knowledge of the exact solution. In Chapter 8 we present some information on the analysis of errors and also discuss several algorithms that are computationally more efficient than the Euler method. However, the best that we can expect, or hope for, from a numerical approximation is that it reflect the behavior of the actual solution. Thus a member of a diverging family of solutions will always be harder to approximate than a member of a converging family. Finally, remember that drawing a direction field is often a helpful first step in understanding the behavior of differential equations and their solutions.

PROBLEMS Many of the problems in this section call for fairly extensive numerical computations. The amount of computing that it is reasonable for you to do depends strongly on the type of computing equipment that you have. A few steps of the requested calculations can be carried out on almost any pocket calculator—or even by hand if necessary. To do more, you will find at least a programmable calculator desirable, and for some problems a computer may be needed.

Remember also that numerical results may vary somewhat depending on how your program is constructed and on how your computer executes arithmetic steps, rounds off, and so forth. Minor variations in the last decimal place may be due to such causes and do not necessarily indicate that something is amiss. Answers in the back of the book are recorded to six digits in most cases, although more digits were retained in the intermediate calculations.

In each of Problems 1 through 4:

(a) Find approximate values of the solution of the given initial value problem at $t = 0.1, 0.2,$ 0.3, and 0.4 using the Euler method with $h = 0.1$.

(b) Repeat part (a) with $h = 0.05$. Compare the results with those found in (a).

(c) Repeat part (a) with $h = 0.025$. Compare the results with those found in (a) and (b).

(d) Find the solution $y = \phi(t)$ of the given problem and evaluate $\phi(t)$ at $t = 0.1, 0.2, 0.3,$ and 0.4. Compare these values with the results of (a), (b), and (c).

1. $y' = 3 + t - y,$ $y(0) = 1$ 2. $y' = 2y - 1,$ $y(0) = 1$
3. $y' = 0.5 - t + 2y,$ $y(0) = 1$ 4. $y' = 3\cos t - 2y,$ $y(0) = 0$

In each of Problems 5 through 10 draw a direction field for the given differential equation and state whether you think that the solutions are converging or diverging.

5. $y' = 5 - 3\sqrt{y}$ 6. $y' = y(3 - ty)$
7. $y' = (4 - ty)/(1 + y^2)$ 8. $y' = -ty + 0.1y^3$
9. $y' = t^2 + y^2$ 10. $y' = (y^2 + 2ty)/(3 + t^2)$

In each of Problems 11 through 14 use Euler's method to find approximate values of the solution of the given initial value problem at $t = 0.5, 1, 1.5, 2, 2.5,$ and 3:

(a) With $h = 0.1$. (b) With $h = 0.05$.
(c) With $h = 0.025$. (d) With $h = 0.01$.

11. $y' = 5 - 3\sqrt{y},$ $y(0) = 2$
12. $y' = y(3 - ty),$ $y(0) = 0.5$
13. $y' = (4 - ty)/(1 + y^2),$ $y(0) = -2$
14. $y' = -ty + 0.1y^3,$ $y(0) = 1$

15. Consider the initial value problem

$$y' = 3t^2/(3y^2 - 4), \qquad y(1) = 0.$$

(a) Use the Euler formula (6) with $h = 0.1$ to obtain approximate values of the solution at $t = 1.2, 1.4, 1.6,$ and 1.8.

(b) Repeat part (a) with $h = 0.05$.

(c) Compare the results of parts (a) and (b). Note that they are reasonably close for $t = 1.2, 1.4,$ and 1.6 but are quite different for $t = 1.8$. Also note (from the differential equation) that the line tangent to the solution is parallel to the y-axis when $y = \pm 2/\sqrt{3} \cong \pm 1.155$. Explain how this might cause such a difference in the calculated values.

16. Consider the initial value problem

$$y' = t^2 + y^2, \qquad y(0) = 1.$$

Use Euler's method with $h = 0.1, 0.05, 0.025,$ and 0.01 to explore the solution of this problem for $0 \le t \le 1$. What is your best estimate of the value of the solution at $t = 0.8$? At $t = 1$? Are your results consistent with the direction field in Problem 9?

17. Consider the initial value problem

$$y' = (y^2 + 2ty)/(3 + t^2), \qquad y(1) = 2.$$

Use Euler's method with $h = 0.1$, 0.05, 0.025, and 0.01 to explore the solution of this problem for $1 \leq t \leq 3$. What is your best estimate of the value of the solution at $t = 2.5$? At $t = 3$? Are your results consistent with the direction field in Problem 10?

 18. Consider the initial value problem

$$y' = -ty + 0.1y^3, \qquad y(0) = \alpha,$$

where α is a given number.

(a) Draw a direction field for the differential equation (or reexamine the one from Problem 8). Observe that there is a critical value of α in the interval $2 \leq \alpha \leq 3$ that separates converging solutions from diverging ones. Call this critical value α_0.

(b) Use Euler's method with $h = 0.01$ to estimate α_0. Do this by restricting α_0 to an interval $[a, b]$, where $b - a = 0.01$.

 19. Consider the initial value problem

$$y' = y^2 - t^2, \qquad y(0) = \alpha,$$

where α is a given number.

(a) Draw a direction field for the differential equation. Observe that there is a critical value of α in the interval $0 \leq \alpha \leq 1$ that separates converging solutions from diverging ones. Call this critical value α_0.

(b) Use Euler's method with $h = 0.01$ to estimate α_0. Do this by restricting α_0 to an interval $[a, b]$, where $b - a = 0.01$.

20. **Convergence of Euler's Method.** It can be shown that, under suitable conditions on f, the numerical approximation generated by the Euler method for the initial value problem $y' = f(t, y)$, $y(t_0) = y_0$ converges to the exact solution as the step size h decreases. This is illustrated by the following example. Consider the initial value problem

$$y' = 1 - t + y, \quad y(t_0) = y_0.$$

(a) Show that the exact solution is $y = \phi(t) = (y_0 - t_0)e^{t-t_0} + t$.

(b) Using the Euler formula, show that

$$y_k = (1 + h)y_{k-1} + h - ht_{k-1}, \qquad k = 1, 2, \ldots.$$

(c) Noting that $y_1 = (1 + h)(y_0 - t_0) + t_1$, show by induction that

$$y_n = (1 + h)^n(y_0 - t_0) + t_n \tag{i}$$

for each positive integer n.

(d) Consider a fixed point $t > t_0$ and for a given n choose $h = (t - t_0)/n$. Then $t_n = t$ for every n. Note also that $h \to 0$ as $n \to \infty$. By substituting for h in Eq. (i) and letting $n \to \infty$, show that $y_n \to \phi(t)$ as $n \to \infty$.
Hint: $\lim_{n \to \infty} (1 + a/n)^n = e^a$.

In each of Problems 21 through 23 use the technique discussed in Problem 20 to show that the approximation obtained by the Euler method converges to the exact solution at any fixed point as $h \to 0$.

21. $y' = y, \qquad y(0) = 1$

22. $y' = 2y - 1, \qquad y(0) = 1 \qquad$ *Hint:* $y_1 = (1 + 2h)/2 + 1/2$

23. $y' = \frac{1}{2} - t + 2y, \qquad y(0) = 1 \qquad$ *Hint:* $y_1 = (1 + 2h) + t_1/2$

2.8 The Existence and Uniqueness Theorem

In this section we discuss the proof of Theorem 2.4.2, the fundamental existence and uniqueness theorem for first order initial value problems. This theorem states that under certain conditions on $f(t, y)$, the initial value problem

$$y' = f(t, y), \qquad y(t_0) = y_0 \tag{1}$$

has a unique solution in some interval containing the point t_0.

In some cases (for example, if the differential equation is linear) the existence of a solution of the initial value problem (1) can be established directly by actually solving the problem and exhibiting a formula for the solution. However, in general, this approach is not feasible because there is no method of solving the differential equation that applies in all cases. Therefore, for the general case, it is necessary to adopt an indirect approach that demonstrates the existence of a solution of Eqs. (1) but usually does not provide a practical means of finding it. The heart of this method is the construction of a sequence of functions that converges to a limit function satisfying the initial value problem, although the members of the sequence individually do not. As a rule, it is impossible to compute explicitly more than a few members of the sequence; therefore the limit function can be determined only in rare cases. Nevertheless, under the restrictions on $f(t, y)$ stated in Theorem 2.4.2, it is possible to show that the sequence in question converges and that the limit function has the desired properties. The argument is fairly intricate and depends, in part, on techniques and results that are usually encountered for the first time in a course on advanced calculus. Consequently, we do not go into all the details of the proof here; we do, however, indicate its main features and point out some of the difficulties that must be overcome.

First of all, we note that it is sufficient to consider the problem in which the initial point (t_0, y_0) is the origin; that is, we consider the problem

$$y' = f(t, y), \qquad y(0) = 0. \tag{2}$$

If some other initial point is given, then we can always make a preliminary change of variables, corresponding to a translation of the coordinate axes, that will take the given point (t_0, y_0) into the origin. The existence and uniqueness theorem can now be stated in the following way.

Theorem 2.8.1 If f and $\partial f / \partial y$ are continuous in a rectangle $R: |t| \leq a$, $|y| \leq b$, then there is some interval $|t| \leq h \leq a$ in which there exists a unique solution $y = \phi(t)$ of the initial value problem (2).

For the method of proof discussed here it is necessary to transform the initial value problem (2) into a more convenient form. If we suppose temporarily that there is a function $y = \phi(t)$ that satisfies the initial value problem, then $f[t, \phi(t)]$ is a continuous function of t only. Hence we can integrate $y' = f(t, y)$ from the initial point $t = 0$ to an arbitrary value of t, obtaining

$$\phi(t) = \int_0^t f[s, \phi(s)] \, ds, \tag{3}$$

where we have made use of the initial condition $\phi(0) = 0$. We also denote the dummy variable of integration by s.

Since Eq. (3) contains an integral of the unknown function ϕ, it is called an **integral equation**. This integral equation is not a formula for the solution of the initial value problem, but it does provide another relation satisfied by any solution of Eqs. (2). Conversely, suppose that there is a continuous function $y = \phi(t)$ that satisfies the integral equation (3); then this function also satisfies the initial value problem (2). To show this, we first substitute zero for t in Eq. (3), which shows that the initial condition is satisfied. Further, since the integrand in Eq. (3) is continuous, it follows from the fundamental theorem of calculus that $\phi'(t) = f[t, \phi(t)]$. Therefore the initial value problem and the integral equation are equivalent in the sense that any solution of one is also a solution of the other. It is more convenient to show that there is a unique solution of the integral equation in a certain interval $|t| \leq h$. The same conclusion will then hold also for the initial value problem.

One method of showing that the integral equation (3) has a unique solution is known as the **method of successive approximations** or Picard's[18] **iteration method**. In using this method, we start by choosing an initial function ϕ_0, either arbitrarily or to approximate in some way the solution of the initial value problem. The simplest choice is

$$\phi_0(t) = 0; \tag{4}$$

then ϕ_0 at least satisfies the initial condition in Eqs. (2), although presumably not the differential equation. The next approximation ϕ_1 is obtained by substituting $\phi_0(s)$ for $\phi(s)$ in the right side of Eq. (3) and calling the result of this operation $\phi_1(t)$. Thus

$$\phi_1(t) = \int_0^t f[s, \phi_0(s)]\, ds. \tag{5}$$

Similarly, ϕ_2 is obtained from ϕ_1:

$$\phi_2(t) = \int_0^t f[s, \phi_1(s)]\, ds, \tag{6}$$

and, in general,

$$\phi_{n+1}(t) = \int_0^t f[s, \phi_n(s)]\, ds. \tag{7}$$

In this manner we generate the sequence of functions $\{\phi_n\} = \phi_0, \phi_1, \ldots, \phi_n, \ldots$. Each member of the sequence satisfies the initial condition, but in general none satisfies the differential equation. However, if at some stage, say for $n = k$, we find that $\phi_{k+1}(t) = \phi_k(t)$, then it follows that ϕ_k is a solution of the integral equation (3). Hence ϕ_k is also a solution of the initial value problem (2), and the sequence is terminated at this point. In general, this does not occur, and it is necessary to consider the entire infinite sequence.

[18]Charles-Émile Picard (1856–1914), except for Henri Poincaré, perhaps the most distinguished French mathematician of his generation, was appointed professor at the Sorbonne before the age of 30. He is known for important theorems in complex variables and algebraic geometry as well as differential equations. A special case of the method of successive approximations was first published by Liouville in 1838. However, the method is usually credited to Picard, who established it in a general and widely applicable form in a series of papers beginning in 1890.

To establish Theorem 2.8.1 we must answer four principal questions:

1. Do all members of the sequence $\{\phi_n\}$ exist, or may the process break down at some stage?
2. Does the sequence converge?
3. What are the properties of the limit function? In particular, does it satisfy the integral equation (3) and hence the initial value problem (2)?
4. Is this the only solution, or may there be others?

We first show how these questions can be answered in a specific and relatively simple example and then comment on some of the difficulties that may be encountered in the general case.

EXAMPLE
1

Solve the initial value problem

$$y' = 2t(1 + y), \qquad y(0) = 0 \tag{8}$$

by the method of successive approximations.

Note first that if $y = \phi(t)$, then the corresponding integral equation is

$$\phi(t) = \int_0^t 2s[1 + \phi(s)]\, ds. \tag{9}$$

If the initial approximation is $\phi_0(t) = 0$, it follows that

$$\phi_1(t) = \int_0^t 2s[1 + \phi_0(s)]\, ds = \int_0^t 2s\, ds = t^2. \tag{10}$$

Similarly,

$$\phi_2(t) = \int_0^t 2s[1 + \phi_1(s)]\, ds = \int_0^t 2s[1 + s^2]\, ds = t^2 + \frac{t^4}{2} \tag{11}$$

and

$$\phi_3(t) = \int_0^t 2s[1 + \phi_2(s)]\, ds = \int_0^t 2s\left[1 + s^2 + \frac{s^4}{2}\right] ds = t^2 + \frac{t^4}{2} + \frac{t^6}{2 \cdot 3}. \tag{12}$$

Equations (10), (11), and (12) suggest that

$$\phi_n(t) = t^2 + \frac{t^4}{2!} + \frac{t^6}{3!} + \cdots + \frac{t^{2n}}{n!} \tag{13}$$

for each $n \geq 1$, and this result can be established by mathematical induction. Equation (13) is certainly true for $n = 1$; see Eq. (10). We must show that if it is true for $n = k$, then it also holds for $n = k + 1$. We have

$$\begin{aligned}
\phi_{k+1}(t) &= \int_0^t 2s[1 + \phi_k(s)]\, ds \\
&= \int_0^t 2s\left(1 + s^2 + \frac{s^4}{2!} + \cdots + \frac{s^{2k}}{k!}\right) ds \\
&= t^2 + \frac{t^4}{2!} + \frac{t^6}{3!} + \cdots + \frac{t^{2k+2}}{(k+1)!},
\end{aligned} \tag{14}$$

and the inductive proof is complete.

A plot of the first four iterates, $\phi_1(t), \ldots, \phi_4(t)$, is shown in Figure 2.8.1. As k increases, the iterates seem to remain close over a gradually increasing interval, suggesting eventual convergence to a limit function.

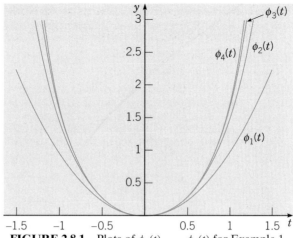

FIGURE 2.8.1 Plots of $\phi_1(t), \ldots, \phi_4(t)$ for Example 1.

It follows from Eq. (13) that $\phi_n(t)$ is the nth partial sum of the infinite series

$$\sum_{k=1}^{\infty} \frac{t^{2k}}{k!} \, ; \tag{15}$$

hence $\lim_{n \to \infty} \phi_n(t)$ exists if and only if the series (15) converges. Applying the ratio test, we see that, for each t,

$$\left| \frac{t^{2k+2}}{(k+1)!} \, \frac{k!}{t^{2k}} \right| = \frac{t^2}{k+1} \to 0 \quad \text{as} \quad k \to \infty. \tag{16}$$

Thus the series (15) converges for all t, and its sum $\phi(t)$ is the limit of the sequence $\{\phi_n(t)\}$. Further, since the series (15) is a Taylor series, it can be differentiated or integrated term by term as long as t remains within the interval of convergence, which in this case is the entire t-axis. Therefore, we can verify by direct computation that $\phi(t) = \sum_{k=1}^{\infty} t^{2k}/k!$ is a solution of the integral equation (9). Alternatively, by substituting $\phi(t)$ for y in Eqs. (8), we can verify that this function satisfies the initial value problem. In this example it is also possible, from the series (15), to identify ϕ in terms of elementary functions, namely, $\phi(t) = e^{t^2} - 1$. However, this is not necessary for the discussion of existence and uniqueness.

Explicit knowledge of $\phi(t)$ does make it possible to visualize the convergence of the sequence of iterates more clearly by plotting $\phi(t) - \phi_k(t)$ for various values of k. Figure 2.8.2 shows this difference for $k = 1, \ldots, 4$. This figure clearly shows the gradually increasing interval over which successive iterates provide a good approximation to the solution of the initial value problem.

Finally, to deal with the question of uniqueness, let us suppose that the initial value problem has two solutions ϕ and ψ. Since ϕ and ψ both satisfy the integral equation (9), we have by subtraction that

$$\phi(t) - \psi(t) = \int_0^t 2s[\phi(s) - \psi(s)] \, ds.$$

Taking absolute values of both sides, we have, if $t > 0$,

$$|\phi(t) - \psi(t)| = \left| \int_0^t 2s[\phi(s) - \psi(s)] \, ds \right| \le \int_0^t 2s|\phi(s) - \psi(s)| \, ds.$$

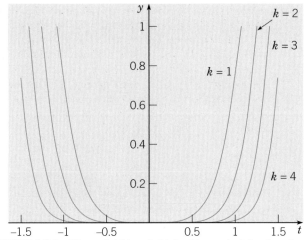

FIGURE 2.8.2 Plots of $\phi(t) - \phi_k(t)$ for Example 1 for $k = 1, \ldots, 4$.

If we restrict t to lie in the interval $0 \leq t \leq A/2$, where A is arbitrary, then $2t \leq A$, and

$$|\phi(t) - \psi(t)| \leq A \int_0^t |\phi(s) - \psi(s)| \, ds. \tag{17}$$

It is now convenient to introduce the function U defined by

$$U(t) = \int_0^t |\phi(s) - \psi(s)| \, ds. \tag{18}$$

Then it follows at once that

$$U(0) = 0, \tag{19}$$

$$U(t) \geq 0, \quad \text{for} \quad t \geq 0. \tag{20}$$

Further, U is differentiable, and $U'(t) = |\phi(t) - \psi(t)|$. Hence, by Eq. (17),

$$U'(t) - AU(t) \leq 0. \tag{21}$$

Multiplying Eq. (21) by the positive quantity e^{-At} gives

$$[e^{-At}U(t)]' \leq 0. \tag{22}$$

Then, upon integrating Eq. (22) from zero to t and using Eq. (19), we obtain

$$e^{-At}U(t) \leq 0 \quad \text{for} \quad t \geq 0.$$

Hence $U(t) \leq 0$ for $t \geq 0$, and in conjunction with Eq. (20), this requires that $U(t) = 0$ for each $t \geq 0$. Thus $U'(t) \equiv 0$, and therefore $\psi(t) \equiv \phi(t)$, which contradicts the original hypothesis. Consequently, there cannot be two different solutions of the initial value problem for $t \geq 0$. A slight modification of this argument leads to the same conclusion for $t \leq 0$.

Returning now to the general problem of solving the integral equation (3), let us consider briefly each of the questions raised earlier:

1. Do all members of the sequence $\{\phi_n\}$ exist? In the example, f and $\partial f/\partial y$ were continuous in the whole ty-plane, and each member of the sequence could be explicitly calculated. In contrast, in the general case, f and $\partial f/\partial y$ are assumed to be continuous only in the rectangle R: $|t| \leq a$, $|y| \leq b$ (see Figure 2.8.3). Furthermore, the members of the sequence cannot as a rule be explicitly determined. The danger is that at some stage, say for $n = k$, the graph of $y = \phi_k(t)$ may contain points that lie outside the rectangle R. Hence at the next stage—in the computation of $\phi_{k+1}(t)$—it would be necessary to evaluate $f(t,y)$ at points where it is not known to be continuous or even to exist. Thus the calculation of $\phi_{k+1}(t)$ might be impossible.

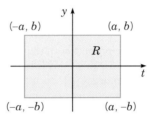

FIGURE 2.8.3 Region of definition for Theorem 2.8.1.

To avoid this danger it may be necessary to restrict t to a smaller interval than $|t| \leq a$. To find such an interval we make use of the fact that a continuous function on a closed bounded region is bounded. Hence f is bounded on R; thus there exists a positive number M such that

$$|f(t,y)| \leq M, \qquad (t,y) \text{ in } R. \qquad (23)$$

We have mentioned before that

$$\phi_n(0) = 0$$

for each n. Since $f[t, \phi_k(t)]$ is equal to $\phi'_{k+1}(t)$, the maximum absolute slope of the graph of the equation $y = \phi_{k+1}(t)$ is M. Since this graph contains the point $(0,0)$, it must lie in the wedge-shaped shaded region in Figure 2.8.4. Hence the point $[t, \phi_{k+1}(t)]$ remains in R at least as long as R contains the wedge-shaped region, which is for $|t| \leq b/M$. We hereafter consider only the rectangle D: $|t| \leq h$, $|y| \leq b$, where h is equal either to a or to b/M, whichever is smaller. With this restriction, all members of the sequence $\{\phi_n(t)\}$ exist. Note that whenever $b/M < a$, then you can try to obtain a larger value of h by finding a better (that is, smaller) bound M for $|f(t, y)|$, if this is possible.

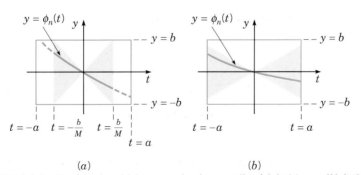

FIGURE 2.8.4 Regions in which successive iterates lie. (a) $b/M < a$; (b) $b/M > a$.

2. Does the sequence $\{\phi_n(t)\}$ converge? As in the example, we can identify $\phi_n(t) = \phi_1(t) + [\phi_2(t) - \phi_1(t)] + \cdots + [\phi_n(t) - \phi_{n-1}(t)]$ as the nth partial sum of the series

$$\phi_1(t) + \sum_{k=1}^{\infty} [\phi_{k+1}(t) - \phi_k(t)]. \tag{24}$$

The convergence of the sequence $\{\phi_n(t)\}$ is established by showing that the series (24) converges. To do this, it is necessary to estimate the magnitude $|\phi_{k+1}(t) - \phi_k(t)|$ of the general term. The argument by which this is done is indicated in Problems 15 through 18 and will be omitted here. Assuming that the sequence converges, we denote the limit function by ϕ, so that

$$\phi(t) = \lim_{n \to \infty} \phi_n(t). \tag{25}$$

3. What are the properties of the limit function ϕ? In the first place, we would like to know that ϕ is continuous. This is not, however, a necessary consequence of the convergence of the sequence $\{\phi_n(t)\}$, even though each member of the sequence is itself continuous. Sometimes a sequence of continuous functions converges to a limit function that is discontinuous. A simple example of this phenomenon is given in Problem 13. One way to show that ϕ is continuous is to show not only that the sequence $\{\phi_n\}$ converges, but also that it converges in a certain manner, known as uniform convergence. We do not take up this question here but note only that the argument referred to in paragraph 2 is sufficient to establish the uniform convergence of the sequence $\{\phi_n\}$ and, hence, the continuity of the limit function ϕ in the interval $|t| \le h$.

Now let us return to Eq. (7),

$$\phi_{n+1}(t) = \int_0^t f[s, \phi_n(s)]\, ds.$$

Allowing n to approach ∞ on both sides, we obtain

$$\phi(t) = \lim_{n \to \infty} \int_0^t f[s, \phi_n(s)]\, ds. \tag{26}$$

We would like to interchange the operations of integrating and taking the limit on the right side of Eq. (26) so as to obtain

$$\phi(t) = \int_0^t \lim_{n \to \infty} f[s, \phi_n(s)]\, ds. \tag{27}$$

In general, such an interchange is not permissible (see Problem 14, for example), but once again, the fact that the sequence $\{\phi_n(t)\}$ converges uniformly is sufficient to allow us to take the limiting operation inside the integral sign. Next, we wish to take the limit inside the function f, which would give

$$\phi(t) = \int_0^t f[s, \lim_{n \to \infty} \phi_n(s)]\, ds \tag{28}$$

and hence

$$\phi(t) = \int_0^t f[s, \phi(s)]\, ds. \tag{29}$$

The statement that $\lim_{n \to \infty} f[s, \phi_n(s)] = f[s, \lim_{n \to \infty} \phi_n(s)]$ is equivalent to the statement that f is continuous in its second variable, which is known by hypothesis. Hence Eq. (29) is valid,

and the function ϕ satisfies the integral equation (3). Thus ϕ is also a solution of the initial value problem (2).

4. Are there other solutions of the integral equation (3) besides $y = \phi(t)$? To show the uniqueness of the solution $y = \phi(t)$, we can proceed much as in the example. First, assume the existence of another solution $y = \psi(t)$. It is then possible to show (see Problem 19) that the difference $\phi(t) - \psi(t)$ satisfies the inequality

$$|\phi(t) - \psi(t)| \le A \int_0^t |\phi(s) - \psi(s)|\, ds \tag{30}$$

for $0 \le t \le h$ and a suitable positive number A. From this point the argument is identical to that given in the example, and we conclude that there is no solution of the initial value problem (2) other than the one generated by the method of successive approximations.

PROBLEMS

In each of Problems 1 and 2 transform the given initial value problem into an equivalent problem with the initial point at the origin.

1. $dy/dt = t^2 + y^2$, $y(1) = 2$ 2. $dy/dt = 1 - y^3$, $y(-1) = 3$

In each of Problems 3 through 6 let $\phi_0(t) = 0$ and use the method of successive approximations to solve the given initial value problem.

(a) Determine $\phi_n(t)$ for an arbitrary value of n.

(b) Plot $\phi_n(t)$ for $n = 1, \ldots, 4$. Observe whether the iterates appear to be converging.

(c) Express $\lim_{n \to \infty} \phi_n(t) = \phi(t)$ in terms of elementary functions; that is, solve the given initial value problem.

(d) Plot $|\phi(t) - \phi_n(t)|$ for $n = 1, \ldots, 4$. For each of $\phi_1(t), \ldots, \phi_4(t)$, estimate the interval in which it is a reasonably good approximation to the actual solution.

3. $y' = 2(y + 1)$, $y(0) = 0$ 4. $y' = -y - 1$, $y(0) = 0$

5. $y' = -y/2 + t$, $y(0) = 0$ 6. $y' = y + 1 - t$, $y(0) = 0$

In each of Problems 7 and 8 let $\phi_0(t) = 0$ and use the method of successive approximations to solve the given initial value problem.

(a) Determine $\phi_n(t)$ for an arbitrary value of n.

(b) Plot $\phi_n(t)$ for $n = 1, \ldots, 4$. Observe whether the iterates appear to be converging.

7. $y' = ty + 1$, $y(0) = 0$ 8. $y' = t^2 y - t$, $y(0) = 0$

In each of Problems 9 and 10 let $\phi_0(t) = 0$ and use the method of successive approximations to approximate the solution of the given initial value problem.

(a) Calculate $\phi_1(t), \ldots, \phi_3(t)$.

(b) Plot $\phi_1(t), \ldots, \phi_3(t)$ and observe whether the iterates appear to be converging.

9. $y' = t^2 + y^2$, $y(0) = 0$ 10. $y' = 1 - y^3$, $y(0) = 0$

In each of Problems 11 and 12 let $\phi_0(t) = 0$ and use the method of successive approximations to approximate the solution of the given initial value problem.

(a) Calculate $\phi_1(t), \ldots, \phi_4(t)$, or (if necessary) Taylor approximations to these iterates. Keep terms up to order six.

(b) Plot the functions you found in part (a) and observe whether they appear to be converging.

11. $y' = -\sin y + 1$, $y(0) = 0$ 12. $y' = (3t^2 + 4t + 2)/2(y - 1)$, $y(0) = 0$

13. Let $\phi_n(x) = x^n$ for $0 \le x \le 1$ and show that

$$\lim_{n \to \infty} \phi_n(x) = \begin{cases} 0, & 0 \le x < 1, \\ 1, & x = 1. \end{cases}$$

This example shows that a sequence of continuous functions may converge to a limit function that is discontinuous.

14. Consider the sequence $\phi_n(x) = 2nxe^{-nx^2}$, $0 \le x \le 1$.

 (a) Show that $\lim\limits_{n \to \infty} \phi_n(x) = 0$ for $0 \le x \le 1$; hence

 $$\int_0^1 \lim_{n \to \infty} \phi_n(x)\, dx = 0.$$

 (b) Show that $\int_0^1 2nxe^{-nx^2}\, dx = 1 - e^{-n}$; hence

 $$\lim_{n \to \infty} \int_0^1 \phi_n(x)\, dx = 1.$$

 Thus, in this example,

 $$\lim_{n \to \infty} \int_a^b \phi_n(x)\, dx \ne \int_a^b \lim_{n \to \infty} \phi_n(x)\, dx,$$

 even though $\lim\limits_{n \to \infty} \phi_n(x)$ exists and is continuous.

In Problems 15 through 18 we indicate how to prove that the sequence $\{\phi_n(t)\}$, defined by Eqs. (4) through (7), converges.

15. If $\partial f / \partial y$ is continuous in the rectangle D, show that there is a positive constant K such that

 $$|f(t, y_1) - f(t, y_2)| \le K|y_1 - y_2|, \tag{i}$$

 where (t, y_1) and (t, y_2) are any two points in D having the same t coordinate. This inequality is known as a Lipschitz[19] condition.

 Hint: Hold t fixed and use the mean value theorem on f as a function of y only. Choose K to be the maximum value of $|\partial f / \partial y|$ in D.

16. If $\phi_{n-1}(t)$ and $\phi_n(t)$ are members of the sequence $\{\phi_n(t)\}$, use the result of Problem 15 to show that

 $$|f[t, \phi_n(t)] - f[t, \phi_{n-1}(t)]| \le K|\phi_n(t) - \phi_{n-1}(t)|.$$

17. (a) Show that if $|t| \le h$, then

 $$|\phi_1(t)| \le M|t|,$$

 where M is chosen so that $|f(t, y)| \le M$ for (t, y) in D.

 (b) Use the results of Problem 16 and part (a) of Problem 17 to show that

 $$|\phi_2(t) - \phi_1(t)| \le \frac{MK|t|^2}{2}.$$

 (c) Show, by mathematical induction, that

 $$|\phi_n(t) - \phi_{n-1}(t)| \le \frac{MK^{n-1}|t|^n}{n!} \le \frac{MK^{n-1}h^n}{n!}.$$

[19]Rudolf Lipschitz (1832–1903), professor at the University of Bonn for many years, worked in several areas of mathematics. The inequality (i) can replace the hypothesis that $\partial f / \partial y$ is continuous in Theorem 2.8.1; this results in a slightly stronger theorem.

18. Note that

$$\phi_n(t) = \phi_1(t) + [\phi_2(t) - \phi_1(t)] + \cdots + [\phi_n(t) - \phi_{n-1}(t)].$$

(a) Show that

$$|\phi_n(t)| \le |\phi_1(t)| + |\phi_2(t) - \phi_1(t)| + \cdots + |\phi_n(t) - \phi_{n-1}(t)|.$$

(b) Use the results of Problem 17 to show that

$$|\phi_n(t)| \le \frac{M}{K}\left[Kh + \frac{(Kh)^2}{2!} + \cdots + \frac{(Kh)^n}{n!}\right].$$

(c) Show that the sum in part (b) converges as $n \to \infty$ and, hence, the sum in part (a) also converges as $n \to \infty$. Conclude therefore that the sequence $\{\phi_n(t)\}$ converges since it is the sequence of partial sums of a convergent infinite series.

19. In this problem we deal with the question of uniqueness of the solution of the integral equation (3),

$$\phi(t) = \int_0^t f[s, \phi(s)]\, ds.$$

(a) Suppose that ϕ and ψ are two solutions of Eq. (3). Show that, for $t \ge 0$,

$$\phi(t) - \psi(t) = \int_0^t \{f[s, \phi(s)] - f[s, \psi(s)]\}\, ds.$$

(b) Show that

$$|\phi(t) - \psi(t)| \le \int_0^t |f[s, \phi(s)] - f[s, \psi(s)]|\, ds.$$

(c) Use the result of Problem 15 to show that

$$|\phi(t) - \psi(t)| \le K \int_0^t |\phi(s) - \psi(s)|\, ds,$$

where K is an upper bound for $|\partial f/\partial y|$ in D. This is the same as Eq. (30), and the rest of the proof may be constructed as indicated in the text.

2.9 First Order Difference Equations

Although a continuous model leading to a differential equation is reasonable and attractive for many problems, there are some cases in which a discrete model may be more natural. For instance, the continuous model of compound interest used in Section 2.3 is only an approximation to the actual discrete process. Similarly, sometimes population growth may be described more accurately by a discrete than by a continuous model. This is true, for example, of species whose generations do not overlap and that propagate at regular intervals, such as at particular times of the calendar year. Then the population y_{n+1} of the species in the year $n + 1$ is some function of n and the population y_n in the preceding year; that is,

$$y_{n+1} = f(n, y_n), \qquad n = 0, 1, 2, \ldots. \tag{1}$$

Equation (1) is called a **first order difference equation**. It is first order because the value of y_{n+1} depends on the value of y_n but not on earlier values y_{n-1}, y_{n-2}, and so forth. As for differential equations, the difference equation (1) is **linear** if f is a linear function of y_n; otherwise, it is **nonlinear**. A **solution** of the difference equation (1) is a sequence of numbers y_0, y_1, y_2, \ldots that satisfy the equation for each n. In addition to the difference equation itself, there may also be an **initial condition**

$$y_0 = \alpha \tag{2}$$

that prescribes the value of the first term of the solution sequence.

We now assume temporarily that the function f in Eq. (1) depends only on y_n, but not on n. In this case

$$y_{n+1} = f(y_n), \qquad n = 0, 1, 2, \ldots . \tag{3}$$

If y_0 is given, then successive terms of the solution can be found from Eq. (3). Thus

$$y_1 = f(y_0),$$

and

$$y_2 = f(y_1) = f[f(y_0)].$$

The quantity $f[f(y_0)]$ is called the second iterate of the difference equation and is sometimes denoted by $f^2(y_0)$. Similarly, the third iterate y_3 is given by

$$y_3 = f(y_2) = f\{f[f(y_0)]\} = f^3(y_0),$$

and so on. In general, the nth iterate y_n is

$$y_n = f(y_{n-1}) = f^n(y_0).$$

This procedure is referred to as iterating the difference equation. It is often of primary interest to determine the behavior of y_n as $n \to \infty$; in particular, does y_n approach a limit, and if so, what is it?

Solutions for which y_n has the same value for all n are called **equilibrium solutions**. They are frequently of special importance, just as in the study of differential equations. If equilibrium solutions exist, one can find them by setting y_{n+1} equal to y_n in Eq. (3) and solving the resulting equation

$$y_n = f(y_n) \tag{4}$$

for y_n.

Linear Equations. Suppose that the population of a certain species in a given region in year $n + 1$, denoted by y_{n+1}, is a positive multiple ρ_n of the population y_n in year n; that is,

$$y_{n+1} = \rho_n y_n, \qquad n = 0, 1, 2, \ldots . \tag{5}$$

Note that the reproduction rate ρ_n may differ from year to year. The difference equation (5) is linear and can easily be solved by iteration. We obtain

$$y_1 = \rho_0 y_0,$$
$$y_2 = \rho_1 y_1 = \rho_1 \rho_0 y_0,$$

and, in general,

$$y_n = \rho_{n-1} \cdots \rho_0 y_0, \qquad n = 1, 2, \ldots . \tag{6}$$

Thus, if the initial population y_0 is given, then the population of each succeeding generation is determined by Eq. (6). Although for a population problem ρ_n is intrinsically positive, the solution (6) is also valid if ρ_n is negative for some or all values of n. Note, however, that if ρ_n is zero for some n, then y_{n+1} and all succeeding values of y are zero; in other words, the species has become extinct.

If the reproduction rate ρ_n has the same value ρ for each n, then the difference equation (5) becomes

$$y_{n+1} = \rho y_n \tag{7}$$

and its solution is

$$y_n = \rho^n y_0. \tag{8}$$

Equation (7) also has an equilibrium solution, namely, $y_n = 0$ for all n, corresponding to the initial value $y_0 = 0$. The limiting behavior of y_n is easy to determine from Eq. (8). In fact,

$$\lim_{n \to \infty} y_n = \begin{cases} 0, & \text{if } |\rho| < 1; \\ y_0, & \text{if } \rho = 1; \\ \text{does not exist}, & \text{otherwise.} \end{cases} \tag{9}$$

In other words, the equilibrium solution $y_n = 0$ is asymptotically stable for $|\rho| < 1$ and unstable if $|\rho| > 1$.

Now we will modify the population model represented by Eq. (5) to include the effect of immigration or emigration. If b_n is the net increase in population in year n due to immigration, then the population in year $n + 1$ is the sum of those due to natural reproduction and those due to immigration. Thus

$$y_{n+1} = \rho y_n + b_n, \qquad n = 0, 1, 2, \ldots, \tag{10}$$

where we are now assuming that the reproduction rate ρ is constant. We can solve Eq. (10) by iteration in the same manner as before. We have

$$y_1 = \rho y_0 + b_0,$$
$$y_2 = \rho(\rho y_0 + b_0) + b_1 = \rho^2 y_0 + \rho b_0 + b_1,$$
$$y_3 = \rho(\rho^2 y_0 + \rho b_0 + b_1) + b_2 = \rho^3 y_0 + \rho^2 b_0 + \rho b_1 + b_2,$$

and so forth. In general, we obtain

$$y_n = \rho^n y_0 + \rho^{n-1} b_0 + \cdots + \rho b_{n-2} + b_{n-1} = \rho^n y_0 + \sum_{j=0}^{n-1} \rho^{n-1-j} b_j. \tag{11}$$

Note that the first term on the right side of Eq. (11) represents the descendants of the original population, while the other terms represent the population in year n resulting from immigration in all preceding years.

In the special case where $b_n = b \neq 0$ for all n, the difference equation is

$$y_{n+1} = \rho y_n + b, \tag{12}$$

and from Eq. (11) its solution is

$$y_n = \rho^n y_0 + (1 + \rho + \rho^2 + \cdots + \rho^{n-1})b. \tag{13}$$

If $\rho \neq 1$, we can write this solution in the more compact form

$$y_n = \rho^n y_0 + \frac{1 - \rho^n}{1 - \rho} b, \tag{14}$$

where again the two terms on the right side are the effects of the original population and of immigration, respectively. Rewriting Eq. (14) as

$$y_n = \rho^n \left(y_0 - \frac{b}{1 - \rho} \right) + \frac{b}{1 - \rho} \tag{15}$$

makes the long-time behavior of y_n more evident. It follows from Eq. (15) that $y_n \rightarrow b/(1 - \rho)$ if $|\rho| < 1$. If $|\rho| > 1$ or if $\rho = -1$ then y_n has no limit unless $y_0 = b/(1 - \rho)$. The quantity $b/(1 - \rho)$, for $\rho \neq 1$, is an equilibrium solution of Eq. (12), as can readily be seen directly from that equation. Of course, Eq. (14) is not valid for $\rho = 1$. To deal with that case, we must return to Eq. (13) and let $\rho = 1$ there. It follows that

$$y_n = y_0 + nb, \tag{16}$$

so in this case y_n becomes unbounded as $n \rightarrow \infty$.

The same model also provides a framework for solving many problems of a financial character. For such problems y_n is the account balance in the nth time period, $\rho_n = 1 + r_n$, where r_n is the interest rate for that period, and b_n is the amount deposited or withdrawn. The following example is typical.

EXAMPLE

1

A recent college graduate takes out a \$10,000 loan to purchase a car. If the interest rate is 12%, what monthly payment is required to pay off the loan in 4 years?

The relevant difference equation is Eq. (12), where y_n is the loan balance outstanding in the nth month, $\rho = 1 + r$, where r is the interest rate per month, and b is the monthly payment. Note that b must be negative and $\rho = 1.01$, corresponding to a monthly interest rate of 1%.

The solution of the difference equation (12) with this value for ρ and the initial condition $y_0 = 10,000$ is given by Eq. (15); that is,

$$y_n = (1.01)^n (10,000 + 100b) - 100b. \tag{17}$$

The payment b needed to pay off the loan in 4 years is found by setting $y_{48} = 0$ and solving for b. This gives

$$b = -100 \frac{(1.01)^{48}}{(1.01)^{48} - 1} = -263.34. \tag{18}$$

The total amount paid on the loan is 48 times b, or \$12,640.32. Of this amount, \$10,000 is repayment of the principal and the remaining \$2640.32 is interest.

Nonlinear Equations. Nonlinear difference equations are much more complicated and have much more varied solutions than linear equations. We will restrict our attention to a single equation, the logistic difference equation

$$y_{n+1} = \rho y_n \left(1 - \frac{y_n}{k} \right), \tag{19}$$

which is analogous to the logistic differential equation

$$\frac{dy}{dt} = ry \left(1 - \frac{y}{K} \right) \tag{20}$$

that was discussed in Section 2.5. Note that if the derivative dy/dt in Eq. (20) is replaced by the difference $(y_{n+1} - y_n)/h$, then Eq. (20) reduces to Eq. (19) with $\rho = 1 + hr$ and $k = (1 + hr)K/hr$. To simplify Eq. (19) a little more, we can scale the variable y_n by introducing the new variable $u_n = y_n/k$. Then Eq. (19) becomes

$$u_{n+1} = \rho u_n (1 - u_n), \tag{21}$$

where ρ is a positive parameter.

We begin our investigation of Eq. (21) by seeking the equilibrium, or constant, solutions. These can be found by setting u_{n+1} equal to u_n in Eq. (21), which corresponds to setting dy/dt equal to zero in Eq. (20). The resulting equation is

$$u_n = \rho u_n - \rho u_n^2, \tag{22}$$

so it follows that the equilibrium solutions of Eq. (21) are

$$u_n = 0, \qquad u_n = \frac{\rho - 1}{\rho}. \tag{23}$$

The next question is whether the equilibrium solutions are asymptotically stable or unstable. That is, for an initial condition near one of the equilibrium solutions, does the resulting solution sequence approach or depart from the equilibrium solution? One way to examine this question is by approximating Eq. (21) by a linear equation in the neighborhood of an equilibrium solution. For example, near the equilibrium solution $u_n = 0$, the quantity u_n^2 is small compared to u_n itself, so we assume that we can neglect the quadratic term in Eq. (21) in comparison with the linear terms. This leaves us with the linear difference equation

$$u_{n+1} = \rho u_n, \tag{24}$$

which is presumably a good approximation to Eq. (21) for u_n sufficiently near zero. However, Eq. (24) is the same as Eq. (7), and we have already concluded, in Eq. (9), that $u_n \to 0$ as $n \to \infty$ if and only if $|\rho| < 1$, or (since ρ must be positive) for $0 < \rho < 1$. Thus the equilibrium solution $u_n = 0$ is asymptotically stable for the linear approximation (24) for this set of ρ values, so we conclude that it is also asymptotically stable for the full nonlinear equation (21). This conclusion is correct, although our argument is not complete. What is lacking is a theorem stating that the solutions of the nonlinear equation (21) resemble those of the linear equation (24) near the equilibrium solution $u_n = 0$. We will not take time to discuss this issue here; the same question is treated for differential equations in Section 9.3.

Now consider the other equilibrium solution $u_n = (\rho - 1)/\rho$. To study solutions in the neighborhood of this point, we write

$$u_n = \frac{\rho - 1}{\rho} + v_n, \tag{25}$$

where we assume that v_n is small. By substituting from Eq. (25) in Eq. (21) and simplifying the resulting equation, we eventually obtain

$$v_{n+1} = (2 - \rho)v_n - \rho v_n^2. \tag{26}$$

Since v_n is small, we again neglect the quadratic term in comparison with the linear terms and thereby obtain the linear equation

$$v_{n+1} = (2 - \rho)v_n. \tag{27}$$

Referring to Eq. (9) once more, we find that $v_n \to 0$ as $n \to \infty$ for $|2 - \rho| < 1$, that is, for $1 < \rho < 3$. Therefore we conclude that, for this range of values of ρ, the equilibrium solution $u_n = (\rho - 1)/\rho$ is asymptotically stable.

Figure 2.9.1 contains the graphs of solutions of Eq. (21) for $\rho = 0.8, \rho = 1.5$, and $\rho = 2.8$, respectively. Observe that the solution converges to zero for $\rho = 0.8$ and to the nonzero equilibrium solution for $\rho = 1.5$ and $\rho = 2.8$. The convergence is monotone for $\rho = 0.8$ and $\rho = 1.5$ and is oscillatory for $\rho = 2.8$. The graphs shown are for particular initial conditions, but the graphs for other initial conditions are similar.

FIGURE 2.9.1 Solutions of $u_{n+1} = \rho u_n(1 - u_n)$: (a) $\rho = 0.8$; (b) $\rho = 1.5$; (c) $\rho = 2.8$.

Another way of displaying the solution of a difference equation is shown in Figure 2.9.2. In each part of this figure the graphs of the parabola $y = \rho x(1 - x)$ and of the straight line $y = x$ are shown. The equilibrium solutions correspond to the points of intersection of these two curves. The piecewise linear graph consisting of successive vertical and horizontal line segments, sometimes called a stairstep diagram, represents the solution sequence. The sequence starts at the point u_0 on the x-axis. The vertical line segment drawn upward to the parabola at u_0 corresponds to the calculation of $\rho u_0(1 - u_0) = u_1$. This value is then transferred from the y-axis to the x-axis; this step is represented by the horizontal line segment from the parabola to the line $y = x$. Then the process is repeated over and over again. Clearly, the sequence converges to the origin in Figure 2.9.2a and to the nonzero equilibrium solution in the other two cases.

To summarize our results so far: The difference equation (21) has two equilibrium solutions, $u_n = 0$ and $u_n = (\rho - 1)/\rho$; the former is asymptotically stable for $0 \le \rho < 1$, and the latter is asymptotically stable for $1 < \rho < 3$. When $\rho = 1$ the two equilibrium solutions coincide at $u = 0$; this solution can be shown to be asymptotically stable. In Figure 2.9.3 the parameter ρ is plotted on the horizontal axis and u on the vertical axis. The equilibrium solutions $u = 0$ and $u = (\rho - 1)/\rho$ are shown. The intervals in which each one is asymptotically stable are indicated by the solid portions of the curves. There is an **exchange of stability** from one equilibrium solution to the other at $\rho = 1$.

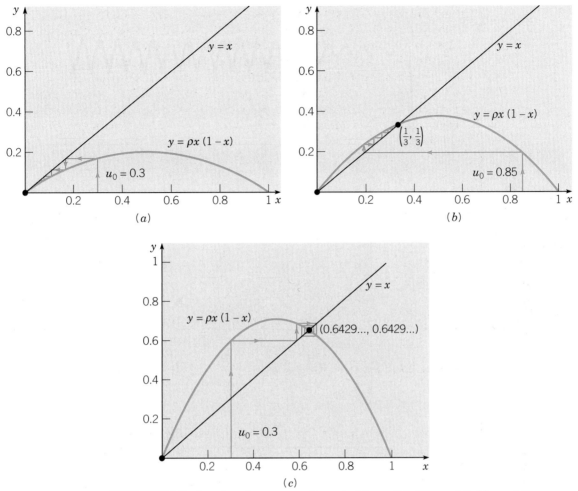

FIGURE 2.9.2 Iterates of $u_{n+1} = \rho u_n(1 - u_n)$. (a) $\rho = 0.8$; (b) $\rho = 1.5$; (c) $\rho = 2.8$.

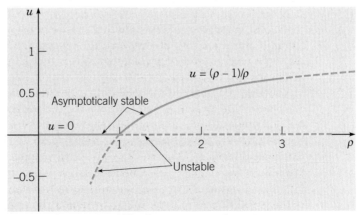

FIGURE 2.9.3 Exchange of stability for $u_{n+1} = \rho u_n(1 - u_n)$.

(a)

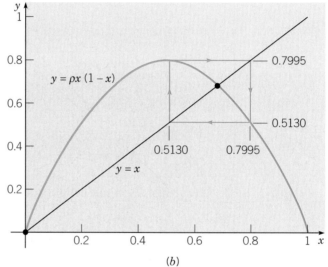

(b)

FIGURE 2.9.4 A solution of $u_{n+1} = \rho u_n(1 - u_n)$ for $\rho = 3.2$; period 2. (a) u_n versus n; (b) a two-cycle.

For $\rho > 3$ neither of the equilibrium solutions is stable, and the solutions of Eq. (21) exhibit increasing complexity as ρ increases. For ρ somewhat greater than 3, the sequence u_n rapidly approaches a steady oscillation of period 2; that is, u_n oscillates back and forth between two distinct values. For $\rho = 3.2$ a solution is shown in Figure 2.9.4. For n greater than about 20, the solution alternates between the values 0.5130 and 0.7995. The graph is drawn for the particular initial condition $u_0 = 0.3$, but it is similar for all other initial values between 0 and 1. Figure 2.9.4b also shows the same steady oscillation as a rectangular path that is traversed repeatedly in the clockwise direction. At about $\rho = 3.449$, each state in the oscillation of period 2 separates into two distinct states, and the solution becomes periodic with period 4; see Figure 2.9.5, which shows a solution of period 4 for $\rho = 3.5$. As ρ increases further, periodic solutions of period 8, 16, ... appear. The appearance of a new solution at a certain parameter value is called a **bifurcation**.

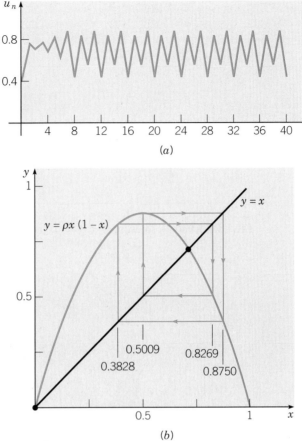

FIGURE 2.9.5 A solution of $u_{n+1} = \rho u_n(1 - u_n)$ for $\rho = 3.5$; period 4. (*a*) u_n versus n; (*b*) a four-cycle.

The ρ-values at which the successive period doublings occur approach a limit that is approximately 3.57. For $\rho > 3.57$ the solutions possess some regularity, but no discernible detailed pattern for most values of ρ. For example, a solution for $\rho = 3.65$ is shown in Figure 2.9.6. It oscillates between approximately 0.3 and 0.9, but its fine structure is unpredictable. The term **chaotic** is used to describe this situation. One of the features of chaotic solutions is extreme sensitivity to the initial conditions. This is illustrated in Figure 2.9.7, where two solutions of Eq. (21) for $\rho = 3.65$ are shown. One solution is the same as that in Figure 2.9.6 and has the initial value $u_0 = 0.3$, while the other solution has the initial value $u_0 = 0.305$. For about 15 iterations the two solutions remain close and are hard to distinguish from each other in the figure. After that, although they continue to wander about in approximately the same set of values, their graphs are quite dissimilar. It would certainly not be possible to use one of these solutions to estimate the value of the other for values of n larger than about 15.

It is only comparatively recently that chaotic solutions of difference and differential equations have become widely known. Equation (20) was one of the first instances

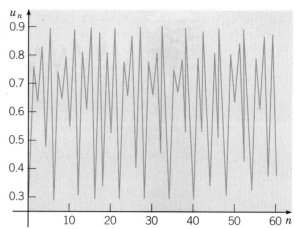

FIGURE 2.9.6 A solution of $u_{n+1} = \rho u_n(1 - u_n)$ for $\rho = 3.65$; a chaotic solution.

FIGURE 2.9.7 Two solutions of $u_{n+1} = \rho u_n(1 - u_n)$ for $\rho = 3.65$; $u_0 = 0.3$ and $u_0 = 0.305$.

of mathematical chaos to be found and studied in detail, by Robert May[20] in 1974. On the basis of his analysis of this equation as a model of the population of certain insect species, May suggested that if the growth rate ρ is too large, then it will be impossible to make effective long-range predictions about these insect populations.

[20]R. M. May, "Biological Populations with Nonoverlapping Generations: Stable Points, Stable Cycles, and Chaos," *Science 186* (1974), pp. 645–647; "Biological Populations Obeying Difference Equations: Stable Points, Stable Cycles, and Chaos," *Journal of Theoretical Biology 51* (1975), pp. 511–524.

The occurrence of chaotic solutions in simple problems has stimulated an enormous amount of research in recent years, but many questions remain unanswered. It is increasingly clear, however, that chaotic solutions are much more common than was suspected at first and that they may be a part of the investigation of a wide range of phenomena.

PROBLEMS

In each of Problems 1 through 6 solve the given difference equation in terms of the initial value y_0. Describe the behavior of the solution as $n \to \infty$.

1. $y_{n+1} = -0.9y_n$

2. $y_{n+1} = \dfrac{n+1}{n+2}y_n$

3. $y_{n+1} = \sqrt{\dfrac{n+3}{n+1}}y_n$

4. $y_{n+1} = (-1)^{n+1}y_n$

5. $y_{n+1} = 0.5y_n + 6$

6. $y_{n+1} = -0.5y_n + 6$

7. Find the effective annual yield of a bank account that pays interest at a rate of 7%, compounded daily; that is, divide the difference between the final and initial balances by the initial balance.

8. An investor deposits $1000 in an account paying interest at a rate of 8% compounded monthly, and also makes additional deposits of $25 per month. Find the balance in the account after 3 years.

9. A certain college graduate borrows $8000 to buy a car. The lender charges interest at an annual rate of 10%. What monthly payment rate is required to pay off the loan in 3 years? Compare your result with that of Problem 9 in Section 2.3.

10. A homebuyer wishes to take out a mortgage of $100,000 for a 30-year period. What monthly payment is required if the interest rate is (a) 9%, (b) 10%, (c) 12%?

11. A homebuyer takes out a mortgage of $100,000 with an interest rate of 9%. What monthly payment is required to pay off the loan in 30 years? In 20 years? What is the total amount paid during the term of the loan in each of these cases?

12. If the interest rate on a 20-year mortgage is fixed at 10% and if a monthly payment of $1000 is the maximum that the buyer can afford, what is the maximum mortgage loan that can be made under these conditions?

13. A homebuyer wishes to finance the purchase with a $95,000 mortgage with a 20-year term. What is the maximum interest rate the buyer can afford if the monthly payment is not to exceed $900?

The Logistic Difference Equation. Problems 14 through 19 deal with the difference equation (21), $u_{n+1} = \rho u_n(1 - u_n)$.

14. Carry out the details in the linear stability analysis of the equilibrium solution $u_n = (\rho - 1)/\rho$. That is, derive the difference equation (26) in the text for the perturbation v_n.

 15. (a) For $\rho = 3.2$ plot or calculate the solution of the logistic equation (21) for several initial conditions, say, $u_0 = 0.2, 0.4, 0.6,$ and 0.8. Observe that in each case the solution approaches a steady oscillation between the same two values. This illustrates that the long-term behavior of the solution is independent of the initial conditions.

(b) Make similar calculations and verify that the nature of the solution for large n is independent of the initial condition for other values of ρ, such as 2.6, 2.8, and 3.4.

16. Assume that $\rho > 1$ in Eq. (21).

 (a) Draw a qualitatively correct stairstep diagram and thereby show that if $u_0 < 0$, then $u_n \to -\infty$ as $n \to \infty$.

 (b) In a similar way, determine what happens as $n \to \infty$ if $u_0 > 1$.

 17. The solutions of Eq. (21) change from convergent sequences to periodic oscillations of period 2 as the parameter ρ passes through the value 3. To see more clearly how this happens, carry out the following calculations.

 (a) Plot or calculate the solution for $\rho = 2.9, 2.95,$ and 2.99, respectively, using an initial value u_0 of your choice in the interval $(0, 1)$. In each case estimate how many iterations are required for the solution to get "very close" to the limiting value. Use any convenient interpretation of what "very close" means in the preceding sentence.

 (b) Plot or calculate the solution for $\rho = 3.01, 3.05,$ and 3.1, respectively, using the same initial condition as in part (a). In each case estimate how many iterations are needed to reach a steady-state oscillation. Also find or estimate the two values in the steady-state oscillation.

18. By calculating or plotting the solution of Eq. (21) for different values of ρ, estimate the value of ρ at which the solution changes from an oscillation of period 2 to one of period 4. In the same way, estimate the value of ρ at which the solution changes from period 4 to period 8.

19. Let ρ_k be the value of ρ at which the solution of Eq. (21) changes from period 2^{k-1} to period 2^k. Thus, as noted in the text, $\rho_1 = 3, \rho_2 \cong 3.449,$ and $\rho_3 \cong 3.544$.

 (a) Using these values of $\rho_1, \rho_2,$ and ρ_3, or those you found in Problem 18, calculate $(\rho_2 - \rho_1)/(\rho_3 - \rho_2)$.

 (b) Let $\delta_n = (\rho_n - \rho_{n-1})/(\rho_{n+1} - \rho_n)$. It has been shown that δ_n approaches a limit δ as $n \to \infty$, where $\delta \cong 4.6692$ is known as the Feigenbaum [21] number. Determine the percentage difference between the limiting value δ and δ_2, as calculated in part (a).

 (c) Assume that $\delta_3 = \delta$ and use this relation to estimate ρ_4, the value of ρ at which solutions of period 16 appear.

 (d) By plotting or calculating solutions near the value of ρ_4 found in part (c), try to detect the appearance of a period 16 solution.

 (e) Observe that

 $$\rho_n = \rho_1 + (\rho_2 - \rho_1) + (\rho_3 - \rho_2) + \cdots + (\rho_n - \rho_{n-1}).$$

 Assuming that $(\rho_4 - \rho_3) = (\rho_3 - \rho_2)\delta^{-1}, (\rho_5 - \rho_4) = (\rho_3 - \rho_2)\delta^{-2}$, and so forth, express ρ_n as a geometric sum. Then find the limit of ρ_n as $n \to \infty$. This is an estimate of the value of ρ at which the onset of chaos occurs in the solution of the logistic equation (21).

[21]This result for the logistic difference equation was discovered by Mitchell Feigenbaum (1944 –) in August 1975, while he was working at the Los Alamos National Laboratory. Within a few weeks he had established that the same limiting value also appears in a large class of period-doubling difference equations. Feigenbaum, who has a doctorate in physics from M.I.T., is now at Rockefeller University.

PROBLEMS

Miscellaneous Problems. One of the difficulties in solving first order equations is that there are several methods of solution, each of which can be used on a certain type of equation. It may take some time to become proficient in matching solution methods with equations. The first 32 of the following problems are presented to give you some practice in identifying the method or methods applicable to a given equation. The remaining problems involve certain types of equations that can be solved by specialized methods.

In each of Problems 1 through 32 solve the given differential equation. If an initial condition is given, also find the solution that satisfies it.

1. $\dfrac{dy}{dx} = \dfrac{x^3 - 2y}{x}$

2. $(x + y)\,dx - (x - y)\,dy = 0$

3. $\dfrac{dy}{dx} = \dfrac{2x + y}{3 + 3y^2 - x}, \qquad y(0) = 0$

4. $(x + e^y)\,dy - dx = 0$

5. $\dfrac{dy}{dx} = -\dfrac{2xy + y^2 + 1}{x^2 + 2xy}$

6. $x\dfrac{dy}{dx} + xy = 1 - y, \qquad y(1) = 0$

7. $\dfrac{dy}{dx} = \dfrac{x}{x^2 y + y^3} \qquad$ *Hint:* Let $u = x^2$.

8. $x\dfrac{dy}{dx} + 2y = \dfrac{\sin x}{x}, \qquad y(2) = 1$

9. $\dfrac{dy}{dx} = -\dfrac{2xy + 1}{x^2 + 2y}$

10. $(3y^2 + 2xy)\,dx - (2xy + x^2)\,dy = 0$

11. $(x^2 + y)\,dx + (x + e^y)\,dy = 0$

12. $\dfrac{dy}{dx} + y = \dfrac{1}{1 + e^x}$

13. $x\,dy - y\,dx = (xy)^{1/2}\,dx$

14. $(x + y)\,dx + (x + 2y)\,dy = 0, \qquad y(2) = 3$

15. $(e^x + 1)\dfrac{dy}{dx} = y - ye^x$

16. $\dfrac{dy}{dx} = \dfrac{x^2 + y^2}{x^2}$

17. $\dfrac{dy}{dx} = e^{2x} + 3y$

18. $(2y + 3x)\,dx = -x\,dy$

19. $x\,dy - y\,dx = 2x^2 y^2\,dy, \qquad y(1) = -2$

20. $y' = e^{x+y}$

21. $xy' = y + xe^{y/x}$

22. $\dfrac{dy}{dx} = \dfrac{x^2 - 1}{y^2 + 1}, \qquad y(-1) = 1$

23. $xy' + y - y^2 e^{2x} = 0$

24. $2\sin y \cos x\,dx + \cos y \sin x\,dy = 0$

25. $\left(2\dfrac{x}{y} - \dfrac{y}{x^2 + y^2}\right)dx + \left(\dfrac{x}{x^2 + y^2} - \dfrac{x^2}{y^2}\right)dy = 0$

26. $(2y + 1)\,dx + \left(\dfrac{x^2 - y}{x}\right)dy = 0$

27. $(\cos 2y - \sin x)\,dx - 2\tan x \sin 2y\,dy = 0$

28. $\dfrac{dy}{dx} = \dfrac{3x^2 - 2y - y^3}{2x + 3xy^2}$

29. $\dfrac{dy}{dx} = \dfrac{2y + \sqrt{x^2 - y^2}}{2x}$

30. $\dfrac{dy}{dx} = \dfrac{y^3}{1 - 2xy^2}, \qquad y(0) = 1$

31. $(x^2 y + xy - y)\,dx + (x^2 y - 2x^2)\,dy = 0$

32. $\dfrac{dy}{dx} = -\dfrac{3x^2 y + y^2}{2x^3 + 3xy}, \qquad y(1) = -2$

33. **Riccati Equations.** The equation

$$\frac{dy}{dt} = q_1(t) + q_2(t)y + q_3(t)y^2$$

is known as a Riccati[22] equation. Suppose that some particular solution y_1 of this equation is known. A more general solution containing one arbitrary constant can be obtained through the substitution

$$y = y_1(t) + \frac{1}{v(t)}.$$

Show that $v(t)$ satisfies the first order *linear* equation

$$\frac{dv}{dt} = -(q_2 + 2q_3 y_1)v - q_3.$$

Note that $v(t)$ will contain a single arbitrary constant.

34. Using the method of Problem 33 and the given particular solution, solve each of the following Riccati equations:

(a) $y' = 1 + t^2 - 2ty + y^2$; $y_1(t) = t$

(b) $y' = -\frac{1}{t^2} - \frac{y}{t} + y^2$; $y_1(t) = \frac{1}{t}$

(c) $\frac{dy}{dt} = \frac{2\cos^2 t - \sin^2 t + y^2}{2\cos t}$; $y_1(t) = \sin t$

35. The propagation of a single action in a large population (for example, drivers turning on headlights at sunset) often depends partly on external circumstances (gathering darkness) and partly on a tendency to imitate others who have already performed the action in question. In this case the proportion $y(t)$ of people who have performed the action can be described[23] by the equation

$$dy/dt = (1 - y)[x(t) + by], \tag{i}$$

where $x(t)$ measures the external stimulus and b is the imitation coefficient.

(a) Observe that Eq. (i) is a Riccati equation and that $y_1(t) = 1$ is one solution. Use the transformation suggested in Problem 33, and find the linear equation satisfied by $v(t)$.

(b) Find $v(t)$ in the case that $x(t) = at$, where a is a constant. Leave your answer in the form of an integral.

Some Special Second Order Equations. Second order equations involve the second derivative of the unknown function and have the general form $y'' = f(t, y, y')$. Usually such equations cannot be solved by methods designed for first order equations. However, there are two types of second order equations that can be transformed into first order equations by a suitable change of variable. The resulting equation can sometimes be solved by the methods presented in this chapter. Problems 36 through 51 deal with these types of equations.

[22] Riccati equations are named for Jacopo Francesco Riccati (1676–1754), a Venetian nobleman, who declined university appointments in Italy, Austria, and Russia to pursue his mathematical studies privately at home. Riccati studied these equations extensively; however, it was Euler (in 1760) who discovered the result stated in this problem.

[23] See Anatol Rapoport, "Contribution to the Mathematical Theory of Mass Behavior: I. The Propagation of Single Acts," *Bulletin of Mathematical Biophysics 14* (1952), pp. 159–169, and John Z. Hearon, "Note on the Theory of Mass Behavior," *Bulletin of Mathematical Biophysics 17* (1955), pp. 7–13.

Equations with the Dependent Variable Missing. For a second order differential equation of the form $y'' = f(t, y')$, the substitution $v = y'$, $v' = y''$ leads to a first order equation of the form $v' = f(t, v)$. If this equation can be solved for v, then y can be obtained by integrating $dy/dt = v$. Note that one arbitrary constant is obtained in solving the first order equation for v, and a second is introduced in the integration for y. In each of Problems 36 through 41 use this substitution to solve the given equation.

36. $t^2 y'' + 2ty' - 1 = 0$, $t > 0$ 37. $ty'' + y' = 1$, $t > 0$

38. $y'' + t(y')^2 = 0$ 39. $2t^2 y'' + (y')^3 = 2ty'$, $t > 0$

40. $y'' + y' = e^{-t}$ 41. $t^2 y'' = (y')^2$, $t > 0$

Equations with the Independent Variable Missing. Consider second order differential equations of the form $y'' = f(y, y')$, in which the independent variable t does not appear explicitly. If we let $v = y'$, then we obtain $dv/dt = f(y, v)$. Since the right side of this equation depends on y and v, rather than on t and v, this equation contains too many variables. However, if we think of y as the independent variable, then by the chain rule, $dv/dt = (dv/dy)(dy/dt) = v(dv/dy)$. Hence the original differential equation can be written as $v(dv/dy) = f(y, v)$. Provided that this first order equation can be solved, we obtain v as a function of y. A relation between y and t results from solving $dy/dt = v(y)$, which is a separable equation. Again, there are two arbitrary constants in the final result. In each of Problems 42 through 47 use this method to solve the given differential equation.

42. $yy'' + (y')^2 = 0$ 43. $y'' + y = 0$

44. $y'' + y(y')^3 = 0$ 45. $2y^2 y'' + 2y(y')^2 = 1$

46. $yy'' - (y')^3 = 0$ 47. $y'' + (y')^2 = 2e^{-y}$

Hint: In Problem 47 the transformed equation is a Bernoulli equation. See Problem 27 in Section 2.4.

In each of Problems 48 through 51 solve the given initial value problem using the methods of Problems 36 through 47.

48. $y'y'' = 2$, $y(0) = 1$, $y'(0) = 2$

49. $y'' - 3y^2 = 0$, $y(0) = 2$, $y'(0) = 4$

50. $(1 + t^2)y'' + 2ty' + 3t^{-2} = 0$, $y(1) = 2$, $y'(1) = -1$

51. $y'y'' - t = 0$, $y(1) = 2$, $y'(1) = 1$

REFERENCES The two books mentioned in Section 2.5 are

Bailey, N. T. J., *The Mathematical Theory of Infectious Diseases and Its Applications* (2nd ed.) (New York: Hafner Press, 1975).

Clark, Colin W., *Mathematical Bioeconomics* (2nd ed.) (New York: Wiley-Interscience, 1990).

A good introduction to population dynamics in general is

Frauenthal, J. C., *Introduction to Population Modeling* (Boston: Birkhauser, 1980).

A fuller discussion of the proof of the fundamental existence and uniqueness theorem can be found in many more advanced books on differential equations. Two that are reasonably accessible to elementary readers are

Coddington, E. A., *An Introduction to Ordinary Differential Equations* (Englewood Cliffs, NJ: Prentice-Hall, 1961; New York: Dover, 1989).

Brauer, F., and Nohel, J., *The Qualitative Theory of Ordinary Differential Equations* (New York: Benjamin, 1969; New York: Dover, 1989).

To use the second initial condition we must first differentiate Eq. (26). This gives $y' = -2c_1e^{-2t} - 3c_2e^{-3t}$. Then, setting $t = 0$ and $y' = 3$, we obtain

$$-2c_1 - 3c_2 = 3. \tag{29}$$

By solving Eqs. (28) and (29) we find that $c_1 = 9$ and $c_2 = -7$. Using these values in the expression (26), we obtain the solution

$$y = 9e^{-2t} - 7e^{-3t} \tag{30}$$

of the initial value problem (27). The graph of the solution is shown in Figure 3.1.1.

FIGURE 3.1.1 Solution of $y'' + 5y' + 6y = 0$, $y(0) = 2$, $y'(0) = 3$.

Find the solution of the initial value problem

$$4y'' - 8y' + 3y = 0, \qquad y(0) = 2, \quad y'(0) = \tfrac{1}{2}. \tag{31}$$

EXAMPLE 3

If $y = e^{rt}$, then the characteristic equation is

$$4r^2 - 8r + 3 = 0$$

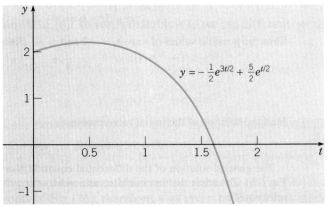

FIGURE 3.1.2 Solution of $4y'' - 8y' + 3y = 0$, $y(0) = 2$, $y'(0) = 0.5$.

and its roots are $r = 3/2$ and $r = 1/2$. Therefore the general solution of the differential equation is

$$y = c_1 e^{3t/2} + c_2 e^{t/2}. \tag{32}$$

Applying the initial conditions, we obtain the following two equations for c_1 and c_2:

$$c_1 + c_2 = 2, \qquad \tfrac{3}{2}c_1 + \tfrac{1}{2}c_2 = \tfrac{1}{2}.$$

The solution of these equations is $c_1 = -\tfrac{1}{2}, c_2 = \tfrac{5}{2}$, and the solution of the initial value problem (31) is

$$y = -\tfrac{1}{2}e^{3t/2} + \tfrac{5}{2}e^{t/2}. \tag{33}$$

Figure 3.1.2 shows the graph of the solution.

EXAMPLE

4

The solution (30) of the initial value problem (27) initially increases (because its initial slope is positive) but eventually approaches zero (because both terms involve negative exponential functions). Therefore the solution must have a maximum point, and the graph in Figure 3.1.1 confirms this. Determine the location of this maximum point.

One can estimate the coordinates of the maximum point from the graph, but to find them more precisely, we seek the point where the solution has a horizontal tangent line. By differentiating the solution (30), $y = 9e^{-2t} - 7e^{-3t}$, with respect to t, we obtain

$$y' = -18e^{-2t} + 21e^{-3t}. \tag{34}$$

Setting y' equal to zero and multiplying by e^{3t}, we find that the critical value t_c satisfies $e^t = 7/6$; hence

$$t_c = \ln(7/6) \cong 0.15415. \tag{35}$$

The corresponding maximum value y_M is given by

$$y_M = 9e^{-2t_c} - 7e^{-3t_c} = \frac{108}{49} \cong 2.20408. \tag{36}$$

In this example the initial slope is 3, but the solution of the given differential equation behaves in a similar way for any other positive initial slope. In Problem 26 you are asked to determine how the coordinates of the maximum point depend on the initial slope.

Returning to the equation $ay'' + by' + cy = 0$ with arbitrary coefficients, recall that when $r_1 \neq r_2$, its general solution (17) is the sum of two exponential functions. Therefore the solution has a relatively simple geometrical behavior: as t increases, the magnitude of the solution either tends to zero (when both exponents are negative) or else grows rapidly (when at least one exponent is positive). These two cases are illustrated by the solutions of Examples 2 and 3, which are shown in Figures 3.1.1 and 3.1.2, respectively. There is also a third case that occurs less often: the solution approaches a constant when one exponent is zero and the other is negative.

In Sections 3.4 and 3.5, respectively, we return to the problem of solving the equation $ay'' + by' + cy = 0$ when the roots of the characteristic equation either are complex conjugates or are real and equal. In the meantime, in Sections 3.2 and 3.3, we provide a systematic discussion of the mathematical structure of the solutions of all second order linear equations.

PROBLEMS

In each of Problems 1 through 8 find the general solution of the given differential equation.

1. $y'' + 2y' - 3y = 0$ 2. $y'' + 3y' + 2y = 0$

3. $6y'' - y' - y = 0$ 4. $2y'' - 3y' + y = 0$

5. $y'' + 5y' = 0$ 6. $4y'' - 9y = 0$

7. $y'' - 9y' + 9y = 0$ 8. $y'' - 2y' - 2y = 0$

In each of Problems 9 through 16 find the solution of the given initial value problem. Sketch the graph of the solution and describe its behavior as t increases.

9. $y'' + y' - 2y = 0,$ $y(0) = 1,$ $y'(0) = 1$

10. $y'' + 4y' + 3y = 0,$ $y(0) = 2,$ $y'(0) = -1$

11. $6y'' - 5y' + y = 0,$ $y(0) = 4,$ $y'(0) = 0$

12. $y'' + 3y' = 0,$ $y(0) = -2,$ $y'(0) = 3$

13. $y'' + 5y' + 3y = 0,$ $y(0) = 1,$ $y'(0) = 0$

14. $2y'' + y' - 4y = 0,$ $y(0) = 0,$ $y'(0) = 1$

15. $y'' + 8y' - 9y = 0,$ $y(1) = 1,$ $y'(1) = 0$

16. $4y'' - y = 0,$ $y(-2) = 1,$ $y'(-2) = -1$

17. Find a differential equation whose general solution is $y = c_1 e^{2t} + c_2 e^{-3t}$.

18. Find a differential equation whose general solution is $y = c_1 e^{-t/2} + c_2 e^{-2t}$.

19. Find the solution of the initial value problem

$$y'' - y = 0, \qquad y(0) = \tfrac{5}{4}, \quad y'(0) = -\tfrac{3}{4}.$$

Plot the solution for $0 \le t \le 2$ and determine its minimum value.

20. Find the solution of the initial value problem

$$2y'' - 3y' + y = 0, \qquad y(0) = 2, \quad y'(0) = \tfrac{1}{2}.$$

Then determine the maximum value of the solution and also find the point where the solution is zero.

21. Solve the initial value problem $y'' - y' - 2y = 0, y(0) = \alpha, y'(0) = 2$. Then find α so that the solution approaches zero as $t \to \infty$.

22. Solve the initial value problem $4y'' - y = 0, y(0) = 2, y'(0) = \beta$. Then find β so that the solution approaches zero as $t \to \infty$.

In each of Problems 23 and 24 determine the values of α, if any, for which all solutions tend to zero as $t \to \infty$; also determine the values of α, if any, for which all (nonzero) solutions become unbounded as $t \to \infty$.

23. $y'' - (2\alpha - 1)y' + \alpha(\alpha - 1)y = 0$

24. $y'' + (3 - \alpha)y' - 2(\alpha - 1)y = 0$

25. Consider the initial value problem

$$2y'' + 3y' - 2y = 0, \qquad y(0) = 1, \quad y'(0) = -\beta,$$

where $\beta > 0$.

(a) Solve the initial value problem.

(b) Plot the solution when $\beta = 1$. Find the coordinates (t_0, y_0) of the minimum point of the solution in this case.

(c) Find the smallest value of β for which the solution has no minimum point.

26. Consider the initial value problem (see Example 4)

$$y'' + 5y' + 6y = 0, \qquad y(0) = 2, \quad y'(0) = \beta,$$

where $\beta > 0$.

(a) Solve the initial value problem.

(b) Determine the coordinates t_m and y_m of the maximum point of the solution as functions of β.

(c) Determine the smallest value of β for which $y_m \geq 4$.

(d) Determine the behavior of t_m and y_m as $\beta \to \infty$.

27. Consider the equation $ay'' + by' + cy = d$, where a, b, c, and d are constants.

(a) Find all equilibrium, or constant, solutions of this differential equation.

(b) Let y_e denote an equilibrium solution, and let $Y = y - y_e$. Thus Y is the deviation of a solution y from an equilibrium solution. Find the differential equation satisfied by Y.

28. Consider the equation $ay'' + by' + cy = 0$, where a, b, and c are constants with $a > 0$. Find conditions on a, b, and c such that the roots of the characteristic equation are:

(a) real, different, and negative.

(b) real with opposite signs.

(c) real, different, and positive.

3.2 Fundamental Solutions of Linear Homogeneous Equations

In the preceding section we showed how to solve some differential equations of the form

$$ay'' + by' + cy = 0,$$

where a, b, and c are constants. Now we build on those results to provide a clearer picture of the structure of the solutions of all second order linear homogeneous equations. In turn, this understanding will assist us in finding the solutions of other problems that we will encounter later.

In developing the theory of linear differential equations, it is helpful to introduce a differential operator notation. Let p and q be continuous functions on an open interval I, that is, for $\alpha < t < \beta$. The cases $\alpha = -\infty$, or $\beta = \infty$, or both, are included. Then, for any function ϕ that is twice differentiable on I, we define the differential operator L by the equation

$$L[\phi] = \phi'' + p\phi' + q\phi. \tag{1}$$

Note that $L[\phi]$ is a function on I. The value of $L[\phi]$ at a point t is

$$L[\phi](t) = \phi''(t) + p(t)\phi'(t) + q(t)\phi(t).$$

For example, if $p(t) = t^2, q(t) = 1 + t$, and $\phi(t) = \sin 3t$, then

$$L[\phi](t) = (\sin 3t)'' + t^2(\sin 3t)' + (1 + t)\sin 3t$$
$$= -9\sin 3t + 3t^2 \cos 3t + (1 + t)\sin 3t.$$

The operator L is often written as $L = D^2 + pD + q$, where D is the derivative operator.

In this section we study the second order linear homogeneous equation $L[\phi](t) = 0$. Since it is customary to use the symbol y to denote $\phi(t)$, we will usually write this equation in the form

$$L[y] = y'' + p(t)y' + q(t)y = 0. \tag{2}$$

With Eq. (2) we associate a set of initial conditions

$$y(t_0) = y_0, \qquad y'(t_0) = y_0', \tag{3}$$

where t_0 is any point in the interval I, and y_0 and y_0' are given real numbers. We would like to know whether the initial value problem (2), (3) always has a solution, and whether it may have more than one solution. We would also like to know whether anything can be said about the form and structure of solutions that might be helpful in finding solutions of particular problems. Answers to these questions are contained in the theorems in this section.

The fundamental theoretical result for initial value problems for second order linear equations is stated in Theorem 3.2.1, which is analogous to Theorem 2.4.1 for first order linear equations. The result applies equally well to nonhomogeneous equations, so the theorem is stated in that form.

Theorem 3.2.1

Consider the initial value problem

$$y'' + p(t)y' + q(t)y = g(t), \qquad y(t_0) = y_0, \quad y'(t_0) = y_0', \tag{4}$$

where p, q, and g are continuous on an open interval I that contains the point t_0. Then there is exactly one solution $y = \phi(t)$ of this problem, and the solution exists throughout the interval I.

We emphasize that the theorem says three things:

1. The initial value problem *has* a solution; in other words, a solution *exists*.
2. The initial value problem has *only one* solution; that is, the solution is *unique*.
3. The solution ϕ is defined *throughout the interval I* where the coefficients are continuous and is at least twice differentiable there.

For some problems some of these assertions are easy to prove. For example, we found in Section 3.1 that the initial value problem

$$y'' - y = 0, \qquad y(0) = 2, \quad y'(0) = -1 \tag{5}$$

has the solution

$$y = \tfrac{1}{2}e^t + \tfrac{3}{2}e^{-t}. \tag{6}$$

The fact that we found a solution certainly establishes that a solution exists for this initial value problem. Further, the solution (6) is twice differentiable, indeed differentiable any number of times, throughout the interval $(-\infty, \infty)$ where the coefficients in the differential equation are continuous. On the other hand, it is not obvious, and is more difficult to show, that the initial value problem (5) has no solutions other

than the one given by Eq. (6). Nevertheless, Theorem 3.2.1 states that this solution is indeed the only solution of the initial value problem (5).

However, for most problems of the form (4), it is not possible to write down a useful expression for the solution. This is a major difference between first order and second order linear equations. Therefore, all parts of the theorem must be proved by general methods that do not involve having such an expression. The proof of Theorem 3.2.1 is fairly difficult, and we do not discuss it here.[2] We will, however, accept Theorem 3.2.1 as true and make use of it whenever necessary.

EXAMPLE 1

Find the longest interval in which the solution of the initial value problem

$$(t^2 - 3t)y'' + ty' - (t + 3)y = 0, \qquad y(1) = 2, \quad y'(1) = 1$$

is certain to exist.

If the given differential equation is written in the form of Eq. (4), then $p(t) = 1/(t - 3)$, $q(t) = -(t + 3)/t(t - 3)$, and $g(t) = 0$. The only points of discontinuity of the coefficients are $t = 0$ and $t = 3$. Therefore, the longest open interval, containing the initial point $t = 1$, in which all the coefficients are continuous is $0 < t < 3$. Thus, this is the longest interval in which Theorem 3.2.1 guarantees that the solution exists.

EXAMPLE 2

Find the unique solution of the initial value problem

$$y'' + p(t)y' + q(t)y = 0, \qquad y(t_0) = 0, \quad y'(t_0) = 0,$$

where p and q are continuous in an open interval I containing t_0.

The function $y = \phi(t) = 0$ for all t in I certainly satisfies the differential equation and initial conditions. By the uniqueness part of Theorem 3.2.1, it is the only solution of the given problem.

Let us now assume that y_1 and y_2 are two solutions of Eq. (2); in other words,

$$L[y_1] = y_1'' + py_1' + qy_1 = 0, \tag{7}$$

and similarly for y_2. Then, just as in the examples in Section 3.1, we can generate more solutions by forming linear combinations of y_1 and y_2. We state this result as a theorem.

Theorem 3.2.2

(Principle of Superposition) If y_1 and y_2 are two solutions of the differential equation (2),

$$L[y] = y'' + p(t)y' + q(t)y = 0,$$

then the linear combination $c_1y_1 + c_2y_2$ is also a solution for any values of the constants c_1 and c_2.

[2]A proof of Theorem 3.2.1 may be found, for example, in Chapter 6, Section 8 of the book by Coddington listed in the references at the end of this chapter.

A special case of Theorem 3.2.2 occurs if either c_1 or c_2 is zero. Then we conclude that any multiple of a solution of Eq. (2) is also a solution.

To prove Theorem 3.2.2 we need only substitute

$$y = c_1 y_1(t) + c_2 y_2(t) \tag{8}$$

for y in Eq. (2). The result is

$$
\begin{aligned}
L[c_1 y_1 + c_2 y_2] &= [c_1 y_1 + c_2 y_2]'' + p[c_1 y_1 + c_2 y_2]' + q[c_1 y_1 + c_2 y_2] \\
&= c_1 y_1'' + c_2 y_2'' + c_1 p y_1' + c_2 p y_2' + c_1 q y_1 + c_2 q y_2 \\
&= c_1 [y_1'' + p y_1' + q y_1] + c_2 [y_2'' + p y_2' + q y_2] \\
&= c_1 L[y_1] + c_2 L[y_2].
\end{aligned}
$$

Since $L[y_1] = 0$ and $L[y_2] = 0$, it follows that $L[c_1 y_1 + c_2 y_2] = 0$ also. Therefore, regardless of the values of c_1 and c_2, y as given by Eq. (8) does satisfy the differential equation (2), and the proof of Theorem 3.2.2 is complete.

Theorem 3.2.2 states that, beginning with only two solutions of Eq. (2), we can construct an infinite family of solutions by means of Eq. (8). The next question is whether all solutions of Eq. (2) are included in Eq. (8) or whether there may be other solutions of a different form. We begin to address this question by examining whether the constants c_1 and c_2 in Eq. (8) can be chosen so as to satisfy the initial conditions (3). These initial conditions require c_1 and c_2 to satisfy the equations

$$
\begin{aligned}
c_1 y_1(t_0) + c_2 y_2(t_0) &= y_0, \\
c_1 y_1'(t_0) + c_2 y_2'(t_0) &= y_0'.
\end{aligned}
\tag{9}
$$

Upon solving Eqs. (9) for c_1 and c_2, we find that

$$c_1 = \frac{y_0 y_2'(t_0) - y_0' y_2(t_0)}{y_1(t_0) y_2'(t_0) - y_1'(t_0) y_2(t_0)}, \qquad c_2 = \frac{-y_0 y_1'(t_0) + y_0' y_1(t_0)}{y_1(t_0) y_2'(t_0) - y_1'(t_0) y_2(t_0)}, \tag{10}$$

or, in terms of determinants,

$$
c_1 = \frac{\begin{vmatrix} y_0 & y_2(t_0) \\ y_0' & y_2'(t_0) \end{vmatrix}}{\begin{vmatrix} y_1(t_0) & y_2(t_0) \\ y_1'(t_0) & y_2'(t_0) \end{vmatrix}}, \qquad
c_2 = \frac{\begin{vmatrix} y_1(t_0) & y_0 \\ y_1'(t_0) & y_0' \end{vmatrix}}{\begin{vmatrix} y_1(t_0) & y_2(t_0) \\ y_1'(t_0) & y_2'(t_0) \end{vmatrix}}. \tag{11}
$$

With these values for c_1 and c_2, the expression (8) satisfies the initial conditions (3) as well as the differential equation (2).

In order for the expressions for c_1 and c_2 in Eqs. (10) or (11) to make sense, it is necessary that the denominators be nonzero. For both c_1 and c_2 the denominator is the same, namely, the determinant

$$W = \begin{vmatrix} y_1(t_0) & y_2(t_0) \\ y_1'(t_0) & y_2'(t_0) \end{vmatrix} = y_1(t_0) y_2'(t_0) - y_1'(t_0) y_2(t_0). \tag{12}$$

The determinant W is called the **Wronskian**[3] **determinant**, or simply the **Wronskian**, of the solutions y_1 and y_2. Sometimes we use the more extended notation $W(y_1, y_2)(t_0)$ to stand for the expression on the right side of Eq. (12), thereby emphasizing that the Wronskian depends on the functions y_1 and y_2, and that it is evaluated at the point t_0. The preceding argument establishes the following result.

Theorem 3.2.3

Suppose that y_1 and y_2 are two solutions of Eq. (2),

$$L[y] = y'' + p(t)y' + q(t)y = 0,$$

and that the Wronskian

$$W = y_1 y_2' - y_1' y_2$$

is not zero at the point t_0 where the initial conditions (3),

$$y(t_0) = y_0, \qquad y'(t_0) = y_0',$$

are assigned. Then there is a choice of the constants c_1, c_2 for which $y = c_1 y_1(t) + c_2 y_2(t)$ satisfies the differential equation (2) and the initial conditions (3).

EXAMPLE 3

In Example 1 of Section 3.1 we found that $y_1(t) = e^{-2t}$ and $y_2(t) = e^{-3t}$ are solutions of the differential equation

$$y'' + 5y' + 6y = 0.$$

Find the Wronskian of y_1 and y_2.

The Wronskian of these two functions is

$$W = \begin{vmatrix} e^{-2t} & e^{-3t} \\ -2e^{-2t} & -3e^{-3t} \end{vmatrix} = -e^{-5t}.$$

Since W is nonzero for all values of t, the functions y_1 and y_2 can be used to construct solutions of the given differential equation, together with initial conditions prescribed at any value of t. One such initial value problem was solved in Example 2 of Section 3.1.

The next theorem justifies the term "general solution" that we introduced in Section 3.1 for the linear combination $c_1 y_1 + c_2 y_2$.

Theorem 3.2.4

If y_1 and y_2 are two solutions of the differential equation (2),

$$L[y] = y'' + p(t)y' + q(t)y = 0,$$

and if there is a point t_0 where the Wronskian of y_1 and y_2 is nonzero, then the family of solutions

$$y = c_1 y_1(t) + c_2 y_2(t)$$

with arbitrary coefficients c_1 and c_2 includes every solution of Eq. (2).

[3]Wronskian determinants are named for Jósef Maria Hoëné-Wronski (1776–1853), who was born in Poland but spent most of his life in France. Wronski was a gifted but troubled man, and his life was marked by frequent heated disputes with other individuals and institutions.

Let ϕ be any solution of Eq. (2). To prove the theorem we must show that ϕ is included in the linear combination $c_1y_1 + c_2y_2$; that is, for some choice of the constants c_1 and c_2, the linear combination is equal to ϕ. Let t_0 be a point where the Wronskian of y_1 and y_2 is nonzero. Then evaluate ϕ and ϕ' at this point and call these values y_0 and y_0', respectively; thus

$$y_0 = \phi(t_0), \qquad y_0' = \phi'(t_0).$$

Next, consider the initial value problem

$$y'' + p(t)y' + q(t)y = 0, \qquad y(t_0) = y_0, \quad y'(t_0) = y_0'. \tag{13}$$

The function ϕ is certainly a solution of this initial value problem. On the other hand, since $W(y_1, y_2)(t_0)$ is nonzero, it is possible (by Theorem 3.2.3) to choose c_1 and c_2 so that $y = c_1y_1(t) + c_2y_2(t)$ is also a solution of the initial value problem (13). In fact, the proper values of c_1 and c_2 are given by Eqs. (10) or (11). The uniqueness part of Theorem 3.2.1 guarantees that these two solutions of the same initial value problem are actually the same function; thus, for the proper choice of c_1 and c_2,

$$\phi(t) = c_1y_1(t) + c_2y_2(t),$$

and therefore ϕ is included in the family of functions of $c_1y_1 + c_2y_2$. Finally, since ϕ is an *arbitrary* solution of Eq. (2), it follows that *every* solution of this equation is included in this family. This completes the proof of Theorem 3.2.4.

Theorem 3.2.4 states that, as long as the Wronskian of y_1 and y_2 is not everywhere zero, the linear combination $c_1y_1 + c_2y_2$ contains all solutions of Eq. (2). It is therefore natural (and we have already done this in the preceding section) to call the expression

$$y = c_1y_1(t) + c_2y_2(t)$$

with arbitrary constant coefficients the **general solution** of Eq. (2). The solutions y_1 and y_2, with a nonzero Wronskian, are said to form a **fundamental set of solutions** of Eq. (2).

We can restate the result of Theorem 3.2.4 in slightly different language: to find the general solution, and therefore all solutions, of an equation of the form (2), we need only find two solutions of the given equation whose Wronskian is nonzero. We did precisely this in several examples in Section 3.1, although there we did not calculate the Wronskians. You should now go back and do that, thereby verifying that all the solutions we called "general solutions" in Section 3.1 do satisfy the necessary Wronskian condition. Alternatively, the following example includes all those mentioned in Section 3.1, as well as many other problems of a similar type.

EXAMPLE

4

Suppose that $y_1(t) = e^{r_1t}$ and $y_2(t) = e^{r_2t}$ are two solutions of an equation of the form (2). Show that they form a fundamental set of solutions if $r_1 \neq r_2$.

We calculate the Wronskian of y_1 and y_2:

$$W = \begin{vmatrix} e^{r_1t} & e^{r_2t} \\ r_1e^{r_1t} & r_2e^{r_2t} \end{vmatrix} = (r_2 - r_1)\exp[(r_1 + r_2)t].$$

Since the exponential function is never zero, and since we are assuming that $r_2 - r_1 \neq 0$, it follows that W is nonzero for every value of t. Consequently, y_1 and y_2 form a fundamental set of solutions.

**EXAMPLE
5**

Show that $y_1(t) = t^{1/2}$ and $y_2(t) = t^{-1}$ form a fundamental set of solutions of

$$2t^2 y'' + 3ty' - y = 0, \qquad t > 0. \tag{14}$$

We will show in Section 5.5 how to solve Eq. (14); see also Problem 38 in Section 3.4. However, at this stage we can verify by direct substitution that y_1 and y_2 are solutions of the differential equation. Since $y_1'(t) = \frac{1}{2}t^{-1/2}$ and $y_1''(t) = -\frac{1}{4}t^{-3/2}$, we have

$$2t^2(-\tfrac{1}{4}t^{-3/2}) + 3t(\tfrac{1}{2}t^{-1/2}) - t^{1/2} = (-\tfrac{1}{2} + \tfrac{3}{2} - 1)t^{1/2} = 0.$$

Similarly, $y_2'(t) = -t^{-2}$ and $y_2''(t) = 2t^{-3}$, so

$$2t^2(2t^{-3}) + 3t(-t^{-2}) - t^{-1} = (4 - 3 - 1)t^{-1} = 0.$$

Next we calculate the Wronskian W of y_1 and y_2:

$$W = \begin{vmatrix} t^{1/2} & t^{-1} \\ \tfrac{1}{2}t^{-1/2} & -t^{-2} \end{vmatrix} = -\tfrac{3}{2}t^{-3/2}. \tag{15}$$

Since $W \neq 0$ for $t > 0$, we conclude that y_1 and y_2 form a fundamental set of solutions there.

In several cases we have been able to find a fundamental set of solutions, and therefore the general solution, of a given differential equation. However, this is often a difficult task, and the question may arise as to whether a differential equation of the form (2) always has a fundamental set of solutions. The following theorem provides an affirmative answer to this question.

Theorem 3.2.5

Consider the differential equation (2),

$$L[y] = y'' + p(t)y' + q(t)y = 0,$$

whose coefficients p and q are continuous on some open interval I. Choose some point t_0 in I. Let y_1 be the solution of Eq. (2) that also satisfies the initial conditions

$$y(t_0) = 1, \qquad y'(t_0) = 0,$$

and let y_2 be the solution of Eq. (2) that satisfies the initial conditions

$$y(t_0) = 0, \qquad y'(t_0) = 1.$$

Then y_1 and y_2 form a fundamental set of solutions of Eq. (2).

First observe that the *existence* of the functions y_1 and y_2 is ensured by the existence part of Theorem 3.2.1. To show that they form a fundamental set of solutions, we need only calculate their Wronskian at t_0:

$$W(y_1, y_2)(t_0) = \begin{vmatrix} y_1(t_0) & y_2(t_0) \\ y_1'(t_0) & y_2'(t_0) \end{vmatrix} = \begin{vmatrix} 1 & 0 \\ 0 & 1 \end{vmatrix} = 1.$$

Since their Wronskian is not zero at the point t_0, the functions y_1 and y_2 do form a fundamental set of solutions, thus completing the proof of Theorem 3.2.5.

Note that the potentially difficult part of this proof, demonstrating the existence of a pair of solutions, is taken care of by reference to Theorem 3.2.1. Note also that Theorem 3.2.5 does not address the question of how to solve the specified initial value

problems so as to find the functions y_1 and y_2 indicated in the theorem. Nevertheless, it may be reassuring to know that a fundamental set of solutions always exists.

EXAMPLE

6

Find the fundamental set of solutions specified by Theorem 3.2.5 for the differential equation

$$y'' - y = 0, \tag{16}$$

using the initial point $t_0 = 0$.

In Section 3.1 we noted that two solutions of Eq. (16) are $y_1(t) = e^t$ and $y_2(t) = e^{-t}$. The Wronskian of these solutions is $W(y_1, y_2)(t) = -2 \neq 0$, so they form a fundamental set of solutions. However, they are not the fundamental solutions indicated by Theorem 3.2.5 because they do not satisfy the initial conditions mentioned in that theorem at the point $t = 0$.

To find the fundamental solutions specified by the theorem, we need to find the solutions satisfying the proper initial conditions. Let us denote by $y_3(t)$ the solution of Eq. (16) that satisfies the initial conditions

$$y(0) = 1, \qquad y'(0) = 0. \tag{17}$$

The general solution of Eq. (16) is

$$y = c_1 e^t + c_2 e^{-t}, \tag{18}$$

and the initial conditions (17) are satisfied if $c_1 = 1/2$ and $c_2 = 1/2$. Thus

$$y_3(t) = \tfrac{1}{2}e^t + \tfrac{1}{2}e^{-t} = \cosh t.$$

Similarly, if $y_4(t)$ satisfies the initial conditions

$$y(0) = 0, \qquad y'(0) = 1, \tag{19}$$

then

$$y_4(t) = \tfrac{1}{2}e^t - \tfrac{1}{2}e^{-t} = \sinh t.$$

Since the Wronskian of y_3 and y_4 is

$$W(y_3, y_4)(t) = \cosh^2 t - \sinh^2 t = 1,$$

these functions also form a fundamental set of solutions, as stated by Theorem 3.2.5. Therefore, the general solution of Eq. (16) can be written as

$$y = k_1 \cosh t + k_2 \sinh t, \tag{20}$$

as well as in the form (18). We have used k_1 and k_2 for the arbitrary constants in Eq. (20) because they are not the same as the constants c_1 and c_2 in Eq. (18). One purpose of this example is to make clear that a given differential equation has more than one fundamental set of solutions; indeed, it has infinitely many. As a rule, you should choose the set that is most convenient.

We can summarize the discussion in this section as follows: to find the general solution of the differential equation

$$y'' + p(t)y' + q(t)y = 0, \qquad \alpha < t < \beta,$$

we must first find two functions y_1 and y_2 that satisfy the differential equation in $\alpha < t < \beta$. Then we must make sure that there is a point in the interval where the

Wronskian W of y_1 and y_2 is nonzero. Under these circumstances y_1 and y_2 form a fundamental set of solutions and the general solution is

$$y = c_1 y_1(t) + c_2 y_2(t),$$

where c_1 and c_2 are arbitrary constants. If initial conditions are prescribed at a point in $\alpha < t < \beta$ where $W \neq 0$, then c_1 and c_2 can be chosen so as to satisfy these conditions.

PROBLEMS

In each of Problems 1 through 6 find the Wronskian of the given pair of functions.

1. e^{2t}, $e^{-3t/2}$
2. $\cos t$, $\sin t$
3. e^{-2t}, te^{-2t}
4. x, xe^x
5. $e^t \sin t$, $e^t \cos t$
6. $\cos^2 \theta$, $1 + \cos 2\theta$

In each of Problems 7 through 12 determine the longest interval in which the given initial value problem is certain to have a unique twice differentiable solution. Do not attempt to find the solution.

7. $ty'' + 3y = t$, $y(1) = 1$, $y'(1) = 2$
8. $(t-1)y'' - 3ty' + 4y = \sin t$, $y(-2) = 2$, $y'(-2) = 1$
9. $t(t-4)y'' + 3ty' + 4y = 2$, $y(3) = 0$, $y'(3) = -1$
10. $y'' + (\cos t)y' + 3(\ln|t|)y = 0$, $y(2) = 3$, $y'(2) = 1$
11. $(x-3)y'' + xy' + (\ln|x|)y = 0$, $y(1) = 0$, $y'(1) = 1$
12. $(x-2)y'' + y' + (x-2)(\tan x)y = 0$, $y(3) = 1$, $y'(3) = 2$
13. Verify that $y_1(t) = t^2$ and $y_2(t) = t^{-1}$ are two solutions of the differential equation $t^2 y'' - 2y = 0$ for $t > 0$. Then show that $c_1 t^2 + c_2 t^{-1}$ is also a solution of this equation for any c_1 and c_2.
14. Verify that $y_1(t) = 1$ and $y_2(t) = t^{1/2}$ are solutions of the differential equation $yy'' + (y')^2 = 0$ for $t > 0$. Then show that $c_1 + c_2 t^{1/2}$ is not, in general, a solution of this equation. Explain why this result does not contradict Theorem 3.2.2.
15. Show that if $y = \phi(t)$ is a solution of the differential equation $y'' + p(t)y' + q(t)y = g(t)$, where $g(t)$ is not always zero, then $y = c\phi(t)$, where c is any constant other than 1, is not a solution. Explain why this result does not contradict the remark following Theorem 3.2.2.
16. Can $y = \sin(t^2)$ be a solution on an interval containing $t = 0$ of an equation $y'' + p(t)y' + q(t)y = 0$ with continuous coefficients? Explain your answer.
17. If the Wronskian W of f and g is $3e^{4t}$, and if $f(t) = e^{2t}$, find $g(t)$.
18. If the Wronskian W of f and g is $t^2 e^t$, and if $f(t) = t$, find $g(t)$.
19. If $W(f,g)$ is the Wronskian of f and g, and if $u = 2f - g$, $v = f + 2g$, find the Wronskian $W(u,v)$ of u and v in terms of $W(f,g)$.
20. If the Wronskian of f and g is $t \cos t - \sin t$, and if $u = f + 3g$, $v = f - g$, find the Wronskian of u and v.

In each of Problems 21 and 22 find the fundamental set of solutions specified by Theorem 3.2.5 for the given differential equation and initial point.

21. $y'' + y' - 2y = 0$, $t_0 = 0$
22. $y'' + 4y' + 3y = 0$, $t_0 = 1$

In each of Problems 23 through 26 verify that the functions y_1 and y_2 are solutions of the given differential equation. Do they constitute a fundamental set of solutions?

23. $y'' + 4y = 0$; $y_1(t) = \cos 2t$, $y_2(t) = \sin 2t$

24. $y'' - 2y' + y = 0$; $y_1(t) = e^t$, $y_2(t) = te^t$

25. $x^2 y'' - x(x+2)y' + (x+2)y = 0$, $x > 0$; $y_1(x) = x$, $y_2(x) = xe^x$

26. $(1 - x\cot x)y'' - xy' + y = 0$, $0 < x < \pi$; $y_1(x) = x$, $y_2(x) = \sin x$

27. Consider the equation $y'' - y' - 2y = 0$.

(a) Show that $y_1(t) = e^{-t}$ and $y_2(t) = e^{2t}$ form a fundamental set of solutions.

(b) Let $y_3(t) = -2e^{2t}$, $y_4(t) = y_1(t) + 2y_2(t)$, and $y_5(t) = 2y_1(t) - 2y_3(t)$. Are $y_3(t)$, $y_4(t)$, and $y_5(t)$ also solutions of the given differential equation?

(c) Determine whether each of the following pairs form a fundamental set of solutions: $[y_1(t), y_3(t)]$; $[y_2(t), y_3(t)]$; $[y_1(t), y_4(t)]$; $[y_4(t), y_5(t)]$.

28. **Exact Equations.** The equation $P(x)y'' + Q(x)y' + R(x)y = 0$ is said to be exact if it can be written in the form $[P(x)y']' + [f(x)y]' = 0$, where $f(x)$ is to be determined in terms of $P(x), Q(x)$, and $R(x)$. The latter equation can be integrated once immediately, resulting in a first order linear equation for y that can be solved as in Section 2.1. By equating the coefficients of the preceding equations and then eliminating $f(x)$, show that a necessary condition for exactness is $P''(x) - Q'(x) + R(x) = 0$. It can be shown that this is also a sufficient condition.

In each of Problems 29 through 32 use the result of Problem 28 to determine whether the given equation is exact. If so, solve the equation.

29. $y'' + xy' + y = 0$ 30. $y'' + 3x^2 y' + xy = 0$

31. $xy'' - (\cos x)y' + (\sin x)y = 0$, $x > 0$ 32. $x^2 y'' + xy' - y = 0$, $x > 0$

33. **The Adjoint Equation.** If a second order linear homogeneous equation is not exact, it can be made exact by multiplying by an appropriate integrating factor $\mu(x)$. Thus we require that $\mu(x)$ be such that $\mu(x)P(x)y'' + \mu(x)Q(x)y' + \mu(x)R(x)y = 0$ can be written in the form $[\mu(x)P(x)y']' + [f(x)y]' = 0$. By equating coefficients in these two equations and eliminating $f(x)$, show that the function μ must satisfy

$$P\mu'' + (2P' - Q)\mu' + (P'' - Q' + R)\mu = 0.$$

This equation is known as the adjoint of the original equation and is important in the advanced theory of differential equations. In general, the problem of solving the adjoint differential equation is as difficult as that of solving the original equation, so only occasionally is it possible to find an integrating factor for a second order equation.

In each of Problems 34 through 36 use the result of Problem 33 to find the adjoint of the given differential equation.

34. $x^2 y'' + xy' + (x^2 - \nu^2)y = 0$, Bessel's equation

35. $(1 - x^2)y'' - 2xy' + \alpha(\alpha + 1)y = 0$, Legendre's equation

36. $y'' - xy = 0$, Airy's equation

37. For the second order linear equation $P(x)y'' + Q(x)y' + R(x)y = 0$, show that the adjoint of the adjoint equation is the original equation.

38. A second order linear equation $P(x)y'' + Q(x)y' + R(x)y = 0$ is said to be self-adjoint if its adjoint is the same as the original equation. Show that a necessary condition for this equation to be self-adjoint is that $P'(x) = Q(x)$. Determine whether each of the equations in Problems 34 through 36 is self-adjoint.

3.3 Linear Independence and the Wronskian

In this section we will relate the ideas of a general solution and a fundamental set of solutions of a linear differential equation to the concept of linear independence, which is central to the study of linear algebra. This relation between differential equations and linear algebra is more significant for higher order equations and for systems of equations, but we will explain it here first in a simpler context. The results presented here will reappear in more general forms in Section 4.1 and Section 7.4.

We will refer to the following basic property of systems of linear homogeneous algebraic equations. Consider the two-by-two system

$$
\begin{aligned}
a_{11}x_1 + a_{12}x_2 &= 0, \\
a_{21}x_1 + a_{22}x_2 &= 0,
\end{aligned}
\tag{1}
$$

and let $\Delta = a_{11}a_{22} - a_{12}a_{21}$ be the corresponding determinant of coefficients. Then $x_1 = 0, x_2 = 0$ is the only solution of the system (1) if and only if $\Delta \neq 0$. Further, the system (1) has nonzero solutions if and only if $\Delta = 0$.

Two functions f and g are said to be **linearly dependent** on an interval I if there exist two constants k_1 and k_2, not both zero, such that

$$
k_1 f(t) + k_2 g(t) = 0
\tag{2}
$$

for all t in I. The functions f and g are said to be **linearly independent** on an interval I if they are not linearly dependent; that is, Eq. (2) holds for all t in I only if $k_1 = k_2 = 0$. In Section 4.1 these definitions are extended to an arbitrary number of functions. Although it may be difficult to determine whether a large set of functions is linearly independent or linearly dependent, it is usually easy to answer this question for a set of only two functions: they are linearly dependent if they are proportional to each other and are linearly independent otherwise. The following examples illustrate these definitions.

EXAMPLE 1

Determine whether the functions $\sin t$ and $\cos(t - \pi/2)$ are linearly independent or linearly dependent on an arbitrary interval.

The given functions are linearly dependent on any interval since

$$
k_1 \sin t + k_2 \cos(t - \pi/2) = 0
$$

for all t if we choose $k_1 = 1$ and $k_2 = -1$.

EXAMPLE 2

Show that the functions e^t and e^{2t} are linearly independent on any interval.

To establish this result we suppose that

$$
k_1 e^t + k_2 e^{2t} = 0
\tag{3}
$$

for all t in the interval; we must then show that $k_1 = k_2 = 0$. Choose two points t_0 and t_1 in the interval, where $t_1 \neq t_0$. Evaluating Eq. (3) at these points, we obtain

$$
\begin{aligned}
k_1 e^{t_0} + k_2 e^{2t_0} &= 0, \\
k_1 e^{t_1} + k_2 e^{2t_1} &= 0.
\end{aligned}
\tag{4}
$$

The determinant of coefficients is

$$e^{t_0} e^{2t_1} - e^{2t_0} e^{t_1} = e^{t_0} e^{t_1} (e^{t_1} - e^{t_0}).$$

Since this determinant is not zero, it follows that the only solution of Eq. (4) is $k_1 = k_2 = 0$. Hence e^t and e^{2t} are linearly independent.

The following theorem relates linear independence and dependence to the Wronskian.

Theorem 3.3.1 If f and g are differentiable functions on an open interval I, and if $W(f,g)(t_0) \neq 0$ for some point t_0 in I, then f and g are linearly independent on I. Moreover, if f and g are linearly dependent on I, then $W(f,g)(t) = 0$ for every t in I.

To prove the first statement in Theorem 3.3.1, consider a linear combination $k_1 f(t) + k_2 g(t)$, and suppose that this expression is zero throughout the interval. Evaluating the expression and its derivative at t_0, we have

$$
\begin{aligned}
k_1 f(t_0) + k_2 g(t_0) &= 0, \\
k_1 f'(t_0) + k_2 g'(t_0) &= 0.
\end{aligned}
\tag{5}
$$

The determinant of coefficients of Eqs. (5) is precisely $W(f,g)(t_0)$, which is not zero by hypothesis. Therefore, the only solution of Eqs. (5) is $k_1 = k_2 = 0$, so f and g are linearly independent.

The second part of Theorem 3.3.1 follows immediately from the first. Let f and g be linearly dependent, and suppose that the conclusion is false—that is, $W(f,g)$ is not everywhere zero in I. Then there is a point t_0 such that $W(f,g)(t_0) \neq 0$; by the first part of Theorem 3.3.1, this implies that f and g are linearly independent, which is a contradiction, thus completing the proof.

We can apply this result to the two functions $f(t) = e^t$ and $g(t) = e^{2t}$ discussed in Example 2. For any point t_0 we have

$$W(f,g)(t_0) = \begin{vmatrix} e^{t_0} & e^{2t_0} \\ e^{t_0} & 2e^{2t_0} \end{vmatrix} = e^{3t_0} \neq 0. \tag{6}$$

The functions e^t and e^{2t} are therefore linearly independent on any interval.

You should be careful not to read too much into Theorem 3.3.1. In particular, two functions f and g may be linearly independent even though $W(f,g)(t) = 0$ for every t in the interval I. This is illustrated in Problem 28.

Now let us examine further the properties of the Wronskian of two solutions of a second order linear homogeneous differential equation. The following theorem, perhaps surprisingly, gives a simple explicit formula for the Wronskian of any two solutions of any such equation, even if the solutions themselves are not known.

Theorem 3.3.2 **(Abel's Theorem)**[4] If y_1 and y_2 are solutions of the differential equation

$$L[y] = y'' + p(t)y' + q(t)y = 0, \tag{7}$$

where p and q are continuous on an open interval I, then the Wronskian $W(y_1, y_2)(t)$ is given by

$$W(y_1, y_2)(t) = c \exp\left[-\int p(t)\,dt\right], \tag{8}$$

where c is a certain constant that depends on y_1 and y_2, but not on t. Further, $W(y_1, y_2)(t)$ either is zero for all t in I (if $c = 0$) or else is never zero in I (if $c \neq 0$).

To prove Abel's theorem, we start by noting that y_1 and y_2 satisfy

$$\begin{aligned} y_1'' + p(t)y_1' + q(t)y_1 &= 0, \\ y_2'' + p(t)y_2' + q(t)y_2 &= 0. \end{aligned} \tag{9}$$

If we multiply the first equation by $-y_2$, multiply the second by y_1, and add the resulting equations, we obtain

$$(y_1 y_2'' - y_1'' y_2) + p(t)(y_1 y_2' - y_1' y_2) = 0. \tag{10}$$

Next, we let $W(t) = W(y_1, y_2)(t)$ and observe that

$$W' = y_1 y_2'' - y_1'' y_2. \tag{11}$$

Then we can write Eq. (10) in the form

$$W' + p(t)W = 0. \tag{12}$$

Equation (12) can be solved immediately since it is both a first order linear equation (Section 2.1) and a separable equation (Section 2.2). Thus

$$W(t) = c \exp\left[-\int p(t)\,dt\right], \tag{13}$$

where c is a constant. The value of c depends on which pair of solutions of Eq. (7) is involved. However, since the exponential function is never zero, $W(t)$ is not zero unless $c = 0$, in which case $W(t)$ is zero for all t, which completes the proof of Theorem 3.3.2.

[4]The result in Theorem 3.3.2 was derived by the Norwegian mathematician Niels Henrik Abel (1802–1829) in 1827 and is known as Abel's formula. Abel also showed that there is no general formula for solving a quintic, or fifth degree, polynomial equation in terms of explicit algebraic operations on the coefficients, thereby resolving a question that had been open since the sixteenth century. His greatest contributions, however, were in analysis, particularly in the study of elliptic functions. Unfortunately, his work was not widely noticed until after his death. The distinguished French mathematician Legendre called it a "monument more lasting than bronze."

Note that the Wronskians of any two fundamental sets of solutions of the same differential equation can differ only by a multiplicative constant, and that the Wronskian of any fundamental set of solutions can be determined, up to a multiplicative constant, without solving the differential equation.

EXAMPLE

3

In Example 5 of Section 3.2 we verified that $y_1(t) = t^{1/2}$ and $y_2(t) = t^{-1}$ are solutions of the equation

$$2t^2 y'' + 3ty' - y = 0, \qquad t > 0. \tag{14}$$

Verify that the Wronskian of y_1 and y_2 is given by Eq. (13).

From the example just cited we know that $W(y_1, y_2)(t) = -(3/2)t^{-3/2}$. To use Eq. (13) we must write the differential equation (14) in the standard form with the coefficient of y'' equal to 1. Thus we obtain

$$y'' + \frac{3}{2t}y' - \frac{1}{2t^2}y = 0,$$

so $p(t) = 3/2t$. Hence

$$W(y_1, y_2)(t) = c \exp\left[-\int \frac{3}{2t}\, dt\right] = c \exp\left(-\frac{3}{2}\ln t\right)$$

$$= c\, t^{-3/2}. \tag{15}$$

Equation (15) gives the Wronskian of any pair of solutions of Eq. (14). For the particular solutions given in this example we must choose $c = -3/2$.

A stronger version of Theorem 3.3.1 can be established if the two functions involved are solutions of a second order linear homogeneous differential equation.

Theorem 3.3.3

Let y_1 and y_2 be the solutions of Eq. (7),

$$L[y] = y'' + p(t)y' + q(t)y = 0,$$

where p and q are continuous on an open interval I. Then y_1 and y_2 are linearly dependent on I if and only if $W(y_1, y_2)(t)$ is zero for all t in I. Alternatively, y_1 and y_2 are linearly independent on I if and only if $W(y_1, y_2)(t)$ is never zero in I.

Of course, we know by Theorem 3.3.2 that $W(y_1, y_2)(t)$ is either everywhere zero or nowhere zero in I. In proving Theorem 3.3.3, observe first that if y_1 and y_2 are linearly dependent, then $W(y_1, y_2)(t)$ is zero for all t in I by Theorem 3.3.1. It remains to prove the converse; that is, if $W(y_1, y_2)(t)$ is zero throughout I, then y_1 and y_2 are linearly dependent. Let t_0 be any point in I; then necessarily $W(y_1, y_2)(t_0) = 0$. Consequently, the system of equations

$$c_1 y_1(t_0) + c_2 y_2(t_0) = 0,$$

$$c_1 y_1'(t_0) + c_2 y_2'(t_0) = 0 \tag{16}$$

for c_1 and c_2 has a nontrivial solution. Using these values of c_1 and c_2, let $\phi(t) = c_1 y_1(t) + c_2 y_2(t)$. Then ϕ is a solution of Eq. (7), and by Eqs. (16), ϕ also satisfies the initial conditions

$$\phi(t_0) = 0, \qquad \phi'(t_0) = 0. \tag{17}$$

Therefore, by the uniqueness part of Theorem 3.2.1, or by Example 2 of Section 3.2, $\phi(t) = 0$ for all t in I. Since $\phi(t) = c_1 y_1(t) + c_2 y_2(t)$ with c_1 and c_2 not both zero, this means that y_1 and y_2 are linearly dependent. The alternative statement of the theorem follows immediately.

We can now summarize the facts about fundamental sets of solutions, Wronskians, and linear independence in the following way. Let y_1 and y_2 be solutions of Eq. (7),

$$y'' + p(t)y' + q(t)y = 0,$$

where p and q are continuous on an open interval I. Then the following four statements are equivalent, in the sense that each one implies the other three:

1. The functions y_1 and y_2 are a fundamental set of solutions on I.
2. The functions y_1 and y_2 are linearly independent on I.
3. $W(y_1, y_2)(t_0) \neq 0$ for some t_0 in I.
4. $W(y_1, y_2)(t) \neq 0$ for all t in I.

It is interesting to note the similarity between second order linear homogeneous differential equations and two-dimensional vector algebra. Two vectors \mathbf{a} and \mathbf{b} are said to be linearly dependent if there are two scalars k_1 and k_2, not both zero, such that $k_1\mathbf{a} + k_2\mathbf{b} = \mathbf{0}$; otherwise, they are said to be linearly independent. Let \mathbf{i} and \mathbf{j} be unit vectors directed along the positive x- and y-axes, respectively. Since $k_1\mathbf{i} + k_2\mathbf{j} = \mathbf{0}$ only if $k_1 = k_2 = 0$, the vectors \mathbf{i} and \mathbf{j} are linearly independent. Further, we know that any vector \mathbf{a} with components a_1 and a_2 can be written as $\mathbf{a} = a_1\mathbf{i} + a_2\mathbf{j}$, that is, as a linear combination of the two linearly independent vectors \mathbf{i} and \mathbf{j}. It is not difficult to show that any vector in two dimensions can be expressed as a linear combination of any two linearly independent two-dimensional vectors (see Problem 14). Such a pair of linearly independent vectors is said to form a basis for the vector space of two-dimensional vectors.

The term **vector space** is also applied to other collections of mathematical objects that obey the same laws of addition and multiplication by scalars that geometric vectors do. For example, it can be shown that the set of functions that are twice differentiable on the open interval I forms a vector space. Similarly, the set V of functions satisfying Eq. (7) also forms a vector space.

Since every member of V can be expressed as a linear combination of two linearly independent members y_1 and y_2, we say that such a pair forms a basis for V. This leads to the conclusion that V is two-dimensional; therefore, it is analogous in many respects to the space of geometric vectors in a plane. Later we find that the set of solutions of an nth order linear homogeneous differential equation forms a vector space of dimension n, and that any set of n linearly independent solutions of the differential equation forms a basis for the space. This connection between differential equations and vectors constitutes a good reason for the study of abstract linear algebra.

PROBLEMS In each of Problems 1 through 8 determine whether the given pair of functions is linearly independent or linearly dependent.

1. $f(t) = t^2 + 5t$, $g(t) = t^2 - 5t$
2. $f(\theta) = \cos 2\theta - 2\cos^2\theta$, $g(\theta) = \cos 2\theta + 2\sin^2\theta$
3. $f(t) = e^{\lambda t}\cos\mu t$, $g(t) = e^{\lambda t}\sin\mu t$, $\mu \neq 0$
4. $f(x) = e^{3x}$, $g(x) = e^{3(x-1)}$
5. $f(t) = 3t - 5$, $g(t) = 9t - 15$
6. $f(t) = t$, $g(t) = t^{-1}$
7. $f(t) = 3t$, $g(t) = |t|$
8. $f(x) = x^3$, $g(x) = |x|^3$

9. The Wronskian of two functions is $W(t) = t\sin^2 t$. Are the functions linearly independent or linearly dependent? Why?

10. The Wronskian of two functions is $W(t) = t^2 - 4$. Are the functions linearly independent or linearly dependent? Why?

11. If the functions y_1 and y_2 are linearly independent solutions of $y'' + p(t)y' + q(t)y = 0$, prove that $c_1 y_1$ and $c_2 y_2$ are also linearly independent solutions, provided that neither c_1 nor c_2 is zero.

12. If the functions y_1 and y_2 are linearly independent solutions of $y'' + p(t)y' + q(t)y = 0$, prove that $y_3 = y_1 + y_2$ and $y_4 = y_1 - y_2$ also form a linearly independent set of solutions. Conversely, if y_3 and y_4 are linearly independent solutions of the differential equation, show that y_1 and y_2 are also.

13. If the functions y_1 and y_2 are linearly independent solutions of $y'' + p(t)y' + q(t)y = 0$, determine under what conditions the functions $y_3 = a_1 y_1 + a_2 y_2$ and $y_4 = b_1 y_1 + b_2 y_2$ also form a linearly independent set of solutions.

14. (a) Prove that any two-dimensional vector can be written as a linear combination of $\mathbf{i} + \mathbf{j}$ and $\mathbf{i} - \mathbf{j}$.

(b) Prove that if the vectors $\mathbf{x} = x_1\mathbf{i} + x_2\mathbf{j}$ and $\mathbf{y} = y_1\mathbf{i} + y_2\mathbf{j}$ are linearly independent, then any vector $\mathbf{z} = z_1\mathbf{i} + z_2\mathbf{j}$ can be expressed as a linear combination of \mathbf{x} and \mathbf{y}. Note that if \mathbf{x} and \mathbf{y} are linearly independent, then $x_1 y_2 - x_2 y_1 \neq 0$. Why?

In each of Problems 15 through 18 find the Wronskian of two solutions of the given differential equation without solving the equation.

15. $t^2 y'' - t(t+2)y' + (t+2)y = 0$
16. $(\cos t)y'' + (\sin t)y' - ty = 0$
17. $x^2 y'' + xy' + (x^2 - \nu^2)y = 0$, Bessel's equation
18. $(1 - x^2)y'' - 2xy' + \alpha(\alpha + 1)y = 0$, Legendre's equation

19. Show that if p is differentiable and $p(t) > 0$, then the Wronskian $W(t)$ of two solutions of $[p(t)y']' + q(t)y = 0$ is $W(t) = c/p(t)$, where c is a constant.

20. If y_1 and y_2 are linearly independent solutions of $ty'' + 2y' + te^t y = 0$ and if $W(y_1, y_2)(1) = 2$, find the value of $W(y_1, y_2)(5)$.

21. If y_1 and y_2 are linearly independent solutions of $t^2 y'' - 2y' + (3 + t)y = 0$ and if $W(y_1, y_2)(2) = 3$, find the value of $W(y_1, y_2)(4)$.

22. If the Wronskian of any two solutions of $y'' + p(t)y' + q(t)y = 0$ is constant, what does this imply about the coefficients p and q?

23. If f, g, and h are differentiable functions, show that $W(fg, fh) = f^2 W(g, h)$.

In Problems 24 through 26 assume that p and q are continuous and that the functions y_1 and y_2 are solutions of the differential equation $y'' + p(t)y' + q(t)y = 0$ on an open interval I.

24. Prove that if y_1 and y_2 are zero at the same point in I, then they cannot be a fundamental set of solutions on that interval.

$y_1 = y_3 - y_2$

$y_2 = y_4 + y$

25. Prove that if y_1 and y_2 have maxima or minima at the same point in I, then they cannot be a fundamental set of solutions on that interval.

26. Prove that if y_1 and y_2 have a common point of inflection t_0 in I, then they cannot be a fundamental set of solutions on I unless both p and q are zero at t_0.

27. Show that t and t^2 are linearly independent on $-1 < t < 1$; indeed, they are linearly independent on every interval. Show also that $W(t, t^2)$ is zero at $t = 0$. What can you conclude from this about the possibility that t and t^2 are solutions of a differential equation $y'' + p(t)y' + q(t)y = 0$? Verify that t and t^2 are solutions of the equation $t^2 y'' - 2ty' + 2y = 0$. Does this contradict your conclusion? Does the behavior of the Wronskian of t and t^2 contradict Theorem 3.3.2?

28. Show that the functions $f(t) = t^2|t|$ and $g(t) = t^3$ are linearly dependent on $0 < t < 1$ and on $-1 < t < 0$ but are linearly independent on $-1 < t < 1$. Although f and g are linearly independent there, show that $W(f, g)$ is zero for all t in $-1 < t < 1$. Hence f and g cannot be solutions of an equation $y'' + p(t)y' + q(t)y = 0$ with p and q continuous on $-1 < t < 1$.

3.4 Complex Roots of the Characteristic Equation

We continue our discussion of the equation

$$ay'' + by' + cy = 0, \tag{1}$$

where a, b, and c are given real numbers. In Section 3.1 we found that if we seek solutions of the form $y = e^{rt}$, then r must be a root of the characteristic equation

$$ar^2 + br + c = 0. \tag{2}$$

If the roots r_1 and r_2 are real and different, which occurs whenever the discriminant $b^2 - 4ac$ is positive, then the general solution of Eq. (1) is

$$y = c_1 e^{r_1 t} + c_2 e^{r_2 t}. \tag{3}$$

Suppose now that $b^2 - 4ac$ is negative. Then the roots of Eq. (2) are conjugate complex numbers; we denote them by

$$r_1 = \lambda + i\mu, \qquad r_2 = \lambda - i\mu, \tag{4}$$

where λ and μ are real. The corresponding expressions for y are

$$y_1(t) = \exp[(\lambda + i\mu)t], \qquad y_2(t) = \exp[(\lambda - i\mu)t]. \tag{5}$$

Our first task is to explore what is meant by these expressions, which involve evaluating the exponential function for a complex exponent. For example, if $\lambda = -1, \mu = 2$, and $t = 3$, then from Eq. (5),

$$y_1(3) = e^{-3+6i}. \tag{6}$$

What does it mean to raise the number e to a complex power? The answer is provided by an important relation known as Euler's formula.

Euler's Formula. To assign a meaning to the expressions in Eqs. (5) we need to give a definition of the complex exponential function. Of course, we want the definition to reduce to the familiar real exponential function when the exponent is real. There

are several ways to accomplish this extension of the exponential function. Here we use a method based on infinite series; an alternative is outlined in Problem 28.

Recall from calculus that the Taylor series for e^t about $t = 0$ is

$$e^t = \sum_{n=0}^{\infty} \frac{t^n}{n!}, \qquad -\infty < t < \infty. \tag{7}$$

If we now assume that we can substitute it for t in Eq. (7), then we have

$$
\begin{aligned}
e^{it} &= \sum_{n=0}^{\infty} \frac{(it)^n}{n!} \\
&= \sum_{n=0}^{\infty} \frac{(-1)^n t^{2n}}{(2n)!} + i \sum_{n=1}^{\infty} \frac{(-1)^{n-1} t^{2n-1}}{(2n-1)!},
\end{aligned} \tag{8}
$$

where we have separated the sum into its real and imaginary parts, making use of the fact that $i^2 = -1$, $i^3 = -i$, $i^4 = 1$, and so forth. The first series in Eq. (8) is precisely the Taylor series for $\cos t$ about $t = 0$, and the second is the Taylor series for $\sin t$ about $t = 0$. Thus we have

$$e^{it} = \cos t + i \sin t. \tag{9}$$

Equation (9) is known as Euler's formula and is an extremely important mathematical relationship. Although our derivation of Eq. (9) is based on the unverified assumption that the series (7) can be used for complex as well as real values of the independent variable, our intention is to use this derivation only to make Eq. (9) seem plausible. We now put matters on a firm foundation by adopting Eq. (9) as the *definition* of e^{it}. In other words, whenever we write e^{it}, we mean the expression on the right side of Eq. (9).

There are some variations of Euler's formula that are also worth noting. If we replace t by $-t$ in Eq. (9) and recall that $\cos(-t) = \cos t$ and $\sin(-t) = -\sin t$, then we have

$$e^{-it} = \cos t - i \sin t. \tag{10}$$

Further, if t is replaced by μt in Eq. (9), then we obtain a generalized version of Euler's formula, namely,

$$e^{i\mu t} = \cos \mu t + i \sin \mu t. \tag{11}$$

Next, we want to extend the definition of the exponential function to arbitrary complex exponents of the form $(\lambda + i\mu)t$. Since we want the usual properties of the exponential function to hold for complex exponents, we certainly want $\exp[(\lambda + i\mu)t]$ to satisfy

$$e^{(\lambda + i\mu)t} = e^{\lambda t} e^{i\mu t}. \tag{12}$$

Then, substituting for $e^{i\mu t}$ from Eq. (11), we obtain

$$
\begin{aligned}
e^{(\lambda + i\mu)t} &= e^{\lambda t}(\cos \mu t + i \sin \mu t) \\
&= e^{\lambda t} \cos \mu t + i e^{\lambda t} \sin \mu t.
\end{aligned} \tag{13}
$$

We now take Eq. (13) as the definition of $\exp[(\lambda + i\mu)t]$. The value of the exponential function with a complex exponent is a complex number whose real and imaginary parts are given by the terms on the right side of Eq. (13). Observe that the real

and imaginary parts of $\exp[(\lambda + i\mu)t]$ are expressed entirely in terms of elementary real-valued functions. For example, the quantity in Eq. (6) has the value

$$e^{-3+6i} = e^{-3}\cos 6 + ie^{-3}\sin 6 \cong 0.0478041 - 0.0139113i.$$

With the definitions (9) and (13) it is straightforward to show that the usual laws of exponents are valid for the complex exponential function. It is also easy to verify that the differentiation formula

$$\frac{d}{dt}(e^{rt}) = re^{rt} \tag{14}$$

also holds for complex values of r.

Real-Valued Solutions. The functions $y_1(t)$ and $y_2(t)$, given by Eqs. (5) and with the meaning expressed by Eq. (13), are solutions of Eq. (1) when the roots of the characteristic equation (2) are complex numbers $\lambda \pm i\mu$. Unfortunately, the solutions y_1 and y_2 are complex-valued functions, whereas in general we would prefer to have real-valued solutions, if possible, because the differential equation itself has real coefficients. Such solutions can be found as a consequence of Theorem 3.2.2, which states that if y_1 and y_2 are solutions of Eq. (1), then any linear combination of y_1 and y_2 is also a solution. In particular, let us form the sum and then the difference of y_1 and y_2. We have

$$y_1(t) + y_2(t) = e^{\lambda t}(\cos \mu t + i\sin \mu t) + e^{\lambda t}(\cos \mu t - i\sin \mu t)$$
$$= 2e^{\lambda t}\cos \mu t$$

and

$$y_1(t) - y_2(t) = e^{\lambda t}(\cos \mu t + i\sin \mu t) - e^{\lambda t}(\cos \mu t - i\sin \mu t)$$
$$= 2ie^{\lambda t}\sin \mu t.$$

Hence, neglecting the constant multipliers 2 and $2i$, respectively, we have obtained a pair of real-valued solutions

$$u(t) = e^{\lambda t}\cos \mu t, \qquad v(t) = e^{\lambda t}\sin \mu t. \tag{15}$$

Observe that u and v are simply the real and imaginary parts, respectively, of y_1.

By direct computation you can show that the Wronskian of u and v is

$$W(u, v)(t) = \mu e^{2\lambda t}. \tag{16}$$

Thus, as long as $\mu \neq 0$, the Wronskian W is not zero, so u and v form a fundamental set of solutions. (Of course, if $\mu = 0$, then the roots are real and the discussion in this section is not applicable.) Consequently, if the roots of the characteristic equation are complex numbers $\lambda \pm i\mu$, with $\mu \neq 0$, then the general solution of Eq. (1) is

$$y = c_1 e^{\lambda t}\cos \mu t + c_2 e^{\lambda t}\sin \mu t, \tag{17}$$

where c_1 and c_2 are arbitrary constants. Note that the solution (17) can be written down as soon as the values of λ and μ are known.

EXAMPLE 1

Find the general solution of

$$y'' + y' + y = 0. \tag{18}$$

The characteristic equation is

$$r^2 + r + 1 = 0,$$

and its roots are

$$r = \frac{-1 \pm (1-4)^{1/2}}{2} = -\frac{1}{2} \pm i\frac{\sqrt{3}}{2}.$$

Thus $\lambda = -1/2$ and $\mu = \sqrt{3}/2$, so the general solution of Eq. (18) is

$$y = c_1 e^{-t/2} \cos(\sqrt{3}t/2) + c_2 e^{-t/2} \sin(\sqrt{3}t/2). \tag{19}$$

EXAMPLE 2

Find the general solution of

$$y'' + 9y = 0. \tag{20}$$

The characteristic equation is $r^2 + 9 = 0$ with the roots $r = \pm 3i$; thus $\lambda = 0$ and $\mu = 3$. The general solution is

$$y = c_1 \cos 3t + c_2 \sin 3t; \tag{21}$$

note that if the real part of the roots is zero, as in this example, then there is no exponential factor in the solution.

EXAMPLE 3

Find the solution of the initial value problem

$$16y'' - 8y' + 145y = 0, \qquad y(0) = -2, \quad y'(0) = 1. \tag{22}$$

The characteristic equation is $16r^2 - 8r + 145 = 0$ and its roots are $r = 1/4 \pm 3i$. Thus the general solution of the differential equation is

$$y = c_1 e^{t/4} \cos 3t + c_2 e^{t/4} \sin 3t. \tag{23}$$

To apply the first initial condition we set $t = 0$ in Eq. (23); this gives

$$y(0) = c_1 = -2.$$

For the second initial condition we must differentiate Eq. (23) and then set $t = 0$. In this way we find that

$$y'(0) = \tfrac{1}{4}c_1 + 3c_2 = 1,$$

from which $c_2 = 1/2$. Using these values of c_1 and c_2 in Eq. (23), we obtain

$$y = -2e^{t/4} \cos 3t + \tfrac{1}{2}e^{t/4} \sin 3t \tag{24}$$

as the solution of the initial value problem (22).

We will discuss the geometrical properties of solutions such as these more fully in Section 3.8, so we will be very brief here. Each of the solutions u and v in Eqs. (15) represents an oscillation, because of the trigonometric factors, and also either grows or decays exponentially, depending on the sign of λ (unless $\lambda = 0$). In Example 1 we have $\lambda = -1/2 < 0$, so solutions are decaying oscillations. The graph of a typical solution of Eq. (18) is shown in Figure 3.4.1. On the other hand, $\lambda = 1/4 > 0$ in Example 3, so solutions of the differential equation (22) are growing oscillations. The graph of the solution (24) of the given initial value problem is shown in Figure 3.4.2.

The intermediate case is illustrated in Example 2, in which $\lambda = 0$. In this case the solution neither grows nor decays exponentially but, rather, oscillates steadily; a typical solution of Eq. (20) is shown in Figure 3.4.3.

FIGURE 3.4.1 A typical solution of $y'' + y' + y = 0$.

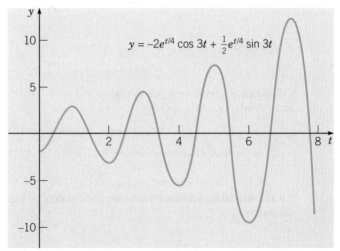

FIGURE 3.4.2 Solution of $16y'' - 8y' + 145y = 0$, $y(0) = -2$, $y'(0) = 1$.

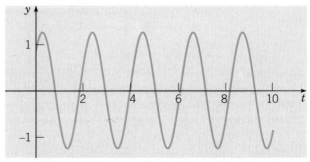

FIGURE 3.4.3 A typical solution of $y'' + 9y = 0$.

PROBLEMS In each of Problems 1 through 6 use Euler's formula to write the given expression in the form
$a + ib$.

1. $\exp(1 + 2i)$ 2. $\exp(2 - 3i)$

3. $e^{i\pi}$ 4. $e^{2-(\pi/2)i}$

5. 2^{1-i} 6. π^{-1+2i}

In each of Problems 7 through 16 find the general solution of the given differential equation.

7. $y'' - 2y' + 2y = 0$ 8. $y'' - 2y' + 6y = 0$

9. $y'' + 2y' - 8y = 0$ 10. $y'' + 2y' + 2y = 0$

11. $y'' + 6y' + 13y = 0$ 12. $4y'' + 9y = 0$

13. $y'' + 2y' + 1.25y = 0$ 14. $9y'' + 9y' - 4y = 0$

15. $y'' + y' + 1.25y = 0$ 16. $y'' + 4y' + 6.25y = 0$

In each of Problems 17 through 22 find the solution of the given initial value problem. Sketch
the graph of the solution and describe its behavior for increasing t.

17. $y'' + 4y = 0$, $y(0) = 0$, $y'(0) = 1$

18. $y'' + 4y' + 5y = 0$, $y(0) = 1$, $y'(0) = 0$

19. $y'' - 2y' + 5y = 0$, $y(\pi/2) = 0$, $y'(\pi/2) = 2$

20. $y'' + y = 0$, $y(\pi/3) = 2$, $y'(\pi/3) = -4$

21. $y'' + y' + 1.25y = 0$, $y(0) = 3$, $y'(0) = 1$

22. $y'' + 2y' + 2y = 0$, $y(\pi/4) = 2$, $y'(\pi/4) = -2$

23. Consider the initial value problem

$$3u'' - u' + 2u = 0, \qquad u(0) = 2, \quad u'(0) = 0.$$

(a) Find the solution $u(t)$ of this problem.

(b) Find the first time at which $|u(t)| = 10$.

24. Consider the initial value problem

$$5u'' + 2u' + 7u = 0, \qquad u(0) = 2, \quad u'(0) = 1.$$

(a) Find the solution $u(t)$ of this problem.

(b) Find the smallest T such that $|u(t)| \le 0.1$ for all $t > T$.

25. Consider the initial value problem

$$y'' + 2y' + 6y = 0, \qquad y(0) = 2, \quad y'(0) = \alpha \ge 0.$$

(a) Find the solution $y(t)$ of this problem.

(b) Find α so that $y = 0$ when $t = 1$.

(c) Find, as a function of α, the smallest positive value of t for which $y = 0$.

(d) Determine the limit of the expression found in part (c) as $\alpha \to \infty$.

26. Consider the initial value problem

$$y'' + 2ay' + (a^2 + 1)y = 0, \qquad y(0) = 1, \quad y'(0) = 0.$$

(a) Find the solution $y(t)$ of this problem.

(b) For $a = 1$ find the smallest T such that $|y(t)| < 0.1$ for $t > T$.

(c) Repeat part (b) for $a = 1/4, 1/2$, and 2.

(d) Using the results of parts (b) and (c), plot T versus a and describe the relation between
T and a.

27. Show that $W(e^{\lambda t} \cos \mu t, e^{\lambda t} \sin \mu t) = \mu e^{2\lambda t}$.

28. In this problem we outline a different derivation of Euler's formula.

 (a) Show that $y_1(t) = \cos t$ and $y_2(t) = \sin t$ are a fundamental set of solutions of $y'' + y = 0$; that is, show that they are solutions and that their Wronskian is not zero.

 (b) Show (formally) that $y = e^{it}$ is also a solution of $y'' + y = 0$. Therefore,

$$e^{it} = c_1 \cos t + c_2 \sin t \tag{i}$$

 for some constants c_1 and c_2. Why is this so?

 (c) Set $t = 0$ in Eq. (i) to show that $c_1 = 1$.

 (d) Assuming that Eq. (14) is true, differentiate Eq. (i) and then set $t = 0$ to conclude that $c_2 = i$. Use the values of c_1 and c_2 in Eq. (i) to arrive at Euler's formula.

29. Using Euler's formula, show that

$$\cos t = (e^{it} + e^{-it})/2, \qquad \sin t = (e^{it} - e^{-it})/2i.$$

30. If e^{rt} is given by Eq. (13), show that $e^{(r_1 + r_2)t} = e^{r_1 t} e^{r_2 t}$ for any complex numbers r_1 and r_2.

31. If e^{rt} is given by Eq. (13), show that

$$\frac{d}{dt} e^{rt} = r e^{rt}$$

 for any complex number r.

32. Let the real-valued functions p and q be continuous on the open interval I, and let $y = \phi(t) = u(t) + iv(t)$ be a complex-valued solution of

$$y'' + p(t)y' + q(t)y = 0, \tag{i}$$

 where u and v are real-valued functions. Show that u and v are also solutions of Eq. (i). *Hint:* Substitute $y = \phi(t)$ in Eq. (i) and separate into real and imaginary parts.

33. If the functions y_1 and y_2 are linearly independent solutions of $y'' + p(t)y' + q(t)y = 0$, show that between consecutive zeros of y_1 there is one and only one zero of y_2. Note that this result is illustrated by the solutions $y_1(t) = \cos t$ and $y_2(t) = \sin t$ of the equation $y'' + y = 0$.

 Hint: Suppose that t_1 and t_2 are two zeros of y_1 between which there are no zeros of y_2. Apply Rolle's theorem to y_1/y_2 to reach a contradiction.

Change of Variables. Often a differential equation with variable coefficients,

$$y'' + p(t)y' + q(t)y = 0, \tag{i}$$

can be put in a more suitable form for finding a solution by making a change of the independent and/or dependent variables. We explore these ideas in Problems 34 through 42. In particular, in Problem 34 we determine conditions under which Eq. (i) can be transformed into a differential equation with constant coefficients and thereby becomes easily solvable. Problems 35 through 42 give specific applications of this procedure.

34. In this problem we determine conditions on p and q that enable Eq. (i) to be transformed into an equation with constant coefficients by a change of the independent variable. Let $x = u(t)$ be the new independent variable, with the relation between x and t to be specified later.

 (a) Show that

$$\frac{dy}{dt} = \frac{dx}{dt}\frac{dy}{dx}, \qquad \frac{d^2y}{dt^2} = \left(\frac{dx}{dt}\right)^2 \frac{d^2y}{dx^2} + \frac{d^2x}{dt^2}\frac{dy}{dx}.$$

(b) Show that the differential equation (i) becomes

$$\left(\frac{dx}{dt}\right)^2 \frac{d^2y}{dx^2} + \left(\frac{d^2x}{dt^2} + p(t)\frac{dx}{dt}\right)\frac{dy}{dx} + q(t)y = 0. \tag{ii}$$

(c) In order for Eq. (ii) to have constant coefficients, the coefficients of d^2y/dx^2 and of y must be proportional. If $q(t) > 0$, then we can choose the constant of proportionality to be 1; hence

$$x = u(t) = \int [q(t)]^{1/2}\, dt. \tag{iii}$$

(d) With x chosen as in part (c), show that the coefficient of dy/dx in Eq. (ii) is also a constant, provided that the expression

$$\frac{q'(t) + 2p(t)q(t)}{2[q(t)]^{3/2}} \tag{iv}$$

is a constant. Thus Eq. (i) can be transformed into an equation with constant coefficients by a change of the independent variable, provided that the function $(q' + 2pq)/q^{3/2}$ is a constant. How must this result be modified if $q(t) < 0$?

In each of Problems 35 through 37 try to transform the given equation into one with constant coefficients by the method of Problem 34. If this is possible, find the general solution of the given equation.

35. $y'' + ty' + e^{-t^2}y = 0, \qquad -\infty < t < \infty$

36. $y'' + 3ty' + t^2y = 0, \qquad -\infty < t < \infty$

37. $ty'' + (t^2 - 1)y' + t^3y = 0, \qquad 0 < t < \infty$

38. **Euler Equations.** An equation of the form

$$t^2 y'' + \alpha t y' + \beta y = 0, \qquad t > 0,$$

where α and β are real constants, is called an Euler equation. Show that the substitution $x = \ln t$ transforms an Euler equation into an equation with constant coefficients. Euler equations are discussed in detail in Section 5.5.

In each of Problems 39 through 42 use the result of Problem 38 to solve the given equation for $t > 0$.

39. $t^2y'' + ty' + y = 0$ 40. $t^2y'' + 4ty' + 2y = 0$

41. $t^2y'' + 3ty' + 1.25y = 0$ 42. $t^2y'' - 4ty' - 6y = 0$

3.5 Repeated Roots; Reduction of Order

In earlier sections we showed how to solve the equation

$$ay'' + by' + cy = 0 \tag{1}$$

when the roots of the characteristic equation

$$ar^2 + br + c = 0 \tag{2}$$

either are real and different or are complex conjugates. Now we consider the third possibility, namely, that the two roots r_1 and r_2 are equal. This case is transitional

between the other two and occurs when the discriminant $b^2 - 4ac$ is zero. Then it follows from the quadratic formula that

$$r_1 = r_2 = -b/2a. \tag{3}$$

The difficulty is immediately apparent; both roots yield the same solution

$$y_1(t) = e^{-bt/2a} \tag{4}$$

of the differential equation (1), and it is not obvious how to find a second solution.

EXAMPLE 1

Solve the differential equation
$$y'' + 4y' + 4y = 0. \tag{5}$$

The characteristic equation is

$$r^2 + 4r + 4 = (r + 2)^2 = 0,$$

so $r_1 = r_2 = -2$. Therefore one solution of Eq. (5) is $y_1(t) = e^{-2t}$. To find the general solution of Eq. (5) we need a second solution that is not a multiple of y_1. This second solution can be found in several ways (see Problems 20 through 22); here we use a method originated by D'Alembert[5] in the eighteenth century. Recall that since $y_1(t)$ is a solution of Eq. (1), so is $cy_1(t)$ for any constant c. The basic idea is to generalize this observation by replacing c by a function $v(t)$ and then trying to determine $v(t)$ so that the product $v(t)y_1(t)$ is also a solution of Eq. (1).

To carry out this program we substitute $y = v(t)y_1(t)$ in Eq. (1) and use the resulting equation to find $v(t)$. Starting with

$$y = v(t)y_1(t) = v(t)e^{-2t}, \tag{6}$$

we have

$$y' = v'(t)e^{-2t} - 2v(t)e^{-2t} \tag{7}$$

and

$$y'' = v''(t)e^{-2t} - 4v'(t)e^{-2t} + 4v(t)e^{-2t}. \tag{8}$$

By substituting the expressions in Eqs. (6), (7), and (8) in Eq. (5) and collecting terms, we obtain

$$[v''(t) - 4v'(t) + 4v(t) + 4v'(t) - 8v(t) + 4v(t)]e^{-2t} = 0,$$

which simplifies to

$$v''(t) = 0. \tag{9}$$

Therefore

$$v'(t) = c_1$$

and

$$v(t) = c_1 t + c_2, \tag{10}$$

[5]Jean d'Alembert (1717–1783), a French mathematician, was a contemporary of Euler and Daniel Bernoulli and is known primarily for his work in mechanics and differential equations. D'Alembert's principle in mechanics and d'Alembert's paradox in hydrodynamics are named for him, and the wave equation first appeared in his paper on vibrating strings in 1747. In his later years he devoted himself primarily to philosophy and to his duties as science editor of Diderot's *Encyclopédie*.

where c_1 and c_2 are arbitrary constants. Finally, substituting for $v(t)$ in Eq. (6), we obtain

$$y = c_1 t e^{-2t} + c_2 e^{-2t}. \tag{11}$$

The second term on the right side of Eq. (11) corresponds to the original solution $y_1(t) = \exp(-2t)$, but the first term arises from a second solution, namely, $y_2(t) = t \exp(-2t)$. These two solutions are obviously not proportional, but we can verify that they are linearly independent by calculating their Wronskian:

$$W(y_1, y_2)(t) = \begin{vmatrix} e^{-2t} & t e^{-2t} \\ -2e^{-2t} & (1-2t)e^{-2t} \end{vmatrix}$$

$$= e^{-4t} - 2te^{-4t} + 2te^{-4t} = e^{-4t} \neq 0.$$

Therefore

$$y_1(t) = e^{-2t}, \qquad y_2(t) = te^{-2t} \tag{12}$$

form a fundamental set of solutions of Eq. (5), and the general solution of that equation is given by Eq. (11). Note that both $y_1(t)$ and $y_2(t)$ tend to zero as $t \to \infty$; consequently, all solutions of Eq. (5) behave in this way. The graph of a typical solution is shown in Figure 3.5.1.

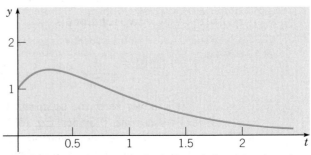

FIGURE 3.5.1 A typical solution of $y'' + 4y' + 4y = 0$.

The procedure used in Example 1 can be extended to a general equation whose characteristic equation has repeated roots. That is, we assume that the coefficients in Eq. (1) satisfy $b^2 - 4ac = 0$, in which case

$$y_1(t) = e^{-bt/2a}$$

is a solution. Then we assume that

$$y = v(t)y_1(t) = v(t)e^{-bt/2a} \tag{13}$$

and substitute for y in Eq. (1) to determine $v(t)$. We have

$$y' = v'(t)e^{-bt/2a} - \frac{b}{2a}v(t)e^{-bt/2a} \tag{14}$$

and

$$y'' = v''(t)e^{-bt/2a} - \frac{b}{a}v'(t)e^{-bt/2a} + \frac{b^2}{4a^2}v(t)e^{-bt/2a}. \tag{15}$$

Then, by substituting in Eq. (1), we obtain

$$\left\{ a\left[v''(t) - \frac{b}{a}v'(t) + \frac{b^2}{4a^2}v(t) \right] + b\left[v'(t) - \frac{b}{2a}v(t) \right] + cv(t) \right\} e^{-bt/2a} = 0. \tag{16}$$

Canceling the factor $\exp(-bt/2a)$, which is nonzero, and rearranging the remaining terms, we find that

$$av''(t) + (-b + b)v'(t) + \left(\frac{b^2}{4a} - \frac{b^2}{2a} + c\right)v(t) = 0. \tag{17}$$

The term involving $v'(t)$ is obviously zero. Further, the coefficient of $v(t)$ is $c - (b^2/4a)$, which is also zero because $b^2 - 4ac = 0$ in the problem that we are considering. Thus, just as in Example 1, Eq. (17) reduces to

$$v''(t) = 0;$$

therefore,

$$v(t) = c_1 t + c_2.$$

Hence, from Eq. (13), we have

$$y = c_1 t e^{-bt/2a} + c_2 e^{-bt/2a}. \tag{18}$$

Thus y is a linear combination of the two solutions

$$y_1(t) = e^{-bt/2a}, \qquad y_2(t) = t e^{-bt/2a}. \tag{19}$$

The Wronskian of these two solutions is

$$W(y_1, y_2)(t) = \begin{vmatrix} e^{-bt/2a} & t e^{-bt/2a} \\ -\dfrac{b}{2a} e^{-bt/2a} & \left(1 - \dfrac{bt}{2a}\right) e^{-bt/2a} \end{vmatrix} = e^{-bt/a}. \tag{20}$$

Since $W(y_1, y_2)(t)$ is never zero, the solutions y_1 and y_2 given by Eq. (19) are a fundamental set of solutions. Further, Eq. (18) is the general solution of Eq. (1) when the roots of the characteristic equation are equal. In other words, in this case, there is one exponential solution corresponding to the repeated root, and a second solution is obtained by multiplying the exponential solution by t.

EXAMPLE 2

Find the solution of the initial value problem

$$y'' - y' + 0.25y = 0, \qquad y(0) = 2, \quad y'(0) = \tfrac{1}{3}. \tag{21}$$

The characteristic equation is

$$r^2 - r + 0.25 = 0,$$

so the roots are $r_1 = r_2 = 1/2$. Thus the general solution of the differential equation is

$$y = c_1 e^{t/2} + c_2 t e^{t/2}. \tag{22}$$

The first initial condition requires that

$$y(0) = c_1 = 2.$$

To satisfy the second initial condition, we first differentiate Eq. (22) and then set $t = 0$. This gives

$$y'(0) = \tfrac{1}{2}c_1 + c_2 = \tfrac{1}{3},$$

so $c_2 = -2/3$. Thus the solution of the initial value problem is

$$y = 2e^{t/2} - \tfrac{2}{3} t e^{t/2}. \tag{23}$$

The graph of this solution is shown in Figure 3.5.2.

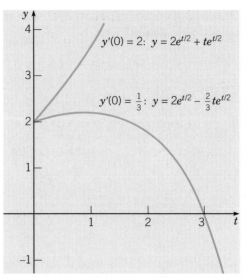

FIGURE 3.5.2 Solutions of $y'' - y' + 0.25y = 0$, $y(0) = 2$, with $y'(0) = 1/3$ and with $y'(0) = 2$, respectively.

Let us now modify the initial value problem (21) by changing the initial slope; to be specific, let the second initial condition be $y'(0) = 2$. The solution of this modified problem is

$$y = 2e^{t/2} + te^{t/2},$$

and its graph is also shown in Figure 3.5.2. The graphs shown in this figure suggest that there is a critical initial slope, with a value between $\frac{1}{3}$ and 2, that separates solutions that grow positively from those that ultimately grow negatively. In Problem 16 you are asked to determine this critical initial slope.

The geometrical behavior of solutions is similar in this case to that when the roots are real and different. If the exponents are either positive or negative, then the magnitude of the solution grows or decays accordingly; the linear factor t has little influence. A decaying solution is shown in Figure 3.5.1 and growing solutions in Figure 3.5.2. However, if the repeated root is zero, then the differential equation is $y'' = 0$ and the general solution is a linear function of t.

Summary. We can now summarize the results that we have obtained for second order linear homogeneous equations with constant coefficients,

$$ay'' + by' + cy = 0. \tag{1}$$

Let r_1 and r_2 be the roots of the corresponding characteristic polynomial

$$ar^2 + br + c = 0. \tag{2}$$

If r_1 and r_2 are real but not equal, then the general solution of the differential equation (1) is

$$y = c_1 e^{r_1 t} + c_2 e^{r_2 t}. \tag{24}$$

If r_1 and r_2 are complex conjugates $\lambda \pm i\mu$, then the general solution is

$$y = c_1 e^{\lambda t} \cos \mu t + c_2 e^{\lambda t} \sin \mu t. \tag{25}$$

If $r_1 = r_2$, then the general solution is

$$y = c_1 e^{r_1 t} + c_2 t e^{r_1 t}. \tag{26}$$

Reduction of Order. It is worth noting that the procedure used earlier in this section for equations with constant coefficients is more generally applicable. Suppose we know one solution $y_1(t)$, not everywhere zero, of

$$y'' + p(t)y' + q(t)y = 0. \tag{27}$$

To find a second solution, let

$$y = v(t)y_1(t); \tag{28}$$

then

$$y' = v'(t)y_1(t) + v(t)y_1'(t)$$

and

$$y'' = v''(t)y_1(t) + 2v'(t)y_1'(t) + v(t)y_1''(t).$$

Substituting for y, y', and y'' in Eq. (27) and collecting terms, we find that

$$y_1 v'' + (2y_1' + py_1)v' + (y_1'' + py_1' + qy_1)v = 0. \tag{29}$$

Since y_1 is a solution of Eq. (27), the coefficient of v in Eq. (29) is zero, so that Eq. (29) becomes

$$y_1 v'' + (2y_1' + py_1)v' = 0. \tag{30}$$

Despite its appearance, Eq. (30) is actually a first order equation for the function v' and can be solved either as a first order linear equation or as a separable equation. Once v' has been found, then v is obtained by an integration. Finally, y is determined from Eq. (28). This procedure is called the method of reduction of order, because the crucial step is the solution of a first order differential equation for v' rather than the original second order equation for y. Although it is possible to write down a formula for $v(t)$, we will instead illustrate how this method works by an example.

EXAMPLE 3

Given that $y_1(t) = t^{-1}$ is a solution of

$$2t^2 y'' + 3ty' - y = 0, \qquad t > 0, \tag{31}$$

find a second linearly independent solution.

We set $y = v(t)t^{-1}$; then

$$y' = v't^{-1} - vt^{-2}, \qquad y'' = v''t^{-1} - 2v't^{-2} + 2vt^{-3}.$$

Substituting for y, y', and y'' in Eq. (31) and collecting terms, we obtain

$$2t^2(v''t^{-1} - 2v't^{-2} + 2vt^{-3}) + 3t(v't^{-1} - vt^{-2}) - vt^{-1}$$
$$= 2tv'' + (-4 + 3)v' + (4t^{-1} - 3t^{-1} - t^{-1})v$$
$$= 2tv'' - v' = 0. \tag{32}$$

Note that the coefficient of v is zero, as it should be; this provides a useful check on our algebra.

Separating the variables in Eq. (32) and solving for $v'(t)$, we find that

$$v'(t) = ct^{1/2};$$

then

$$v(t) = \tfrac{2}{3}ct^{3/2} + k.$$

It follows that

$$y = \tfrac{2}{3}ct^{1/2} + kt^{-1}, \tag{33}$$

where c and k are arbitrary constants. The second term on the right side of Eq. (33) is a multiple of $y_1(t)$ and can be dropped, but the first term provides a new independent solution. Neglecting the arbitrary multiplicative constant, we have $y_2(t) = t^{1/2}$.

PROBLEMS

In each of Problems 1 through 10 find the general solution of the given differential equation.

1. $y'' - 2y' + y = 0$ 2. $9y'' + 6y' + y = 0$
3. $4y'' - 4y' - 3y = 0$ 4. $4y'' + 12y' + 9y = 0$
5. $y'' - 2y' + 10y = 0$ 6. $y'' - 6y' + 9y = 0$
7. $4y'' + 17y' + 4y = 0$ 8. $16y'' + 24y' + 9y = 0$
9. $25y'' - 20y' + 4y = 0$ 10. $2y'' + 2y' + y = 0$

In each of Problems 11 through 14 solve the given initial value problem. Sketch the graph of the solution and describe its behavior for increasing t.

11. $9y'' - 12y' + 4y = 0$, $y(0) = 2$, $y'(0) = -1$
12. $y'' - 6y' + 9y = 0$, $y(0) = 0$, $y'(0) = 2$
13. $9y'' + 6y' + 82y = 0$, $y(0) = -1$, $y'(0) = 2$
14. $y'' + 4y' + 4y = 0$, $y(-1) = 2$, $y'(-1) = 1$

 15. Consider the initial value problem

$$4y'' + 12y' + 9y = 0, \qquad y(0) = 1, \quad y'(0) = -4.$$

(a) Solve the initial value problem and plot its solution for $0 \le t \le 5$.

(b) Determine where the solution has the value zero.

(c) Determine the coordinates (t_0, y_0) of the minimum point.

(d) Change the second initial condition to $y'(0) = b$ and find the solution as a function of b. Then find the critical value of b that separates solutions that always remain positive from those that eventually become negative.

16. Consider the following modification of the initial value problem in Example 2:

$$y'' - y' + 0.25y = 0, \qquad y(0) = 2, \quad y'(0) = b.$$

Find the solution as a function of b and then determine the critical value of b that separates solutions that grow positively from those that eventually grow negatively.

17. Consider the initial value problem

$$4y'' + 4y' + y = 0, \qquad y(0) = 1, \quad y'(0) = 2.$$

(a) Solve the initial value problem and plot the solution.

(b) Determine the coordinates (t_M, y_M) of the maximum point.

(c) Change the second initial condition to $y'(0) = b > 0$ and find the solution as a function of b.

(d) Find the coordinates (t_M, y_M) of the maximum point in terms of b. Describe the dependence of t_M and y_M on b as b increases.

18. Consider the initial value problem

$$9y'' + 12y' + 4y = 0, \qquad y(0) = a > 0, \quad y'(0) = -1.$$

(a) Solve the initial value problem.

(b) Find the critical value of a that separates solutions that become negative from those that are always positive.

19. If the roots of the characteristic equation are real, show that a solution of $ay'' + by' + cy = 0$ can take on the value zero at most once.

Problems 20 through 22 indicate other ways of finding the second solution when the characteristic equation has repeated roots.

20. (a) Consider the equation $y'' + 2ay' + a^2y = 0$. Show that the roots of the characteristic equation are $r_1 = r_2 = -a$, so that one solution of the equation is e^{-at}.

(b) Use Abel's formula [Eq. (8) of Section 3.3] to show that the Wronskian of any two solutions of the given equation is

$$W(t) = y_1(t)y_2'(t) - y_1'(t)y_2(t) = c_1 e^{-2at},$$

where c_1 is a constant.

(c) Let $y_1(t) = e^{-at}$ and use the result of part (b) to obtain a differential equation satisfied by a second solution $y_2(t)$. By solving this equation, show that $y_2(t) = te^{-at}$.

21. Suppose that r_1 and r_2 are roots of $ar^2 + br + c = 0$ and that $r_1 \neq r_2$; then $\exp(r_1 t)$ and $\exp(r_2 t)$ are solutions of the differential equation $ay'' + by' + cy = 0$. Show that $\phi(t; r_1, r_2) = [\exp(r_2 t) - \exp(r_1 t)]/(r_2 - r_1)$ is also a solution of the equation for $r_2 \neq r_1$. Then think of r_1 as fixed and use L'Hospital's rule to evaluate the limit of $\phi(t; r_1, r_2)$ as $r_2 \to r_1$, thereby obtaining the second solution in the case of equal roots.

22. (a) If $ar^2 + br + c = 0$ has equal roots r_1, show that

$$L[e^{rt}] = a(e^{rt})'' + b(e^{rt})' + ce^{rt} = a(r - r_1)^2 e^{rt}. \qquad (i)$$

Since the right side of Eq. (i) is zero when $r = r_1$, it follows that $\exp(r_1 t)$ is a solution of $L[y] = ay'' + by' + cy = 0$.

(b) Differentiate Eq. (i) with respect to r and interchange differentiation with respect to r and with respect to t, thus showing that

$$\frac{\partial}{\partial r} L[e^{rt}] = L\left[\frac{\partial}{\partial r} e^{rt}\right] = L[te^{rt}] = ate^{rt}(r - r_1)^2 + 2ae^{rt}(r - r_1). \qquad (ii)$$

Since the right side of Eq. (ii) is zero when $r = r_1$, conclude that $t\exp(r_1 t)$ is also a solution of $L[y] = 0$.

In each of Problems 23 through 30 use the method of reduction of order to find a second solution of the given differential equation.

23. $t^2 y'' - 4ty' + 6y = 0$, $t > 0$; $y_1(t) = t^2$

24. $t^2 y'' + 2ty' - 2y = 0$, $t > 0$; $y_1(t) = t$

25. $t^2 y'' + 3ty' + y = 0$, $t > 0$; $y_1(t) = t^{-1}$

26. $t^2 y'' - t(t + 2)y' + (t + 2)y = 0$, $t > 0$; $y_1(t) = t$

27. $xy'' - y' + 4x^3 y = 0$, $x > 0$; $y_1(x) = \sin x^2$

28. $(x - 1)y'' - xy' + y = 0$, $x > 1$; $y_1(x) = e^x$

29. $x^2 y'' - (x - 0.1875)y = 0$, $x > 0$; $y_1(x) = x^{1/4} e^{2\sqrt{x}}$

30. $x^2 y'' + xy' + (x^2 - 0.25)y = 0, \quad x > 0; \qquad y_1(x) = x^{-1/2} \sin x$

31. The differential equation

$$xy'' - (x + N)y' + Ny = 0,$$

where N is a nonnegative integer, has been discussed by several authors.[6] One reason why it is interesting is that it has an exponential solution and a polynomial solution.

(a) Verify that one solution is $y_1(x) = e^x$.

(b) Show that a second solution has the form $y_2(x) = ce^x \int x^N e^{-x}\, dx$. Calculate $y_2(x)$ for $N = 1$ and $N = 2$; convince yourself that, with $c = -1/N!$,

$$y_2(x) = 1 + \frac{x}{1!} + \frac{x^2}{2!} + \cdots + \frac{x^N}{N!}.$$

Note that $y_2(x)$ is exactly the first $N + 1$ terms in the Taylor series about $x = 0$ for e^x, that is, for $y_1(x)$.

32. The differential equation

$$y'' + \delta(xy' + y) = 0$$

arises in the study of the turbulent flow of a uniform stream past a circular cylinder. Verify that $y_1(x) = \exp(-\delta x^2/2)$ is one solution and then find the general solution in the form of an integral.

33. The method of Problem 20 can be extended to second order equations with variable coefficients. If y_1 is a known nonvanishing solution of $y'' + p(t)y' + q(t)y = 0$, show that a second solution y_2 satisfies $(y_2/y_1)' = W(y_1, y_2)/y_1^2$, where $W(y_1, y_2)$ is the Wronskian of y_1 and y_2. Then use Abel's formula [Eq. (8) of Section 3.3] to determine y_2.

In each of Problems 34 through 37 use the method of Problem 33 to find a second independent solution of the given equation.

34. $t^2 y'' + 3ty' + y = 0, \quad t > 0; \qquad y_1(t) = t^{-1}$

35. $ty'' - y' + 4t^3 y = 0, \quad t > 0; \qquad y_1(t) = \sin(t^2)$

36. $(x - 1)y'' - xy' + y = 0, \quad x > 1; \qquad y_1(x) = e^x$

37. $x^2 y'' + xy' + (x^2 - 0.25)y = 0, \quad x > 0; \qquad y_1(x) = x^{-1/2} \sin x$

Behavior of Solutions as $t \to \infty$. Problems 38 through 40 are concerned with the behavior of solutions as $t \to \infty$.

38. If a, b, and c are positive constants, show that all solutions of $ay'' + by' + cy = 0$ approach zero as $t \to \infty$.

39. (a) If $a > 0$ and $c > 0$, but $b = 0$, show that the result of Problem 38 is no longer true, but that all solutions are bounded as $t \to \infty$.

(b) If $a > 0$ and $b > 0$, but $c = 0$, show that the result of Problem 38 is no longer true, but that all solutions approach a constant that depends on the initial conditions as $t \to \infty$. Determine this constant for the initial conditions $y(0) = y_0, y'(0) = y_0'$.

40. Show that $y = \sin t$ is a solution of

$$y'' + (k \sin^2 t)y' + (1 - k \cos t \sin t)y = 0$$

[6]T. A. Newton, "On Using a Differential Equation to Generate Polynomials," *American Mathematical Monthly 81* (1974), pp. 592–601. Also see the references given there.

for any value of the constant k. If $0 < k < 2$, show that $1 - k \cos t \sin t > 0$ and $k \sin^2 t \geq 0$. Thus observe that even though the coefficients of this variable-coefficient differential equation are nonnegative (and the coefficient of y' is zero only at the points $t = 0, \pi, 2\pi, \ldots$), it has a solution that does not approach zero as $t \to \infty$. Compare this situation with the result of Problem 38. Thus we observe a not unusual situation in the theory of differential equations: equations that are apparently very similar can have quite different properties.

Euler Equations. Use the substitution introduced in Problem 38 in Section 3.4 to solve each of the equations in Problems 41 and 42.

41. $t^2 y'' - 3ty' + 4y = 0$, $t > 0$

42. $t^2 y'' + 2ty' + 0.25y = 0$, $t > 0$

3.6 Nonhomogeneous Equations; Method of Undetermined Coefficients

We now return to the nonhomogeneous equation

$$L[y] = y'' + p(t)y' + q(t)y = g(t), \tag{1}$$

where p, q, and g are given (continuous) functions on the open interval I. The equation

$$L[y] = y'' + p(t)y' + q(t)y = 0, \tag{2}$$

in which $g(t) = 0$ and p and q are the same as in Eq. (1), is called the homogeneous equation corresponding to Eq. (1). The following two results describe the structure of solutions of the nonhomogeneous equation (1) and provide a basis for constructing its general solution.

Theorem 3.6.1 If Y_1 and Y_2 are two solutions of the nonhomogeneous equation (1), then their difference $Y_1 - Y_2$ is a solution of the corresponding homogeneous equation (2). If, in addition, y_1 and y_2 are a fundamental set of solutions of Eq. (2), then

$$Y_1(t) - Y_2(t) = c_1 y_1(t) + c_2 y_2(t), \tag{3}$$

where c_1 and c_2 are certain constants.

To prove this result, note that Y_1 and Y_2 satisfy the equations

$$L[Y_1](t) = g(t), \qquad L[Y_2](t) = g(t). \tag{4}$$

Subtracting the second of these equations from the first, we have

$$L[Y_1](t) - L[Y_2](t) = g(t) - g(t) = 0. \tag{5}$$

However,

$$L[Y_1] - L[Y_2] = L[Y_1 - Y_2],$$

so Eq. (5) becomes

$$L[Y_1 - Y_2](t) = 0. \tag{6}$$

Equation (6) states that $Y_1 - Y_2$ is a solution of Eq. (2). Finally, since all solutions of Eq. (2) can be expressed as linear combinations of a fundamental set of solutions by Theorem 3.2.4, it follows that the solution $Y_1 - Y_2$ can be so written. Hence Eq. (3) holds and the proof is complete.

Theorem 3.6.2

The general solution of the nonhomogeneous equation (1) can be written in the form

$$y = \phi(t) = c_1 y_1(t) + c_2 y_2(t) + Y(t), \tag{7}$$

where y_1 and y_2 are a fundamental set of solutions of the corresponding homogeneous equation (2), c_1 and c_2 are arbitrary constants, and Y is some specific solution of the nonhomogeneous equation (1).

The proof of Theorem 3.6.2 follows quickly from the preceding theorem. Note that Eq. (3) holds if we identify Y_1 with an arbitrary solution ϕ of Eq. (1) and Y_2 with the specific solution Y. From Eq. (3) we thereby obtain

$$\phi(t) - Y(t) = c_1 y_1(t) + c_2 y_2(t), \tag{8}$$

which is equivalent to Eq. (7). Since ϕ is an arbitrary solution of Eq. (1), the expression on the right side of Eq. (7) includes all solutions of Eq. (1); thus it is natural to call it the general solution of Eq. (1).

In somewhat different words, Theorem 3.6.2 states that to solve the nonhomogeneous equation (1), we must do three things:

1. Find the general solution $c_1 y_1(t) + c_2 y_2(t)$ of the corresponding homogeneous equation. This solution is frequently called the complementary solution and may be denoted by $y_c(t)$.
2. Find some single solution $Y(t)$ of the nonhomogeneous equation. Often this solution is referred to as a particular solution.
3. Add together the functions found in the two preceding steps.

We have already discussed how to find $y_c(t)$, at least when the homogeneous equation (2) has constant coefficients. Therefore, in the remainder of this section and in the next, we will focus on finding a particular solution $Y(t)$ of the nonhomogeneous equation (1). There are two methods that we wish to discuss. They are known as the method of undetermined coefficients and the method of variation of parameters, respectively. Each has some advantages and some possible shortcomings.

Method of Undetermined Coefficients. The method of undetermined coefficients requires that we make an initial assumption about the form of the particular solution $Y(t)$, but with the coefficients left unspecified. We then substitute the assumed expression into Eq. (1) and attempt to determine the coefficients so as to satisfy that equation. If we are successful, then we have found a solution of the differential equation (1) and can use it for the particular solution $Y(t)$. If we cannot determine the coefficients, then this means that there is no solution of the form that we assumed. In this case we may modify the initial assumption and try again.

The main advantage of the method of undetermined coefficients is that it is straightforward to execute once the assumption is made as to the form of $Y(t)$. Its major limitation is that it is useful primarily for equations for which we can easily write

down the correct form of the particular solution in advance. For this reason, this method is usually used only for problems in which the homogeneous equation has constant coefficients and the nonhomogeneous term is restricted to a relatively small class of functions. In particular, we consider only nonhomogeneous terms that consist of polynomials, exponential functions, sines, and cosines. Despite this limitation, the method of undetermined coefficients is useful for solving many problems that have important applications. However, the algebraic details may become tedious, and a computer algebra system can be very helpful in practical applications. We will illustrate the method of undetermined coefficients by several simple examples and then summarize some rules for using it.

EXAMPLE 1

Find a particular solution of

$$y'' - 3y' - 4y = 3e^{2t}. \tag{9}$$

We seek a function Y such that the combination $Y''(t) - 3Y'(t) - 4Y(t)$ is equal to $3e^{2t}$. Since the exponential function reproduces itself through differentiation, the most plausible way to achieve the desired result is to assume that $Y(t)$ is some multiple of e^{2t}, that is,

$$Y(t) = Ae^{2t},$$

where the coefficient A is yet to be determined. To find A we calculate

$$Y'(t) = 2Ae^{2t}, \qquad Y''(t) = 4Ae^{2t},$$

and substitute for y, y', and y'' in Eq. (9). We obtain

$$(4A - 6A - 4A)e^{2t} = 3e^{2t}.$$

Hence $-6Ae^{2t}$ must equal $3e^{2t}$, so $A = -1/2$. Thus a particular solution is

$$Y(t) = -\tfrac{1}{2}e^{2t}. \tag{10}$$

EXAMPLE 2

Find a particular solution of

$$y'' - 3y' - 4y = 2\sin t. \tag{11}$$

By analogy with Example 1, let us first assume that $Y(t) = A\sin t$, where A is a constant to be determined. On substituting in Eq. (11) and rearranging the terms, we obtain

$$-5A\sin t - 3A\cos t = 2\sin t,$$

or

$$(2 + 5A)\sin t + 3A\cos t = 0. \tag{12}$$

The functions $\sin t$ and $\cos t$ are linearly independent, so Eq. (12) can hold on an interval only if the coefficients $2 + 5A$ and $3A$ are both zero. These contradictory requirements mean that there is no choice of the constant A that makes Eq. (12) true for all t. Thus we conclude that our assumption concerning $Y(t)$ is inadequate. The appearance of the cosine term in Eq. (12) suggests that we modify our original assumption to include a cosine term in $Y(t)$; that is,

$$Y(t) = A\sin t + B\cos t,$$

where A and B are to be determined. Then

$$Y'(t) = A\cos t - B\sin t, \qquad Y''(t) = -A\sin t - B\cos t.$$

By substituting these expressions for y, y', and y'' in Eq. (11) and collecting terms, we obtain

$$(-A + 3B - 4A)\sin t + (-B - 3A - 4B)\cos t = 2\sin t. \tag{13}$$

To satisfy Eq. (13) we must match the coefficients of $\sin t$ and $\cos t$ on each side of the equation; thus A and B must satisfy the equations

$$-5A + 3B = 2, \qquad -3A - 5B = 0.$$

Hence $A = -5/17$ and $B = 3/17$, so a particular solution of Eq. (11) is

$$Y(t) = -\tfrac{5}{17}\sin t + \tfrac{3}{17}\cos t.$$

The method illustrated in the preceding examples can also be used when the right side of the equation is a polynomial. Thus, to find a particular solution of

$$y'' - 3y' - 4y = 4t^2 - 1, \tag{14}$$

we initially assume that $Y(t)$ is a polynomial of the same degree as the nonhomogeneous term, that is, $Y(t) = At^2 + Bt + C$.

To summarize our conclusions up to this point: if the nonhomogeneous term $g(t)$ in Eq. (1) is an exponential function $e^{\alpha t}$, then assume that $Y(t)$ is proportional to the same exponential function; if $g(t)$ is $\sin \beta t$ or $\cos \beta t$, then assume that $Y(t)$ is a linear combination of $\sin \beta t$ and $\cos \beta t$; if $g(t)$ is a polynomial, then assume that $Y(t)$ is a polynomial of like degree. The same principle extends to the case where $g(t)$ is a product of any two, or all three, of these types of functions, as the next example illustrates.

EXAMPLE 3

Find a particular solution of

$$y'' - 3y' - 4y = -8e^t \cos 2t. \tag{15}$$

In this case we assume that $Y(t)$ is the product of e^t and a linear combination of $\cos 2t$ and $\sin 2t$, that is,

$$Y(t) = Ae^t \cos 2t + Be^t \sin 2t.$$

The algebra is more tedious in this example, but it follows that

$$Y'(t) = (A + 2B)e^t \cos 2t + (-2A + B)e^t \sin 2t$$

and

$$Y''(t) = (-3A + 4B)e^t \cos 2t + (-4A - 3B)e^t \sin 2t.$$

By substituting these expressions in Eq. (15), we find that A and B must satisfy

$$10A + 2B = 8, \qquad 2A - 10B = 0.$$

Hence $A = 10/13$ and $B = 2/13$; therefore a particular solution of Eq. (15) is

$$Y(t) = \tfrac{10}{13}e^t \cos 2t + \tfrac{2}{13}e^t \sin 2t.$$

Now suppose that $g(t)$ is the sum of two terms, $g(t) = g_1(t) + g_2(t)$, and suppose that Y_1 and Y_2 are solutions of the equations

$$ay'' + by' + cy = g_1(t) \tag{16}$$

and

$$ay'' + by' + cy = g_2(t), \tag{17}$$

respectively. Then $Y_1 + Y_2$ is a solution of the equation

$$ay'' + by' + cy = g(t). \tag{18}$$

To prove this statement, substitute $Y_1(t) + Y_2(t)$ for y in Eq. (18) and make use of Eqs. (16) and (17). A similar conclusion holds if $g(t)$ is the sum of any finite number of terms. The practical significance of this result is that for an equation whose nonhomogeneous function $g(t)$ can be expressed as a sum, one can consider instead several simpler equations and then add the results together. The following example is an illustration of this procedure.

EXAMPLE 4

Find a particular solution of

$$y'' - 3y' - 4y = 3e^{2t} + 2\sin t - 8e^t \cos 2t. \tag{19}$$

By splitting up the right side of Eq. (19), we obtain the three equations

$$y'' - 3y' - 4y = 3e^{2t},$$
$$y'' - 3y' - 4y = 2\sin t,$$

and

$$y'' - 3y' - 4y = -8e^t \cos 2t.$$

Solutions of these three equations have been found in Examples 1, 2, and 3, respectively. Therefore a particular solution of Eq. (19) is their sum, namely,

$$Y(t) = -\tfrac{1}{2}e^{2t} + \tfrac{3}{17}\cos t - \tfrac{5}{17}\sin t + \tfrac{10}{13}e^t \cos 2t + \tfrac{2}{13}e^t \sin 2t.$$

The procedure illustrated in these examples enables us to solve a fairly large class of problems in a reasonably efficient manner. However, there is one difficulty that sometimes occurs. The next example illustrates how it arises.

EXAMPLE 5

Find a particular solution of

$$y'' - 3y' - 4y = 2e^{-t}. \tag{20}$$

Proceeding as in Example 1, we assume that $Y(t) = Ae^{-t}$. By substituting in Eq. (20), we then obtain

$$(A + 3A - 4A)e^{-t} = 2e^{-t}. \tag{21}$$

Since the left side of Eq. (21) is zero, there is no choice of A that satisfies this equation. Therefore, there is no particular solution of Eq. (20) of the assumed form. The reason for this possibly unexpected result becomes clear if we solve the homogeneous equation

$$y'' - 3y' - 4y = 0 \tag{22}$$

that corresponds to Eq. (20). A fundamental set of solutions of Eq. (22) is $y_1(t) = e^{-t}$ and $y_2(t) = e^{4t}$. Thus our assumed particular solution of Eq. (20) is actually a solution of the homogeneous equation (22); consequently, it cannot possibly be a solution of the nonhomogeneous equation (20). To find a solution of Eq. (20) we must therefore consider functions of a somewhat different form.

At this stage we have several possible alternatives. One is simply to try to guess the proper form of the particular solution of Eq. (20). Another is to solve this equation in some different way and then to use the result to guide our assumptions if this situation arises again in the future; see Problems 27 and 33 for other solution methods. Still another possibility is to seek a simpler equation where this difficulty occurs and to use its solution to suggest how we might proceed with Eq. (20). Adopting the latter approach, we look for a first order equation analogous to Eq. (20). One possibility is

$$y' + y = 2e^{-t}. \tag{23}$$

If we try to find a particular solution of Eq. (23) of the form Ae^{-t}, we will fail because e^{-t} is a solution of the homogeneous equation $y' + y = 0$. However, from Section 2.1 we already know how to solve Eq. (23). An integrating factor is $\mu(t) = e^t$, and by multiplying by $\mu(t)$ and then integrating both sides, we obtain the solution

$$y = 2te^{-t} + ce^{-t}. \tag{24}$$

The second term on the right side of Eq. (24) is the general solution of the homogeneous equation $y' + y = 0$, but the first term is a solution of the full nonhomogeneous equation (23). Observe that it involves the exponential factor e^{-t} multiplied by the factor t. This is the clue that we were looking for.

We now return to Eq. (20) and assume a particular solution of the form $Y(t) = Ate^{-t}$. Then

$$Y'(t) = Ae^{-t} - Ate^{-t}, \qquad Y''(t) = -2Ae^{-t} + Ate^{-t}. \tag{25}$$

Substituting these expressions for y, y', and y'' in Eq. (20), we obtain $-5A = 2$, so $A = -2/5$. Thus a particular solution of Eq. (20) is

$$Y(t) = -\tfrac{2}{5}te^{-t}. \tag{26}$$

The outcome of Example 5 suggests a modification of the principle stated previously: if the assumed form of the particular solution duplicates a solution of the corresponding homogeneous equation, then modify the assumed particular solution by multiplying it by t. Occasionally, this modification will be insufficient to remove all duplication with the solutions of the homogeneous equation, in which case it is necessary to multiply by t a second time. For a second order equation, it will never be necessary to carry the process further than this.

Summary. We now summarize the steps involved in finding the solution of an initial value problem consisting of a nonhomogeneous equation of the form

$$ay'' + by' + cy = g(t), \tag{27}$$

where the coefficients a, b, and c are constants, together with a given set of initial conditions:

1. Find the general solution of the corresponding homogeneous equation.
2. Make sure that the function $g(t)$ in Eq. (27) belongs to the class of functions discussed in this section; that is, be sure it involves nothing more than exponential functions, sines, cosines, polynomials, or sums or products of such functions. If this is not the case, use the method of variation of parameters (discussed in the next section).
3. If $g(t) = g_1(t) + \cdots + g_n(t)$, that is, if $g(t)$ is a sum of n terms, then form n subproblems, each of which contains only one of the terms $g_1(t), \ldots, g_n(t)$. The ith subproblem consists of the equation

$$ay'' + by' + cy = g_i(t),$$

where i runs from 1 to n.

4. For the ith subproblem assume a particular solution $Y_i(t)$ consisting of the appropriate exponential function, sine, cosine, polynomial, or combination thereof. If there is any duplication in the assumed form of $Y_i(t)$ with the solutions of the homogeneous equation (found in step 1), then multiply $Y_i(t)$ by t, or (if necessary) by t^2, so as to remove the duplication. See Table 3.6.1.

TABLE 3.6.1 The Particular Solution of $ay'' + by' + cy = g_i(t)$

$g_i(t)$	$Y_i(t)$
$P_n(t) = a_0 t^n + a_1 t^{n-1} + \cdots + a_n$	$t^s(A_0 t^n + A_1 t^{n-1} + \cdots + A_n)$
$P_n(t)e^{\alpha t}$	$t^s(A_0 t^n + A_1 t^{n-1} + \cdots + A_n)e^{\alpha t}$
$P_n(t)e^{\alpha t} \begin{cases} \sin \beta t \\ \cos \beta t \end{cases}$	$t^s[(A_0 t^n + A_1 t^{n-1} + \cdots + A_n)e^{\alpha t} \cos \beta t$ $+ (B_0 t^n + B_1 t^{n-1} + \cdots + B_n)e^{\alpha t} \sin \beta t]$

Notes. Here s is the smallest nonnegative integer ($s = 0, 1$, or 2) that will ensure that no term in $Y_i(t)$ is a solution of the corresponding homogeneous equation. Equivalently, for the three cases, s is the number of times 0 is a root of the characteristic equation, α is a root of the characteristic equation, and $\alpha + i\beta$ is a root of the characteristic equation, respectively.

5. Find a particular solution $Y_i(t)$ for each of the subproblems. Then the sum $Y_1(t) + \cdots + Y_n(t)$ is a particular solution of the full nonhomogeneous equation (27).
6. Form the sum of the general solution of the homogeneous equation (step 1) and the particular solution of the nonhomogeneous equation (step 5). This is the general solution of the nonhomogeneous equation.
7. Use the initial conditions to determine the values of the arbitrary constants remaining in the general solution.

For some problems this entire procedure is easy to carry out by hand, but in many cases it requires considerable algebra. Once you understand clearly how the method works, a computer algebra system can be of great assistance in executing the details.

The method of undetermined coefficients is self-correcting in the sense that if one assumes too little for $Y(t)$, then a contradiction is soon reached that usually points the way to the modification that is needed in the assumed form. On the other hand, if one

assumes too many terms, then some unnecessary work is done and some coefficients turn out to be zero, but at least the correct answer is obtained.

Proof of the Method of Undetermined Coefficients. In the preceding discussion we have described the method of undetermined coefficients on the basis of several examples. To prove that the procedure always works as stated, we now give a general argument, in which we consider several cases corresponding to different forms for the nonhomogeneous term $g(t)$.

$g(t) = P_n(t) = a_0 t^n + a_1 t^{n-1} + \cdots + a_n$. In this case Eq. (27) becomes

$$ay'' + by' + cy = a_0 t^n + a_1 t^{n-1} + \cdots + a_n. \tag{28}$$

To obtain a particular solution we assume that

$$Y(t) = A_0 t^n + A_1 t^{n-1} + \cdots + A_{n-2} t^2 + A_{n-1} t + A_n. \tag{29}$$

Substituting in Eq. (28), we obtain

$$a[n(n-1)A_0 t^{n-2} + \cdots + 2A_{n-2}] + b(nA_0 t^{n-1} + \cdots + A_{n-1})$$
$$+ c(A_0 t^n + A_1 t^{n-1} + \cdots + A_n) = a_0 t^n + \cdots + a_n. \tag{30}$$

Equating the coefficients of like powers of t gives

$$cA_0 = a_0,$$
$$cA_1 + nbA_0 = a_1,$$
$$\vdots$$
$$cA_n + bA_{n-1} + 2aA_{n-2} = a_n.$$

Provided that $c \neq 0$, the solution of the first equation is $A_0 = a_0/c$, and the remaining equations determine A_1, \ldots, A_n successively. If $c = 0$ but $b \neq 0$, then the polynomial on the left side of Eq. (30) is of degree $n - 1$, and we cannot satisfy Eq. (30). To be sure that $aY''(t) + bY'(t)$ is a polynomial of degree n, we must choose $Y(t)$ to be a polynomial of degree $n + 1$. Hence we assume that

$$Y(t) = t(A_0 t^n + \cdots + A_n).$$

There is no constant term in this expression for $Y(t)$, but there is no need to include such a term since a constant is a solution of the homogeneous equation when $c = 0$. Since $b \neq 0$, we have $A_0 = a_0/b(n+1)$, and the other coefficients A_1, \ldots, A_n can be determined similarly. If both c and b are zero, we assume that

$$Y(t) = t^2(A_0 t^n + \cdots + A_n).$$

The term $aY''(t)$ gives rise to a term of degree n, and we can proceed as before. Again the constant and linear terms in $Y(t)$ are omitted, since in this case they are both solutions of the homogeneous equation.

$g(t) = e^{\alpha t} P_n(t)$. The problem of determining a particular solution of

$$ay'' + by' + cy = e^{\alpha t} P_n(t) \tag{31}$$

can be reduced to the preceding case by a substitution. Let

$$Y(t) = e^{\alpha t} u(t);$$

then

$$Y'(t) = e^{\alpha t}[u'(t) + \alpha u(t)]$$

and

$$Y''(t) = e^{\alpha t}[u''(t) + 2\alpha u'(t) + \alpha^2 u(t)].$$

Substituting for y, y', and y'' in Eq. (31), canceling the factor $e^{\alpha t}$, and collecting terms, we obtain

$$au''(t) + (2a\alpha + b)u'(t) + (a\alpha^2 + b\alpha + c)u(t) = P_n(t). \tag{32}$$

The determination of a particular solution of Eq. (32) is precisely the same problem, except for the names of the constants, as solving Eq. (28). Therefore, if $a\alpha^2 + b\alpha + c$ is not zero, we assume that $u(t) = A_0 t^n + \cdots + A_n$; hence a particular solution of Eq. (31) is of the form

$$Y(t) = e^{\alpha t}(A_0 t^n + A_1 t^{n-1} + \cdots + A_n). \tag{33}$$

On the other hand, if $a\alpha^2 + b\alpha + c$ is zero but $2a\alpha + b$ is not, we must take $u(t)$ to be of the form $t(A_0 t^n + \cdots + A_n)$. The corresponding form for $Y(t)$ is t times the expression on the right side of Eq. (33). Note that if $a\alpha^2 + b\alpha + c$ is zero, then $e^{\alpha t}$ is a solution of the homogeneous equation. If both $a\alpha^2 + b\alpha + c$ and $2a\alpha + b$ are zero (and this implies that both $e^{\alpha t}$ and $te^{\alpha t}$ are solutions of the homogeneous equation), then the correct form for $u(t)$ is $t^2(A_0 t^n + \cdots + A_n)$. Hence $Y(t)$ is t^2 times the expression on the right side of Eq. (33).

$g(t) = e^{\alpha t} P_n(t) \cos \beta t$ or $e^{\alpha t} P_n(t) \sin \beta t$. These two cases are similar, so we consider only the latter. We can reduce this problem to the preceding one by noting that, as a consequence of Euler's formula, $\sin \beta t = (e^{i\beta t} - e^{-i\beta t})/2i$. Hence $g(t)$ is of the form

$$g(t) = P_n(t) \frac{e^{(\alpha+i\beta)t} - e^{(\alpha-i\beta)t}}{2i},$$

and we should choose

$$Y(t) = e^{(\alpha+i\beta)t}(A_0 t^n + \cdots + A_n) + e^{(\alpha-i\beta)t}(B_0 t^n + \cdots + B_n),$$

or, equivalently,

$$Y(t) = e^{\alpha t}(A_0 t^n + \cdots + A_n) \cos \beta t + e^{\alpha t}(B_0 t^n + \cdots + B_n) \sin \beta t.$$

Usually, the latter form is preferred. If $\alpha \pm i\beta$ satisfy the characteristic equation corresponding to the homogeneous equation, we must, of course, multiply each of the polynomials by t to increase their degrees by one.

If the nonhomogeneous function involves both $\cos \beta t$ and $\sin \beta t$, it is usually convenient to treat these terms together, since each one individually may give rise to the same form for a particular solution. For example, if $g(t) = t \sin t + 2 \cos t$, the form for $Y(t)$ would be

$$Y(t) = (A_0 t + A_1) \sin t + (B_0 t + B_1) \cos t,$$

provided that $\sin t$ and $\cos t$ are not solutions of the homogeneous equation.

PROBLEMS In each of Problems 1 through 12 find the general solution of the given differential equation.

1. $y'' - 2y' - 3y = 3e^{2t}$ 2. $y'' + 2y' + 5y = 3\sin 2t$

3. $y'' - 2y' - 3y = -3te^{-t}$ 4. $y'' + 2y' = 3 + 4\sin 2t$

5. $y'' + 9y = t^2 e^{3t} + 6$ 6. $y'' + 2y' + y = 2e^{-t}$

7. $2y'' + 3y' + y = t^2 + 3\sin t$ 8. $y'' + y = 3\sin 2t + t\cos 2t$

9. $u'' + \omega_0^2 u = \cos \omega t, \qquad \omega^2 \neq \omega_0^2$ 10. $u'' + \omega_0^2 u = \cos \omega_0 t$

11. $y'' + y' + 4y = 2\sinh t \qquad$ *Hint:* $\sinh t = (e^t - e^{-t})/2$

12. $y'' - y' - 2y = \cosh 2t \qquad$ *Hint:* $\cosh t = (e^t + e^{-t})/2$

In each of Problems 13 through 18 find the solution of the given initial value problem.

13. $y'' + y' - 2y = 2t, \qquad y(0) = 0, \quad y'(0) = 1$

14. $y'' + 4y = t^2 + 3e^t, \qquad y(0) = 0, \quad y'(0) = 2$

15. $y'' - 2y' + y = te^t + 4, \qquad y(0) = 1, \quad y'(0) = 1$

16. $y'' - 2y' - 3y = 3te^{2t}, \qquad y(0) = 1, \quad y'(0) = 0$

17. $y'' + 4y = 3\sin 2t, \qquad y(0) = 2, \quad y'(0) = -1$

18. $y'' + 2y' + 5y = 4e^{-t}\cos 2t, \qquad y(0) = 1, \quad y'(0) = 0$

In each of Problems 19 through 26:

(a) Determine a suitable form for $Y(t)$ if the method of undetermined coefficients is to be used.

(b) Use a computer algebra system to find a particular solution of the given equation.

19. $y'' + 3y' = 2t^4 + t^2 e^{-3t} + \sin 3t$

20. $y'' + y = t(1 + \sin t)$

21. $y'' - 5y' + 6y = e^t \cos 2t + e^{2t}(3t + 4)\sin t$

22. $y'' + 2y' + 2y = 3e^{-t} + 2e^{-t}\cos t + 4e^{-t}t^2 \sin t$

23. $y'' - 4y' + 4y = 2t^2 + 4te^{2t} + t\sin 2t$

24. $y'' + 4y = t^2 \sin 2t + (6t + 7)\cos 2t$

25. $y'' + 3y' + 2y = e^t(t^2 + 1)\sin 2t + 3e^{-t}\cos t + 4e^t$

26. $y'' + 2y' + 5y = 3te^{-t}\cos 2t - 2te^{-2t}\cos t$

27. Consider the equation

$$y'' - 3y' - 4y = 2e^{-t} \tag{i}$$

from Example 5. Recall that $y_1(t) = e^{-t}$ and $y_2(t) = e^{4t}$ are solutions of the corresponding homogeneous equation. Adapting the method of reduction of order (Section 3.5), seek a solution of the nonhomogeneous equation of the form $Y(t) = v(t)y_1(t) = v(t)e^{-t}$, where $v(t)$ is to be determined.

(a) Substitute $Y(t)$, $Y'(t)$, and $Y''(t)$ into Eq. (i) and show that $v(t)$ must satisfy $v'' - 5v' = 2$.

(b) Let $w(t) = v'(t)$ and show that $w(t)$ must satisfy $w' - 5w = 2$. Solve this equation for $w(t)$.

(c) Integrate $w(t)$ to find $v(t)$ and then show that

$$Y(t) = -\tfrac{2}{5}te^{-t} + \tfrac{1}{5}c_1 e^{4t} + c_2 e^{-t}.$$

The first term on the right side is the desired particular solution of the nonhomogeneous equation. Note that it is a product of t and e^{-t}.

28. Determine the general solution of

$$y'' + \lambda^2 y = \sum_{m=1}^{N} a_m \sin m\pi t,$$

where $\lambda > 0$ and $\lambda \neq m\pi$ for $m = 1, \ldots, N$.

29. In many physical problems the nonhomogeneous term may be specified by different formulas in different time periods. As an example, determine the solution $y = \phi(t)$ of

$$y'' + y = \begin{cases} t, & 0 \leq t \leq \pi, \\ \pi e^{\pi - t}, & t > \pi, \end{cases}$$

satisfying the initial conditions $y(0) = 0$ and $y'(0) = 1$. Assume that y and y' are also continuous at $t = \pi$. Plot the nonhomogeneous term and the solution as functions of time. *Hint:* First solve the initial value problem for $t \leq \pi$; then solve for $t > \pi$, determining the constants in the latter solution from the continuity conditions at $t = \pi$.

30. Follow the instructions in Problem 29 to solve the differential equation

$$y'' + 2y' + 5y = \begin{cases} 1, & 0 \leq t \leq \pi/2, \\ 0, & t > \pi/2 \end{cases}$$

with the initial conditions $y(0) = 0$ and $y'(0) = 0$.

Behavior of Solutions as $t \to \infty$. In Problems 31 and 32 we continue the discussion started with Problems 38 through 40 of Section 3.5. Consider the differential equation

$$ay'' + by' + cy = g(t), \tag{i}$$

where $a, b,$ and c are positive.

31. If $Y_1(t)$ and $Y_2(t)$ are solutions of Eq. (i), show that $Y_1(t) - Y_2(t) \to 0$ as $t \to \infty$. Is this result true if $b = 0$?

32. If $g(t) = d,$ a constant, show that every solution of Eq. (i) approaches d/c as $t \to \infty$. What happens if $c = 0$? What if $b = 0$ also?

33. In this problem we indicate an alternative procedure[7] for solving the differential equation

$$y'' + by' + cy = (D^2 + bD + c)y = g(t), \tag{i}$$

where b and c are constants, and D denotes differentiation with respect to t. Let r_1 and r_2 be the zeros of the characteristic polynomial of the corresponding homogeneous equation. These roots may be real and different, real and equal, or conjugate complex numbers.

(a) Verify that Eq. (i) can be written in the factored form

$$(D - r_1)(D - r_2)y = g(t),$$

where $r_1 + r_2 = -b$ and $r_1 r_2 = c$.

(b) Let $u = (D - r_2)y$. Then show that the solution of Eq (i) can be found by solving the following two first order equations:

$$(D - r_1)u = g(t), \qquad (D - r_2)y = u(t).$$

[7]R. S. Luthar, "Another Approach to a Standard Differential Equation," *Two Year College Mathematics Journal 10* (1979), pp. 200–201; also see D. C. Sandell and F. M. Stein, "Factorization of Operators of Second Order Linear Homogeneous Ordinary Differential Equations," *Two Year College Mathematics Journal 8* (1977), pp. 132–141, for a more general discussion of factoring operators.

In each of Problems 34 through 37 use the method of Problem 33 to solve the given differential equation.

34. $y'' - 3y' - 4y = 3e^{2t}$ (see Example 1)
35. $2y'' + 3y' + y = t^2 + 3\sin t$ (see Problem 7)
36. $y'' + 2y' + y = 2e^{-t}$ (see Problem 6)
37. $y'' + 2y' = 3 + 4\sin 2t$ (see Problem 4)

3.7 Variation of Parameters

In this section we describe another method of finding a particular solution of a nonhomogeneous equation. This method, known as **variation of parameters**, is due to Lagrange and complements the method of undetermined coefficients rather well. The main advantage of variation of parameters is that it is a *general method*; in principle at least, it can be applied to any equation, and it requires no detailed assumptions about the form of the solution. In fact, later in this section we use this method to derive a formula for a particular solution of an arbitrary second order linear nonhomogeneous differential equation. On the other hand, the method of variation of parameters eventually requires that we evaluate certain integrals involving the nonhomogeneous term in the differential equation, and this may present difficulties. Before looking at this method in the general case, we illustrate its use in an example.

EXAMPLE 1

Find a particular solution of

$$y'' + 4y = 3\csc t. \tag{1}$$

Observe that this problem is not a good candidate for the method of undetermined coefficients, as described in Section 3.6, because the nonhomogeneous term $g(t) = 3\csc t$ involves a quotient (rather than a sum or a product) of $\sin t$ or $\cos t$. Therefore, we need a different approach. Observe also that the homogeneous equation corresponding to Eq. (1) is

$$y'' + 4y = 0, \tag{2}$$

and that the general solution of Eq. (2) is

$$y_c(t) = c_1 \cos 2t + c_2 \sin 2t. \tag{3}$$

The basic idea in the method of variation of parameters is to replace the constants c_1 and c_2 in Eq. (3) by functions $u_1(t)$ and $u_2(t)$, respectively, and then to determine these functions so that the resulting expression

$$y = u_1(t)\cos 2t + u_2(t)\sin 2t \tag{4}$$

is a solution of the nonhomogeneous equation (1).

To determine u_1 and u_2 we need to substitute for y from Eq. (4) in Eq. (1). However, even without carrying out this substitution, we can anticipate that the result will be a single equation involving some combination of u_1, u_2, and their first two derivatives. Since there is only one equation and two unknown functions, we can expect that there are many possible choices of u_1 and u_2 that will meet our needs. Alternatively, we may be able to impose a second condition of our own choosing, thereby obtaining two equations for the two unknown functions u_1 and

u_2. We will soon show (following Lagrange) that it is possible to choose this second condition in a way that makes the computation markedly more efficient.

Returning now to Eq. (4), we differentiate it and rearrange the terms, thereby obtaining

$$y' = -2u_1(t)\sin 2t + 2u_2(t)\cos 2t + u_1'(t)\cos 2t + u_2'(t)\sin 2t. \tag{5}$$

Keeping in mind the possibility of choosing a second condition on u_1 and u_2, let us require the sum of the last two terms on the right side of Eq. (5) to be zero; that is, we require that

$$u_1'(t)\cos 2t + u_2'(t)\sin 2t = 0. \tag{6}$$

It then follows from Eq. (5) that

$$y' = -2u_1(t)\sin 2t + 2u_2(t)\cos 2t. \tag{7}$$

Although the ultimate effect of the condition (6) is not yet clear, at the very least it has simplified the expression for y'. Further, by differentiating Eq. (7), we obtain

$$y'' = -4u_1(t)\cos 2t - 4u_2(t)\sin 2t - 2u_1'(t)\sin 2t + 2u_2'(t)\cos 2t. \tag{8}$$

Then, substituting for y and y'' in Eq. (1) from Eqs. (4) and (8), respectively, we find that u_1 and u_2 must satisfy

$$-2u_1'(t)\sin 2t + 2u_2'(t)\cos 2t = 3\csc t. \tag{9}$$

Summarizing our results to this point, we want to choose u_1 and u_2 so as to satisfy Eqs. (6) and (9). These equations can be viewed as a pair of linear *algebraic* equations for the two unknown quantities $u_1'(t)$ and $u_2'(t)$. Equations (6) and (9) can be solved in various ways. For example, solving Eq. (6) for $u_2'(t)$, we have

$$u_2'(t) = -u_1'(t)\frac{\cos 2t}{\sin 2t}. \tag{10}$$

Then, substituting for $u_2'(t)$ in Eq. (9) and simplifying, we obtain

$$u_1'(t) = -\frac{3\csc t\sin 2t}{2} = -3\cos t. \tag{11}$$

Further, putting this expression for $u_1'(t)$ back in Eq. (10) and using the double-angle formulas, we find that

$$u_2'(t) = \frac{3\cos t\cos 2t}{\sin 2t} = \frac{3(1 - 2\sin^2 t)}{2\sin t} = \frac{3}{2}\csc t - 3\sin t. \tag{12}$$

Having obtained $u_1'(t)$ and $u_2'(t)$, we next integrate so as to find $u_1(t)$ and $u_2(t)$. The result is

$$u_1(t) = -3\sin t + c_1 \tag{13}$$

and

$$u_2(t) = \tfrac{3}{2}\ln|\csc t - \cot t| + 3\cos t + c_2. \tag{14}$$

On substituting these expressions in Eq. (4), we have

$$y = -3\sin t\cos 2t + \tfrac{3}{2}\ln|\csc t - \cot t|\sin 2t + 3\cos t\sin 2t$$
$$+ c_1\cos 2t + c_2\sin 2t.$$

Finally, by using the double-angle formulas once more, we obtain

$$y = 3\sin t + \tfrac{3}{2}\ln|\csc t - \cot t|\sin 2t + c_1\cos 2t + c_2\sin 2t. \tag{15}$$

The terms in Eq. (15) involving the arbitrary constants c_1 and c_2 are the general solution of the corresponding homogeneous equation, while the remaining terms are a particular solution of the nonhomogeneous equation (1). Thus Eq. (15) is the general solution of Eq. (1).

In the preceding example the method of variation of parameters worked well in determining a particular solution, and hence the general solution, of Eq. (1). The next question is whether this method can be applied effectively to an arbitrary equation. Therefore we consider

$$y'' + p(t)y' + q(t)y = g(t), \qquad (16)$$

where p, q, and g are given continuous functions. As a starting point, we assume that we know the general solution

$$y_c(t) = c_1 y_1(t) + c_2 y_2(t) \qquad (17)$$

of the corresponding homogeneous equation

$$y'' + p(t)y' + q(t)y = 0. \qquad (18)$$

This is a major assumption because so far we have shown how to solve Eq. (18) only if it has constant coefficients. If Eq. (18) has coefficients that depend on t, then usually the methods described in Chapter 5 must be used to obtain $y_c(t)$.

The crucial idea, as illustrated in Example 1, is to replace the constants c_1 and c_2 in Eq. (17) by functions $u_1(t)$ and $u_2(t)$, respectively; this gives

$$y = u_1(t)y_1(t) + u_2(t)y_2(t). \qquad (19)$$

Then we try to determine $u_1(t)$ and $u_2(t)$ so that the expression in Eq. (19) is a solution of the nonhomogeneous equation (16) rather than the homogeneous equation (18). Thus we differentiate Eq. (19), obtaining

$$y' = u_1'(t)y_1(t) + u_1(t)y_1'(t) + u_2'(t)y_2(t) + u_2(t)y_2'(t). \qquad (20)$$

As in Example 1, we now set the terms involving $u_1'(t)$ and $u_2'(t)$ in Eq. (20) equal to zero; that is, we require that

$$u_1'(t)y_1(t) + u_2'(t)y_2(t) = 0. \qquad (21)$$

Then, from Eq. (20), we have

$$y' = u_1(t)y_1'(t) + u_2(t)y_2'(t). \qquad (22)$$

Further, by differentiating again, we obtain

$$y'' = u_1'(t)y_1'(t) + u_1(t)y_1''(t) + u_2'(t)y_2'(t) + u_2(t)y_2''(t). \qquad (23)$$

Now we substitute for y, y', and y'' in Eq. (16) from Eqs. (19), (22), and (23), respectively. After rearranging the terms in the resulting equation, we find that

$$u_1(t)[y_1''(t) + p(t)y_1'(t) + q(t)y_1(t)]$$
$$+ u_2(t)[y_2''(t) + p(t)y_2'(t) + q(t)y_2(t)]$$
$$+ u_1'(t)y_1'(t) + u_2'(t)y_2'(t) = g(t). \qquad (24)$$

Each of the expressions in square brackets in Eq. (24) is zero because both y_1 and y_2 are solutions of the homogeneous equation (18). Therefore Eq. (24) reduces to

$$u_1'(t)y_1'(t) + u_2'(t)y_2'(t) = g(t). \qquad (25)$$

Equations (21) and (25) form a system of two linear algebraic equations for the derivatives $u_1'(t)$ and $u_2'(t)$ of the unknown functions. They correspond exactly to Eqs. (6) and (9) in Example 1.

By solving the system (21), (25) we obtain

$$u_1'(t) = -\frac{y_2(t)g(t)}{W(y_1, y_2)(t)}, \qquad u_2'(t) = \frac{y_1(t)g(t)}{W(y_1, y_2)(t)}, \tag{26}$$

where $W(y_1, y_2)$ is the Wronskian of y_1 and y_2. Note that division by W is permissible since y_1 and y_2 are a fundamental set of solutions, and therefore their Wronskian is nonzero. By integrating Eqs. (26) we find the desired functions $u_1(t)$ and $u_2(t)$, namely,

$$u_1(t) = -\int \frac{y_2(t)g(t)}{W(y_1, y_2)(t)} \, dt + c_1, \qquad u_2(t) = \int \frac{y_1(t)g(t)}{W(y_1, y_2)(t)} \, dt + c_2. \tag{27}$$

If the integrals in Eqs. (27) can be evaluated in terms of elementary functions, then we substitute the results in Eq. (19), thereby obtaining the general solution of Eq. (16). More generally, the solution can always be expressed in terms of integrals, as stated in the following theorem.

Theorem 3.7.1

If the functions $p, q,$ and g are continuous on an open interval I, and if the functions y_1 and y_2 are linearly independent solutions of the homogeneous equation (18) corresponding to the nonhomogeneous equation (16),

$$y'' + p(t)y' + q(t)y = g(t),$$

then a particular solution of Eq. (16) is

$$Y(t) = -y_1(t) \int_{t_0}^t \frac{y_2(s)g(s)}{W(y_1, y_2)(s)} \, ds + y_2(t) \int_{t_0}^t \frac{y_1(s)g(s)}{W(y_1, y_2)(s)} \, ds, \tag{28}$$

where t_0 is any conveniently chosen point in I. The general solution is

$$y = c_1 y_1(t) + c_2 y_2(t) + Y(t), \tag{29}$$

as prescribed by Theorem 3.6.2.

By examining the expression (28) and reviewing the process by which we derived it, we can see that there may be two major difficulties in using the method of variation of parameters. As we have mentioned earlier, one is the determination of $y_1(t)$ and $y_2(t)$, a fundamental set of solutions of the homogeneous equation (18), when the coefficients in that equation are not constants. The other possible difficulty is in the evaluation of the integrals appearing in Eq. (28). This depends entirely on the nature of the functions $y_1, y_2,$ and g. In using Eq. (28), be sure that the differential equation is exactly in the form (16); otherwise, the nonhomogeneous term $g(t)$ will not be correctly identified.

A major advantage of the method of variation of parameters is that Eq. (28) provides an expression for the particular solution $Y(t)$ in terms of an arbitrary forcing function $g(t)$. This expression is a good starting point if you wish to investigate the effect of variations in the forcing function, or if you wish to analyze the response of a system to a number of different forcing functions.

PROBLEMS

In each of Problems 1 through 4 use the method of variation of parameters to find a particular solution of the given differential equation. Then check your answer by using the method of undetermined coefficients.

1. $y'' - 5y' + 6y = 2e^t$

2. $y'' - y' - 2y = 2e^{-t}$

3. $y'' + 2y' + y = 3e^{-t}$

4. $4y'' - 4y' + y = 16e^{t/2}$

In each of Problems 5 through 12 find the general solution of the given differential equation. In Problems 11 and 12, g is an arbitrary continuous function.

5. $y'' + y = \tan t, \qquad 0 < t < \pi/2$

6. $y'' + 9y = 9\sec^2 3t, \qquad 0 < t < \pi/6$

7. $y'' + 4y' + 4y = t^{-2}e^{-2t}, \qquad t > 0$

8. $y'' + 4y = 3\csc 2t, \qquad 0 < t < \pi/2$

9. $4y'' + y = 2\sec(t/2), \qquad -\pi < t < \pi$

10. $y'' - 2y' + y = e^t/(1 + t^2)$

11. $y'' - 5y' + 6y = g(t)$

12. $y'' + 4y = g(t)$

In each of Problems 13 through 20 verify that the given functions y_1 and y_2 satisfy the corresponding homogeneous equation; then find a particular solution of the given nonhomogeneous equation. In Problems 19 and 20, g is an arbitrary continuous function.

13. $t^2 y'' - 2y = 3t^2 - 1, \quad t > 0; \qquad y_1(t) = t^2, \quad y_2(t) = t^{-1}$

14. $t^2 y'' - t(t + 2)y' + (t + 2)y = 2t^3, \quad t > 0; \qquad y_1(t) = t, \quad y_2(t) = te^t$

15. $ty'' - (1 + t)y' + y = t^2 e^{2t}, \quad t > 0; \qquad y_1(t) = 1 + t, \quad y_2(t) = e^t$

16. $(1 - t)y'' + ty' - y = 2(t - 1)^2 e^{-t}, \quad 0 < t < 1; \qquad y_1(t) = e^t, \quad y_2(t) = t$

17. $x^2 y'' - 3xy' + 4y = x^2 \ln x, \quad x > 0; \qquad y_1(x) = x^2, \quad y_2(x) = x^2 \ln x$

18. $x^2 y'' + xy' + (x^2 - 0.25)y = 3x^{3/2}\sin x, \quad x > 0;$
$\qquad y_1(x) = x^{-1/2}\sin x, \quad y_2(x) = x^{-1/2}\cos x$

19. $(1 - x)y'' + xy' - y = g(x), \quad 0 < x < 1; \qquad y_1(x) = e^x, \quad y_2(x) = x$

20. $x^2 y'' + xy' + (x^2 - 0.25)y = g(x), \quad x > 0; \qquad y_1(x) = x^{-1/2}\sin x, \quad y_2(x) = x^{-1/2}\cos x$

21. Show that the solution of the initial value problem

$$L[y] = y'' + p(t)y' + q(t)y = g(t), \qquad y(t_0) = y_0, \quad y'(t_0) = y_0' \qquad (i)$$

can be written as $y = u(t) + v(t)$, where u and v are solutions of the two initial value problems

$$L[u] = 0, \qquad u(t_0) = y_0, \quad u'(t_0) = y_0', \qquad (ii)$$

$$L[v] = g(t), \qquad v(t_0) = 0, \quad v'(t_0) = 0, \qquad (iii)$$

respectively. In other words, the nonhomogeneities in the differential equation and in the initial conditions can be dealt with separately. Observe that u is easy to find if a fundamental set of solutions of $L[u] = 0$ is known.

22. By choosing the lower limit of integration in Eq. (28) in the text as the initial point t_0, show that $Y(t)$ becomes

$$Y(t) = \int_{t_0}^t \frac{y_1(s)y_2(t) - y_1(t)y_2(s)}{y_1(s)y_2'(s) - y_1'(s)y_2(s)} g(s)\, ds.$$

Show that $Y(t)$ is a solution of the initial value problem

$$L[y] = g(t), \qquad y(t_0) = 0, \quad y'(t_0) = 0.$$

Thus Y can be identified with v in Problem 21.

23. (a) Use the result of Problem 22 to show that the solution of the initial value problem

$$y'' + y = g(t), \qquad y(t_0) = 0, \quad y'(t_0) = 0 \tag{i}$$

is

$$y = \int_{t_0}^{t} \sin(t - s)g(s)\, ds. \tag{ii}$$

(b) Use the result of Problem 21 to find the solution of the initial value problem

$$y'' + y = g(t), \qquad y(0) = y_0, \quad y'(0) = y_0'.$$

24. Use the result of Problem 22 to find the solution of the initial value problem

$$L[y] = (D - a)(D - b)y = g(t), \qquad y(t_0) = 0, \quad y'(t_0) = 0,$$

where a and b are real numbers with $a \neq b$.

25. Use the result of Problem 22 to find the solution of the initial value problem

$$L[y] = [D^2 - 2\lambda D + (\lambda^2 + \mu^2)]y = g(t), \qquad y(t_0) = 0, \quad y'(t_0) = 0.$$

Note that the roots of the characteristic equation are $\lambda \pm i\mu$.

26. Use the result of Problem 22 to find the solution of the initial value problem

$$L[y] = (D - a)^2 y = g(t), \qquad y(t_0) = 0, \quad y'(t_0) = 0,$$

where a is any real number.

27. By combining the results of Problems 24 through 26, show that the solution of the initial value problem

$$L[y] = (D^2 + bD + c)y = g(t), \qquad y(t_0) = 0, \quad y'(t_0) = 0,$$

where b and c are constants, has the form

$$y = \phi(t) = \int_{t_0}^{t} K(t - s)g(s)\, ds. \tag{i}$$

The function K depends only on the solutions y_1 and y_2 of the corresponding homogeneous equation and is independent of the nonhomogeneous term. Once K is determined, all nonhomogeneous problems involving the same differential operator L are reduced to the evaluation of an integral. Note also that although K depends on both t and s, only the combination $t - s$ appears, so K is actually a function of a single variable. When we think of $g(t)$ as the input to the problem and of $\phi(t)$ as the output, it follows from Eq. (i) that the output depends on the input over the entire interval from the initial point t_0 to the current value t. The integral in Eq. (i) is called the **convolution** of K and g, and K is referred to as the **kernel**.

28. The method of reduction of order (Section 3.5) can also be used for the nonhomogeneous equation

$$y'' + p(t)y' + q(t)y = g(t), \tag{i}$$

provided one solution y_1 of the corresponding homogeneous equation is known. Let $y = v(t)y_1(t)$ and show that y satisfies Eq. (i) if v is a solution of

$$y_1(t)v'' + [2y_1'(t) + p(t)y_1(t)]v' = g(t). \tag{ii}$$

Equation (ii) is a first order linear equation for v'. Solving this equation, integrating the result, and then multiplying by $y_1(t)$ lead to the general solution of Eq. (i).

In each of Problems 29 through 32 use the method outlined in Problem 28 to solve the given differential equation.

29. $t^2 y'' - 2t y' + 2y = 4t^2, \quad t > 0; \qquad y_1(t) = t$

30. $t^2 y'' + 7t y' + 5y = t, \quad t > 0; \qquad y_1(t) = t^{-1}$

31. $t y'' - (1 + t) y' + y = t^2 e^{2t}, \quad t > 0; \qquad y_1(t) = 1 + t$ \quad (see Problem 15)

32. $(1 - t) y'' + t y' - y = 2(t - 1)^2 e^{-t}, \quad 0 < t < 1; \qquad y_1(t) = e^t$ \quad (see Problem 16)

3.8 Mechanical and Electrical Vibrations

One of the reasons why second order linear equations with constant coefficients are worth studying is that they serve as mathematical models of some important physical processes. Two important areas of application are in the fields of mechanical and electrical oscillations. For example, the motion of a mass on a vibrating spring, the torsional oscillations of a shaft with a flywheel, the flow of electric current in a simple series circuit, and many other physical problems are all described by the solution of an initial value problem of the form

$$ay'' + by' + cy = g(t), \qquad y(0) = y_0, \quad y'(0) = y_0'. \tag{1}$$

This illustrates a fundamental relationship between mathematics and physics: *Many physical problems may have the same mathematical model.* Thus, once we know how to solve the initial value problem (1), it is only necessary to make appropriate interpretations of the constants a, b, and c, and of the functions y and g, to obtain solutions of different physical problems.

We will study the motion of a mass on a spring in detail because an understanding of the behavior of this simple system is the first step in the investigation of more complex vibrating systems. Further, the principles involved are common to many problems. Consider a mass m hanging on the end of a vertical spring of original length l, as shown in Figure 3.8.1. The mass causes an elongation L of the spring in the downward (positive) direction. There are two forces acting at the point where the mass is attached to the spring; see Figure 3.8.2. The gravitational force, or weight of the mass, acts downward and has magnitude mg, where g is the acceleration due to gravity. There is also a force F_s, due to the spring, that acts upward. If we assume that the elongation L of the spring is small, the spring force is very nearly proportional to L; this is known as Hooke's[8] law. Thus we write $F_s = -kL$, where the constant of proportionality k is called the spring constant, and the minus sign is due to the fact that the spring force acts in the upward (negative) direction. Since the mass is in equilibrium, the two forces balance each other, which means that

$$mg - kL = 0. \tag{2}$$

[8]Robert Hooke (1635–1703) was an English scientist with wide-ranging interests. His most important book, *Micrographia*, was published in 1665 and described a variety of microscopical observations. Hooke first published his law of elastic behavior in 1676 as *ceiiinosssttuv*; in 1678 he gave the interpretation *ut tensio sic vis*, which means, roughly, "as the force so is the displacement."

For a given weight $w = mg$, one can measure L and then use Eq. (2) to determine k. Note that k has the units of force/length.

FIGURE 3.8.1 A spring–mass system.

FIGURE 3.8.2 Force diagram for a spring–mass system.

In the corresponding dynamic problem, we are interested in studying the motion of the mass when it is acted on by an external force or is initially displaced. Let $u(t)$, measured positive downward, denote the displacement of the mass from its equilibrium position at time t; see Figure 3.8.1. Then $u(t)$ is related to the forces acting on the mass through Newton's law of motion,

$$mu''(t) = f(t), \tag{3}$$

where u'' is the acceleration of the mass and f is the net force acting on the mass. Observe that both u and f are functions of time. In determining f there are four separate forces that must be considered:

1. The weight $w = mg$ of the mass always acts downward.
2. The spring force F_s is assumed to be proportional to the total elongation $L + u$ of the spring and always acts to restore the spring to its natural position. If $L + u > 0$, then the spring is extended, and the spring force is directed upward. In this case

$$F_s = -k(L + u). \tag{4}$$

On the other hand, if $L + u < 0$, then the spring is compressed a distance $|L + u|$, and the spring force, which is now directed downward, is given by $F_s = k|L + u|$. However, when $L + u < 0$, it follows that $|L + u| = -(L + u)$, so F_s is again given by Eq. (4). Thus, regardless of the position of the mass, the force exerted by the spring is always expressed by Eq. (4).

3. The damping or resistive force F_d always acts in the direction opposite to the direction of motion of the mass. This force may arise from several sources: resistance from the air or other medium in which the mass moves, internal energy dissipation due to the extension or compression of the spring, friction between the mass and the guides (if any) that constrain its motion to one dimension, or a mechanical device (dashpot) that imparts a resistive force to the mass. In any case, we assume that the resistive force is proportional to the speed $|du/dt|$ of the mass; this is usually referred to as viscous damping. If $du/dt > 0$, then u is increasing, so the mass is moving downward. Then F_d is directed upward and is given by

$$F_d(t) = -\gamma u'(t), \tag{5}$$

where γ is a positive constant of proportionality known as the damping constant. On the other hand, if $du/dt < 0$, then u is decreasing, the mass is moving upward, and F_d is directed downward. In this case, $F_d = \gamma |u'(t)|$; since $|u'(t)| = -u'(t)$, it follows that $F_d(t)$ is again given by Eq. (5). Thus, regardless of the direction of motion of the mass, the damping force is always expressed by Eq. (5).

The damping force may be rather complicated, and the assumption that it is modeled adequately by Eq. (5) may be open to question. Some dashpots do behave as Eq. (5) states, and if the other sources of dissipation are small, it may be possible to neglect them altogether or to adjust the damping constant γ to approximate them. An important benefit of the assumption (5) is that it leads to a linear (rather than a nonlinear) differential equation. In turn, this means that a thorough analysis of the system is straightforward, as we will show in this section and the next.

4. An applied external force $F(t)$ is directed downward or upward as $F(t)$ is positive or negative. This could be a force due to the motion of the mount to which the spring is attached, or it could be a force applied directly to the mass. Often the external force is periodic.

Taking account of these forces, we can now rewrite Newton's law (3) as

$$\begin{aligned}mu''(t) &= mg + F_s(t) + F_d(t) + F(t) \\ &= mg - k[L + u(t)] - \gamma u'(t) + F(t).\end{aligned} \tag{6}$$

Since $mg - kL = 0$ by Eq. (2), it follows that the equation of motion of the mass is

$$mu''(t) + \gamma u'(t) + ku(t) = F(t), \tag{7}$$

where the constants m, γ, and k are positive. Note that Eq. (7) has the same form as Eq. (1).

It is important to understand that Eq. (7) is only an approximate equation for the displacement $u(t)$. In particular, both Eqs. (4) and (5) should be viewed as approximations for the spring force and the damping force, respectively. In our derivation we have also neglected the mass of the spring in comparison with the mass of the attached body.

The complete formulation of the vibration problem requires that we specify two initial conditions, namely, the initial position u_0 and the initial velocity v_0 of the mass:

$$u(0) = u_0, \qquad u'(0) = v_0. \tag{8}$$

It follows from Theorem 3.2.1 that these conditions give a mathematical problem that has a unique solution. This is consistent with our physical intuition that if the mass is set in motion with a given initial displacement and velocity, then its position will be

determined uniquely at all future times. The position of the mass is given (approximately) by the solution of Eq. (7) subject to the prescribed initial conditions (8).

**EXAMPLE
1**

A mass weighing 4 lb stretches a spring 2 in. Suppose that the mass is displaced an additional 6 in. in the positive direction and then released. The mass is in a medium that exerts a viscous resistance of 6 lb when the mass has a velocity of 3 ft/sec. Under the assumptions discussed in this section, formulate the initial value problem that governs the motion of the mass.

The required initial value problem consists of the differential equation (7) and initial conditions (8), so our task is to determine the various constants that appear in these equations. The first step is to choose the units of measurement. Based on the statement of the problem, it is natural to use the English rather than the metric system of units. The only time unit mentioned is the second, so we will measure t in seconds. On the other hand, both the foot and the inch appear in the statement as units of length. It is immaterial which one we use, but having made a choice, we must be consistent. To be definite, let us measure the displacement u in feet.

Since nothing is said in the statement of the problem about an external force, we assume that $F(t) = 0$. To determine m note that

$$m = \frac{w}{g} = \frac{4 \text{ lb}}{32 \text{ ft/ sec}^2} = \frac{1}{8} \frac{\text{lb-sec}^2}{\text{ft}}.$$

The damping coefficient γ is determined from the statement that $\gamma u'$ is equal to 6 lb when u' is 3 ft/sec. Therefore

$$\gamma = \frac{6 \text{ lb}}{3 \text{ ft/sec}} = 2 \frac{\text{lb-sec}}{\text{ft}}.$$

The spring constant k is found from the statement that the mass stretches the spring by 2 in., or 1/6 ft. Thus

$$k = \frac{4 \text{ lb}}{1/6 \text{ ft}} = 24 \frac{\text{lb}}{\text{ft}}.$$

Consequently, Eq. (7) becomes

$$\tfrac{1}{8}u'' + 2u' + 24u = 0,$$

or

$$u'' + 16u' + 192u = 0. \tag{9}$$

The initial conditions are

$$u(0) = \tfrac{1}{2}, \qquad u'(0) = 0. \tag{10}$$

The second initial condition is implied by the word "released" in the statement of the problem, which we interpret to mean that the mass is set in motion with no initial velocity.

Undamped Free Vibrations. If there is no external force, then $F(t) = 0$ in Eq. (7). Let us also suppose that there is no damping, so that $\gamma = 0$; this is an idealized configuration of the system, seldom (if ever) completely attainable in practice. However, if the actual damping is very small, then the assumption of no damping may yield satisfactory results over short to moderate time intervals. In this case the equation of motion (7) reduces to

$$mu'' + ku = 0. \tag{11}$$

The general solution of Eq. (11) is

$$u = A \cos \omega_0 t + B \sin \omega_0 t, \tag{12}$$

where

$$\omega_0^2 = k/m. \tag{13}$$

The arbitrary constants A and B can be determined if initial conditions of the form (8) are given.

In discussing the solution of Eq. (11) it is convenient to rewrite Eq. (12) in the form

$$u = R \cos(\omega_0 t - \delta), \tag{14}$$

or

$$u = R \cos \delta \cos \omega_0 t + R \sin \delta \sin \omega_0 t. \tag{15}$$

By comparing Eq. (15) with Eq. (12), we find that $A, B, R,$ and δ are related by the equations

$$A = R \cos \delta, \qquad B = R \sin \delta. \tag{16}$$

Thus

$$R = \sqrt{A^2 + B^2}, \qquad \tan \delta = B/A. \tag{17}$$

In calculating δ, we must take care to choose the correct quadrant; this can be done by checking the signs of $\cos \delta$ and $\sin \delta$ in Eqs. (16).

The graph of Eq. (14), or the equivalent Eq. (12), for a typical set of initial conditions is shown in Figure 3.8.3. The graph is a displaced cosine wave that describes a periodic, or simple harmonic, motion of the mass. The **period** of the motion is

$$T = \frac{2\pi}{\omega_0} = 2\pi \left(\frac{m}{k}\right)^{1/2}. \tag{18}$$

The circular frequency $\omega_0 = \sqrt{k/m}$, measured in radians per unit time, is called the **natural frequency** of the vibration. The maximum displacement R of the mass from equilibrium is the **amplitude** of the motion. The dimensionless parameter δ is called the **phase**, or phase angle, and measures the displacement of the wave from its normal position corresponding to $\delta = 0$.

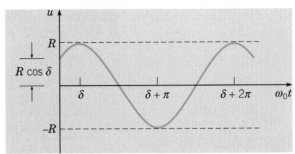

FIGURE 3.8.3 Simple harmonic motion; $u = R \cos(\omega_0 t - \delta)$.

Note that the motion described by Eq. (14) has a constant amplitude that does not diminish with time. This reflects the fact that, in the absence of damping, there is no way for the system to dissipate the energy imparted to it by the initial displacement and velocity. Further, for a given mass m and spring constant k, the system always vibrates at the same frequency ω_0, regardless of the initial conditions. However, the initial conditions do help to determine the amplitude of the motion. Finally, observe from Eq. (18) that T increases as m increases, so larger masses vibrate more slowly. On the other hand, T decreases as k increases, which means that stiffer springs cause the system to vibrate more rapidly.

EXAMPLE 2

Suppose that a mass weighing 10 lb stretches a spring 2 in. If the mass is displaced an additional 2 in. and is then set in motion with an initial upward velocity of 1 ft/sec, determine the position of the mass at any later time. Also determine the period, amplitude, and phase of the motion.

The spring constant is $k = 10\,\text{lb}/2\,\text{in.} = 60\,\text{lb/ft}$, and the mass is $m = w/g = 10/32$ lb-sec^2/ft. Hence the equation of motion reduces to

$$u'' + 192u = 0, \tag{19}$$

and the general solution is

$$u = A\cos(8\sqrt{3}t) + B\sin(8\sqrt{3}t).$$

The solution satisfying the initial conditions $u(0) = 1/6$ ft and $u'(0) = -1$ ft/sec is

$$u = \frac{1}{6}\cos(8\sqrt{3}t) - \frac{1}{8\sqrt{3}}\sin(8\sqrt{3}t). \tag{20}$$

The natural frequency is $\omega_0 = \sqrt{192} \cong 13.856$ rad/sec, so the period is $T = 2\pi/\omega_0 \cong 0.45345$ sec. The amplitude R and phase δ are found from Eqs. (17). We have

$$R^2 = \frac{1}{36} + \frac{1}{192} = \frac{19}{576}, \quad \text{so} \quad R \cong 0.18162 \text{ ft.}$$

The second of Eqs. (17) yields $\tan\delta = -\sqrt{3}/4$. There are two solutions of this equation, one in the second quadrant and one in the fourth. In the present problem $\cos\delta > 0$ and $\sin\delta < 0$, so δ is in the fourth quadrant, namely,

$$\delta = -\arctan(\sqrt{3}/4) \cong -0.40864 \text{ rad.}$$

The graph of the solution (20) is shown in Figure 3.8.4.

FIGURE 3.8.4 An undamped free vibration; $u'' + 192u = 0, \quad u(0) = 1/6, \quad u'(0) = -1.$

Damped Free Vibrations. If we include the effect of damping, the differential equation governing the motion of the mass is

$$mu'' + \gamma u' + ku = 0. \tag{21}$$

We are especially interested in examining the effect of variations in the damping coefficient γ for given values of the mass m and spring constant k. The roots of the corresponding characteristic equation are

$$r_1, r_2 = \frac{-\gamma \pm \sqrt{\gamma^2 - 4km}}{2m} = \frac{\gamma}{2m}\left(-1 \pm \sqrt{1 - \frac{4km}{\gamma^2}}\right). \tag{22}$$

Depending on the sign of $\gamma^2 - 4km$, the solution u has one of the following forms:

$$\gamma^2 - 4km > 0, \qquad u = Ae^{r_1 t} + Be^{r_2 t}; \tag{23}$$

$$\gamma^2 - 4km = 0, \qquad u = (A + Bt)e^{-\gamma t/2m}; \tag{24}$$

$$\gamma^2 - 4km < 0, \quad u = e^{-\gamma t/2m}(A\cos\mu t + B\sin\mu t), \quad \mu = \frac{(4km - \gamma^2)^{1/2}}{2m} > 0. \tag{25}$$

Since m, γ, and k are positive, $\gamma^2 - 4km$ is always less than γ^2. Hence, if $\gamma^2 - 4km \geq 0$, then the values of r_1 and r_2 given by Eq. (22) are *negative*. If $\gamma^2 - 4km < 0$, then the values of r_1 and r_2 are complex, but with *negative* real part. Thus, in all cases, the solution u tends to zero as $t \to \infty$; this occurs regardless of the values of the arbitrary constants A and B, that is, regardless of the initial conditions. This confirms our intuitive expectation, namely, that damping gradually dissipates the energy initially imparted to the system, and consequently the motion dies out with increasing time.

The most important case is the third one, which occurs when the damping is small. If we let $A = R\cos\delta$ and $B = R\sin\delta$ in Eq. (25), then we obtain

$$u = Re^{-\gamma t/2m}\cos(\mu t - \delta). \tag{26}$$

The displacement u lies between the curves $u = \pm Re^{-\gamma t/2m}$; hence it resembles a cosine wave whose amplitude decreases as t increases. A typical example is sketched in Figure 3.8.5. The motion is called a damped oscillation or a damped vibration. The amplitude factor R depends on m, γ, k, and the initial conditions.

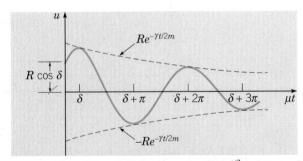

FIGURE 3.8.5 Damped vibration; $u = Re^{-\gamma t/2m}\cos(\mu t - \delta)$.

Although the motion is not periodic, the parameter μ determines the frequency with which the mass oscillates back and forth; consequently, μ is called the **quasi frequency**. By comparing μ with the frequency ω_0 of undamped motion, we find that

$$\frac{\mu}{\omega_0} = \frac{(4km - \gamma^2)^{1/2}/2m}{\sqrt{k/m}} = \left(1 - \frac{\gamma^2}{4km}\right)^{1/2} \cong 1 - \frac{\gamma^2}{8km}. \tag{27}$$

The last approximation is valid when $\gamma^2/4km$ is small; we refer to this situation as "small damping." Thus the effect of small damping is to reduce slightly the frequency of the oscillation. By analogy with Eq. (18), the quantity $T_d = 2\pi/\mu$ is called the **quasi period**. It is the time between successive maxima or successive minima of the position of the mass, or between successive passages of the mass through its equilibrium position while going in the same direction. The relation between T_d and T is given by

$$\frac{T_d}{T} = \frac{\omega_0}{\mu} = \left(1 - \frac{\gamma^2}{4km}\right)^{-1/2} \cong \left(1 + \frac{\gamma^2}{8km}\right), \tag{28}$$

where again the last approximation is valid when $\gamma^2/4km$ is small. Thus small damping increases the quasi period.

Equations (27) and (28) reinforce the significance of the dimensionless ratio $\gamma^2/4km$. It is not the magnitude of γ alone that determines whether damping is large or small, but the magnitude of γ^2 compared to $4km$. When $\gamma^2/4km$ is small, then damping has a small effect on the quasi frequency and quasi period of the motion. On the other hand, if we want to study the detailed motion of the mass for all time, then we can *never* neglect the damping force, no matter how small.

As $\gamma^2/4km$ increases, the quasi frequency μ decreases and the quasi period T_d increases. In fact, $\mu \to 0$ and $T_d \to \infty$ as $\gamma \to 2\sqrt{km}$. As indicated by Eqs. (23), (24), and (25), the nature of the solution changes as γ passes through the value $2\sqrt{km}$. This value is known as **critical damping**, while for larger values of γ the motion is said to be **overdamped**. In these cases, given by Eqs. (24) and (23), respectively, the mass creeps back to its equilibrium position but does not oscillate about it, as for small γ. Two typical examples of critically damped motion are shown in Figure 3.8.6, and the situation is discussed further in Problems 21 and 22.

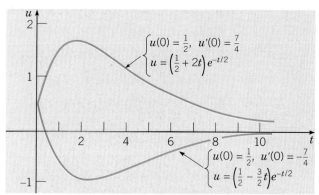

FIGURE 3.8.6 Critically damped motions: $u'' + u' + 0.25u = 0$; $u = (A + Bt)e^{-t/2}$.

EXAMPLE

3

The motion of a certain spring–mass system is governed by the differential equation

$$u'' + 0.125u' + u = 0, \tag{29}$$

where u is measured in feet and t in seconds. If $u(0) = 2$ and $u'(0) = 0$, determine the position of the mass at any time. Find the quasi frequency and the quasi period, as well as the time at which the mass first passes through its equilibrium position. Also find the time τ such that $|u(t)| < 0.1$ for all $t > \tau$.

The solution of Eq. (29) is

$$u = e^{-t/16}\left[A\cos\frac{\sqrt{255}}{16}t + B\sin\frac{\sqrt{255}}{16}t\right].$$

To satisfy the initial conditions we must choose $A = 2$ and $B = 2/\sqrt{255}$; hence the solution of the initial value problem is

$$u = e^{-t/16}\left(2\cos\frac{\sqrt{255}}{16}t + \frac{2}{\sqrt{255}}\sin\frac{\sqrt{255}}{16}t\right)$$

$$= \frac{32}{\sqrt{255}}e^{-t/16}\cos\left(\frac{\sqrt{255}}{16}t - \delta\right), \tag{30}$$

where $\tan\delta = 1/\sqrt{255}$, so $\delta \cong 0.06254$. The displacement of the mass as a function of time is shown in Figure 3.8.7. For purposes of comparison, we also show the motion if the damping term is neglected.

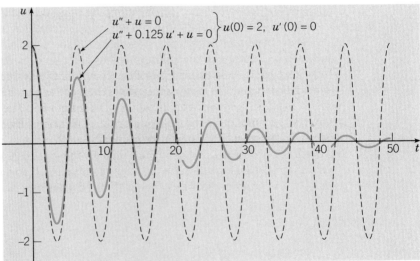

FIGURE 3.8.7 Vibration with small damping (solid curve) and with no damping (dashed curve).

The quasi frequency is $\mu = \sqrt{255}/16 \cong 0.998$ and the quasi period is $T_d = 2\pi/\mu \cong 6.295$ sec. These values differ only slightly from the corresponding values (1 and 2π, respectively) for the undamped oscillation. This is evident also from the graphs in Figure 3.8.7, which rise and fall almost together. The damping coefficient is small in this example, only one-sixteenth

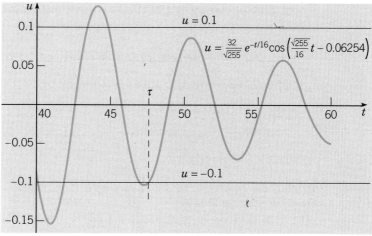

FIGURE 3.8.8 Solution of Example 3; determination of τ.

of the critical value, in fact. Nevertheless, the amplitude of the oscillation is reduced rather rapidly. Figure 3.8.8 shows the graph of the solution for $40 \leq t \leq 60$, together with the graphs of $u = \pm 0.1$. From the graph it appears that τ is about 47.5, and by a more precise calculation we find that $\tau \cong 47.5149$ sec.

To find the time at which the mass first passes through its equilibrium position, we refer to Eq. (30) and set $\sqrt{255}t/16 - \delta$ equal to $\pi/2$, the smallest positive zero of the cosine function. Then, by solving for t, we obtain

$$t = \frac{16}{\sqrt{255}}\left(\frac{\pi}{2} + \delta\right) \cong 1.637 \,\text{sec}.$$

Electric Circuits. A second example of the occurrence of second order linear differential equations with constant coefficients is their use as a model of the flow of electric current in the simple series circuit shown in Figure 3.8.9. The current I, measured in amperes, is a function of time t. The resistance R (ohms), the capacitance C (farads), and the inductance L (henrys) are all positive and are assumed to be known constants. The impressed voltage E (volts) is a given function of time. Another physical quantity that enters the discussion is the total charge Q (coulombs) on the capacitor at time t. The relation between charge Q and current I is

$$I = dQ/dt. \tag{31}$$

FIGURE 3.8.9 A simple electric circuit.

The flow of current in the circuit is governed by Kirchhoff's[9] second law: *In a closed circuit the impressed voltage is equal to the sum of the voltage drops in the rest of the circuit.*

According to the elementary laws of electricity, we know that

The voltage drop across the resistor is IR.

The voltage drop across the capacitor is Q/C.

The voltage drop across the inductor is $L dI/dt$.

Hence, by Kirchhoff's law,

$$L\frac{dI}{dt} + RI + \frac{1}{C}Q = E(t). \tag{32}$$

The units have been chosen so that 1 volt = 1 ohm · 1 ampere = 1 coulomb/1 farad = 1 henry · 1 ampere/1 second.

Substituting for I from Eq. (31), we obtain the differential equation

$$LQ'' + RQ' + \frac{1}{C}Q = E(t) \tag{33}$$

for the charge Q. The initial conditions are

$$Q(t_0) = Q_0, \qquad Q'(t_0) = I(t_0) = I_0. \tag{34}$$

Thus we must know the charge on the capacitor and the current in the circuit at some initial time t_0.

Alternatively, we can obtain a differential equation for the current I by differentiating Eq. (33) with respect to t, and then substituting for dQ/dt from Eq. (31). The result is

$$LI'' + RI' + \frac{1}{C}I = E'(t), \tag{35}$$

with the initial conditions

$$I(t_0) = I_0, \qquad I'(t_0) = I_0'. \tag{36}$$

From Eq. (32) it follows that

$$I_0' = \frac{E(t_0) - RI_0 - (1/C)Q_0}{L}. \tag{37}$$

Hence I_0' is also determined by the initial charge and current, which are physically measurable quantities.

The most important conclusion from this discussion is that the flow of current in the circuit is described by an initial value problem of precisely the same form as the one that describes the motion of a spring–mass system. This is a good example of the unifying role of mathematics: Once you know how to solve second order

[9]Gustav Kirchhoff (1824–1887), professor at Breslau, Heidelberg, and Berlin, was one of the leading physicists of the nineteenth century. He discovered the basic laws of electric circuits about 1845 while still a student at Königsberg. He is also famous for fundamental work in electromagnetic absorption and emission and was one of the founders of spectroscopy.

linear equations with constant coefficients, you can interpret the results in terms of mechanical vibrations, electric circuits, or any other physical situation that leads to the same problem.

PROBLEMS

In each of Problems 1 through 4 determine ω_0, R, and δ so as to write the given expression in the form $u = R\cos(\omega_0 t - \delta)$.

1. $u = 3\cos 2t + 4\sin 2t$

2. $u = -\cos t + \sqrt{3}\sin t$

3. $u = 4\cos 3t - 2\sin 3t$

4. $u = -2\cos \pi t - 3\sin \pi t$

5. A mass weighing 2 lb stretches a spring 6 in. If the mass is pulled down an additional 3 in. and then released, and if there is no damping, determine the position u of the mass at any time t. Plot u versus t. Find the frequency, period, and amplitude of the motion.

6. A mass of 100 g stretches a spring 5 cm. If the mass is set in motion from its equilibrium position with a downward velocity of 10 cm/sec, and if there is no damping, determine the position u of the mass at any time t. When does the mass first return to its equilibrium position?

7. A mass weighing 3 lb stretches a spring 3 in. If the mass is pushed upward, contracting the spring a distance of 1 in., and then set in motion with a downward velocity of 2 ft/sec, and if there is no damping, find the position u of the mass at any time t. Determine the frequency, period, amplitude, and phase of the motion.

8. A series circuit has a capacitor of 0.25×10^{-6} farad and an inductor of 1 henry. If the initial charge on the capacitor is 10^{-6} coulomb and there is no initial current, find the charge Q on the capacitor at any time t.

9. A mass of 20 g stretches a spring 5 cm. Suppose that the mass is also attached to a viscous damper with a damping constant of 400 dyne-sec/cm. If the mass is pulled down an additional 2 cm and then released, find its position u at any time t. Plot u versus t. Determine the quasi frequency and the quasi period. Determine the ratio of the quasi period to the period of the corresponding undamped motion. Also find the time τ such that $|u(t)| < 0.05$ cm for all $t > \tau$.

10. A mass weighing 16 lb stretches a spring 3 in. The mass is attached to a viscous damper with a damping constant of 2 lb-sec/ft. If the mass is set in motion from its equilibrium position with a downward velocity of 3 in./sec, find its position u at any time t. Plot u versus t. Determine when the mass first returns to its equilibrium position. Also find the time τ such that $|u(t)| < 0.01$ in. for all $t > \tau$.

11. A spring is stretched 10 cm by a force of 3 newtons. A mass of 2 kg is hung from the spring and is also attached to a viscous damper that exerts a force of 3 newtons when the velocity of the mass is 5 m/sec. If the mass is pulled down 5 cm below its equilibrium position and given an initial downward velocity of 10 cm/sec, determine its position u at any time t. Find the quasi frequency μ and the ratio of μ to the natural frequency of the corresponding undamped motion.

12. A series circuit has a capacitor of 10^{-5} farad, a resistor of 3×10^2 ohms, and an inductor of 0.2 henry. The initial charge on the capacitor is 10^{-6} coulomb and there is no initial current. Find the charge Q on the capacitor at any time t.

13. A certain vibrating system satisfies the equation $u'' + \gamma u' + u = 0$. Find the value of the damping coefficient γ for which the quasi period of the damped motion is 50% greater than the period of the corresponding undamped motion.

14. Show that the period of motion of an undamped vibration of a mass hanging from a vertical spring is $2\pi\sqrt{L/g}$, where L is the elongation of the spring due to the mass and g is the acceleration due to gravity.

15. Show that the solution of the initial value problem

$$mu'' + \gamma u' + ku = 0, \qquad u(t_0) = u_0, \quad u'(t_0) = u'_0$$

can be expressed as the sum $u = v + w$, where v satisfies the initial conditions $v(t_0) = u_0, v'(t_0) = 0$, w satisfies the initial conditions $w(t_0) = 0, w'(t_0) = u'_0$, and both v and w satisfy the same differential equation as u. This is another instance of superposing solutions of simpler problems to obtain the solution of a more general problem.

16. Show that $A \cos \omega_0 t + B \sin \omega_0 t$ can be written in the form $r \sin(\omega_0 t - \theta)$. Determine r and θ in terms of A and B. If $R \cos(\omega_0 t - \delta) = r \sin(\omega_0 t - \theta)$, determine the relationship among R, r, δ, and θ.

17. A mass weighing 8 lb stretches a spring 1.5 in. The mass is also attached to a damper with coefficient γ. Determine the value of γ for which the system is critically damped; be sure to give the units for γ.

18. If a series circuit has a capacitor of $C = 0.8 \times 10^{-6}$ farad and an inductor of $L = 0.2$ henry, find the resistance R so that the circuit is critically damped.

19. Assume that the system described by the equation $mu'' + \gamma u' + ku = 0$ is either critically damped or overdamped. Show that the mass can pass through the equilibrium position at most once, regardless of the initial conditions.
Hint: Determine all possible values of t for which $u = 0$.

20. Assume that the system described by the equation $mu'' + \gamma u' + ku = 0$ is critically damped and that the initial conditions are $u(0) = u_0, u'(0) = v_0$. If $v_0 = 0$, show that $u \to 0$ as $t \to \infty$ but that u is never zero. If u_0 is positive, determine a condition on v_0 that will ensure that the mass passes through its equilibrium position after it is released.

21. **Logarithmic Decrement.** (a) For the damped oscillation described by Eq. (26), show that the time between successive maxima is $T_d = 2\pi/\mu$.

(b) Show that the ratio of the displacements at two successive maxima is given by $\exp(\gamma T_d/2m)$. Observe that this ratio does not depend on which pair of maxima is chosen. The natural logarithm of this ratio is called the logarithmic decrement and is denoted by Δ.

(c) Show that $\Delta = \pi \gamma/m\mu$. Since m, μ, and Δ are quantities that can be measured easily for a mechanical system, this result provides a convenient and *practical* method for determining the damping constant of the system, which is more difficult to measure directly. In particular, for the motion of a vibrating mass in a viscous fluid, the damping constant depends on the viscosity of the fluid; for simple geometric shapes the form of this dependence is known, and the preceding relation allows the experimental determination of the viscosity. This is one of the most accurate ways of determining the viscosity of a gas at high pressure.

22. Referring to Problem 21, find the logarithmic decrement of the system in Problem 10.

23. For the system in Problem 17 suppose that $\Delta = 3$ and $T_d = 0.3$ sec. Referring to Problem 21, determine the value of the damping coefficient γ.

24. The position of a certain spring–mass system satisfies the initial value problem

$$\tfrac{3}{2} u'' + ku = 0, \qquad u(0) = 2, \quad u'(0) = v.$$

If the period and amplitude of the resulting motion are observed to be π and 3, respectively, determine the values of k and v.

25. Consider the initial value problem

$$u'' + \gamma u' + u = 0, \qquad u(0) = 2, \quad u'(0) = 0.$$

We wish to explore how long a time interval is required for the solution to become "negligible" and how this interval depends on the damping coefficient γ. To be more precise,

let us seek the time τ such that $|u(t)| < 0.01$ for all $t > \tau$. Note that critical damping for this problem occurs for $\gamma = 2$.

(a) Let $\gamma = 0.25$ and determine τ, or at least estimate it fairly accurately from a plot of the solution.

(b) Repeat part (a) for several other values of γ in the interval $0 < \gamma < 1.5$. Note that τ steadily decreases as γ increases for γ in this range.

(c) Create a graph of τ versus γ by plotting the pairs of values found in parts (a) and (b). Is the graph a smooth curve?

(d) Repeat part (b) for values of γ between 1.5 and 2. Show that τ continues to decrease until γ reaches a certain critical value γ_0, after which τ increases. Find γ_0 and the corresponding minimum value of τ to two decimal places.

(e) Another way to proceed is to write the solution of the initial value problem in the form (26). Neglect the cosine factor and consider only the exponential factor and the amplitude R. Then find an expression for τ as a function of γ. Compare the approximate results obtained in this way with the values determined in parts (a), (b), and (d).

26. Consider the initial value problem

$$mu'' + \gamma u' + ku = 0, \qquad u(0) = u_0, \quad u'(0) = v_0.$$

Assume that $\gamma^2 < 4km$.

(a) Solve the initial value problem.

(b) Write the solution in the form $u(t) = R \exp(-\gamma t/2m) \cos(\mu t - \delta)$. Determine R in terms of m, γ, k, u_0, and v_0.

(c) Investigate the dependence of R on the damping coefficient γ for fixed values of the other parameters.

27. A cubic block of side l and mass density ρ per unit volume is floating in a fluid of mass density ρ_0 per unit volume, where $\rho_0 > \rho$. If the block is slightly depressed and then released, it oscillates in the vertical direction. Assuming that the viscous damping of the fluid and air can be neglected, derive the differential equation of motion and determine the period of the motion.

Hint: Use Archimedes' principle: An object that is completely or partially submerged in a fluid is acted on by an upward (buoyant) force equal to the weight of the displaced fluid.

28. The position of a certain undamped spring–mass system satisfies the initial value problem

$$u'' + 2u = 0, \qquad u(0) = 0, \quad u'(0) = 2.$$

(a) Find the solution of this initial value problem.

(b) Plot u versus t and u' versus t on the same axes.

(c) Plot u' versus u; that is, plot $u(t)$ and $u'(t)$ parametrically with t as the parameter. This plot is known as a phase plot, and the uu'-plane is called the phase plane. Observe that a closed curve in the phase plane corresponds to a periodic solution $u(t)$. What is the direction of motion on the phase plot as t increases?

29. The position of a certain spring–mass system satisfies the initial value problem

$$u'' + \tfrac{1}{4}u' + 2u = 0, \qquad u(0) = 0, \quad u'(0) = 2.$$

(a) Find the solution of this initial value problem.

(b) Plot u versus t and u' versus t on the same axes.

(c) Plot u' versus u in the phase plane (see Problem 28). Identify several corresponding points on the curves in parts (b) and (c). What is the direction of motion on the phase plot as t increases?

30. In the absence of damping the motion of a spring–mass system satisfies the initial value problem

$$mu'' + ku = 0, \qquad u(0) = a, \quad u'(0) = b.$$

(a) Show that the kinetic energy initially imparted to the mass is $mb^2/2$ and that the potential energy initially stored in the spring is $ka^2/2$, so that initially the total energy in the system is $(ka^2 + mb^2)/2$.

(b) Solve the given initial value problem.

(c) Using the solution in part (b), determine the total energy in the system at any time t. Your result should confirm the principle of conservation of energy for this system.

31. Suppose that a mass m slides without friction on a horizontal surface. The mass is attached to a spring with spring constant k, as shown in Figure 3.8.10, and is also subject to viscous air resistance with coefficient γ. Show that the displacement $u(t)$ of the mass from its equilibrium position satisfies Eq. (21). How does the derivation of the equation of motion in this case differ from the derivation given in the text?

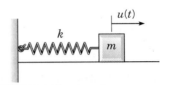

FIGURE 3.8.10 A spring–mass system.

32. In the spring–mass system of Problem 31, suppose that the spring force is not given by Hooke's law but instead satisfies the relation

$$F_s = -(ku + \epsilon u^3),$$

where $k > 0$ and ϵ is small but may be of either sign. The spring is called a hardening spring if $\epsilon > 0$ and a softening spring if $\epsilon < 0$. Why are these terms appropriate?

(a) Show that the displacement $u(t)$ of the mass from its equilibrium position satisfies the differential equation

$$mu'' + \gamma u' + ku + \epsilon u^3 = 0.$$

Suppose that the initial conditions are

$$u(0) = 0, \quad u'(0) = 1.$$

In the remainder of this problem assume that $m = 1, k = 1$, and $\gamma = 0$.

(b) Find $u(t)$ when $\epsilon = 0$ and also determine the amplitude and period of the motion.

(c) Let $\epsilon = 0.1$. Plot a numerical approximation to the solution. Does the motion appear to be periodic? Estimate the amplitude and period.

(d) Repeat part (c) for $\epsilon = 0.2$ and $\epsilon = 0.3$.

(e) Plot your estimated values of the amplitude A and the period T versus ϵ. Describe the way in which A and T, respectively, depend on ϵ.

(f) Repeat parts (c), (d), and (e) for negative values of ϵ.

3.9 Forced Vibrations

We will now investigate the situation in which a periodic external force is applied to a spring–mass system. The behavior of this simple system models that of many oscillatory systems with an external force due, for example, to a motor attached to the system. We will first consider the case in which damping is present and will look later at the idealized special case in which there is assumed to be no damping.

Forced Vibrations with Damping. Suppose that the external force is given by $F_0 \cos \omega t$, where F_0 and ω are positive constants representing the amplitude and frequency, respectively, of the force. Then the equation of motion is

$$mu'' + \gamma u' + ku = F_0 \cos \omega t, \tag{1}$$

where m, γ, and k are the mass, damping coefficient, and spring constant of the spring–mass system. The general solution of Eq. (1) must be of the form

$$u = c_1 u_1(t) + c_2 u_2(t) + A \cos \omega t + B \sin \omega t = u_c(t) + U(t). \tag{2}$$

The first two terms on the right side of Eq. (2) form the general solution $u_c(t)$ of the homogeneous equation corresponding to Eq. (1), while the latter two terms are a particular solution $U(t)$ of the full nonhomogeneous equation. The coefficients A and B can be found, as usual, by substituting these terms into the differential equation (1), while the arbitrary constants c_1 and c_2 are available to satisfy initial conditions, if any are prescribed. The solutions $u_1(t)$ and $u_2(t)$ of the homogeneous equation depend on the roots r_1 and r_2 of the characteristic equation $mr^2 + \gamma r + k = 0$. Since m, γ, and k are all positive, it follows that r_1 and r_2 either are real and negative or are complex conjugates with negative real part. In either case, both $u_1(t)$ and $u_2(t)$ approach zero as $t \to \infty$. Since $u_c(t)$ dies out as t increases, it is called the **transient solution**. In many applications, it is of little importance and (depending on the value of γ) may well be undetectable after only a few seconds.

The remaining terms in Eq. (2), namely $U(t) = A \cos \omega t + B \sin \omega t$, do not die out as t increases but persist indefinitely, or as long as the external force is applied. They represent a steady oscillation with the same frequency as the external force and are called the **steady-state solution** or the **forced response**. The transient solution enables us to satisfy whatever initial conditions may be imposed; with increasing time, the energy put into the system by the initial displacement and velocity is dissipated through the damping force, and the motion then becomes the response of the system to the external force. Without damping, the effect of the initial conditions would persist for all time.

It is convenient to express $U(t)$ as a single trigonometric term rather than as a sum of two terms. Recall that we did this for other similar expressions in Section 3.8. Thus we write

$$U(t) = R \cos(\omega t - \delta). \tag{3}$$

The amplitude R and phase δ depend directly on A and B and indirectly on the parameters in the differential equation (1). It is possible to show, by straightforward but somewhat lengthy algebraic computations, that

$$R = \frac{F_0}{\Delta}, \qquad \cos \delta = \frac{m(\omega_0^2 - \omega^2)}{\Delta}, \qquad \sin \delta = \frac{\gamma \omega}{\Delta}, \tag{4}$$

where

$$\Delta = \sqrt{m^2(\omega_0^2 - \omega^2)^2 + \gamma^2\omega^2} \quad \text{and} \quad \omega_0^2 = k/m. \tag{5}$$

We now investigate how the amplitude R of the steady-state oscillation depends on the frequency ω of the external force. For low-frequency excitation, that is, as $\omega \to 0$, it follows from Eqs. (4) and (5) that $R \to F_0/k$. At the other extreme, for very high-frequency excitation, Eqs. (4) and (5) imply that $R \to 0$ as $\omega \to \infty$. At an intermediate value of ω the amplitude may have a maximum. To find this maximum point, we can differentiate R with respect to ω and set the result equal to zero. In this way we find that the maximum amplitude occurs when $\omega = \omega_{max}$, where

$$\omega_{max}^2 = \omega_0^2 - \frac{\gamma^2}{2m^2} = \omega_0^2\left(1 - \frac{\gamma^2}{2mk}\right). \tag{6}$$

Note that $\omega_{max} < \omega_0$ and that ω_{max} is close to ω_0 when γ is small. The maximum value of R is

$$R_{max} = \frac{F_0}{\gamma\omega_0\sqrt{1 - (\gamma^2/4mk)}} \cong \frac{F_0}{\gamma\omega_0}\left(1 + \frac{\gamma^2}{8mk}\right), \tag{7}$$

where the last expression is an approximation for small γ. If $\gamma^2/mk > 2$, then ω_{max} as given by Eq. (6) is imaginary; in this case the maximum value of R occurs for $\omega = 0$, and R is a monotone decreasing function of ω. Recall that critical damping occurs when $\gamma^2/mk = 4$.

For small γ it follows from Eq. (7) that $R_{max} \cong F_0/\gamma\omega_0$. Thus, for lightly damped systems, the amplitude R of the forced response when ω is near ω_0 is quite large even for relatively small external forces, and the smaller the value of γ, the more pronounced is this effect. This phenomenon is known as **resonance**, and it is often an important design consideration. Resonance can be either good or bad, depending on the circumstances. It must be taken very seriously in the design of structures, such as buildings and bridges, where it can produce instabilities that might lead to the catastrophic failure of the structure. On the other hand, resonance can be put to good use in the design of instruments, such as seismographs, that are intended to detect weak periodic incoming signals.

Figure 3.9.1 contains some representative graphs of Rk/F_0 versus ω/ω_0 for several values of $\Gamma = \gamma^2/mk$. The quantity Rk/F_0 is the ratio of the amplitude R of the forced response to F_0/k, the static displacement of the spring produced by a force F_0. The graph corresponding to $\Gamma = 0.015625$ is included because this is the value of Γ that occurs in Example 1 below. Note particularly the sharp peak in the curve corresponding to $\Gamma = 0.015625$ near $\omega/\omega_0 = 1$. The limiting case as $\Gamma \to 0$ is also shown. It follows from Eqs. (4) and (5) that $R \to F_0/m|\omega_0^2 - \omega^2|$ as $\gamma \to 0$ and hence Rk/F_0 is asymptotic to the vertical line $\omega = \omega_0$, as shown in the figure. As the damping in the system increases, the peak response gradually diminishes.

Figure 3.9.1 also illustrates the usefulness of dimensionless variables. You can easily verify that each of the quantities Rk/F_0, ω/ω_0, and Γ is dimensionless. The importance of this observation is that the number of significant parameters in the problem has been reduced to three rather than the five that appear in Eq. (1). Thus only one family of curves, of which a few are shown in Figure 3.9.1, describes the response-versus-frequency behavior of all systems governed by Eq. (1).

FIGURE 3.9.1 Forced vibration with damping: amplitude of steady-state response versus frequency of driving force; $\Gamma = \gamma^2/mk$.

The phase angle δ also depends in an interesting way on ω. For ω near zero, it follows from Eqs. (4) and (5) that $\cos\delta \cong 1$ and $\sin\delta \cong 0$. Thus $\delta \cong 0$, and the response is nearly in phase with the excitation, meaning that they rise and fall together and, in particular, assume their respective maxima nearly together and their respective minima nearly together. For $\omega = \omega_0$ we find that $\cos\delta = 0$ and $\sin\delta = 1$, so $\delta = \pi/2$. In this case the response lags behind the excitation by $\pi/2$; that is, the peaks of the response occur $\pi/2$ later than the peaks of the excitation, and similarly for the valleys. Finally, for ω very large, we have $\cos\delta \cong -1$ and $\sin\delta \cong 0$. Thus $\delta \cong \pi$, so that the response is nearly out of phase with the excitation; this means that the response is minimum when the excitation is maximum, and vice versa. Figure 3.9.2 shows

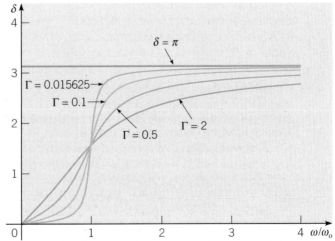

FIGURE 3.9.2 Forced vibration with damping: phase of steady-state response versus frequency of driving force; $\Gamma = \gamma^2/mk$.

the graphs of δ versus ω/ω_0 for several values of Γ. For small damping, the phase transition from near $\delta = 0$ to near $\delta = \pi$ occurs rather abruptly, whereas for larger values of the damping parameter, the transition takes place more gradually.

Consider the initial value problem

$$u'' + 0.125u' + u = 3\cos\omega t, \qquad u(0) = 2, \quad u'(0) = 0. \tag{8}$$

Show plots of the solution for different values of the forcing frequency ω, and compare them with corresponding plots of the forcing function.

For this system we have $\omega_0 = 1$ and $\Gamma = 1/64 = 0.015625$. Its unforced motion was discussed in Example 3 of Section 3.8, and Figure 3.8.7 shows the graph of the solution of the unforced problem. Figures 3.9.3, 3.9.4, and 3.9.5 show the solution of the forced problem (8) for $\omega = 0.3$, $\omega = 1$, and $\omega = 2$, respectively. The graph of the corresponding forcing function is also shown in each figure. In this example the static displacement, F_0/k, is equal to 3.

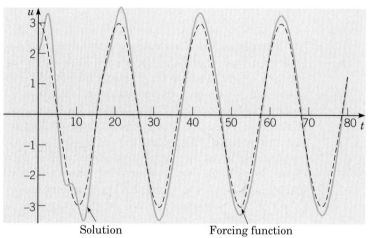

FIGURE 3.9.3 A forced vibration with damping; solution of $u'' + 0.125u' + u = 3\cos 0.3t$, $u(0) = 2$, $u'(0) = 0$.

Figure 3.9.3 shows the low-frequency case, $\omega/\omega_0 = 0.3$. After the initial transient response is substantially damped out, the remaining steady-state response is essentially in phase with the excitation, and the amplitude of the response is somewhat larger than the static displacement. To be specific, $R \cong 3.2939$ and $\delta \cong 0.041185$.

The resonant case, $\omega/\omega_0 = 1$, is shown in Figure 3.9.4. Here the amplitude of the steady-state response is eight times the static displacement, and the figure also shows the predicted phase lag of $\pi/2$ relative to the external force.

The case of comparatively high-frequency excitation is shown in Figure 3.9.5. Observe that the amplitude of the steady forced response is approximately one-third the static displacement and that the phase difference between the excitation and response is approximately π. More precisely, we find that $R \cong 0.99655$ and that $\delta \cong 3.0585$.

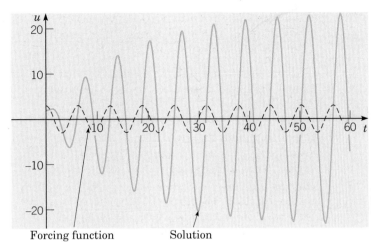

FIGURE 3.9.4 A forced vibration with damping; solution of $u'' + 0.125u' + u = 3\cos t$, $u(0) = 2$, $u'(0) = 0$.

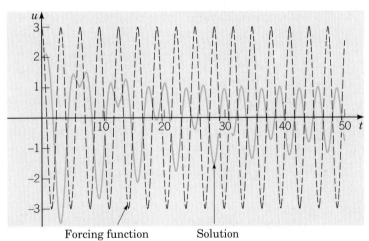

FIGURE 3.9.5 A forced vibration with damping; solution of $u'' + 0.125u' + u = 3\cos 2t$, $u(0) = 2$, $u'(0) = 0$.

Forced Vibrations Without Damping. We now assume that $\gamma = 0$ in Eq. (1), thereby obtaining the equation of motion of an undamped forced oscillator,

$$mu'' + ku = F_0 \cos \omega t. \tag{9}$$

The form of the general solution of Eq. (9) is different, depending on whether the forcing frequency ω is different from or equal to the natural frequency $\omega_0 = \sqrt{k/m}$ of the unforced system. First consider the case $\omega \neq \omega_0$; then the general solution

of Eq. (9) is

$$u = c_1 \cos \omega_0 t + c_2 \sin \omega_0 t + \frac{F_0}{m(\omega_0^2 - \omega^2)} \cos \omega t. \tag{10}$$

The constants c_1 and c_2 are determined by the initial conditions. The resulting motion is, in general, the sum of two periodic motions of different frequencies (ω_0 and ω) and amplitudes.

It is particularly interesting to suppose that the mass is initially at rest, so that the initial conditions are $u(0) = 0$ and $u'(0) = 0$. Then the energy driving the system comes entirely from the external force, with no contribution from the initial conditions. In this case it turns out that the constants c_1 and c_2 in Eq. (10) are given by

$$c_1 = -\frac{F_0}{m(\omega_0^2 - \omega^2)}, \qquad c_2 = 0, \tag{11}$$

and the solution of Eq. (9) is

$$u = \frac{F_0}{m(\omega_0^2 - \omega^2)} (\cos \omega t - \cos \omega_0 t). \tag{12}$$

This is the sum of two periodic functions of different periods but the same amplitude. Making use of the trigonometric identities for $\cos(A \pm B)$ with $A = (\omega_0 + \omega)t/2$ and $B = (\omega_0 - \omega)t/2$, we can write Eq. (12) in the form

$$u = \left[\frac{2F_0}{m(\omega_0^2 - \omega^2)} \sin \frac{(\omega_0 - \omega)t}{2} \right] \sin \frac{(\omega_0 + \omega)t}{2}. \tag{13}$$

If $|\omega_0 - \omega|$ is small, then $\omega_0 + \omega$ is much greater than $|\omega_0 - \omega|$. Consequently, $\sin(\omega_0 + \omega)t/2$ is a rapidly oscillating function compared to $\sin(\omega_0 - \omega)t/2$. Thus the motion is a rapid oscillation with frequency $(\omega_0 + \omega)/2$ but with a slowly varying sinusoidal amplitude

$$\frac{2F_0}{m|\omega_0^2 - \omega^2|} \left| \sin \frac{(\omega_0 - \omega)t}{2} \right|.$$

This type of motion, possessing a periodic variation of amplitude, exhibits what is called a **beat**. For example, such a phenomenon occurs in acoustics when two tuning forks of nearly equal frequency are excited simultaneously. In this case the periodic variation of amplitude is quite apparent to the unaided ear. In electronics, the variation of the amplitude with time is called **amplitude modulation**.

**EXAMPLE
2**

Solve the initial value problem

$$u'' + u = 0.5 \cos 0.8t, \qquad u(0) = 0, \quad u'(0) = 0, \tag{14}$$

and plot the solution.

In this case $\omega_0 = 1, \omega = 0.8$, and $F_0 = 0.5$, so from Eq. (13) the solution of the given problem is

$$u = 2.77778 \sin 0.1t \sin 0.9t. \tag{15}$$

A graph of this solution is shown in Figure 3.9.6. The amplitude variation has a slow frequency of 0.1 and a corresponding slow period of 20π. Note that a half-period of 10π corresponds to

a single cycle of increasing and then decreasing amplitude. The displacement of the spring–mass system oscillates with a relatively fast frequency of 0.9, which is only slightly less than the natural frequency ω_0.

Now imagine that the forcing frequency ω is further increased, say to $\omega = 0.9$. Then the slow frequency is halved to 0.05, and the corresponding slow half-period is doubled to 20π. The multiplier 2.7778 also increases substantially, to 5.2632. However, the fast frequency is only marginally increased, to 0.95. Can you visualize what happens as ω takes on values closer and closer to the natural frequency $\omega_0 = 1$?

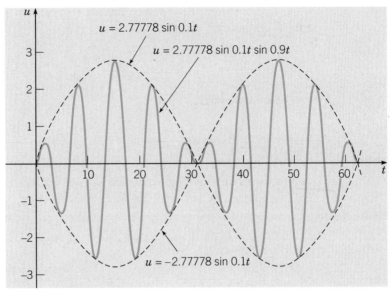

FIGURE 3.9.6 A beat; solution of $u'' + u = 0.5\cos 0.8t$, $u(0) = 0$, $u'(0) = 0$; $u = 2.77778\sin 0.1t\sin 0.9t$.

Now let us return to Eq. (9) and consider the case of resonance, where $\omega = \omega_0$; that is, the frequency of the forcing function is the same as the natural frequency of the system. Then the nonhomogeneous term $F_0\cos\omega t$ is a solution of the homogeneous equation. In this case the solution of Eq. (9) is

$$u = c_1\cos\omega_0 t + c_2\sin\omega_0 t + \frac{F_0}{2m\omega_0}t\sin\omega_0 t. \tag{16}$$

Because of the term $t\sin\omega_0 t$, the solution (16) predicts that the motion will become unbounded as $t \to \infty$ regardless of the values of c_1 and c_2; see Figure 3.9.7 for a typical example. Of course, in reality, unbounded oscillations do not occur. As soon as u becomes large, the mathematical model on which Eq. (9) is based is no longer valid, since the assumption that the spring force depends linearly on the displacement requires that u be small. As we have seen, if damping is included in the model, the predicted motion remains bounded; however, the response to the input function $F_0\cos\omega t$ may be quite large if the damping is small and ω is close to ω_0.

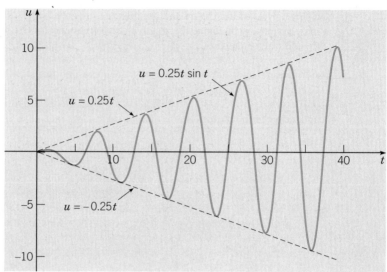

FIGURE 3.9.7 Resonance; solution of $u'' + u = 0.5\cos t$, $u(0) = 0$, $u'(0) = 0$;
$u = 0.25t\sin t$.

PROBLEMS

In each of Problems 1 through 4 write the given expression as a product of two trigonometric functions of different frequencies.

1. $\cos 9t - \cos 7t$ 2. $\sin 7t - \sin 6t$

3. $\cos \pi t + \cos 2\pi t$ 4. $\sin 3t + \sin 4t$

5. A mass weighing 4 lb stretches a spring 1.5 in. The mass is displaced 2 in. in the positive direction from its equilibrium position and released with no initial velocity. Assuming that there is no damping and that the mass is acted on by an external force of $2\cos 3t$ lb, formulate the initial value problem describing the motion of the mass.

6. A mass of 5 kg stretches a spring 10 cm. The mass is acted on by an external force of $10\sin(t/2)$ N (newtons) and moves in a medium that imparts a viscous force of 2 N when the speed of the mass is 4 cm/sec. If the mass is set in motion from its equilibrium position with an initial velocity of 3 cm/sec, formulate the initial value problem describing the motion of the mass.

7. (a) Find the solution of Problem 5.

(b) Plot the graph of the solution.

(c) If the given external force is replaced by a force $4\sin \omega t$ of frequency ω, find the value of ω for which resonance occurs.

8. (a) Find the solution of the initial value problem in Problem 6.

(b) Identify the transient and steady-state parts of the solution.

(c) Plot the graph of the steady-state solution.

(d) If the given external force is replaced by a force of $2\cos \omega t$ of frequency ω, find the value of ω for which the amplitude of the forced response is maximum.

9. If an undamped spring–mass system with a mass that weighs 6 lb and a spring constant 1 lb/in. is suddenly set in motion at $t = 0$ by an external force of $4\cos 7t$ lb, determine the position of the mass at any time and draw a graph of the displacement versus t.

10. A mass that weighs 8 lb stretches a spring 6 in. The system is acted on by an external force of $8 \sin 8t$ lb. If the mass is pulled down 3 in. and then released, determine the position of the mass at any time. Determine the first four times at which the velocity of the mass is zero.

11. A spring is stretched 6 in. by a mass that weighs 8 lb. The mass is attached to a dashpot mechanism that has a damping constant of 0.25 lb-sec/ft and is acted on by an external force of $4 \cos 2t$ lb.

 (a) Determine the steady-state response of this system.

 (b) If the given mass is replaced by a mass m, determine the value of m for which the amplitude of the steady-state response is maximum.

12. A spring–mass system has a spring constant of 3 N/m. A mass of 2 kg is attached to the spring, and the motion takes place in a viscous fluid that offers a resistance numerically equal to the magnitude of the instantaneous velocity. If the system is driven by an external force of $(3 \cos 3t - 2 \sin 3t)$ N, determine the steady-state response. Express your answer in the form $R \cos(\omega t - \delta)$.

13. Furnish the details in determining when the steady-state response given by Eq. (3) is maximum; that is, show that ω_{max}^2 and R_{max} are given by Eqs. (6) and (7), respectively.

14. Find the velocity of the steady-state response given by Eq. (3). Then show that the velocity is maximum when $\omega = \omega_0$.

15. Find the solution of the initial value problem

$$u'' + u = F(t), \qquad u(0) = 0, \quad u'(0) = 0,$$

where

$$F(t) = \begin{cases} F_0 t, & 0 \le t \le \pi, \\ F_0 (2\pi - t), & \pi < t \le 2\pi, \\ 0, & 2\pi < t. \end{cases}$$

Hint: Treat each time interval separately, and match the solutions in the different intervals by requiring that u and u' be continuous functions of t.

16. A series circuit has a capacitor of 0.25×10^{-6} farad, a resistor of 5×10^3 ohms, and an inductor of 1 henry. The initial charge on the capacitor is zero. If a 12-volt battery is connected to the circuit and the circuit is closed at $t = 0$, determine the charge on the capacitor at $t = 0.001$ sec, at $t = 0.01$ sec, and at any time t. Also determine the limiting charge as $t \to \infty$.

17. Consider a vibrating system described by the initial value problem

$$u'' + \tfrac{1}{4}u' + 2u = 2\cos \omega t, \qquad u(0) = 0, \quad u'(0) = 2.$$

 (a) Determine the steady-state part of the solution of this problem.

 (b) Find the amplitude A of the steady-state solution in terms of ω.

 (c) Plot A versus ω.

 (d) Find the maximum value of A and the frequency ω for which it occurs.

18. Consider the forced but undamped system described by the initial value problem

$$u'' + u = 3\cos \omega t, \qquad u(0) = 0, \quad u'(0) = 0.$$

 (a) Find the solution $u(t)$ for $\omega \ne 1$.

 (b) Plot the solution $u(t)$ versus t for $\omega = 0.7$, $\omega = 0.8$, and $\omega = 0.9$. Describe how the response $u(t)$ changes as ω varies in this interval. What happens as ω takes on values closer and closer to 1? Note that the natural frequency of the unforced system is $\omega_0 = 1$.

19. Consider the vibrating system described by the initial value problem

$$u'' + u = 3\cos\omega t, \qquad u(0) = 1, \quad u'(0) = 1.$$

(a) Find the solution for $\omega \neq 1$.

(b) Plot the solution $u(t)$ versus t for $\omega = 0.7, \omega = 0.8$, and $\omega = 0.9$. Compare the results with those of Problem 18; that is, describe the effect of the nonzero initial conditions.

20. For the initial value problem in Problem 18 plot u' versus u for $\omega = 0.7$, $\omega = 0.8$, and $\omega = 0.9$; that is, draw the phase plot of the solution for these values of ω. Use a t interval that is long enough so the phase plot appears as a closed curve. Mark your curve with arrows to show the direction in which it is traversed as t increases.

Problems 21 through 23 deal with the initial value problem

$$u'' + 0.125u' + 4u = F(t), \qquad u(0) = 2, \quad u'(0) = 0.$$

In each of these problems:

(a) Plot the given forcing function $F(t)$ versus t, and also plot the solution $u(t)$ versus t on the same set of axes. Use a t interval that is long enough so the initial transients are substantially eliminated. Observe the relation between the amplitude and phase of the forcing term and the amplitude and phase of the response. Note that $\omega_0 = \sqrt{k/m} = 2$.

(b) Draw the phase plot of the solution; that is, plot u' versus u.

21. $F(t) = 3\cos(t/4)$

22. $F(t) = 3\cos 2t$

23. $F(t) = 3\cos 6t$

24. A spring–mass system with a hardening spring (Problem 32 of Section 3.8) is acted on by a periodic external force. In the absence of damping, suppose that the displacement of the mass satisfies the initial value problem

$$u'' + u + \tfrac{1}{5}u^3 = \cos\omega t, \qquad u(0) = 0, \quad u'(0) = 0.$$

(a) Let $\omega = 1$ and plot a computer-generated solution of the given problem. Does the system exhibit a beat?

(b) Plot the solution for several values of ω between $1/2$ and 2. Describe how the solution changes as ω increases.

25. Suppose that the system of Problem 24 is modified to include a damping term and that the resulting initial value problem is

$$u'' + \tfrac{1}{5}u' + u + \tfrac{1}{5}u^3 = \cos\omega t, \qquad u(0) = 0, \quad u'(0) = 0.$$

(a) Plot a computer-generated solution of the given problem for several values of ω between $1/2$ and 2, and estimate the amplitude R of the steady response in each case.

(b) Using the data from part (a), plot the graph of R versus ω. For what frequency ω is the amplitude greatest?

(c) Compare the results of parts (a) and (b) with the corresponding results for the linear spring.

REFERENCES

Coddington, E. A., *An Introduction to Ordinary Differential Equations* (Englewood Cliffs, NJ: Prentice-Hall, 1961; New York: Dover, 1989).

There are many books on mechanical vibrations and electric circuits. One that deals with both is

Close, C. M., and Frederick, D. K., *Modeling and Analysis of Dynamic Systems* (3rd ed.) (New York: Wiley, 2001).

A classic book on mechanical vibrations is

Den Hartog, J. P., *Mechanical Vibrations* (4th ed.) (New York: McGraw-Hill, 1956; New York; Dover, 1985).

A more recent intermediate-level book is

Thomson, W. T., *Theory of Vibrations with Applications* (5th ed.) (Englewood Cliffs, NJ: Prentice-Hall, 1997).

An elementary book on electrical circuits is

Bobrow, L. S., *Elementary Linear Circuit Analysis* (New York: Oxford University Press, 1996).

The coefficients in Eq. (17) differ only slightly from those in Eq. (16), but the exponentially growing term, even with the relatively small coefficient of 1/32, completely dominates the solution by the time t is larger than about 4 or 5. This is clearly seen in Figure 4.2.3, which shows the graphs of the two solutions (16) and (17).

FIGURE 4.2.2 A plot of the solution (16).

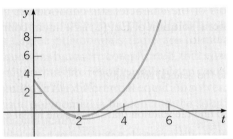

FIGURE 4.2.3 Plots of the solutions (16) (light curve) and (17) (heavy curve).

Repeated Roots. If the roots of the characteristic equation are not distinct—that is, if some of the roots are repeated—then the solution (5) is clearly not the general solution of Eq. (1). Recall that if r_1 is a repeated root for the second order linear equation $a_0 y'' + a_1 y' + a_2 y = 0$, then two linearly independent solutions are $e^{r_1 t}$ and $t e^{r_1 t}$. For an equation of order n, if a root of $Z(r) = 0$, say $r = r_1$, has multiplicity s (where $s \leq n$), then

$$e^{r_1 t}, \quad t e^{r_1 t}, \quad t^2 e^{r_1 t}, \quad \ldots, \quad t^{s-1} e^{r_1 t} \tag{18}$$

are corresponding solutions of Eq. (1); see Problem 41 for a proof of this statement.

If a complex root $\lambda + i\mu$ is repeated s times, the complex conjugate $\lambda - i\mu$ is also repeated s times. Corresponding to these $2s$ complex-valued solutions, we can find $2s$ real-valued solutions by noting that the real and imaginary parts of $e^{(\lambda+i\mu)t}$, $t e^{(\lambda+i\mu)t}, \ldots, t^{s-1} e^{(\lambda+i\mu)t}$ are also linearly independent solutions:

$$e^{\lambda t} \cos \mu t, \quad e^{\lambda t} \sin \mu t, \quad t e^{\lambda t} \cos \mu t, \quad t e^{\lambda t} \sin \mu t,$$
$$\ldots, \quad t^{s-1} e^{\lambda t} \cos \mu t, \quad t^{s-1} e^{\lambda t} \sin \mu t.$$

Hence the general solution of Eq. (1) can always be expressed as a linear combination of n real-valued solutions. Consider the following example.

**EXAMPLE
3**

Find the general solution of

$$y^{(4)} + 2y'' + y = 0. \tag{19}$$

The characteristic equation is

$$r^4 + 2r^2 + 1 = (r^2 + 1)(r^2 + 1) = 0.$$

The roots are $r = i, i, -i, -i$, and the general solution of Eq. (19) is

$$y = c_1 \cos t + c_2 \sin t + c_3 t \cos t + c_4 t \sin t.$$

In determining the roots of the characteristic equation, it may be necessary to compute the cube roots, the fourth roots, or even higher roots of a (possibly complex) number. This can usually be done most conveniently by using Euler's formula $e^{it} = \cos t + i \sin t$ and the algebraic laws given in Section 3.4. This is illustrated in the following example.

**EXAMPLE
4**

Find the general solution of

$$y^{(4)} + y = 0. \tag{20}$$

The characteristic equation is

$$r^4 + 1 = 0.$$

To solve the equation we must compute the fourth roots of -1. Now -1, thought of as a complex number, is $-1 + 0i$. It has magnitude 1 and polar angle π. Thus

$$-1 = \cos \pi + i \sin \pi = e^{i\pi}.$$

Moreover, the angle is determined only up to a multiple of 2π. Thus

$$-1 = \cos(\pi + 2m\pi) + i \sin(\pi + 2m\pi) = e^{i(\pi + 2m\pi)},$$

where m is zero or any positive or negative integer. Thus

$$(-1)^{1/4} = e^{i(\pi/4 + m\pi/2)} = \cos\left(\frac{\pi}{4} + \frac{m\pi}{2}\right) + i \sin\left(\frac{\pi}{4} + \frac{m\pi}{2}\right).$$

The four fourth roots of -1 are obtained by setting $m = 0, 1, 2$, and 3; they are

$$\frac{1+i}{\sqrt{2}}, \quad \frac{-1+i}{\sqrt{2}}, \quad \frac{-1-i}{\sqrt{2}}, \quad \frac{1-i}{\sqrt{2}}.$$

It is easy to verify that, for any other value of m, we obtain one of these four roots. For example, corresponding to $m = 4$, we obtain $(1 + i)/\sqrt{2}$. The general solution of Eq. (20) is

$$y = e^{t/\sqrt{2}}\left(c_1 \cos \frac{t}{\sqrt{2}} + c_2 \sin \frac{t}{\sqrt{2}}\right) + e^{-t/\sqrt{2}}\left(c_3 \cos \frac{t}{\sqrt{2}} + c_4 \sin \frac{t}{\sqrt{2}}\right). \tag{21}$$

In conclusion, we note that the problem of finding all the roots of a polynomial equation may not be entirely straightforward, even with computer assistance. For instance, it may be difficult to determine whether two roots are equal or merely very close together. Recall that the form of the general solution is different in these two cases.

If the constants a_0, a_1, \ldots, a_n in Eq. (1) are complex numbers, the solution of Eq. (1) is still of the form (4). In this case, however, the roots of the characteristic equation

are, in general, complex numbers, and it is no longer true that the complex conjugate of a root is also a root. The corresponding solutions are complex-valued.

PROBLEMS

In each of Problems 1 through 6 express the given complex number in the form $R(\cos\theta + i\sin\theta) = Re^{i\theta}$.

1. $1 + i$
2. $-1 + \sqrt{3}i$
3. -3
4. $-i$
5. $\sqrt{3} - i$
6. $-1 - i$

In each of Problems 7 through 10 follow the procedure illustrated in Example 4 to determine the indicated roots of the given complex number.

7. $1^{1/3}$
8. $(1 - i)^{1/2}$
9. $1^{1/4}$
10. $[2(\cos\pi/3 + i\sin\pi/3)]^{1/2}$

In each of Problems 11 through 28 find the general solution of the given differential equation.

11. $y''' - y'' - y' + y = 0$
12. $y''' - 3y'' + 3y' - y = 0$
13. $2y''' - 4y'' - 2y' + 4y = 0$
14. $y^{(4)} - 4y''' + 4y'' = 0$
15. $y^{(6)} + y = 0$
16. $y^{(4)} - 5y'' + 4y = 0$
17. $y^{(6)} - 3y^{(4)} + 3y'' - y = 0$
18. $y^{(6)} - y'' = 0$
19. $y^{(5)} - 3y^{(4)} + 3y''' - 3y'' + 2y' = 0$
20. $y^{(4)} - 8y' = 0$
21. $y^{(8)} + 8y^{(4)} + 16y = 0$
22. $y^{(4)} + 2y'' + y = 0$
23. $y''' - 5y'' + 3y' + y = 0$
24. $y''' + 5y'' + 6y' + 2y = 0$
25. $18y''' + 21y'' + 14y' + 4y = 0$
26. $y^{(4)} - 7y''' + 6y'' + 30y' - 36y = 0$
27. $12y^{(4)} + 31y''' + 75y'' + 37y' + 5y = 0$
28. $y^{(4)} + 6y''' + 17y'' + 22y' + 14y = 0$

In each of Problems 29 through 36 find the solution of the given initial value problem, and plot its graph. How does the solution behave as $t \to \infty$?

29. $y''' + y' = 0$; $y(0) = 0$, $y'(0) = 1$, $y''(0) = 2$
30. $y^{(4)} + y = 0$; $y(0) = 0$, $y'(0) = 0$, $y''(0) = -1$, $y'''(0) = 0$
31. $y^{(4)} - 4y''' + 4y'' = 0$; $y(1) = -1$, $y'(1) = 2$, $y''(1) = 0$, $y'''(1) = 0$
32. $y''' - y'' + y' - y = 0$; $y(0) = 2$, $y'(0) = -1$, $y''(0) = -2$
33. $2y^{(4)} - y''' - 9y'' + 4y' + 4y = 0$; $y(0) = -2$, $y'(0) = 0$, $y''(0) = -2$, $y'''(0) = 0$
34. $4y''' + y' + 5y = 0$; $y(0) = 2$, $y'(0) = 1$, $y''(0) = -1$
35. $6y''' + 5y'' + y' = 0$; $y(0) = -2$, $y'(0) = 2$, $y''(0) = 0$
36. $y^{(4)} + 6y''' + 17y'' + 22y' + 14y = 0$; $y(0) = 1$, $y'(0) = -2$, $y''(0) = 0$, $y'''(0) = 3$

37. Show that the general solution of $y^{(4)} - y = 0$ can be written as

$$y = c_1 \cos t + c_2 \sin t + c_3 \cosh t + c_4 \sinh t.$$

Determine the solution satisfying the initial conditions $y(0) = 0$, $y'(0) = 0$, $y''(0) = 1$, $y'''(0) = 1$. Why is it convenient to use the solutions $\cosh t$ and $\sinh t$ rather than e^t and e^{-t}?

38. Consider the equation $y^{(4)} - y = 0$.

(a) Use Abel's formula [Problem 20(d) of Section 4.1] to find the Wronskian of a fundamental set of solutions of the given equation.

(b) Determine the Wronskian of the solutions $e^t, e^{-t}, \cos t$, and $\sin t$.

(c) Determine the Wronskian of the solutions $\cosh t, \sinh t, \cos t$, and $\sin t$.

39. Consider the spring-mass system, shown in Figure 4.2.4, consisting of two unit masses suspended from springs with spring constants 3 and 2, respectively. Assume that there is no damping in the system.

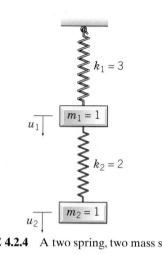

FIGURE 4.2.4 A two spring, two mass system.

(a) Show that the displacements u_1 and u_2 of the masses from their respective equilibrium positions satisfy the equations

$$u_1'' + 5u_1 = 2u_2, \qquad u_2'' + 2u_2 = 2u_1. \tag{i}$$

(b) Solve the first of Eqs. (i) for u_2 and substitute into the second equation, thereby obtaining the following fourth order equation for u_1:

$$u_1^{(4)} + 7u_1'' + 6u_1 = 0. \tag{ii}$$

Find the general solution of Eq. (ii).

(c) Suppose that the initial conditions are

$$u_1(0) = 1, \qquad u_1'(0) = 0, \qquad u_2(0) = 2, \qquad u_2'(0) = 0. \tag{iii}$$

Use the first of Eqs. (i) and the initial conditions (iii) to obtain values for $u_1''(0)$ and $u_1'''(0)$. Then show that the solution of Eq. (ii) that satisfies the four initial conditions on u_1 is $u_1(t) = \cos t$. Show that the corresponding solution u_2 is $u_2(t) = 2\cos t$.

(d) Now suppose that the initial conditions are

$$u_1(0) = -2, \qquad u_1'(0) = 0, \qquad u_2(0) = 1, \qquad u_2'(0) = 0. \tag{iv}$$

Proceed as in part (c) to show that the corresponding solutions are $u_1(t) = -2\cos\sqrt{6}\,t$ and $u_2(t) = \cos\sqrt{6}\,t$.

(e) Observe that the solutions obtained in parts (c) and (d) describe two distinct modes of vibration. In the first, the frequency of the motion is 1, and the two masses move in

phase, both moving up or down together. The second motion has frequency $\sqrt{6}$, and the masses move out of phase with each other, one moving down while the other is moving up, and vice versa. For other initial conditions, the motion of the masses is a combination of these two modes.

40. In this problem we outline one way to show that if r_1, \ldots, r_n are all real and different, then $e^{r_1 t}, \ldots, e^{r_n t}$ are linearly independent on $-\infty < t < \infty$. To do this, we consider the linear relation

$$c_1 e^{r_1 t} + \cdots + c_n e^{r_n t} = 0, \qquad -\infty < t < \infty \tag{i}$$

and show that all the constants are zero.

(a) Multiply Eq. (i) by $e^{-r_1 t}$ and differentiate with respect to t, thereby obtaining

$$c_2 (r_2 - r_1) e^{(r_2 - r_1)t} + \cdots + c_n (r_n - r_1) e^{(r_n - r_1)t} = 0.$$

(b) Multiply the result of part (a) by $e^{-(r_2 - r_1)t}$ and differentiate with respect to t to obtain

$$c_3 (r_3 - r_2)(r_3 - r_1) e^{(r_3 - r_2)t} + \cdots + c_n (r_n - r_2)(r_n - r_1) e^{(r_n - r_2)t} = 0.$$

(c) Continue the procedure from parts (a) and (b), eventually obtaining

$$c_n (r_n - r_{n-1}) \cdots (r_n - r_1) e^{(r_n - r_{n-1})t} = 0.$$

Hence $c_n = 0$, and therefore

$$c_1 e^{r_1 t} + \cdots + c_{n-1} e^{r_{n-1} t} = 0.$$

(d) Repeat the preceding argument to show that $c_{n-1} = 0$. In a similar way it follows that $c_{n-2} = \cdots = c_1 = 0$. Thus the functions $e^{r_1 t}, \ldots, e^{r_n t}$ are linearly independent.

41. In this problem we indicate one way to show that if $r = r_1$ is a root of multiplicity s of the characteristic polynomial $Z(r)$, then $e^{r_1 t}, te^{r_1 t}, \ldots, t^{s-1} e^{r_1 t}$ are solutions of Eq. (1). This problem extends to nth order equations the method given in Problem 22 of Section 3.5 for second order equations. We start from Eq. (2) in the text,

$$L[e^{rt}] = e^{rt} Z(r), \tag{i}$$

and differentiate repeatedly with respect to r, setting $r = r_1$ after each differentiation.

(a) Observe that if r_1 is a root of multiplicity s, then $Z(r) = (r - r_1)^s q(r)$, where $q(r)$ is a polynomial of degree $n - s$ and $q(r_1) \neq 0$. Show that $Z(r_1), Z'(r_1), \ldots, Z^{(s-1)}(r_1)$ are all zero, but $Z^{(s)}(r_1) \neq 0$.

(b) By differentiating Eq. (i) repeatedly with respect to r, show that

$$\frac{\partial}{\partial r} L[e^{rt}] = L\left[\frac{\partial}{\partial r} e^{rt}\right] = L[te^{rt}],$$

$$\vdots$$

$$\frac{\partial^{s-1}}{\partial r^{s-1}} L[e^{rt}] = L[t^{s-1} e^{rt}].$$

(c) Show that $e^{r_1 t}, te^{r_1 t}, \ldots, t^{s-1} e^{r_1 t}$ are solutions of Eq. (1).

4.3 The Method of Undetermined Coefficients

A particular solution Y of the nonhomogeneous nth order linear equation with constant coefficients

$$L[y] = a_0 y^{(n)} + a_1 y^{(n-1)} + \cdots + a_{n-1} y' + a_n y = g(t) \qquad (1)$$

can be obtained by the method of undetermined coefficients, provided that $g(t)$ is of an appropriate form. Although the method of undetermined coefficients is not as general as the method of variation of parameters described in the next section, it is usually much easier to use when it is applicable.

Just as for the second order linear equation, when the constant coefficient linear differential operator L is applied to a polynomial $A_0 t^m + A_1 t^{m-1} + \cdots + A_m$, an exponential function $e^{\alpha t}$, a sine function $\sin \beta t$, or a cosine function $\cos \beta t$, the result is a polynomial, an exponential function, or a linear combination of sine and cosine functions, respectively. Hence, if $g(t)$ is a sum of polynomials, exponentials, sines, and cosines, or products of such functions, we can expect that it is possible to find $Y(t)$ by choosing a suitable combination of polynomials, exponentials, and so forth, multiplied by a number of undetermined constants. The constants are then determined by substituting the assumed expression into Eq. (1).

The main difference in using this method for higher order equations stems from the fact that roots of the characteristic polynomial equation may have multiplicity greater than 2. Consequently, terms proposed for the nonhomogeneous part of the solution may need to be multiplied by higher powers of t to make them different from terms in the solution of the corresponding homogeneous equation. The following examples illustrate this. In these examples we have omitted numerous straightforward algebraic steps, because our main goal is to show how to arrive at the correct form for the assumed solution.

EXAMPLE 1

Find the general solution of

$$y''' - 3y'' + 3y' - y = 4e^t. \qquad (2)$$

The characteristic polynomial for the homogeneous equation corresponding to Eq. (2) is

$$r^3 - 3r^2 + 3r - 1 = (r - 1)^3,$$

so the general solution of the homogeneous equation is

$$y_c(t) = c_1 e^t + c_2 t e^t + c_3 t^2 e^t. \qquad (3)$$

To find a particular solution $Y(t)$ of Eq. (2), we start by assuming that $Y(t) = Ae^t$. However, since e^t, te^t, and $t^2 e^t$ are all solutions of the homogeneous equation, we must multiply this initial choice by t^3. Thus our final assumption is that $Y(t) = At^3 e^t$, where A is an undetermined coefficient. To find the correct value for A, we differentiate $Y(t)$ three times, substitute for y and its derivatives in Eq. (2), and collect terms in the resulting equation. In this way we obtain

$$6Ae^t = 4e^t.$$

Thus $A = \frac{2}{3}$ and the particular solution is

$$Y(t) = \frac{2}{3}t^3 e^t. \qquad (4)$$

The general solution of Eq. (2) is the sum of $y_c(t)$ from Eq. (3) and $Y(t)$ from Eq. (4).

EXAMPLE 2

Find a particular solution of the equation

$$y^{(4)} + 2y'' + y = 3\sin t - 5\cos t. \tag{5}$$

The general solution of the homogeneous equation was found in Example 3 of Section 4.2; it is

$$y_c(t) = c_1 \cos t + c_2 \sin t + c_3 t \cos t + c_4 t \sin t, \tag{6}$$

corresponding to the roots $r = i$, i, $-i$, and $-i$ of the characteristic equation. Our initial assumption for a particular solution is $Y(t) = A \sin t + B \cos t$, but we must multiply this choice by t^2 to make it different from all solutions of the homogeneous equation. Thus our final assumption is

$$Y(t) = At^2 \sin t + Bt^2 \cos t.$$

Next, we differentiate $Y(t)$ four times, substitute into the differential equation (4), and collect terms, obtaining finally

$$-8A \sin t - 8B \cos t = 3 \sin t - 5 \cos t.$$

Thus $A = -\frac{3}{8}$, $B = \frac{5}{8}$, and the particular solution of Eq. (4) is

$$Y(t) = -\frac{3}{8}t^2 \sin t + \frac{5}{8}t^2 \cos t. \tag{7}$$

If $g(t)$ is a sum of several terms, it may be easier in practice to compute separately the particular solution corresponding to each term in $g(t)$. As for the second order equation, the particular solution of the complete problem is the sum of the particular solutions of the individual component problems. This is illustrated in the following example.

EXAMPLE 3

Find a particular solution of

$$y''' - 4y' = t + 3\cos t + e^{-2t}. \tag{8}$$

First we solve the homogeneous equation. The characteristic equation is $r^3 - 4r = 0$, and the roots are $0, \pm 2$; hence

$$y_c(t) = c_1 + c_2 e^{2t} + c_3 e^{-2t}.$$

We can write a particular solution of Eq. (8) as the sum of particular solutions of the differential equations

$$y''' - 4y' = t, \qquad y''' - 4y' = 3\cos t, \qquad y''' - 4y' = e^{-2t}.$$

Our initial choice for a particular solution $Y_1(t)$ of the first equation is $A_0 t + A_1$, but since a constant is a solution of the homogeneous equation, we multiply by t. Thus

$$Y_1(t) = t(A_0 t + A_1).$$

For the second equation we choose

$$Y_2(t) = B \cos t + C \sin t,$$

and there is no need to modify this initial choice since $\cos t$ and $\sin t$ are not solutions of the homogeneous equation. Finally, for the third equation, since e^{-2t} is a solution of the homogeneous equation, we assume that

$$Y_3(t) = Ete^{-2t}.$$

The constants are determined by substituting into the individual differential equations; they are $A_0 = -\frac{1}{8}$, $A_1 = 0$, $B = 0$, $C = -\frac{3}{5}$, and $E = \frac{1}{8}$. Hence a particular solution of Eq. (8) is

$$Y(t) = -\tfrac{1}{8}t^2 - \tfrac{3}{5}\sin t + \tfrac{1}{8}te^{-2t}. \tag{9}$$

You should keep in mind that the amount of algebra required to calculate the coefficients may be quite substantial for higher order equations, especially if the nonhomogeneous term is even moderately complicated. A computer algebra system can be extremely helpful in executing these algebraic calculations.

The method of undetermined coefficients can be used whenever it is possible to guess the correct form for $Y(t)$. However, this is usually impossible for differential equations not having constant coefficients, or for nonhomogeneous terms other than the type described previously. For more complicated problems we can use the method of variation of parameters, which is discussed in the next section.

PROBLEMS

In each of Problems 1 through 8 determine the general solution of the given differential equation.

1. $y''' - y'' - y' + y = 2e^{-t} + 3$
2. $y^{(4)} - y = 3t + \cos t$
3. $y''' + y'' + y' + y = e^{-t} + 4t$
4. $y''' - y' = 2\sin t$
5. $y^{(4)} - 4y'' = t^2 + e^t$
6. $y^{(4)} + 2y'' + y = 3 + \cos 2t$
7. $y^{(6)} + y''' = t$
8. $y^{(4)} + y''' = \sin 2t$

In each of Problems 9 through 12 find the solution of the given initial value problem. Then plot a graph of the solution.

9. $y''' + 4y' = t;$ $y(0) = y'(0) = 0,$ $y''(0) = 1$
10. $y^{(4)} + 2y'' + y = 3t + 4;$ $y(0) = y'(0) = 0,$ $y''(0) = y'''(0) = 1$
11. $y''' - 3y'' + 2y' = t + e^t;$ $y(0) = 1,$ $y'(0) = -\frac{1}{4},$ $y''(0) = -\frac{3}{2}$
12. $y^{(4)} + 2y''' + y'' + 8y' - 12y = 12\sin t - e^{-t};$ $y(0) = 3,$ $y'(0) = 0,$
 $y''(0) = -1,$ $y'''(0) = 2$

In each of Problems 13 through 18 determine a suitable form for $Y(t)$ if the method of undetermined coefficients is to be used. Do not evaluate the constants.

13. $y''' - 2y'' + y' = t^3 + 2e^t$
14. $y''' - y' = te^{-t} + 2\cos t$
15. $y^{(4)} - 2y'' + y = e^t + \sin t$
16. $y^{(4)} + 4y'' = \sin 2t + te^t + 4$
17. $y^{(4)} - y''' - y'' + y' = t^2 + 4 + t\sin t$
18. $y^{(4)} + 2y''' + 2y'' = 3e^t + 2te^{-t} + e^{-t}\sin t$

19. Consider the nonhomogeneous nth order linear differential equation

$$a_0 y^{(n)} + a_1 y^{(n-1)} + \cdots + a_n y = g(t), \tag{i}$$

where a_0, \ldots, a_n are constants. Verify that if $g(t)$ is of the form

$$e^{\alpha t}(b_0 t^m + \cdots + b_m),$$

then the substitution $y = e^{\alpha t}u(t)$ reduces Eq. (i) to the form

$$k_0 u^{(n)} + k_1 u^{(n-1)} + \cdots + k_n u = b_0 t^m + \cdots + b_m, \tag{ii}$$

where k_0, \ldots, k_n are constants. Determine k_0 and k_n in terms of the a's and α. Thus the problem of determining a particular solution of the original equation is reduced to the simpler problem of determining a particular solution of an equation with constant coefficients and a polynomial for the nonhomogeneous term.

Method of Annihilators. In Problems 20 through 22 we consider another way of arriving at the proper form of $Y(t)$ for use in the method of undetermined coefficients. The procedure is based on the observation that exponential, polynomial, or sinusoidal terms (or sums and products of such terms) can be viewed as solutions of certain linear homogeneous differential equations with constant coefficients. It is convenient to use the symbol D for d/dt. Then, for example, e^{-t} is a solution of $(D+1)y = 0$; the differential operator $D+1$ is said to *annihilate*, or to be an *annihilator* of, e^{-t}. Similarly, $D^2 + 4$ is an annihilator of $\sin 2t$ or $\cos 2t$, $(D-3)^2 = D^2 - 6D + 9$ is an annihilator of e^{3t} or te^{3t}, and so forth.

20. Show that linear differential operators with constant coefficients obey the commutative law. That is, show that

$$(D-a)(D-b)f = (D-b)(D-a)f$$

for any twice differentiable function f and any constants a and b. The result extends at once to any finite number of factors.

21. Consider the problem of finding the form of a particular solution $Y(t)$ of

$$(D-2)^3(D+1)Y = 3e^{2t} - te^{-t}, \qquad \text{(i)}$$

where the left side of the equation is written in a form corresponding to the factorization of the characteristic polynomial.

(a) Show that $D-2$ and $(D+1)^2$, respectively, are annihilators of the terms on the right side of Eq. (i), and that the combined operator $(D-2)(D+1)^2$ annihilates both terms on the right side of Eq. (i) simultaneously.

(b) Apply the operator $(D-2)(D+1)^2$ to Eq. (i) and use the result of Problem 20 to obtain

$$(D-2)^4(D+1)^3 Y = 0. \qquad \text{(ii)}$$

Thus Y is a solution of the homogeneous equation (ii). By solving Eq. (ii), show that

$$Y(t) = c_1 e^{2t} + c_2 t e^{2t} + c_3 t^2 e^{2t} + c_4 t^3 e^{2t} + c_5 e^{-t} + c_6 t e^{-t} + c_7 t^2 e^{-t}, \qquad \text{(iii)}$$

where c_1, \ldots, c_7 are constants, as yet undetermined.

(c) Observe that e^{2t}, te^{2t}, $t^2 e^{2t}$, and e^{-t} are solutions of the homogeneous equation corresponding to Eq. (i); hence these terms are not useful in solving the nonhomogeneous equation. Therefore, choose c_1, c_2, c_3, and c_5 to be zero in Eq. (iii), so that

$$Y(t) = c_4 t^3 e^{2t} + c_6 t e^{-t} + c_7 t^2 e^{-t}. \qquad \text{(iv)}$$

This is the form of the particular solution Y of Eq. (i). The values of the coefficients c_4, c_6, and c_7 can be found by substituting from Eq. (iv) in the differential equation (i).

Summary. Suppose that

$$L(D)y = g(t), \qquad \text{(v)}$$

where $L(D)$ is a linear differential operator with constant coefficients, and $g(t)$ is a sum or product of exponential, polynomial, or sinusoidal terms. To find the form of a particular solution of Eq. (v), you can proceed as follows:

(a) Find a differential operator $H(D)$ with constant coefficients that annihilates $g(t)$, that is, an operator such that $H(D)g(t) = 0$.

(b) Apply $H(D)$ to Eq. (v), obtaining

$$H(D)L(D)y = 0, \tag{vi}$$

which is a homogeneous equation of higher order.

(c) Solve Eq. (vi).

(d) Eliminate from the solution found in step (c) the terms that also appear in the solution of $L(D)y = 0$. The remaining terms constitute the correct form of a particular solution of Eq. (v).

22. Use the method of annihilators to find the form of a particular solution $Y(t)$ for each of the equations in Problems 13 through 18. Do not evaluate the coefficients.

4.4 The Method of Variation of Parameters

The method of variation of parameters for determining a particular solution of the nonhomogeneous nth order linear differential equation

$$L[y] = y^{(n)} + p_1(t)y^{(n-1)} + \cdots + p_{n-1}(t)y' + p_n(t)y = g(t) \tag{1}$$

is a direct extension of the method for the second order differential equation (see Section 3.7). As before, to use the method of variation of parameters, it is first necessary to solve the corresponding homogeneous differential equation. In general, this may be difficult unless the coefficients are constants. However, the method of variation of parameters is still more general than the method of undetermined coefficients in that it leads to an expression for the particular solution for *any* continuous function g, whereas the method of undetermined coefficients is restricted in practice to a limited class of functions g.

Suppose then that we know a fundamental set of solutions y_1, y_2, \ldots, y_n of the homogeneous equation. Then the general solution of the homogeneous equation is

$$y_c(t) = c_1 y_1(t) + c_2 y_2(t) + \cdots + c_n y_n(t). \tag{2}$$

The method of variation of parameters for determining a particular solution of Eq. (1) rests on the possibility of determining n functions u_1, u_2, \ldots, u_n such that $Y(t)$ is of the form

$$Y(t) = u_1(t)y_1(t) + u_2(t)y_2(t) + \cdots + u_n(t)y_n(t). \tag{3}$$

Since we have n functions to determine, we will have to specify n conditions. One of these is clearly that Y satisfy Eq. (1). The other $n - 1$ conditions are chosen so as to make the calculations as simple as possible. Since we can hardly expect a simplification in determining Y if we must solve high order differential equations for u_1, \ldots, u_n, it is natural to impose conditions to suppress the terms that lead to higher derivatives of u_1, \ldots, u_n. From Eq. (3) we obtain

$$Y' = (u_1 y_1' + u_2 y_2' + \cdots + u_n y_n') + (u_1' y_1 + u_2' y_2 + \cdots + u_n' y_n), \tag{4}$$

where we have omitted the independent variable t on which each function in Eq. (4) depends. Thus the first condition that we impose is that

$$u_1'y_1 + u_2'y_2 + \cdots + u_n'y_n = 0. \tag{5}$$

Continuing this process in a similar manner through $n - 1$ derivatives of Y gives

$$Y^{(m)} = u_1 y_1^{(m)} + u_2 y_2^{(m)} + \cdots + u_n y_n^{(m)}, \qquad m = 0, 1, 2, \ldots, n - 1, \tag{6}$$

and the following $n - 1$ conditions on the functions u_1, \ldots, u_n:

$$u_1'y_1^{(m-1)} + u_2'y_2^{(m-1)} + \cdots + u_n'y_n^{(m-1)} = 0, \qquad m = 1, 2, \ldots, n - 1. \tag{7}$$

The nth derivative of Y is

$$Y^{(n)} = (u_1 y_1^{(n)} + \cdots + u_n y_n^{(n)}) + (u_1'y_1^{(n-1)} + \cdots + u_n'y_n^{(n-1)}). \tag{8}$$

Finally, we impose the condition that Y must be a solution of Eq. (1). On substituting for the derivatives of Y from Eqs. (6) and (8), collecting terms, and making use of the fact that $L[y_i] = 0, i = 1, 2, \ldots, n$, we obtain

$$u_1'y_1^{(n-1)} + u_2'y_2^{(n-1)} + \cdots + u_n'y_n^{(n-1)} = g. \tag{9}$$

Equation (9), coupled with the $n - 1$ equations (7), gives n simultaneous linear nonhomogeneous algebraic equations for u_1', u_2', \ldots, u_n':

$$\begin{aligned}
y_1 u_1' + y_2 u_2' + \cdots + y_n u_n' &= 0, \\
y_1' u_1' + y_2' u_2' + \cdots + y_n' u_n' &= 0, \\
y_1'' u_1' + y_2'' u_2' + \cdots + y_n'' u_n' &= 0, \\
&\ \ \vdots \\
y_1^{(n-1)} u_1' + \cdots + y_n^{(n-1)} u_n' &= g.
\end{aligned} \tag{10}$$

The system (10) is a linear algebraic system for the unknown quantities u_1', \ldots, u_n'. By solving this system and then integrating the resulting expressions, you can obtain the coefficients u_1, \ldots, u_n. A sufficient condition for the existence of a solution of the system of equations (10) is that the determinant of coefficients is nonzero for each value of t. However, the determinant of coefficients is precisely $W(y_1, y_2, \ldots, y_n)$, and it is nowhere zero since y_1, \ldots, y_n are linearly independent solutions of the homogeneous equation. Hence it is possible to determine u_1', \ldots, u_n'. Using Cramer's[3] rule, we can write the solution of the system of equations (10) in the form

$$u_m'(t) = \frac{g(t) W_m(t)}{W(t)}, \qquad m = 1, 2, \ldots, n. \tag{11}$$

[3]Cramer's rule is credited to Gabriel Cramer (1704–1752), professor at the Académie de Calvin in Geneva, who published it in a general form (but without proof) in 1750. For small systems the result had been known earlier.

Here $W(t) = W(y_1, y_2, \ldots, y_n)(t)$, and W_m is the determinant obtained from W by replacing the mth column by the column $(0, 0, \ldots, 0, 1)$. With this notation a particular solution of Eq. (1) is given by

$$Y(t) = \sum_{m=1}^{n} y_m(t) \int_{t_0}^{t} \frac{g(s) W_m(s)}{W(s)} \, ds, \tag{12}$$

where t_0 is arbitrary. Although the procedure is straightforward, the algebraic computations involved in determining $Y(t)$ from Eq. (12) become more and more complicated as n increases. In some cases the calculations may be simplified to some extent by using Abel's identity (Problem 20 of Section 4.1),

$$W(t) = W(y_1, \ldots, y_n)(t) = c \exp\left[-\int p_1(t) \, dt \right].$$

The constant c can be determined by evaluating W at some convenient point.

EXAMPLE 1

Given that $y_1(t) = e^t$, $y_2(t) = te^t$, and $y_3(t) = e^{-t}$ are solutions of the homogeneous equation corresponding to

$$y''' - y'' - y' + y = g(t), \tag{13}$$

determine a particular solution of Eq. (13) in terms of an integral.

We use Eq. (12). First, we have

$$W(t) = W(e^t, te^t, e^{-t})(t) = \begin{vmatrix} e^t & te^t & e^{-t} \\ e^t & (t+1)e^t & -e^{-t} \\ e^t & (t+2)e^t & e^{-t} \end{vmatrix}.$$

Factoring e^t from each of the first two columns and e^{-t} from the third column, we obtain

$$W(t) = e^t \begin{vmatrix} 1 & t & 1 \\ 1 & t+1 & -1 \\ 1 & t+2 & 1 \end{vmatrix}.$$

Then, by subtracting the first row from the second and third rows, we have

$$W(t) = e^t \begin{vmatrix} 1 & t & 1 \\ 0 & 1 & -2 \\ 0 & 2 & 0 \end{vmatrix}.$$

Finally, evaluating the latter determinant by minors associated with the first column, we find that

$$W(t) = 4e^t.$$

Next,

$$W_1(t) = \begin{vmatrix} 0 & te^t & e^{-t} \\ 0 & (t+1)e^t & -e^{-t} \\ 1 & (t+2)e^t & e^{-t} \end{vmatrix}.$$

Using minors associated with the first column, we obtain

$$W_1(t) = \begin{vmatrix} te^t & e^{-t} \\ (t+1)e^t & -e^{-t} \end{vmatrix} = -2t - 1.$$

In a similar way

$$W_2(t) = \begin{vmatrix} e^t & 0 & e^{-t} \\ e^t & 0 & -e^{-t} \\ e^t & 1 & e^{-t} \end{vmatrix} = -\begin{vmatrix} e^t & e^{-t} \\ e^t & -e^{-t} \end{vmatrix} = 2,$$

and

$$W_3(t) = \begin{vmatrix} e^t & te^t & 0 \\ e^t & (t+1)e^t & 0 \\ e^t & (t+2)e^t & 1 \end{vmatrix} = \begin{vmatrix} e^t & te^t \\ e^t & (t+1)e^t \end{vmatrix} = e^{2t}.$$

Substituting these results in Eq. (12), we have

$$Y(t) = e^t \int_{t_0}^t \frac{g(s)(-1-2s)}{4e^s}\,ds + te^t \int_{t_0}^t \frac{g(s)(2)}{4e^s}\,ds + e^{-t}\int_{t_0}^t \frac{g(s)e^{2s}}{4e^s}\,ds$$

$$= \frac{1}{4}\int_{t_0}^t \left\{e^{t-s}[-1 + 2(t-s)] + e^{-(t-s)}\right\}g(s)\,ds.$$

PROBLEMS

In each of Problems 1 through 6 use the method of variation of parameters to determine the general solution of the given differential equation.

1. $y''' + y' = \tan t$, $0 < t < \pi$
2. $y''' - y' = t$
3. $y''' - 2y'' - y' + 2y = e^{4t}$
4. $y''' + y' = \sec t$, $-\pi/2 < t < \pi/2$
5. $y''' - y'' + y' - y = e^{-t}\sin t$
6. $y^{(4)} + 2y'' + y = \sin t$

In each of Problems 7 and 8 find the general solution of the given differential equation. Leave your answer in terms of one or more integrals.

7. $y''' - y'' + y' - y = \sec t$, $-\pi/2 < t < \pi/2$
8. $y''' - y' = \csc t$, $0 < t < \pi$

In each of Problems 9 through 12 find the solution of the given initial value problem. Then plot a graph of the solution.

9. $y''' + y' = \sec t$; $y(0) = 2$, $y'(0) = 1$, $y''(0) = -2$
10. $y^{(4)} + 2y'' + y = \sin t$; $y(0) = 2$, $y'(0) = 0$, $y''(0) = -1$, $y'''(0) = 1$
11. $y''' - y'' + y' - y = \sec t$; $y(0) = 2$, $y'(0) = -1$, $y''(0) = 1$
12. $y''' - y' = \csc t$; $y(\pi/2) = 2$, $y'(\pi/2) = 1$, $y''(\pi/2) = -1$
13. Given that x, x^2, and $1/x$ are solutions of the homogeneous equation corresponding to

$$x^3 y''' + x^2 y'' - 2xy' + 2y = 2x^4, \qquad x > 0,$$

determine a particular solution.

14. Find a formula involving integrals for a particular solution of the differential equation

$$y''' - y'' + y' - y = g(t).$$

15. Find a formula involving integrals for a particular solution of the differential equation

$$y^{(4)} - y = g(t).$$

Hint: The functions $\sin t, \cos t, \sinh t$, and $\cosh t$ form a fundamental set of solutions of the homogeneous equation.

16. Find a formula involving integrals for a particular solution of the differential equation

$$y''' - 3y'' + 3y' - y = g(t).$$

If $g(t) = t^{-2}e^t$, determine $Y(t)$.

17. Find a formula involving integrals for a particular solution of the differential equation

$$x^3 y''' - 3x^2 y'' + 6xy' - 6y = g(x), \qquad x > 0.$$

Hint: Verify that x, x^2, and x^3 are solutions of the homogeneous equation.

REFERENCES Coddington, E. A., *An Introduction to Ordinary Differential Equations* (Englewood Cliffs, NJ: Prentice-Hall, 1961; New York: Dover, 1989).

Ince, E. L., *Ordinary Differential Equations* (London: Longmans, Green, 1927; New York: Dover, 1953).

Series Solutions of Second Order Linear Equations

Finding the general solution of a linear differential equation depends on determining a fundamental set of solutions of the homogeneous equation. So far, we have given a systematic procedure for constructing fundamental solutions only if the equation has constant coefficients. To deal with the much larger class of equations that have variable coefficients, it is necessary to extend our search for solutions beyond the familiar elementary functions of calculus. The principal tool that we need is the representation of a given function by a power series. The basic idea is similar to that in the method of undetermined coefficients: we assume that the solutions of a given differential equation have power series expansions, and then we attempt to determine the coefficients so as to satisfy the differential equation.

5.1 Review of Power Series

In this chapter we discuss the use of power series to construct fundamental sets of solutions of second order linear differential equations whose coefficients are functions of the independent variable. We begin by summarizing very briefly the pertinent results about power series that we need. Readers who are familiar with power series may go on to Section 5.2. Those who need more details than are presented here should consult a book on calculus.

1. A power series $\sum\limits_{n=0}^{\infty} a_n(x - x_0)^n$ is said to converge at a point x if

$$\lim_{m \to \infty} \sum_{n=0}^{m} a_n(x - x_0)^n$$

exists for that x. The series certainly converges for $x = x_0$; it may converge for all x, or it may converge for some values of x and not for others.

2. The series $\sum\limits_{n=0}^{\infty} a_n(x - x_0)^n$ is said to converge absolutely at a point x if the series

$$\sum_{n=0}^{\infty} |a_n(x - x_0)^n| = \sum_{n=0}^{\infty} |a_n||x - x_0|^n$$

converges. It can be shown that if the series converges absolutely, then the series also converges; however, the converse is not necessarily true.

3. One of the most useful tests for the absolute convergence of a power series is the ratio test. If $a_n \neq 0$, and if, for a fixed value of x,

$$\lim_{n \to \infty} \left| \frac{a_{n+1}(x - x_0)^{n+1}}{a_n(x - x_0)^n} \right| = |x - x_0| \lim_{n \to \infty} \left| \frac{a_{n+1}}{a_n} \right| = |x - x_0|L,$$

then the power series converges absolutely at that value of x if $|x - x_0|L < 1$ and diverges if $|x - x_0|L > 1$. If $|x - x_0|L = 1$, the test is inconclusive.

EXAMPLE
1

For which values of x does the power series

$$\sum_{n=1}^{\infty} (-1)^{n+1} n(x - 2)^n$$

converge?

To test for convergence we use the ratio test. We have

$$\lim_{n \to \infty} \left| \frac{(-1)^{n+2}(n + 1)(x - 2)^{n+1}}{(-1)^{n+1} n(x - 2)^n} \right| = |x - 2| \lim_{n \to \infty} \frac{n + 1}{n} = |x - 2|.$$

According to statement 3 the series converges absolutely for $|x - 2| < 1$, or $1 < x < 3$, and diverges for $|x - 2| > 1$. The values of x corresponding to $|x - 2| = 1$ are $x = 1$ and $x = 3$. The series diverges for each of these values of x since the nth term of the series does not approach zero as $n \to \infty$.

4. If the power series $\sum\limits_{n=0}^{\infty} a_n(x - x_0)^n$ converges at $x = x_1$, it converges absolutely for $|x - x_0| < |x_1 - x_0|$; and if it diverges at $x = x_1$, it diverges for $|x - x_0| > |x_1 - x_0|$.

5. There is a nonnegative number ρ, called the **radius of convergence**, such that $\sum\limits_{n=0}^{\infty} a_n(x - x_0)^n$ converges absolutely for $|x - x_0| < \rho$ and diverges for $|x - x_0| > \rho$. For a series that converges only at x_0, we define ρ to be zero; for a series that converges for all x, we say that ρ is infinite. If $\rho > 0$, then the interval $|x - x_0| < \rho$ is called the **interval of convergence**; it is indicated by the hatched lines in Figure 5.1.1. The series may either converge or diverge when $|x - x_0| = \rho$.

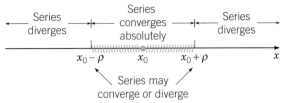

FIGURE 5.1.1 The interval of convergence of a power series.

<div style="float:left">

EXAMPLE

2

</div>

Determine the radius of convergence of the power series

$$\sum_{n=1}^{\infty} \frac{(x+1)^n}{n2^n}.$$

We apply the ratio test:

$$\lim_{n\to\infty} \left| \frac{(x+1)^{n+1}}{(n+1)2^{n+1}} \frac{n2^n}{(x+1)^n} \right| = \frac{|x+1|}{2} \lim_{n\to\infty} \frac{n}{n+1} = \frac{|x+1|}{2}.$$

Thus the series converges absolutely for $|x+1| < 2$, or $-3 < x < 1$, and diverges for $|x+1| > 2$. The radius of convergence of the power series is $\rho = 2$. Finally, we check the endpoints of the interval of convergence. At $x = 1$ the series becomes the harmonic series

$$\sum_{n=1}^{\infty} \frac{1}{n},$$

which diverges. At $x = -3$ we have

$$\sum_{n=1}^{\infty} \frac{(-3+1)^n}{n2^n} = \sum_{n=1}^{\infty} \frac{(-1)^n}{n},$$

which converges but does not converge absolutely. The series is said to converge conditionally at $x = -3$. To summarize, the given power series converges for $-3 \le x < 1$ and diverges otherwise. It converges absolutely for $-3 < x < 1$ and has a radius of convergence 2.

Suppose that $\sum_{n=0}^{\infty} a_n(x-x_0)^n$ and $\sum_{n=0}^{\infty} b_n(x-x_0)^n$ converge to $f(x)$ and $g(x)$, respectively, for $|x - x_0| < \rho, \rho > 0$.

6. The series can be added or subtracted termwise, and

$$f(x) \pm g(x) = \sum_{n=0}^{\infty} (a_n \pm b_n)(x - x_0)^n;$$

the resulting series converges at least for $|x - x_0| < \rho$.

7. The series can be formally multiplied, and

$$f(x)g(x) = \left[\sum_{n=0}^{\infty} a_n(x-x_0)^n \right] \left[\sum_{n=0}^{\infty} b_n(x-x_0)^n \right] = \sum_{n=0}^{\infty} c_n(x-x_0)^n,$$

where $c_n = a_0 b_n + a_1 b_{n-1} + \cdots + a_n b_0$. The resulting series converges at least for $|x - x_0| < \rho$.

Further, if $g(x_0) \neq 0$, the series can be formally divided, and

$$\frac{f(x)}{g(x)} = \sum_{n=0}^{\infty} d_n (x - x_0)^n.$$

In most cases the coefficients d_n can be most easily obtained by equating coefficients in the equivalent relation

$$\sum_{n=0}^{\infty} a_n (x - x_0)^n = \left[\sum_{n=0}^{\infty} d_n (x - x_0)^n \right] \left[\sum_{n=0}^{\infty} b_n (x - x_0)^n \right]$$

$$= \sum_{n=0}^{\infty} \left(\sum_{k=0}^{n} d_k b_{n-k} \right) (x - x_0)^n.$$

In the case of division the radius of convergence of the resulting power series may be less than ρ.

8. The function f is continuous and has derivatives of all orders for $|x - x_0| < \rho$. Further, f', f'', ... can be computed by differentiating the series termwise; that is,

$$f'(x) = a_1 + 2a_2(x - x_0) + \cdots + na_n(x - x_0)^{n-1} + \cdots$$

$$= \sum_{n=1}^{\infty} na_n (x - x_0)^{n-1},$$

$$f''(x) = 2a_2 + 6a_3(x - x_0) + \cdots + n(n-1)a_n(x - x_0)^{n-2} + \cdots$$

$$= \sum_{n=2}^{\infty} n(n-1)a_n (x - x_0)^{n-2},$$

and so forth, and each of the series converges absolutely for $|x - x_0| < \rho$.

9. The value of a_n is given by

$$a_n = \frac{f^{(n)}(x_0)}{n!}.$$

The series is called the Taylor[1] series for the function f about $x = x_0$.

10. If $\sum_{n=0}^{\infty} a_n(x - x_0)^n = \sum_{n=0}^{\infty} b_n(x - x_0)^n$ for each x in some open interval with center x_0, then $a_n = b_n$ for $n = 0, 1, 2, 3, \ldots$. In particular, if $\sum_{n=0}^{\infty} a_n(x - x_0)^n = 0$ for each such x, then $a_0 = a_1 = \cdots = a_n = \cdots = 0$.

A function f that has a Taylor series expansion about $x = x_0$,

$$f(x) = \sum_{n=0}^{\infty} \frac{f^{(n)}(x_0)}{n!} (x - x_0)^n,$$

with a radius of convergence $\rho > 0$, is said to be **analytic** at $x = x_0$. All of the familiar functions of calculus are analytic except perhaps at certain easily recognized points.

[1] Brook Taylor (1685–1731) was the leading English mathematician in the generation following Newton. In 1715 he published a general statement of the expansion theorem that is named for him, a result that is fundamental in all branches of analysis. He was also one of the founders of the calculus of finite differences and was the first to recognize the existence of singular solutions of differential equations.

For example, $\sin x$ and e^x are analytic everywhere, $1/x$ is analytic except at $x = 0$, and $\tan x$ is analytic except at odd multiples of $\pi/2$. According to statements 6 and 7, if f and g are analytic at x_0, then $f \pm g, f \cdot g$, and f/g [provided that $g(x_0) \neq 0$] are also analytic at $x = x_0$. In many respects the natural context for the use of power series is the complex plane. The methods and results of this chapter nearly always can be directly extended to differential equations in which the independent and dependent variables are complex valued.

Shift of Index of Summation. The index of summation in an infinite series is a dummy parameter just as the integration variable in a definite integral is a dummy variable. Thus it is immaterial which letter is used for the index of summation. For example,

$$\sum_{n=0}^{\infty} \frac{2^n x^n}{n!} = \sum_{j=0}^{\infty} \frac{2^j x^j}{j!}.$$

Just as we make changes of the variable of integration in a definite integral, we find it convenient to make changes of summation indices in calculating series solutions of differential equations. We illustrate by several examples how to shift the summation index.

EXAMPLE 3

Write $\sum_{n=2}^{\infty} a_n x^n$ as a series whose first term corresponds to $n = 0$ rather than $n = 2$.

Let $m = n - 2$; then $n = m + 2$, and $n = 2$ corresponds to $m = 0$. Hence

$$\sum_{n=2}^{\infty} a_n x^n = \sum_{m=0}^{\infty} a_{m+2} x^{m+2}. \tag{1}$$

By writing out the first few terms of each of these series, you can verify that they contain precisely the same terms. Finally, in the series on the right side of Eq. (1), we can replace the dummy index m by n, obtaining

$$\sum_{n=2}^{\infty} a_n x^n = \sum_{n=0}^{\infty} a_{n+2} x^{n+2}. \tag{2}$$

In effect, we have shifted the index upward by 2 and have compensated by starting to count at a level 2 lower than originally.

EXAMPLE 4

Write the series

$$\sum_{n=2}^{\infty} (n+2)(n+1) a_n (x - x_0)^{n-2} \tag{3}$$

as a series whose generic term involves $(x - x_0)^n$ rather than $(x - x_0)^{n-2}$.

Again, we shift the index by 2 so that n is replaced by $n + 2$ and start counting 2 lower. We obtain

$$\sum_{n=0}^{\infty} (n+4)(n+3) a_{n+2} (x - x_0)^n. \tag{4}$$

You can readily verify that the terms in the series (3) and (4) are exactly the same.

EXAMPLE
5

Write the expression

$$x^2 \sum_{n=0}^{\infty} (r+n)a_n x^{r+n-1} \tag{5}$$

as a series whose generic term involves x^{r+n}.

First, take the x^2 inside the summation, obtaining

$$\sum_{n=0}^{\infty} (r+n)a_n x^{r+n+1}. \tag{6}$$

Next, shift the index down by 1 and start counting 1 higher. Thus

$$\sum_{n=0}^{\infty} (r+n)a_n x^{r+n+1} = \sum_{n=1}^{\infty} (r+n-1)a_{n-1} x^{r+n}. \tag{7}$$

Again, you can easily verify that the two series in Eq. (7) are identical and that both are exactly the same as the expression (5).

EXAMPLE
6

Assume that

$$\sum_{n=1}^{\infty} n a_n x^{n-1} = \sum_{n=0}^{\infty} a_n x^n \tag{8}$$

for all x, and determine what this implies about the coefficients a_n.

We want to use statement 10 to equate corresponding coefficients in the two series. In order to do this, we must first rewrite Eq. (8) so that the series display the same power of x in their generic terms. For instance, in the series on the left side of Eq. (8), we can replace n by $n+1$ and start counting 1 lower. Thus Eq. (8) becomes

$$\sum_{n=0}^{\infty} (n+1)a_{n+1} x^n = \sum_{n=0}^{\infty} a_n x^n. \tag{9}$$

According to statement 10, we conclude that

$$(n+1)a_{n+1} = a_n, \qquad n = 0, 1, 2, 3, \ldots$$

or

$$a_{n+1} = \frac{a_n}{n+1}, \qquad n = 0, 1, 2, 3, \ldots \tag{10}$$

Hence, choosing successive values of n in Eq. (10), we have

$$a_1 = a_0, \qquad a_2 = \frac{a_1}{2} = \frac{a_0}{2}, \qquad a_3 = \frac{a_2}{3} = \frac{a_0}{3!},$$

and so forth. In general,

$$a_n = \frac{a_0}{n!}, \qquad n = 1, 2, 3, \ldots. \tag{11}$$

Thus the relation (8) determines all the following coefficients in terms of a_0. Finally, using the coefficients given by Eq. (11), we obtain

$$\sum_{n=0}^{\infty} a_n x^n = a_0 \sum_{n=0}^{\infty} \frac{x^n}{n!} = a_0 e^x,$$

where we have followed the usual convention that $0! = 1$.

PROBLEMS In each of Problems 1 through 8 determine the radius of convergence of the given power series.

1. $\displaystyle\sum_{n=0}^{\infty} (x-3)^n$

2. $\displaystyle\sum_{n=0}^{\infty} \frac{n}{2^n} x^n$

3. $\displaystyle\sum_{n=0}^{\infty} \frac{x^{2n}}{n!}$

4. $\displaystyle\sum_{n=0}^{\infty} 2^n x^n$

5. $\displaystyle\sum_{n=1}^{\infty} \frac{(2x+1)^n}{n^2}$

6. $\displaystyle\sum_{n=1}^{\infty} \frac{(x-x_0)^n}{n}$

7. $\displaystyle\sum_{n=1}^{\infty} \frac{(-1)^n n^2 (x+2)^n}{3^n}$

8. $\displaystyle\sum_{n=1}^{\infty} \frac{n! x^n}{n^n}$

In each of Problems 9 through 16 determine the Taylor series about the point x_0 for the given function. Also determine the radius of convergence of the series.

9. $\sin x$, $x_0 = 0$

10. e^x, $x_0 = 0$

11. x, $x_0 = 1$

12. x^2, $x_0 = -1$

13. $\ln x$, $x_0 = 1$

14. $\dfrac{1}{1+x}$, $x_0 = 0$

15. $\dfrac{1}{1-x}$, $x_0 = 0$

16. $\dfrac{1}{1-x}$, $x_0 = 2$

17. Given that $y = \displaystyle\sum_{n=0}^{\infty} n x^n$, compute y' and y'' and write out the first four terms of each series as well as the coefficient of x^n in the general term.

18. Given that $y = \displaystyle\sum_{n=0}^{\infty} a_n x^n$, compute y' and y'' and write out the first four terms of each series as well as the coefficient of x^n in the general term. Show that if $y'' = y$, then the coefficients a_0 and a_1 are arbitrary, and determine a_2 and a_3 in terms of a_0 and a_1. Show that $a_{n+2} = a_n/(n+2)(n+1)$, $n = 0,1,2,3,\ldots$.

In each of Problems 19 and 20 verify the given equation.

19. $\displaystyle\sum_{n=0}^{\infty} a_n (x-1)^{n+1} = \sum_{n=1}^{\infty} a_{n-1}(x-1)^n$

20. $\displaystyle\sum_{k=0}^{\infty} a_{k+1} x^k + \sum_{k=0}^{\infty} a_k x^{k+1} = a_1 + \sum_{k=1}^{\infty} (a_{k+1} + a_{k-1}) x^k$

In each of Problems 21 through 27 rewrite the given expression as a sum whose generic term involves x^n.

21. $\displaystyle\sum_{n=2}^{\infty} n(n-1) a_n x^{n-2}$

22. $\displaystyle\sum_{n=0}^{\infty} a_n x^{n+2}$

23. $x \displaystyle\sum_{n=1}^{\infty} n a_n x^{n-1} + \sum_{k=0}^{\infty} a_k x^k$

24. $(1-x^2) \displaystyle\sum_{n=2}^{\infty} n(n-1) a_n x^{n-2}$

25. $\displaystyle\sum_{m=2}^{\infty} m(m-1) a_m x^{m-2} + x \sum_{k=1}^{\infty} k a_k x^{k-1}$

26. $\displaystyle\sum_{n=1}^{\infty} n a_n x^{n-1} + x \sum_{n=0}^{\infty} a_n x^n$

27. $x \displaystyle\sum_{n=2}^{\infty} n(n-1) a_n x^{n-2} + \sum_{n=0}^{\infty} a_n x^n$

28. Determine the a_n so that the equation

$$\sum_{n=1}^{\infty} n a_n x^{n-1} + 2 \sum_{n=0}^{\infty} a_n x^n = 0$$

is satisfied. Try to identify the function represented by the series $\sum_{n=0}^{\infty} a_n x^n$.

5.2 Series Solutions Near an Ordinary Point, Part I

In Chapter 3 we described methods of solving second order linear differential equations with constant coefficients. We now consider methods of solving second order linear equations when the coefficients are functions of the independent variable. In this chapter we will denote the independent variable by x. It is sufficient to consider the homogeneous equation

$$P(x)\frac{d^2y}{dx^2} + Q(x)\frac{dy}{dx} + R(x)y = 0, \tag{1}$$

since the procedure for the corresponding nonhomogeneous equation is similar.

Many problems in mathematical physics lead to equations of the form (1) having polynomial coefficients; examples include the Bessel equation

$$x^2 y'' + x y' + (x^2 - v^2)y = 0,$$

where v is a constant, and the Legendre equation

$$(1 - x^2)y'' - 2xy' + \alpha(\alpha + 1)y = 0,$$

where α is a constant. For this reason, as well as to simplify the algebraic computations, we primarily consider the case in which the functions P, Q, and R are polynomials. However, as we will see, the method of solution is also applicable when P, Q, and R are general analytic functions.

For the present, then, suppose that P, Q, and R are polynomials and that they have no common factors. Suppose also that we wish to solve Eq. (1) in the neighborhood of a point x_0. The solution of Eq. (1) in an interval containing x_0 is closely associated with the behavior of P in that interval.

A point x_0 such that $P(x_0) \neq 0$ is called an **ordinary point**. Since P is continuous, it follows that there is an interval about x_0 in which $P(x)$ is never zero. In that interval we can divide Eq. (1) by $P(x)$ to obtain

$$y'' + p(x)y' + q(x)y = 0, \tag{2}$$

where $p(x) = Q(x)/P(x)$ and $q(x) = R(x)/P(x)$ are continuous functions. Hence, according to the existence and uniqueness Theorem 3.2.1, there exists in that interval a unique solution of Eq. (1) that also satisfies the initial conditions $y(x_0) = y_0$, $y'(x_0) = y'_0$ for arbitrary values of y_0 and y'_0. In this and the following section we discuss the solution of Eq. (1) in the neighborhood of an ordinary point.

On the other hand, if $P(x_0) = 0$, then x_0 is called a **singular point** of Eq. (1). In this case at least one of $Q(x_0)$ and $R(x_0)$ is not zero. Consequently, at least one of

the coefficients p and q in Eq. (2) becomes unbounded as $x \to x_0$, and therefore Theorem 3.2.1 does not apply in this case. Sections 5.4 through 5.8 deal with finding solutions of Eq. (1) in the neighborhood of a singular point.

We now take up the problem of solving Eq. (1) in the neighborhood of an ordinary point x_0. We look for solutions of the form

$$y = a_0 + a_1(x - x_0) + \cdots + a_n(x - x_0)^n + \cdots = \sum_{n=0}^{\infty} a_n(x - x_0)^n, \qquad (3)$$

and assume that the series converges in the interval $|x - x_0| < \rho$ for some $\rho > 0$. While at first sight it may appear unattractive to seek a solution in the form of a power series, this is actually a convenient and useful form for a solution. Within their intervals of convergence, power series behave very much like polynomials and are easy to manipulate both analytically and numerically. Indeed, even if we can obtain a solution in terms of elementary functions, such as exponential or trigonometric functions, we are likely to need a power series or some equivalent expression if we want to evaluate them numerically or to plot their graphs.

The most practical way to determine the coefficients a_n is to substitute the series (3) and its derivatives for y, y', and y'' in Eq. (1). The following examples illustrate this process. The operations, such as differentiation, that are involved in the procedure are justified so long as we stay within the interval of convergence. The differential equations in these examples are also of considerable importance in their own right.

EXAMPLE 1

Find a series solution of the equation

$$y'' + y = 0, \qquad -\infty < x < \infty. \qquad (4)$$

As we know, two linearly independent solutions of this equation are $\sin x$ and $\cos x$, so series methods are not needed to solve this equation. However, this example illustrates the use of power series in a relatively simple case. For Eq. (4), $P(x) = 1, Q(x) = 0$, and $R(x) = 1$; hence every point is an ordinary point.

We look for a solution in the form of a power series about $x_0 = 0$,

$$y = a_0 + a_1 x + a_2 x^2 + \cdots + a_n x^n + \cdots = \sum_{n=0}^{\infty} a_n x^n, \qquad (5)$$

and assume that the series converges in some interval $|x| < \rho$.

Differentiating Eq. (5) term by term yields

$$y' = a_1 + 2a_2 x + \cdots + na_n x^{n-1} + \cdots = \sum_{n=1}^{\infty} na_n x^{n-1}, \qquad (6)$$

$$y'' = 2a_2 + \cdots + n(n-1)a_n x^{n-2} + \cdots = \sum_{n=2}^{\infty} n(n-1)a_n x^{n-2}. \qquad (7)$$

Substituting the series (5) and (7) for y and y'' in Eq. (4) gives

$$\sum_{n=2}^{\infty} n(n-1)a_n x^{n-2} + \sum_{n=0}^{\infty} a_n x^n = 0.$$

To combine the two series, we need to rewrite at least one of them so that both series display the same generic term. Thus, in the first sum, we shift the index of summation by replacing n by $n + 2$ and starting the sum at 0 rather than 2. We obtain

$$\sum_{n=0}^{\infty} (n + 2)(n + 1)a_{n+2}x^n + \sum_{n=0}^{\infty} a_n x^n = 0$$

or

$$\sum_{n=0}^{\infty} [(n + 2)(n + 1)a_{n+2} + a_n]x^n = 0.$$

For this equation to be satisfied for all x, the coefficient of each power of x must be zero; hence we conclude that

$$(n + 2)(n + 1)a_{n+2} + a_n = 0, \qquad n = 0, 1, 2, 3, \ldots. \tag{8}$$

Equation (8) is referred to as a **recurrence relation**. The successive coefficients can be evaluated one by one by writing the recurrence relation first for $n = 0$, then for $n = 1$, and so forth. In this example Eq. (8) relates each coefficient to the second one before it. Thus the even-numbered coefficients (a_0, a_2, a_4, \ldots) and the odd-numbered ones (a_1, a_3, a_5, \ldots) are determined separately. For the even-numbered coefficients we have

$$a_2 = -\frac{a_0}{2 \cdot 1} = -\frac{a_0}{2!}, \qquad a_4 = -\frac{a_2}{4 \cdot 3} = +\frac{a_0}{4!}, \qquad a_6 = -\frac{a_4}{6 \cdot 5} = -\frac{a_0}{6!}, \ldots.$$

These results suggest that in general, if $n = 2k$, then

$$a_n = a_{2k} = \frac{(-1)^k}{(2k)!}a_0, \qquad k = 1, 2, 3, \ldots. \tag{9}$$

We can prove Eq. (9) by mathematical induction. First, observe that it is true for $k = 1$. Next, assume that it is true for an arbitrary value of k and consider the case $k + 1$. We have

$$a_{2k+2} = -\frac{a_{2k}}{(2k + 2)(2k + 1)} = -\frac{(-1)^k}{(2k + 2)(2k + 1)(2k)!}a_0 = \frac{(-1)^{k+1}}{(2k + 2)!}a_0.$$

Hence Eq. (9) is also true for $k + 1$, and consequently it is true for all positive integers k.
 Similarly, for the odd-numbered coefficients

$$a_3 = -\frac{a_1}{2 \cdot 3} = -\frac{a_1}{3!}, \qquad a_5 = -\frac{a_3}{5 \cdot 4} = +\frac{a_1}{5!}, \qquad a_7 = -\frac{a_5}{7 \cdot 6} = -\frac{a_1}{7!}, \ldots,$$

and in general, if $n = 2k + 1$, then[2]

$$a_n = a_{2k+1} = \frac{(-1)^k}{(2k + 1)!}a_1, \qquad k = 1, 2, 3, \ldots. \tag{10}$$

[2]The result given in Eq. (10) and other similar formulas in this chapter can be proved by an induction argument resembling the one just given for Eq. (9). We assume that the results are plausible and omit the inductive argument hereafter.

Substituting these coefficients into Eq. (5), we have

$$y = a_0 + a_1 x - \frac{a_0}{2!}x^2 - \frac{a_1}{3!}x^3 + \frac{a_0}{4!}x^4 + \frac{a_1}{5!}x^5$$

$$+ \cdots + \frac{(-1)^n a_0}{(2n)!}x^{2n} + \frac{(-1)^n a_1}{(2n+1)!}x^{2n+1} + \cdots$$

$$= a_0\left[1 - \frac{x^2}{2!} + \frac{x^4}{4!} + \cdots + \frac{(-1)^n}{(2n)!}x^{2n} + \cdots\right]$$

$$+ a_1\left[x - \frac{x^3}{3!} + \frac{x^5}{5!} + \cdots + \frac{(-1)^n}{(2n+1)!}x^{2n+1} + \cdots\right]$$

$$= a_0\sum_{n=0}^{\infty}\frac{(-1)^n}{(2n)!}x^{2n} + a_1\sum_{n=0}^{\infty}\frac{(-1)^n}{(2n+1)!}x^{2n+1}. \tag{11}$$

Now that we have formally obtained two series solutions of Eq. (4), we can test them for convergence. Using the ratio test, it is easy to show that each of the series in Eq. (11) converges for all x, and this justifies retroactively all the steps used in obtaining the solutions. Indeed, we recognize that the first series in Eq. (11) is exactly the Taylor series for $\cos x$ about $x = 0$ and that the second is the Taylor series for $\sin x$ about $x = 0$. Thus, as expected, we obtain the solution $y = a_0\cos x + a_1\sin x$.

Notice that no conditions are imposed on a_0 and a_1; hence they are arbitrary. From Eqs. (5) and (6) we see that y and y' evaluated at $x = 0$ are a_0 and a_1, respectively. Since the initial conditions $y(0)$ and $y'(0)$ can be chosen arbitrarily, it follows that a_0 and a_1 should be arbitrary until specific initial conditions are stated.

Figures 5.2.1 and 5.2.2 show how the partial sums of the series in Eq. (11) approximate $\cos x$ and $\sin x$. As the number of terms increases, the interval over which the approximation is satisfactory becomes longer, and for each x in this interval the accuracy of the approximation improves. However, you should always remember that a truncated power series provides only a local approximation of the solution in a neighborhood of the initial point $x = 0$; it cannot adequately represent the solution for large $|x|$.

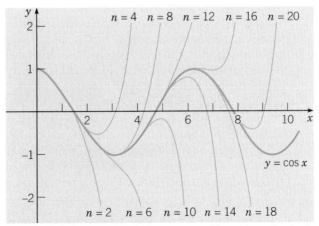

FIGURE 5.2.1 Polynomial approximations to $\cos x$. The value of n is the degree of the approximating polynomial.

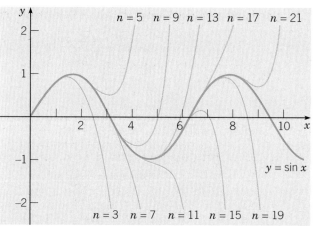

FIGURE 5.2.2 Polynomial approximations to $\sin x$. The value of n is the degree of the approximating polynomial.

In Example 1 we knew from the start that $\sin x$ and $\cos x$ form a fundamental set of solutions of Eq. (4). However, if we had not known this and had simply solved Eq. (4) using series methods, we would still have obtained the solution (11). In recognition of the fact that the differential equation (4) often occurs in applications, we might decide to give the two solutions of Eq. (11) special names, perhaps

$$C(x) = \sum_{n=0}^{\infty} \frac{(-1)^n}{(2n)!} x^{2n}, \qquad S(x) = \sum_{n=0}^{\infty} \frac{(-1)^n}{(2n+1)!} x^{2n+1}.$$

Then we might ask what properties these functions have. For instance, it follows at once from the series expansions that $C(0) = 1$, $S(0) = 0$, $C(-x) = C(x)$, and $S(-x) = -S(x)$. It is also easy to show that

$$S'(x) = C(x), \qquad C'(x) = -S(x).$$

Moreover, by calculating with the infinite series,[3] we can show that the functions $C(x)$ and $S(x)$ have all the usual analytical and algebraic properties of the cosine and sine functions, respectively.

Although you probably first saw the sine and cosine functions defined in a more elementary manner in terms of right triangles, it is interesting that these functions can be defined as solutions of a certain simple second order linear differential equation. To be precise, the function $\sin x$ can be defined as the unique solution of the initial value problem $y'' + y = 0$, $y(0) = 0$, $y'(0) = 1$; similarly, $\cos x$ can be defined as the unique solution of the initial value problem $y'' + y = 0$, $y(0) = 1$, $y'(0) = 0$. Many other functions that are important in mathematical physics are also defined as solutions of certain initial value problems. For most of these functions there is no simpler or more elementary way to approach them.

[3]Such an analysis is given in Section 24 of K. Knopp, *Theory and Applications of Infinite Series* (New York: Hafner, 1951).

<div style="float:left">**EXAMPLE**

2</div>

Find a series solution in powers of x of Airy's[4] equation

$$y'' - xy = 0, \qquad -\infty < x < \infty. \tag{12}$$

For this equation $P(x) = 1, Q(x) = 0$, and $R(x) = -x$; hence every point is an ordinary point. We assume that

$$y = \sum_{n=0}^{\infty} a_n x^n \tag{13}$$

and that the series converges in some interval $|x| < \rho$. The series for y'' is given by Eq. (7); as explained in the preceding example, we can rewrite it as

$$y'' = \sum_{n=0}^{\infty} (n+2)(n+1)a_{n+2}x^n. \tag{14}$$

Substituting the series (13) and (14) for y and y'' in Eq. (12), we obtain

$$\sum_{n=0}^{\infty} (n+2)(n+1)a_{n+2}x^n = x \sum_{n=0}^{\infty} a_n x^n = \sum_{n=0}^{\infty} a_n x^{n+1}. \tag{15}$$

Next, we shift the index of summation in the series on the right side of this equation by replacing n by $n - 1$ and starting the summation at 1 rather than zero. Thus we have

$$2 \cdot 1 a_2 + \sum_{n=1}^{\infty} (n+2)(n+1)a_{n+2}x^n = \sum_{n=1}^{\infty} a_{n-1}x^n.$$

Again, for this equation to be satisfied for all x, it is necessary that the coefficients of like powers of x be equal; hence $a_2 = 0$, and we obtain the recurrence relation

$$(n+2)(n+1)a_{n+2} = a_{n-1} \qquad \text{for} \quad n = 1, 2, 3, \ldots. \tag{16}$$

Since a_{n+2} is given in terms of a_{n-1}, the a's are determined in steps of three. Thus a_0 determines a_3, which in turn determines a_6, \ldots; a_1 determines a_4, which in turn determines a_7, \ldots; and a_2 determines a_5, which in turn determines a_8, \ldots. Since $a_2 = 0$, we immediately conclude that $a_5 = a_8 = a_{11} = \cdots = 0$.

For the sequence $a_0, a_3, a_6, a_9, \ldots$ we set $n = 1, 4, 7, 10, \ldots$ in the recurrence relation:

$$a_3 = \frac{a_0}{2 \cdot 3}, \qquad a_6 = \frac{a_3}{5 \cdot 6} = \frac{a_0}{2 \cdot 3 \cdot 5 \cdot 6}, \qquad a_9 = \frac{a_6}{8 \cdot 9} = \frac{a_0}{2 \cdot 3 \cdot 5 \cdot 6 \cdot 8 \cdot 9}, \ldots.$$

These results suggest the general formula

$$a_{3n} = \frac{a_0}{2 \cdot 3 \cdot 5 \cdot 6 \cdots (3n-1)(3n)}, \qquad n \geq 4.$$

For the sequence $a_1, a_4, a_7, a_{10}, \ldots$, we set $n = 2, 5, 8, 11, \ldots$ in the recurrence relation:

$$a_4 = \frac{a_1}{3 \cdot 4}, \qquad a_7 = \frac{a_4}{6 \cdot 7} = \frac{a_1}{3 \cdot 4 \cdot 6 \cdot 7}, \qquad a_{10} = \frac{a_7}{9 \cdot 10} = \frac{a_1}{3 \cdot 4 \cdot 6 \cdot 7 \cdot 9 \cdot 10}, \ldots.$$

[4]Sir George Biddell Airy (1801–1892), an English astronomer and mathematician, was director of the Greenwich Observatory from 1835 to 1881. One reason why Airy's equation is of interest is that for x negative the solutions are oscillatory, similar to trigonometric functions, and for x positive they are monotonic, similar to hyperbolic functions. Can you explain why it is reasonable to expect such behavior?

In general, we have

$$a_{3n+1} = \frac{a_1}{3 \cdot 4 \cdot 6 \cdot 7 \cdots (3n)(3n+1)}, \qquad n \geq 4.$$

Thus the general solution of Airy's equation is

$$y = a_0 \left[1 + \frac{x^3}{2 \cdot 3} + \frac{x^6}{2 \cdot 3 \cdot 5 \cdot 6} + \cdots + \frac{x^{3n}}{2 \cdot 3 \cdots (3n-1)(3n)} + \cdots \right]$$

$$+ a_1 \left[x + \frac{x^4}{3 \cdot 4} + \frac{x^7}{3 \cdot 4 \cdot 6 \cdot 7} + \cdots + \frac{x^{3n+1}}{3 \cdot 4 \cdots (3n)(3n+1)} + \cdots \right]. \tag{17}$$

Having obtained these two series solutions, we can now investigate their convergence. Because of the rapid growth of the denominators of the terms in the series (17), we might expect these series to have a large radius of convergence. Indeed, it is easy to use the ratio test to show that both these series converge for all x; see Problem 20.

Assuming for the moment that the series do converge for all x, let y_1 and y_2 denote the functions defined by the expressions in the first and second sets of brackets, respectively, in Eq. (17). Then, by choosing first $a_0 = 1$, $a_1 = 0$ and then $a_0 = 0$, $a_1 = 1$, it follows that y_1 and y_2 are individually solutions of Eq. (12). Notice that y_1 satisfies the initial conditions $y_1(0) = 1$, $y_1'(0) = 0$ and that y_2 satisfies the initial conditions $y_2(0) = 0$, $y_2'(0) = 1$. Thus $W(y_1, y_2)(0) = 1 \neq 0$, and consequently y_1 and y_2 are linearly independent. Hence the general solution of Airy's equation is

$$y = a_0 y_1(x) + a_1 y_2(x), \qquad -\infty < x < \infty.$$

In Figures 5.2.3 and 5.2.4, respectively, we show the graphs of the solutions y_1 and y_2 of Airy's equation, as well as graphs of several partial sums of the two series in Eq. (17). Again, the partial sums provide local approximations to the solutions in a neighborhood of the origin. Although the quality of the approximation improves as the number of terms increases, no polynomial can adequately represent y_1 and y_2 for large $|x|$. A practical way to estimate the interval in which a given partial sum is reasonably accurate is to compare the graphs of that partial sum and the next one, obtained by including one more term. As soon as the graphs

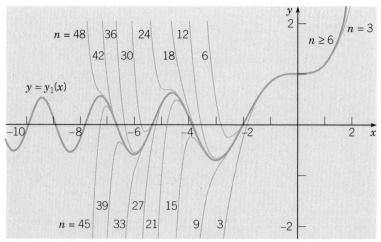

FIGURE 5.2.3 Polynomial approximations to the solution $y_1(x)$ of Airy's equation. The value of n is the degree of the approximating polynomial.

begin to separate noticeably, we can be confident that the original partial sum is no longer accurate. For example, in Figure 5.2.3 the graphs for $n = 24$ and $n = 27$ begin to separate at about $x = -9/2$. Thus, beyond this point, the partial sum of degree 24 is worthless as an approximation to the solution.

FIGURE 5.2.4 Polynomial approximations to the solution $y_2(x)$ of Airy's equation. The value of n is the degree of the approximating polynomial.

Observe that both y_1 and y_2 are monotone for $x > 0$ and oscillatory for $x < 0$. One can also see from the figures that the oscillations are not uniform but, rather, decay in amplitude and increase in frequency as the distance from the origin increases. In contrast to Example 1, the solutions y_1 and y_2 of Airy's equation are not elementary functions that you have already encountered in calculus. However, because of their importance in some physical applications, these functions have been extensively studied, and their properties are well known to applied mathematicians and scientists.

EXAMPLE

3

Find a solution of Airy's equation in powers of $x - 1$.

The point $x = 1$ is an ordinary point of Eq. (12), and thus we look for a solution of the form

$$y = \sum_{n=0}^{\infty} a_n(x - 1)^n,$$

where we assume that the series converges in some interval $|x - 1| < \rho$. Then

$$y' = \sum_{n=1}^{\infty} na_n(x - 1)^{n-1} = \sum_{n=0}^{\infty} (n + 1)a_{n+1}(x - 1)^n,$$

and

$$y'' = \sum_{n=2}^{\infty} n(n - 1)a_n(x - 1)^{n-2} = \sum_{n=0}^{\infty} (n + 2)(n + 1)a_{n+2}(x - 1)^n.$$

Substituting for y and y'' in Eq. (12), we obtain

$$\sum_{n=0}^{\infty}(n+2)(n+1)a_{n+2}(x-1)^n = x\sum_{n=0}^{\infty}a_n(x-1)^n. \tag{18}$$

Now to equate the coefficients of like powers of $(x-1)$ we must express x, the coefficient of y in Eq. (12), in powers of $x-1$; that is, we write $x = 1 + (x-1)$. Note that this is precisely the Taylor series for x about $x = 1$. Then Eq. (18) takes the form

$$\sum_{n=0}^{\infty}(n+2)(n+1)a_{n+2}(x-1)^n = [1+(x-1)]\sum_{n=0}^{\infty}a_n(x-1)^n$$

$$= \sum_{n=0}^{\infty}a_n(x-1)^n + \sum_{n=0}^{\infty}a_n(x-1)^{n+1}.$$

Shifting the index of summation in the second series on the right gives

$$\sum_{n=0}^{\infty}(n+2)(n+1)a_{n+2}(x-1)^n = \sum_{n=0}^{\infty}a_n(x-1)^n + \sum_{n=1}^{\infty}a_{n-1}(x-1)^n.$$

Equating coefficients of like powers of $x-1$, we obtain

$$2a_2 = a_0,$$
$$(3\cdot2)a_3 = a_1 + a_0,$$
$$(4\cdot3)a_4 = a_2 + a_1,$$
$$(5\cdot4)a_5 = a_3 + a_2,$$
$$\vdots$$

The general recurrence relation is

$$(n+2)(n+1)a_{n+2} = a_n + a_{n-1} \qquad \text{for} \quad n \geq 1. \tag{19}$$

Solving for the first few coefficients a_n in terms of a_0 and a_1, we find that

$$a_2 = \frac{a_0}{2}, \qquad a_3 = \frac{a_1}{6} + \frac{a_0}{6}, \qquad a_4 = \frac{a_2}{12} + \frac{a_1}{12} = \frac{a_0}{24} + \frac{a_1}{12}, \qquad a_5 = \frac{a_3}{20} + \frac{a_2}{20} = \frac{a_0}{30} + \frac{a_1}{120}.$$

Hence

$$y = a_0\left[1 + \frac{(x-1)^2}{2} + \frac{(x-1)^3}{6} + \frac{(x-1)^4}{24} + \frac{(x-1)^5}{30} + \cdots\right]$$

$$+ a_1\left[(x-1) + \frac{(x-1)^3}{6} + \frac{(x-1)^4}{12} + \frac{(x-1)^5}{120} + \cdots\right]. \tag{20}$$

In general, when the recurrence relation has more than two terms, as in Eq. (19), the determination of a formula for a_n in terms a_0 and a_1 will be fairly complicated, if not impossible. In this example such a formula is not readily apparent. Lacking such a formula, we cannot test the two series in Eq. (20) for convergence by direct methods such as the ratio test. However, we shall see in Section 5.3 that even without knowing the formula for a_n, it is possible to establish that the series in Eq. (20) converge for all x. Further, they define functions y_3 and y_4 that are linearly independent solutions of the Airy equation (12). Thus

$$y = a_0 y_3(x) + a_1 y_4(x)$$

is the general solution of Airy's equation for $-\infty < x < \infty$.

It is worth emphasizing, as we saw in Example 3, that if we look for a solution of Eq. (1) of the form $y = \sum\limits_{n=0}^{\infty} a_n(x - x_0)^n$, then the coefficients $P(x)$, $Q(x)$, and $R(x)$ in Eq. (1) must also be expressed in powers of $x - x_0$. Alternatively, we can make the change of variable $x - x_0 = t$, obtaining a new differential equation for y as a function of t, and then look for solutions of this new equation of the form $\sum\limits_{n=0}^{\infty} a_n t^n$.

When we have finished the calculations, we replace t by $x - x_0$ (see Problem 19).

In Examples 2 and 3 we have found two sets of solutions of Airy's equation. The functions y_1 and y_2 defined by the series in Eq. (17) are linearly independent solutions of Eq. (12) for all x, and this is also true for the functions y_3 and y_4 defined by the series in Eq. (20). According to the general theory of second order linear equations each of the first two functions can be expressed as a linear combination of the latter two functions and vice versa—a result that is certainly not obvious from an examination of the series alone.

Finally, we emphasize that it is not particularly important if, as in Example 3, we are unable to determine the general coefficient a_n in terms of a_0 and a_1. What is essential is that we can determine *as many coefficients as we want*. Thus we can find as many terms in the two series solutions as we want, even if we cannot determine the general term. While the task of calculating several coefficients in a power series solution is not difficult, it can be tedious. A symbolic manipulation package can be very helpful here; some are able to find a specified number of terms in a power series solution in response to a single command. With a suitable graphics package one can also produce plots such as those shown in the figures in this section.

PROBLEMS

In each of Problems 1 through 14 solve the given differential equation by means of a power series about the given point x_0. Find the recurrence relation; also find the first four terms in each of two linearly independent solutions (unless the series terminates sooner). If possible, find the general term in each solution.

1. $y'' - y = 0$, $x_0 = 0$

2. $y'' - xy' - y = 0$, $x_0 = 0$

3. $y'' - xy' - y = 0$, $x_0 = 1$

4. $y'' + k^2 x^2 y = 0$, $x_0 = 0$, k a constant

5. $(1 - x)y'' + y = 0$, $x_0 = 0$

6. $(2 + x^2)y'' - xy' + 4y = 0$, $x_0 = 0$

7. $y'' + xy' + 2y = 0$, $x_0 = 0$

8. $xy'' + y' + xy = 0$, $x_0 = 1$

9. $(1 + x^2)y'' - 4xy' + 6y = 0$, $x_0 = 0$

10. $(4 - x^2)y'' + 2y = 0$, $x_0 = 0$

11. $(3 - x^2)y'' - 3xy' - y = 0$, $x_0 = 0$

12. $(1 - x)y'' + xy' - y = 0$, $x_0 = 0$

13. $2y'' + xy' + 3y = 0$, $x_0 = 0$

14. $2y'' + (x + 1)y' + 3y = 0$, $x_0 = 2$

In each of Problems 15 through 18:

(a) Find the first five nonzero terms in the solution of the given initial value problem.

(b) Plot the four-term and the five-term approximations to the solution on the same axes.

(c) From the plot in part (b) estimate the interval in which the four-term approximation is reasonably accurate.

15. $y'' - xy' - y = 0$, $y(0) = 2$, $y'(0) = 1$; see Problem 2

16. $(2 + x^2)y'' - xy' + 4y = 0$, $y(0) = -1$, $y'(0) = 3$; see Problem 6

17. $y'' + xy' + 2y = 0,$ $y(0) = 4,$ $y'(0) = -1;$ see Problem 7

18. $(1 - x)y'' + xy' - y = 0,$ $y(0) = -3,$ $y'(0) = 2;$ see Problem 12

19. By making the change of variable $x - 1 = t$ and assuming that y has a Taylor series in powers of t, find two linearly independent series solutions of

$$y'' + (x - 1)^2 y' + (x^2 - 1)y = 0$$

in powers of $x - 1$. Show that you obtain the same result directly by assuming that y has a Taylor series in powers of $x - 1$ and also expressing the coefficient $x^2 - 1$ in powers of $x - 1$.

20. Show directly, using the ratio test, that the two series solutions of Airy's equation about $x = 0$ converge for all x; see Eq. (17) of the text.

21. **The Hermite Equation.** The equation

$$y'' - 2xy' + \lambda y = 0, \qquad -\infty < x < \infty,$$

where λ is a constant, is known as the Hermite[5] equation. It is an important equation in mathematical physics.

(a) Find the first four terms in each of two linearly independent solutions about $x = 0$.

(b) Observe that if λ is a nonnegative even integer, then one or the other of the series solutions terminates and becomes a polynomial. Find the polynomial solutions for $\lambda = 0$, 2, 4, 6, 8, and 10. Note that each polynomial is determined only up to a multiplicative constant.

(c) The Hermite polynomial $H_n(x)$ is defined as the polynomial solution of the Hermite equation with $\lambda = 2n$ for which the coefficient of x^n is 2^n. Find $H_0(x), \ldots, H_5(x)$.

22. Consider the initial value problem $y' = \sqrt{1 - y^2}, y(0) = 0$.

(a) Show that $y = \sin x$ is the solution of this initial value problem.

(b) Look for a solution of the initial value problem in the form of a power series about $x = 0$. Find the coefficients up to the term in x^3 in this series.

In each of Problems 23 through 28 plot several partial sums in a series solution of the given initial value problem about $x = 0$, thereby obtaining graphs analogous to those in Figures 5.2.1 through 5.2.4.

23. $y'' - xy' - y = 0,$ $y(0) = 1,$ $y'(0) = 0;$ see Problem 2

24. $(2 + x^2)y'' - xy' + 4y = 0,$ $y(0) = 1,$ $y'(0) = 0;$ see Problem 6

25. $y'' + xy' + 2y = 0,$ $y(0) = 0,$ $y'(0) = 1;$ see Problem 7

26. $(4 - x^2)y'' + 2y = 0,$ $y(0) = 0,$ $y'(0) = 1;$ see Problem 10

27. $y'' + x^2 y = 0,$ $y(0) = 1,$ $y'(0) = 0;$ see Problem 4

28. $(1 - x)y'' + xy' - 2y = 0,$ $y(0) = 0,$ $y'(0) = 1$

[5] Charles Hermite (1822–1901) was an influential French analyst and algebraist. He introduced the Hermite functions in 1864 and showed in 1873 that e is a transcendental number (that is, e is not a root of any polynomial equation with rational coefficients). His name is also associated with Hermitian matrices (see Section 7.3), some of whose properties he discovered.

5.3 Series Solutions Near an Ordinary Point, Part II

In the preceding section we considered the problem of finding solutions of

$$P(x)y'' + Q(x)y' + R(x)y = 0, \tag{1}$$

where P, Q, and R are polynomials, in the neighborhood of an ordinary point x_0. Assuming that Eq. (1) does have a solution $y = \phi(x)$ and that ϕ has a Taylor series

$$y = \phi(x) = \sum_{n=0}^{\infty} a_n(x - x_0)^n, \tag{2}$$

which converges for $|x - x_0| < \rho$, where $\rho > 0$, we found that the a_n can be determined by directly substituting the series (2) for y in Eq. (1).

Let us now consider how we might justify the statement that if x_0 is an ordinary point of Eq. (1), then there exist solutions of the form (2). We also consider the question of the radius of convergence of such a series. In doing this we are led to a generalization of the definition of an ordinary point.

Suppose, then, that there is a solution of Eq. (1) of the form (2). By differentiating Eq. (2) m times and setting x equal to x_0, it follows that

$$m!a_m = \phi^{(m)}(x_0).$$

Hence, to compute a_n in the series (2), we must show that we can determine $\phi^{(n)}(x_0)$ for $n = 0, 1, 2, \ldots$ from the differential equation (1).

Suppose that $y = \phi(x)$ is a solution of Eq. (1) satisfying the initial conditions $y(x_0) = y_0$, $y'(x_0) = y_0'$. Then $a_0 = y_0$ and $a_1 = y_0'$. If we are solely interested in finding a solution of Eq. (1) without specifying any initial conditions, then a_0 and a_1 remain arbitrary. To determine $\phi^{(n)}(x_0)$ and the corresponding a_n for $n = 2, 3, \ldots$, we turn to Eq. (1). Since ϕ is a solution of Eq. (1), we have

$$P(x)\phi''(x) + Q(x)\phi'(x) + R(x)\phi(x) = 0.$$

For the interval about x_0 for which P is nonvanishing, we can write this equation in the form

$$\phi''(x) = -p(x)\phi'(x) - q(x)\phi(x), \tag{3}$$

where $p(x) = Q(x)/P(x)$ and $q(x) = R(x)/P(x)$. Setting x equal to x_0 in Eq. (3) gives

$$\phi''(x_0) = -p(x_0)\phi'(x_0) - q(x_0)\phi(x_0).$$

Hence a_2 is given by

$$2!a_2 = \phi''(x_0) = -p(x_0)a_1 - q(x_0)a_0. \tag{4}$$

To determine a_3 we differentiate Eq. (3) and then set x equal to x_0, obtaining

$$3!a_3 = \phi'''(x_0) = -[p\phi'' + (p' + q)\phi' + q'\phi]\Big|_{x=x_0}$$

$$= -2!p(x_0)a_2 - [p'(x_0) + q(x_0)]a_1 - q'(x_0)a_0. \tag{5}$$

Substituting for a_2 from Eq. (4) gives a_3 in terms of a_1 and a_0. Since P, Q, and R are polynomials and $P(x_0) \neq 0$, all the derivatives of p and q exist at x_0. Hence, we can

continue to differentiate Eq. (3) indefinitely, determining after each differentiation the successive coefficients a_4, a_5, \ldots by setting x equal to x_0.

Notice that the important property that we used in determining the a_n was that we could compute infinitely many derivatives of the functions p and q. It might seem reasonable to relax our assumption that the functions p and q are ratios of polynomials and simply require that they be infinitely differentiable in the neighborhood of x_0. Unfortunately, this condition is too weak to ensure that we can prove the convergence of the resulting series expansion for $y = \phi(x)$. What is needed is to assume that the functions p and q are *analytic* at x_0; that is, they have Taylor series expansions that converge to them in some interval about the point x_0:

$$p(x) = p_0 + p_1(x - x_0) + \cdots + p_n(x - x_0)^n + \cdots = \sum_{n=0}^{\infty} p_n(x - x_0)^n, \qquad (6)$$

$$q(x) = q_0 + q_1(x - x_0) + \cdots + q_n(x - x_0)^n + \cdots = \sum_{n=0}^{\infty} q_n(x - x_0)^n. \qquad (7)$$

With this idea in mind, we can generalize the definitions of an ordinary point and a singular point of Eq. (1) as follows: If the functions $p = Q/P$ and $q = R/P$ are analytic at x_0, then the point x_0 is said to be an **ordinary point** of the differential equation (1); otherwise, it is a **singular point**.

Now let us turn to the question of the interval of convergence of the series solution. One possibility is actually to compute the series solution for each problem and then to apply one of the tests for convergence of an infinite series to determine its radius of convergence. However, the question can be answered at once for a wide class of problems by the following theorem.

Theorem 5.3.1 If x_0 is an ordinary point of the differential equation (1),

$$P(x)y'' + Q(x)y' + R(x)y = 0,$$

that is, if $p = Q/P$ and $q = R/P$ are analytic at x_0, then the general solution of Eq. (1) is

$$y = \sum_{n=0}^{\infty} a_n(x - x_0)^n = a_0 y_1(x) + a_1 y_2(x), \qquad (8)$$

where a_0 and a_1 are arbitrary, and y_1 and y_2 are linearly independent series solutions that are analytic at x_0. Further, the radius of convergence for each of the series solutions y_1 and y_2 is at least as large as the minimum of the radii of convergence of the series for p and q.

Notice from the form of the series solution that $y_1(x) = 1 + b_2(x - x_0)^2 + \cdots$ and $y_2(x) = (x - x_0) + c_2(x - x_0)^2 + \cdots$. Hence y_1 is the solution satisfying the initial conditions $y_1(x_0) = 1$, $y_1'(x_0) = 0$, and y_2 is the solution satisfying the initial conditions $y_2(x_0) = 0$, $y_2'(x_0) = 1$. Also note that although calculating the coefficients by successively differentiating the differential equation is excellent in theory, it is usually not a practical computational procedure. Rather, one should substitute the

series (2) for y in the differential equation (1) and determine the coefficients so that the differential equation is satisfied, as in the examples in the preceding section.

We will not prove this theorem, which in a slightly more general form was established by Fuchs.[6] What is important for our purposes is that there is a series solution of the form (2) and that the radius of convergence of the series solution cannot be less than the smaller of the radii of convergence of the series for p and q; hence we need only determine these.

This can be done in either of two ways. Again, one possibility is simply to compute the power series for p and q and then to determine the radii of convergence by using one of the convergence tests for infinite series. However, there is an easier way when P, Q, and R are polynomials. It is shown in the theory of functions of a complex variable that the ratio of two polynomials, say Q/P, has a convergent power series expansion about a point $x = x_0$ if $P(x_0) \neq 0$. Further, if we assume that any factors common to Q and P have been canceled, then the radius of convergence of the power series for Q/P about the point x_0 is precisely the distance from x_0 to the nearest zero of P. In determining this distance we must remember that $P(x) = 0$ may have complex roots, and these must also be considered.

EXAMPLE 1

What is the radius of convergence of the Taylor series for $(1 + x^2)^{-1}$ about $x = 0$?

One way to proceed is to find the Taylor series in question, namely

$$\frac{1}{1 + x^2} = 1 - x^2 + x^4 - x^6 + \cdots + (-1)^n x^{2n} + \cdots.$$

Then it can be verified by the ratio test that $\rho = 1$. Another approach is to note that the zeros of $1 + x^2$ are $x = \pm i$. Since the distance in the complex plane from 0 to i or to $-i$ is 1, the radius of convergence of the power series about $x = 0$ is 1.

EXAMPLE 2

What is the radius of convergence of the Taylor series for $(x^2 - 2x + 2)^{-1}$ about $x = 0$? About $x = 1$?

First notice that

$$x^2 - 2x + 2 = 0$$

has solutions $x = 1 \pm i$. The distance in the complex plane from $x = 0$ to either $x = 1 + i$ or $x = 1 - i$ is $\sqrt{2}$; hence the radius of convergence of the Taylor series expansion $\sum\limits_{n=0}^{\infty} a_n x^n$ about $x = 0$ is $\sqrt{2}$.

The distance in the complex plane from $x = 1$ to either $x = 1 + i$ or $x = 1 - i$ is 1; hence the radius of convergence of the Taylor series expansion $\sum\limits_{n=0}^{\infty} b_n (x - 1)^n$ about $x = 1$ is 1.

[6]Immanuel Lazarus Fuchs (1833–1902) was a student and later a professor at the University of Berlin. He proved the result of Theorem 5.3.1 in 1866. His most important research was on singular points of linear differential equations. He recognized the significance of regular singular points (Section 5.4), and equations whose only singularities, including the point at infinity, are regular singular points are known as Fuchsian equations.

According to Theorem 5.3.1 the series solutions of the Airy equation in Examples 2 and 3 of the preceding section converge for all values of x and $x - 1$, respectively, since in each problem $P(x) = 1$ and hence is never zero.

A series solution may converge for a wider range of x than indicated by Theorem 5.3.1, so the theorem actually gives only a lower bound on the radius of convergence of the series solution. This is illustrated by the Legendre polynomial solution of the Legendre equation given in the next example.

EXAMPLE 3

Determine a lower bound for the radius of convergence of series solutions about $x = 0$ for the Legendre equation

$$(1 - x^2)y'' - 2xy' + \alpha(\alpha + 1)y = 0,$$

where α is a constant.

Note that $P(x) = 1 - x^2$, $Q(x) = -2x$, and $R(x) = \alpha(\alpha + 1)$ are polynomials, and that the zeros of P, namely $x = \pm 1$, are a distance 1 from $x = 0$. Hence a series solution of the form $\sum_{n=0}^{\infty} a_n x^n$ converges for $|x| < 1$ at least, and possibly for larger values of x. Indeed, it can be shown that if α is a positive integer, one of the series solutions terminates after a finite number of terms and hence converges not just for $|x| < 1$ but for all x. For example, if $\alpha = 1$, the polynomial solution is $y = x$. See Problems 22 through 29 at the end of this section for a more comprehensive discussion of the Legendre equation.

EXAMPLE 4

Determine a lower bound for the radius of convergence of series solutions of the differential equation

$$(1 + x^2)y'' + 2xy' + 4x^2 y = 0 \tag{9}$$

about the point $x = 0$; about the point $x = -\frac{1}{2}$.

Again P, Q, and R are polynomials, and P has zeros at $x = \pm i$. The distance in the complex plane from 0 to $\pm i$ is 1, and from $-\frac{1}{2}$ to $\pm i$ is $\sqrt{1 + \frac{1}{4}} = \sqrt{5}/2$. Hence in the first case the series $\sum_{n=0}^{\infty} a_n x^n$ converges at least for $|x| < 1$, and in the second case the series $\sum_{n=0}^{\infty} b_n \left(x + \frac{1}{2}\right)^n$ converges at least for $|x + \frac{1}{2}| < \sqrt{5}/2$.

An interesting observation that we can make about Eq. (9) follows from Theorems 3.2.1 and 5.3.1. Suppose that initial conditions $y(0) = y_0$ and $y'(0) = y_0'$ are given. Since $1 + x^2 \neq 0$ for all x, we know from Theorem 3.2.1 that there exists a unique solution of the initial value problem on $-\infty < x < \infty$. On the other hand, Theorem 5.3.1 only guarantees a series solution of the form $\sum_{n=0}^{\infty} a_n x^n$ (with $a_0 = y_0, a_1 = y_0'$) for $-1 < x < 1$. The unique solution on the interval $-\infty < x < \infty$ may not have a power series about $x = 0$ that converges for all x.

EXAMPLE 5

Can we determine a series solution about $x = 0$ for the differential equation

$$y'' + (\sin x)y' + (1 + x^2)y = 0,$$

and if so, what is the radius of convergence?

For this differential equation, $p(x) = \sin x$ and $q(x) = 1 + x^2$. Recall from calculus that $\sin x$ has a Taylor series expansion about $x = 0$ that converges for all x. Further, q also has a Taylor series expansion about $x = 0$, namely $q(x) = 1 + x^2$, that converges for all x. Thus

there is a series solution of the form $y = \sum\limits_{n=0}^{\infty} a_n x^n$ with a_0 and a_1 arbitrary, and the series converges for all x.

PROBLEMS

In each of Problems 1 through 4 determine $\phi''(x_0)$, $\phi'''(x_0)$, and $\phi^{(4)}(x_0)$ for the given point x_0 if $y = \phi(x)$ is a solution of the given initial value problem.

1. $y'' + xy' + y = 0$; $y(0) = 1$, $y'(0) = 0$

2. $y'' + (\sin x)y' + (\cos x)y = 0$; $y(0) = 0$, $y'(0) = 1$

3. $x^2 y'' + (1 + x)y' + 3(\ln x)y = 0$; $y(1) = 2$, $y'(1) = 0$

4. $y'' + x^2 y' + (\sin x)y = 0$; $y(0) = a_0$, $y'(0) = a_1$

In each of Problems 5 through 8 determine a lower bound for the radius of convergence of series solutions about each given point x_0 for the given differential equation.

5. $y'' + 4y' + 6xy = 0$; $x_0 = 0$, $x_0 = 4$

6. $(x^2 - 2x - 3)y'' + xy' + 4y = 0$; $x_0 = 4$, $x_0 = -4$, $x_0 = 0$

7. $(1 + x^3)y'' + 4xy' + y = 0$; $x_0 = 0$, $x_0 = 2$

8. $xy'' + y = 0$; $x_0 = 1$

9. Determine a lower bound for the radius of convergence of series solutions about the given x_0 for each of the differential equations in Problems 1 through 14 of Section 5.2.

10. **The Chebyshev Equation.** The Chebyshev[7] differential equation is

$$(1 - x^2)y'' - xy' + \alpha^2 y = 0,$$

where α is a constant.

(a) Determine two linearly independent solutions in powers of x for $|x| < 1$.

(b) Show that if α is a nonnegative integer n, then there is a polynomial solution of degree n. These polynomials, when properly normalized, are called the Chebyshev polynomials. They are very useful in problems that require a polynomial approximation to a function defined on $-1 \le x \le 1$.

(c) Find a polynomial solution for each of the cases $\alpha = n = 0, 1, 2, 3$.

For each of the differential equations in Problems 11 through 14 find the first four nonzero terms in each of two linearly independent power series solutions about the origin. What do you expect the radius of convergence to be for each solution?

11. $y'' + (\sin x)y = 0$ 12. $e^x y'' + xy = 0$

13. $(\cos x)y'' + xy' - 2y = 0$ 14. $e^{-x}y'' + \ln(1 + x)y' - xy = 0$

15. Suppose that you are told that x and x^2 are solutions of a differential equation $P(x)y'' + Q(x)y' + R(x)y = 0$. Can you say whether the point $x = 0$ is an ordinary point or a singular point?

Hint: Use Theorem 3.2.1, and note the values of x and x^2 at $x = 0$.

[7]Pafnuty L. Chebyshev (1821–1894), professor at Petersburg University for 35 years and the most influential nineteenth-century Russian mathematician, founded the so-called Petersburg school, which produced a long line of distinguished mathematicians. His study of Chebyshev polynomials began about 1854 as part of an investigation of the approximation of functions by polynomials. Chebyshev is also known for his work in number theory and probability.

First Order Equations. The series methods discussed in this section are directly applicable to the first order linear differential equation $P(x)y' + Q(x)y = 0$ at a point x_0, if the function $p = Q/P$ has a Taylor series expansion about that point. Such a point is called an ordinary point, and further, the radius of convergence of the series $y = \sum_{n=0}^{\infty} a_n(x - x_0)^n$ is at least as large as the radius of convergence of the series for Q/P. In each of Problems 16 through 21 solve the given differential equation by a series in powers of x and verify that a_0 is arbitrary in each case. Problems 20 and 21 involve nonhomogeneous differential equations to which series methods can be easily extended. Where possible, compare the series solution with the solution obtained by using the methods of Chapter 2.

16. $y' - y = 0$ 17. $y' - xy = 0$

18. $y' = e^{x^2}y$, three terms only 19. $(1 - x)y' = y$

20. $y' - y = x^2$ 21. $y' + xy = 1 + x$

The Legendre Equation. Problems 22 through 29 deal with the Legendre[8] equation

$$(1 - x^2)y'' - 2xy' + \alpha(\alpha + 1)y = 0.$$

As indicated in Example 3, the point $x = 0$ is an ordinary point of this equation, and the distance from the origin to the nearest zero of $P(x) = 1 - x^2$ is 1. Hence the radius of convergence of series solutions about $x = 0$ is at least 1. Also notice that we need to consider only $\alpha > -1$ because if $\alpha \leq -1$, then the substitution $\alpha = -(1 + \gamma)$, where $\gamma \geq 0$, leads to the Legendre equation $(1 - x^2)y'' - 2xy' + \gamma(\gamma + 1)y = 0$.

22. Show that two linearly independent solutions of the Legendre equation for $|x| < 1$ are

$$y_1(x) = 1 - \frac{\alpha(\alpha + 1)}{2!}x^2 + \frac{\alpha(\alpha - 2)(\alpha + 1)(\alpha + 3)}{4!}x^4$$

$$+ \sum_{m=3}^{\infty}(-1)^m\frac{\alpha \cdots (\alpha - 2m + 2)(\alpha + 1)\cdots(\alpha + 2m - 1)}{(2m)!}x^{2m},$$

$$y_2(x) = x - \frac{(\alpha - 1)(\alpha + 2)}{3!}x^3 + \frac{(\alpha - 1)(\alpha - 3)(\alpha + 2)(\alpha + 4)}{5!}x^5$$

$$+ \sum_{m=3}^{\infty}(-1)^m\frac{(\alpha - 1)\cdots(\alpha - 2m + 1)(\alpha + 2)\cdots(\alpha + 2m)}{(2m + 1)!}x^{2m+1}.$$

23. Show that, if α is zero or a positive even integer $2n$, the series solution y_1 reduces to a polynomial of degree $2n$ containing only even powers of x. Find the polynomials corresponding to $\alpha = 0, 2$, and 4. Show that, if α is a positive odd integer $2n + 1$, the series solution y_2 reduces to a polynomial of degree $2n + 1$ containing only odd powers of x. Find the polynomials corresponding to $\alpha = 1, 3$, and 5.

24. The Legendre polynomial $P_n(x)$ is defined as the polynomial solution of the Legendre equation with $\alpha = n$ that also satisfies the condition $P_n(1) = 1$.

 (a) Using the results of Problem 23, find the Legendre polynomials $P_0(x), \ldots, P_5(x)$.

 (b) Plot the graphs of $P_0(x), \ldots, P_5(x)$ for $-1 \leq x \leq 1$.

 (c) Find the zeros of $P_0(x), \ldots, P_5(x)$.

[8]Adrien-Marie Legendre (1752–1833) held various positions in the French Académie des Sciences from 1783 onward. His primary work was in the fields of elliptic functions and number theory. The Legendre functions, solutions of Legendre's equation, first appeared in 1784 in his study of the attraction of spheroids.

25. It can be shown that the general formula for $P_n(x)$ is

$$P_n(x) = \frac{1}{2^n} \sum_{k=0}^{[n/2]} \frac{(-1)^k (2n-2k)!}{k!(n-k)!(n-2k)!} x^{n-2k},$$

where $[n/2]$ denotes the greatest integer less than or equal to $n/2$. By observing the form of $P_n(x)$ for n even and n odd, show that $P_n(-1) = (-1)^n$.

26. The Legendre polynomials play an important role in mathematical physics. For example, in solving Laplace's equation (the potential equation) in spherical coordinates, we encounter the equation

$$\frac{d^2 F(\varphi)}{d\varphi^2} + \cot \varphi \frac{dF(\varphi)}{d\varphi} + n(n+1)F(\varphi) = 0, \qquad 0 < \varphi < \pi,$$

where n is a positive integer. Show that the change of variable $x = \cos \varphi$ leads to the Legendre equation with $\alpha = n$ for $y = f(x) = F(\arccos x)$.

27. Show that for $n = 0, 1, 2, 3$, the corresponding Legendre polynomial is given by

$$P_n(x) = \frac{1}{2^n n!} \frac{d^n}{dx^n} (x^2 - 1)^n.$$

This formula, known as Rodrigues' formula,[9] is true for all positive integers n.

28. Show that the Legendre equation can also be written as

$$[(1-x^2)y']' = -\alpha(\alpha+1)y.$$

Then it follows that

$$[(1-x^2)P_n'(x)]' = -n(n+1)P_n(x) \quad \text{and} \quad [(1-x^2)P_m'(x)]' = -m(m+1)P_m(x).$$

By multiplying the first equation by $P_m(x)$ and the second equation by $P_n(x)$, integrating by parts, and then subtracting one equation from the other, show that

$$\int_{-1}^{1} P_n(x)P_m(x)\, dx = 0 \qquad \text{if} \quad n \neq m.$$

This property of the Legendre polynomials is known as the orthogonality property. If $m = n$, it can be shown that the value of the preceding integral is $2/(2n+1)$.

29. Given a polynomial f of degree n, it is possible to express f as a linear combination of $P_0, P_1, P_2, \ldots, P_n$:

$$f(x) = \sum_{k=0}^{n} a_k P_k(x).$$

Using the result of Problem 28, show that

$$a_k = \frac{2k+1}{2} \int_{-1}^{1} f(x)P_k(x)\, dx.$$

[9] Olinde Rodrigues (1794–1851) published this result as part of his doctoral thesis from the École Normale in Paris in 1816. He later became a banker and social reformer.

5.4 Regular Singular Points

In this section we will consider the equation

$$P(x)y'' + Q(x)y' + R(x)y = 0 \qquad (1)$$

in the neighborhood of a singular point x_0. Recall that if the functions P, Q, and R are polynomials having no common factors, the singular points of Eq. (1) are the points for which $P(x) = 0$.

EXAMPLE 1

Determine the singular points and ordinary points of the Bessel equation of order ν

$$x^2 y'' + xy' + (x^2 - \nu^2)y = 0. \qquad (2)$$

The point $x = 0$ is a singular point since $P(x) = x^2$ is zero there. All other points are ordinary points of Eq. (2).

EXAMPLE 2

Determine the singular points and ordinary points of the Legendre equation

$$(1 - x^2)y'' - 2xy' + \alpha(\alpha + 1)y = 0, \qquad (3)$$

where α is a constant.

The singular points are the zeros of $P(x) = 1 - x^2$, namely, the points $x = \pm 1$. All other points are ordinary points.

Unfortunately, if we attempt to use the methods of the preceding two sections to solve Eq. (1) in the neighborhood of a singular point x_0, we find that these methods fail. This is because the solution of Eq. (1) is often not analytic at x_0 and consequently cannot be represented by a Taylor series in powers of $x - x_0$. Instead, we must use a more general series expansion.

Since the singular points of a differential equation are usually few in number, we might ask whether we can simply ignore them, especially since we already know how to construct solutions about ordinary points. However, this is not feasible because the singular points determine the principal features of the solution to a much larger extent than one might at first suspect. In the neighborhood of a singular point the solution often becomes large in magnitude or experiences rapid changes in magnitude. Thus the behavior of a physical system modeled by a differential equation frequently is most interesting in the neighborhood of a singular point. Often geometric singularities in a physical problem, such as corners or sharp edges, lead to singular points in the corresponding differential equation. Thus, although at first we might want to avoid the few points where a differential equation is singular, it is precisely at these points that it is necessary to study the solution most carefully.

As an alternative to analytical methods, one can consider the use of numerical methods, which are discussed in Chapter 8. However, these methods are ill suited for the study of solutions near a singular point. Thus, even if one adopts a numerical approach, it is advantageous to combine it with the analytical methods of this chapter in order to examine the behavior of solutions near singular points.

Without any additional information about the behavior of Q/P and R/P in the neighborhood of the singular point, it is impossible to describe the behavior of the solutions of Eq. (1) near $x = x_0$. It may be that there are two linearly independent solutions of Eq. (1) that remain bounded as $x \to x_0$; or there may be only one, with the other becoming unbounded as $x \to x_0$; or they may both become unbounded as $x \to x_0$. Examples 3, 4, and 5 below illustrate these possibilities. If Eq. (1) has solutions that become unbounded as $x \to x_0$, it is often important to determine how these solutions behave as $x \to x_0$. For example, does $y \to \infty$ in the same manner as $(x - x_0)^{-1}$ or $|x - x_0|^{-1/2}$, or in some other manner?

EXAMPLE 3

The differential equation

$$x^2 y'' - 2y = 0 \tag{4}$$

has a singular point at $x = 0$. It can be easily verified by direct substitution that $y_1(x) = x^2$ and $y_2(x) = 1/x$ are linearly independent solutions of Eq. (4) for $x > 0$ or $x < 0$. Thus, in any interval not containing the origin, the general solution of Eq. (4) is $y = c_1 x^2 + c_2 x^{-1}$. The only solution of Eq. (4) that is bounded as $x \to 0$ is $y = c_1 x^2$. Indeed, this solution is analytic at the origin even though if Eq. (4) is put in the standard form, $y'' - (2/x^2)y = 0$, the function $q(x) = -2/x^2$ is not analytic at $x = 0$ and Theorem 5.3.1 is not applicable. On the other hand, notice that the solution $y_2(x) = x^{-1}$ does not have a Taylor series expansion about the origin (is not analytic at $x = 0$); therefore, the method of Section 5.2 would fail in this case.

EXAMPLE 4

The differential equation

$$x^2 y'' - 2xy' + 2y = 0 \tag{5}$$

also has a singular point at $x = 0$. It can be verified that $y_1(x) = x$ and $y_2(x) = x^2$ are linearly independent solutions of Eq. (5), and both are analytic at $x = 0$. Nevertheless, it is still not proper to pose an initial value problem with initial conditions at $x = 0$. It is impossible to satisfy arbitrary initial conditions at $x = 0$, since any linear combination of x and x^2 is zero at $x = 0$.

EXAMPLE 5

The differential equation

$$x^2 y'' + 5xy' + 3y = 0 \tag{6}$$

also has a singular point at $x = 0$. You can verify that $y_1(x) = 1/x$ and $y_2(x) = 1/x^3$ are linearly independent solutions of Eq. (6) and that neither is analytic at $x = 0$. Indeed, every (nonzero) solution of Eq. (6) becomes unbounded as $x \to 0$.

Our goal is to extend the method already developed for solving Eq. (1) near an ordinary point so that it also applies to the neighborhood of a singular point x_0. To do this in a reasonably simple manner, it is necessary to restrict ourselves to cases in which the singularities in the functions Q/P and R/P at $x = x_0$ are not too severe— that is, to what we might call "weak singularities." At this stage it is not clear exactly what is an acceptable singularity. However, as we develop the method of solution,

you will see that the appropriate conditions (see also Section 5.7, Problem 21) to distinguish "weak singularities" are

$$\lim_{x \to x_0} (x - x_0) \frac{Q(x)}{P(x)} \quad \text{is finite} \tag{7}$$

and

$$\lim_{x \to x_0} (x - x_0)^2 \frac{R(x)}{P(x)} \quad \text{is finite.} \tag{8}$$

This means that the singularity in Q/P can be no worse than $(x - x_0)^{-1}$ and the singularity in R/P can be no worse than $(x - x_0)^{-2}$. Such a point is called a **regular singular point** of Eq. (1). For more general functions than polynomials, x_0 is a regular singular point of Eq. (1) if it is a singular point and if both[10]

$$(x - x_0) \frac{Q(x)}{P(x)} \quad \text{and} \quad (x - x_0)^2 \frac{R(x)}{P(x)} \tag{9}$$

have convergent Taylor series about x_0—that is, if the functions in Eq. (9) are analytic at $x = x_0$. Equations (7) and (8) imply that this will be the case when P, Q, and R are polynomials. Any singular point of Eq. (1) that is not a regular singular point is called an **irregular singular point** of Eq. (1).

In the following sections we discuss how to solve Eq. (1) in the neighborhood of a regular singular point. A discussion of the solutions of differential equations in the neighborhood of irregular singular points is more complicated and may be found in more advanced books.

EXAMPLE

6

In Example 2 we observed that the singular points of the Legendre equation

$$(1 - x^2)y'' - 2xy' + \alpha(\alpha + 1)y = 0$$

are $x = \pm 1$. Determine whether these singular points are regular or irregular singular points.

We consider the point $x = 1$ first and also observe that when we divide by $1 - x^2$, the coefficients of y' and y are $-2x/(1 - x^2)$ and $\alpha(\alpha + 1)/(1 - x^2)$, respectively. Thus we calculate

$$\lim_{x \to 1}(x - 1)\frac{-2x}{1 - x^2} = \lim_{x \to 1} \frac{(x - 1)(-2x)}{(1 - x)(1 + x)} = \lim_{x \to 1} \frac{2x}{1 + x} = 1$$

and

$$\lim_{x \to 1}(x - 1)^2 \frac{\alpha(\alpha + 1)}{1 - x^2} = \lim_{x \to 1} \frac{(x - 1)^2 \alpha(\alpha + 1)}{(1 - x)(1 + x)}$$

$$= \lim_{x \to 1} \frac{(x - 1)(-\alpha)(\alpha + 1)}{1 + x} = 0.$$

Since these limits are finite, the point $x = 1$ is a regular singular point. It can be shown in a similar manner that $x = -1$ is also a regular singular point.

[10]The functions given in Eq. (9) may not be defined at x_0, in which case their values at x_0 are to be assigned as their limits as $x \to x_0$.

EXAMPLE 7

Determine the singular points of the differential equation

$$2x(x-2)^2 y'' + 3xy' + (x-2)y = 0$$

and classify them as regular or irregular.

Dividing the differential equation by $2x(x-2)^2$, we have

$$y'' + \frac{3}{2(x-2)^2}y' + \frac{1}{2x(x-2)}y = 0,$$

so $p(x) = Q(x)/P(x) = 3/2(x-2)^2$ and $q(x) = R(x)/P(x) = 1/2x(x-2)$. The singular points are $x = 0$ and $x = 2$. Consider $x = 0$. We have

$$\lim_{x \to 0} xp(x) = \lim_{x \to 0} x\frac{3}{2(x-2)^2} = 0,$$

$$\lim_{x \to 0} x^2 q(x) = \lim_{x \to 0} x^2 \frac{1}{2x(x-2)} = 0.$$

Since these limits are finite, $x = 0$ is a regular singular point. For $x = 2$ we have

$$\lim_{x \to 2}(x-2)p(x) = \lim_{x \to 2}(x-2)\frac{3}{2(x-2)^2} = \lim_{x \to 2}\frac{3}{2(x-2)},$$

so the limit does not exist; hence $x = 2$ is an irregular singular point.

EXAMPLE 8

Determine the singular points of

$$\left(x - \frac{\pi}{2}\right)^2 y'' + (\cos x)y' + (\sin x)y = 0$$

and classify them as regular or irregular.

The only singular point is $x = \pi/2$. To study it we consider the functions

$$\left(x - \frac{\pi}{2}\right)p(x) = \left(x - \frac{\pi}{2}\right)\frac{Q(x)}{P(x)} = \frac{\cos x}{x - \pi/2}$$

and

$$\left(x - \frac{\pi}{2}\right)^2 q(x) = \left(x - \frac{\pi}{2}\right)^2 \frac{R(x)}{P(x)} = \sin x.$$

Starting from the Taylor series for $\cos x$ about $x = \pi/2$, we find that

$$\frac{\cos x}{x - \pi/2} = -1 + \frac{(x - \pi/2)^2}{3!} - \frac{(x - \pi/2)^4}{5!} + \cdots,$$

which converges for all x. Similarly, $\sin x$ is analytic at $x = \pi/2$. Therefore, we conclude that $\pi/2$ is a regular singular point for this equation.

PROBLEMS

In each of Problems 1 through 18 find all singular points of the given equation and determine whether each one is regular or irregular.

1. $xy'' + (1-x)y' + xy = 0$
2. $x^2(1-x)^2 y'' + 2xy' + 4y = 0$
3. $x^2(1-x)y'' + (x-2)y' - 3xy = 0$
4. $x^2(1-x^2)y'' + (2/x)y' + 4y = 0$
5. $(1-x^2)^2 y'' + x(1-x)y' + (1+x)y = 0$

6. $x^2y'' + xy' + (x^2 - v^2)y = 0,$ Bessel equation

7. $(x + 3)y'' - 2xy' + (1 - x^2)y = 0$

8. $x(1 - x^2)^3 y'' + (1 - x^2)^2 y' + 2(1 + x)y = 0$

9. $(x + 2)^2(x - 1)y'' + 3(x - 1)y' - 2(x + 2)y = 0$

10. $x(3 - x)y'' + (x + 1)y' - 2y = 0$

11. $(x^2 + x - 2)y'' + (x + 1)y' + 2y = 0$ 12. $xy'' + e^x y' + (3\cos x)y = 0$

13. $y'' + (\ln|x|)y' + 3xy = 0$ 14. $x^2y'' + 2(e^x - 1)y' + (e^{-x}\cos x)y = 0$

15. $x^2y'' - 3(\sin x)y' + (1 + x^2)y = 0$ 16. $xy'' + y' + (\cot x)y = 0$

17. $(\sin x)y'' + xy' + 4y = 0$ 18. $(x\sin x)y'' + 3y' + xy = 0$

In each of Problems 19 and 20 show that the point $x = 0$ is a regular singular point. In each problem try to find solutions of the form $\sum_{n=0}^{\infty} a_n x^n$. Show that there is only one nonzero solution of this form in Problem 19 and that there are no nonzero solutions of this form in Problem 20. Thus in neither case can the general solution be found in this manner. This is typical of equations with singular points.

19. $2xy'' + 3y' + xy = 0$

20. $2x^2y'' + 3xy' - (1 + x)y = 0$

21. **Singularities at Infinity.** The definitions of an ordinary point and a regular singular point given in the preceding sections apply only if the point x_0 is finite. In more advanced work in differential equations it is often necessary to consider the point at infinity. This is done by making the change of variable $\xi = 1/x$ and studying the resulting equation at $\xi = 0$. Show that, for the differential equation $P(x)y'' + Q(x)y' + R(x)y = 0$, the point at infinity is an ordinary point if

$$\frac{1}{P(1/\xi)}\left[\frac{2P(1/\xi)}{\xi} - \frac{Q(1/\xi)}{\xi^2}\right] \quad \text{and} \quad \frac{R(1/\xi)}{\xi^4 P(1/\xi)}$$

have Taylor series expansions about $\xi = 0$. Show also that the point at infinity is a regular singular point if at least one of the above functions does not have a Taylor series expansion, but both

$$\frac{\xi}{P(1/\xi)}\left[\frac{2P(1/\xi)}{\xi} - \frac{Q(1/\xi)}{\xi^2}\right] \quad \text{and} \quad \frac{R(1/\xi)}{\xi^2 P(1/\xi)}$$

do have such expansions.

In each of Problems 22 through 27 use the results of Problem 21 to determine whether the point at infinity is an ordinary point, a regular singular point, or an irregular singular point of the given differential equation.

22. $y'' + y = 0$

23. $x^2y'' + xy' - 4y = 0$

24. $(1 - x^2)y'' - 2xy' + \alpha(\alpha + 1)y = 0,$ Legendre equation

25. $x^2y'' + xy' + (x^2 - v^2)y = 0,$ Bessel equation

26. $y'' - 2xy' + \lambda y = 0,$ Hermite equation

27. $y'' - xy = 0,$ Airy equation

5.5 Euler Equations

A relatively simple differential equation that has a regular singular point is the **Euler equation,**[11]

$$L[y] = x^2 y'' + \alpha x y' + \beta y = 0, \tag{1}$$

where α and β are real constants. It is easy to show that $x = 0$ is a regular singular point of Eq. (1). Because the solution of the Euler equation is typical of the solutions of all differential equations with a regular singular point, it is worthwhile considering this equation in detail before discussing the more general problem.

In any interval not including the origin, Eq. (1) has a general solution of the form $y = c_1 y_1(x) + c_2 y_2(x)$, where y_1 and y_2 are linearly independent. For convenience we first consider the interval $x > 0$, extending our results later to the interval $x < 0$. First, we note that $(x^r)' = r x^{r-1}$ and $(x^r)'' = r(r-1)x^{r-2}$. Hence, if we assume that Eq. (1) has a solution of the form

$$y = x^r, \tag{2}$$

then we obtain

$$
\begin{aligned}
L[x^r] &= x^2 (x^r)'' + \alpha x (x^r)' + \beta x^r \\
&= x^r [r(r-1) + \alpha r + \beta].
\end{aligned} \tag{3}
$$

If r is a root of the quadratic equation

$$F(r) = r(r-1) + \alpha r + \beta = 0, \tag{4}$$

then $L[x^r]$ is zero, and $y = x^r$ is a solution of Eq. (1). The roots of Eq. (4) are

$$r_1, r_2 = \frac{-(\alpha - 1) \pm \sqrt{(\alpha - 1)^2 - 4\beta}}{2}, \tag{5}$$

and $F(r) = (r - r_1)(r - r_2)$. As for second order linear equations with constant coefficients, it is necessary to consider separately the cases in which the roots are real and different, real but equal, and complex conjugates. Indeed, the entire discussion in this section is similar to the treatment of second order linear equations with constant coefficients in Chapter 3, with e^{rx} replaced by x^r; see also Problem 23.

Real, Distinct Roots. If $F(r) = 0$ has real roots r_1 and r_2, with $r_1 \neq r_2$, then $y_1(x) = x^{r_1}$ and $y_2(x) = x^{r_2}$ are solutions of Eq. (1). Since $W(x^{r_1}, x^{r_2}) = (r_2 - r_1)x^{r_1 + r_2 - 1}$ is nonvanishing for $r_1 \neq r_2$ and $x > 0$, it follows that the general solution of Eq. (1) is

$$y = c_1 x^{r_1} + c_2 x^{r_2}, \qquad x > 0. \tag{6}$$

Note that if r is not a rational number, then x^r is defined by $x^r = e^{r \ln x}$.

[11]This equation is sometimes called the Cauchy-Euler equation or the equidimensional equation. It was studied by Euler about 1740, but its solution was known to Johann Bernoulli before 1700.

EXAMPLE 1

Solve
$$2x^2 y'' + 3xy' - y = 0, \qquad x > 0. \tag{7}$$

Substituting $y = x^r$ in Eq. (7) gives
$$x^r[2r(r-1) + 3r - 1] = x^r(2r^2 + r - 1) = x^r(2r - 1)(r + 1) = 0.$$

Hence $r_1 = \frac{1}{2}$ and $r_2 = -1$, so the general solution of Eq. (7) is
$$y = c_1 x^{1/2} + c_2 x^{-1}, \qquad x > 0. \tag{8}$$

Equal Roots. If the roots r_1 and r_2 are equal, then we obtain only one solution $y_1(x) = x^{r_1}$ of the assumed form. A second solution can be obtained by the method of reduction of order, but for the purpose of our future discussion we consider an alternative method. Since $r_1 = r_2$, it follows that $F(r) = (r - r_1)^2$. Thus in this case not only does $F(r_1) = 0$ but also $F'(r_1) = 0$. This suggests differentiating Eq. (3) with respect to r and then setting r equal to r_1. Differentiating Eq. (3) with respect to r gives
$$\frac{\partial}{\partial r} L[x^r] = \frac{\partial}{\partial r} [x^r F(r)].$$

Substituting for $F(r)$, interchanging differentiation with respect to x and with respect to r, and noting that $\partial(x^r)/\partial r = x^r \ln x$, we obtain
$$L[x^r \ln x] = (r - r_1)^2 x^r \ln x + 2(r - r_1)x^r. \tag{9}$$

The right side of Eq. (9) is zero for $r = r_1$; consequently,
$$y_2(x) = x^{r_1} \ln x, \qquad x > 0 \tag{10}$$

is a second solution of Eq. (1). It is easy to show that $W(x^{r_1}, x^{r_1} \ln x) = x^{2r_1 - 1}$. Hence x^{r_1} and $x^{r_1} \ln x$ are linearly independent for $x > 0$, and the general solution of Eq. (1) is
$$y = (c_1 + c_2 \ln x)x^{r_1}, \qquad x > 0. \tag{11}$$

EXAMPLE 2

Solve
$$x^2 y'' + 5xy' + 4y = 0, \qquad x > 0. \tag{12}$$

Substituting $y = x^r$ in Eq. (12) gives
$$x^r[r(r-1) + 5r + 4] = x^r(r^2 + 4r + 4) = 0.$$

Hence $r_1 = r_2 = -2$, and
$$y = x^{-2}(c_1 + c_2 \ln x), \qquad x > 0. \tag{13}$$

Complex Roots. Finally, suppose that the roots r_1 and r_2 are complex conjugates, say, $r_1 = \lambda + i\mu$ and $r_2 = \lambda - i\mu$, with $\mu \neq 0$. We must now explain what is meant by x^r when r is complex. Remembering that
$$x^r = e^{r \ln x} \tag{14}$$

when $x > 0$ and r is real, we can use this equation to *define* x^r when r is complex. Then

$$x^{\lambda+i\mu} = e^{(\lambda+i\mu)\ln x} = e^{\lambda \ln x}e^{i\mu \ln x} = x^\lambda e^{i\mu \ln x}$$
$$= x^\lambda[\cos(\mu \ln x) + i\sin(\mu \ln x)], \qquad x > 0. \qquad (15)$$

With this definition of x^r for complex values of r, it can be verified that the usual laws of algebra and the differential calculus hold, and hence x^{r_1} and x^{r_2} are indeed solutions of Eq. (1). The general solution of Eq. (1) is

$$y = c_1 x^{\lambda+i\mu} + c_2 x^{\lambda-i\mu}. \qquad (16)$$

The disadvantage of this expression is that the functions $x^{\lambda+i\mu}$ and $x^{\lambda-i\mu}$ are complex-valued. Recall that we had a similar situation for the second order differential equation with constant coefficients when the roots of the characteristic equation were complex. In the same way as we did then, we observe that the real and imaginary parts of $x^{\lambda+i\mu}$, namely

$$x^\lambda \cos(\mu \ln x) \quad \text{and} \quad x^\lambda \sin(\mu \ln x), \qquad (17)$$

are also solutions of Eq. (1). A straightforward calculation shows that

$$W[x^\lambda \cos(\mu \ln x), x^\lambda \sin(\mu \ln x)] = \mu x^{2\lambda-1}.$$

Hence these solutions are also linearly independent for $x > 0$, and the general solution of Eq. (1) is

$$y = c_1 x^\lambda \cos(\mu \ln x) + c_2 x^\lambda \sin(\mu \ln x), \qquad x > 0. \qquad (18)$$

EXAMPLE 3

Solve

$$x^2 y'' + xy' + y = 0. \qquad (19)$$

Substituting $y = x^r$ in Eq. (19) gives

$$x^r[r(r-1) + r + 1] = x^r(r^2 + 1) = 0.$$

Hence $r = \pm i$, and the general solution is

$$y = c_1 \cos(\ln x) + c_2 \sin(\ln x), \qquad x > 0. \qquad (20)$$

Now let us consider the qualitative behavior of the solutions of Eq. (1) near the singular point $x = 0$. This depends entirely on the nature of the exponents r_1 and r_2. First, if r is real and positive, then $x^r \to 0$ as x tends to zero through positive values. On the other hand, if r is real and negative, then x^r becomes unbounded. Finally, if $r = 0$, then $x^r = 1$. Figure 5.5.1 shows these possibilities for various values of r. If r is complex, then a typical solution is $x^\lambda \cos(\mu \ln x)$. This function becomes unbounded or approaches zero if λ is negative or positive, respectively, and also oscillates more and more rapidly as $x \to 0$. This behavior is shown in Figures 5.5.2 and 5.5.3 for selected values of λ and μ. If $\lambda = 0$, the oscillation is of constant amplitude. Finally, if there are repeated roots, then one solution is of the form $x^r \ln x$, which tends to zero if $r > 0$ and becomes unbounded if $r \leq 0$. An example of each case is shown in Figure 5.5.4.

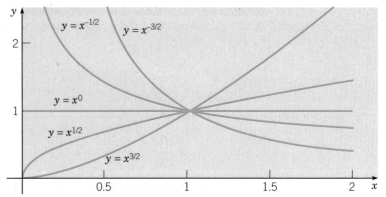

FIGURE 5.5.1 Solutions of an Euler equation; real roots.

FIGURE 5.5.2 Solution of an Euler equation; complex roots with negative real part.

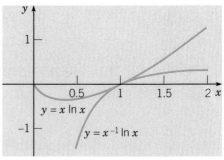

FIGURE 5.5.3 Solution of an Euler equa-
tion; complex roots with positive real part.

FIGURE 5.5.4 Solutions of an Euler equa-
tion; repeated roots.

 The extension of the solutions of Eq. (1) into the interval $x < 0$ can be carried out in a relatively straightforward manner. The difficulty lies in understanding what is meant by x^r when x is negative and r is not an integer; similarly, $\ln x$ has not been defined for $x < 0$. The solutions of the Euler equation that we have given for $x > 0$ can be shown to be valid for $x < 0$, but in general they are complex-valued. Thus in Example 1 the solution $x^{1/2}$ is imaginary for $x < 0$.

It is always possible to obtain real-valued solutions of the Euler equation (1) in the interval $x < 0$ by making the following change of variable. Let $x = -\xi$, where $\xi > 0$, and let $y = u(\xi)$. Then we have

$$\frac{dy}{dx} = \frac{du}{d\xi}\frac{d\xi}{dx} = -\frac{du}{d\xi}, \qquad \frac{d^2y}{dx^2} = \frac{d}{d\xi}\left(-\frac{du}{d\xi}\right)\frac{d\xi}{dx} = \frac{d^2u}{d\xi^2}. \tag{21}$$

Thus Eq. (1), for $x < 0$, takes the form

$$\xi^2\frac{d^2u}{d\xi^2} + \alpha\xi\frac{du}{d\xi} + \beta u = 0, \qquad \xi > 0. \tag{22}$$

But this is exactly the problem that we have just solved; from Eqs. (6), (11), and (18) we have

$$u(\xi) = \begin{cases} c_1\xi^{r_1} + c_2\xi^{r_2} \\ (c_1 + c_2\ln\xi)\xi^{r_1} \\ c_1\xi^\lambda\cos(\mu\ln\xi) + c_2\xi^\lambda\sin(\mu\ln\xi), \end{cases} \tag{23}$$

depending on whether the zeros of $F(r) = r(r-1) + \alpha r + \beta$ are real and different, real and equal, or complex conjugates. To obtain u in terms of x, we replace ξ by $-x$ in Eqs. (23).

We can combine the results for $x > 0$ and $x < 0$ by recalling that $|x| = x$ when $x > 0$ and that $|x| = -x$ when $x < 0$. Thus we need only replace x by $|x|$ in Eqs. (6), (11), and (18) to obtain real-valued solutions valid in any interval not containing the origin (also see Problems 30 and 31). These results are summarized in the following theorem.

Theorem 5.5.1

The general solution of the Euler equation (1),

$$x^2y'' + \alpha xy' + \beta y = 0,$$

in any interval not containing the origin is determined by the roots r_1 and r_2 of the equation

$$F(r) = r(r-1) + \alpha r + \beta = 0.$$

If the roots are real and different, then

$$y = c_1|x|^{r_1} + c_2|x|^{r_2}. \tag{24}$$

If the roots are real and equal, then

$$y = (c_1 + c_2\ln|x|)|x|^{r_1}. \tag{25}$$

If the roots are complex conjugates, then

$$y = |x|^\lambda\left[c_1\cos(\mu\ln|x|) + c_2\sin(\mu\ln|x|)\right], \tag{26}$$

where $r_1, r_2 = \lambda \pm i\mu$.

The solutions of an Euler equation of the form

$$(x - x_0)^2 y'' + \alpha(x - x_0)y' + \beta y = 0 \tag{27}$$

are similar to those given in Theorem 5.5.1. If one looks for solutions of the form $y = (x - x_0)^r$, then the general solution is given by Eq. (24), Eq. (25), or Eq. (26) with x replaced by $x - x_0$. Alternatively, we can reduce Eq. (27) to the form of Eq. (1) by making the change of independent variable $t = x - x_0$.

The situation for a general second order differential equation with a regular singular point is analogous to that for an Euler equation. We consider that problem in the next section.

PROBLEMS

In each of Problems 1 through 12 determine the general solution of the given differential equation that is valid in any interval not including the singular point.

1. $x^2 y'' + 4xy' + 2y = 0$
2. $(x + 1)^2 y'' + 3(x + 1)y' + 0.75y = 0$
3. $x^2 y'' - 3xy' + 4y = 0$
4. $x^2 y'' + 3xy' + 5y = 0$
5. $x^2 y'' - xy' + y = 0$
6. $(x - 1)^2 y'' + 8(x - 1)y' + 12y = 0$
7. $x^2 y'' + 6xy' - y = 0$
8. $2x^2 y'' - 4xy' + 6y = 0$
9. $x^2 y'' - 5xy' + 9y = 0$
10. $(x - 2)^2 y'' + 5(x - 2)y' + 8y = 0$
11. $x^2 y'' + 2xy' + 4y = 0$
12. $x^2 y'' - 4xy' + 4y = 0$

In each of Problems 13 through 16 find the solution of the given initial value problem. Plot the graph of the solution and describe how the solution behaves as $x \to 0$.

13. $2x^2 y'' + xy' - 3y = 0$, $y(1) = 1$, $y'(1) = 4$
14. $4x^2 y'' + 8xy' + 17y = 0$, $y(1) = 2$, $y'(1) = -3$
15. $x^2 y'' - 3xy' + 4y = 0$, $y(-1) = 2$, $y'(-1) = 3$
16. $x^2 y'' + 3xy' + 5y = 0$, $y(1) = 1$, $y'(1) = -1$

17. Find all values of α for which all solutions of $x^2 y'' + \alpha xy' + (5/2)y = 0$ approach zero as $x \to 0$.

18. Find all values of β for which all solutions of $x^2 y'' + \beta y = 0$ approach zero as $x \to 0$.

19. Find γ so that the solution of the initial value problem $x^2 y'' - 2y = 0, y(1) = 1, y'(1) = \gamma$ is bounded as $x \to 0$.

20. Find all values of α for which all solutions of $x^2 y'' + \alpha xy' + (5/2)y = 0$ approach zero as $x \to \infty$.

21. Consider the Euler equation $x^2 y'' + \alpha xy' + \beta y = 0$. Find conditions on α and β so that:
 (a) All solutions approach zero as $x \to 0$.
 (b) All solutions are bounded as $x \to 0$.
 (c) All solutions approach zero as $x \to \infty$.
 (d) All solutions are bounded as $x \to \infty$.
 (e) All solutions are bounded both as $x \to 0$ and as $x \to \infty$.

22. Using the method of reduction of order, show that if r_1 is a repeated root of $r(r - 1) + \alpha r + \beta = 0$, then x^{r_1} and $x^{r_1} \ln x$ are solutions of $x^2 y'' + \alpha xy' + \beta y = 0$ for $x > 0$.

23. **Transformation to a Constant Coefficient Equation.** The Euler equation $x^2 y'' + \alpha xy' + \beta y = 0$ can be reduced to an equation with constant coefficients by a change of the independent variable. Let $x = e^z$, or $z = \ln x$, and consider only the interval $x > 0$.

(a) Show that

$$\frac{dy}{dx} = \frac{1}{x}\frac{dy}{dz} \quad \text{and} \quad \frac{d^2y}{dx^2} = \frac{1}{x^2}\frac{d^2y}{dz^2} - \frac{1}{x^2}\frac{dy}{dz}.$$

(b) Show that the Euler equation becomes

$$\frac{d^2y}{dz^2} + (\alpha - 1)\frac{dy}{dz} + \beta y = 0.$$

Letting r_1 and r_2 denote the roots of $r^2 + (\alpha - 1)r + \beta = 0$, show that:

(c) If r_1 and r_2 are real and different, then

$$y = c_1 e^{r_1 z} + c_2 e^{r_2 z} = c_1 x^{r_1} + c_2 x^{r_2}.$$

(d) If r_1 and r_2 are real and equal, then

$$y = (c_1 + c_2 z)e^{r_1 z} = (c_1 + c_2 \ln x)x^{r_1}.$$

(e) If r_1 and r_2 are complex conjugates, $r_1 = \lambda + i\mu$, then

$$y = e^{\lambda z}[c_1 \cos(\mu z) + c_2 \sin(\mu z)] = x^\lambda [c_1 \cos(\mu \ln x) + c_2 \sin(\mu \ln x)].$$

In each of Problems 24 through 29 use the method of Problem 23 to solve the given equation for $x > 0$.

24. $x^2 y'' - 2y = 0$ 25. $x^2 y'' - 3xy' + 4y = \ln x$

26. $x^2 y'' + 7xy' + 5y = x$ 27. $x^2 y'' - 2xy' + 2y = 3x^2 + 2\ln x$

28. $x^2 y'' + xy' + 4y = \sin(\ln x)$ 29. $3x^2 y'' + 12xy' + 9y = 0$

30. Show that if $L[y] = x^2 y'' + \alpha xy' + \beta y$, then

$$L[(-x)^r] = (-x)^r F(r)$$

for all $x < 0$, where $F(r) = r(r - 1) + \alpha r + \beta$. Hence conclude that if $r_1 \neq r_2$ are roots of $F(r) = 0$, then linearly independent solutions of $L[y] = 0$ for $x < 0$ are $(-x)^{r_1}$ and $(-x)^{r_2}$.

31. Suppose that x^{r_1} and x^{r_2} are solutions of an Euler equation for $x > 0$, where $r_1 \neq r_2$, and r_1 is an integer. According to Eq. (24), the general solution in any interval not containing the origin is $y = c_1|x|^{r_1} + c_2|x|^{r_2}$. Show that the general solution can also be written as $y = k_1 x^{r_1} + k_2|x|^{r_2}$.

Hint: Show by a proper choice of constants that the expressions are identical for $x > 0$, and by a different choice of constants that they are identical for $x < 0$.

5.6 Series Solutions Near a Regular Singular Point, Part I

We now consider the question of solving the general second order linear equation

$$P(x)y'' + Q(x)y' + R(x)y = 0 \tag{1}$$

in the neighborhood of a regular singular point $x = x_0$. For convenience we assume that $x_0 = 0$. If $x_0 \neq 0$, we can transform the equation into one for which the regular singular point is at the origin by letting $x - x_0$ equal t.

The fact that $x = 0$ is a regular singular point of Eq. (1) means that $xQ(x)/P(x) = xp(x)$ and $x^2 R(x)/P(x) = x^2 q(x)$ have finite limits as $x \to 0$ and are analytic at

$x = 0$. Thus they have convergent power series expansions of the form

$$xp(x) = \sum_{n=0}^{\infty} p_n x^n, \qquad x^2 q(x) = \sum_{n=0}^{\infty} q_n x^n, \tag{2}$$

on some interval $|x| < \rho$ about the origin, where $\rho > 0$. To make the quantities $xp(x)$ and $x^2 q(x)$ appear in Eq. (1), it is convenient to divide Eq. (1) by $P(x)$ and then to multiply by x^2, obtaining

$$x^2 y'' + x[xp(x)]y' + [x^2 q(x)]y = 0, \tag{3}$$

or

$$x^2 y'' + x(p_0 + p_1 x + \cdots + p_n x^n + \cdots)y'$$
$$+ (q_0 + q_1 x + \cdots + q_n x^n + \cdots)y = 0. \tag{4}$$

If all of the coefficients p_n and q_n are zero, except possibly

$$p_0 = \lim_{x \to 0} \frac{xQ(x)}{P(x)} \quad \text{and} \quad q_0 = \lim_{x \to 0} \frac{x^2 R(x)}{P(x)}, \tag{5}$$

then Eq. (4) reduces to the Euler equation

$$x^2 y'' + p_0 xy' + q_0 y = 0, \tag{6}$$

which was discussed in the preceding section. In general, of course, some of the p_n and q_n, $n \geq 1$, are not zero. However, the essential character of solutions of Eq. (4) is identical to that of solutions of the Euler equation (6). The presence of the terms $p_1 x + \cdots + p_n x^n + \cdots$ and $q_1 x + \cdots + q_n x^n + \cdots$ merely complicates the calculations.

We restrict our discussion primarily to the interval $x > 0$. The interval $x < 0$ can be treated, just as for the Euler equation, by making the change of variable $x = -\xi$ and then solving the resulting equation for $\xi > 0$.

Since the coefficients in Eq. (4) are "Euler coefficients" times power series, it is natural to seek solutions in the form of "Euler solutions" times power series. Thus we assume that

$$y = x^r(a_0 + a_1 x + \cdots + a_n x^n + \cdots) = x^r \sum_{n=0}^{\infty} a_n x^n = \sum_{n=0}^{\infty} a_n x^{r+n}, \tag{7}$$

where $a_0 \neq 0$. In other words, r is the exponent of the first term in the series, and a_0 is its coefficient. As part of the solution we have to determine:

1. The values of r for which Eq. (1) has a solution of the form (7).

2. The recurrence relation for the coefficients a_n.

3. The radius of convergence of the series $\sum_{n=0}^{\infty} a_n x^n$.

The general theory was constructed by Frobenius[12] and is fairly complicated. Rather than trying to present this theory, we simply assume in this and the next two sections that there does exist a solution of the stated form. In particular, we assume that any power series in an expression for a solution has a nonzero radius of convergence and concentrate on showing how to determine the coefficients in such a series. To illustrate the method of Frobenius we first consider an example.

E X A M P L E

1

Solve the differential equation

$$2x^2 y'' - xy' + (1 + x)y = 0. \tag{8}$$

It is easy to show that $x = 0$ is a regular singular point of Eq. (8). Further, $xp(x) = -1/2$ and $x^2 q(x) = (1 + x)/2$. Thus $p_0 = -1/2, q_0 = 1/2, q_1 = 1/2$, and all other p's and q's are zero. Then, from Eq. (6), the Euler equation corresponding to Eq. (8) is

$$2x^2 y'' - xy' + y = 0. \tag{9}$$

To solve Eq. (8) we assume that there is a solution of the form (7). Then y' and y'' are given by

$$y' = \sum_{n=0}^{\infty} a_n(r + n)x^{r+n-1} \tag{10}$$

and

$$y'' = \sum_{n=0}^{\infty} a_n(r + n)(r + n - 1)x^{r+n-2}. \tag{11}$$

By substituting the expressions for y, y', and y'' in Eq. (8), we obtain

$$2x^2 y'' - xy' + (1 + x)y = \sum_{n=0}^{\infty} 2a_n(r + n)(r + n - 1)x^{r+n}$$

$$- \sum_{n=0}^{\infty} a_n(r + n)x^{r+n} + \sum_{n=0}^{\infty} a_n x^{r+n} + \sum_{n=0}^{\infty} a_n x^{r+n+1}. \tag{12}$$

The last term in Eq. (12) can be written as $\sum_{n=1}^{\infty} a_{n-1}x^{r+n}$, so by combining the terms in Eq. (12), we obtain

$$2x^2 y'' - xy' + (1 + x)y = a_0[2r(r - 1) - r + 1]x^r$$

$$+ \sum_{n=1}^{\infty} \left\{ [2(r + n)(r + n - 1) - (r + n) + 1]a_n + a_{n-1} \right\} x^{r+n} = 0. \tag{13}$$

[12]Ferdinand Georg Frobenius (1849–1917) was (like Fuchs) a student and eventually a professor at the University of Berlin. He showed how to construct series solutions about regular singular points in 1874. His most distinguished work, however, was in algebra, where he was one of the foremost early developers of group theory.

If Eq. (13) is to be satisfied for all x, the coefficient of each power of x in Eq. (13) must be zero. From the coefficient of x^r we obtain, since $a_0 \neq 0$,

$$2r(r-1) - r + 1 = 2r^2 - 3r + 1 = (r-1)(2r-1) = 0. \tag{14}$$

Equation (14) is called the **indicial equation** for Eq. (8). Note that it is exactly the polynomial equation we would obtain for the Euler equation (9) associated with Eq. (8). The roots of the indicial equation are

$$r_1 = 1, \qquad r_2 = 1/2. \tag{15}$$

These values of r are called the **exponents at the singularity** for the regular singular point $x = 0$. They determine the qualitative behavior of the solution (7) in the neighborhood of the singular point.

Now we return to Eq. (13) and set the coefficient of x^{r+n} equal to zero. This gives the relation

$$[2(r+n)(r+n-1) - (r+n) + 1]a_n + a_{n-1} = 0 \tag{16}$$

or

$$a_n = -\frac{a_{n-1}}{2(r+n)^2 - 3(r+n) + 1}$$

$$= -\frac{a_{n-1}}{[(r+n)-1][2(r+n)-1]}, \qquad n \geq 1. \tag{17}$$

For each root r_1 and r_2 of the indicial equation, we use the recurrence relation (17) to determine a set of coefficients a_1, a_2, \ldots. For $r = r_1 = 1$, Eq. (17) becomes

$$a_n = -\frac{a_{n-1}}{(2n+1)n}, \qquad n \geq 1.$$

Thus

$$a_1 = -\frac{a_0}{3 \cdot 1},$$

$$a_2 = -\frac{a_1}{5 \cdot 2} = \frac{a_0}{(3 \cdot 5)(1 \cdot 2)},$$

and

$$a_3 = -\frac{a_2}{7 \cdot 3} = -\frac{a_0}{(3 \cdot 5 \cdot 7)(1 \cdot 2 \cdot 3)}.$$

In general, we have

$$a_n = \frac{(-1)^n}{[3 \cdot 5 \cdot 7 \cdots (2n+1)]n!} a_0, \qquad n \geq 4. \tag{18}$$

Multiplying the numerator and denominator of the right side of Eq. (18) by $2 \cdot 4 \cdot 6 \cdots 2n = 2^n n!$, we can rewrite a_n as

$$a_n = \frac{(-1)^n 2^n}{(2n+1)!} a_0, \qquad n \geq 1.$$

Hence, if we omit the constant multiplier a_0, one solution of Eq. (8) is

$$y_1(x) = x \left[1 + \sum_{n=1}^{\infty} \frac{(-1)^n 2^n}{(2n+1)!} x^n \right], \qquad x > 0. \tag{19}$$

To determine the radius of convergence of the series in Eq. (19) we use the ratio test:

$$\lim_{n \to \infty} \left| \frac{a_{n+1}x^{n+1}}{a_n x^n} \right| = \lim_{n \to \infty} \frac{2|x|}{(2n+2)(2n+3)} = 0$$

for all x. Thus the series converges for all x.

Corresponding to the second root $r = r_2 = \frac{1}{2}$, we proceed similarly. From Eq. (17) we have

$$a_n = -\frac{a_{n-1}}{2n\left(n - \frac{1}{2}\right)} = -\frac{a_{n-1}}{n(2n-1)}, \qquad n \geq 1.$$

Hence

$$a_1 = -\frac{a_0}{1 \cdot 1},$$

$$a_2 = -\frac{a_1}{2 \cdot 3} = \frac{a_0}{(1 \cdot 2)(1 \cdot 3)},$$

$$a_3 = -\frac{a_2}{3 \cdot 5} = -\frac{a_0}{(1 \cdot 2 \cdot 3)(1 \cdot 3 \cdot 5)},$$

and, in general,

$$a_n = \frac{(-1)^n}{n![1 \cdot 3 \cdot 5 \cdots (2n-1)]} a_0, \qquad n \geq 4. \tag{20}$$

Just as in the case of the first root r_1, we multiply the numerator and denominator by $2 \cdot 4 \cdot 6 \cdots 2n = 2^n n!$. Then we have

$$a_n = \frac{(-1)^n 2^n}{(2n)!} a_0, \qquad n \geq 1.$$

Again omitting the constant multiplier a_0, we obtain the second solution

$$y_2(x) = x^{1/2} \left[1 + \sum_{n=1}^{\infty} \frac{(-1)^n 2^n}{(2n)!} x^n \right], \qquad x > 0. \tag{21}$$

As before, we can show that the series in Eq. (21) converges for all x. Since the leading terms in the series solutions y_1 and y_2 are x and $x^{1/2}$, respectively, it follows that the solutions are linearly independent. Hence the general solution of Eq. (8) is

$$y = c_1 y_1(x) + c_2 y_2(x), \qquad x > 0.$$

The preceding example illustrates that if $x = 0$ is a regular singular point, then sometimes there are two solutions of the form (7) in the neighborhood of this point. Similarly, if there is a regular singular point at $x = x_0$, then there may be two solutions of the form

$$y = (x - x_0)^r \sum_{n=0}^{\infty} a_n (x - x_0)^n \tag{22}$$

that are valid near $x = x_0$. However, just as an Euler equation may not have two solutions of the form $y = x^r$, so a more general equation with a regular singular point may not have two solutions of the form (7) or (22). In particular, we show in the next section that if the roots r_1 and r_2 of the indicial equation are equal, or differ by an integer, then the second solution normally has a more complicated structure. In all cases, though, it is possible to find at least one solution of the form (7) or (22); if r_1 and r_2 differ by an integer, this solution corresponds to the larger value of r. If there

is only one such solution, then the second solution involves a logarithmic term, just as for the Euler equation when the roots of the characteristic equation are equal. The method of reduction of order or some other procedure can be invoked to determine the second solution in such cases. This is discussed in Sections 5.7 and 5.8.

If the roots of the indicial equation are complex, then they cannot be equal or differ by an integer, so there are always two solutions of the form (7) or (22). Of course, these solutions are complex-valued functions of x. However, as for the Euler equation, it is possible to obtain real-valued solutions by taking the real and imaginary parts of the complex solutions.

Finally, we mention a practical point. If P, Q, and R are polynomials, it is often much better to work directly with Eq. (1) than with Eq. (3). This avoids the necessity of expressing $xQ(x)/P(x)$ and $x^2R(x)/P(x)$ as power series. For example, it is more convenient to consider the equation

$$x(1 + x)y'' + 2y' + xy = 0$$

than to write it in the form

$$x^2y'' + \frac{2x}{1+x}y' + \frac{x^2}{1+x}y = 0,$$

which would entail expanding $2x/(1 + x)$ and $x^2/(1 + x)$ in power series.

PROBLEMS

In each of Problems 1 through 10 show that the given differential equation has a regular singular point at $x = 0$. Determine the indicial equation, the recurrence relation, and the roots of the indicial equation. Find the series solution ($x > 0$) corresponding to the larger root. If the roots are unequal and do not differ by an integer, find the series solution corresponding to the smaller root also.

1. $2xy'' + y' + xy = 0$
2. $x^2y'' + xy' + \left(x^2 - \frac{1}{9}\right)y = 0$
3. $xy'' + y = 0$
4. $xy'' + y' - y = 0$
5. $3x^2y'' + 2xy' + x^2y = 0$
6. $x^2y'' + xy' + (x - 2)y = 0$
7. $xy'' + (1 - x)y' - y = 0$
8. $2x^2y'' + 3xy' + (2x^2 - 1)y = 0$
9. $x^2y'' - x(x + 3)y' + (x + 3)y = 0$
10. $x^2y'' + \left(x^2 + \frac{1}{4}\right)y = 0$

11. The Legendre equation of order α is

$$(1 - x^2)y'' - 2xy' + \alpha(\alpha + 1)y = 0.$$

The solution of this equation near the ordinary point $x = 0$ was discussed in Problems 22 and 23 of Section 5.3. In Example 6 of Section 5.4 it was shown that $x = \pm 1$ are regular singular points. Determine the indicial equation and its roots for the point $x = 1$. Find a series solution in powers of $x - 1$ for $x - 1 > 0$.
Hint: Write $1 + x = 2 + (x - 1)$ and $x = 1 + (x - 1)$. Alternatively, make the change of variable $x - 1 = t$ and determine a series solution in powers of t.

12. The Chebyshev equation is

$$(1 - x^2)y'' - xy' + \alpha^2y = 0,$$

where α is a constant; see Problem 10 of Section 5.3.

(a) Show that $x = 1$ and $x = -1$ are regular singular points, and find the exponents at each of these singularities.

(b) Find two linearly independent solutions about $x = 1$.

13. The Laguerre[13] differential equation is

$$xy'' + (1 - x)y' + \lambda y = 0.$$

Show that $x = 0$ is a regular singular point. Determine the indicial equation, its roots, the recurrence relation, and one solution $(x > 0)$. Show that if $\lambda = m$, a positive integer, this solution reduces to a polynomial. When properly normalized, this polynomial is known as the Laguerre polynomial, $L_m(x)$.

14. The Bessel equation of order zero is

$$x^2 y'' + xy' + x^2 y = 0.$$

Show that $x = 0$ is a regular singular point, that the roots of the indicial equation are $r_1 = r_2 = 0$, and that one solution for $x > 0$ is

$$J_0(x) = 1 + \sum_{n=1}^{\infty} \frac{(-1)^n x^{2n}}{2^{2n} (n!)^2}.$$

Show that the series converges for all x. The function J_0 is known as the Bessel function of the first kind of order zero.

15. Referring to Problem 14, use the method of reduction of order to show that the second solution of the Bessel equation of order zero contains a logarithmic term.
Hint: If $y_2(x) = J_0(x)v(x)$, then

$$y_2(x) = J_0(x) \int \frac{dx}{x[J_0(x)]^2}.$$

Find the first term in the series expansion of $1/x[J_0(x)]^2$.

16. The Bessel equation of order one is

$$x^2 y'' + xy' + (x^2 - 1)y = 0.$$

(a) Show that $x = 0$ is a regular singular point, that the roots of the indicial equation are $r_1 = 1$ and $r_2 = -1$, and that one solution for $x > 0$ is

$$J_1(x) = \frac{x}{2} \sum_{n=0}^{\infty} \frac{(-1)^n x^{2n}}{(n+1)! \, n! \, 2^{2n}}.$$

Show that the series converges for all x. The function J_1 is known as the Bessel function of the first kind of order one.

(b) Show that it is impossible to determine a second solution of the form

$$x^{-1} \sum_{n=0}^{\infty} b_n x^n, \qquad x > 0.$$

[13]Edmond Nicolas Laguerre (1834–1886), a French geometer and analyst, studied the polynomials named for him about 1879.

5.7 Series Solutions Near a Regular Singular Point, Part II

Now let us consider the general problem of determining a solution of the equation

$$L[y] = x^2 y'' + x[xp(x)]y' + [x^2 q(x)]y = 0, \tag{1}$$

where

$$xp(x) = \sum_{n=0}^{\infty} p_n x^n, \qquad x^2 q(x) = \sum_{n=0}^{\infty} q_n x^n, \tag{2}$$

and both series converge in an interval $|x| < \rho$ for some $\rho > 0$. The point $x = 0$ is a regular singular point, and the corresponding Euler equation is

$$x^2 y'' + p_0 xy' + q_0 y = 0. \tag{3}$$

We seek a solution of Eq. (1) for $x > 0$ and assume that it has the form

$$y = \phi(r, x) = x^r \sum_{n=0}^{\infty} a_n x^n = \sum_{n=0}^{\infty} a_n x^{r+n}, \tag{4}$$

where $a_0 \neq 0$, and we have written $y = \phi(r, x)$ to emphasize that ϕ depends on r as well as x. It follows that

$$y' = \sum_{n=0}^{\infty} (r+n) a_n x^{r+n-1}, \qquad y'' = \sum_{n=0}^{\infty} (r+n)(r+n-1) a_n x^{r+n-2}. \tag{5}$$

Then, substituting from Eqs. (2), (4), and (5) in Eq. (1) gives

$$a_0 r(r-1)x^r + a_1(r+1)rx^{r+1} + \cdots + a_n(r+n)(r+n-1)x^{r+n} + \cdots$$
$$+ (p_0 + p_1 x + \cdots + p_n x^n + \cdots)$$
$$\times [a_0 r x^r + a_1(r+1)x^{r+1} + \cdots + a_n(r+n)x^{r+n} + \cdots]$$
$$+ (q_0 + q_1 x + \cdots + q_n x^n + \cdots)$$
$$\times (a_0 x^r + a_1 x^{r+1} + \cdots + a_n x^{r+n} + \cdots) = 0.$$

Multiplying the infinite series together and then collecting terms, we obtain

$$a_0 F(r)x^r + [a_1 F(r+1) + a_0(p_1 r + q_1)]x^{r+1}$$
$$+ \left\{ a_2 F(r+2) + a_0(p_2 r + q_2) + a_1[p_1(r+1) + q_1] \right\} x^{r+2}$$
$$+ \cdots + \left\{ a_n F(r+n) + a_0(p_n r + q_n) + a_1[p_{n-1}(r+1) + q_{n-1}] \right.$$
$$+ \cdots + a_{n-1}[p_1(r+n-1) + q_1] \right\} x^{r+n} + \cdots = 0,$$

or, in a more compact form,

$$L[\phi](r, x) = a_0 F(r)x^r$$
$$+ \sum_{n=1}^{\infty} \left\{ F(r+n)a_n + \sum_{k=0}^{n-1} a_k[(r+k)p_{n-k} + q_{n-k}] \right\} x^{r+n} = 0, \tag{6}$$

where

$$F(r) = r(r-1) + p_0 r + q_0. \qquad (7)$$

For Eq. (6) to be satisfied identically, the coefficient of each power of x must be zero.

Since $a_0 \neq 0$, the term involving x^r yields the equation $F(r) = 0$. This equation is called the **indicial equation**; note that it is exactly the equation we would obtain in looking for solutions $y = x^r$ of the Euler equation (3). Let us denote the roots of the indicial equation by r_1 and r_2 with $r_1 \geq r_2$ if the roots are real. If the roots are complex, the designation of the roots is immaterial. Only for these values of r can we expect to find solutions of Eq. (1) of the form (4). The roots r_1 and r_2 are called the **exponents at the singularity**; they determine the qualitative nature of the solution in the neighborhood of the singular point.

Setting the coefficient of x^{r+n} in Eq. (6) equal to zero gives the **recurrence relation**

$$F(r+n)a_n + \sum_{k=0}^{n-1} a_k[(r+k)p_{n-k} + q_{n-k}] = 0, \qquad n \geq 1. \qquad (8)$$

Equation (8) shows that, in general, a_n depends on the value of r and all the preceding coefficients $a_0, a_1, \ldots, a_{n-1}$. It also shows that we can successively compute $a_1, a_2, \ldots, a_n, \ldots$ in terms of a_0 and the coefficients in the series for $xp(x)$ and $x^2q(x)$, provided that $F(r+1), F(r+2), \ldots, F(r+n), \ldots$ are not zero. The only values of r for which $F(r) = 0$ are $r = r_1$ and $r = r_2$; since $r_1 \geq r_2$, it follows that $r_1 + n$ is not equal to r_1 or r_2 for $n \geq 1$. Consequently, $F(r_1 + n) \neq 0$ for $n \geq 1$. Hence we can always determine one solution of Eq. (1) in the form (4), namely

$$y_1(x) = x^{r_1}\left[1 + \sum_{n=1}^{\infty} a_n(r_1)x^n\right], \qquad x > 0. \qquad (9)$$

Here we have introduced the notation $a_n(r_1)$ to indicate that a_n has been determined from Eq. (8) with $r = r_1$. To specify the arbitrary constant in the solution we have taken a_0 to be 1.

If r_2 is not equal to r_1, and $r_1 - r_2$ is not a positive integer, then $r_2 + n$ is not equal to r_1 for any value of $n \geq 1$; hence $F(r_2 + n) \neq 0$, and we can also obtain a second solution

$$y_2(x) = x^{r_2}\left[1 + \sum_{n=1}^{\infty} a_n(r_2)x^n\right], \qquad x > 0. \qquad (10)$$

Just as for the series solutions about ordinary points discussed in Section 5.3, the series in Eqs. (9) and (10) converge at least in the interval $|x| < \rho$ where the series for both $xp(x)$ and $x^2q(x)$ converge. Within their radii of convergence, the power series $1 + \sum_{n=1}^{\infty} a_n(r_1)x^n$ and $1 + \sum_{n=1}^{\infty} a_n(r_2)x^n$ define functions that are analytic at $x = 0$. Thus the singular behavior, if there is any, of the solutions y_1 and y_2 is due to the factors x^{r_1} and x^{r_2} that multiply these two analytic functions. Next, to obtain real-valued solutions for $x < 0$, we can make the substitution $x = -\xi$ with $\xi > 0$. As we might expect from our discussion of the Euler equation, it turns out that we need only replace x^{r_1} in Eq. (9) and x^{r_2} in Eq. (10) by $|x|^{r_1}$ and $|x|^{r_2}$, respectively. Finally, note that if r_1 and r_2 are complex numbers, then they are necessarily complex conjugates

and $r_2 \neq r_1 + N$. Thus, in this case we can always find two series solutions of the form (4); however, they are complex-valued functions of x. Real-valued solutions can be obtained by taking the real and imaginary parts of the complex-valued solutions. The exceptional cases in which $r_1 = r_2$ or $r_1 - r_2 = N$, where N is a positive integer, require more discussion and will be considered later in this section.

It is important to realize that r_1 and r_2, the exponents at the singular point, are easy to find and that they determine the qualitative behavior of the solutions. To calculate r_1 and r_2 it is only necessary to solve the quadratic indicial equation

$$r(r-1) + p_0 r + q_0 = 0, \tag{11}$$

whose coefficients are given by

$$p_0 = \lim_{x \to 0} x p(x), \qquad q_0 = \lim_{x \to 0} x^2 q(x). \tag{12}$$

Note that these are exactly the limits that must be evaluated in order to classify the singularity as a regular singular point; thus they have usually been determined at an earlier stage of the investigation.

Further, if $x = 0$ is a regular singular point of the equation

$$P(x)y'' + Q(x)y' + R(x)y = 0, \tag{13}$$

where the functions P, Q, and R are polynomials, then $xp(x) = xQ(x)/P(x)$ and $x^2 q(x) = x^2 R(x)/P(x)$. Thus

$$p_0 = \lim_{x \to 0} x \frac{Q(x)}{P(x)}, \qquad q_0 = \lim_{x \to 0} x^2 \frac{R(x)}{P(x)}. \tag{14}$$

Finally, the radii of convergence for the series in Eqs. (9) and (10) are at least equal to the distance from the origin to the nearest zero of P other than $x = 0$ itself.

EXAMPLE
1

Discuss the nature of the solutions of the equation

$$2x(1+x)y'' + (3+x)y' - xy = 0$$

near the singular points.

This equation is of the form (13) with $P(x) = 2x(1+x)$, $Q(x) = 3+x$, and $R(x) = -x$. The points $x = 0$ and $x = -1$ are the only singular points. The point $x = 0$ is a regular singular point, since

$$\lim_{x \to 0} x \frac{Q(x)}{P(x)} = \lim_{x \to 0} x \frac{3+x}{2x(1+x)} = \frac{3}{2},$$

$$\lim_{x \to 0} x^2 \frac{R(x)}{P(x)} = \lim_{x \to 0} x^2 \frac{-x}{2x(1+x)} = 0.$$

Further, from Eq. (14), $p_0 = \frac{3}{2}$ and $q_0 = 0$. Thus the indicial equation is $r(r-1) + \frac{3}{2}r = 0$, and the roots are $r_1 = 0$, $r_2 = -\frac{1}{2}$. Since these roots are not equal and do not differ by an integer, there are two linearly independent solutions of the form

$$y_1(x) = 1 + \sum_{n=1}^{\infty} a_n(0)x^n \quad \text{and} \quad y_2(x) = |x|^{-1/2}\left[1 + \sum_{n=1}^{\infty} a_n\left(-\tfrac{1}{2}\right)x^n\right]$$

for $0 < |x| < \rho$. A lower bound for the radius of convergence of each series is 1, the distance from $x = 0$ to $x = -1$, the other zero of $P(x)$. Note that the solution y_1 is bounded as $x \to 0$, indeed is analytic there, and that the second solution y_2 is unbounded as $x \to 0$.

The point $x = -1$ is also a regular singular point, since

$$\lim_{x \to -1} (x + 1) \frac{Q(x)}{P(x)} = \lim_{x \to -1} \frac{(x+1)(3+x)}{2x(1+x)} = -1,$$

$$\lim_{x \to -1} (x + 1)^2 \frac{R(x)}{P(x)} = \lim_{x \to -1} \frac{(x+1)^2(-x)}{2x(1+x)} = 0.$$

In this case $p_0 = -1, q_0 = 0$, so the indicial equation is $r(r - 1) - r = 0$. The roots of the indicial equation are $r_1 = 2$ and $r_2 = 0$. Corresponding to the larger root there is a solution of the form

$$y_1(x) = (x + 1)^2 \left[1 + \sum_{n=1}^{\infty} a_n(2)(x + 1)^n \right].$$

The series converges at least for $|x + 1| < 1$, and y_1 is an analytic function there. Since the two roots differ by a positive integer, there may or may not be a second solution of the form

$$y_2(x) = 1 + \sum_{n=1}^{\infty} a_n(0)(x + 1)^n.$$

We cannot say more without further analysis.

Observe that no complicated calculations were required to discover the information about the solutions presented in this example. All that was needed was to evaluate a few limits and solve two quadratic equations.

We now consider the cases in which the roots of the indicial equation are equal, or differ by a positive integer, $r_1 - r_2 = N$. As we have shown earlier, there is always one solution of the form (9) corresponding to the larger root r_1 of the indicial equation. By analogy with the Euler equation, we might expect that if $r_1 = r_2$, then the second solution contains a logarithmic term. This may also be true if the roots differ by an integer.

Equal Roots. The method of finding the second solution is essentially the same as the one we used in finding the second solution of the Euler equation (see Section 5.5) when the roots of the indicial equation were equal. We consider r to be a continuous variable and determine a_n as a function of r by solving the recurrence relation (8). For this choice of $a_n(r)$ for $n \geq 1$, Eq. (6) reduces to

$$L[\phi](r, x) = a_0 F(r)x^r = a_0(r - r_1)^2 x^r, \tag{15}$$

since r_1 is a repeated root of $F(r)$. Setting $r = r_1$ in Eq. (15), we find that $L[\phi](r_1, x) = 0$; hence, as we already know, $y_1(x)$ given by Eq. (9) is one solution of Eq. (1). But more important, it also follows from Eq. (15), just as for the Euler equation, that

$$L\left[\frac{\partial \phi}{\partial r} \right] (r_1, x) = a_0 \frac{\partial}{\partial r} [x^r (r - r_1)^2] \Big|_{r=r_1}$$

$$= a_0[(r - r_1)^2 x^r \ln x + 2(r - r_1)x^r] \Big|_{r=r_1} = 0. \tag{16}$$

Hence, a second solution of Eq. (1) is

$$y_2(x) = \left. \frac{\partial \phi(r,x)}{\partial r} \right|_{r=r_1} = \left. \frac{\partial}{\partial r} \left\{ x^r \left[a_0 + \sum_{n=1}^{\infty} a_n(r)x^n \right] \right\} \right|_{r=r_1}$$

$$= (x^{r_1} \ln x) \left[a_0 + \sum_{n=1}^{\infty} a_n(r_1)x^n \right] + x^{r_1} \sum_{n=1}^{\infty} a_n'(r_1)x^n$$

$$= y_1(x) \ln x + x^{r_1} \sum_{n=1}^{\infty} a_n'(r_1)x^n, \qquad x > 0, \tag{17}$$

where $a_n'(r_1)$ denotes da_n/dr evaluated at $r = r_1$.

It may turn out that it is difficult to determine $a_n(r)$ as a function of r from the recurrence relation (8) and then to differentiate the resulting expression with respect to r. An alternative is simply to assume that y has the *form* of Eq. (17). That is, assume that

$$y = y_1(x) \ln x + x^{r_1} \sum_{n=1}^{\infty} b_n x^n, \qquad x > 0, \tag{18}$$

where $y_1(x)$ has already been found. The coefficients b_n are calculated, as usual, by substituting into the differential equation, collecting terms, and setting the coefficient of each power of x equal to zero. A third possibility is to use the method of reduction of order to find $y_2(x)$ once $y_1(x)$ is known.

Roots Differing by an Integer. For this case the derivation of the second solution is considerably more complicated and will not be given here. The form of this solution is stated in Eq. (24) in the following theorem. The coefficients $c_n(r_2)$ in Eq. (24) are given by

$$c_n(r_2) = \left. \frac{d}{dr}[(r - r_2)a_n(r)] \right|_{r=r_2}, \qquad n = 1, 2, \ldots, \tag{19}$$

where $a_n(r)$ is determined from the recurrence relation (8) with $a_0 = 1$. Further, the coefficient a in Eq. (24) is

$$a = \lim_{r \to r_2} (r - r_2)a_N(r). \tag{20}$$

If $a_N(r_2)$ is finite, then $a = 0$ and there is no logarithmic term in y_2. A full derivation of formulas (19) and (20) may be found in Coddington (Chapter 4).

In practice, the best way to determine whether a is zero in the second solution is simply to try to compute the a_n corresponding to the root r_2 and to see whether it is possible to determine $a_N(r_2)$. If so, there is no further problem. If not, we must use the form (24) with $a \neq 0$.

When $r_1 - r_2 = N$, there are again three ways to find a second solution. First, we can calculate a and $c_n(r_2)$ directly by substituting the expression (24) for y in Eq. (1). Second, we can calculate $c_n(r_2)$ and a of Eq. (24) using the formulas (19) and (20). If this is the planned procedure, then in calculating the solution corresponding to $r = r_1$, be sure to obtain the general formula for $a_n(r)$ rather than just $a_n(r_1)$. The third alternative is to use the method of reduction of order.

Theorem 5.7.1 Consider the differential equation (1),

$$x^2 y'' + x[xp(x)]y' + [x^2 q(x)]y = 0,$$

where $x = 0$ is a regular singular point. Then $xp(x)$ and $x^2 q(x)$ are analytic at $x = 0$ with convergent power series expansions

$$xp(x) = \sum_{n=0}^{\infty} p_n x^n, \qquad x^2 q(x) = \sum_{n=0}^{\infty} q_n x^n$$

for $|x| < \rho$, where $\rho > 0$ is the minimum of the radii of convergence of the power series for $xp(x)$ and $x^2 q(x)$. Let r_1 and r_2 be the roots of the indicial equation

$$F(r) = r(r-1) + p_0 r + q_0 = 0,$$

with $r_1 \geq r_2$ if r_1 and r_2 are real. Then in either the interval $-\rho < x < 0$ or the interval $0 < x < \rho$, there exists a solution of the form

$$y_1(x) = |x|^{r_1}\left[1 + \sum_{n=1}^{\infty} a_n(r_1)x^n\right], \tag{21}$$

where the $a_n(r_1)$ are given by the recurrence relation (8) with $a_0 = 1$ and $r = r_1$.

If $r_1 - r_2$ is not zero or a positive integer, then in either the interval $-\rho < x < 0$ or the interval $0 < x < \rho$, there exists a second linearly independent solution of the form

$$y_2(x) = |x|^{r_2}\left[1 + \sum_{n=1}^{\infty} a_n(r_2)x^n\right]. \tag{22}$$

The $a_n(r_2)$ are also determined by the recurrence relation (8) with $a_0 = 1$ and $r = r_2$. The power series in Eqs. (21) and (22) converge at least for $|x| < \rho$.

If $r_1 = r_2$, then the second solution is

$$y_2(x) = y_1(x)\ln|x| + |x|^{r_1}\sum_{n=1}^{\infty} b_n(r_1)x^n. \tag{23}$$

If $r_1 - r_2 = N$, a positive integer, then

$$y_2(x) = ay_1(x)\ln|x| + |x|^{r_2}\left[1 + \sum_{n=1}^{\infty} c_n(r_2)x^n\right]. \tag{24}$$

The coefficients $a_n(r_1), b_n(r_1), c_n(r_2)$, and the constant a can be determined by substituting the form of the series solutions for y in Eq. (1). The constant a may turn out to be zero, in which case there is no logarithmic term in the solution (24). Each of the series in Eqs. (23) and (24) converges at least for $|x| < \rho$ and defines a function that is analytic in some neighborhood of $x = 0$.

PROBLEMS In each of Problems 1 through 12 find all the regular singular points of the given differential
 equation. Determine the indicial equation and the exponents at the singularity for each regular
 singular point.

1. $xy'' + 2xy' + 6e^x y = 0$ 2. $x^2 y'' - x(2 + x)y' + (2 + x^2)y = 0$

3. $x(x - 1)y'' + 6x^2 y' + 3y = 0$ 4. $y'' + 4xy' + 6y = 0$

5. $x^2 y'' + 3(\sin x)y' - 2y = 0$ 6. $2x(x + 2)y'' + y' - xy = 0$

7. $x^2 y'' + \frac{1}{2}(x + \sin x)y' + y = 0$ 8. $(x + 1)^2 y'' + 3(x^2 - 1)y' + 3y = 0$

9. $x^2(1 - x)y'' - (1 + x)y' + 2xy = 0$ 10. $(x - 2)^2(x + 2)y'' + 2xy' + 3(x - 2)y = 0$

11. $(4 - x^2)y'' + 2xy' + 3y = 0$ 12. $x(x + 3)^2 y'' - 2(x + 3)y' - xy = 0$

In each of Problems 13 through 17:

(a) Show that $x = 0$ is a regular singular point of the given differential equation.

(b) Find the exponents at the singular point $x = 0$.

(c) Find the first three nonzero terms in each of two linearly independent solutions about
$x = 0$.

13. $xy'' + y' - y = 0$

14. $xy'' + 2xy' + 6e^x y = 0$; see Problem 1

15. $x(x - 1)y'' + 6x^2 y' + 3y = 0$; see Problem 3

16. $xy'' + y = 0$

17. $x^2 y'' + (\sin x)y' - (\cos x)y = 0$

18. Show that

$$(\ln x)y'' + \tfrac{1}{2}y' + y = 0$$

has a regular singular point at $x = 1$. Determine the roots of the indicial equation at $x = 1$.
Determine the first three nonzero terms in the series $\sum\limits_{n=0}^{\infty} a_n(x - 1)^{r+n}$ corresponding to the
larger root. Take $x - 1 > 0$. What would you expect the radius of convergence of the
series to be?

19. In several problems in mathematical physics it is necessary to study the differential equa-
tion

$$x(1 - x)y'' + [\gamma - (1 + \alpha + \beta)x]y' - \alpha\beta y = 0, \qquad \text{(i)}$$

where α, β, and γ are constants. This equation is known as the **hypergeometric equation**.

(a) Show that $x = 0$ is a regular singular point and that the roots of the indicial equation
are 0 and $1 - \gamma$.

(b) Show that $x = 1$ is a regular singular point and that the roots of the indicial equation
are 0 and $\gamma - \alpha - \beta$.

(c) Assuming that $1 - \gamma$ is not a positive integer, show that, in the neighborhood of $x = 0$,
one solution of (i) is

$$y_1(x) = 1 + \frac{\alpha\beta}{\gamma \cdot 1!}x + \frac{\alpha(\alpha + 1)\beta(\beta + 1)}{\gamma(\gamma + 1)2!}x^2 + \cdots.$$

What would you expect the radius of convergence of this series to be?

(d) Assuming that $1 - \gamma$ is not an integer or zero, show that a second solution for $0 < x < 1$ is

$$y_2(x) = x^{1-\gamma} \left[1 + \frac{(\alpha - \gamma + 1)(\beta - \gamma + 1)}{(2 - \gamma)1!} x \right.$$

$$\left. + \frac{(\alpha - \gamma + 1)(\alpha - \gamma + 2)(\beta - \gamma + 1)(\beta - \gamma + 2)}{(2 - \gamma)(3 - \gamma)2!} x^2 + \cdots \right].$$

(e) Show that the point at infinity is a regular singular point and that the roots of the indicial equation are α and β. See Problem 21 of Section 5.4.

20. Consider the differential equation

$$x^3 y'' + \alpha x y' + \beta y = 0,$$

where α and β are real constants and $\alpha \neq 0$.

(a) Show that $x = 0$ is an irregular singular point.

(b) By attempting to determine a solution of the form $\sum_{n=0}^{\infty} a_n x^{r+n}$, show that the indicial equation for r is linear and that, consequently, there is only one formal solution of the assumed form.

(c) Show that if $\beta/\alpha = -1, 0, 1, 2, \ldots$, then the formal series solution terminates and therefore is an actual solution. For other values of β/α, show that the formal series solution has a zero radius of convergence and so does not represent an actual solution in any interval.

21. Consider the differential equation

$$y'' + \frac{\alpha}{x^s} y' + \frac{\beta}{x^t} y = 0, \tag{i}$$

where $\alpha \neq 0$ and $\beta \neq 0$ are real numbers, and s and t are positive integers that for the moment are arbitrary.

(a) Show that if $s > 1$ or $t > 2$, then the point $x = 0$ is an irregular singular point.

(b) Try to find a solution of Eq. (i) of the form

$$y = \sum_{n=0}^{\infty} a_n x^{r+n}, \qquad x > 0. \tag{ii}$$

Show that if $s = 2$ and $t = 2$, then there is only one possible value of r for which there is a formal solution of Eq. (i) of the form (ii).

(c) Show that if $s = 1$ and $t = 3$, then there are no solutions of Eq. (i) of the form (ii).

(d) Show that the maximum values of s and t for which the indicial equation is quadratic in r [and hence we can hope to find two solutions of the form (ii)] are $s = 1$ and $t = 2$. These are precisely the conditions that distinguish a "weak singularity," or a regular singular point, from an irregular singular point, as we defined them in Section 5.4.

As a note of caution, we point out that although it is sometimes possible to obtain a formal series solution of the form (ii) at an irregular singular point, the series may not have a positive radius of convergence. See Problem 20 for an example.

5.8 Bessel's Equation

In this section we consider three special cases of Bessel's[14] equation,

$$x^2 y'' + x y' + (x^2 - \nu^2) y = 0, \tag{1}$$

where ν is a constant, which illustrate the theory discussed in Section 5.7. It is easy to show that $x = 0$ is a regular singular point. For simplicity we consider only the case $x > 0$.

Bessel Equation of Order Zero. This example illustrates the situation in which the roots of the indicial equation are equal. Setting $\nu = 0$ in Eq. (1) gives

$$L[y] = x^2 y'' + x y' + x^2 y = 0. \tag{2}$$

Substituting

$$y = \phi(r, x) = a_0 x^r + \sum_{n=1}^{\infty} a_n x^{r+n}, \tag{3}$$

we obtain

$$L[\phi](r, x) = \sum_{n=0}^{\infty} a_n [(r+n)(r+n-1) + (r+n)] x^{r+n} + \sum_{n=0}^{\infty} a_n x^{r+n+2}$$

$$= a_0 [r(r-1) + r] x^r + a_1 [(r+1)r + (r+1)] x^{r+1}$$

$$+ \sum_{n=2}^{\infty} \left\{ a_n [(r+n)(r+n-1) + (r+n)] + a_{n-2} \right\} x^{r+n} = 0. \tag{4}$$

The roots of the indicial equation $F(r) = r(r-1) + r = 0$ are $r_1 = 0$ and $r_2 = 0$; hence we have the case of equal roots. The recurrence relation is

$$a_n(r) = -\frac{a_{n-2}(r)}{(r+n)(r+n-1) + (r+n)} = -\frac{a_{n-2}(r)}{(r+n)^2}, \qquad n \geq 2. \tag{5}$$

To determine $y_1(x)$ we set r equal to 0. Then, from Eq. (4), it follows that for the coefficient of x^{r+1} to be zero we must choose $a_1 = 0$. Hence, from Eq. (5), $a_3 = a_5 = a_7 = \cdots = 0$. Further,

$$a_n(0) = -a_{n-2}(0)/n^2, \qquad n = 2, 4, 6, 8, \ldots,$$

or, letting $n = 2m$, we obtain

$$a_{2m}(0) = -a_{2m-2}(0)/(2m)^2, \qquad m = 1, 2, 3, \ldots.$$

[14]Friedrich Wilhelm Bessel (1784–1846) embarked on a career in business as a youth but soon became interested in astronomy and mathematics. He was appointed director of the observatory at Königsberg in 1810 and held this position until his death. His study of planetary perturbations led him in 1824 to make the first systematic analysis of the solutions, known as Bessel functions, of Eq. (1). He is also famous for making, in 1838, the first accurate determination of the distance from the earth to a star.

Thus

$$a_2(0) = -\frac{a_0}{2^2}, \qquad a_4(0) = \frac{a_0}{2^4 2^2}, \qquad a_6(0) = -\frac{a_0}{2^6(3 \cdot 2)^2},$$

and, in general,

$$a_{2m}(0) = \frac{(-1)^m a_0}{2^{2m}(m!)^2}, \qquad m = 1, 2, 3, \ldots. \tag{6}$$

Hence

$$y_1(x) = a_0 \left[1 + \sum_{m=1}^{\infty} \frac{(-1)^m x^{2m}}{2^{2m}(m!)^2} \right], \qquad x > 0. \tag{7}$$

The function in brackets is known as the *Bessel function of the first kind of order zero* and is denoted by $J_0(x)$. It follows from Theorem 5.7.1 that the series converges for all x and that J_0 is analytic at $x = 0$. Some of the important properties of J_0 are discussed in the problems. Figure 5.8.1 shows the graphs of $y = J_0(x)$ and some of the partial sums of the series (7).

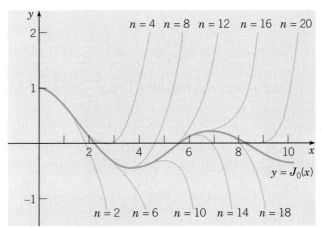

FIGURE 5.8.1 Polynomial approximations to $J_0(x)$. The value of n is the degree of the approximating polynomial.

To determine $y_2(x)$ we will calculate $a_n'(0)$.[15] First we note from the coefficient of x^{r+1} in Eq. (4) that $(r+1)^2 a_1(r) = 0$. It follows that not only does $a_1(0) = 0$ but also $a_1'(0) = 0$. It is easy to deduce from the recurrence relation (5) that $a_3'(0) = a_5'(0) = \cdots = a_{2n+1}'(0) = \cdots = 0$; hence we need only compute $a_{2m}'(0), m = 1, 2, 3, \ldots$. From Eq. (5) we have

$$a_{2m}(r) = -a_{2m-2}(r)/(r + 2m)^2, \qquad m = 1, 2, 3, \ldots.$$

[15]Problem 10 outlines an alternative procedure, in which we simply substitute the form (23) of Section 5.7 in Eq. (2) and then determine the b_n.

By solving this recurrence relation we obtain

$$a_2(r) = -\frac{a_0}{(r+2)^2}, \qquad a_4(r) = \frac{a_0}{(r+2)^2(r+4)^2},$$

and, in general,

$$a_{2m}(r) = \frac{(-1)^m a_0}{(r+2)^2 \cdots (r+2m)^2}, \qquad m \geq 3. \tag{8}$$

The computation of $a'_{2m}(r)$ can be carried out most conveniently by noting that if

$$f(x) = (x - \alpha_1)^{\beta_1}(x - \alpha_2)^{\beta_2}(x - \alpha_3)^{\beta_3} \cdots (x - \alpha_n)^{\beta_n},$$

and if x is not equal to $\alpha_1, \alpha_2, \ldots, \alpha_n$, then

$$\frac{f'(x)}{f(x)} = \frac{\beta_1}{x - \alpha_1} + \frac{\beta_2}{x - \alpha_2} + \cdots + \frac{\beta_n}{x - \alpha_n}.$$

Applying this result to $a_{2m}(r)$ from Eq. (8), we find that

$$\frac{a'_{2m}(r)}{a_{2m}(r)} = -2\left(\frac{1}{r+2} + \frac{1}{r+4} + \cdots + \frac{1}{r+2m}\right),$$

and setting r equal to 0, we obtain

$$a'_{2m}(0) = -2\left[\frac{1}{2} + \frac{1}{4} + \cdots + \frac{1}{2m}\right] a_{2m}(0).$$

Substituting for $a_{2m}(0)$ from Eq. (6), and letting

$$H_m = 1 + \frac{1}{2} + \frac{1}{3} + \cdots + \frac{1}{m}, \tag{9}$$

we obtain, finally,

$$a'_{2m}(0) = -H_m \frac{(-1)^m a_0}{2^{2m}(m!)^2}, \qquad m = 1, 2, 3, \ldots.$$

The second solution of the Bessel equation of order zero is found by setting $a_0 = 1$ and substituting for $y_1(x)$ and $a'_{2m}(0) = b_{2m}(0)$ in Eq. (23) of Section 5.7. We obtain

$$y_2(x) = J_0(x) \ln x + \sum_{m=1}^{\infty} \frac{(-1)^{m+1} H_m}{2^{2m}(m!)^2} x^{2m}, \qquad x > 0. \tag{10}$$

Instead of y_2, the second solution is usually taken to be a certain linear combination of J_0 and y_2. It is known as the Bessel function of the second kind of order zero

and is denoted by Y_0. Following Copson (Chapter 12), we define[16]

$$Y_0(x) = \frac{2}{\pi}[y_2(x) + (\gamma - \ln 2)J_0(x)]. \tag{11}$$

Here γ is a constant known as the Euler–Máscheroni[17] constant; it is defined by the equation

$$\gamma = \lim_{n \to \infty} (H_n - \ln n) \cong 0.5772. \tag{12}$$

Substituting for $y_2(x)$ in Eq. (11), we obtain

$$Y_0(x) = \frac{2}{\pi}\left[\left(\gamma + \ln \frac{x}{2}\right) J_0(x) + \sum_{m=1}^{\infty} \frac{(-1)^{m+1} H_m}{2^{2m}(m!)^2} x^{2m}\right], \qquad x > 0. \tag{13}$$

The general solution of the Bessel equation of order zero for $x > 0$ is

$$y = c_1 J_0(x) + c_2 Y_0(x).$$

Note that $J_0(x) \to 1$ as $x \to 0$ and that $Y_0(x)$ has a logarithmic singularity at $x = 0$; that is, $Y_0(x)$ behaves as $(2/\pi) \ln x$ when $x \to 0$ through positive values. Thus, if we are interested in solutions of Bessel's equation of order zero that are finite at the origin, which is often the case, we must discard Y_0. The graphs of the functions J_0 and Y_0 are shown in Figure 5.8.2.

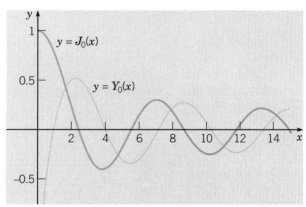

FIGURE 5.8.2 The Bessel functions J_0 and Y_0.

It is interesting to note from Figure 5.8.2 that for x large, both $J_0(x)$ and $Y_0(x)$ are oscillatory. Such a behavior might be anticipated from the original equation; indeed

[16]Other authors use other definitions for Y_0. The present choice for Y_0 is also known as the Weber function, after Heinrich Weber (1842–1913), who taught at several German universities.

[17]Lorenzo Máscheroni (1750–1800) was an Italian priest and professor at the University of Pavia. He correctly calculated the first 19 decimal places of γ in 1790.

it is true for the solutions of the Bessel equation of order v. If we divide Eq. (1) by x^2, we obtain

$$y'' + \frac{1}{x}y' + \left(1 - \frac{v^2}{x^2}\right)y = 0.$$

For x very large it is reasonable to suspect that the terms $(1/x)y'$ and $(v^2/x^2)y$ are small and hence can be neglected. If this is true, then the Bessel equation of order v can be approximated by

$$y'' + y = 0.$$

The solutions of this equation are $\sin x$ and $\cos x$; thus we might anticipate that the solutions of Bessel's equation for large x are similar to linear combinations of $\sin x$ and $\cos x$. This is correct insofar as the Bessel functions are oscillatory; however, it is only partly correct. For x large the functions J_0 and Y_0 also decay as x increases; thus the equation $y'' + y = 0$ does not provide an adequate approximation to the Bessel equation for large x, and a more delicate analysis is required. In fact, it is possible to show that

$$J_0(x) \cong \left(\frac{2}{\pi x}\right)^{1/2} \cos\left(x - \frac{\pi}{4}\right) \qquad \text{as} \quad x \to \infty, \tag{14}$$

and that

$$Y_0(x) \cong \left(\frac{2}{\pi x}\right)^{1/2} \sin\left(x - \frac{\pi}{4}\right) \qquad \text{as} \quad x \to \infty. \tag{15}$$

These asymptotic approximations, as $x \to \infty$, are actually very good. For example, Figure 5.8.3 shows that the asymptotic approximation (14) to $J_0(x)$ is reasonably accurate for all $x \geq 1$. Thus to approximate $J_0(x)$ over the entire range from zero to infinity, one can use two or three terms of the series (7) for $x \leq 1$ and the asymptotic approximation (14) for $x \geq 1$.

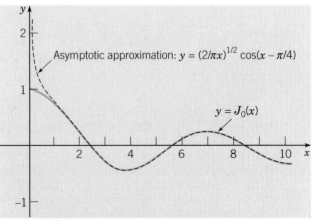

FIGURE 5.8.3 Asymptotic approximation to $J_0(x)$.

Bessel Equation of Order One-Half. This example illustrates the situation in which the roots of the indicial equation differ by a positive integer but there is no logarithmic term in the second solution. Setting $\nu = \frac{1}{2}$ in Eq. (1) gives

$$L[y] = x^2 y'' + x y' + \left(x^2 - \tfrac{1}{4}\right) y = 0. \tag{16}$$

When we substitute the series (3) for $y = \phi(r, x)$, we obtain

$$
\begin{aligned}
L[\phi](r, x) &= \sum_{n=0}^{\infty} \left[(r+n)(r+n-1) + (r+n) - \tfrac{1}{4}\right] a_n x^{r+n} + \sum_{n=0}^{\infty} a_n x^{r+n+2} \\
&= \left(r^2 - \tfrac{1}{4}\right) a_0 x^r + \left[(r+1)^2 - \tfrac{1}{4}\right] a_1 x^{r+1} \\
&\quad + \sum_{n=2}^{\infty} \left\{\left[(r+n)^2 - \tfrac{1}{4}\right] a_n + a_{n-2}\right\} x^{r+n} = 0.
\end{aligned}
\tag{17}
$$

The roots of the indicial equation are $r_1 = \frac{1}{2}$, $r_2 = -\frac{1}{2}$; hence the roots differ by an integer. The recurrence relation is

$$\left[(r+n)^2 - \tfrac{1}{4}\right] a_n = -a_{n-2}, \qquad n \geq 2. \tag{18}$$

Corresponding to the larger root $r_1 = \frac{1}{2}$, we find from the coefficient of x^{r+1} in Eq. (17) that $a_1 = 0$. Hence, from Eq. (18), $a_3 = a_5 = \cdots = a_{2n+1} = \cdots = 0$. Further, for $r = \frac{1}{2}$,

$$a_n = -\frac{a_{n-2}}{n(n+1)}, \qquad n = 2, 4, 6 \ldots,$$

or, letting $n = 2m$, we obtain

$$a_{2m} = -\frac{a_{2m-2}}{2m(2m+1)}, \qquad m = 1, 2, 3, \ldots.$$

By solving this recurrence relation we find that

$$a_2 = -\frac{a_0}{3!}, \qquad a_4 = \frac{a_0}{5!}, \ldots$$

and, in general,

$$a_{2m} = \frac{(-1)^m a_0}{(2m+1)!}, \qquad m = 1, 2, 3, \ldots.$$

Hence, taking $a_0 = 1$, we obtain

$$y_1(x) = x^{1/2} \left[1 + \sum_{m=1}^{\infty} \frac{(-1)^m x^{2m}}{(2m+1)!}\right] = x^{-1/2} \sum_{m=0}^{\infty} \frac{(-1)^m x^{2m+1}}{(2m+1)!}, \qquad x > 0. \tag{19}$$

The power series in Eq. (19) is precisely the Taylor series for $\sin x$; hence one solution of the Bessel equation of order one-half is $x^{-1/2} \sin x$. The Bessel function of the first kind of order one-half, $J_{1/2}$, is defined as $(2/\pi)^{1/2} y_1$. Thus

$$J_{1/2}(x) = \left(\frac{2}{\pi x}\right)^{1/2} \sin x, \qquad x > 0. \tag{20}$$

Corresponding to the root $r_2 = -\frac{1}{2}$, it is possible that we may have difficulty in computing a_1 since $N = r_1 - r_2 = 1$. However, from Eq. (17) for $r = -\frac{1}{2}$, the coefficients of x^r and x^{r+1} are both zero regardless of the choice of a_0 and a_1. Hence a_0 and a_1 can be chosen arbitrarily. From the recurrence relation (18) we obtain a set of even-numbered coefficients corresponding to a_0 and a set of odd-numbered coefficients corresponding to a_1. Thus no logarithmic term is needed to obtain a second solution in this case. It is left as an exercise to show that, for $r = -\frac{1}{2}$,

$$a_{2n} = \frac{(-1)^n a_0}{(2n)!}, \qquad a_{2n+1} = \frac{(-1)^n a_1}{(2n+1)!}, \qquad n = 1, 2, \ldots.$$

Hence

$$y_2(x) = x^{-1/2} \left[a_0 \sum_{n=0}^{\infty} \frac{(-1)^n x^{2n}}{(2n)!} + a_1 \sum_{n=0}^{\infty} \frac{(-1)^n x^{2n+1}}{(2n+1)!} \right]$$

$$= a_0 \frac{\cos x}{x^{1/2}} + a_1 \frac{\sin x}{x^{1/2}}, \qquad x > 0. \tag{21}$$

The constant a_1 simply introduces a multiple of $y_1(x)$. The second linearly independent solution of the Bessel equation of order one-half is usually taken to be the solution for which $a_0 = (2/\pi)^{1/2}$ and $a_1 = 0$. It is denoted by $J_{-1/2}$. Then

$$J_{-1/2}(x) = \left(\frac{2}{\pi x} \right)^{1/2} \cos x, \qquad x > 0. \tag{22}$$

The general solution of Eq. (16) is $y = c_1 J_{1/2}(x) + c_2 J_{-1/2}(x)$.

By comparing Eqs. (20) and (22) with Eqs. (14) and (15), we see that, except for a phase shift of $\pi/4$, the functions $J_{-1/2}$ and $J_{1/2}$ resemble J_0 and Y_0, respectively, for large x. The graphs of $J_{1/2}$ and $J_{-1/2}$ are shown in Figure 5.8.4.

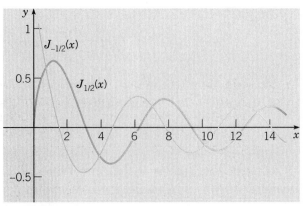

FIGURE 5.8.4 The Bessel functions $J_{1/2}26$ and $J_{-1/2}$.

Bessel Equation of Order One. This example illustrates the situation in which the roots of the indicial equation differ by a positive integer and the second solution involves a logarithmic term. Setting $\nu = 1$ in Eq. (1) gives

$$L[y] = x^2 y'' + xy' + (x^2 - 1)y = 0. \tag{23}$$

If we substitute the series (3) for $y = \phi(r, x)$ and collect terms as in the preceding cases, we obtain

$$L[\phi](r, x) = a_0(r^2 - 1)x^r + a_1[(r+1)^2 - 1]x^{r+1}$$

$$+ \sum_{n=2}^{\infty} \left\{ [(r+n)^2 - 1] a_n + a_{n-2} \right\} x^{r+n} = 0. \tag{24}$$

The roots of the indicial equation are $r_1 = 1$ and $r_2 = -1$. The recurrence relation is

$$[(r+n)^2 - 1]a_n(r) = -a_{n-2}(r), \qquad n \geq 2. \tag{25}$$

Corresponding to the larger root $r = 1$, the recurrence relation becomes

$$a_n = -\frac{a_{n-2}}{(n+2)n}, \qquad n = 2, 3, 4, \ldots.$$

We also find from the coefficient of x^{r+1} in Eq. (24) that $a_1 = 0$; hence from the recurrence relation, $a_3 = a_5 = \cdots = 0$. For even values of n, let $n = 2m$; then

$$a_{2m} = -\frac{a_{2m-2}}{(2m+2)(2m)} = -\frac{a_{2m-2}}{2^2(m+1)m}, \qquad m = 1, 2, 3, \ldots.$$

By solving this recurrence relation we obtain

$$a_{2m} = \frac{(-1)^m a_0}{2^{2m}(m+1)!m!}, \qquad m = 1, 2, 3, \ldots. \tag{26}$$

The Bessel function of the first kind of order one, denoted by J_1, is obtained by choosing $a_0 = 1/2$. Hence

$$J_1(x) = \frac{x}{2} \sum_{m=0}^{\infty} \frac{(-1)^m x^{2m}}{2^{2m}(m+1)!m!}. \tag{27}$$

The series converges absolutely for all x, so the function J_1 is analytic everywhere.

In determining a second solution of Bessel's equation of order one, we illustrate the method of direct substitution. The calculation of the general term in Eq. (28) below is rather complicated, but the first few coefficients can be found fairly easily. According to Theorem 5.7.1, we assume that

$$y_2(x) = aJ_1(x) \ln x + x^{-1} \left[1 + \sum_{n=1}^{\infty} c_n x^n \right], \qquad x > 0. \tag{28}$$

Computing $y_2'(x)$, $y_2''(x)$, substituting in Eq. (23), and making use of the fact that J_1 is a solution of Eq. (23) give

$$2axJ_1'(x) + \sum_{n=0}^{\infty} [(n-1)(n-2)c_n + (n-1)c_n - c_n]x^{n-1} + \sum_{n=0}^{\infty} c_n x^{n+1} = 0, \quad (29)$$

where $c_0 = 1$. Substituting for $J_1(x)$ from Eq. (27), shifting the indices of summation in the two series, and carrying out several steps of algebra give

$$-c_1 + [0 \cdot c_2 + c_0]x + \sum_{n=2}^{\infty} [(n^2-1)c_{n+1} + c_{n-1}]x^n$$

$$= -a\left[x + \sum_{m=1}^{\infty} \frac{(-1)^m(2m+1)x^{2m+1}}{2^{2m}(m+1)!\,m!}\right]. \quad (30)$$

From Eq. (30) we observe first that $c_1 = 0$, and $a = -c_0 = -1$. Further, since there are only odd powers of x on the right, the coefficient of each even power of x on the left must be zero. Thus, since $c_1 = 0$, we have $c_3 = c_5 = \cdots = 0$. Corresponding to the odd powers of x, we obtain the recurrence relation [let $n = 2m + 1$ in the series on the left side of Eq. (30)]

$$[(2m+1)^2 - 1]c_{2m+2} + c_{2m} = \frac{(-1)^m(2m+1)}{2^{2m}(m+1)!\,m!}, \quad m = 1,2,3,\ldots. \quad (31)$$

When we set $m = 1$ in Eq. (31), we obtain

$$(3^2 - 1)c_4 + c_2 = (-1)3/(2^2 \cdot 2!).$$

Notice that c_2 can be selected *arbitrarily*, and then this equation determines c_4. Also notice that in the equation for the coefficient of x, c_2 appeared multiplied by 0, and that equation was used to determine a. That c_2 is arbitrary is not surprising, since c_2 is the coefficient of x in the expression $x^{-1}\left[1 + \sum_{n=1}^{\infty} c_n x^n\right]$. Consequently, c_2 simply generates a multiple of J_1, and y_2 is determined only up to an additive multiple of J_1. In accordance with the usual practice, we choose $c_2 = 1/2^2$. Then we obtain

$$c_4 = \frac{-1}{2^4 \cdot 2}\left[\frac{3}{2} + 1\right] = \frac{-1}{2^4 2!}\left[\left(1 + \frac{1}{2}\right) + 1\right]$$

$$= \frac{(-1)}{2^4 \cdot 2!}(H_2 + H_1).$$

It is possible to show that the solution of the recurrence relation (31) is

$$c_{2m} = \frac{(-1)^{m+1}(H_m + H_{m-1})}{2^{2m}m!(m-1)!}, \quad m = 1,2,\ldots$$

with the understanding that $H_0 = 0$. Thus

$$y_2(x) = -J_1(x)\ln x + \frac{1}{x}\left[1 - \sum_{m=1}^{\infty}\frac{(-1)^m(H_m + H_{m-1})}{2^{2m}m!(m-1)!}x^{2m}\right], \qquad x > 0. \qquad (32)$$

The calculation of $y_2(x)$ using the alternative procedure [see Eqs. (19) and (20) of Section 5.7] in which we determine the $c_n(r_2)$ is slightly easier. In particular, the latter procedure yields the general formula for c_{2m} without the necessity of solving a recurrence relation of the form (31) (see Problem 11). In this regard, the reader may also wish to compare the calculations of the second solution of Bessel's equation of order zero in the text and in Problem 10.

The second solution of Eq. (23), the Bessel function of the second kind of order one, Y_1, is usually taken to be a certain linear combination of J_1 and y_2. Following Copson (Chapter 12), Y_1 is defined as

$$Y_1(x) = \frac{2}{\pi}[-y_2(x) + (\gamma - \ln 2)J_1(x)], \qquad (33)$$

where γ is defined in Eq. (12). The general solution of Eq. (23) for $x > 0$ is

$$y = c_1J_1(x) + c_2Y_1(x).$$

Notice that although J_1 is analytic at $x = 0$, the second solution Y_1 becomes unbounded in the same manner as $1/x$ as $x \to 0$. The graphs of J_1 and Y_1 are shown in Figure 5.8.5.

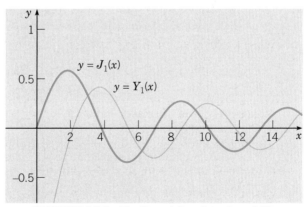

FIGURE 5.8.5 The Bessel functions J_1 and Y_1.

PROBLEMS

In each of Problems 1 through 4 show that the given differential equation has a regular singular point at $x = 0$, and determine two linearly independent solutions for $x > 0$.

1. $x^2y'' + 2xy' + xy = 0$
2. $x^2y'' + 3xy' + (1 + x)y = 0$
3. $x^2y'' + xy' + 2xy = 0$
4. $x^2y'' + 4xy' + (2 + x)y = 0$
5. Find two linearly independent solutions of the Bessel equation of order $\frac{3}{2}$,

$$x^2y'' + xy' + \left(x^2 - \tfrac{9}{4}\right)y = 0, \qquad x > 0.$$

6. Show that the Bessel equation of order one-half,

$$x^2 y'' + x y' + \left(x^2 - \tfrac{1}{4}\right) y = 0, \qquad x > 0,$$

can be reduced to the equation

$$v'' + v = 0$$

by the change of dependent variable $y = x^{-1/2} v(x)$. From this conclude that $y_1(x) = x^{-1/2} \cos x$ and $y_2(x) = x^{-1/2} \sin x$ are solutions of the Bessel equation of order one-half.

7. Show directly that the series for $J_0(x)$, Eq. (7), converges absolutely for all x.

8. Show directly that the series for $J_1(x)$, Eq. (27), converges absolutely for all x and that $J_0'(x) = -J_1(x)$.

9. Consider the Bessel equation of order v,

$$x^2 y'' + x y' + (x^2 - v^2) y = 0, \qquad x > 0,$$

where v is real and positive.

(a) Show that $x = 0$ is a regular singular point and that the roots of the indicial equation are v and $-v$.

(b) Corresponding to the larger root v, show that one solution is

$$y_1(x) = x^v \left[1 - \frac{1}{1!(1+v)} \left(\frac{x}{2}\right)^2 + \frac{1}{2!(1+v)(2+v)} \left(\frac{x}{2}\right)^4 \right.$$
$$\left. + \sum_{m=3}^{\infty} \frac{(-1)^m}{m!(1+v)\cdots(m+v)} \left(\frac{x}{2}\right)^{2m} \right].$$

(c) If $2v$ is not an integer, show that a second solution is

$$y_2(x) = x^{-v} \left[1 - \frac{1}{1!(1-v)} \left(\frac{x}{2}\right)^2 + \frac{1}{2!(1-v)(2-v)} \left(\frac{x}{2}\right)^4 \right.$$
$$\left. + \sum_{m=3}^{\infty} \frac{(-1)^m}{m!(1-v)\cdots(m-v)} \left(\frac{x}{2}\right)^{2m} \right].$$

Note that $y_1(x) \to 0$ as $x \to 0$, and that $y_2(x)$ is unbounded as $x \to 0$.

(d) Verify by direct methods that the power series in the expressions for $y_1(x)$ and $y_2(x)$ converge absolutely for all x. Also verify that y_2 is a solution provided only that v is not an integer.

10. In this section we showed that one solution of Bessel's equation of order zero,

$$L[y] = x^2 y'' + x y' + x^2 y = 0,$$

is J_0, where $J_0(x)$ is given by Eq. (7) with $a_0 = 1$. According to Theorem 5.7.1, a second solution has the form $(x > 0)$

$$y_2(x) = J_0(x) \ln x + \sum_{n=1}^{\infty} b_n x^n.$$

(a) Show that

$$L[y_2](x) = \sum_{n=2}^{\infty} n(n-1) b_n x^n + \sum_{n=1}^{\infty} n b_n x^n + \sum_{n=1}^{\infty} b_n x^{n+2} + 2 x J_0'(x). \tag{i}$$

(b) Substituting the series representation for $J_0(x)$ in Eq. (i), show that

$$b_1 x + 2^2 b_2 x^2 + \sum_{n=3}^{\infty} (n^2 b_n + b_{n-2}) x^n = -2 \sum_{n=1}^{\infty} \frac{(-1)^n 2n x^{2n}}{2^{2n} (n!)^2}. \tag{ii}$$

(c) Note that only even powers of x appear on the right side of Eq. (ii). Show that $b_1 = b_3 = b_5 = \cdots = 0, b_2 = 1/2^2 (1!)^2$, and that

$$(2n)^2 b_{2n} + b_{2n-2} = -2(-1)^n (2n)/2^{2n} (n!)^2, \qquad n = 2, 3, 4, \ldots.$$

Deduce that

$$b_4 = -\frac{1}{2^2 \, 4^2} \left(1 + \frac{1}{2} \right) \quad \text{and} \quad b_6 = \frac{1}{2^2 \, 4^2 \, 6^2} \left(1 + \frac{1}{2} + \frac{1}{3} \right).$$

The general solution of the recurrence relation is $b_{2n} = (-1)^{n+1} H_n / 2^{2n} (n!)^2$. Substituting for b_n in the expression for $y_2(x)$, we obtain the solution given in Eq. (10).

11. Find a second solution of Bessel's equation of order one by computing the $c_n(r_2)$ and a of Eq. (24) of Section 5.7 according to the formulas (19) and (20) of that section. Some guidelines along the way of this calculation are the following. First, use Eq. (24) of this section to show that $a_1(-1)$ and $a'_1(-1)$ are 0. Then show that $c_1(-1) = 0$ and, from the recurrence relation, that $c_n(-1) = 0$ for $n = 3, 5, \ldots$. Finally, use Eq. (25) to show that

$$a_2(r) = -\frac{a_0}{(r+1)(r+3)}, \qquad a_4(r) = \frac{a_0}{(r+1)(r+3)(r+3)(r+5)},$$

and that

$$a_{2m}(r) = \frac{(-1)^m a_0}{(r+1) \cdots (r+2m-1)(r+3) \cdots (r+2m+1)}, \qquad m \geq 3.$$

Then show that

$$c_{2m}(-1) = (-1)^{m+1} (H_m + H_{m-1}) / 2^{2m} m! (m-1)!, \qquad m \geq 1.$$

12. By a suitable change of variables it is sometimes possible to transform another differential equation into a Bessel equation. For example, show that a solution of

$$x^2 y'' + \left(\alpha^2 \beta^2 x^{2\beta} + \tfrac{1}{4} - \nu^2 \beta^2 \right) y = 0, \qquad x > 0$$

is given by $y = x^{1/2} f(\alpha x^\beta)$, where $f(\xi)$ is a solution of the Bessel equation of order ν.

13. Using the result of Problem 12, show that the general solution of the Airy equation

$$y'' - xy = 0, \qquad x > 0$$

is $y = x^{1/2} [c_1 f_1(\tfrac{2}{3} i x^{3/2}) + c_2 f_2(\tfrac{2}{3} i x^{3/2})]$, where $f_1(\xi)$ and $f_2(\xi)$ are linearly independent solutions of the Bessel equation of order one-third.

14. It can be shown that J_0 has infinitely many zeros for $x > 0$. In particular, the first three zeros are approximately 2.405, 5.520, and 8.653 (see Figure 5.8.1). Let $\lambda_j, j = 1, 2, 3, \ldots$, denote the zeros of J_0; it follows that

$$J_0(\lambda_j x) = \begin{cases} 1, & x = 0, \\ 0, & x = 1. \end{cases}$$

Verify that $y = J_0(\lambda_j x)$ satisfies the differential equation

$$y'' + \frac{1}{x} y' + \lambda_j^2 y = 0, \qquad x > 0.$$

Hence show that

$$\int_0^1 x J_0(\lambda_i x) J_0(\lambda_j x)\, dx = 0 \qquad \text{if} \quad \lambda_i \neq \lambda_j.$$

This important property of $J_0(\lambda_i x)$, known as the orthogonality property, is useful in solving boundary value problems.

Hint: Write the differential equation for $J_0(\lambda_i x)$. Multiply it by $x J_0(\lambda_j x)$ and subtract it from $x J_0(\lambda_i x)$ times the differential equation for $J_0(\lambda_j x)$. Then integrate from 0 to 1.

REFERENCES

Coddington, E. A., *An Introduction to Ordinary Differential Equations* (Englewood Cliffs, NJ: Prentice-Hall, 1961; New York: Dover, 1989).

Copson, E. T., *An Introduction to the Theory of Functions of a Complex Variable* (Oxford: Oxford University Press, 1935).

Proofs of Theorems 5.3.1 and 5.7.1 can be found in intermediate or advanced books; for example, see Chapters 3 and 4 of Coddington or Chapters 3 and 4 of

Rainville, E. D., *Intermediate Differential Equations* (2nd ed.) (New York: Macmillan, 1964).

Also see these texts for a discussion of the point at infinity, which was mentioned in Problem 21 of Section 5.4. The behavior of solutions near an irregular singular point is an even more advanced topic; a brief discussion can be found in Chapter 5 of

Coddington, E. A., and Levinson, N., *Theory of Ordinary Differential Equations* (New York: McGraw-Hill, 1955).

Fuller discussions of the Bessel equation, the Legendre equation, and many of the other named equations can be found in advanced books on differential equations, methods of applied mathematics, and special functions. A text dealing with special functions such as the Legendre polynomials and the Bessel functions is

Hochstadt, H., *Special Functions of Mathematical Physics* (New York: Holt, 1961).

An excellent compilation of formulas, graphs, and tables of Bessel functions, Legendre functions, and other special functions of mathematical physics may be found in

Abramowitz, M., and Stegun, I. A. (eds.), *Handbook of Mathematical Functions* (New York: Dover, 1965); originally published by the National Bureau of Standards, Washington, DC, 1964.

The Laplace Transform

Many practical engineering problems involve mechanical or electrical systems acted on by discontinuous or impulsive forcing terms. For such problems the methods described in Chapter 3 are often rather awkward to use. Another method that is especially well suited to these problems, although useful much more generally, is based on the Laplace transform. In this chapter we describe how this important method works, emphasizing problems typical of those that arise in engineering applications.

6.1 Definition of the Laplace Transform

Improper Integrals. Since the Laplace transform involves an integral from zero to infinity, a knowledge of improper integrals of this type is necessary to appreciate the subsequent development of the properties of the transform. We provide a brief review of such improper integrals here. If you are already familiar with improper integrals, you may wish to skip over this review. On the other hand, if improper integrals are new to you, then you should probably consult a calculus book, where many more details and examples will be found.

An improper integral over an unbounded interval is defined as a limit of integrals over finite intervals; thus

$$\int_a^\infty f(t)\,dt = \lim_{A\to\infty} \int_a^A f(t)\,dt, \tag{1}$$

where A is a positive real number. If the integral from a to A exists for each $A > a$, and if the limit as $A \to \infty$ exists, then the improper integral is said to **converge** to that limiting value. Otherwise the integral is said to **diverge**, or to fail to exist. The following examples illustrate both possibilities.

EXAMPLE 1

Let $f(t) = e^{ct}$, $t \geq 0$, where c is a real nonzero constant. Then

$$\int_0^{\infty} e^{ct}\, dt = \lim_{A \to \infty} \int_0^A e^{ct}\, dt = \lim_{A \to \infty} \frac{e^{ct}}{c}\bigg|_0^A$$

$$= \lim_{A \to \infty} \frac{1}{c}(e^{cA} - 1).$$

It follows that the improper integral converges to the value $-1/c$ if $c < 0$ and diverges if $c > 0$. If $c = 0$, the integrand $f(t)$ is the constant function with value 1, and the integral again diverges.

EXAMPLE 2

Let $f(t) = 1/t$, $t \geq 1$. Then

$$\int_1^{\infty} \frac{dt}{t} = \lim_{A \to \infty} \int_1^A \frac{dt}{t} = \lim_{A \to \infty} \ln A.$$

Since $\lim_{A \to \infty} \ln A = \infty$, the improper integral diverges.

EXAMPLE 3

Let $f(t) = t^{-p}$, $t \geq 1$, where p is a real constant and $p \neq 1$; the case $p = 1$ was considered in Example 2. Then

$$\int_1^{\infty} t^{-p}\, dt = \lim_{A \to \infty} \int_1^A t^{-p}\, dt = \lim_{A \to \infty} \frac{1}{1-p}(A^{1-p} - 1).$$

As $A \to \infty$, $A^{1-p} \to 0$ if $p > 1$, but $A^{1-p} \to \infty$ if $p < 1$. Hence $\int_1^{\infty} t^{-p}\, dt$ converges to the value $1/(p-1)$ for $p > 1$ but (incorporating the result of Example 2) diverges for $p \leq 1$. These results are analogous to those for the infinite series $\sum_{n=1}^{\infty} n^{-p}$.

Before discussing the possible existence of $\int_a^{\infty} f(t)\, dt$, it is helpful to define certain terms. A function f is said to be **piecewise continuous** on an interval $\alpha \leq t \leq \beta$ if the interval can be partitioned by a finite number of points $\alpha = t_0 < t_1 < \cdots < t_n = \beta$ so that

1. f is continuous on each open subinterval $t_{i-1} < t < t_i$.
2. f approaches a finite limit as the endpoints of each subinterval are approached from within the subinterval.

In other words, f is piecewise continuous on $\alpha \leq t \leq \beta$ if it is continuous there except for a finite number of jump discontinuities. If f is piecewise continuous on $\alpha \leq t \leq \beta$ for every $\beta > \alpha$, then f is said to be piecewise continuous on $t \geq \alpha$. An example of a piecewise continuous function is shown in Figure 6.1.1.

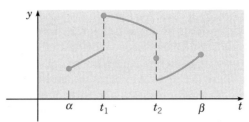

FIGURE 6.1.1 A piecewise continuous function.

If f is piecewise continuous on the interval $a \leq t \leq A$, then it can be shown that $\int_a^A f(t)\,dt$ exists. Hence, if f is piecewise continuous for $t \geq a$, then $\int_a^A f(t)\,dt$ exists for each $A > a$. However, piecewise continuity is not enough to ensure convergence of the improper integral $\int_a^\infty f(t)\,dt$, as the preceding examples show.

If f cannot be integrated easily in terms of elementary functions, the definition of convergence of $\int_a^\infty f(t)\,dt$ may be difficult to apply. Frequently, the most convenient way to test the convergence or divergence of an improper integral is by the following comparison theorem, which is analogous to a similar theorem for infinite series.

Theorem 6.1.1 If f is piecewise continuous for $t \geq a$, if $|f(t)| \leq g(t)$ when $t \geq M$ for some positive constant M, and if $\int_M^\infty g(t)\,dt$ converges, then $\int_a^\infty f(t)\,dt$ also converges. On the other hand, if $f(t) \geq g(t) \geq 0$ for $t \geq M$, and if $\int_M^\infty g(t)\,dt$ diverges, then $\int_a^\infty f(t)\,dt$ also diverges.

The proof of this result from the calculus will not be given here. It is made plausible, however, by comparing the areas represented by $\int_M^\infty g(t)\,dt$ and $\int_M^\infty |f(t)|\,dt$. The functions most useful for comparison purposes are e^{ct} and t^{-p}, which we considered in Examples 1, 2, and 3.

The Laplace Transform. Among the tools that are very useful for solving linear differential equations are **integral transforms**. An integral transform is a relation of the form

$$F(s) = \int_\alpha^\beta K(s,t) f(t)\,dt, \tag{2}$$

where $K(s,t)$ is a given function, called the **kernel** of the transformation, and the limits of integration α and β are also given. It is possible that $\alpha = -\infty$ or $\beta = \infty$, or both. The relation (2) transforms the function f into another function F, which is called the transform of f. The general idea in using an integral transform to solve a

differential equation is as follows: Use the relation (2) to transform a problem for an unknown function f into a simpler problem for F, then solve this simpler problem to find F, and finally recover the desired function f from its transform F. This last step is known as "inverting the transform."

There are several integral transforms that are useful in applied mathematics, but in this chapter we consider only the Laplace[1] transform. This transform is defined in the following way. Let $f(t)$ be given for $t \geq 0$, and suppose that f satisfies certain conditions to be stated a little later. Then the Laplace transform of f, which we will denote by $\mathcal{L}\{f(t)\}$ or by $F(s)$, is defined by the equation

$$\mathcal{L}\{f(t)\} = F(s) = \int_0^\infty e^{-st} f(t)\, dt, \tag{3}$$

whenever this improper integral converges. The Laplace transform makes use of the kernel $K(s, t) = e^{-st}$. Since the solutions of linear differential equations with constant coefficients are based on the exponential function, the Laplace transform is particularly useful for such equations.

In general, the parameter s may be complex, and the full power of the Laplace transform becomes available only when we regard $F(s)$ as a function of a complex variable. However, for the problems discussed here, it is sufficient to consider only real values of s. The Laplace transform F of a function f exists if f satisfies certain conditions, such as those stated in the following theorem.

Theorem 6.1.2

Suppose that

1. f is piecewise continuous on the interval $0 \leq t \leq A$ for any positive A.
2. $|f(t)| \leq Ke^{at}$ when $t \geq M$. In this inequality, K, a, and M are real constants, K and M necessarily positive.

Then the Laplace transform $\mathcal{L}\{f(t)\} = F(s)$, defined by Eq. (3), exists for $s > a$.

To establish this theorem we must show that the integral in Eq. (3) converges for $s > a$. Splitting the improper integral into two parts, we have

$$\int_0^\infty e^{-st} f(t)\, dt = \int_0^M e^{-st} f(t)\, dt + \int_M^\infty e^{-st} f(t)\, dt. \tag{4}$$

The first integral on the right side of Eq. (4) exists by hypothesis (1) of the theorem; hence the existence of $F(s)$ depends on the convergence of the second integral. By hypothesis (2) we have, for $t \geq M$,

$$|e^{-st} f(t)| \leq Ke^{-st} e^{at} = Ke^{(a-s)t},$$

[1]The Laplace transform is named for the eminent French mathematician P. S. Laplace, who studied the relation (3) in 1782. However, the techniques described in this chapter were not developed until a century or more later. We owe them mainly to Oliver Heaviside (1850–1925), an innovative but unconventional English electrical engineer, who made significant contributions to the development and application of electromagnetic theory.

and thus, by Theorem 6.1.1, $F(s)$ exists provided that $\int_M^\infty e^{(a-s)t}\,dt$ converges. Referring to Example 1 with c replaced by $a - s$, we see that this latter integral converges when $a - s < 0$, which establishes Theorem 6.1.2.

In this chapter (except in Section 6.5) we deal almost exclusively with functions that satisfy the conditions of Theorem 6.1.2. Such functions are described as piecewise continuous and of **exponential order** as $t \to \infty$. The Laplace transforms of some important elementary functions are given in the following examples.

**EXAMPLE
4**

Let $f(t) = 1$, $t \geq 0$. Then, as in Example 1,

$$\mathcal{L}\{1\} = \int_0^\infty e^{-st}\,dt = -\lim_{A \to \infty} \left.\frac{e^{-st}}{s}\right|_0^A = \frac{1}{s}, \qquad s > 0.$$

**EXAMPLE
5**

Let $f(t) = e^{at}$, $t \geq 0$. Then, again referring to Example 1,

$$\mathcal{L}\{e^{at}\} = \int_0^\infty e^{-st} e^{at}\,dt = \int_0^\infty e^{-(s-a)t}\,dt$$

$$= \frac{1}{s - a}, \qquad s > a.$$

**EXAMPLE
6**

Let $f(t) = \sin at$, $t \geq 0$. Then

$$\mathcal{L}\{\sin at\} = F(s) = \int_0^\infty e^{-st} \sin at\,dt, \qquad s > 0.$$

Since

$$F(s) = \lim_{A \to \infty} \int_0^A e^{-st} \sin at\,dt,$$

upon integrating by parts, we obtain

$$F(s) = \lim_{A \to \infty} \left[-\left.\frac{e^{-st} \cos at}{a}\right|_0^A - \frac{s}{a} \int_0^A e^{-st} \cos at\,dt \right]$$

$$= \frac{1}{a} - \frac{s}{a} \int_0^\infty e^{-st} \cos at\,dt.$$

A second integration by parts then yields

$$F(s) = \frac{1}{a} - \frac{s^2}{a^2} \int_0^\infty e^{-st} \sin at\,dt$$

$$= \frac{1}{a} - \frac{s^2}{a^2} F(s).$$

Hence, solving for $F(s)$, we have

$$F(s) = \frac{a}{s^2 + a^2}, \qquad s > 0.$$

Now let us suppose that f_1 and f_2 are two functions whose Laplace transforms exist for $s > a_1$ and $s > a_2$, respectively. Then, for s greater than the maximum of a_1 and a_2,

$$\mathcal{L}\{c_1 f_1(t) + c_2 f_2(t)\} = \int_0^\infty e^{-st}[c_1 f_1(t) + c_2 f_2(t)]\, dt$$

$$= c_1 \int_0^\infty e^{-st} f_1(t)\, dt + c_2 \int_0^\infty e^{-st} f_2(t)\, dt;$$

hence

$$\mathcal{L}\{c_1 f_1(t) + c_2 f_2(t)\} = c_1 \mathcal{L}\{f_1(t)\} + c_2 \mathcal{L}\{f_2(t)\}. \tag{5}$$

Equation (5) states that the Laplace transform is a **linear operator**, and we make frequent use of this property later. The sum in Eq. (5) can be readily extended to an arbitrary number of terms.

EXAMPLE 7

Find the Laplace transform of $f(t) = 5e^{-2t} - 3\sin 4t,\ t \geq 0$.
 Using Eq. (5), we write

$$\mathcal{L}\{f(t)\} = 5\mathcal{L}\{e^{-2t}\} - 3\mathcal{L}\{\sin 4t\}.$$

Then, from Examples 5 and 6, we obtain

$$\mathcal{L}\{f(t)\} = \frac{5}{s+2} - \frac{12}{s^2 + 16}, \qquad s > 0.$$

PROBLEMS

In each of Problems 1 through 4 sketch the graph of the given function. In each case determine whether f is continuous, piecewise continuous, or neither on the interval $0 \leq t \leq 3$.

1. $f(t) = \begin{cases} t^2, & 0 \leq t \leq 1 \\ 2+t, & 1 < t \leq 2 \\ 6-t, & 2 < t \leq 3 \end{cases}$

2. $f(t) = \begin{cases} t^2, & 0 \leq t \leq 1 \\ (t-1)^{-1}, & 1 < t \leq 2 \\ 1, & 2 < t \leq 3 \end{cases}$

3. $f(t) = \begin{cases} t^2, & 0 \leq t \leq 1 \\ 1, & 1 < t \leq 2 \\ 3-t, & 2 < t \leq 3 \end{cases}$

4. $f(t) = \begin{cases} t, & 0 \leq t \leq 1 \\ 3-t, & 1 < t \leq 2 \\ 1, & 2 < t \leq 3 \end{cases}$

5. Find the Laplace transform of each of the following functions:
 (a) t
 (b) t^2
 (c) t^n, where n is a positive integer

6. Find the Laplace transform of $f(t) = \cos at$, where a is a real constant.

Recall that $\cosh bt = (e^{bt} + e^{-bt})/2$ and $\sinh bt = (e^{bt} - e^{-bt})/2$. In each of Problems 7 through 10 find the Laplace transform of the given function; a and b are real constants.

7. $\cosh bt$

8. $\sinh bt$

9. $e^{at} \cosh bt$

10. $e^{at} \sinh bt$

In each of Problems 11 through 14 recall that $\cos bt = (e^{ibt} + e^{-ibt})/2$ and $\sin bt = (e^{ibt} - e^{-ibt})/2i$. Assuming that the necessary elementary integration formulas extend to this case, find the Laplace transform of the given function; a and b are real constants.

11. $\sin bt$

12. $\cos bt$

13. $e^{at} \sin bt$

14. $e^{at} \cos bt$

In each of Problems 15 through 20, using integration by parts, find the Laplace transform of the given function; n is a positive integer and a is a real constant.

15. te^{at}

16. $t \sin at$

17. $t \cosh at$

18. $t^n e^{at}$

19. $t^2 \sin at$

20. $t^2 \sinh at$

In each of Problems 21 through 24 determine whether the given integral converges or diverges.

21. $\displaystyle\int_0^\infty (t^2 + 1)^{-1}\, dt$

22. $\displaystyle\int_0^\infty te^{-t}\, dt$

23. $\displaystyle\int_1^\infty t^{-2}e^t\, dt$

24. $\displaystyle\int_0^\infty e^{-t}\cos t\, dt$

25. Suppose that f and f' are continuous for $t \geq 0$ and of exponential order as $t \to \infty$. Show by integration by parts that if $F(s) = \mathcal{L}\{f(t)\}$, then $\lim\limits_{s\to\infty} F(s) = 0$. The result is actually true under less restrictive conditions, such as those of Theorem 6.1.2.

26. **The Gamma Function.** The gamma function is denoted by $\Gamma(p)$ and is defined by the integral

$$\Gamma(p + 1) = \int_0^\infty e^{-x}x^p\, dx. \tag{i}$$

The integral converges as $x \to \infty$ for all p. For $p < 0$ it is also improper because the integrand becomes unbounded as $x \to 0$. However, the integral can be shown to converge at $x = 0$ for $p > -1$.

(a) Show that, for $p > 0$,

$$\Gamma(p + 1) = p\Gamma(p).$$

(b) Show that $\Gamma(1) = 1$.

(c) If p is a positive integer n, show that

$$\Gamma(n + 1) = n!.$$

Since $\Gamma(p)$ is also defined when p is not an integer, this function provides an extension of the factorial function to nonintegral values of the independent variable. Note that it is also consistent to define $0! = 1$.

(d) Show that, for $p > 0$,

$$p(p + 1)(p + 2)\cdots(p + n - 1) = \Gamma(p + n)/\Gamma(p).$$

Thus $\Gamma(p)$ can be determined for all positive values of p if $\Gamma(p)$ is known in a single interval of unit length, say, $0 < p \leq 1$. It is possible to show that $\Gamma\left(\frac{1}{2}\right) = \sqrt{\pi}$. Find $\Gamma\left(\frac{3}{2}\right)$ and $\Gamma\left(\frac{11}{2}\right)$.

27. Consider the Laplace transform of t^p, where $p > -1$.

 (a) Referring to Problem 26, show that

$$\mathcal{L}\{t^p\} = \int_0^\infty e^{-st} t^p \, dt = \frac{1}{s^{p+1}} \int_0^\infty e^{-x} x^p \, dx$$

$$= \Gamma(p+1)/s^{p+1}, \qquad s > 0.$$

 (b) Let p be a positive integer n in (a); show that

$$\mathcal{L}\{t^n\} = n!/s^{n+1}, \qquad s > 0.$$

 (c) Show that

$$\mathcal{L}\{t^{-1/2}\} = \frac{2}{\sqrt{s}} \int_0^\infty e^{-x^2} \, dx, \qquad s > 0.$$

 It is possible to show that

$$\int_0^\infty e^{-x^2} \, dx = \frac{\sqrt{\pi}}{2};$$

 hence

$$\mathcal{L}\{t^{-1/2}\} = \sqrt{\pi/s}, \qquad s > 0.$$

 (d) Show that

$$\mathcal{L}\{t^{1/2}\} = \sqrt{\pi}/(2s^{3/2}), \qquad s > 0.$$

6.2 Solution of Initial Value Problems

In this section we show how the Laplace transform can be used to solve initial value problems for linear differential equations with constant coefficients. The usefulness of the Laplace transform in this connection rests primarily on the fact that the transform of f' is related in a simple way to the transform of f. The relationship is expressed in the following theorem.

Theorem 6.2.1 Suppose that f is continuous and f' is piecewise continuous on any interval $0 \le t \le A$. Suppose further that there exist constants K, a, and M such that $|f(t)| \le Ke^{at}$ for $t \ge M$. Then $\mathcal{L}\{f'(t)\}$ exists for $s > a$, and moreover

$$\mathcal{L}\{f'(t)\} = s\mathcal{L}\{f(t)\} - f(0). \tag{1}$$

To prove this theorem, we consider the integral

$$\int_0^A e^{-st} f'(t) \, dt.$$

If f' has points of discontinuity in the interval $0 \leq t \leq A$, let them be denoted by t_1, t_2, \ldots, t_n. Then we can write this integral as

$$\int_0^A e^{-st} f'(t)\, dt = \int_0^{t_1} e^{-st} f'(t)\, dt + \int_{t_1}^{t_2} e^{-st} f'(t)\, dt + \cdots + \int_{t_n}^A e^{-st} f'(t)\, dt.$$

Integrating each term on the right by parts yields

$$\int_0^A e^{-st} f'(t)\, dt = e^{-st} f(t) \Big|_0^{t_1} + e^{-st} f(t) \Big|_{t_1}^{t_2} + \cdots + e^{-st} f(t) \Big|_{t_n}^A$$

$$+ s \left[\int_0^{t_1} e^{-st} f(t)\, dt + \int_{t_1}^{t_2} e^{-st} f(t)\, dt + \cdots + \int_{t_n}^A e^{-st} f(t)\, dt \right].$$

Since f is continuous, the contributions of the integrated terms at t_1, t_2, \ldots, t_n cancel. Combining the integrals gives

$$\int_0^A e^{-st} f'(t)\, dt = e^{-sA} f(A) - f(0) + s \int_0^A e^{-st} f(t)\, dt.$$

As $A \to \infty$, $e^{-sA} f(A) \to 0$ whenever $s > a$. Hence, for $s > a$,

$$\mathcal{L}\{f'(t)\} = s\mathcal{L}\{f(t)\} - f(0),$$

which establishes the theorem.

If f' and f'' satisfy the same conditions that are imposed on f and f', respectively, in Theorem 6.2.1, then it follows that the Laplace transform of f'' also exists for $s > a$ and is given by

$$\mathcal{L}\{f''(t)\} = s^2 \mathcal{L}\{f(t)\} - sf(0) - f'(0). \tag{2}$$

Indeed, provided the function f and its derivatives satisfy suitable conditions, an expression for the transform of the nth derivative $f^{(n)}$ can be derived by successive applications of this theorem. The result is given in the following corollary.

Corollary 6.2.2 Suppose that the functions $f, f', \ldots, f^{(n-1)}$ are continuous and that $f^{(n)}$ is piecewise continuous on any interval $0 \leq t \leq A$. Suppose further that there exist constants K, a, and M such that $|f(t)| \leq Ke^{at}$, $|f'(t)| \leq Ke^{at}, \ldots, |f^{(n-1)}(t)| \leq Ke^{at}$ for $t \geq M$. Then $\mathcal{L}\{f^{(n)}(t)\}$ exists for $s > a$ and is given by

$$\mathcal{L}\{f^{(n)}(t)\} = s^n \mathcal{L}\{f(t)\} - s^{n-1} f(0) - \cdots - sf^{(n-2)}(0) - f^{(n-1)}(0). \tag{3}$$

We now show how the Laplace transform can be used to solve initial value problems. It is most useful for problems involving nonhomogeneous differential equations, as we will demonstrate in later sections of this chapter. However, we begin by looking at some homogeneous equations, which are a bit simpler. For example, consider the differential equation

$$y'' - y' - 2y = 0 \tag{4}$$

and the initial conditions

$$y(0) = 1, \qquad y'(0) = 0. \tag{5}$$

This problem is easily solved by the methods of Section 3.1. The characteristic equation is

$$r^2 - r - 2 = (r - 2)(r + 1) = 0, \tag{6}$$

and consequently the general solution of Eq. (4) is

$$y = c_1 e^{-t} + c_2 e^{2t}. \tag{7}$$

To satisfy the initial conditions (5) we must have $c_1 + c_2 = 1$ and $-c_1 + 2c_2 = 0$; hence $c_1 = \frac{2}{3}$ and $c_2 = \frac{1}{3}$, so that the solution of the initial value problem (4) and (5) is

$$y = \phi(t) = \tfrac{2}{3} e^{-t} + \tfrac{1}{3} e^{2t}. \tag{8}$$

Now let us solve the same problem by using the Laplace transform. To do this we must assume that the problem has a solution $y = \phi(t)$, which with its first two derivatives satisfies the conditions of Corollary 6.2.2. Then, taking the Laplace transform of the differential equation (4), we obtain

$$\mathcal{L}\{y''\} - \mathcal{L}\{y'\} - 2\mathcal{L}\{y\} = 0, \tag{9}$$

where we have used the linearity of the transform to write the transform of a sum as the sum of the separate transforms. Upon using the corollary to express $\mathcal{L}\{y''\}$ and $\mathcal{L}\{y'\}$ in terms of $\mathcal{L}\{y\}$, we find that Eq. (9) becomes

$$s^2 \mathcal{L}\{y\} - sy(0) - y'(0) - [s\mathcal{L}\{y\} - y(0)] - 2\mathcal{L}\{y\} = 0,$$

or

$$(s^2 - s - 2)Y(s) + (1 - s)y(0) - y'(0) = 0, \tag{10}$$

where $Y(s) = \mathcal{L}\{y\}$. Substituting for $y(0)$ and $y'(0)$ in Eq. (10) from the initial conditions (5), and then solving for $Y(s)$, we obtain

$$Y(s) = \frac{s - 1}{s^2 - s - 2} = \frac{s - 1}{(s - 2)(s + 1)}. \tag{11}$$

We have thus obtained an expression for the Laplace transform $Y(s)$ of the solution $y = \phi(t)$ of the given initial value problem. To determine the function ϕ, we must find the function whose Laplace transform is $Y(s)$, as given by Eq. (11).

This can be done most easily by expanding the right side of Eq. (11) in partial fractions. Thus we write

$$Y(s) = \frac{s - 1}{(s - 2)(s + 1)} = \frac{a}{s - 2} + \frac{b}{s + 1} = \frac{a(s + 1) + b(s - 2)}{(s - 2)(s + 1)}, \tag{12}$$

where the coefficients a and b are to be determined. By equating numerators of the second and fourth members of Eq. (12), we obtain

$$s - 1 = a(s + 1) + b(s - 2),$$

an equation that must hold for all s. In particular, if we set $s = 2$, then it follows that $a = \frac{1}{3}$. Similarly, if we set $s = -1$, then we find that $b = \frac{2}{3}$. By substituting these

values for a and b, respectively, we have

$$Y(s) = \frac{1/3}{s-2} + \frac{2/3}{s+1}. \tag{13}$$

Finally, if we use the result of Example 5 of Section 6.1, it follows that $\frac{1}{3}e^{2t}$ has the transform $\frac{1}{3}(s-2)^{-1}$; similarly, $\frac{2}{3}e^{-t}$ has the transform $\frac{2}{3}(s+1)^{-1}$. Hence, by the linearity of the Laplace transform,

$$y = \phi(t) = \tfrac{1}{3}e^{2t} + \tfrac{2}{3}e^{-t}$$

has the transform (13) and is therefore the solution of the initial value problem (4), (5). Observe that it does satisfy the conditions of Corollary 6.2.2, as we assumed initially. Of course, this is the same solution that we obtained earlier.

The same procedure can be applied to the general second order linear equation with constant coefficients,

$$ay'' + by' + cy = f(t). \tag{14}$$

Assuming that the solution $y = \phi(t)$ satisfies the conditions of Corollary 6.2.2 for $n = 2$, we can take the transform of Eq. (14) and thereby obtain

$$a[s^2 Y(s) - sy(0) - y'(0)] + b[sY(s) - y(0)] + cY(s) = F(s), \tag{15}$$

where $F(s)$ is the transform of $f(t)$. By solving Eq. (15) for $Y(s)$, we find that

$$Y(s) = \frac{(as+b)y(0) + ay'(0)}{as^2 + bs + c} + \frac{F(s)}{as^2 + bs + c}. \tag{16}$$

The problem is then solved, provided that we can find the function $y = \phi(t)$ whose transform is $Y(s)$.

Even at this early stage of our discussion we can point out some of the essential features of the transform method. In the first place, the transform $Y(s)$ of the unknown function $y = \phi(t)$ is found by solving an *algebraic equation* rather than a *differential equation*, Eq. (10) rather than Eq. (4), or in general Eq. (15) rather than Eq. (14). This is the key to the usefulness of Laplace transforms for solving linear, constant coefficient, ordinary differential equations—the problem is reduced from a differential equation to an algebraic one. Next, the solution satisfying given initial conditions is automatically found, so that the task of determining appropriate values for the arbitrary constants in the general solution does not arise. Further, as indicated in Eq. (15), nonhomogeneous equations are handled in exactly the same way as homogeneous ones; it is not necessary to solve the corresponding homogeneous equation first. Finally, the method can be applied in the same way to higher order equations, as long as we assume that the solution satisfies the conditions of the corollary for the appropriate value of n.

Observe that the polynomial $as^2 + bs + c$ in the denominator on the right side of Eq. (16) is precisely the characteristic polynomial associated with Eq. (14). Since the use of a partial fraction expansion of $Y(s)$ to determine $\phi(t)$ requires us to factor this polynomial, the use of Laplace transforms does not avoid the necessity of finding roots of the characteristic equation. For equations of higher than second order this may be a difficult algebraic problem, particularly if the roots are irrational or complex.

The main difficulty that occurs in solving initial value problems by the transform technique lies in the problem of determining the function $y = \phi(t)$ corresponding to the transform $Y(s)$. This problem is known as the inversion problem for the Laplace transform; $\phi(t)$ is called the inverse transform corresponding to $Y(s)$, and the process of finding $\phi(t)$ from $Y(s)$ is known as inverting the transform. We also use the notation $\mathcal{L}^{-1}\{Y(s)\}$ to denote the inverse transform of $Y(s)$. There is a general formula for the inverse Laplace transform, but its use requires a knowledge of the theory of functions of a complex variable, and we do not consider it in this book. However, it is still possible to develop many important properties of the Laplace transform, and to solve many interesting problems, without the use of complex variables.

In solving the initial value problem (4), (5), we did not consider the question of whether there may be functions other than the one given by Eq. (8) that also have the transform (13). In fact, it can be shown that if f is a continuous function with the Laplace transform F, then there is no other continuous function having the same transform. In other words, there is essentially a one-to-one correspondence between functions and their Laplace transforms. This fact suggests the compilation of a table, such as Table 6.2.1, giving the transforms of functions frequently encountered, and vice versa. The entries in the second column of Table 6.2.1 are the transforms of those in the first column. Perhaps more important, the functions in the first column are the inverse transforms of those in the second column. Thus, for example, if the transform of the solution of a differential equation is known, the solution itself can often be found merely by looking it up in the table. Some of the entries in Table 6.2.1 have been used as examples, or appear as problems in Section 6.1, while others will be developed later in the chapter. The third column of the table indicates where the derivation of the given transforms may be found. Although Table 6.2.1 is sufficient for the examples and problems in this book, much larger tables are also available (see the list of references at the end of the chapter). Transforms and inverse transforms can also be readily obtained electronically by using a computer algebra system.

Frequently, a Laplace transform $F(s)$ is expressible as a sum of several terms,

$$F(s) = F_1(s) + F_2(s) + \cdots + F_n(s). \tag{17}$$

Suppose that $f_1(t) = \mathcal{L}^{-1}\{F_1(s)\}, \ldots, f_n(t) = \mathcal{L}^{-1}\{F_n(s)\}$. Then the function

$$f(t) = f_1(t) + \cdots + f_n(t)$$

has the Laplace transform $F(s)$. By the uniqueness property stated previously, no other continuous function f has the same transform. Thus

$$\mathcal{L}^{-1}\{F(s)\} = \mathcal{L}^{-1}\{F_1(s)\} + \cdots + \mathcal{L}^{-1}\{F_n(s)\}; \tag{18}$$

that is, the inverse Laplace transform is also a linear operator.

In many problems it is convenient to make use of this property by decomposing a given transform into a sum of functions whose inverse transforms are already known or can be found in the table. Partial fraction expansions are particularly useful in this connection, and a general result covering many cases is given in Problem 38. Other useful properties of Laplace transforms are derived later in this chapter.

As further illustrations of the technique of solving initial value problems by means of the Laplace transform and partial fraction expansions, consider the following examples.

TABLE 6.2.1 Elementary Laplace Transforms

$f(t) = \mathcal{L}^{-1}\{F(s)\}$	$F(s) = \mathcal{L}\{f(t)\}$	Notes		
1. 1	$\dfrac{1}{s}, \quad s > 0$	Sec. 6.1; Ex. 4		
2. e^{at}	$\dfrac{1}{s-a}, \quad s > a$	Sec. 6.1; Ex. 5		
3. $t^n, \quad n = $ positive integer	$\dfrac{n!}{s^{n+1}}, \quad s > 0$	Sec. 6.1; Prob. 27		
4. $t^p, \quad p > -1$	$\dfrac{\Gamma(p+1)}{s^{p+1}}, \quad s > 0$	Sec. 6.1; Prob. 27		
5. $\sin at$	$\dfrac{a}{s^2 + a^2}, \quad s > 0$	Sec. 6.1; Ex. 6		
6. $\cos at$	$\dfrac{s}{s^2 + a^2}, \quad s > 0$	Sec. 6.1; Prob. 6		
7. $\sinh at$	$\dfrac{a}{s^2 - a^2}, \quad s >	a	$	Sec. 6.1; Prob. 8
8. $\cosh at$	$\dfrac{s}{s^2 - a^2}, \quad s >	a	$	Sec. 6.1; Prob. 7
9. $e^{at}\sin bt$	$\dfrac{b}{(s-a)^2 + b^2}, \quad s > a$	Sec. 6.1; Prob. 13		
10. $e^{at}\cos bt$	$\dfrac{s-a}{(s-a)^2 + b^2}, \quad s > a$	Sec. 6.1; Prob. 14		
11. $t^n e^{at}, \quad n = $ positive integer	$\dfrac{n!}{(s-a)^{n+1}}, \quad s > a$	Sec. 6.1; Prob. 18		
12. $u_c(t)$	$\dfrac{e^{-cs}}{s}, \quad s > 0$	Sec. 6.3		
13. $u_c(t)f(t-c)$	$e^{-cs}F(s)$	Sec. 6.3		
14. $e^{ct}f(t)$	$F(s-c)$	Sec. 6.3		
15. $f(ct)$	$\dfrac{1}{c}F\left(\dfrac{s}{c}\right), \quad c > 0$	Sec. 6.3; Prob. 19		
16. $\displaystyle\int_0^t f(t-\tau)g(\tau)\,d\tau$	$F(s)G(s)$	Sec. 6.6		
17. $\delta(t-c)$	e^{-cs}	Sec. 6.5		
18. $f^{(n)}(t)$	$s^n F(s) - s^{n-1}f(0) - \cdots - f^{(n-1)}(0)$	Sec. 6.2		
19. $(-t)^n f(t)$	$F^{(n)}(s)$	Sec. 6.2; Prob. 28		

EXAMPLE 1

Find the solution of the differential equation

$$y'' + y = \sin 2t, \tag{19}$$

satisfying the initial conditions

$$y(0) = 2, \qquad y'(0) = 1. \tag{20}$$

We assume that this initial value problem has a solution $y = \phi(t)$, which with its first two derivatives satisfies the conditions of Corollary 6.2.2. Then, taking the Laplace transform of the differential equation, we have

$$s^2 Y(s) - sy(0) - y'(0) + Y(s) = 2/(s^2 + 4),$$

where the transform of $\sin 2t$ has been obtained from line 5 of Table 6.2.1. Substituting for $y(0)$ and $y'(0)$ from the initial conditions and solving for $Y(s)$, we obtain

$$Y(s) = \frac{2s^3 + s^2 + 8s + 6}{(s^2 + 1)(s^2 + 4)}. \tag{21}$$

Using partial fractions, we can write $Y(s)$ in the form

$$Y(s) = \frac{as + b}{s^2 + 1} + \frac{cs + d}{s^2 + 4} = \frac{(as + b)(s^2 + 4) + (cs + d)(s^2 + 1)}{(s^2 + 1)(s^2 + 4)}. \tag{22}$$

By expanding the numerator on the right side of Eq. (22) and equating it to the numerator in Eq. (21), we find that

$$2s^3 + s^2 + 8s + 6 = (a + c)s^3 + (b + d)s^2 + (4a + c)s + (4b + d)$$

for all s. Then, comparing coefficients of like powers of s, we have

$$a + c = 2, \qquad b + d = 1,$$
$$4a + c = 8, \qquad 4b + d = 6.$$

Consequently, $a = 2, c = 0, b = \frac{5}{3}$, and $d = -\frac{2}{3}$, from which it follows that

$$Y(s) = \frac{2s}{s^2 + 1} + \frac{5/3}{s^2 + 1} - \frac{2/3}{s^2 + 4}. \tag{23}$$

From lines 5 and 6 of Table 6.2.1, the solution of the given initial value problem is

$$y = \phi(t) = 2\cos t + \tfrac{5}{3}\sin t - \tfrac{1}{3}\sin 2t. \tag{24}$$

EXAMPLE 2

Find the solution of the initial value problem

$$y^{(4)} - y = 0, \tag{25}$$

$$y(0) = 0, \qquad y'(0) = 1, \qquad y''(0) = 0, \qquad y'''(0) = 0. \tag{26}$$

In this problem we need to assume that the solution $y = \phi(t)$ satisfies the conditions of Corollary 6.2.2 for $n = 4$. The Laplace transform of the differential equation (25) is

$$s^4 Y(s) - s^3 y(0) - s^2 y'(0) - sy''(0) - y'''(0) - Y(s) = 0.$$

Then, using the initial conditions (26) and solving for $Y(s)$, we have

$$Y(s) = \frac{s^2}{s^4 - 1}. \tag{27}$$

A partial fraction expansion of $Y(s)$ is

$$Y(s) = \frac{as+b}{s^2-1} + \frac{cs+d}{s^2+1},$$

and it follows that

$$(as+b)(s^2+1) + (cs+d)(s^2-1) = s^2 \tag{28}$$

for all s. By setting $s = 1$ and $s = -1$, respectively, in Eq. (28), we obtain the pair of equations

$$2(a+b) = 1, \qquad 2(-a+b) = 1,$$

and therefore $a = 0$ and $b = \frac{1}{2}$. If we set $s = 0$ in Eq. (28), then $b - d = 0$, so $d = \frac{1}{2}$. Finally, equating the coefficients of the cubic terms on each side of Eq. (28), we find that $a + c = 0$, so $c = 0$. Thus

$$Y(s) = \frac{1/2}{s^2-1} + \frac{1/2}{s^2+1}, \tag{29}$$

and from lines 7 and 5 of Table 6.2.1, the solution of the initial value problem (25), (26) is

$$y = \phi(t) = \frac{\sinh t + \sin t}{2}. \tag{30}$$

The most important elementary applications of the Laplace transform are in the study of mechanical vibrations and in the analysis of electric circuits; the governing equations were derived in Section 3.8. A vibrating spring–mass system has the equation of motion

$$m\frac{d^2u}{dt^2} + \gamma\frac{du}{dt} + ku = F(t), \tag{31}$$

where m is the mass, γ the damping coefficient, k the spring constant, and $F(t)$ the applied external force. The equation that describes an electric circuit containing an inductance L, a resistance R, and a capacitance C (an LRC circuit) is

$$L\frac{d^2Q}{dt^2} + R\frac{dQ}{dt} + \frac{1}{C}Q = E(t), \tag{32}$$

where $Q(t)$ is the charge on the capacitor and $E(t)$ is the applied voltage. In terms of the current $I(t) = dQ(t)/dt$, we can differentiate Eq. (32) and write

$$L\frac{d^2I}{dt^2} + R\frac{dI}{dt} + \frac{1}{C}I = \frac{dE}{dt}(t). \tag{33}$$

Suitable initial conditions on u, Q, or I must also be prescribed.

We have noted previously, in Section 3.8, that Eq. (31) for the spring–mass system and Eq. (32) or (33) for the electric circuit are identical mathematically, differing only in the interpretation of the constants and variables appearing in them. There are other physical problems that also lead to the same differential equation. Thus, once the mathematical problem is solved, its solution can be interpreted in terms of whichever corresponding physical problem is of immediate interest.

In the problem lists following this and other sections in this chapter are numerous initial value problems for second order linear differential equations with constant coefficients. Many can be interpreted as models of particular physical systems, but usually we do not point this out explicitly.

PROBLEMS

In each of Problems 1 through 10 find the inverse Laplace transform of the given function.

1. $\dfrac{3}{s^2 + 4}$

2. $\dfrac{4}{(s-1)^3}$

3. $\dfrac{2}{s^2 + 3s - 4}$

4. $\dfrac{3s}{s^2 - s - 6}$

5. $\dfrac{2s + 2}{s^2 + 2s + 5}$

6. $\dfrac{2s - 3}{s^2 - 4}$

7. $\dfrac{2s + 1}{s^2 - 2s + 2}$

8. $\dfrac{8s^2 - 4s + 12}{s(s^2 + 4)}$

9. $\dfrac{1 - 2s}{s^2 + 4s + 5}$

10. $\dfrac{2s - 3}{s^2 + 2s + 10}$

In each of Problems 11 through 23 use the Laplace transform to solve the given initial value problem.

11. $y'' - y' - 6y = 0;$ $y(0) = 1,$ $y'(0) = -1$

12. $y'' + 3y' + 2y = 0;$ $y(0) = 1,$ $y'(0) = 0$

13. $y'' - 2y' + 2y = 0;$ $y(0) = 0,$ $y'(0) = 1$

14. $y'' - 4y' + 4y = 0;$ $y(0) = 1,$ $y'(0) = 1$

15. $y'' - 2y' + 4y = 0;$ $y(0) = 2,$ $y'(0) = 0$

16. $y'' + 2y' + 5y = 0;$ $y(0) = 2,$ $y'(0) = -1$

17. $y^{(4)} - 4y''' + 6y'' - 4y' + y = 0;$ $y(0) = 0,$ $y'(0) = 1,$ $y''(0) = 0,$ $y'''(0) = 1$

18. $y^{(4)} - y = 0;$ $y(0) = 1,$ $y'(0) = 0,$ $y''(0) = 1,$ $y'''(0) = 0$

19. $y^{(4)} - 4y = 0;$ $y(0) = 1,$ $y'(0) = 0,$ $y''(0) = -2,$ $y'''(0) = 0$

20. $y'' + \omega^2 y = \cos 2t, \quad \omega^2 \neq 4;$ $y(0) = 1,$ $y'(0) = 0$

21. $y'' - 2y' + 2y = \cos t;$ $y(0) = 1,$ $y'(0) = 0$

22. $y'' - 2y' + 2y = e^{-t};$ $y(0) = 0,$ $y'(0) = 1$

23. $y'' + 2y' + y = 4e^{-t};$ $y(0) = 2,$ $y'(0) = -1$

In each of Problems 24 through 26 find the Laplace transform $Y(s) = \mathcal{L}\{y\}$ of the solution of the given initial value problem. A method of determining the inverse transform is developed in Section 6.3.

24. $y'' + 4y = \begin{cases} 1, & 0 \le t < \pi, \\ 0, & \pi \le t < \infty; \end{cases}$ $y(0) = 1,$ $y'(0) = 0$

25. $y'' + y = \begin{cases} t, & 0 \le t < 1, \\ 0, & 1 \le t < \infty; \end{cases}$ $y(0) = 0,$ $y'(0) = 0$

26. $y'' + 4y = \begin{cases} t, & 0 \le t < 1, \\ 1, & 1 \le t < \infty; \end{cases}$ $y(0) = 0,$ $y'(0) = 0$

27. The Laplace transforms of certain functions can be found conveniently from their Taylor series expansions.

(a) Using the Taylor series for $\sin t$,

$$\sin t = \sum_{n=0}^{\infty} \frac{(-1)^n t^{2n+1}}{(2n+1)!},$$

and assuming that the Laplace transform of this series can be computed term by term, verify that

$$\mathcal{L}\{\sin t\} = \frac{1}{s^2+1}, \qquad s > 1.$$

(b) Let

$$f(t) = \begin{cases} (\sin t)/t, & t \neq 0, \\ 1, & t = 0. \end{cases}$$

Find the Taylor series for f about $t = 0$. Assuming that the Laplace transform of this function can be computed term by term, verify that

$$\mathcal{L}\{f(t)\} = \arctan(1/s), \qquad s > 1.$$

(c) The Bessel function of the first kind of order zero, J_0, has the Taylor series (see Section 5.8)

$$J_0(t) = \sum_{n=0}^{\infty} \frac{(-1)^n t^{2n}}{2^{2n}(n!)^2}.$$

Assuming that the following Laplace transforms can be computed term by term, verify that

$$\mathcal{L}\{J_0(t)\} = (s^2+1)^{-1/2}, \qquad s > 1,$$

and

$$\mathcal{L}\{J_0(\sqrt{t})\} = s^{-1} e^{-1/4s}, \qquad s > 0.$$

Problems 28 through 36 are concerned with differentiation of the Laplace transform.

28. Let

$$F(s) = \int_0^{\infty} e^{-st} f(t)\, dt.$$

It is possible to show that as long as f satisfies the conditions of Theorem 6.1.2, it is legitimate to differentiate under the integral sign with respect to the parameter s when $s > a$.

(a) Show that $F'(s) = \mathcal{L}\{-tf(t)\}$.

(b) Show that $F^{(n)}(s) = \mathcal{L}\{(-t)^n f(t)\}$; hence differentiating the Laplace transform corresponds to multiplying the original function by $-t$.

In each of Problems 29 through 34 use the result of Problem 28 to find the Laplace transform of the given function; a and b are real numbers and n is a positive integer.

29. te^{at}	30. $t^2 \sin bt$
31. t^n	32. $t^n e^{at}$
33. $te^{at} \sin bt$	34. $te^{at} \cos bt$

35. Consider Bessel's equation of order zero,

$$ty'' + y' + ty = 0.$$

Recall from Section 5.4 that $t = 0$ is a regular singular point for this equation, and therefore solutions may become unbounded as $t \to 0$. However, let us try to determine whether there are any solutions that remain finite at $t = 0$ and have finite derivatives there. Assuming that there is such a solution $y = \phi(t)$, let $Y(s) = \mathcal{L}\{\phi(t)\}$.

(a) Show that $Y(s)$ satisfies

$$(1 + s^2)Y'(s) + sY(s) = 0.$$

(b) Show that $Y(s) = c(1 + s^2)^{-1/2}$, where c is an arbitrary constant.

(c) Writing $(1 + s^2)^{-1/2} = s^{-1}(1 + s^{-2})^{-1/2}$, expanding in a binomial series valid for $s > 1$, and assuming that it is permissible to take the inverse transform term by term, show that

$$y = c \sum_{n=0}^{\infty} \frac{(-1)^n t^{2n}}{2^{2n}(n!)^2} = cJ_0(t),$$

where J_0 is the Bessel function of the first kind of order zero. Note that $J_0(0) = 1$ and that J_0 has finite derivatives of all orders at $t = 0$. It was shown in Section 5.8 that the second solution of this equation becomes unbounded as $t \to 0$.

36. For each of the following initial value problems use the results of Problem 28 to find the differential equation satisfied by $Y(s) = \mathcal{L}\{\phi(t)\}$, where $y = \phi(t)$ is the solution of the given initial value problem.

(a) $y'' - ty = 0$; $y(0) = 1$, $y'(0) = 0$ (Airy's equation)

(b) $(1 - t^2)y'' - 2ty' + \alpha(\alpha + 1)y = 0$; $y(0) = 0$, $y'(0) = 1$ (Legendre's equation)

Note that the differential equation for $Y(s)$ is of first order in part (a), but of second order in part (b). This is due to the fact that t appears at most to the first power in the equation of part (a), whereas it appears to the second power in that of part (b). This illustrates that the Laplace transform is not often useful in solving differential equations with variable coefficients, unless all the coefficients are at most linear functions of the independent variable.

37. Suppose that

$$g(t) = \int_0^t f(\tau)\,d\tau.$$

If $G(s)$ and $F(s)$ are the Laplace transforms of $g(t)$ and $f(t)$, respectively, show that

$$G(s) = F(s)/s.$$

38. In this problem we show how a general partial fraction expansion can be used to calculate many inverse Laplace transforms. Suppose that

$$F(s) = P(s)/Q(s),$$

where $Q(s)$ is a polynomial of degree n with distinct zeros r_1, \ldots, r_n, and $P(s)$ is a polynomial of degree less than n. In this case it is possible to show that $P(s)/Q(s)$ has a partial fraction expansion of the form

$$\frac{P(s)}{Q(s)} = \frac{A_1}{s - r_1} + \cdots + \frac{A_n}{s - r_n}, \tag{i}$$

where the coefficients A_1, \ldots, A_n must be determined.

(a) Show that

$$A_k = P(r_k)/Q'(r_k), \qquad k = 1, \ldots, n. \tag{ii}$$

Hint: One way to do this is to multiply Eq. (i) by $s - r_k$ and then to take the limit as $s \to r_k$.

(b) Show that

$$\mathcal{L}^{-1}\{F(s)\} = \sum_{k=1}^{n} \frac{P(r_k)}{Q'(r_k)} e^{r_k t}. \tag{iii}$$

6.3 Step Functions

In Section 6.2 we outlined the general procedure involved in solving initial value problems by means of the Laplace transform. Some of the most interesting elementary applications of the transform method occur in the solution of linear differential equations with discontinuous or impulsive forcing functions. Equations of this type frequently arise in the analysis of the flow of current in electric circuits or the vibrations of mechanical systems. In this section and the following ones we develop some additional properties of the Laplace transform that are useful in the solution of such problems. Unless a specific statement is made to the contrary, all functions appearing below will be assumed to be piecewise continuous and of exponential order, so that their Laplace transforms exist, at least for s sufficiently large.

To deal effectively with functions having jump discontinuities, it is very helpful to introduce a function known as the **unit step function** or **Heaviside function**. This function will be denoted by u_c and is defined by

$$u_c(t) = \begin{cases} 0, & t < c, \\ 1, & t \geq c, \end{cases} \qquad c \geq 0. \tag{1}$$

The graph of $y = u_c(t)$ is shown in Figure 6.3.1. The step can also be negative. For instance, Figure 6.3.2 shows the graph $y = 1 - u_c(t)$.

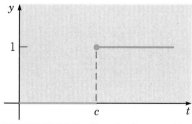

FIGURE 6.3.1 Graph of $y = u_c(t)$.

FIGURE 6.3.2 Graph of $y = 1 - u_c(t)$.

$$u_c = \begin{cases} 0 & t < c \\ 1 & t \geq 1 \end{cases}$$

EXAMPLE 1

Sketch the graph of $y = h(t)$, where

$$h(t) = u_\pi(t) - u_{2\pi}(t), \qquad t \geq 0.$$

From the definition of $u_c(t)$ in Eq. (1), we have

$$h(t) = \begin{cases} 0 - 0 = 0, & 0 \leq t < \pi, \\ 1 - 0 = 1, & \pi \leq t < 2\pi, \\ 1 - 1 = 0, & 2\pi \leq t < \infty. \end{cases}$$

Thus the equation $y = h(t)$ has the graph shown in Figure 6.3.3. This function can be thought of as a rectangular pulse.

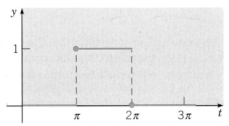

FIGURE 6.3.3 Graph of $y = u_\pi(t) - u_{2\pi}(t)$.

The Laplace transform of u_c is easily determined:

$$\mathcal{L}\{u_c(t)\} = \int_0^\infty e^{-st} u_c(t)\, dt = \int_c^\infty e^{-st}\, dt$$

$$= \frac{e^{-cs}}{s}, \qquad s > 0. \tag{2}$$

For a given function f defined for $t \geq 0$, we will often want to consider the related function g defined by

$$y = g(t) = \begin{cases} 0, & t < c, \\ f(t-c), & t \geq c, \end{cases}$$

which represents a translation of f a distance c in the positive t direction; see Figure 6.3.4. In terms of the unit step function we can write $g(t)$ in the convenient form

$$g(t) = u_c(t)f(t-c).$$

(a) (b)

FIGURE 6.3.4 A translation of the given function. (a) $y = f(t)$; (b) $y = u_c(t)f(t-c)$.

The unit step function is particularly important in transform use because of the following relation between the transform of $f(t)$ and that of its translation $u_c(t)f(t-c)$.

Theorem 6.3.1 If $F(s) = \mathcal{L}\{f(t)\}$ exists for $s > a \geq 0$, and if c is a positive constant, then

$$\mathcal{L}\{u_c(t)f(t-c)\} = e^{-cs}\mathcal{L}\{f(t)\} = e^{-cs}F(s), \qquad s > a. \tag{3}$$

Conversely, if $f(t) = \mathcal{L}^{-1}\{F(s)\}$, then

$$u_c(t)f(t-c) = \mathcal{L}^{-1}\{e^{-cs}F(s)\}. \tag{4}$$

Theorem 6.3.1 simply states that the translation of $f(t)$ a distance c in the positive t direction corresponds to the multiplication of $F(s)$ by e^{-cs}. To prove Theorem 6.3.1, it is sufficient to compute the transform of $u_c(t)f(t-c)$:

$$\mathcal{L}\{u_c(t)f(t-c)\} = \int_0^\infty e^{-st}u_c(t)f(t-c)\,dt$$

$$= \int_c^\infty e^{-st}f(t-c)\,dt.$$

Introducing a new integration variable $\xi = t - c$, we have

$$\mathcal{L}\{u_c(t)f(t-c)\} = \int_0^\infty e^{-(\xi+c)s}f(\xi)\,d\xi = e^{-cs}\int_0^\infty e^{-s\xi}f(\xi)\,d\xi$$

$$= e^{-cs}F(s).$$

Thus Eq. (3) is established; Eq. (4) follows by taking the inverse transform of both sides of Eq. (3).

A simple example of this theorem occurs if we take $f(t) = 1$. Recalling that $\mathcal{L}\{1\} = 1/s$, we immediately have from Eq. (3) that $\mathcal{L}\{u_c(t)\} = e^{-cs}/s$. This result agrees with that of Eq. (2). Examples 2 and 3 illustrate further how Theorem 6.3.1 can be used in the calculation of transforms and inverse transforms.

EXAMPLE 2

If the function f is defined by

$$f(t) = \begin{cases} \sin t, & 0 \leq t < \pi/4, \\ \sin t + \cos(t - \pi/4), & t \geq \pi/4, \end{cases}$$

find $\mathcal{L}\{f(t)\}$. The graph of $y = f(t)$ is shown in Figure 6.3.5.

Note that $f(t) = \sin t + g(t)$, where

$$g(t) = \begin{cases} 0, & t < \pi/4, \\ \cos(t - \pi/4), & t \geq \pi/4. \end{cases}$$

Thus

$$g(t) = u_{\pi/4}(t)\cos(t - \pi/4),$$

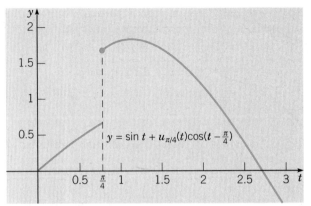

FIGURE 6.3.5 Graph of the function in Example 2.

and

$$\mathcal{L}\{f(t)\} = \mathcal{L}\{\sin t\} + \mathcal{L}\{u_{\pi/4}(t)\cos(t - \pi/4)\}$$
$$= \mathcal{L}\{\sin t\} + e^{-\pi s/4}\mathcal{L}\{\cos t\}.$$

Introducing the transforms of $\sin t$ and $\cos t$, we obtain

$$\mathcal{L}\{f(t)\} = \frac{1}{s^2 + 1} + e^{-\pi s/4}\frac{s}{s^2 + 1} = \frac{1 + se^{-\pi s/4}}{s^2 + 1}.$$

You should compare this method with the calculation of $\mathcal{L}\{f(t)\}$ directly from the definition.

EXAMPLE
3

Find the inverse transform of

$$F(s) = \frac{1 - e^{-2s}}{s^2}.$$

From the linearity of the inverse transform we have

$$f(t) = \mathcal{L}^{-1}\{F(s)\} = \mathcal{L}^{-1}\left\{\frac{1}{s^2}\right\} - \mathcal{L}^{-1}\left\{\frac{e^{-2s}}{s^2}\right\}$$
$$= t - u_2(t)(t - 2).$$

The function f may also be written as

$$f(t) = \begin{cases} t, & 0 \le t < 2, \\ 2, & t \ge 2. \end{cases}$$

The following theorem contains another very useful property of Laplace transforms that is somewhat analogous to that given in Theorem 6.3.1.

Theorem 6.3.2 If $F(s) = \mathcal{L}\{f(t)\}$ exists for $s > a \geq 0$, and if c is a constant, then

$$\mathcal{L}\{e^{ct}f(t)\} = F(s-c), \qquad s > a + c. \tag{5}$$

Conversely, if $f(t) = \mathcal{L}^{-1}\{F(s)\}$, then

$$e^{ct}f(t) = \mathcal{L}^{-1}\{F(s-c)\}. \tag{6}$$

According to Theorem 6.3.2, multiplication of $f(t)$ by e^{ct} results in a translation of the transform $F(s)$ a distance c in the positive s direction, and conversely. The proof of this theorem requires merely the evaluation of $\mathcal{L}\{e^{ct}f(t)\}$. Thus

$$\mathcal{L}\{e^{ct}f(t)\} = \int_0^\infty e^{-st}e^{ct}f(t)\,dt = \int_0^\infty e^{-(s-c)t}f(t)\,dt$$

$$= F(s-c),$$

which is Eq. (5). The restriction $s > a + c$ follows from the observation that, according to hypothesis (ii) of Theorem 6.1.2, $|f(t)| \leq Ke^{at}$; hence $|e^{ct}f(t)| \leq Ke^{(a+c)t}$. Equation (6) follows by taking the inverse transform of Eq. (5), and the proof is complete.

The principal application of Theorem 6.3.2 is in the evaluation of certain inverse transforms, as illustrated by Example 4.

EXAMPLE 4

Find the inverse transform of

$$G(s) = \frac{1}{s^2 - 4s + 5}.$$

By completing the square in the denominator we can write

$$G(s) = \frac{1}{(s-2)^2 + 1} = F(s-2),$$

where $F(s) = (s^2 + 1)^{-1}$. Since $\mathcal{L}^{-1}\{F(s)\} = \sin t$, it follows from Theorem 6.3.2 that

$$g(t) = \mathcal{L}^{-1}\{G(s)\} = e^{2t}\sin t.$$

The results of this section are often useful in solving differential equations, particularly those that have discontinuous forcing functions. The next section is devoted to examples illustrating this fact.

PROBLEMS

In each of Problems 1 through 6 sketch the graph of the given function on the interval $t \geq 0$.

1. $u_1(t) + 2u_3(t) - 6u_4(t)$

2. $(t-3)u_2(t) - (t-2)u_3(t)$

3. $f(t-\pi)u_\pi(t)$, where $f(t) = t^2$

4. $f(t-3)u_3(t)$, where $f(t) = \sin t$

5. $f(t-1)u_2(t)$, where $f(t) = 2t$

6. $(t-1)u_1(t) - 2(t-2)u_2(t) + (t-3)u_3(t)$

In each of Problems 7 through 12 find the Laplace transform of the given function.

7. $f(t) = \begin{cases} 0, & t < 2 \\ (t-2)^2, & t \geq 2 \end{cases}$

8. $f(t) = \begin{cases} 0, & t < 1 \\ t^2 - 2t + 2, & t \geq 1 \end{cases}$

9. $f(t) = \begin{cases} 0, & t < \pi \\ t - \pi, & \pi \leq t < 2\pi \\ 0, & t \geq 2\pi \end{cases}$

10. $f(t) = u_1(t) + 2u_3(t) - 6u_4(t)$

11. $f(t) = (t-3)u_2(t) - (t-2)u_3(t)$

12. $f(t) = t - u_1(t)(t-1), \qquad t \geq 0$

In each of Problems 13 through 18 find the inverse Laplace transform of the given function.

13. $F(s) = \dfrac{3!}{(s-2)^4}$

14. $F(s) = \dfrac{e^{-2s}}{s^2 + s - 2}$

15. $F(s) = \dfrac{2(s-1)e^{-2s}}{s^2 - 2s + 2}$

16. $F(s) = \dfrac{2e^{-2s}}{s^2 - 4}$

17. $F(s) = \dfrac{(s-2)e^{-s}}{s^2 - 4s + 3}$

18. $F(s) = \dfrac{e^{-s} + e^{-2s} - e^{-3s} - e^{-4s}}{s}$

19. Suppose that $F(s) = \mathcal{L}\{f(t)\}$ exists for $s > a \geq 0$.
 (a) Show that if c is a positive constant, then

$$\mathcal{L}\{f(ct)\} = \frac{1}{c}F\left(\frac{s}{c}\right), \qquad s > ca.$$

 (b) Show that if k is a positive constant, then

$$\mathcal{L}^{-1}\{F(ks)\} = \frac{1}{k}f\left(\frac{t}{k}\right).$$

 (c) Show that if a and b are constants with $a > 0$, then

$$\mathcal{L}^{-1}\{F(as+b)\} = \frac{1}{a}e^{-bt/a}f\left(\frac{t}{a}\right).$$

In each of Problems 20 through 23 use the results of Problem 19 to find the inverse Laplace transform of the given function.

20. $F(s) = \dfrac{2^{n+1}n!}{s^{n+1}}$

21. $F(s) = \dfrac{2s+1}{4s^2 + 4s + 5}$

22. $F(s) = \dfrac{1}{9s^2 - 12s + 3}$

23. $F(s) = \dfrac{e^2 e^{-4s}}{2s - 1}$

In each of Problems 24 through 27 find the Laplace transform of the given function. In Problem 27 assume that term-by-term integration of the infinite series is permissible.

24. $f(t) = \begin{cases} 1, & 0 \leq t < 1 \\ 0, & t \geq 1 \end{cases}$

25. $f(t) = \begin{cases} 1, & 0 \leq t < 1 \\ 0, & 1 \leq t < 2 \\ 1, & 2 \leq t < 3 \\ 0, & t \geq 3 \end{cases}$

26. $f(t) = 1 - u_1(t) + \cdots + u_{2n}(t) - u_{2n+1}(t) = 1 + \displaystyle\sum_{k=1}^{2n+1}(-1)^k u_k(t)$

27. $f(t) = 1 + \sum_{k=1}^{\infty} (-1)^k u_k(t).$ See Figure 6.3.6.

FIGURE 6.3.6 A square wave.

28. Let f satisfy $f(t + T) = f(t)$ for all $t \geq 0$ and for some fixed positive number $T; f$ is said to be periodic with period T on $0 \leq t < \infty$. Show that

$$\mathcal{L}\{f(t)\} = \frac{\int_0^T e^{-st} f(t)\, dt}{1 - e^{-sT}}.$$

In each of Problems 29 through 32, use the result of Problem 28 to find the Laplace transform of the given function.

29. $f(t) = \begin{cases} 1, & 0 \leq t < 1, \\ 0, & 1 \leq t < 2; \end{cases}$

 $f(t + 2) = f(t).$

 Compare with Problem 27.

30. $f(t) = \begin{cases} 1, & 0 \leq t < 1, \\ -1, & 1 \leq t < 2; \end{cases}$

 $f(t + 2) = f(t).$

 See Figure 6.3.7.

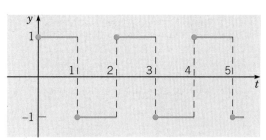

FIGURE 6.3.7 A square wave.

31. $f(t) = t,$ $0 \leq t < 1;$

 $f(t + 1) = f(t).$

 See Figure 6.3.8.

32. $f(t) = \sin t,$ $0 \leq t < \pi;$

 $f(t + \pi) = f(t).$

 See Figure 6.3.9.

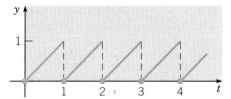

FIGURE 6.3.8 A sawtooth wave.

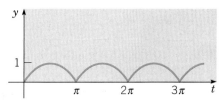

FIGURE 6.3.9 A rectified sine wave.

33. (a) If $f(t) = 1 - u_1(t)$, find $\mathcal{L}\{f(t)\}$; compare with Problem 24. Sketch the graph of $y = f(t)$.

(b) Let $g(t) = \int_0^t f(\xi)\, d\xi$, where the function f is defined in part (a). Sketch the graph of $y = g(t)$ and find $\mathcal{L}\{g(t)\}$.

(c) Let $h(t) = g(t) - u_1(t)g(t-1)$, where g is defined in part (b). Sketch the graph of $y = h(t)$ and find $\mathcal{L}\{h(t)\}$.

34. Consider the function p defined by

$$p(t) = \begin{cases} t, & 0 \leq t < 1, \\ 2 - t, & 1 \leq t < 2; \end{cases} \qquad p(t+2) = p(t).$$

(a) Sketch the graph of $y = p(t)$.

(b) Find $\mathcal{L}\{p(t)\}$ by noting that p is the periodic extension of the function h in Problem 33(c) and then using the result of Problem 28.

(c) Find $\mathcal{L}\{p(t)\}$ by noting that

$$p(t) = \int_0^t f(t)\, dt,$$

where f is the function in Problem 30, and then using Theorem 6.2.1.

6.4 Differential Equations with Discontinuous Forcing Functions

In this section we turn our attention to some examples in which the nonhomogeneous term, or forcing function, is discontinuous.

EXAMPLE 1

Find the solution of the differential equation

$$2y'' + y' + 2y = g(t), \tag{1}$$

where

$$g(t) = u_5(t) - u_{20}(t) = \begin{cases} 1, & 5 \leq t < 20, \\ 0, & 0 \leq t < 5 \quad \text{and} \quad t \geq 20. \end{cases} \tag{2}$$

Assume that the initial conditions are

$$y(0) = 0, \qquad y'(0) = 0. \tag{3}$$

This problem governs the charge on the capacitor in a simple electric circuit with a unit voltage pulse for $5 \leq t < 20$. Alternatively, y may represent the response of a damped oscillator subject to the applied force $g(t)$.

The Laplace transform of Eq. (1) is

$$2s^2 Y(s) - 2sy(0) - 2y'(0) + sY(s) - y(0) + 2Y(s) = \mathcal{L}\{u_5(t)\} - \mathcal{L}\{u_{20}(t)\}$$

$$= (e^{-5s} - e^{-20s})/s.$$

Introducing the initial values (3) and solving for $Y(s)$, we obtain

$$Y(s) = \frac{e^{-5s} - e^{-20s}}{s(2s^2 + s + 2)}. \tag{4}$$

To find $y = \phi(t)$ it is convenient to write $Y(s)$ as

$$Y(s) = (e^{-5s} - e^{-20s})H(s), \tag{5}$$

where

$$H(s) = \frac{1}{s(2s^2 + s + 2)}. \tag{6}$$

Then, if $h(t) = \mathcal{L}^{-1}\{H(s)\}$, we have

$$y = \phi(t) = u_5(t)h(t - 5) - u_{20}(t)h(t - 20). \tag{7}$$

Observe that we have used Theorem 6.3.1 to write the inverse transforms of $e^{-5s}H(s)$ and $e^{-20s}H(s)$, respectively. Finally, to determine $h(t)$ we use the partial fraction expansion of $H(s)$:

$$H(s) = \frac{a}{s} + \frac{bs + c}{2s^2 + s + 2}. \tag{8}$$

Upon determining the coefficients, we find that $a = \frac{1}{2}, b = -1$, and $c = -\frac{1}{2}$. Thus

$$H(s) = \frac{1/2}{s} - \frac{s + \frac{1}{2}}{2s^2 + s + 2}$$

$$= \frac{1/2}{s} - \left(\frac{1}{2}\right)\frac{\left(s + \frac{1}{4}\right) + \frac{1}{4}}{\left(s + \frac{1}{4}\right)^2 + \frac{15}{16}}, \tag{9}$$

so that, by referring to lines 9 and 10 of Table 6.2.1, we obtain

$$h(t) = \frac{1}{2} - \frac{1}{2}\left[e^{-t/4}\cos(\sqrt{15}\,t/4) + (\sqrt{15}/15)e^{-t/4}\sin(\sqrt{15}\,t/4)\right]. \tag{10}$$

In Figure 6.4.1 the graph of $y = \phi(t)$ from Eqs. (7) and (10) shows that the solution consists of three distinct parts. For $0 < t < 5$ the differential equation is

$$2y'' + y' + 2y = 0 \tag{11}$$

and the initial conditions are given by Eq. (3). Since the initial conditions impart no energy to the system, and since there is no external forcing, the system remains at rest; that is, $y = 0$ for $0 < t < 5$. This can be confirmed by solving Eq. (11) subject to the initial conditions (3). In particular, evaluating the solution and its derivative at $t = 5$, or, more precisely, as t approaches 5 from below, we have

$$y(5) = 0, \qquad y'(5) = 0. \tag{12}$$

Once $t > 5$, the differential equation becomes

$$2y'' + y' + 2y = 1, \tag{13}$$

whose solution is the sum of a constant (the response to the constant forcing function) and a damped oscillation (the solution of the corresponding homogeneous equation). The plot in Figure 6.4.1 shows this behavior clearly for the interval $5 \le t \le 20$. An expression for this portion of the solution can be found by solving the differential equation (13) subject to the initial conditions (12). Finally, for $t > 20$ the differential equation becomes Eq. (11) again,

and the initial conditions are obtained by evaluating the solution of Eqs. (13), (12) and its derivative at $t = 20$. These values are

$$y(20) \cong 0.50162, \qquad y'(20) \cong 0.01125. \tag{14}$$

The initial value problem (11), (14) contains no external forcing, so its solution is a damped oscillation about $y = 0$, as can be seen in Figure 6.4.1.

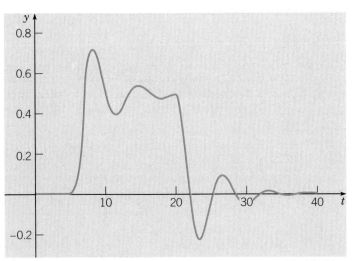

FIGURE 6.4.1 Solution of the initial value problem (1), (2), (3).

Although it may be helpful to visualize the solution shown in Figure 6.4.1 as composed of solutions of three separate initial value problems in three separate intervals, it is somewhat tedious to find the solution by solving these separate problems. Laplace transform methods provide a much more convenient and elegant approach to this problem and to others having discontinuous forcing functions.

The effect of the discontinuity in the forcing function can be seen if we examine the solution $\phi(t)$ of Example 1 more closely. According to the existence and uniqueness Theorem 3.2.1, the solution ϕ and its first two derivatives are continuous except possibly at the points $t = 5$ and $t = 20$, where g is discontinuous. This can also be seen at once from Eq. (7). One can also show by direct computation from Eq. (7) that ϕ and ϕ' are continuous even at $t = 5$ and $t = 20$. However, if we calculate ϕ'', we find that

$$\lim_{t \to 5-} \phi''(t) = 0, \qquad \lim_{t \to 5+} \phi''(t) = 1/2.$$

Consequently, $\phi''(t)$ has a jump of $1/2$ at $t = 5$. In a similar way one can show that $\phi''(t)$ has a jump of $-1/2$ at $t = 20$. Thus the jump in the forcing term $g(t)$ at these points is balanced by a corresponding jump in the highest order term $2y''$ on the left side of the equation.

Consider now the general second order linear equation

$$y'' + p(t)y' + q(t)y = g(t), \tag{15}$$

where p and q are continuous on some interval $\alpha < t < \beta$, but g is only piecewise continuous there. If $y = \psi(t)$ is a solution of Eq. (15), then ψ and ψ' are continuous on $\alpha < t < \beta$, but ψ'' has jump discontinuities at the same points as g. Similar remarks apply to higher order equations; the highest derivative of the solution appearing in the differential equation has jump discontinuities at the same points as the forcing function, but the solution itself and its lower derivatives are continuous even at those points.

EXAMPLE 2

Describe the qualitative nature of the solution of the initial value problem

$$y'' + 4y = g(t), \tag{16}$$

$$y(0) = 0, \qquad y'(0) = 0, \tag{17}$$

where

$$g(t) = \begin{cases} 0, & 0 \le t < 5, \\ (t-5)/5, & 5 \le t < 10, \\ 1, & t \ge 10, \end{cases} \tag{18}$$

and then find the solution.

In this example the forcing function has the graph shown in Figure 6.4.2 and is known as ramp loading. It is relatively easy to identify the general form of the solution. For $t < 5$ the solution is simply $y = 0$. On the other hand, for $t > 10$ the solution has the form

$$y = c_1 \cos 2t + c_2 \sin 2t + 1/4. \tag{19}$$

The constant $1/4$ is a particular solution of the nonhomogeneous equation, while the other two terms are the general solution of the corresponding homogeneous equation. Thus the solution (19) is a simple harmonic oscillation about $y = 1/4$. Similarly, in the intermediate range $5 < t < 10$, the solution is an oscillation about a certain linear function. In an engineering context, for example, we might be interested in knowing the amplitude of the eventual steady oscillation.

To solve the problem it is convenient to write

$$g(t) = \left[u_5(t)(t-5) - u_{10}(t)(t-10) \right]/5, \tag{20}$$

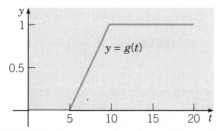

FIGURE 6.4.2 Ramp loading; $y = g(t)$ from Eq. (18).

as you may verify. Then we take the Laplace transform of the differential equation and use the initial conditions, thereby obtaining

$$(s^2 + 4)Y(s) = (e^{-5s} - e^{-10s})/5s^2,$$

or

$$Y(s) = (e^{-5s} - e^{-10s})H(s)/5, \tag{21}$$

where

$$H(s) = \frac{1}{s^2(s^2 + 4)} \tag{22}$$

Thus the solution of the initial value problem (16), (17), (18) is

$$y = \phi(t) = \left[u_5(t)h(t - 5) - u_{10}(t)h(t - 10)\right]/5, \tag{23}$$

where $h(t)$ is the inverse transform of $H(s)$. The partial fraction expansion of $H(s)$ is

$$H(s) = \frac{1/4}{s^2} - \frac{1/4}{s^2 + 4}, \tag{24}$$

and it then follows from lines 3 and 5 of Table 6.2.1 that

$$h(t) = \tfrac{1}{4}t - \tfrac{1}{8}\sin 2t. \tag{25}$$

The graph of $y = \phi(t)$ is shown in Figure 6.4.3. Observe that it has the qualitative form that we indicated earlier. To find the amplitude of the eventual steady oscillation, it is sufficient to locate one of the maximum or minimum points for $t > 10$. Setting the derivative of the solution (23) equal to zero, we find that the first maximum is located approximately at $(10.642, 0.2979)$, so the amplitude of the oscillation is approximately 0.0479.

FIGURE 6.4.3 Solution of the initial value problem (16), (17), (18).

Note that in this example the forcing function g is continuous but g' is discontinuous at $t = 5$ and $t = 10$. It follows that the solution ϕ and its first two derivatives are continuous everywhere, but ϕ''' has discontinuities at $t = 5$ and at $t = 10$ that match the discontinuities in g' at those points.

PROBLEMS

In each of Problems 1 through 13 find the solution of the given initial value problem. Draw the graphs of the solution and of the forcing function; explain how they are related.

1. $y'' + y = f(t)$; $y(0) = 0$, $y'(0) = 1$; $f(t) = \begin{cases} 1, & 0 \le t < \pi/2 \\ 0, & \pi/2 \le t < \infty \end{cases}$

2. $y'' + 2y' + 2y = h(t)$; $y(0) = 0$, $y'(0) = 1$; $h(t) = \begin{cases} 1, & \pi \le t < 2\pi \\ 0, & 0 \le t < \pi \text{ and } t \ge 2\pi \end{cases}$

3. $y'' + 4y = \sin t - u_{2\pi}(t)\sin(t - 2\pi)$; $y(0) = 0$, $y'(0) = 0$

4. $y'' + 4y = \sin t + u_{\pi}(t)\sin(t - \pi)$; $y(0) = 0$, $y'(0) = 0$

5. $y'' + 3y' + 2y = f(t)$; $y(0) = 0$, $y'(0) = 0$; $f(t) = \begin{cases} 1, & 0 \le t < 10 \\ 0, & t \ge 10 \end{cases}$

6. $y'' + 3y' + 2y = u_2(t)$; $y(0) = 0$, $y'(0) = 1$

7. $y'' + y = u_{3\pi}(t)$; $y(0) = 1$, $y'(0) = 0$

8. $y'' + y' + \frac{5}{4}y = t - u_{\pi/2}(t)(t - \pi/2)$; $y(0) = 0$, $y'(0) = 0$

9. $y'' + y = g(t)$; $y(0) = 0$, $y'(0) = 1$; $g(t) = \begin{cases} t/2, & 0 \le t < 6 \\ 3, & t \ge 6 \end{cases}$

10. $y'' + y' + \frac{5}{4}y = g(t)$; $y(0) = 0$, $y'(0) = 0$; $g(t) = \begin{cases} \sin t, & 0 \le t < \pi \\ 0, & t \ge \pi \end{cases}$

11. $y'' + 4y = u_{\pi}(t) - u_{3\pi}(t)$; $y(0) = 0$, $y'(0) = 0$

12. $y^{(4)} - y = u_1(t) - u_2(t)$; $y(0) = 0$, $y'(0) = 0$, $y''(0) = 0$, $y'''(0) = 0$

13. $y^{(4)} + 5y'' + 4y = 1 - u_{\pi}(t)$; $y(0) = 0$, $y'(0) = 0$, $y''(0) = 0$, $y'''(0) = 0$

14. Find an expression involving $u_c(t)$ for a function f that ramps up from zero at $t = t_0$ to the value h at $t = t_0 + k$.

15. Find an expression involving $u_c(t)$ for a function g that ramps up from zero at $t = t_0$ to the value h at $t = t_0 + k$ and then ramps back down to zero at $t = t_0 + 2k$.

16. A certain spring–mass system satisfies the initial value problem

$$u'' + \tfrac{1}{4}u' + u = kg(t), \qquad u(0) = 0, \quad u'(0) = 0,$$

where $g(t) = u_{3/2}(t) - u_{5/2}(t)$ and $k > 0$ is a parameter.

(a) Sketch the graph of $g(t)$. Observe that it is a pulse of unit magnitude extending over one time unit.

(b) Solve the initial value problem.

(c) Plot the solution for $k = 1/2, k = 1$, and $k = 2$. Describe the principal features of the solution and how they depend on k.

(d) Find, to two decimal places, the smallest value of k for which the solution $u(t)$ reaches the value 2.

(e) Suppose $k = 2$. Find the time τ after which $|u(t)| < 0.1$ for all $t > \tau$.

17. Modify the problem in Example 2 of this section by replacing the given forcing function $g(t)$ by

$$f(t) = \left[u_5(t)(t - 5) - u_{5+k}(t)(t - 5 - k) \right]/k.$$

(a) Sketch the graph of $f(t)$ and describe how it depends on k. For what value of k is $f(t)$ identical to $g(t)$ in the example?

only graph of forcing function required

(b) Solve the initial value problem

$$y'' + 4y = f(t), \qquad y(0) = 0, \quad y'(0) = 0.$$

(c) The solution in part (b) depends on k, but for sufficiently large t the solution is always a simple harmonic oscillation about $y = 1/4$. Try to decide how the amplitude of this eventual oscillation depends on k. Then confirm your conclusion by plotting the solution for a few different values of k.

18. Consider the initial value problem

$$y'' + \tfrac{1}{3}y' + 4y = f_k(t), \qquad y(0) = 0, \quad y'(0) = 0,$$

where

$$f_k(t) = \begin{cases} 1/2k, & 4 - k \le t < 4 + k \\ 0, & 0 \le t < 4 - k \quad \text{and} \quad t \ge 4 + k \end{cases}$$

and $0 < k < 4$.

(a) Sketch the graph of $f_k(t)$. Observe that the area under the graph is independent of k. If $f_k(t)$ represents a force, this means that the product of the magnitude of the force and the time interval during which it acts does not depend on k.

(b) Write $f_k(t)$ in terms of the unit step function and then solve the given initial value problem.

(c) Plot the solution for $k = 2$, $k = 1$, and $k = \tfrac{1}{2}$. Describe how the solution depends on k.

Resonance and Beats. In Section 3.9 we observed that an undamped harmonic oscillator (such as a spring–mass system) with a sinusoidal forcing term experiences resonance if the frequency of the forcing term is the same as the natural frequency. If the forcing frequency is slightly different from the natural frequency, then the system exhibits a beat. In Problems 19 through 23 we explore the effect of some nonsinusoidal periodic forcing functions.

19. Consider the initial value problem

$$y'' + y = f(t), \qquad y(0) = 0, \quad y'(0) = 0,$$

where

$$f(t) = u_0(t) + 2 \sum_{k=1}^{n} (-1)^k u_{k\pi}(t).$$

(a) Draw the graph of $f(t)$ on an interval such as $0 \le t \le 6\pi$.

(b) Find the solution of the initial value problem.

(c) Let $n = 15$ and plot the graph of the solution for $0 \le t \le 60$. Describe the solution and explain why it behaves as it does.

(d) Investigate how the solution changes as n increases. What happens as $n \to \infty$?

20. Consider the initial value problem

$$y'' + 0.1y' + y = f(t), \qquad y(0) = 0, \quad y'(0) = 0,$$

where $f(t)$ is the same as in Problem 19.

(a) Plot the graph of the solution. Use a large enough value of n and a long enough t-interval so that the transient part of the solution has become negligible and the steady state is clearly shown.

(b) Estimate the amplitude and frequency of the steady-state part of the solution.

(c) Compare the results of part (b) with those from Section 3.9 for a sinusoidally forced oscillator.

21. Consider the initial value problem

$$y'' + y = g(t), \qquad y(0) = 0, \quad y'(0) = 0,$$

where

$$g(t) = u_0(t) + \sum_{k=1}^{n}(-1)^k u_{k\pi}(t).$$

(a) Draw the graph of $g(t)$ on an interval such as $0 \le t \le 6\pi$. Compare the graph with that of $f(t)$ in Problem 19(a).

(b) Find the solution of the initial value problem.

(c) Let $n = 15$ and plot the graph of the solution for $0 \le t \le 60$. Describe the solution and explain why it behaves as it does. Compare it with the solution of Problem 19.

(d) Investigate how the solution changes as n increases. What happens as $n \to \infty$?

22. Consider the initial value problem

$$y'' + 0.1y' + y = g(t), \qquad y(0) = 0, \quad y'(0) = 0,$$

where $g(t)$ is the same as in Problem 21.

(a) Plot the graph of the solution. Use a large enough value of n and a long enough t-interval so that the transient part of the solution has become negligible and the steady state is clearly shown.

(b) Estimate the amplitude and frequency of the steady-state part of the solution.

(c) Compare the results of part (b) with those from Problem 20 and from Section 3.9 for a sinusoidally forced oscillator.

23. Consider the initial value problem

$$y'' + y = h(t), \qquad y(0) = 0, \quad y'(0) = 0,$$

where

$$f(t) = u_0(t) + 2\sum_{k=1}^{n}(-1)^k u_{11k/4}(t).$$

Observe that this problem is identical to Problem 19 except that the frequency of the forcing term has been increased somewhat.

(a) Find the solution of this initial value problem.

(b) Let $n \ge 33$ and plot the solution for $0 \le t \le 90$ or longer. Your plot should show a clearly recognizable beat.

(c) From the graph in part (b) estimate the "slow period" and the "fast period" for this oscillator.

(d) For a sinusoidally forced oscillator, it was shown in Section 3.9 that the "slow frequency" is given by $|\omega - \omega_0|/2$, where ω_0 is the natural frequency of the system and ω is the forcing frequency. Similarly, the "fast frequency" is $(\omega + \omega_0)/2$. Use these expressions to calculate the "fast period" and the "slow period" for the oscillator in this problem. How well do the results compare with your estimates from part (c)?

6.5 Impulse Functions

In some applications it is necessary to deal with phenomena of an impulsive nature—for example, voltages or forces of large magnitude that act over very short time intervals. Such problems often lead to differential equations of the form

$$ay'' + by' + cy = g(t), \tag{1}$$

where $g(t)$ is large during a short interval $t_0 - \tau < t < t_0 + \tau$ and is otherwise zero. The integral $I(\tau)$, defined by

$$I(\tau) = \int_{t_0-\tau}^{t_0+\tau} g(t)\, dt, \tag{2}$$

or, since $g(t) = 0$ outside of the interval $(t_0 - \tau, t_0 + \tau)$,

$$I(\tau) = \int_{-\infty}^{\infty} g(t)\, dt, \tag{3}$$

is a measure of the strength of the forcing function. In a mechanical system, where $g(t)$ is a force, $I(\tau)$ is the total **impulse** of the force $g(t)$ over the time interval $(t_0 - \tau, t_0 + \tau)$. Similarly, if y is the current in an electric circuit and $g(t)$ is the time derivative of the voltage, then $I(\tau)$ represents the total voltage impressed on the circuit during the interval $(t_0 - \tau, t_0 + \tau)$.

In particular, let us suppose that t_0 is zero and that $g(t)$ is given by

$$g(t) = d_\tau(t) = \begin{cases} 1/2\tau, & -\tau < t < \tau, \\ 0, & t \le -\tau \quad \text{or} \quad t \ge \tau, \end{cases} \tag{4}$$

where τ is a small positive constant (see Figure 6.5.1). According to Eq. (2) or (3) it follows immediately that in this case $I(\tau) = 1$ independent of the value of τ, as long as $\tau \ne 0$. Now let us idealize the forcing function d_τ by prescribing it to act over shorter and shorter time intervals; that is, we require that $\tau \to 0$, as indicated in Figure 6.5.2. As a result of this limiting operation, we obtain

$$\lim_{\tau \to 0} d_\tau(t) = 0, \qquad t \ne 0. \tag{5}$$

Further, since $I(\tau) = 1$ for each $\tau \ne 0$, it follows that

$$\lim_{\tau \to 0} I(\tau) = 1. \tag{6}$$

FIGURE 6.5.1 Graph of $y = d_\tau(t)$.

FIGURE 6.5.2 Graphs of $y = d_\tau(t)$ as $\tau \to 0$.

Equations (5) and (6) can be used to define an idealized **unit impulse function** δ, which imparts an impulse of magnitude one at $t = 0$ but is zero for all values of t other than zero. That is, the "function" δ is defined to have the properties

$$\delta(t) = 0, \qquad t \neq 0; \tag{7}$$

$$\int_{-\infty}^{\infty} \delta(t)\, dt = 1. \tag{8}$$

There is no ordinary function of the kind studied in elementary calculus that satisfies both of Eqs. (7) and (8). The "function" δ, defined by those equations, is an example of what are known as generalized functions; it is usually called the Dirac[2] **delta function**. Since $\delta(t)$ corresponds to a unit impulse at $t = 0$, a unit impulse at an arbitrary point $t = t_0$ is given by $\delta(t - t_0)$. From Eqs. (7) and (8) it follows that

$$\delta(t - t_0) = 0, \qquad t \neq t_0; \tag{9}$$

$$\int_{-\infty}^{\infty} \delta(t - t_0)\, dt = 1. \tag{10}$$

The delta function does not satisfy the conditions of Theorem 6.1.2, but its Laplace transform can nevertheless be formally defined. Since $\delta(t)$ is defined as the limit of $d_\tau(t)$ as $\tau \to 0$, it is natural to define the Laplace transform of δ as a similar limit of the transform of d_τ. In particular, we will assume that $t_0 > 0$ and define $\mathcal{L}\{\delta(t - t_0)\}$ by the equation

$$\mathcal{L}\{\delta(t - t_0)\} = \lim_{\tau \to 0} \mathcal{L}\{d_\tau(t - t_0)\}. \tag{11}$$

[2]Paul A. M. Dirac (1902–1984), English mathematical physicist, received his Ph.D. from Cambridge in 1926 and was professor of mathematics there until 1969. He was awarded the Nobel Prize in 1933 (with Erwin Schrödinger) for fundamental work in quantum mechanics. His most celebrated result was the relativistic equation for the electron, published in 1928. From this equation he predicted the existence of an "anti-electron," or positron, which was first observed in 1932. Following his retirement from Cambridge, Dirac moved to the United States and held a research professorship at Florida State University.

To evaluate the limit in Eq. (11), we first observe that if $\tau < t_0$, which must eventually be the case as $\tau \to 0$, then $t_0 - \tau > 0$. Since $d_\tau(t - t_0)$ is nonzero only in the interval from $t_0 - \tau$ to $t_0 + \tau$, we have

$$\mathcal{L}\{d_\tau(t - t_0)\} = \int_0^\infty e^{-st} d_\tau(t - t_0)\, dt$$

$$= \int_{t_0 - \tau}^{t_0 + \tau} e^{-st} d_\tau(t - t_0)\, dt.$$

Substituting for $d_\tau(t - t_0)$ from Eq. (4), we obtain

$$\mathcal{L}\{d_\tau(t - t_0)\} = \frac{1}{2\tau} \int_{t_0 - \tau}^{t_0 + \tau} e^{-st}\, dt = -\frac{1}{2s\tau} e^{-st} \Big|_{t = t_0 - \tau}^{t = t_0 + \tau}$$

$$= \frac{1}{2s\tau} e^{-st_0}(e^{s\tau} - e^{-s\tau})$$

or

$$\mathcal{L}\{d_\tau(t - t_0)\} = \frac{\sinh s\tau}{s\tau} e^{-st_0}. \tag{12}$$

The quotient $(\sinh s\tau)/s\tau$ is indeterminate as $\tau \to 0$, but its limit can be evaluated by L'Hospital's rule. We obtain

$$\lim_{\tau \to 0} \frac{\sinh s\tau}{s\tau} = \lim_{\tau \to 0} \frac{s \cosh s\tau}{s} = 1.$$

Then from Eq. (11) it follows that

$$\mathcal{L}\{\delta(t - t_0)\} = e^{-st_0}. \tag{13}$$

Equation (13) defines $\mathcal{L}\{\delta(t - t_0)\}$ for any $t_0 > 0$. We extend this result, to allow t_0 to be zero, by letting $t_0 \to 0$ on the right side of Eq. (13); thus

$$\mathcal{L}\{\delta(t)\} = \lim_{t_0 \to 0} e^{-st_0} = 1. \tag{14}$$

In a similar way it is possible to define the integral of the product of the delta function and any continuous function f. We have

$$\int_{-\infty}^\infty \delta(t - t_0) f(t)\, dt = \lim_{\tau \to 0} \int_{-\infty}^\infty d_\tau(t - t_0) f(t)\, dt. \tag{15}$$

Using the definition (4) of $d_\tau(t)$ and the mean value theorem for integrals, we find that

$$\int_{-\infty}^\infty d_\tau(t - t_0) f(t)\, dt = \frac{1}{2\tau} \int_{t_0 - \tau}^{t_0 + \tau} f(t)\, dt$$

$$= \frac{1}{2\tau} \cdot 2\tau \cdot f(t^*) = f(t^*),$$

where $t_0 - \tau < t^* < t_0 + \tau$. Hence $t^* \to t_0$ as $\tau \to 0$, and it follows from Eq. (15) that

$$\int_{-\infty}^{\infty} \delta(t - t_0) f(t) \, dt = f(t_0). \tag{16}$$

It is often convenient to introduce the delta function when working with impulse problems and to operate formally on it as though it were a function of the ordinary kind. This is illustrated in the example below. It is important to realize, however, that the ultimate justification of such procedures must rest on a careful analysis of the limiting operations involved. Such a rigorous mathematical theory has been developed, but we do not discuss it here.

EXAMPLE 1

Find the solution of the initial value problem

$$2y'' + y' + 2y = \delta(t - 5), \tag{17}$$

$$y(0) = 0, \qquad y'(0) = 0. \tag{18}$$

This initial value problem arises from the study of the same electrical circuit or mechanical oscillator as in Example 1 of Section 6.4. The only difference is in the forcing term.

To solve the given problem we take the Laplace transform of the differential equation and use the initial conditions, obtaining

$$(2s^2 + s + 2) Y(s) = e^{-5s}.$$

Thus

$$Y(s) = \frac{e^{-5s}}{2s^2 + s + 2} = \frac{e^{-5s}}{2} \frac{1}{\left(s + \frac{1}{4}\right)^2 + \frac{15}{16}}. \tag{19}$$

By Theorem 6.3.2 or from line 9 of Table 6.2.1,

$$\mathcal{L}^{-1} \left\{ \frac{1}{\left(s + \frac{1}{4}\right)^2 + \frac{15}{16}} \right\} = \frac{4}{\sqrt{15}} e^{-t/4} \sin \frac{\sqrt{15}}{4} t. \tag{20}$$

Hence, by Theorem 6.3.1, we have

$$y = \mathcal{L}^{-1}\{Y(s)\} = \frac{2}{\sqrt{15}} u_5(t) e^{-(t-5)/4} \sin \frac{\sqrt{15}}{4} (t - 5), \tag{21}$$

which is the formal solution of the given problem. It is also possible to write y in the form

$$y = \begin{cases} 0, & t < 5, \\ \dfrac{2}{\sqrt{15}} e^{-(t-5)/4} \sin \dfrac{\sqrt{15}}{4} (t - 5), & t \geq 5. \end{cases} \tag{22}$$

The graph of Eq. (22) is shown in Figure 6.5.3. Since the initial conditions at $t = 0$ are homogeneous and there is no external excitation until $t = 5$, there is no response in the interval $0 < t < 5$. The impulse at $t = 5$ produces a decaying oscillation that persists indefinitely. The response is continuous at $t = 5$ despite the singularity in the forcing function at that point. However, the first derivative of the solution has a jump discontinuity at $t = 5$, and the second derivative has an infinite discontinuity there. This is required by the differential equation (17), since a singularity on one side of the equation must be balanced by a corresponding singularity on the other side.

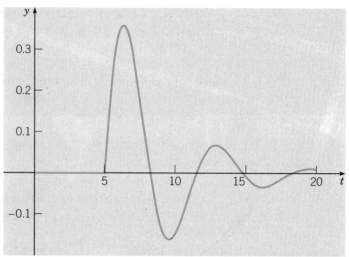

FIGURE 6.5.3 Solution of the initial value problem (17), (18).

PROBLEMS

In each of Problems 1 through 12 find the solution of the given initial value problem and draw its graph.

1. $y'' + 2y' + 2y = \delta(t - \pi)$; $y(0) = 1$, $y'(0) = 0$
2. $y'' + 4y = \delta(t - \pi) - \delta(t - 2\pi)$; $y(0) = 0$, $y'(0) = 0$
3. $y'' + 3y' + 2y = \delta(t - 5) + u_{10}(t)$; $y(0) = 0$, $y'(0) = 1/2$
4. $y'' - y = -20\delta(t - 3)$; $y(0) = 1$, $y'(0) = 0$
5. $y'' + 2y' + 3y = \sin t + \delta(t - 3\pi)$; $y(0) = 0$, $y'(0) = 0$
6. $y'' + 4y = \delta(t - 4\pi)$; $y(0) = 1/2$, $y'(0) = 0$
7. $y'' + y = \delta(t - 2\pi)\cos t$; $y(0) = 0$, $y'(0) = 1$
8. $y'' + 4y = 2\delta(t - \pi/4)$; $y(0) = 0$, $y'(0) = 0$
9. $y'' + y = u_{\pi/2}(t) + 3\delta(t - 3\pi/2) - u_{2\pi}(t)$; $y(0) = 0$, $y'(0) = 0$
10. $2y'' + y' + 4y = \delta(t - \pi/6)\sin t$; $y(0) = 0$, $y'(0) = 0$
11. $y'' + 2y' + 2y = \cos t + \delta(t - \pi/2)$; $y(0) = 0$, $y'(0) = 0$
12. $y^{(4)} - y = \delta(t - 1)$; $y(0) = 0$, $y'(0) = 0$, $y''(0) = 0$, $y'''(0) = 0$

13. Consider again the system in Example 1 of this section, in which an oscillation is excited by a unit impulse at $t = 5$. Suppose that it is desired to bring the system to rest again after exactly one cycle—that is, when the response first returns to equilibrium moving in the positive direction.

(a) Determine the impulse $k\delta(t - t_0)$ that should be applied to the system in order to accomplish this objective. Note that k is the magnitude of the impulse and t_0 is the time of its application.

(b) Solve the resulting initial value problem and plot its solution to confirm that it behaves in the specified manner.

14. Consider the initial value problem

$$y'' + \gamma y' + y = \delta(t - 1), \qquad y(0) = 0, \quad y'(0) = 0,$$

where γ is the damping coefficient (or resistance).

3, 5, 7, 9,
12, 15

Only 7

you need
to graph
solution

(a) Let $\gamma = \frac{1}{2}$. Find the solution of the initial value problem and plot its graph.

(b) Find the time t_1 at which the solution attains its maximum value. Also find the maximum value y_1 of the solution.

(c) Let $\gamma = \frac{1}{4}$ and repeat parts (a) and (b).

(d) Determine how t_1 and y_1 vary as γ decreases. What are the values of t_1 and y_1 when $\gamma = 0$?

 15. Consider the initial value problem

$$y'' + \gamma y' + y = k\delta(t-1), \qquad y(0) = 0, \quad y'(0) = 0,$$

where k is the magnitude of an impulse at $t = 1$ and γ is the damping coefficient (or resistance).

(a) Let $\gamma = \frac{1}{2}$. Find the value of k for which the response has a peak value of 2; call this value k_1.

(b) Repeat part (a) for $\gamma = \frac{1}{4}$.

(c) Determine how k_1 varies as γ decreases. What is the value of k_1 when $\gamma = 0$?

 16. Consider the initial value problem

$$y'' + y = f_k(t), \qquad y(0) = 0, \quad y'(0) = 0,$$

where $f_k(t) = [u_{4-k}(t) - u_{4+k}(t)]/2k$ with $0 < k \le 1$.

(a) Find the solution $y = \phi(t, k)$ of the initial value problem.

(b) Calculate $\lim_{k \to 0} \phi(t, k)$ from the solution found in part (a).

(c) Observe that $\lim_{k \to 0} f_k(t) = \delta(t-4)$. Find the solution $\phi_0(t)$ of the given initial value problem with $f_k(t)$ replaced by $\delta(t-4)$. Is it true that $\phi_0(t) = \lim_{k \to 0} \phi(t, k)$?

(d) Plot $\phi(t, 1/2)$, $\phi(t, 1/4)$, and $\phi_0(t)$ on the same axes. Describe the relation between $\phi(t, k)$ and $\phi_0(t)$.

Problems 17 through 22 deal with the effect of a sequence of impulses on an undamped oscillator. Suppose that

$$y'' + y = f(t), \qquad y(0) = 0, \quad y'(0) = 0.$$

For each of the following choices for $f(t)$:

(a) Try to predict the nature of the solution without solving the problem.

(b) Test your prediction by finding the solution and drawing its graph.

(c) Determine what happens after the sequence of impulses ends.

17. $f(t) = \sum_{k=1}^{20} \delta(t - k\pi)$ 18. $f(t) = \sum_{k=1}^{20} (-1)^{k+1} \delta(t - k\pi)$

19. $f(t) = \sum_{k=1}^{20} \delta(t - k\pi/2)$ 20. $f(t) = \sum_{k=1}^{20} (-1)^{k+1} \delta(t - k\pi/2)$

21. $f(t) = \sum_{k=1}^{15} \delta[t - (2k-1)\pi]$ 22. $f(t) = \sum_{k=1}^{40} (-1)^{k+1} \delta(t - 11k/4)$

23. The position of a certain lightly damped oscillator satisfies the initial value problem

$$y'' + 0.1y' + y = \sum_{k=1}^{20} (-1)^{k+1} \delta(t - k\pi), \qquad y(0) = 0, \quad y'(0) = 0.$$

Observe that, except for the damping term, this problem is the same as Problem 18.

(a) Try to predict the nature of the solution without solving the problem.

(b) Test your prediction by finding the solution and drawing its graph.

(c) Determine what happens after the sequence of impulses ends.

24. Proceed as in Problem 23 for the oscillator satisfying

$$y'' + 0.1y' + y = \sum_{k=1}^{15} \delta[t - (2k-1)\pi], \qquad y(0) = 0, \quad y'(0) = 0.$$

Observe that, except for the damping term, this problem is the same as Problem 21.

25. (a) By the method of variation of parameters, show that the solution of the initial value problem

$$y'' + 2y' + 2y = f(t); \qquad y(0) = 0, \quad y'(0) = 0$$

is

$$y = \int_0^t e^{-(t-\tau)} f(\tau) \sin(t - \tau) \, d\tau.$$

(b) Show that if $f(t) = \delta(t - \pi)$, then the solution of part (a) reduces to

$$y = u_\pi(t) e^{-(t-\pi)} \sin(t - \pi).$$

(c) Use a Laplace transform to solve the given initial value problem with $f(t) = \delta(t - \pi)$ and confirm that the solution agrees with the result of part (b).

6.6 The Convolution Integral

Sometimes it is possible to identify a Laplace transform $H(s)$ as the product of two other transforms $F(s)$ and $G(s)$, the latter transforms corresponding to known functions f and g, respectively. In this event, we might anticipate that $H(s)$ would be the transform of the product of f and g. However, this is not the case; in other words, the Laplace transform cannot be commuted with ordinary multiplication. On the other hand, if an appropriately defined "generalized product" is introduced, then the situation changes, as stated in the following theorem.

Theorem 6.6.1 If $F(s) = \mathcal{L}\{f(t)\}$ and $G(s) = \mathcal{L}\{g(t)\}$ both exist for $s > a \geq 0$, then

$$H(s) = F(s)G(s) = \mathcal{L}\{h(t)\}, \qquad s > a, \tag{1}$$

where

$$h(t) = \int_0^t f(t - \tau)g(\tau) \, d\tau = \int_0^t f(\tau)g(t - \tau) \, d\tau. \tag{2}$$

The function h is known as the convolution of f and g; the integrals in Eq. (2) are known as convolution integrals.

The equality of the two integrals in Eq. (2) follows by making the change of variable $t - \tau = \xi$ in the first integral. Before giving the proof of this theorem, let us make some observations about the convolution integral. According to this theorem, the transform of the convolution of two functions, rather than the transform of their ordinary product, is given by the product of the separate transforms. It is conventional to emphasize that the convolution integral can be thought of as a "generalized product" by writing

$$h(t) = (f * g)(t). \tag{3}$$

In particular, the notation $(f * g)(t)$ serves to indicate the first integral appearing in Eq. (2).

The convolution $f * g$ has many of the properties of ordinary multiplication. For example, it is relatively simple to show that

$$f * g = g * f \qquad \text{(commutative law)} \tag{4}$$

$$f * (g_1 + g_2) = f * g_1 + f * g_2 \qquad \text{(distributive law)} \tag{5}$$

$$(f * g) * h = f * (g * h) \qquad \text{(associative law)} \tag{6}$$

$$f * 0 = 0 * f = 0. \tag{7}$$

The proofs of these properties are left to the reader. However, there are other properties of ordinary multiplication that the convolution integral does not have. For example, it is not true in general that $f * 1$ is equal to f. To see this, note that

$$(f * 1)(t) = \int_0^t f(t - \tau) \cdot 1 \, d\tau = \int_0^t f(t - \tau) \, d\tau.$$

If, for example, $f(t) = \cos t$, then

$$(f * 1)(t) = \int_0^t \cos(t - \tau) \, d\tau = -\sin(t - \tau) \Big|_{\tau=0}^{\tau=t}$$

$$= -\sin 0 + \sin t$$

$$= \sin t.$$

Clearly, $(f * 1)(t) \neq f(t)$. Similarly, it may not be true that $f * f$ is nonnegative. See Problem 3 for an example.

Convolution integrals arise in various applications in which the behavior of the system at time t depends not only on its state at time t but also on its past history. Systems of this kind are sometimes called hereditary systems and occur in such diverse fields as neutron transport, viscoelasticity, and population dynamics.

Turning now to the proof of Theorem 6.6.1, we note first that if

$$F(s) = \int_0^\infty e^{-s\xi} f(\xi) \, d\xi$$

and

$$G(s) = \int_0^\infty e^{-s\tau} g(\tau) \, d\tau,$$

then

$$F(s)G(s) = \int_0^\infty e^{-s\xi} f(\xi)\, d\xi \int_0^\infty e^{-s\tau} g(\tau)\, d\tau. \tag{8}$$

Since the integrand of the first integral does not depend on the integration variable of the second, we can write $F(s)G(s)$ as an iterated integral,

$$F(s)G(s) = \int_0^\infty e^{-s\tau} g(\tau) \left[\int_0^\infty e^{-s\xi} f(\xi)\, d\xi \right] d\tau$$

$$= \int_0^\infty g(\tau) \left[\int_0^\infty e^{-s(\xi+\tau)} f(\xi)\, d\xi \right] d\tau. \tag{9}$$

The latter integral can be put into a more convenient form by introducing a change of variable. Let $\xi = t - \tau$, for fixed τ, so that $d\xi = dt$. Further, $\xi = 0$ corresponds to $t = \tau$ and $\xi = \infty$ corresponds to $t = \infty$; then the integral with respect to ξ in Eq. (9) is transformed into one with respect to t:

$$F(s)G(s) = \int_0^\infty g(\tau) \left[\int_\tau^\infty e^{-st} f(t - \tau)\, dt \right] d\tau. \tag{10}$$

The iterated integral on the right side of Eq. (10) is carried out over the shaded wedge-shaped region extending to infinity in the $t\tau$-plane shown in Figure 6.6.1. Assuming that the order of integration can be reversed, we finally obtain

$$F(s)G(s) = \int_0^\infty e^{-st} \left[\int_0^t f(t - \tau)g(\tau)\, d\tau \right] dt, \tag{11}$$

or

$$F(s)G(s) = \int_0^\infty e^{-st} h(t)\, dt$$

$$= \mathcal{L}\{h(t)\}, \tag{12}$$

where $h(t)$ is defined by Eq. (2). This completes the proof of Theorem 6.6.1.

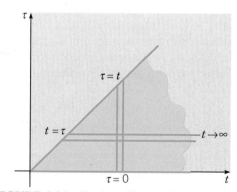

FIGURE 6.6.1 Region of integration in $F(s)G(s)$.

**EXAMPLE
1**

Find the inverse transform of

$$H(s) = \frac{a}{s^2(s^2 + a^2)}. \tag{13}$$

It is convenient to think of $H(s)$ as the product of s^{-2} and $a/(s^2 + a^2)$, which, according to lines 3 and 5 of Table 6.2.1, are the transforms of t and $\sin at$, respectively. Hence, by Theorem 6.6.1, the inverse transform of $H(s)$ is

$$h(t) = \int_0^t (t - \tau) \sin a\tau \, d\tau = \frac{at - \sin at}{a^2}. \tag{14}$$

You can verify that the same result is obtained if $h(t)$ is written in the alternative form

$$h(t) = \int_0^t \tau \sin a(t - \tau) \, d\tau,$$

which confirms Eq. (2) in this case. Of course, $h(t)$ can also be found by expanding $H(s)$ in partial fractions.

**EXAMPLE
2**

Find the solution of the initial value problem

$$y'' + 4y = g(t), \tag{15}$$

$$y(0) = 3, \qquad y'(0) = -1. \tag{16}$$

By taking the Laplace transform of the differential equation and using the initial conditions, we obtain

$$s^2 Y(s) - 3s + 1 + 4Y(s) = G(s),$$

or

$$Y(s) = \frac{3s - 1}{s^2 + 4} + \frac{G(s)}{s^2 + 4}. \tag{17}$$

Observe that the first and second terms on the right side of Eq. (17) contain the dependence of $Y(s)$ on the initial conditions and forcing function, respectively. It is convenient to write $Y(s)$ in the form

$$Y(s) = 3\frac{s}{s^2 + 4} - \frac{1}{2}\frac{2}{s^2 + 4} + \frac{1}{2}\frac{2}{s^2 + 4}G(s). \tag{18}$$

Then, using lines 5 and 6 of Table 6.2.1 and Theorem 6.6.1, we obtain

$$y = 3\cos 2t - \tfrac{1}{2}\sin 2t + \tfrac{1}{2}\int_0^t \sin 2(t - \tau)g(\tau) \, d\tau. \tag{19}$$

If a specific forcing function g is given, then the integral in Eq. (19) can be evaluated (by numerical means, if necessary).

Example 2 illustrates the power of the convolution integral as a tool for writing the solution of an initial value problem in terms of an integral. In fact, it is possible to proceed in much the same way in more general problems. Consider the problem consisting of the differential equation

$$ay'' + by' + cy = g(t), \tag{20}$$

where $a, b,$ and c are real constants and g is a given function, together with the initial conditions

$$y(0) = y_0, \qquad y'(0) = y'_0. \tag{21}$$

The transform approach yields some important insights concerning the structure of the solution of any problem of this type.

The initial value problem (20), (21) is often referred to as an input–output problem. The coefficients $a, b,$ and c describe the properties of some physical system, and $g(t)$ is the input to the system. The values y_0 and y'_0 describe the initial state, and the solution y is the output at time t.

By taking the Laplace transform of Eq. (20) and using the initial conditions (21), we obtain

$$(as^2 + bs + c)Y(s) - (as + b)y_0 - ay'_0 = G(s).$$

If we let

$$\Phi(s) = \frac{(as + b)y_0 + ay'_0}{as^2 + bs + c}, \qquad \Psi(s) = \frac{G(s)}{as^2 + bs + c}, \tag{22}$$

then we can write

$$Y(s) = \Phi(s) + \Psi(s). \tag{23}$$

Consequently,

$$y = \phi(t) + \psi(t), \tag{24}$$

where $\phi(t) = \mathcal{L}^{-1}\{\Phi(s)\}$ and $\psi(t) = \mathcal{L}^{-1}\{\Psi(s)\}$. Observe that $\phi(t)$ is the solution of the initial value problem

$$ay'' + by' + cy = 0, \qquad y(0) = y_0, \quad y'(0) = y'_0, \tag{25}$$

obtained from Eqs. (20) and (21) by setting $g(t)$ equal to zero. Similarly, $\psi(t)$ is the solution of

$$ay'' + by' + cy = g(t), \qquad y(0) = 0, \quad y'(0) = 0, \tag{26}$$

in which the initial values y_0 and y'_0 are each replaced by zero.

Once specific values of $a, b,$ and c are given, we can find $\phi(t) = \mathcal{L}^{-1}\{\Phi(s)\}$ by using Table 6.2.1, possibly in conjunction with a translation or a partial fraction expansion. To find $\psi(t) = \mathcal{L}^{-1}\{\Psi(s)\}$, it is convenient to write $\Psi(s)$ as

$$\Psi(s) = H(s)G(s), \tag{27}$$

where $H(s) = (as^2 + bs + c)^{-1}$. The function H is known as the **transfer function**[3] and depends only on the properties of the system under consideration; that is, $H(s)$ is determined entirely by the coefficients $a, b,$ and c. On the other hand, $G(s)$ depends

[3]This terminology arises from the fact that $H(s)$ is the ratio of the transforms of the output and the input of the problem (26).

only on the external excitation $g(t)$ that is applied to the system. By the convolution theorem we can write

$$\psi(t) = \mathcal{L}^{-1}\{H(s)G(s)\} = \int_0^t h(t-\tau)g(\tau)\,d\tau, \tag{28}$$

where $h(t) = \mathcal{L}^{-1}\{H(s)\}$, and $g(t)$ is the given forcing function.

To obtain a better understanding of the significance of $h(t)$, we consider the case in which $G(s) = 1$; consequently, $g(t) = \delta(t)$ and $\Psi(s) = H(s)$. This means that $y = h(t)$ is the solution of the initial value problem

$$ay'' + by' + cy = \delta(t), \qquad y(0) = 0, \quad y'(0) = 0, \tag{29}$$

obtained from Eq. (26) by replacing $g(t)$ by $\delta(t)$. Thus $h(t)$ is the response of the system to a unit impulse applied at $t = 0$, and it is natural to call $h(t)$ the **impulse response** of the system. Equation (28) then says that $\psi(t)$ is the convolution of the impulse response and the forcing function.

Referring to Example 2, we note that in that case, the transfer function is $H(s) = 1/(s^2 + 4)$ and the impulse response is $h(t) = (\sin 2t)/2$. Also, the first two terms on the right side of Eq. (19) constitute the function $\phi(t)$, the solution of the corresponding homogeneous equation that satisfies the given initial conditions.

PROBLEMS

1. Establish the commutative, distributive, and associative properties of the convolution integral.

 (a) $f * g = g * f$

 (b) $f * (g_1 + g_2) = f * g_1 + f * g_2$

 (c) $f * (g * h) = (f * g) * h$

2. Find an example different from the one in the text showing that $(f * 1)(t)$ need not be equal to $f(t)$.

3. Show, by means of the example $f(t) = \sin t$, that $f * f$ is not necessarily nonnegative.

In each of Problems 4 through 7 find the Laplace transform of the given function.

4. $f(t) = \int_0^t (t - \tau)^2 \cos 2\tau \, d\tau$

5. $f(t) = \int_0^t e^{-(t-\tau)} \sin \tau \, d\tau$

6. $f(t) = \int_0^t (t - \tau)e^\tau \, d\tau$

7. $f(t) = \int_0^t \sin(t - \tau) \cos \tau \, d\tau$

In each of Problems 8 through 11 find the inverse Laplace transform of the given function by using the convolution theorem.

8. $F(s) = \dfrac{1}{s^4(s^2 + 1)}$

9. $F(s) = \dfrac{s}{(s + 1)(s^2 + 4)}$

10. $F(s) = \dfrac{1}{(s + 1)^2(s^2 + 4)}$

11. $F(s) = \dfrac{G(s)}{s^2 + 1}$

12. (a) If $f(t) = t^m$ and $g(t) = t^n$, where m and n are positive integers, show that

$$f * g = t^{m+n+1} \int_0^1 u^m (1 - u)^n \, du.$$

5, 7, 9, 10, 13, 17

y
 online

(b) Use the convolution theorem to show that

$$\int_0^1 u^m (1 - u)^n \, du = \frac{m! \, n!}{(m + n + 1)!}.$$

(c) Extend the result of part (b) to the case where m and n are positive numbers but not necessarily integers.

In each of Problems 13 through 20 express the solution of the given initial value problem in terms of a convolution integral.

13. $y'' + \omega^2 y = g(t);$ $y(0) = 0, \quad y'(0) = 1$

14. $y'' + 2y' + 2y = \sin \alpha t;$ $y(0) = 0, \quad y'(0) = 0$

15. $4y'' + 4y' + 17y = g(t);$ $y(0) = 0, \quad y'(0) = 0$

16. $y'' + y' + \frac{5}{4} y = 1 - u_\pi(t);$ $y(0) = 1, \quad y'(0) = -1$

17. $y'' + 4y' + 4y = g(t);$ $y(0) = 2, \quad y'(0) = -3$

18. $y'' + 3y' + 2y = \cos \alpha t;$ $y(0) = 1, \quad y'(0) = 0$

19. $y^{(4)} - y = g(t);$ $y(0) = 0, \quad y'(0) = 0, \quad y''(0) = 0, \quad y'''(0) = 0$

20. $y^{(4)} + 5y'' + 4y = g(t);$ $y(0) = 1, \quad y'(0) = 0, \quad y''(0) = 0, \quad y'''(0) = 0$

21. Consider the equation

$$\phi(t) + \int_0^t k(t - \xi) \phi(\xi) \, d\xi = f(t),$$

in which f and k are known functions, and ϕ is to be determined. Since the unknown function ϕ appears under an integral sign, the given equation is called an **integral equation**; in particular, it belongs to a class of integral equations known as Volterra integral equations. Take the Laplace transform of the given integral equation and obtain an expression for $\mathcal{L}\{\phi(t)\}$ in terms of the transforms $\mathcal{L}\{f(t)\}$ and $\mathcal{L}\{k(t)\}$ of the given functions f and k. The inverse transform of $\mathcal{L}\{\phi(t)\}$ is the solution of the original integral equation.

22. Consider the Volterra integral equation (see Problem 21)

$$\phi(t) + \int_0^t (t - \xi) \phi(\xi) \, d\xi = \sin 2t. \tag{i}$$

(a) Solve the integral equation (i) by using the Laplace transform.

(b) By differentiating Eq. (i) twice, show that $\phi(t)$ satisfies the differential equation

$$\phi''(t) + \phi(t) = -4 \sin 2t.$$

Show also that the initial conditions are

$$\phi(0) = 0, \qquad \phi'(0) = 2.$$

(c) Solve the initial value problem in part (b) and verify that the solution is the same as the one in part (a).

In each of Problems 23 through 25:

(a) Solve the given Volterra integral equation by using the Laplace transform.

(b) Convert the integral equation into an initial value problem, as in Problem 22(b).

(c) Solve the initial value problem in part (b) and verify that the solution is the same as the one in part (a).

23. $\phi(t) + \int_0^t (t - \xi)\phi(\xi)\,d\xi = 1$ 24. $\phi(t) - \int_0^t (t - \xi)\phi(\xi)\,d\xi = 1$

25. $\phi(t) + 2\int_0^t \cos(t - \xi)\phi(\xi)\,d\xi = e^{-t}$

There are also equations, known as **integro-differential equations**, in which both derivatives and integrals of the unknown function appear. In each of Problems 26 through 28:

(a) Solve the given integro-differential equation by using the Laplace transform.

(b) By differentiating the integro-differential equation a sufficient number of times, convert it into an initial value problem.

(c) Solve the initial value problem in part (b) and verify that the solution is the same as the one in part (a).

26. $\phi'(t) + \int_0^t (t - \xi)\phi(\xi)\,d\xi = t, \qquad \phi(0) = 0$

27. $\phi'(t) - \frac{1}{2}\int_0^t (t - \xi)^2\phi(\xi)\,d\xi = -t, \qquad \phi(0) = 1$

28. $\phi'(t) + \phi(t) = \int_0^t \sin(t - \xi)\phi(\xi)\,d\xi, \qquad \phi(0) = 1$

29. **The Tautochrone.** A problem of interest in the history of mathematics is that of finding the tautochrone[4]—the curve down which a particle will slide freely under gravity alone, reaching the bottom in the same time regardless of its starting point on the curve. This problem arose in the construction of a clock pendulum whose period is independent of the amplitude of its motion. The tautochrone was found by Christian Huygens (1629–1695) in 1673 by geometrical methods, and later by Leibniz and Jakob Bernoulli using analytical arguments. Bernoulli's solution (in 1690) was one of the first occasions on which a differential equation was explicitly solved.

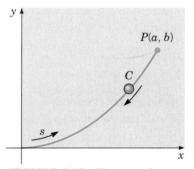

FIGURE 6.6.2 The tautochrone.

The geometric configuration is shown in Figure 6.6.2. The starting point $P(a, b)$ is joined to the terminal point $(0, 0)$ by the arc C. Arc length s is measured from the origin, and $f(y)$ denotes the rate of change of s with respect to y:

$$f(y) = \frac{ds}{dy} = \left[1 + \left(\frac{dx}{dy}\right)^2\right]^{1/2}. \tag{i}$$

[4]The word "tautochrone" comes from the Greek words *tauto*, meaning same, and *chronos*, meaning time.

Then it follows from the principle of conservation of energy that the time $T(b)$ required for a particle to slide from P to the origin is

$$T(b) = \frac{1}{\sqrt{2g}} \int_0^b \frac{f(y)}{\sqrt{b-y}} \, dy. \tag{ii}$$

(a) Assume that $T(b) = T_0$, a constant, for each b. By taking the Laplace transform of Eq. (ii) in this case, and using the convolution theorem, show that

$$F(s) = \sqrt{\frac{2g}{\pi}} \frac{T_0}{\sqrt{s}}; \tag{iii}$$

then show that

$$f(y) = \frac{\sqrt{2g}}{\pi} \frac{T_0}{\sqrt{y}}. \tag{iv}$$

Hint: See Problem 27 of Section 6.1.

(b) Combining Eqs. (i) and (iv), show that

$$\frac{dx}{dy} = \sqrt{\frac{2\alpha - y}{y}}, \tag{v}$$

where $\alpha = gT_0^2/\pi^2$.

(c) Use the substitution $y = 2\alpha \sin^2(\theta/2)$ to solve Eq. (v), and show that

$$x = \alpha(\theta + \sin\theta), \qquad y = \alpha(1 - \cos\theta). \tag{vi}$$

Equations (vi) can be identified as parametric equations of a cycloid. Thus the tautochrone is an arc of a cycloid.

REFERENCES The books listed below contain additional information on the Laplace transform and its applications:

Churchill, R. V., *Operational Mathematics* (3rd ed.) (New York: McGraw-Hill, 1971).

Doetsch, G., *Introduction to the Theory and Application of the Laplace Transform* (trans. W. Nader) (New York: Springer, 1974).

Kaplan, W., *Operational Methods for Linear Systems* (Reading, MA: Addison-Wesley, 1962).

Kuhfittig, P. K. F., *Introduction to the Laplace Transform* (New York: Plenum, 1978).

Miles, J. W., *Integral Transforms in Applied Mathematics* (London: Cambridge University Press, 1971).

Rainville, E. D., *The Laplace Transform: An Introduction* (New York: Macmillan, 1963).

Each of the books just mentioned contains a table of transforms. Extensive tables are also available; see, for example:

Erdelyi, A. (ed.), *Tables of Integral Transforms* (Vol. 1) (New York: McGraw-Hill, 1954).

Roberts, G. E., and Kaufman, H., *Table of Laplace Transforms* (Philadelphia: Saunders, 1966).

A further discussion of generalized functions can be found in

Lighthill, M. J., *Fourier Analysis and Generalized Functions* (London: Cambridge University Press, 1958).

Systems of First Order Linear Equations

There are many physical problems that involve a number of separate elements linked together in some manner. For example, electrical networks have this character, as do some problems in mechanics and in other fields. In these and similar cases, the corresponding mathematical problem consists of a *system* of two or more differential equations, which can always be written as first order equations. In this chapter we focus on systems of first order *linear* equations, utilizing some of the elementary aspects of linear algebra to unify the presentation.

7.1 Introduction

Systems of simultaneous ordinary differential equations arise naturally in problems involving several dependent variables, each of which is a function of the same single independent variable. We will denote the independent variable by t, and will let x_1, x_2, x_3, \ldots represent dependent variables that are functions of t. Differentiation with respect to t will be denoted by a prime.

For example, consider the spring–mass system in Figure 7.1.1. The two masses move on a frictionless surface under the influence of external forces $F_1(t)$ and $F_2(t)$, and they are also constrained by the three springs whose constants are $k_1, k_2,$ and k_3, respectively. Using arguments similar to those in Section 3.8, we find the following

355

equations for the coordinates x_1 and x_2 of the two masses:

$$m_1\frac{d^2x_1}{dt^2} = k_2(x_2 - x_1) - k_1x_1 + F_1(t)$$

$$= -(k_1 + k_2)x_1 + k_2x_2 + F_1(t),$$

$$m_2\frac{d^2x_2}{dt^2} = -k_3x_2 - k_2(x_2 - x_1) + F_2(t)$$

$$= k_2x_1 - (k_2 + k_3)x_2 + F_2(t). \tag{1}$$

A derivation of Eqs. (1) is outlined in Problem 17.

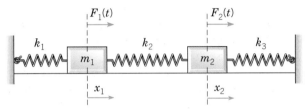

FIGURE 7.1.1 A two-mass, three-spring system.

Next, consider the parallel LRC circuit shown in Figure 7.1.2. Let V be the voltage drop across the capacitor and I the current through the inductor. Then, referring to Section 3.8 and to Problem 19 of this section, we can show that the voltage and current are described by the system of equations

$$\frac{dI}{dt} = \frac{V}{L},$$

$$\frac{dV}{dt} = -\frac{I}{C} - \frac{V}{RC}, \tag{2}$$

where L is the inductance, C is the capacitance, and R is the resistance.

FIGURE 7.1.2 A parallel LRC circuit.

As a final example, we mention the predator–prey problem, one of the fundamental problems of mathematical ecology, which is discussed in more detail in Section 9.5. Let $H(t)$ and $P(t)$ denote the populations at time t of two species, one of which (P) preys on the other (H). For example, $P(t)$ and $H(t)$ may be the numbers of foxes and rabbits, respectively, in a woods, or the numbers of bass and redear (which are eaten

by bass) in a pond. Without the prey the predators will decrease, and without the predator the prey will increase. A mathematical model showing how an ecological balance can be maintained when both are present was proposed about 1925 by Lotka and Volterra. The model consists of the system of differential equations

$$dH/dt = a_1 H - b_1 HP,$$
$$dP/dt = -a_2 P + b_2 HP, \tag{3}$$

known as the predator–prey equations. In Eqs. (3) the coefficient a_1 is the birth rate of the population H; similarly, a_2 is the death rate of the population P. The HP terms in the two equations model the interaction of the two populations. The number of encounters between predator and prey is assumed to be proportional to the product of the populations. Since any such encounter tends to be good for the predator and bad for the prey, the sign of the HP term is negative in the first equation and positive in the second. The coefficients b_1 and b_2 are the coefficients of interaction between predator and prey.

Another reason why systems of first order equations are important is that equations of higher order can always be transformed into such systems. This is usually required if a numerical approach is planned, because almost all codes for generating approximate numerical solutions of differential equations are written for systems of first order equations. The following example illustrates how easy it is to make the transformation.

EXAMPLE 1

The motion of a certain spring–mass system (see Example 3 of Section 3.8) is described by the second order differential equation

$$u'' + 0.125u' + u = 0. \tag{4}$$

Rewrite this equation as a system of first order equations.

Let $x_1 = u$ and $x_2 = u'$. Then it follows that $x_1' = x_2$. Further, $u'' = x_2'$. Then, by substituting for u, u', and u'' in Eq. (4), we obtain

$$x_2' + 0.125x_2 + x_1 = 0.$$

Thus x_1 and x_2 satisfy the following system of two first order differential equations:

$$x_1' = x_2,$$
$$x_2' = -x_1 - 0.125x_2. \tag{5}$$

The general equation of motion of a spring–mass system,

$$mu'' + \gamma u' + ku = F(t), \tag{6}$$

can be transformed into a system of first order equations in the same manner. If we let $x_1 = u$ and $x_2 = u'$, and proceed as in Example 1, we quickly obtain the system

$$x_1' = x_2,$$
$$x_2' = -(k/m)x_1 - (\gamma/m)x_2 + F(t)/m. \tag{7}$$

To transform an arbitrary nth order equation

$$y^{(n)} = F(t, y, y', \ldots, y^{(n-1)}) \tag{8}$$

into a system of n first order equations, we extend the method of Example 1 by introducing the variables x_1, x_2, \ldots, x_n defined by

$$x_1 = y, \quad x_2 = y', \quad x_3 = y'', \quad \ldots, \quad x_n = y^{(n-1)}. \tag{9}$$

It then follows immediately that

$$\begin{aligned} x_1' &= x_2, \\ x_2' &= x_3, \\ &\;\;\vdots \\ x_{n-1}' &= x_n, \end{aligned} \tag{10}$$

and, from Eq. (8),

$$x_n' = F(t, x_1, x_2, \ldots, x_n). \tag{11}$$

Equations (10) and (11) are a special case of the more general system

$$\begin{aligned} x_1' &= F_1(t, x_1, x_2, \ldots, x_n), \\ x_2' &= F_2(t, x_1, x_2, \ldots, x_n), \\ &\;\;\vdots \\ x_n' &= F_n(t, x_1, x_2, \ldots, x_n). \end{aligned} \tag{12}$$

In a similar way the system (1) can be reduced to a system of four first order equations of the form (12), and the systems (2) and (3) are already in this form. In fact, systems of the form (12) include almost all cases of interest, so much of the more advanced theory of differential equations is devoted to such systems.

The system (12) is said to have a **solution** on the interval I: $\alpha < t < \beta$ if there exists a set of n functions

$$x_1 = \phi_1(t), \quad x_2 = \phi_2(t), \quad \ldots, \quad x_n = \phi_n(t), \tag{13}$$

that are differentiable at all points in the interval I and that satisfy the system of equations (12) at all points in this interval. In addition to the given system of differential equations, there may also be given initial conditions of the form

$$x_1(t_0) = x_1^0, \quad x_2(t_0) = x_2^0, \quad \ldots, \quad x_n(t_0) = x_n^0, \tag{14}$$

where t_0 is a specified value of t in I, and x_1^0, \ldots, x_n^0 are prescribed numbers. The differential equations (12) and the initial conditions (14) together form an initial value problem.

A solution (13) can be viewed as a set of parametric equations in an n-dimensional space. For a given value of t, Eqs. (13) give values for the coordinates x_1, \ldots, x_n of a point in the space. As t changes, the coordinates in general also change. The collection of points corresponding to $\alpha < t < \beta$ form a curve in the space. It is often helpful to think of the curve as the trajectory, or path, of a particle moving in accordance with the system of differential equations (12). The initial conditions (14) determine the starting point of the moving particle.

The following conditions on F_1, F_2, \ldots, F_n, which are easily checked in specific problems, are sufficient to ensure that the initial value problem (12), (14) has a unique solution. Theorem 7.1.1 is analogous to Theorem 2.4.2, the existence and uniqueness theorem for a single first order equation.

Theorem 7.1.1

Let each of the functions F_1, \ldots, F_n and the partial derivatives $\partial F_1/\partial x_1, \ldots, \partial F_1/\partial x_n,$ $\ldots, \partial F_n/\partial x_1, \ldots, \partial F_n/\partial x_n$ be continuous in a region R of $tx_1 x_2 \cdots x_n$-space defined by $\alpha < t < \beta$, $\alpha_1 < x_1 < \beta_1, \ldots, \alpha_n < x_n < \beta_n$, and let the point $(t_0, x_1^0, x_2^0, \ldots, x_n^0)$ be in R. Then there is an interval $|t - t_0| < h$ in which there exists a unique solution $x_1 = \phi_1(t), \ldots, x_n = \phi_n(t)$ of the system of differential equations (12) that also satisfies the initial conditions (14).

The proof of this theorem can be constructed by generalizing the argument in Section 2.8, but we do not give it here. However, note that, in the hypotheses of the theorem, nothing is said about the partial derivatives of F_1, \ldots, F_n with respect to the independent variable t. Also, in the conclusion, the length $2h$ of the interval in which the solution exists is not specified exactly, and in some cases it may be very short. Finally, the same result can be established on the basis of somewhat weaker but more complicated hypotheses, so the theorem as stated is not the most general one known, and the given conditions are sufficient, but not necessary, for the conclusion to hold.

If each of the functions F_1, \ldots, F_n in Eqs. (12) is a linear function of the dependent variables x_1, \ldots, x_n, then the system of equations is said to be **linear**; otherwise, it is **nonlinear**. Thus the most general system of n first order linear equations has the form

$$
\begin{aligned}
x_1' &= p_{11}(t)x_1 + \cdots + p_{1n}(t)x_n + g_1(t), \\
x_2' &= p_{21}(t)x_1 + \cdots + p_{2n}(t)x_n + g_2(t), \\
&\ \vdots \\
x_n' &= p_{n1}(t)x_1 + \cdots + p_{nn}(t)x_n + g_n(t).
\end{aligned}
\tag{15}
$$

If each of the functions $g_1(t), \ldots, g_n(t)$ is zero for all t in the interval I, then the system (15) is said to be **homogeneous**; otherwise, it is **nonhomogeneous**. Observe that the systems (1) and (2) are both linear, but the system (3) is nonlinear. The system (1) is nonhomogeneous unless $F_1(t) = F_2(t) = 0$, while the system (2) is homogeneous. For the linear system (15), the existence and uniqueness theorem is simpler and also has a stronger conclusion. It is analogous to Theorems 2.4.1 and 3.2.1.

Theorem 7.1.2

If the functions $p_{11}, p_{12}, \ldots, p_{nn}, g_1, \ldots, g_n$ are continuous on an open interval $I: \alpha < t < \beta$, then there exists a unique solution $x_1 = \phi_1(t), \ldots, x_n = \phi_n(t)$ of the system (15) that also satisfies the initial conditions (14), where t_0 is any point in I, and x_1^0, \ldots, x_n^0 are any prescribed numbers. Moreover, the solution exists throughout the interval I.

Note that, in contrast to the situation for a nonlinear system, the existence and uniqueness of the solution of a linear system are guaranteed throughout the interval

in which the hypotheses are satisfied. Furthermore, for a linear system the initial values x_1^0, \ldots, x_n^0 at $t = t_0$ are completely arbitrary, whereas in the nonlinear case the initial point must lie in the region R defined in Theorem 7.1.1.

The rest of this chapter is devoted to systems of linear first order equations (nonlinear systems are included in the discussion in Chapters 8 and 9). Our presentation makes use of matrix notation and assumes that you have some familiarity with the properties of matrices. The basic facts about matrices are summarized in Sections 7.2 and 7.3, and some more advanced material is reviewed as needed in later sections.

PROBLEMS

In each of Problems 1 through 4 transform the given equation into a system of first order equations.

1. $u'' + 0.5u' + 2u = 0$

2. $u'' + 0.5u' + 2u = 3\sin t$

3. $t^2 u'' + tu' + (t^2 - 0.25)u = 0$

4. $u^{(4)} - u = 0$

In each of Problems 5 and 6 transform the given initial value problem into an initial value problem for two first order equations.

5. $u'' + 0.25u' + 4u = 2\cos 3t, \qquad u(0) = 1, \quad u'(0) = -2$

6. $u'' + p(t)u' + q(t)u = g(t), \qquad u(0) = u_0, \quad u'(0) = u_0'$

7. Systems of first order equations can sometimes be transformed into a single equation of higher order. Consider the system

$$x_1' = -2x_1 + x_2, \qquad x_2' = x_1 - 2x_2.$$

(a) Solve the first equation for x_2 and substitute into the second equation, thereby obtaining a second order equation for x_1. Solve this equation for x_1 and then determine x_2 also.

(b) Find the solution of the given system that also satisfies the initial conditions $x_1(0) = 2$, $x_2(0) = 3$.

(c) Sketch the curve, for $t \geq 0$, given parametrically by the expressions for x_1 and x_2 obtained in part (b).

In each of Problems 8 through 12 proceed as in Problem 7 to transform the given system into a single equation of second order. Then find x_1 and x_2 that also satisfy the given initial conditions. Finally, sketch the graph of the solution in the $x_1 x_2$-plane for $t \geq 0$.

8. $x_1' = 3x_1 - 2x_2, \qquad x_1(0) = 3$
$x_2' = 2x_1 - 2x_2, \qquad x_2(0) = \frac{1}{2}$

9. $x_1' = 1.25x_1 + 0.75x_2, \qquad x_1(0) = -2$
$x_2' = 0.75x_1 + 1.25x_2, \qquad x_2(0) = 1$

10. $x_1' = x_1 - 2x_2, \qquad x_1(0) = -1$
$x_2' = 3x_1 - 4x_2, \qquad x_2(0) = 2$

11. $x_1' = 2x_2, \qquad x_1(0) = 3$
$x_2' = -2x_1, \qquad x_2(0) = 4$

12. $x_1' = -0.5x_1 + 2x_2, \qquad x_1(0) = -2$
$x_2' = -2x_1 - 0.5x_2, \qquad x_2(0) = 2$

13. Transform Eqs. (2) for the parallel circuit into a single second order equation.

14. Show that if a_{11}, a_{12}, a_{21}, and a_{22} are constants with a_{12} and a_{21} not both zero, and if the functions g_1 and g_2 are differentiable, then the initial value problem

$$x_1' = a_{11}x_1 + a_{12}x_2 + g_1(t), \qquad x_1(0) = x_1^0$$
$$x_2' = a_{21}x_1 + a_{22}x_2 + g_2(t), \qquad x_2(0) = x_2^0$$

can be transformed into an initial value problem for a single second order equation. Can the same procedure be carried out if a_{11}, \ldots, a_{22} are functions of t?

15. Consider the linear homogeneous system

$$x' = p_{11}(t)x + p_{12}(t)y,$$

$$y' = p_{21}(t)x + p_{22}(t)y.$$

Show that if $x = x_1(t)$, $y = y_1(t)$ and $x = x_2(t)$, $y = y_2(t)$ are two solutions of the given system, then $x = c_1 x_1(t) + c_2 x_2(t)$, $y = c_1 y_1(t) + c_2 y_2(t)$ is also a solution for any constants c_1 and c_2. This is the principle of superposition.

16. Let $x = x_1(t)$, $y = y_1(t)$ and $x = x_2(t)$, $y = y_2(t)$ be any two solutions of the linear nonhomogeneous system

$$x' = p_{11}(t)x + p_{12}(t)y + g_1(t),$$

$$y' = p_{21}(t)x + p_{22}(t)y + g_2(t).$$

Show that $x = x_1(t) - x_2(t)$, $y = y_1(t) - y_2(t)$ is a solution of the corresponding homogeneous system.

17. Equations (1) can be derived by drawing a free-body diagram showing the forces acting on each mass. Figure 7.1.3a shows the situation when the displacements x_1 and x_2 of the two masses are both positive (to the right) and $x_2 > x_1$. Then springs 1 and 2 are elongated and spring 3 is compressed, giving rise to forces as shown in Figure 7.1.3b. Use Newton's law ($F = ma$) to derive Eqs. (1).

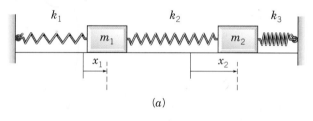

(a)

(b)

FIGURE 7.1.3 (a) The displacements x_1 and x_2 are both positive. (b) The free-body diagram for the spring–mass system.

18. Transform the system (1) into a system of first order equations by letting $y_1 = x_1, y_2 = x_2$, $y_3 = x_1'$, and $y_4 = x_2'$.

Electric Circuits. The theory of electric circuits, such as that shown in Figure 7.1.2, consisting of inductors, resistors, and capacitors, is based on Kirchhoff's laws: (1) The net flow of current through each node (or junction) is zero, and (2) the net voltage drop around each closed loop is zero. In addition to Kirchhoff's laws, we also have the relation between the current I in

amperes through each circuit element and the voltage drop V in volts across the element:

$$V = RI, \qquad R = \text{resistance in ohms};$$

$$C\frac{dV}{dt} = I, \qquad C = \text{capacitance in farads}[1];$$

$$L\frac{dI}{dt} = V, \qquad L = \text{inductance in henrys}.$$

Kirchhoff's laws and the current–voltage relation for each circuit element provide a system of algebraic and differential equations from which the voltage and current throughout the circuit can be determined. Problems 19 through 21 illustrate the procedure just described.

19. Consider the circuit shown in Figure 7.1.2. Let I_1, I_2, and I_3 be the currents through the capacitor, resistor, and inductor, respectively. Likewise, let V_1, V_2, and V_3 be the corresponding voltage drops. The arrows denote the arbitrarily chosen directions in which currents and voltage drops will be taken to be positive.

 (a) Applying Kirchhoff's second law to the upper loop in the circuit, show that

 $$V_1 - V_2 = 0. \tag{i}$$

 In a similar way, show that

 $$V_2 - V_3 = 0. \tag{ii}$$

 (b) Applying Kirchhoff's first law to either node in the circuit, show that

 $$I_1 + I_2 + I_3 = 0. \tag{iii}$$

 (c) Use the current–voltage relation through each element in the circuit to obtain the equations

 $$CV_1' = I_1, \qquad V_2 = RI_2, \qquad LI_3' = V_3. \tag{iv}$$

 (d) Eliminate V_2, V_3, I_1, and I_2 among Eqs. (i) through (iv) to obtain

 $$CV_1' = -I_3 - \frac{V_1}{R}, \qquad LI_3' = V_1. \tag{v}$$

 Observe that if we omit the subscripts in Eqs. (v), then we have the system (2) of this section.

20. Consider the circuit shown in Figure 7.1.4. Use the method outlined in Problem 19 to show that the current I through the inductor and the voltage V across the capacitor satisfy the system of differential equations

 $$\frac{dI}{dt} = -I - V, \qquad \frac{dV}{dt} = 2I - V.$$

[1] Actual capacitors typically have capacitances measured in microfarads. We use farad as the unit for numerical convenience.

FIGURE 7.1.4 The circuit in Problem 20.

21. Consider the circuit shown in Figure 7.1.5. Use the method outlined in Problem 19 to show that the current I through the inductor and the voltage V across the capacitor satisfy the system of differential equations

$$L\frac{dI}{dt} = -R_1 I - V, \qquad C\frac{dV}{dt} = I - \frac{V}{R_2}.$$

FIGURE 7.1.5 The circuit in Problem 21.

22. Consider the two interconnected tanks shown in Figure 7.1.6. Tank 1 initially contains 30 gal of water and 25 oz of salt, and Tank 2 initially contains 20 gal of water and 15 oz of salt. Water containing 1 oz/gal of salt flows into Tank 1 at a rate of 1.5 gal/min. The mixture flows from Tank 1 to Tank 2 at a rate of 3 gal/min. Water containing 3 oz/gal of salt also flows into Tank 2 at a rate of 1 gal/min (from the outside). The mixture drains from Tank 2 at a rate of 4 gal/min, of which some flows back into Tank 1 at a rate of 1.5 gal/min, while the remainder leaves the system.

(a) Let $Q_1(t)$ and $Q_2(t)$, respectively, be the amount of salt in each tank at time t. Write down differential equations and initial conditions that model the flow process. Observe that the system of differential equations is nonhomogeneous.

(b) Find the values of Q_1 and Q_2 for which the system is in equilibrium, that is, does not change with time. Let Q_1^E and Q_2^E be the equilibrium values. Can you predict which tank will approach its equilibrium state more rapidly?

(c) Let $x_1 = Q_1(t) - Q_1^E$ and $x_2 = Q_2(t) - Q_2^E$. Determine an initial value problem for x_1 and x_2. Observe that the system of equations for x_1 and x_2 is homogeneous.

23. Consider two interconnected tanks similar to those in Figure 7.1.6. Tank 1 initially contains 60 gal of water and Q_1^0 oz of salt, and Tank 2 initially contains 100 gal of water and Q_2^0 oz of salt. Water containing q_1 oz/gal of salt flows into Tank 1 at a rate of 3 gal/min. The mixture in Tank 1 flows out at a rate of 4 gal/min, of which half flows into Tank 2 while

FIGURE 7.1.6　Two interconnected tanks (Problem 22).

the remainder leaves the system. Water containing q_2 oz/gal of salt also flows into Tank 2 from the outside at the rate of 1 gal/min. The mixture in Tank 2 leaves the tank at a rate of 3 gal/min, of which some flows back into Tank 1 at a rate of 1 gal/min, while the rest leaves the system.

(a) Draw a diagram that depicts the flow process described above. Let $Q_1(t)$ and $Q_2(t)$, respectively, be the amount of salt in each tank at time t. Write down differential equations and initial conditions for Q_1 and Q_2 that model the flow process.

(b) Find the equilibrium values Q_1^E and Q_2^E in terms of the concentrations q_1 and q_2.

(c) Is it possible (by adjusting q_1 and q_2) to obtain $Q_1^E = 60$ and $Q_2^E = 50$ as an equilibrium state?

(d) Describe which equilibrium states are possible for this system for various values of q_1 and q_2.

7.2 Review of Matrices

For both theoretical and computational reasons, it is advisable to bring some of the results of matrix theory[2] to bear on the initial value problem for a system of linear differential equations. For reference purposes this section and the next are devoted to a brief summary of the facts that will be needed later. More details can be found

[2]The properties of matrices were first extensively explored in 1858 in a paper by the English algebraist Arthur Cayley (1821–1895), although the word "matrix" was introduced by his good friend James Sylvester (1814–1897) in 1850. Cayley did some of his best mathematical work while practicing law from 1849 to 1863; he then became professor of mathematics at Cambridge, a position he held for the rest of his life. After Cayley's groundbreaking work, the development of matrix theory proceeded rapidly, with significant contributions by Charles Hermite, Georg Frobenius, and Camille Jordan, among others.

in any elementary book on linear algebra. We assume, however, that you are familiar with determinants and how to evaluate them.

We designate matrices by boldfaced capitals $\mathbf{A}, \mathbf{B}, \mathbf{C}, \ldots$, occasionally using bold-faced Greek capitals $\boldsymbol{\Phi}, \boldsymbol{\Psi}, \ldots$. A matrix \mathbf{A} consists of a rectangular array of numbers, or elements, arranged in m rows and n columns—that is,

$$
\mathbf{A} = \begin{pmatrix} a_{11} & a_{12} & \cdots & a_{1n} \\ a_{21} & a_{22} & \cdots & a_{2n} \\ \vdots & \vdots & & \vdots \\ a_{m1} & a_{m2} & \cdots & a_{mn} \end{pmatrix}. \tag{1}
$$

We speak of \mathbf{A} as an $m \times n$ matrix. Although later in the chapter we will often assume that the elements of certain matrices are real numbers, in this section we assume that the elements of matrices may be complex numbers. The element lying in the ith row and jth column is designated by a_{ij}, the first subscript identifying its row and the second its column. Sometimes the notation (a_{ij}) is used to denote the matrix whose generic element is a_{ij}.

Associated with each matrix \mathbf{A} is the matrix \mathbf{A}^T, which is known as the **transpose** of \mathbf{A} and is obtained from \mathbf{A} by interchanging the rows and columns of \mathbf{A}. Thus, if $\mathbf{A} = (a_{ij})$, then $\mathbf{A}^T = (a_{ji})$. Also, we will denote by \bar{a}_{ij} the complex conjugate of a_{ij}, and by $\overline{\mathbf{A}}$ the matrix obtained from \mathbf{A} by replacing each element a_{ij} by its conjugate \bar{a}_{ij}. The matrix $\overline{\mathbf{A}}$ is called the **conjugate** of \mathbf{A}. It will also be necessary to consider the transpose of the conjugate matrix $\overline{\mathbf{A}}^T$. This matrix is called the **adjoint** of \mathbf{A} and will be denoted by \mathbf{A}^*.

For example, let
$$
\mathbf{A} = \begin{pmatrix} 3 & 2 - i \\ 4 + 3i & -5 + 2i \end{pmatrix}.
$$
Then
$$
\mathbf{A}^T = \begin{pmatrix} 3 & 4 + 3i \\ 2 - i & -5 + 2i \end{pmatrix}, \qquad \overline{\mathbf{A}} = \begin{pmatrix} 3 & 2 + i \\ 4 - 3i & -5 - 2i \end{pmatrix},
$$
$$
\mathbf{A}^* = \begin{pmatrix} 3 & 4 - 3i \\ 2 + i & -5 - 2i \end{pmatrix}.
$$

We are particularly interested in two somewhat special kinds of matrices: square matrices, which have the same number of rows and columns—that is, $m = n$; and vectors (or column vectors), which can be thought of as $n \times 1$ matrices, or matrices having only one column. Square matrices having n rows and n columns are said to be of order n. We denote (column) vectors by boldfaced lower-case letters, $\mathbf{x}, \mathbf{y}, \boldsymbol{\xi}$, $\boldsymbol{\eta}, \ldots$. The transpose \mathbf{x}^T of an $n \times 1$ column vector is a $1 \times n$ row vector—that is, the matrix consisting of one row whose elements are the same as the elements in the corresponding positions of \mathbf{x}.

Properties of Matrices.
1. **Equality.** Two $m \times n$ matrices \mathbf{A} and \mathbf{B} are said to be equal if all corresponding elements are equal—that is, if $a_{ij} = b_{ij}$ for each i and j.

2. **Zero.** The symbol $\mathbf{0}$ will be used to denote the matrix (or vector) each of whose elements is zero.

3. **Addition.** The sum of two $m \times n$ matrices **A** and **B** is defined as the matrix obtained by adding corresponding elements:

$$\mathbf{A} + \mathbf{B} = (a_{ij}) + (b_{ij}) = (a_{ij} + b_{ij}). \tag{2}$$

With this definition, it follows that matrix addition is commutative and associative, so that

$$\mathbf{A} + \mathbf{B} = \mathbf{B} + \mathbf{A}, \qquad \mathbf{A} + (\mathbf{B} + \mathbf{C}) = (\mathbf{A} + \mathbf{B}) + \mathbf{C}. \tag{3}$$

4. **Multiplication by a Number.** The product of a matrix **A** by a complex number α is defined as follows:

$$\alpha \mathbf{A} = \alpha(a_{ij}) = (\alpha a_{ij}). \tag{4}$$

The distributive laws

$$\alpha(\mathbf{A} + \mathbf{B}) = \alpha \mathbf{A} + \alpha \mathbf{B}, \qquad (\alpha + \beta)\mathbf{A} = \alpha \mathbf{A} + \beta \mathbf{A} \tag{5}$$

are satisfied for this type of multiplication. In particular, the negative of **A**, denoted by $-\mathbf{A}$, is defined by

$$-\mathbf{A} = (-1)\mathbf{A}. \tag{6}$$

5. **Subtraction.** The difference $\mathbf{A} - \mathbf{B}$ of two $m \times n$ matrices is defined by

$$\mathbf{A} - \mathbf{B} = \mathbf{A} + (-\mathbf{B}). \tag{7}$$

Thus

$$\mathbf{A} - \mathbf{B} = (a_{ij}) - (b_{ij}) = (a_{ij} - b_{ij}), \tag{8}$$

which is similar to Eq. (2).

6. **Multiplication.** The product **AB** of two matrices is defined whenever the number of columns in the first factor is the same as the number of rows in the second. If **A** and **B** are $m \times n$ and $n \times r$ matrices, respectively, then the product $\mathbf{C} = \mathbf{AB}$ is an $m \times r$ matrix. The element in the ith row and jth column of **C** is found by multiplying each element of the ith row of **A** by the corresponding element of the jth column of **B** and then adding the resulting products. In symbols,

$$c_{ij} = \sum_{k=1}^{n} a_{ik} b_{kj}. \tag{9}$$

By direct calculation, it can be shown that matrix multiplication satisfies the associative law

$$(\mathbf{AB})\mathbf{C} = \mathbf{A}(\mathbf{BC}) \tag{10}$$

and the distributive law

$$\mathbf{A}(\mathbf{B} + \mathbf{C}) = \mathbf{AB} + \mathbf{AC}. \tag{11}$$

However, in general, matrix multiplication is not commutative. For both products **AB** and **BA** to exist and to be of the same size, it is necessary that **A** and **B** be square matrices of the same order. Even in that case the two products are usually unequal, so that, in general,

$$\mathbf{AB} \neq \mathbf{BA}. \tag{12}$$

EXAMPLE 1

To illustrate the multiplication of matrices, and also the fact that matrix multiplication is not necessarily commutative, consider the matrices

$$\mathbf{A} = \begin{pmatrix} 1 & -2 & 1 \\ 0 & 2 & -1 \\ 2 & 1 & 1 \end{pmatrix}, \qquad \mathbf{B} = \begin{pmatrix} 2 & 1 & -1 \\ 1 & -1 & 0 \\ 2 & -1 & 1 \end{pmatrix}.$$

From the definition of multiplication given in Eq. (9) we have

$$\mathbf{AB} = \begin{pmatrix} 2-2+2 & 1+2-1 & -1+0+1 \\ 0+2-2 & 0-2+1 & 0+0-1 \\ 4+1+2 & 2-1-1 & -2+0+1 \end{pmatrix}$$

$$= \begin{pmatrix} 2 & 2 & 0 \\ 0 & -1 & -1 \\ 7 & 0 & -1 \end{pmatrix}.$$

Similarly, we find that

$$\mathbf{BA} = \begin{pmatrix} 0 & -3 & 0 \\ 1 & -4 & 2 \\ 4 & -5 & 4 \end{pmatrix}.$$

Clearly, $\mathbf{AB} \neq \mathbf{BA}$.

7. **Multiplication of Vectors.** There are several ways of forming a product of two vectors \mathbf{x} and \mathbf{y}, each with n components. One is a direct extension to n dimensions of the familiar dot product from physics and calculus; we denote it by $\mathbf{x}^T\mathbf{y}$ and write

$$\mathbf{x}^T\mathbf{y} = \sum_{i=1}^{n} x_i y_i. \tag{13}$$

The result of Eq. (13) is a (complex) number, and it follows directly from Eq. (13) that

$$\mathbf{x}^T\mathbf{y} = \mathbf{y}^T\mathbf{x}, \qquad \mathbf{x}^T(\mathbf{y}+\mathbf{z}) = \mathbf{x}^T\mathbf{y} + \mathbf{x}^T\mathbf{z}, \qquad (\alpha\mathbf{x})^T\mathbf{y} = \alpha(\mathbf{x}^T\mathbf{y}) = \mathbf{x}^T(\alpha\mathbf{y}). \tag{14}$$

There is another vector product that is also defined for any two vectors having the same number of components. This product, denoted by (\mathbf{x}, \mathbf{y}), is called the **scalar** or **inner product**, and is defined by

$$(\mathbf{x}, \mathbf{y}) = \sum_{i=1}^{n} x_i \bar{y}_i, \tag{15}$$

The scalar product is also a (complex) number, and by comparing Eqs. (13) and (15) we see that

$$(\mathbf{x}, \mathbf{y}) = \mathbf{x}^T\bar{\mathbf{y}}. \tag{16}$$

Thus, if all the elements of \mathbf{y} are real, then the two products (13) and (15) are identical. From Eq. (15) it follows that

$$(\mathbf{x}, \mathbf{y}) = \overline{(\mathbf{y}, \mathbf{x})}, \qquad (\mathbf{x}, \mathbf{y} + \mathbf{z}) = (\mathbf{x}, \mathbf{y}) + (\mathbf{x}, \mathbf{z}),$$
$$(\alpha\mathbf{x}, \mathbf{y}) = \alpha(\mathbf{x}, \mathbf{y}), \qquad (\mathbf{x}, \alpha\mathbf{y}) = \overline{\alpha}(\mathbf{x}, \mathbf{y}). \tag{17}$$

Note that even if the vector \mathbf{x} has elements with nonzero imaginary parts, the scalar product of \mathbf{x} with itself yields a nonnegative real number,

$$(\mathbf{x}, \mathbf{x}) = \sum_{i=1}^{n} x_i \overline{x}_i = \sum_{i=1}^{n} |x_i|^2. \tag{18}$$

The nonnegative quantity $(\mathbf{x}, \mathbf{x})^{1/2}$, often denoted by $\|\mathbf{x}\|$, is called the **length**, or **magnitude**, of \mathbf{x}. If $(\mathbf{x}, \mathbf{y}) = 0$, then the two vectors \mathbf{x} and \mathbf{y} are said to be **orthogonal**. For example, the unit vectors $\mathbf{i}, \mathbf{j}, \mathbf{k}$ of three-dimensional vector geometry form an orthogonal set. On the other hand, if some of the elements of \mathbf{x} are not real, then the product

$$\mathbf{x}^T\mathbf{x} = \sum_{i=1}^{n} x_i^2 \tag{19}$$

may not be a real number.

For example, let

$$\mathbf{x} = \begin{pmatrix} i \\ -2 \\ 1+i \end{pmatrix}, \qquad \mathbf{y} = \begin{pmatrix} 2-i \\ i \\ 3 \end{pmatrix}.$$

Then

$$\mathbf{x}^T\mathbf{y} = (i)(2 - i) + (-2)(i) + (1 + i)(3) = 4 + 3i,$$
$$(\mathbf{x}, \mathbf{y}) = (i)(2 + i) + (-2)(-i) + (1 + i)(3) = 2 + 7i,$$
$$\mathbf{x}^T\mathbf{x} = (i)^2 + (-2)^2 + (1 + i)^2 = 3 + 2i,$$
$$(\mathbf{x}, \mathbf{x}) = (i)(-i) + (-2)(-2) + (1 + i)(1 - i) = 7.$$

8. **Identity.** The multiplicative identity, or simply the identity matrix \mathbf{I}, is given by

$$\mathbf{I} = \begin{pmatrix} 1 & 0 & \cdots & 0 \\ 0 & 1 & \cdots & 0 \\ \vdots & \vdots & & \vdots \\ 0 & 0 & \cdots & 1 \end{pmatrix}. \tag{20}$$

From the definition of matrix multiplication we have

$$\mathbf{AI} = \mathbf{IA} = \mathbf{A} \tag{21}$$

for any (square) matrix \mathbf{A}. Hence the commutative law does hold for square matrices if one of the matrices is the identity.

9. **Inverse.** The square matrix \mathbf{A} is said to be **nonsingular** or **invertible** if there is another matrix \mathbf{B} such that $\mathbf{AB} = \mathbf{I}$ and $\mathbf{BA} = \mathbf{I}$, where \mathbf{I} is the identity. If there is such a \mathbf{B}, it can be shown that there is only one. It is called the multiplicative inverse, or simply the inverse, of \mathbf{A}, and we write $\mathbf{B} = \mathbf{A}^{-1}$. Then

$$\mathbf{AA}^{-1} = \mathbf{A}^{-1}\mathbf{A} = \mathbf{I}. \tag{22}$$

Matrices that do not have an inverse are called **singular** or **noninvertible**.

There are various ways to compute \mathbf{A}^{-1} from \mathbf{A}, assuming that it exists. One way involves the use of determinants. Associated with each element a_{ij} of a given matrix is the minor M_{ij}, which is the determinant of the matrix obtained by deleting the ith row and jth column of the original matrix—that is, the row and column containing a_{ij}. Also associated with each element a_{ij} is the cofactor C_{ij} defined by the equation

$$C_{ij} = (-1)^{i+j} M_{ij}. \tag{23}$$

If $\mathbf{B} = \mathbf{A}^{-1}$, then it can be shown that the general element b_{ij} is given by

$$b_{ij} = \frac{C_{ji}}{\det \mathbf{A}}. \tag{24}$$

Although Eq. (24) is not an efficient way[3] to calculate \mathbf{A}^{-1}, it does suggest a condition that \mathbf{A} must satisfy for it to have an inverse. In fact, the condition is both necessary and sufficient: \mathbf{A} is nonsingular if and only if $\det \mathbf{A} \neq 0$. If $\det \mathbf{A} = 0$, then \mathbf{A} is singular.

Another and usually better way to compute \mathbf{A}^{-1} is by means of elementary row operations. There are three such operations:

1. Interchange of two rows.
2. Multiplication of a row by a nonzero scalar.
3. Addition of any multiple of one row to another row.

The transformation of a matrix by a sequence of elementary row operations is referred to as **row reduction** or **Gaussian[4] elimination**. Any nonsingular matrix \mathbf{A} can be transformed into the identity \mathbf{I} by a systematic sequence of these operations. It is possible to show that if the same sequence of operations is then performed on \mathbf{I}, it is transformed into \mathbf{A}^{-1}. It is most efficient to perform the sequence of operations on both matrices at the same time by forming the augmented matrix $\mathbf{A} \mid \mathbf{I}$. The following example illustrates the calculation of an inverse matrix in this way.

[3]For large n the number of multiplications required to evaluate \mathbf{A}^{-1} by Eq. (24) is proportional to $n!$. If one uses more efficient methods, such as the row reduction procedure described later, the number of multiplications is proportional only to n^3. Even for small values of n (such as $n = 4$), determinants are not an economical tool in calculating inverses, and row reduction methods are preferred.

[4]Carl Friedrich Gauss (1777–1855) spent most of his life at Göttingen and made major contributions to many areas of mathematics, including number theory, algebra, nonEuclidean and differential geometry, and analysis, as well as to more applied fields such as geodesy, statistics, and celestial mechanics. He is generally considered to be among the half-dozen best mathematicians of all time.

Find the inverse of

EXAMPLE

2

$$A = \begin{pmatrix} 1 & -1 & -1 \\ 3 & -1 & 2 \\ 2 & 2 & 3 \end{pmatrix}.$$

We begin by forming the augmented matrix $\mathbf{A} \,|\, \mathbf{I}$:

$$\mathbf{A} \,|\, \mathbf{I} = \begin{pmatrix} 1 & -1 & -1 & | & 1 & 0 & 0 \\ 3 & -1 & 2 & | & 0 & 1 & 0 \\ 2 & 2 & 3 & | & 0 & 0 & 1 \end{pmatrix}.$$

The matrix \mathbf{A} can be transformed into \mathbf{I} by the following sequence of operations, and at the same time, \mathbf{I} is transformed into \mathbf{A}^{-1}. The result of each step appears below the statement.

(a) Obtain zeros in the off-diagonal positions in the first column by adding (-3) times the first row to the second row and adding (-2) times the first row to the third row.

$$\begin{pmatrix} 1 & -1 & -1 & | & 1 & 0 & 0 \\ 0 & 2 & 5 & | & -3 & 1 & 0 \\ 0 & 4 & 5 & | & -2 & 0 & 1 \end{pmatrix}$$

(b) Obtain a 1 in the diagonal position in the second column by multiplying the second row by $\frac{1}{2}$.

$$\begin{pmatrix} 1 & -1 & -1 & | & 1 & 0 & 0 \\ 0 & 1 & \frac{5}{2} & | & -\frac{3}{2} & \frac{1}{2} & 0 \\ 0 & 4 & 5 & | & -2 & 0 & 1 \end{pmatrix}$$

(c) Obtain zeros in the off-diagonal positions in the second column by adding the second row to the first row and adding (-4) times the second row to the third row.

$$\begin{pmatrix} 1 & 0 & \frac{3}{2} & | & -\frac{1}{2} & \frac{1}{2} & 0 \\ 0 & 1 & \frac{5}{2} & | & -\frac{3}{2} & \frac{1}{2} & 0 \\ 0 & 0 & -5 & | & 4 & -2 & 1 \end{pmatrix}$$

(d) Obtain a 1 in the diagonal position in the third column by multiplying the third row by $(-\frac{1}{5})$.

$$\begin{pmatrix} 1 & 0 & \frac{3}{2} & | & -\frac{1}{2} & \frac{1}{2} & 0 \\ 0 & 1 & \frac{5}{2} & | & -\frac{3}{2} & \frac{1}{2} & 0 \\ 0 & 0 & 1 & | & -\frac{4}{5} & \frac{2}{5} & -\frac{1}{5} \end{pmatrix}$$

(e) Obtain zeros in the off-diagonal positions in the third column by adding $(-\frac{3}{2})$ times the third row to the first row and adding $(-\frac{5}{2})$ times the third row to the second row.

$$\begin{pmatrix} 1 & 0 & 0 & | & \frac{7}{10} & -\frac{1}{10} & \frac{3}{10} \\ 0 & 1 & 0 & | & \frac{1}{2} & -\frac{1}{2} & \frac{1}{2} \\ 0 & 0 & 1 & | & -\frac{4}{5} & \frac{2}{5} & -\frac{1}{5} \end{pmatrix}$$

The last of these matrices is $\mathbf{I} \,|\, \mathbf{A}^{-1}$, a fact that can be verified by direct multiplication with the original matrix \mathbf{A}.

This example was made slightly simpler by the fact that the original matrix **A** had a 1 in the upper left corner ($a_{11} = 1$). If this is not the case, then the first step is to produce a 1 there by multiplying the first row by $1/a_{11}$, as long as $a_{11} \neq 0$. If $a_{11} = 0$, then the first row must be interchanged with some other row to bring a nonzero element into the upper left position before proceeding.

Matrix Functions. We sometimes need to consider vectors or matrices whose elements are functions of a real variable t. We write

$$\mathbf{x}(t) = \begin{pmatrix} x_1(t) \\ \vdots \\ x_n(t) \end{pmatrix}, \qquad \mathbf{A}(t) = \begin{pmatrix} a_{11}(t) & \cdots & a_{1n}(t) \\ \vdots & & \vdots \\ a_{n1}(t) & \cdots & a_{nn}(t) \end{pmatrix}, \tag{25}$$

respectively.

The matrix $\mathbf{A}(t)$ is said to be continuous at $t = t_0$ or on an interval $\alpha < t < \beta$ if each element of **A** is a continuous function at the given point or on the given interval. Similarly, $\mathbf{A}(t)$ is said to be differentiable if each of its elements is differentiable, and its derivative $d\mathbf{A}/dt$ is defined by

$$\frac{d\mathbf{A}}{dt} = \left(\frac{da_{ij}}{dt} \right); \tag{26}$$

that is, each element of $d\mathbf{A}/dt$ is the derivative of the corresponding element of **A**. In the same way, the integral of a matrix function is defined as

$$\int_a^b \mathbf{A}(t)\, dt = \left(\int_a^b a_{ij}(t)\, dt \right). \tag{27}$$

For example, if

$$\mathbf{A}(t) = \begin{pmatrix} \sin t & t \\ 1 & \cos t \end{pmatrix},$$

then

$$\mathbf{A}'(t) = \begin{pmatrix} \cos t & 1 \\ 0 & -\sin t \end{pmatrix}, \qquad \int_0^\pi \mathbf{A}(t)\, dt = \begin{pmatrix} 2 & \pi^2/2 \\ \pi & 0 \end{pmatrix}.$$

Many of the rules of elementary calculus extend easily to matrix functions; in particular,

$$\frac{d}{dt}(\mathbf{CA}) = \mathbf{C}\frac{d\mathbf{A}}{dt}, \qquad \text{where } \mathbf{C} \text{ is a constant matrix;} \tag{28}$$

$$\frac{d}{dt}(\mathbf{A} + \mathbf{B}) = \frac{d\mathbf{A}}{dt} + \frac{d\mathbf{B}}{dt}; \tag{29}$$

$$\frac{d}{dt}(\mathbf{AB}) = \mathbf{A}\frac{d\mathbf{B}}{dt} + \frac{d\mathbf{A}}{dt}\mathbf{B}. \tag{30}$$

In Eqs. (28) and (30), care must be taken in each term to avoid interchanging the order of multiplication. The definitions expressed by Eqs. (26) and (27) also apply as special cases to vectors.

PROBLEMS

1. If $\mathbf{A} = \begin{pmatrix} 1 & -2 & 0 \\ 3 & 2 & -1 \\ -2 & 1 & 3 \end{pmatrix}$ and $\mathbf{B} = \begin{pmatrix} 4 & -2 & 3 \\ -1 & 5 & 0 \\ 6 & 1 & 2 \end{pmatrix}$, find

 (a) $2\mathbf{A} + \mathbf{B}$ (b) $\mathbf{A} - 4\mathbf{B}$
 (c) \mathbf{AB} (d) \mathbf{BA}

2. If $\mathbf{A} = \begin{pmatrix} 1+i & -1+2i \\ 3+2i & 2-i \end{pmatrix}$ and $\mathbf{B} = \begin{pmatrix} i & 3 \\ 2 & -2i \end{pmatrix}$, find

 (a) $\mathbf{A} - 2\mathbf{B}$ (b) $3\mathbf{A} + \mathbf{B}$
 (c) \mathbf{AB} (d) \mathbf{BA}

3. If $\mathbf{A} = \begin{pmatrix} -2 & 1 & 2 \\ 1 & 0 & -3 \\ 2 & -1 & 1 \end{pmatrix}$ and $\mathbf{B} = \begin{pmatrix} 1 & 2 & 3 \\ 3 & -1 & -1 \\ -2 & 1 & 0 \end{pmatrix}$, find

 (a) \mathbf{A}^T (b) \mathbf{B}^T
 (c) $\mathbf{A}^T + \mathbf{B}^T$ (d) $(\mathbf{A} + \mathbf{B})^T$

4. If $\mathbf{A} = \begin{pmatrix} 3-2i & 1+i \\ 2-i & -2+3i \end{pmatrix}$, find

 (a) \mathbf{A}^T (b) $\overline{\mathbf{A}}$ (c) \mathbf{A}^*

5. If $\mathbf{A} = \begin{pmatrix} 3 & 2 & -1 \\ 2 & -1 & 2 \\ 1 & 2 & 1 \end{pmatrix}$ and $\mathbf{B} = \begin{pmatrix} 2 & 1 & -1 \\ -2 & 3 & 3 \\ 1 & 0 & 2 \end{pmatrix}$, verify that $2(\mathbf{A} + \mathbf{B}) = 2\mathbf{A} + 2\mathbf{B}$.

6. If $\mathbf{A} = \begin{pmatrix} 1 & -2 & 0 \\ 3 & 2 & -1 \\ -2 & 0 & 3 \end{pmatrix}$, $\mathbf{B} = \begin{pmatrix} 2 & 1 & -1 \\ -2 & 3 & 3 \\ 1 & 0 & 2 \end{pmatrix}$, and $\mathbf{C} = \begin{pmatrix} 2 & 1 & 0 \\ 1 & 2 & 2 \\ 0 & 1 & -1 \end{pmatrix}$, verify that

 (a) $(\mathbf{AB})\mathbf{C} = \mathbf{A}(\mathbf{BC})$ (b) $(\mathbf{A} + \mathbf{B}) + \mathbf{C} = \mathbf{A} + (\mathbf{B} + \mathbf{C})$
 (c) $\mathbf{A}(\mathbf{B} + \mathbf{C}) = \mathbf{AB} + \mathbf{AC}$

7. Prove each of the following laws of matrix algebra:
 (a) $\mathbf{A} + \mathbf{B} = \mathbf{B} + \mathbf{A}$ (b) $\mathbf{A} + (\mathbf{B} + \mathbf{C}) = (\mathbf{A} + \mathbf{B}) + \mathbf{C}$
 (c) $\alpha(\mathbf{A} + \mathbf{B}) = \alpha\mathbf{A} + \alpha\mathbf{B}$ (d) $(\alpha + \beta)\mathbf{A} = \alpha\mathbf{A} + \beta\mathbf{A}$
 (e) $\mathbf{A}(\mathbf{BC}) = (\mathbf{AB})\mathbf{C}$ (f) $\mathbf{A}(\mathbf{B} + \mathbf{C}) = \mathbf{AB} + \mathbf{AC}$

8. If $\mathbf{x} = \begin{pmatrix} 2 \\ 3i \\ 1-i \end{pmatrix}$ and $\mathbf{y} = \begin{pmatrix} -1+i \\ 2 \\ 3-i \end{pmatrix}$, find

 (a) $\mathbf{x}^T\mathbf{y}$ (b) $\mathbf{y}^T\mathbf{y}$
 (c) (\mathbf{x}, \mathbf{y}) (d) (\mathbf{y}, \mathbf{y})

9. If $\mathbf{x} = \begin{pmatrix} 1-2i \\ i \\ 2 \end{pmatrix}$ and $\mathbf{y} = \begin{pmatrix} 2 \\ 3-i \\ 1+2i \end{pmatrix}$, show that

 (a) $\mathbf{x}^T\mathbf{y} = \mathbf{y}^T\mathbf{x}$ (b) $(\mathbf{x},\mathbf{y}) = \overline{(\mathbf{y}, \mathbf{x})}$

In each of Problems 10 through 19 either compute the inverse of the given matrix, or else show that it is singular.

10. $\begin{pmatrix} 1 & 4 \\ -2 & 3 \end{pmatrix}$

11. $\begin{pmatrix} 3 & -1 \\ 6 & 2 \end{pmatrix}$

12. $\begin{pmatrix} 1 & 2 & 3 \\ 2 & 4 & 5 \\ 3 & 5 & 6 \end{pmatrix}$

13. $\begin{pmatrix} 1 & 1 & -1 \\ 2 & -1 & 1 \\ 1 & 1 & 2 \end{pmatrix}$

14. $\begin{pmatrix} 1 & 2 & 1 \\ -2 & 1 & 8 \\ 1 & -2 & -7 \end{pmatrix}$

15. $\begin{pmatrix} 2 & 1 & 0 \\ 0 & 2 & 1 \\ 0 & 0 & 2 \end{pmatrix}$

16. $\begin{pmatrix} 1 & -1 & -1 \\ 2 & 1 & 0 \\ 3 & -2 & 1 \end{pmatrix}$

17. $\begin{pmatrix} 2 & 3 & 1 \\ -1 & 2 & 1 \\ 4 & -1 & -1 \end{pmatrix}$

18. $\begin{pmatrix} 1 & 0 & 0 & -1 \\ 0 & -1 & 1 & 0 \\ -1 & 0 & 1 & 0 \\ 0 & 1 & -1 & 1 \end{pmatrix}$

19. $\begin{pmatrix} 1 & -1 & 2 & 0 \\ -1 & 2 & -4 & 2 \\ 1 & 0 & 1 & 3 \\ -2 & 2 & 0 & -1 \end{pmatrix}$

20. Prove that if there are two matrices **B** and **C** such that $\mathbf{AB} = \mathbf{I}$ and $\mathbf{AC} = \mathbf{I}$, then $\mathbf{B} = \mathbf{C}$. This shows that a matrix **A** can have only one inverse.

21. If $\mathbf{A}(t) = \begin{pmatrix} e^t & 2e^{-t} & e^{2t} \\ 2e^t & e^{-t} & -e^{2t} \\ -e^t & 3e^{-t} & 2e^{2t} \end{pmatrix}$ and $\mathbf{B}(t) = \begin{pmatrix} 2e^t & e^{-t} & 3e^{2t} \\ -e^t & 2e^{-t} & e^{2t} \\ 3e^t & -e^{-t} & -e^{2t} \end{pmatrix}$, find

(a) $\mathbf{A} + 3\mathbf{B}$

(b) \mathbf{AB}

(c) $d\mathbf{A}/dt$

(d) $\int_0^1 \mathbf{A}(t)\,dt$

In each of Problems 22 through 24 verify that the given vector satisfies the given differential equation.

22. $\mathbf{x}' = \begin{pmatrix} 3 & -2 \\ 2 & -2 \end{pmatrix} \mathbf{x}, \qquad \mathbf{x} = \begin{pmatrix} 4 \\ 2 \end{pmatrix} e^{2t}$

23. $\mathbf{x}' = \begin{pmatrix} 2 & -1 \\ 3 & -2 \end{pmatrix} \mathbf{x} + \begin{pmatrix} 1 \\ -1 \end{pmatrix} e^t, \qquad \mathbf{x} = \begin{pmatrix} 1 \\ 0 \end{pmatrix} e^t + 2 \begin{pmatrix} 1 \\ 1 \end{pmatrix} te^t$

24. $\mathbf{x}' = \begin{pmatrix} 1 & 1 & 1 \\ 2 & 1 & -1 \\ 0 & -1 & 1 \end{pmatrix} \mathbf{x}, \qquad \mathbf{x} = \begin{pmatrix} 6 \\ -8 \\ -4 \end{pmatrix} e^{-t} + 2 \begin{pmatrix} 0 \\ 1 \\ -1 \end{pmatrix} e^{2t}$

In each of Problems 25 and 26 verify that the given matrix satisfies the given differential equation.

25. $\boldsymbol{\Psi}' = \begin{pmatrix} 1 & 1 \\ 4 & -2 \end{pmatrix} \boldsymbol{\Psi}, \qquad \boldsymbol{\Psi}(t) = \begin{pmatrix} e^{-3t} & e^{2t} \\ -4e^{-3t} & e^{2t} \end{pmatrix}$

26. $\mathbf{\Psi}' = \begin{pmatrix} 1 & -1 & 4 \\ 3 & 2 & -1 \\ 2 & 1 & -1 \end{pmatrix} \mathbf{\Psi}, \qquad \mathbf{\Psi}(t) = \begin{pmatrix} e^t & e^{-2t} & e^{3t} \\ -4e^t & -e^{-2t} & 2e^{3t} \\ -e^t & -e^{-2t} & e^{3t} \end{pmatrix}$

7.3 Systems of Linear Algebraic Equations; Linear Independence, Eigenvalues, Eigenvectors

In this section we review some results from linear algebra that are important for the solution of systems of linear differential equations. Some of these results are easily proved and others are not; since we are interested simply in summarizing some useful information in compact form, we give no indication of proofs in either case. All the results in this section depend on some basic facts about the solution of systems of linear algebraic equations.

Systems of Linear Algebraic Equations. A set of n simultaneous linear algebraic equations in n variables,

$$a_{11}x_1 + a_{12}x_2 + \cdots + a_{1n}x_n = b_1,$$
$$\vdots$$
$$a_{n1}x_1 + a_{n2}x_2 + \cdots + a_{nn}x_n = b_n,$$

(1)

can be written as

$$\mathbf{Ax} = \mathbf{b}, \tag{2}$$

where the $n \times n$ matrix \mathbf{A} and the vector \mathbf{b} are given, and the components of \mathbf{x} are to be determined. If $\mathbf{b} = \mathbf{0}$, the system is said to be **homogeneous**; otherwise, it is **nonhomogeneous**.

If the coefficient matrix \mathbf{A} is nonsingular—that is, if $\det \mathbf{A}$ is not zero—then there is a unique solution of the system (2). Since \mathbf{A} is nonsingular, \mathbf{A}^{-1} exists, and the solution can be found by multiplying each side of Eq. (2) on the left by \mathbf{A}^{-1}; thus

$$\mathbf{x} = \mathbf{A}^{-1}\mathbf{b}. \tag{3}$$

In particular, the homogeneous problem $\mathbf{Ax} = \mathbf{0}$, corresponding to $\mathbf{b} = \mathbf{0}$ in Eq. (2), has only the trivial solution $\mathbf{x} = \mathbf{0}$.

On the other hand, if \mathbf{A} is singular—that is, if $\det \mathbf{A}$ is zero—then solutions of Eq. (2) either do not exist, or do exist but are not unique. Since \mathbf{A} is singular, \mathbf{A}^{-1} does not exist, so Eq. (3) is no longer valid. The homogeneous system

$$\mathbf{Ax} = \mathbf{0} \tag{4}$$

has (infinitely many) nonzero solutions in addition to the trivial solution. The situation for the nonhomogeneous system (2) is more complicated. This system has no solution unless the vector \mathbf{b} satisfies a certain further condition. This condition is that

$$(\mathbf{b}, \mathbf{y}) = 0, \tag{5}$$

for all vectors \mathbf{y} satisfying $\mathbf{A}^*\mathbf{y} = \mathbf{0}$, where \mathbf{A}^* is the adjoint of \mathbf{A}. If condition (5) is met, then the system (2) has (infinitely many) solutions. Each of these solutions has the form

$$\mathbf{x} = \mathbf{x}^{(0)} + \boldsymbol{\xi}, \tag{6}$$

where $\mathbf{x}^{(0)}$ is a particular solution of Eq. (2), and $\boldsymbol{\xi}$ is any solution of the homogeneous system (4). Note the resemblance between Eq. (6) and the solution of a nonhomogeneous linear differential equation. The proofs of some of the preceding statements are outlined in Problems 25 through 29.

The results in the preceding paragraph are important as a means of classifying the solutions of linear systems. However, for solving particular systems it is generally best to use row reduction to transform the system into a much simpler one from which the solution(s), if there are any, can be written down easily. To do this efficiently, we can form the augmented matrix

$$\mathbf{A} \mid \mathbf{b} = \begin{pmatrix} a_{11} & \cdots & a_{1n} & \mid & b_1 \\ \vdots & & \vdots & \mid & \vdots \\ a_{n1} & \cdots & a_{nn} & \mid & b_n \end{pmatrix} \tag{7}$$

by adjoining the vector \mathbf{b} to the coefficient matrix \mathbf{A} as an additional column. The dashed line replaces the equals sign and is said to partition the augmented matrix. We now perform row operations on the augmented matrix so as to transform \mathbf{A} into a triangular matrix—that is, a matrix whose elements below the main diagonal are all zero. Once this is done, it is easy to see whether the system has solutions, and to find them if it does. Observe that elementary row operations on the augmented matrix (7) correspond to legitimate operations on the equations in the system (1). The following examples illustrate the process.

EXAMPLE 1

Solve the system of equations

$$\begin{aligned} x_1 - 2x_2 + 3x_3 &= 7, \\ -x_1 + x_2 - 2x_3 &= -5, \\ 2x_1 - x_2 - x_3 &= 4. \end{aligned} \tag{8}$$

The augmented matrix for the system (8) is

$$\begin{pmatrix} 1 & -2 & 3 & \mid & 7 \\ -1 & 1 & -2 & \mid & -5 \\ 2 & -1 & -1 & \mid & 4 \end{pmatrix} \tag{9}$$

We now perform row operations on the matrix (9) with a view to introducing zeros in the lower left part of the matrix. Each step is described and the result recorded below.

(a) Add the first row to the second row and add (-2) times the first row to the third row.

$$\begin{pmatrix} 1 & -2 & 3 & \mid & 7 \\ 0 & -1 & 1 & \mid & 2 \\ 0 & 3 & -7 & \mid & -10 \end{pmatrix}$$

(b) Multiply the second row by -1.

$$\begin{pmatrix} 1 & -2 & 3 & | & 7 \\ 0 & 1 & -1 & | & -2 \\ 0 & 3 & -7 & | & -10 \end{pmatrix}$$

(c) Add (-3) times the second row to the third row.

$$\begin{pmatrix} 1 & -2 & 3 & | & 7 \\ 0 & 1 & -1 & | & -2 \\ 0 & 0 & -4 & | & -4 \end{pmatrix}$$

(d) Divide the third row by -4.

$$\begin{pmatrix} 1 & -2 & 3 & | & 7 \\ 0 & 1 & -1 & | & -2 \\ 0 & 0 & 1 & | & 1 \end{pmatrix}$$

The matrix obtained in this manner corresponds to the system of equations

$$\begin{aligned} x_1 - 2x_2 + 3x_3 &= 7, \\ x_2 - x_3 &= -2, \\ x_3 &= 1, \end{aligned} \tag{10}$$

which is equivalent to the original system (8). Note that the coefficients in Eqs. (10) form a triangular matrix. From the last of Eqs. (10) we have $x_3 = 1$, from the second equation $x_2 = -2 + x_3 = -1$, and from the first equation $x_1 = 7 + 2x_2 - 3x_3 = 2$. Thus we obtain

$$\mathbf{x} = \begin{pmatrix} 2 \\ -1 \\ 1 \end{pmatrix},$$

which is the solution of the given system (8). Incidentally, since the solution is unique, we conclude that the coefficient matrix is nonsingular.

EXAMPLE 2

Discuss solutions of the system

$$\begin{aligned} x_1 - 2x_2 + 3x_3 &= b_1, \\ -x_1 + x_2 - 2x_3 &= b_2, \\ 2x_1 - x_2 + 3x_3 &= b_3 \end{aligned} \tag{11}$$

for various values of $b_1, b_2,$ and b_3.

Observe that the coefficients in the system (11) are the same as those in the system (8) except for the coefficient of x_3 in the third equation. The augmented matrix for the system (11) is

$$\begin{pmatrix} 1 & -2 & 3 & | & b_1 \\ -1 & 1 & -2 & | & b_2 \\ 2 & -1 & 3 & | & b_3 \end{pmatrix}. \tag{12}$$

By performing steps (a), (b), and (c) as in Example 1, we transform the matrix (12) into

$$\begin{pmatrix} 1 & -2 & 3 & | & b_1 \\ 0 & 1 & -1 & | & -b_1 - b_2 \\ 0 & 0 & 0 & | & b_1 + 3b_2 + b_3 \end{pmatrix}. \tag{13}$$

The equation corresponding to the third row of the matrix (13) is

$$b_1 + 3b_2 + b_3 = 0; \tag{14}$$

thus the system (11) has no solution unless the condition (14) is satisfied by b_1, b_2, and b_3. It is possible to show that this condition is just Eq. (5) for the system (11).

Let us now assume that $b_1 = 2$, $b_2 = 1$, and $b_3 = -5$, in which case Eq. (14) is satisfied. Then the first two rows of the matrix (13) correspond to the equations

$$\begin{aligned} x_1 - 2x_2 + 3x_3 &= 2, \\ x_2 - x_3 &= -3. \end{aligned} \tag{15}$$

To solve the system (15) we can choose one of the unknowns arbitrarily and then solve for the other two. If we let $x_3 = \alpha$, where α is arbitrary, it then follows that

$$x_2 = \alpha - 3,$$
$$x_1 = 2(\alpha - 3) - 3\alpha + 2 = -\alpha - 4.$$

If we write the solution in vector notation, we have

$$\mathbf{x} = \begin{pmatrix} -\alpha - 4 \\ \alpha - 3 \\ \alpha \end{pmatrix} = \alpha \begin{pmatrix} -1 \\ 1 \\ 1 \end{pmatrix} + \begin{pmatrix} -4 \\ -3 \\ 0 \end{pmatrix}. \tag{16}$$

It is easy to verify that the second term on the right side of Eq. (16) is a solution of the nonhomogeneous system (11) and that the first term is the most general solution of the homogeneous system corresponding to (11).

Row reduction is also useful in solving homogeneous systems and systems in which the number of equations is different from the number of unknowns.

Linear Independence. A set of k vectors $\mathbf{x}^{(1)}, \ldots, \mathbf{x}^{(k)}$ is said to be **linearly dependent** if there exists a set of (complex) numbers c_1, \ldots, c_k, at least one of which is nonzero, such that

$$c_1 \mathbf{x}^{(1)} + \cdots + c_k \mathbf{x}^{(k)} = \mathbf{0}. \tag{17}$$

In other words, $\mathbf{x}^{(1)}, \ldots, \mathbf{x}^{(k)}$ are linearly dependent if there is a linear relation among them. On the other hand, if the only set c_1, \ldots, c_k for which Eq. (17) is satisfied is $c_1 = c_2 = \cdots = c_k = 0$, then $\mathbf{x}^{(1)}, \ldots, \mathbf{x}^{(k)}$ are said to be **linearly independent**.

Consider now a set of n vectors, each of which has n components. Let $x_{ij} = x_i^{(j)}$ be the ith component of the vector $\mathbf{x}^{(j)}$, and let $\mathbf{X} = (x_{ij})$. Then Eq. (17) can be written as

$$\begin{pmatrix} x_1^{(1)} c_1 + \cdots + x_1^{(n)} c_n \\ \vdots \qquad\qquad \vdots \\ x_n^{(1)} c_1 + \cdots + x_n^{(n)} c_n \end{pmatrix} = \begin{pmatrix} x_{11} c_1 + \cdots + x_{1n} c_n \\ \vdots \qquad\qquad \vdots \\ x_{n1} c_1 + \cdots + x_{nn} c_n \end{pmatrix} = \mathbf{Xc} = \mathbf{0}. \tag{18}$$

If $\det \mathbf{X} \neq 0$, then the only solution of Eq. (18) is $\mathbf{c} = \mathbf{0}$, but if $\det \mathbf{X} = 0$, there are nonzero solutions. Thus the set of vectors $\mathbf{x}^{(1)}, \ldots, \mathbf{x}^{(n)}$ is linearly independent if and only if $\det \mathbf{X} \neq 0$.

EXAMPLE 3

Determine whether the vectors

$$\mathbf{x}^{(1)} = \begin{pmatrix} 1 \\ 2 \\ -1 \end{pmatrix}, \qquad \mathbf{x}^{(2)} = \begin{pmatrix} 2 \\ 1 \\ 3 \end{pmatrix}, \qquad \mathbf{x}^{(3)} = \begin{pmatrix} -4 \\ 1 \\ -11 \end{pmatrix} \tag{19}$$

are linearly independent or linearly dependent. If they are linearly dependent, find a linear relation among them.

To determine whether $\mathbf{x}^{(1)}$, $\mathbf{x}^{(2)}$, and $\mathbf{x}^{(3)}$ are linearly dependent, we seek constants c_1, c_2, and c_3 such that

$$c_1\mathbf{x}^{(1)} + c_2\mathbf{x}^{(2)} + c_3\mathbf{x}^{(3)} = \mathbf{0}. \tag{20}$$

Equation (20) can also be written in the form

$$\begin{pmatrix} 1 & 2 & -4 \\ 2 & 1 & 1 \\ -1 & 3 & -11 \end{pmatrix} \begin{pmatrix} c_1 \\ c_2 \\ c_3 \end{pmatrix} = \begin{pmatrix} 0 \\ 0 \\ 0 \end{pmatrix} \tag{21}$$

and solved by means of elementary row operations starting from the augmented matrix

$$\begin{pmatrix} 1 & 2 & -4 & | & 0 \\ 2 & 1 & 1 & | & 0 \\ -1 & 3 & -11 & | & 0 \end{pmatrix}. \tag{22}$$

We proceed as in Examples 1 and 2.

(a) Add (-2) times the first row to the second row, and add the first row to the third row.

$$\begin{pmatrix} 1 & 2 & -4 & | & 0 \\ 0 & -3 & 9 & | & 0 \\ 0 & 5 & -15 & | & 0 \end{pmatrix}$$

(b) Divide the second row by -3; then add (-5) times the second row to the third row.

$$\begin{pmatrix} 1 & 2 & -4 & | & 0 \\ 0 & 1 & -3 & | & 0 \\ 0 & 0 & 0 & | & 0 \end{pmatrix}$$

Thus we obtain the equivalent system

$$\begin{aligned} c_1 + 2c_2 - 4c_3 &= 0, \\ c_2 - 3c_3 &= 0. \end{aligned} \tag{23}$$

From the second of Eqs. (23) we have $c_2 = 3c_3$, and from the first we obtain $c_1 = 4c_3 - 2c_2 = -2c_3$. Thus we have solved for c_1 and c_2 in terms of c_3, with the latter remaining arbitrary. If we choose $c_3 = -1$ for convenience, then $c_1 = 2$ and $c_2 = -3$. In this case the relation (20) becomes

$$2\mathbf{x}^{(1)} - 3\mathbf{x}^{(2)} - \mathbf{x}^{(3)} = \mathbf{0},$$

and the given vectors are clearly linearly dependent.

Alternatively, we can compute $\det(x_{ij})$, whose columns are the components of $\mathbf{x}^{(1)}$, $\mathbf{x}^{(2)}$, and $\mathbf{x}^{(3)}$, respectively. Thus

$$\det(x_{ij}) = \begin{vmatrix} 1 & 2 & -4 \\ 2 & 1 & 1 \\ -1 & 3 & -11 \end{vmatrix}$$

and direct calculation shows that it is zero. Hence $\mathbf{x}^{(1)}$, $\mathbf{x}^{(2)}$, and $\mathbf{x}^{(3)}$ are linearly dependent. However, if the coefficients c_1, c_2, and c_3 are required, we still need to solve Eq. (20) to find them.

Frequently, it is useful to think of the columns (or rows) of a matrix \mathbf{A} as vectors. These column (or row) vectors are linearly independent if and only if $\det \mathbf{A} \neq 0$. Further, if $\mathbf{C} = \mathbf{AB}$, then it can be shown that $\det \mathbf{C} = (\det \mathbf{A})(\det \mathbf{B})$. Therefore, if the columns (or rows) of both \mathbf{A} and \mathbf{B} are linearly independent, then the columns (or rows) of \mathbf{C} are also linearly independent.

Now let us extend the concepts of linear dependence and independence to a set of vector functions $\mathbf{x}^{(1)}(t), \ldots, \mathbf{x}^{(k)}(t)$ defined on an interval $\alpha < t < \beta$. The vectors $\mathbf{x}^{(1)}(t), \ldots, \mathbf{x}^{(k)}(t)$ are said to be linearly dependent on $\alpha < t < \beta$ if there exists a set of constants c_1, \ldots, c_k, not all of which are zero, such that $c_1\mathbf{x}^{(1)}(t) + \cdots + c_k\mathbf{x}^{(k)}(t) = \mathbf{0}$ for all t in the interval. Otherwise, $\mathbf{x}^{(1)}(t), \ldots, \mathbf{x}^{(k)}(t)$ are said to be linearly independent. Note that if $\mathbf{x}^{(1)}(t), \ldots, \mathbf{x}^{(k)}(t)$ are linearly dependent on an interval, they are linearly dependent at each point in the interval. However, if $\mathbf{x}^{(1)}(t), \ldots, \mathbf{x}^{(k)}(t)$ are linearly independent on an interval, they may or may not be linearly independent at each point; they may, in fact, be linearly dependent at each point, but with different sets of constants at different points. See Problem 14 for an example.

Eigenvalues and Eigenvectors. The equation

$$\mathbf{Ax} = \mathbf{y} \tag{24}$$

can be viewed as a linear transformation that maps (or transforms) a given vector \mathbf{x} into a new vector \mathbf{y}. Vectors that are transformed into multiples of themselves are important in many applications.[5] To find such vectors, we set $\mathbf{y} = \lambda\mathbf{x}$, where λ is a scalar proportionality factor, and seek solutions of the equation

$$\mathbf{Ax} = \lambda\mathbf{x}, \tag{25}$$

or

$$(\mathbf{A} - \lambda\mathbf{I})\mathbf{x} = \mathbf{0}. \tag{26}$$

The latter equation has nonzero solutions if and only if λ is chosen so that

$$\det(\mathbf{A} - \lambda\mathbf{I}) = 0. \tag{27}$$

[5]For example, this problem is encountered in finding the principal axes of stress or strain in an elastic body, and in finding the modes of free vibration in a conservative system with a finite number of degrees of freedom.

Values of λ that satisfy Eq. (27) are called **eigenvalues** of the matrix **A**, and the nonzero solutions of Eq. (25) or (26) that are obtained by using such a value of λ are called the **eigenvectors** corresponding to that eigenvalue.

If **A** is a 2×2 matrix, then Eq. (26) has the form

$$\begin{pmatrix} a_{11} - \lambda & a_{12} \\ a_{21} & a_{22} - \lambda \end{pmatrix} \begin{pmatrix} x_1 \\ x_2 \end{pmatrix} = \begin{pmatrix} 0 \\ 0 \end{pmatrix} \tag{28}$$

and Eq. (27) becomes

$$(a_{11} - \lambda)(a_{22} - \lambda) - a_{12}a_{21} = 0. \tag{29}$$

The following example illustrates how eigenvalues and eigenvectors are found.

EXAMPLE 4

Find the eigenvalues and eigenvectors of the matrix

$$\mathbf{A} = \begin{pmatrix} 3 & -1 \\ 4 & -2 \end{pmatrix}. \tag{30}$$

The eigenvalues λ and eigenvectors **x** satisfy the equation $(\mathbf{A} - \lambda\mathbf{I})\mathbf{x} = \mathbf{0}$, or

$$\begin{pmatrix} 3 - \lambda & -1 \\ 4 & -2 - \lambda \end{pmatrix} \begin{pmatrix} x_1 \\ x_2 \end{pmatrix} = \begin{pmatrix} 0 \\ 0 \end{pmatrix}. \tag{31}$$

The eigenvalues are the roots of the equation

$$\det(\mathbf{A} - \lambda\mathbf{I}) = \begin{vmatrix} 3 - \lambda & -1 \\ 4 & -2 - \lambda \end{vmatrix} = \lambda^2 - \lambda - 2 = 0. \tag{32}$$

Thus the eigenvalues are $\lambda_1 = 2$ and $\lambda_2 = -1$.

To find the eigenvectors, we return to Eq. (31) and replace λ by each of the eigenvalues in turn. For $\lambda = 2$ we have

$$\begin{pmatrix} 1 & -1 \\ 4 & -4 \end{pmatrix} \begin{pmatrix} x_1 \\ x_2 \end{pmatrix} = \begin{pmatrix} 0 \\ 0 \end{pmatrix}. \tag{33}$$

Hence each row of this vector equation leads to the condition $x_1 - x_2 = 0$, so x_1 and x_2 are equal but their value is not determined. If $x_1 = c$, then $x_2 = c$ also, and the eigenvector $\mathbf{x}^{(1)}$ is

$$\mathbf{x}^{(1)} = c \begin{pmatrix} 1 \\ 1 \end{pmatrix}, \qquad c \neq 0. \tag{34}$$

Usually, we will drop the arbitrary constant c when finding eigenvectors; thus, instead of Eq. (34), we write

$$\mathbf{x}^{(1)} = \begin{pmatrix} 1 \\ 1 \end{pmatrix} \tag{35}$$

and remember that any nonzero multiple of this vector is also an eigenvector. We say that $\mathbf{x}^{(1)}$ is the eigenvector corresponding to the eigenvalue $\lambda_1 = 2$.

Now, setting $\lambda = -1$ in Eq. (31), we obtain

$$\begin{pmatrix} 4 & -1 \\ 4 & -1 \end{pmatrix} \begin{pmatrix} x_1 \\ x_2 \end{pmatrix} = \begin{pmatrix} 0 \\ 0 \end{pmatrix}. \tag{36}$$

Again we obtain a single condition on x_1 and x_2, namely, $4x_1 - x_2 = 0$. Thus the eigenvector corresponding to the eigenvalue $\lambda_2 = -1$ is

$$\mathbf{x}^{(2)} = \begin{pmatrix} 1 \\ 4 \end{pmatrix} \tag{37}$$

or any nonzero multiple of this vector.

As Example 4 illustrates, eigenvectors are determined only up to an arbitrary nonzero multiplicative constant; if this constant is specified in some way, then the eigenvectors are said to be **normalized**. In Example 4, we set the constant equal to 1, but any other nonzero value could also have been used. Sometimes it is convenient to normalize an eigenvector \mathbf{x} by choosing the constant so that $(\mathbf{x}, \mathbf{x}) = 1$.

Equation (27) is a polynomial equation of degree n in λ, so there are n eigenvalues $\lambda_1, \ldots, \lambda_n$, some of which may be repeated. If a given eigenvalue appears m times as a root of Eq. (27), then that eigenvalue is said to have **algebraic multiplicity** m. Each eigenvalue has at least one associated eigenvector, and an eigenvalue of algebraic multiplicity m may have q linearly independent eigenvectors. The number q is called the **geometric multiplicity** of the eigenvalue, and it is possible to show that

$$1 \leq q \leq m. \tag{38}$$

Further, examples demonstrate that q may be any integer in this interval. If each eigenvalue of \mathbf{A} is **simple** (has algebraic multiplicity one), then each eigenvalue also has geometric multiplicity one. Then it is possible to show that the n eigenvectors of \mathbf{A}, one for each eigenvalue, are linearly independent. On the other hand, if \mathbf{A} has one or more repeated eigenvalues, then there may be fewer than n linearly independent eigenvectors associated with \mathbf{A}, since for a repeated eigenvalue we may have $q < m$. As we will see in Section 7.8, this fact may lead to complications later on in the solution of systems of differential equations.

EXAMPLE 5

Find the eigenvalues and eigenvectors of the matrix

$$\mathbf{A} = \begin{pmatrix} 0 & 1 & 1 \\ 1 & 0 & 1 \\ 1 & 1 & 0 \end{pmatrix}. \tag{39}$$

The eigenvalues λ and eigenvectors \mathbf{x} satisfy the equation $(\mathbf{A} - \lambda \mathbf{I})\mathbf{x} = \mathbf{0}$, or

$$\begin{pmatrix} -\lambda & 1 & 1 \\ 1 & -\lambda & 1 \\ 1 & 1 & -\lambda \end{pmatrix} \begin{pmatrix} x_1 \\ x_2 \\ x_3 \end{pmatrix} = \begin{pmatrix} 0 \\ 0 \\ 0 \end{pmatrix}. \tag{40}$$

The eigenvalues are the roots of the equation

$$\det(\mathbf{A} - \lambda \mathbf{I}) = \begin{vmatrix} -\lambda & 1 & 1 \\ 1 & -\lambda & 1 \\ 1 & 1 & -\lambda \end{vmatrix} = -\lambda^3 + 3\lambda + 2 = 0. \tag{41}$$

The roots of Eq. (41) are $\lambda_1 = 2, \lambda_2 = -1$, and $\lambda_3 = -1$. Thus 2 is a simple eigenvalue, and -1 is an eigenvalue of algebraic multiplicity 2, or a double eigenvalue.

To find the eigenvector $\mathbf{x}^{(1)}$ corresponding to the eigenvalue λ_1, we substitute $\lambda = 2$ in Eq. (40); this gives the system

$$
\begin{pmatrix} -2 & 1 & 1 \\ 1 & -2 & 1 \\ 1 & 1 & -2 \end{pmatrix} \begin{pmatrix} x_1 \\ x_2 \\ x_3 \end{pmatrix} = \begin{pmatrix} 0 \\ 0 \\ 0 \end{pmatrix}. \tag{42}
$$

We can reduce this to the equivalent system

$$
\begin{pmatrix} 2 & -1 & -1 \\ 0 & 1 & -1 \\ 0 & 0 & 0 \end{pmatrix} \begin{pmatrix} x_1 \\ x_2 \\ x_3 \end{pmatrix} = \begin{pmatrix} 0 \\ 0 \\ 0 \end{pmatrix} \tag{43}
$$

by elementary row operations. Solving this system yields the eigenvector

$$
\mathbf{x}^{(1)} = \begin{pmatrix} 1 \\ 1 \\ 1 \end{pmatrix}. \tag{44}
$$

For $\lambda = -1$, Eqs. (40) reduce immediately to the single equation

$$
x_1 + x_2 + x_3 = 0. \tag{45}
$$

Thus values for two of the quantities x_1, x_2, x_3 can be chosen arbitrarily, and the third is determined from Eq. (45). For example, if $x_1 = 1$ and $x_2 = 0$, then $x_3 = -1$, and

$$
\mathbf{x}^{(2)} = \begin{pmatrix} 1 \\ 0 \\ -1 \end{pmatrix} \tag{46}
$$

is an eigenvector. Any nonzero multiple of $\mathbf{x}^{(2)}$ is also an eigenvector, but a second independent eigenvector can be found by making another choice of x_1 and x_2—for instance, $x_1 = 0$ and $x_2 = 1$. Again $x_3 = -1$ and

$$
\mathbf{x}^{(3)} = \begin{pmatrix} 0 \\ 1 \\ -1 \end{pmatrix} \tag{47}
$$

is an eigenvector linearly independent of $\mathbf{x}^{(2)}$. Therefore, in this example, two linearly independent eigenvectors are associated with the double eigenvalue.

An important special class of matrices, called **self-adjoint** or **Hermitian** matrices, are those for which $\mathbf{A}^* = \mathbf{A}$; that is, $\bar{a}_{ji} = a_{ij}$. Hermitian matrices include as a subclass real symmetric matrices—that is, matrices that have real elements and for which $\mathbf{A}^T = \mathbf{A}$. The eigenvalues and eigenvectors of Hermitian matrices always have the following useful properties:

1. All eigenvalues are real.
2. There always exists a full set of n linearly independent eigenvectors, regardless of the algebraic multiplicities of the eigenvalues.
3. If $\mathbf{x}^{(1)}$ and $\mathbf{x}^{(2)}$ are eigenvectors that correspond to different eigenvalues, then $(\mathbf{x}^{(1)}, \mathbf{x}^{(2)}) = 0$. Thus, if all eigenvalues are simple, then the associated eigenvectors form an orthogonal set of vectors.

4. Corresponding to an eigenvalue of algebraic multiplicity m, it is possible to choose m eigenvectors that are mutually orthogonal. Thus the full set of n eigenvectors can always be chosen to be orthogonal as well as linearly independent.

Example 5 above involves a real symmetric matrix and illustrates properties 1, 2, and 3, but the choice we have made for $\mathbf{x}^{(2)}$ and $\mathbf{x}^{(3)}$ does not illustrate property 4. However, it is always possible to choose an $\mathbf{x}^{(2)}$ and $\mathbf{x}^{(3)}$ so that $(\mathbf{x}^{(2)}, \mathbf{x}^{(3)}) = 0$. For example, in Example 5 we could have chosen

$$\mathbf{x}^{(2)} = \begin{pmatrix} 1 \\ 0 \\ -1 \end{pmatrix}, \qquad \mathbf{x}^{(3)} = \begin{pmatrix} 1 \\ -2 \\ 1 \end{pmatrix}$$

as the eigenvectors associated with the eigenvalue $\lambda = -1$. These eigenvectors are orthogonal to each other as well as to the eigenvector $\mathbf{x}^{(1)}$ that corresponds to the eigenvalue $\lambda = 2$. The proofs of statements 1 and 3 above are outlined in Problems 32 and 33.

PROBLEMS

In each of Problems 1 through 5 either solve the given set of equations, or else show that there is no solution.

1. $\begin{aligned} x_1 \quad\;\; - x_3 &= 0 \\ 3x_1 + x_2 + x_3 &= 1 \\ -x_1 + x_2 + 2x_3 &= 2 \end{aligned}$

2. $\begin{aligned} x_1 + 2x_2 - x_3 &= 1 \\ 2x_1 + x_2 + x_3 &= 1 \\ x_1 - x_2 + 2x_3 &= 1 \end{aligned}$

3. $\begin{aligned} x_1 + 2x_2 - x_3 &= 2 \\ 2x_1 + x_2 + x_3 &= 1 \\ x_1 - x_2 + 2x_3 &= -1 \end{aligned}$

4. $\begin{aligned} x_1 + 2x_2 - x_3 &= 0 \\ 2x_1 + x_2 + x_3 &= 0 \\ x_1 - x_2 + 2x_3 &= 0 \end{aligned}$

5. $\begin{aligned} x_1 \quad\;\; - x_3 &= 0 \\ 3x_1 + x_2 + x_3 &= 0 \\ -x_1 + x_2 + 2x_3 &= 0 \end{aligned}$

In each of Problems 6 through 10 determine whether the members of the given set of vectors are linearly independent. If they are linearly dependent, find a linear relation among them. The vectors are written as row vectors to save space but may be considered as column vectors; that is, the transposes of the given vectors may be used instead of the vectors themselves.

6. $\mathbf{x}^{(1)} = (1,1,0), \qquad \mathbf{x}^{(2)} = (0,1,1), \qquad \mathbf{x}^{(3)} = (1,0,1)$

7. $\mathbf{x}^{(1)} = (2,1,0), \qquad \mathbf{x}^{(2)} = (0,1,0), \qquad \mathbf{x}^{(3)} = (-1,2,0)$

8. $\mathbf{x}^{(1)} = (1,2,2,3), \qquad \mathbf{x}^{(2)} = (-1,0,3,1), \qquad \mathbf{x}^{(3)} = (-2,-1,1,0), \qquad \mathbf{x}^{(4)} = (-3,0,-1,3)$

9. $\mathbf{x}^{(1)} = (1,2,-1,0), \qquad \mathbf{x}^{(2)} = (2,3,1,-1), \qquad \mathbf{x}^{(3)} = (-1,0,2,2), \qquad \mathbf{x}^{(4)} = (3,-1,1,3)$

10. $\mathbf{x}^{(1)} = (1,2,-2), \qquad \mathbf{x}^{(2)} = (3,1,0), \qquad \mathbf{x}^{(3)} = (2,-1,1), \qquad \mathbf{x}^{(4)} = (4,3,-2)$

11. Suppose that each of the vectors $\mathbf{x}^{(1)}, \ldots, \mathbf{x}^{(m)}$ has n components, where $n < m$. Show that $\mathbf{x}^{(1)}, \ldots, \mathbf{x}^{(m)}$ are linearly dependent.

In each of Problems 12 and 13 determine whether the members of the given set of vectors are linearly independent for $-\infty < t < \infty$. If they are linearly dependent, find the linear relation among them. As in Problems 6 through 10, the vectors are written as row vectors to save space.

12. $\mathbf{x}^{(1)}(t) = (e^{-t}, 2e^{-t})$, $\mathbf{x}^{(2)}(t) = (e^{-t}, e^{-t})$, $\mathbf{x}^{(3)}(t) = (3e^{-t}, 0)$

13. $\mathbf{x}^{(1)}(t) = (2\sin t, \sin t)$, $\mathbf{x}^{(2)}(t) = (\sin t, 2\sin t)$

14. Let

$$\mathbf{x}^{(1)}(t) = \begin{pmatrix} e^t \\ te^t \end{pmatrix}, \qquad \mathbf{x}^{(2)}(t) = \begin{pmatrix} 1 \\ t \end{pmatrix}.$$

Show that $\mathbf{x}^{(1)}(t)$ and $\mathbf{x}^{(2)}(t)$ are linearly dependent at each point in the interval $0 \le t \le 1$. Nevertheless, show that $\mathbf{x}^{(1)}(t)$ and $\mathbf{x}^{(2)}(t)$ are linearly independent on $0 \le t \le 1$.

In each of Problems 15 through 24 find all eigenvalues and eigenvectors of the given matrix.

15. $\begin{pmatrix} 5 & -1 \\ 3 & 1 \end{pmatrix}$

16. $\begin{pmatrix} 3 & -2 \\ 4 & -1 \end{pmatrix}$

17. $\begin{pmatrix} -2 & 1 \\ 1 & -2 \end{pmatrix}$

18. $\begin{pmatrix} 1 & i \\ -i & 1 \end{pmatrix}$

19. $\begin{pmatrix} 1 & \sqrt{3} \\ \sqrt{3} & -1 \end{pmatrix}$

20. $\begin{pmatrix} -3 & 3/4 \\ -5 & 1 \end{pmatrix}$

21. $\begin{pmatrix} 1 & 0 & 0 \\ 2 & 1 & -2 \\ 3 & 2 & 1 \end{pmatrix}$

22. $\begin{pmatrix} 3 & 2 & 2 \\ 1 & 4 & 1 \\ -2 & -4 & -1 \end{pmatrix}$

23. $\begin{pmatrix} 11/9 & -2/9 & 8/9 \\ -2/9 & 2/9 & 10/9 \\ 8/9 & 10/9 & 5/9 \end{pmatrix}$

24. $\begin{pmatrix} 3 & 2 & 4 \\ 2 & 0 & 2 \\ 4 & 2 & 3 \end{pmatrix}$

Problems 25 through 29 deal with the problem of solving $\mathbf{Ax} = \mathbf{b}$ when $\det \mathbf{A} = 0$.

25. Suppose that, for a given matrix \mathbf{A}, there is a nonzero vector \mathbf{x} such that $\mathbf{Ax} = \mathbf{0}$. Show that there is also a nonzero vector \mathbf{y} such that $\mathbf{A}^*\mathbf{y} = \mathbf{0}$.

26. Show that $(\mathbf{Ax}, \mathbf{y}) = (\mathbf{x}, \mathbf{A}^*\mathbf{y})$ for any vectors \mathbf{x} and \mathbf{y}.

27. Suppose that $\det \mathbf{A} = 0$ and that $\mathbf{Ax} = \mathbf{b}$ has solutions. Show that $(\mathbf{b}, \mathbf{y}) = 0$, where \mathbf{y} is any solution of $\mathbf{A}^*\mathbf{y} = \mathbf{0}$. Verify that this statement is true for the set of equations in Example 2.
Hint: Use the result of Problem 26.

28. Suppose that $\det \mathbf{A} = 0$ and that $\mathbf{x} = \mathbf{x}^{(0)}$ is a solution of $\mathbf{Ax} = \mathbf{b}$. Show that if $\boldsymbol{\xi}$ is a solution of $\mathbf{A}\boldsymbol{\xi} = \mathbf{0}$ and α is any constant, then $\mathbf{x} = \mathbf{x}^{(0)} + \alpha\boldsymbol{\xi}$ is also a solution of $\mathbf{Ax} = \mathbf{b}$.

29. Suppose that $\det \mathbf{A} = 0$ and that \mathbf{y} is a solution of $\mathbf{A}^*\mathbf{y} = \mathbf{0}$. Show that if $(\mathbf{b}, \mathbf{y}) = 0$ for every such \mathbf{y}, then $\mathbf{Ax} = \mathbf{b}$ has solutions. Note that this is the converse of Problem 27; the form of the solution is given by Problem 28.

30. Prove that $\lambda = 0$ is an eigenvalue of \mathbf{A} if and only if \mathbf{A} is singular.

31. Prove that if \mathbf{A} is Hermitian, then $(\mathbf{Ax}, \mathbf{y}) = (\mathbf{x}, \mathbf{Ay})$, where \mathbf{x} and \mathbf{y} are any vectors.

32. In this problem we show that the eigenvalues of a Hermitian matrix \mathbf{A} are real. Let \mathbf{x} be an eigenvector corresponding to the eigenvalue λ.
 (a) Show that $(\mathbf{A}\mathbf{x}, \mathbf{x}) = (\mathbf{x}, \mathbf{A}\mathbf{x})$. *Hint:* See Problem 31.
 (b) Show that $\lambda(\mathbf{x}, \mathbf{x}) = \bar{\lambda}(\mathbf{x}, \mathbf{x})$. *Hint:* Recall that $\mathbf{A}\mathbf{x} = \lambda\mathbf{x}$.
 (c) Show that $\lambda = \bar{\lambda}$; that is, the eigenvalue λ is real.

33. Show that if λ_1 and λ_2 are eigenvalues of a Hermitian matrix \mathbf{A}, and if $\lambda_1 \neq \lambda_2$, then the corresponding eigenvectors $\mathbf{x}^{(1)}$ and $\mathbf{x}^{(2)}$ are orthogonal.
 Hint: Use the results of Problems 31 and 32 to show that $(\lambda_1 - \lambda_2)(\mathbf{x}^{(1)}, \mathbf{x}^{(2)}) = 0$.

7.4 Basic Theory of Systems of First Order Linear Equations

The general theory of a system of n first order linear equations

$$x_1' = p_{11}(t)x_1 + \cdots + p_{1n}(t)x_n + g_1(t),$$
$$\vdots \tag{1}$$
$$x_n' = p_{n1}(t)x_1 + \cdots + p_{nn}(t)x_n + g_n(t)$$

closely parallels that of a single linear equation of nth order. The discussion in this section therefore follows the same general lines as that in Sections 3.2, 3.3, and 4.1. To discuss the system (1) most effectively, we write it in matrix notation. That is, we consider $x_1 = \phi_1(t), \ldots, x_n = \phi_n(t)$ to be components of a vector $\mathbf{x} = \boldsymbol{\phi}(t)$; similarly, $g_1(t), \ldots, g_n(t)$ are components of a vector $\mathbf{g}(t)$, and $p_{11}(t), \ldots, p_{nn}(t)$ are elements of an $n \times n$ matrix $\mathbf{P}(t)$. Equation (1) then takes the form

$$\mathbf{x}' = \mathbf{P}(t)\mathbf{x} + \mathbf{g}(t). \tag{2}$$

The use of vectors and matrices not only saves a great deal of space and facilitates calculations but also emphasizes the similarity between systems of equations and single (scalar) equations.

A vector $\mathbf{x} = \boldsymbol{\phi}(t)$ is said to be a solution of Eq. (2) if its components satisfy the system of equations (1). Throughout this section we assume that \mathbf{P} and \mathbf{g} are continuous on some interval $\alpha < t < \beta$; that is, each of the scalar functions $p_{11}, \ldots, p_{nn}, g_1, \ldots, g_n$ is continuous there. According to Theorem 7.1.2, this is sufficient to guarantee the existence of solutions of Eq. (2) on the interval $\alpha < t < \beta$.

It is convenient to consider first the homogeneous equation

$$\mathbf{x}' = \mathbf{P}(t)\mathbf{x} \tag{3}$$

obtained from Eq. (2) by setting $\mathbf{g}(t) = \mathbf{0}$. Once the homogeneous equation has been solved, there are several methods that can be used to solve the nonhomogeneous equation (2); this is taken up in Section 7.9. We use the notation

$$\mathbf{x}^{(1)}(t) = \begin{pmatrix} x_{11}(t) \\ x_{21}(t) \\ \vdots \\ x_{n1}(t) \end{pmatrix}, \quad \ldots, \quad \mathbf{x}^{(k)}(t) = \begin{pmatrix} x_{1k}(t) \\ x_{2k}(t) \\ \vdots \\ x_{nk}(t) \end{pmatrix}, \quad \ldots \tag{4}$$

to designate specific solutions of the system (3). Note that $x_{ij}(t) = x_i^{(j)}(t)$ refers to the ith component of the jth solution $\mathbf{x}^{(j)}(t)$. The main facts about the structure of solutions of the system (3) are stated in Theorems 7.4.1 to 7.4.4. They closely resemble the corresponding theorems in Sections 3.2, 3.3, and 4.1; some of the proofs are left to you as exercises.

Theorem 7.4.1 If the vector functions $\mathbf{x}^{(1)}$ and $\mathbf{x}^{(2)}$ are solutions of the system (3), then the linear combination $c_1\mathbf{x}^{(1)} + c_2\mathbf{x}^{(2)}$ is also a solution for any constants c_1 and c_2.

This is the principle of superposition; it is proved simply by differentiating $c_1\mathbf{x}^{(1)} + c_2\mathbf{x}^{(2)}$ and using the fact that $\mathbf{x}^{(1)}$ and $\mathbf{x}^{(2)}$ satisfy Eq. (3). By repeated application of Theorem 7.4.1 we reach the conclusion that if $\mathbf{x}^{(1)}, \ldots, \mathbf{x}^{(k)}$ are solutions of Eq. (3), then

$$\mathbf{x} = c_1\mathbf{x}^{(1)}(t) + \cdots + c_k\mathbf{x}^{(k)}(t) \tag{5}$$

is also a solution for any constants c_1, \ldots, c_k. As an example, it can be verified that

$$\mathbf{x}^{(1)}(t) = \begin{pmatrix} e^{3t} \\ 2e^{3t} \end{pmatrix} = \begin{pmatrix} 1 \\ 2 \end{pmatrix} e^{3t}, \qquad \mathbf{x}^{(2)}(t) = \begin{pmatrix} e^{-t} \\ -2e^{-t} \end{pmatrix} = \begin{pmatrix} 1 \\ -2 \end{pmatrix} e^{-t} \tag{6}$$

satisfy the equation

$$\mathbf{x}' = \begin{pmatrix} 1 & 1 \\ 4 & 1 \end{pmatrix} \mathbf{x}. \tag{7}$$

According to Theorem 7.4.1,

$$\mathbf{x} = c_1 \begin{pmatrix} 1 \\ 2 \end{pmatrix} e^{3t} + c_2 \begin{pmatrix} 1 \\ -2 \end{pmatrix} e^{-t}$$

$$= c_1\mathbf{x}^{(1)}(t) + c_2\mathbf{x}^{(2)}(t) \tag{8}$$

also satisfies Eq. (7).

As we indicated previously, by repeatedly applying Theorem 7.4.1, it follows that every finite linear combination of solutions of Eq. (3) is also a solution. The question that now arises is whether all solutions of Eq. (3) can be found in this way. By analogy with previous cases, it is reasonable to expect that for a system of the form (3) of nth order, it is sufficient to form linear combinations of n properly chosen solutions. Therefore let $\mathbf{x}^{(1)}, \ldots, \mathbf{x}^{(n)}$ be n solutions of the nth order system (3), and consider the matrix $\mathbf{X}(t)$ whose columns are the vectors $\mathbf{x}^{(1)}(t), \ldots, \mathbf{x}^{(n)}(t)$:

$$\mathbf{X}(t) = \begin{pmatrix} x_{11}(t) & \cdots & x_{1n}(t) \\ \vdots & & \vdots \\ x_{n1}(t) & \cdots & x_{nn}(t) \end{pmatrix}. \tag{9}$$

Recall from Section 7.3 that the columns of $\mathbf{X}(t)$ are linearly independent for a given value of t if and only if $\det \mathbf{X} \neq 0$ for that value of t. This determinant is called the Wronskian of the n solutions $\mathbf{x}^{(1)}, \ldots, \mathbf{x}^{(n)}$ and is also denoted by $W[\mathbf{x}^{(1)}, \ldots, \mathbf{x}^{(n)}]$; that is,

$$W[\mathbf{x}^{(1)}, \ldots, \mathbf{x}^{(n)}](t) = \det \mathbf{X}(t). \tag{10}$$

The solutions $\mathbf{x}^{(1)}, \ldots, \mathbf{x}^{(n)}$ are then linearly independent at a point if and only if $W[\,\mathbf{x}^{(1)}, \ldots, \mathbf{x}^{(n)}]$ is not zero there.

Theorem 7.4.2

If the vector functions $\mathbf{x}^{(1)}, \ldots, \mathbf{x}^{(n)}$ are linearly independent solutions of the system (3) for each point in the interval $\alpha < t < \beta$, then each solution $\mathbf{x} = \boldsymbol{\phi}(t)$ of the system (3) can be expressed as a linear combination of $\mathbf{x}^{(1)}, \ldots, \mathbf{x}^{(n)}$,

$$\boldsymbol{\phi}(t) = c_1 \mathbf{x}^{(1)}(t) + \cdots + c_n \mathbf{x}^{(n)}(t), \tag{11}$$

in exactly one way.

Before proving Theorem 7.4.2, note that according to Theorem 7.4.1 all expressions of the form (11) are solutions of the system (3), while by Theorem 7.4.2 all solutions of Eq. (3) can be written in the form (11). If the constants c_1, \ldots, c_n are thought of as arbitrary, then Eq. (11) includes all solutions of the system (3), and it is customary to call it the **general solution.** Any set of solutions $\mathbf{x}^{(1)}, \ldots, \mathbf{x}^{(n)}$ of Eq. (3), which is linearly independent at each point in the interval $\alpha < t < \beta$, is said to be a **fundamental set of solutions** for that interval.

To prove Theorem 7.4.2 we will show, given any solution $\boldsymbol{\phi}$ of Eq. (3), that $\boldsymbol{\phi}(t) = c_1 \mathbf{x}^{(1)}(t) + \cdots + c_n \mathbf{x}^{(n)}(t)$ for suitable values of c_1, \ldots, c_n. Let $t = t_0$ be some point in the interval $\alpha < t < \beta$ and let $\boldsymbol{\xi} = \boldsymbol{\phi}(t_0)$. We now wish to determine whether there is any solution of the form $\mathbf{x} = c_1 \mathbf{x}^{(1)}(t) + \cdots + c_n \mathbf{x}^{(n)}(t)$ that also satisfies the same initial condition $\mathbf{x}(t_0) = \boldsymbol{\xi}$. That is, we wish to know whether there are values of c_1, \ldots, c_n such that

$$c_1 \mathbf{x}^{(1)}(t_0) + \cdots + c_n \mathbf{x}^{(n)}(t_0) = \boldsymbol{\xi}, \tag{12}$$

or, in scalar form,

$$c_1 x_{11}(t_0) + \cdots + c_n x_{1n}(t_0) = \xi_1,$$
$$\vdots \tag{13}$$
$$c_1 x_{n1}(t_0) + \cdots + c_n x_{nn}(t_0) = \xi_n.$$

The necessary and sufficient condition that Eqs. (13) possess a unique solution c_1, \ldots, c_n is precisely the nonvanishing of the determinant of coefficients, which is the Wronskian $W[\,\mathbf{x}^{(1)}, \ldots, \mathbf{x}^{(n)}]$ evaluated at $t = t_0$. The hypothesis that $\mathbf{x}^{(1)}, \ldots, \mathbf{x}^{(n)}$ are linearly independent throughout $\alpha < t < \beta$ guarantees that $W[\mathbf{x}^{(1)}, \ldots, \mathbf{x}^{(n)}]$ is not zero at $t = t_0$, and therefore there is a (unique) solution of Eq. (3) of the form $\mathbf{x} = c_1 \mathbf{x}^{(1)}(t) + \cdots + c_n \mathbf{x}^{(n)}(t)$ that also satisfies the initial condition (12). By the uniqueness part of Theorem 7.1.2, this solution is identical to $\boldsymbol{\phi}(t)$, and hence $\boldsymbol{\phi}(t) = c_1 \mathbf{x}^{(1)}(t) + \cdots + c_n \mathbf{x}^{(n)}(t)$, as was to be proved.

Theorem 7.4.3

If $\mathbf{x}^{(1)}, \ldots, \mathbf{x}^{(n)}$ are solutions of Eq. (3) on the interval $\alpha < t < \beta$, then in this interval $W[\mathbf{x}^{(1)}, \ldots, \mathbf{x}^{(n)}]$ either is identically zero or else never vanishes.

The significance of Theorem 7.4.3 lies in the fact that it relieves us of the necessity of examining $W[\mathbf{x}^{(1)}, \ldots, \mathbf{x}^{(n)}]$ at all points in the interval of interest, and enables us

to determine whether $\mathbf{x}^{(1)}, \ldots, \mathbf{x}^{(n)}$ form a fundamental set of solutions merely by evaluating their Wronskian at any convenient point in the interval.

Theorem 7.4.3 is proved by first establishing that the Wronskian of $\mathbf{x}^{(1)}, \ldots, \mathbf{x}^{(n)}$ satisfies the differential equation (see Problem 2)

$$\frac{dW}{dt} = (p_{11} + p_{22} + \cdots + p_{nn})W. \tag{14}$$

Hence W is an exponential function, and the conclusion of the theorem follows immediately. The expression for W obtained by solving Eq. (14) is known as Abel's formula; note the analogy with Eq. (8) of Section 3.3.

Alternatively, Theorem 7.4.3 can be established by showing that if n solutions $\mathbf{x}^{(1)}, \ldots, \mathbf{x}^{(n)}$ of Eq. (3) are linearly dependent at one point $t = t_0$, then they must be linearly dependent at each point in $\alpha < t < \beta$ (see Problem 8). Consequently, if $\mathbf{x}^{(1)}, \ldots, \mathbf{x}^{(n)}$ are linearly independent at one point, they must be linearly independent at each point in the interval.

The next theorem states that the system (3) always has at least one fundamental set of solutions.

Theorem 7.4.4 Let

$$\mathbf{e}^{(1)} = \begin{pmatrix} 1 \\ 0 \\ 0 \\ \vdots \\ 0 \end{pmatrix}, \quad \mathbf{e}^{(2)} = \begin{pmatrix} 0 \\ 1 \\ 0 \\ \vdots \\ 0 \end{pmatrix}, \quad \ldots, \quad \mathbf{e}^{(n)} = \begin{pmatrix} 0 \\ 0 \\ \vdots \\ 0 \\ 1 \end{pmatrix};$$

further let $\mathbf{x}^{(1)}, \ldots, \mathbf{x}^{(n)}$ be the solutions of the system (3) that satisfy the initial conditions

$$\mathbf{x}^{(1)}(t_0) = \mathbf{e}^{(1)}, \quad \ldots, \quad \mathbf{x}^{(n)}(t_0) = \mathbf{e}^{(n)}, \tag{15}$$

respectively, where t_0 is any point in $\alpha < t < \beta$. Then $\mathbf{x}^{(1)}, \ldots, \mathbf{x}^{(n)}$ form a fundamental set of solutions of the system (3).

To prove this theorem, note that the existence and uniqueness of the solutions $\mathbf{x}^{(1)}, \ldots, \mathbf{x}^{(n)}$ mentioned in Theorem 7.4.4 are ensured by Theorem 7.1.2. It is not hard to see that the Wronskian of these solutions is equal to 1 when $t = t_0$; therefore $\mathbf{x}^{(1)}, \ldots, \mathbf{x}^{(n)}$ are a fundamental set of solutions.

Once one fundamental set of solutions has been found, other sets can be generated by forming (independent) linear combinations of the first set. For theoretical purposes the set given by Theorem 7.4.4 is usually the simplest.

To summarize, any set of n linearly independent solutions of the system (3) constitutes a fundamental set of solutions. Under the conditions given in this section, such fundamental sets always exist, and every solution of the system (3) can be represented as a linear combination of any fundamental set of solutions.

PROBLEMS

1. Prove the statement following Theorem 7.4.1 for an arbitrary value of the integer k.

2. In this problem we outline a proof of Theorem 7.4.3 in the case $n = 2$. Let $\mathbf{x}^{(1)}$ and $\mathbf{x}^{(2)}$ be solutions of Eq. (3) for $\alpha < t < \beta$, and let W be the Wronskian of $\mathbf{x}^{(1)}$ and $\mathbf{x}^{(2)}$.

 (a) Show that

$$
\frac{dW}{dt} = \begin{vmatrix} \dfrac{dx_1^{(1)}}{dt} & \dfrac{dx_1^{(2)}}{dt} \\ x_2^{(1)} & x_2^{(2)} \end{vmatrix} + \begin{vmatrix} x_1^{(1)} & x_1^{(2)} \\ \dfrac{dx_2^{(1)}}{dt} & \dfrac{dx_2^{(2)}}{dt} \end{vmatrix}.
$$

 (b) Using Eq. (3), show that

$$
\frac{dW}{dt} = (p_{11} + p_{22})W.
$$

 (c) Find $W(t)$ by solving the differential equation obtained in part (b). Use this expression to obtain the conclusion stated in Theorem 7.4.3.

 (d) Prove Theorem 7.4.3 for an arbitrary value of n by generalizing the procedure of parts (a), (b), and (c).

3. Show that the Wronskians of two fundamental sets of solutions of the system (3) can differ at most by a multiplicative constant.
 Hint: Use Eq. (14).

4. If $x_1 = y$ and $x_2 = y'$, then the second order equation

$$
y'' + p(t)y' + q(t)y = 0 \tag{i}
$$

 corresponds to the system

$$
x_1' = x_2,
$$
$$
x_2' = -q(t)x_1 - p(t)x_2. \tag{ii}
$$

 Show that if $\mathbf{x}^{(1)}$ and $\mathbf{x}^{(2)}$ are a fundamental set of solutions of Eqs. (ii), and if $y^{(1)}$ and $y^{(2)}$ are a fundamental set of solutions of Eq. (i), then $W[y^{(1)}, y^{(2)}] = cW[\mathbf{x}^{(1)}, \mathbf{x}^{(2)}]$, where c is a nonzero constant.
 Hint: $y^{(1)}(t)$ and $y^{(2)}(t)$ must be linear combinations of $x_{11}(t)$ and $x_{12}(t)$.

5. Show that the general solution of $\mathbf{x}' = \mathbf{P}(t)\mathbf{x} + \mathbf{g}(t)$ is the sum of any particular solution $\mathbf{x}^{(p)}$ of this equation and the general solution $\mathbf{x}^{(c)}$ of the corresponding homogeneous equation.

6. Consider the vectors $\mathbf{x}^{(1)}(t) = \begin{pmatrix} t \\ 1 \end{pmatrix}$ and $\mathbf{x}^{(2)}(t) = \begin{pmatrix} t^2 \\ 2t \end{pmatrix}$.

 (a) Compute the Wronskian of $\mathbf{x}^{(1)}$ and $\mathbf{x}^{(2)}$.

 (b) In what intervals are $\mathbf{x}^{(1)}$ and $\mathbf{x}^{(2)}$ linearly independent?

 (c) What conclusion can be drawn about the coefficients in the system of homogeneous differential equations satisfied by $\mathbf{x}^{(1)}$ and $\mathbf{x}^{(2)}$?

 (d) Find this system of equations and verify the conclusions of part (c).

7. Consider the vectors $\mathbf{x}^{(1)}(t) = \begin{pmatrix} t^2 \\ 2t \end{pmatrix}$ and $\mathbf{x}^{(2)}(t) = \begin{pmatrix} e^t \\ e^t \end{pmatrix}$, and answer the same questions as in Problem 6.

The following two problems indicate an alternative derivation of Theorem 7.4.2.

8. Let $\mathbf{x}^{(1)}, \ldots, \mathbf{x}^{(m)}$ be solutions of $\mathbf{x}' = \mathbf{P}(t)\mathbf{x}$ on the interval $\alpha < t < \beta$. Assume that \mathbf{P} is continuous and let t_0 be an arbitrary point in the given interval. Show that $\mathbf{x}^{(1)}, \ldots, \mathbf{x}^{(m)}$ are linearly dependent for $\alpha < t < \beta$ if (and only if) $\mathbf{x}^{(1)}(t_0), \ldots, \mathbf{x}^{(m)}(t_0)$ are linearly dependent. In other words $\mathbf{x}^{(1)}, \ldots, \mathbf{x}^{(m)}$ are linearly dependent on the interval (α, β) if they are linearly dependent at any point in it.
 Hint: There are constants c_1, \ldots, c_m such that $c_1\mathbf{x}^{(1)}(t_0) + \cdots + c_m\mathbf{x}^{(m)}(t_0) = \mathbf{0}$. Let $\mathbf{z}(t) = c_1\mathbf{x}^{(1)}(t) + \cdots + c_m\mathbf{x}^{(m)}(t)$, and use the uniqueness theorem to show that $\mathbf{z}(t) = \mathbf{0}$ for each t in $\alpha < t < \beta$.

9. Let $\mathbf{x}^{(1)}, \ldots, \mathbf{x}^{(n)}$ be linearly independent solutions of $\mathbf{x}' = \mathbf{P}(t)\mathbf{x}$, where \mathbf{P} is continuous on $\alpha < t < \beta$.

 (a) Show that any solution $\mathbf{x} = \mathbf{z}(t)$ can be written in the form

 $$\mathbf{z}(t) = c_1\mathbf{x}^{(1)}(t) + \cdots + c_n\mathbf{x}^{(n)}(t)$$

 for suitable constants c_1, \ldots, c_n.
 Hint: Use the result of Problem 11 of Section 7.3, and also Problem 8 above.

 (b) Show that the expression for the solution $\mathbf{z}(t)$ in part (a) is unique; that is, if $\mathbf{z}(t) = k_1\mathbf{x}^{(1)}(t) + \cdots + k_n\mathbf{x}^{(n)}(t)$, then $k_1 = c_1, \ldots, k_n = c_n$.
 Hint: Show that $(k_1 - c_1)\mathbf{x}^{(1)}(t) + \cdots + (k_n - c_n)\mathbf{x}^{(n)}(t) = \mathbf{0}$ for each t in $\alpha < t < \beta$, and use the linear independence of $\mathbf{x}^{(1)}, \ldots, \mathbf{x}^{(n)}$.

7.5 Homogeneous Linear Systems with Constant Coefficients

We will concentrate most of our attention on systems of homogeneous linear equations with constant coefficients—that is, systems of the form

$$\mathbf{x}' = \mathbf{A}\mathbf{x}, \tag{1}$$

where \mathbf{A} is a constant $n \times n$ matrix. Unless stated otherwise, we will assume further that all the elements of \mathbf{A} are real (rather than complex) numbers.

If $n = 1$, then the system reduces to a single first order equation

$$\frac{dx}{dt} = ax, \tag{2}$$

whose solution is $x = ce^{at}$. In Section 2.5 we noted that $x = 0$ is the only equilibrium solution if $a \neq 0$. Other solutions approach $x = 0$ if $a < 0$, and in this case we say that $x = 0$ is an asymptotically stable equilibrium solution. On the other hand, if $a > 0$, then $x = 0$ is unstable, since other solutions depart from it. For higher order systems, the situation is somewhat analogous but more complicated. Equilibrium solutions are found by solving $\mathbf{A}\mathbf{x} = \mathbf{0}$. We assume that $\det \mathbf{A} \neq 0$, so $\mathbf{x} = \mathbf{0}$ is the only equilibrium solution. An important question is whether other solutions approach this equilibrium solution or depart from it as t increases; in other words, is $\mathbf{x} = \mathbf{0}$ asymptotically stable or unstable? Or are there still other possibilities?

The case $n = 2$ is particularly important and lends itself to visualization in the x_1x_2-plane, called the **phase plane**. By evaluating $\mathbf{A}\mathbf{x}$ at a large number of points

and plotting the resulting vectors, we obtain a direction field of tangent vectors to solutions of the system of differential equations. A qualitative understanding of the behavior of solutions can usually be gained from a direction field. More precise information results from including in the plot some solution curves, or trajectories. A plot that shows a representative sample of trajectories for a given system is called a **phase portrait**. Examples of direction fields and phase portraits appear later in this section.

To construct the general solution of the system (1), we proceed by analogy to the treatment of second order linear equations in Section 3.1. Thus we seek solutions of Eq. (1) of the form

$$\mathbf{x} = \boldsymbol{\xi} e^{rt}, \tag{3}$$

where the exponent r and the constant vector $\boldsymbol{\xi}$ are to be determined. Substituting from Eq. (3) for \mathbf{x} in the system (1) gives

$$r\boldsymbol{\xi} e^{rt} = \mathbf{A}\boldsymbol{\xi} e^{rt}.$$

Upon canceling the nonzero scalar factor e^{rt}, we obtain $\mathbf{A}\boldsymbol{\xi} = r\boldsymbol{\xi}$, or

$$(\mathbf{A} - r\mathbf{I})\boldsymbol{\xi} = \mathbf{0}, \tag{4}$$

where \mathbf{I} is the $n \times n$ identity matrix. Thus, to solve the system of differential equations (1), we must solve the system of algebraic equations (4). This latter problem is precisely the one that determines the eigenvalues and eigenvectors of the matrix \mathbf{A}. Therefore the vector \mathbf{x} given by Eq. (3) is a solution of Eq. (1), provided that r is an eigenvalue and $\boldsymbol{\xi}$ an associated eigenvector of the coefficient matrix \mathbf{A}.

The following two examples illustrate the solution procedure in the case of 2×2 coefficient matrices. We also show how to construct the corresponding phase portraits. Later in the section we return to a further discussion of the general $n \times n$ system.

EXAMPLE

1

Consider the system

$$\mathbf{x}' = \begin{pmatrix} 1 & 1 \\ 4 & 1 \end{pmatrix} \mathbf{x}. \tag{5}$$

Plot a direction field and determine the qualitative behavior of solutions. Then find the general solution and draw several trajectories.

A direction field for this system is shown in Figure 7.5.1. From this figure it is easy to see that a typical solution departs from the neighborhood of the origin and ultimately has a slope of approximately 2 in either the first or the third quadrant.

To find solutions explicitly, we assume that $\mathbf{x} = \boldsymbol{\xi} e^{rt}$ and substitute for \mathbf{x} in Eq. (5). We are led to the system of algebraic equations

$$\begin{pmatrix} 1 - r & 1 \\ 4 & 1 - r \end{pmatrix} \begin{pmatrix} \xi_1 \\ \xi_2 \end{pmatrix} = \begin{pmatrix} 0 \\ 0 \end{pmatrix}. \tag{6}$$

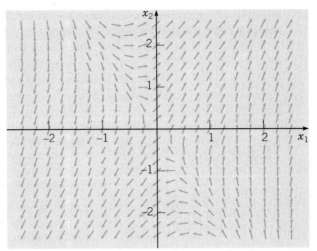

FIGURE 7.5.1 Direction field for the system (5).

Equations (6) have a nontrivial solution if and only if the determinant of coefficients is zero. Thus allowable values of r are found from the equation

$$\begin{vmatrix} 1-r & 1 \\ 4 & 1-r \end{vmatrix} = (1-r)^2 - 4$$

$$= r^2 - 2r - 3 = 0. \tag{7}$$

Equation (7) has the roots $r_1 = 3$ and $r_2 = -1$; these are the eigenvalues of the coefficient matrix in Eq. (5). If $r = 3$, then the system (6) reduces to the single equation

$$-2\xi_1 + \xi_2 = 0. \tag{8}$$

Thus $\xi_2 = 2\xi_1$, and the eigenvector corresponding to $r_1 = 3$ can be taken as

$$\boldsymbol{\xi}^{(1)} = \begin{pmatrix} 1 \\ 2 \end{pmatrix}. \tag{9}$$

Similarly, corresponding to $r_2 = -1$, we find that $\xi_2 = -2\xi_1$, so the eigenvector is

$$\boldsymbol{\xi}^{(2)} = \begin{pmatrix} 1 \\ -2 \end{pmatrix}. \tag{10}$$

The corresponding solutions of the differential equation are

$$\mathbf{x}^{(1)}(t) = \begin{pmatrix} 1 \\ 2 \end{pmatrix} e^{3t}, \qquad \mathbf{x}^{(2)}(t) = \begin{pmatrix} 1 \\ -2 \end{pmatrix} e^{-t}. \tag{11}$$

The Wronskian of these solutions is

$$W[\mathbf{x}^{(1)}, \mathbf{x}^{(2)}](t) = \begin{vmatrix} e^{3t} & e^{-t} \\ 2e^{3t} & -2e^{-t} \end{vmatrix} = -4e^{2t}, \tag{12}$$

which is never zero. Hence the solutions $\mathbf{x}^{(1)}$ and $\mathbf{x}^{(2)}$ form a fundamental set, and the general solution of the system (5) is

$$\mathbf{x} = c_1 \mathbf{x}^{(1)}(t) + c_2 \mathbf{x}^{(2)}(t)$$

$$= c_1 \begin{pmatrix} 1 \\ 2 \end{pmatrix} e^{3t} + c_2 \begin{pmatrix} 1 \\ -2 \end{pmatrix} e^{-t}, \tag{13}$$

where c_1 and c_2 are arbitrary constants.

To visualize the solution (13) it is helpful to consider its graph in the $x_1 x_2$-plane for various values of the constants c_1 and c_2. We start with $\mathbf{x} = c_1 \mathbf{x}^{(1)}(t)$ or, in scalar form,

$$x_1 = c_1 e^{3t}, \qquad x_2 = 2c_1 e^{3t}.$$

By eliminating t between these two equations, we see that this solution lies on the straight line $x_2 = 2x_1$; see Figure 7.5.2a. This is the line through the origin in the direction of the eigenvector $\boldsymbol{\xi}^{(1)}$. If we look on the solution as the trajectory of a moving particle, then the particle is in the first quadrant when $c_1 > 0$ and in the third quadrant when $c_1 < 0$. In either case the particle departs from the origin as t increases. Next consider $\mathbf{x} = c_2 \mathbf{x}^{(2)}(t)$, or

$$x_1 = c_2 e^{-t}, \qquad x_2 = -2c_2 e^{-t}.$$

This solution lies on the line $x_2 = -2x_1$, whose direction is determined by the eigenvector $\boldsymbol{\xi}^{(2)}$. The solution is in the fourth quadrant when $c_2 > 0$ and in the second quadrant when $c_2 < 0$, as shown in Figure 7.5.2a. In both cases the particle moves toward the origin as t increases. The solution (13) is a combination of $\mathbf{x}^{(1)}(t)$ and $\mathbf{x}^{(2)}(t)$. For large t the term $c_1 \mathbf{x}^{(1)}(t)$ is dominant and the term $c_2 \mathbf{x}^{(2)}(t)$ becomes negligible. Thus all solutions for which $c_1 \neq 0$ are asymptotic to the line $x_2 = 2x_1$ as $t \to \infty$. Similarly, all solutions for which $c_2 \neq 0$ are asymptotic to the line $x_2 = -2x_1$ as $t \to -\infty$. The graphs of several solutions are shown in Figure 7.5.2a. The pattern of trajectories in this figure is typical of all second order systems $\mathbf{x}' = \mathbf{A}\mathbf{x}$ for which the eigenvalues are real and of opposite signs. The origin is called a **saddle point** in this case. Saddle points are always unstable because almost all trajectories depart from them as t increases.

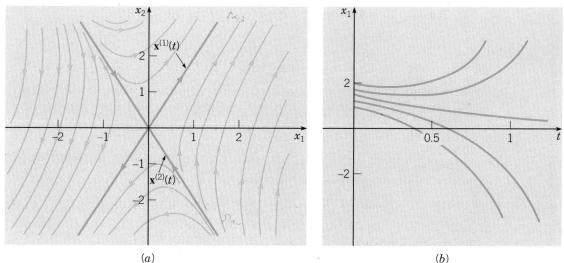

(a) (b)

FIGURE 7.5.2 (a) Trajectories of the system (5); the origin is a saddle point. (b) Plots of x_1 versus t for the system (5).

In the preceding paragraph we have described how to draw by hand a qualitatively correct sketch of the trajectories of a system such as Eq. (5), once the eigenvalues and eigenvectors have been determined. However, to produce a detailed and accurate drawing, such as Figure 7.5.2a and other figures that appear later in this chapter, a computer is extremely helpful, if not indispensable.

As an alternative to Figure 7.5.2a, one can also plot x_1 or x_2 as a function of t; some typical plots of x_1 versus t are shown in Figure 7.5.2b, and those of x_2 versus t are similar. For certain initial conditions it follows that $c_1 = 0$ in Eq. (13), so that $x_1 = c_2 e^{-t}$ and $x_1 \to 0$ as $t \to \infty$. One such graph is shown in Figure 7.5.2b, corresponding to a trajectory that approaches the origin in Figure 7.5.2a. For most initial conditions, however, $c_1 \neq 0$ and x_1 is given by $x_1 = c_1 e^{3t} + c_2 e^{-t}$. Then the presence of the positive exponential term causes x_1 to grow exponentially in magnitude as t increases. Several graphs of this type are shown in Figure 7.5.2b, corresponding to trajectories that depart from the neighborhood of the origin in Figure 7.5.2a. It is important to understand the relation between parts (a) and (b) of Figure 7.5.2 and other similar figures that appear later, since one may want to visualize solutions either in the $x_1 x_2$-plane or as functions of the independent variable t.

EXAMPLE

2

Consider the system

$$\mathbf{x}' = \begin{pmatrix} -3 & \sqrt{2} \\ \sqrt{2} & -2 \end{pmatrix} \mathbf{x}. \tag{14}$$

Draw a direction field for this system; then find its general solution and plot several trajectories in the phase plane.

The direction field for the system (14) in Figure 7.5.3 shows clearly that all solutions approach the origin. To find the solutions, we assume that $\mathbf{x} = \boldsymbol{\xi} e^{rt}$; then we obtain the algebraic system

$$\begin{pmatrix} -3 - r & \sqrt{2} \\ \sqrt{2} & -2 - r \end{pmatrix} \begin{pmatrix} \xi_1 \\ \xi_2 \end{pmatrix} = \begin{pmatrix} 0 \\ 0 \end{pmatrix}. \tag{15}$$

The eigenvalues satisfy

$$(-3 - r)(-2 - r) - 2 = r^2 + 5r + 4$$
$$= (r + 1)(r + 4) = 0, \tag{16}$$

so $r_1 = -1$ and $r_2 = -4$. For $r = -1$, Eq. (15) becomes

$$\begin{pmatrix} -2 & \sqrt{2} \\ \sqrt{2} & -1 \end{pmatrix} \begin{pmatrix} \xi_1 \\ \xi_2 \end{pmatrix} = \begin{pmatrix} 0 \\ 0 \end{pmatrix}. \tag{17}$$

Hence $\xi_2 = \sqrt{2}\, \xi_1$, and the eigenvector $\boldsymbol{\xi}^{(1)}$ corresponding to the eigenvalue $r_1 = -1$ can be taken as

$$\boldsymbol{\xi}^{(1)} = \begin{pmatrix} 1 \\ \sqrt{2} \end{pmatrix}. \tag{18}$$

Similarly, corresponding to the eigenvalue $r_2 = -4$ we have $\xi_1 = -\sqrt{2}\, \xi_2$, so the eigenvector is

$$\boldsymbol{\xi}^{(2)} = \begin{pmatrix} -\sqrt{2} \\ 1 \end{pmatrix}. \tag{19}$$

Thus a fundamental set of solutions of the system (14) is

$$\mathbf{x}^{(1)}(t) = \begin{pmatrix} 1 \\ \sqrt{2} \end{pmatrix} e^{-t}, \qquad \mathbf{x}^{(2)}(t) = \begin{pmatrix} -\sqrt{2} \\ 1 \end{pmatrix} e^{-4t}, \tag{20}$$

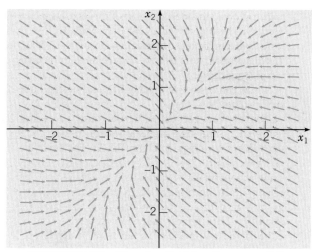

FIGURE 7.5.3 Direction field for the system (14).

and the general solution is

$$\mathbf{x} = c_1 \mathbf{x}^{(1)}(t) + c_2 \mathbf{x}^{(2)} = c_1 \begin{pmatrix} 1 \\ \sqrt{2} \end{pmatrix} e^{-t} + c_2 \begin{pmatrix} -\sqrt{2} \\ 1 \end{pmatrix} e^{-4t}. \tag{21}$$

Graphs of the solution (21) for several values of c_1 and c_2 are shown in Figure 7.5.4a. The solution $\mathbf{x}^{(1)}(t)$ approaches the origin along the line $x_2 = \sqrt{2}\,x_1$, and the solution $\mathbf{x}^{(2)}(t)$ approaches the origin along the line $x_1 = -\sqrt{2}\,x_2$. The directions of these lines are determined by the eigenvectors $\boldsymbol{\xi}^{(1)}$ and $\boldsymbol{\xi}^{(2)}$, respectively. In general, we have a combination of these two fundamental solutions. As $t \to \infty$, the solution $\mathbf{x}^{(2)}(t)$ is negligible compared to $\mathbf{x}^{(1)}(t)$. Thus, unless $c_1 = 0$, the solution (21) approaches the origin tangent to the line $x_2 = \sqrt{2}x_1$. The pattern of trajectories shown in Figure 7.5.4a is typical of all second order systems $\mathbf{x}' = \mathbf{Ax}$ for

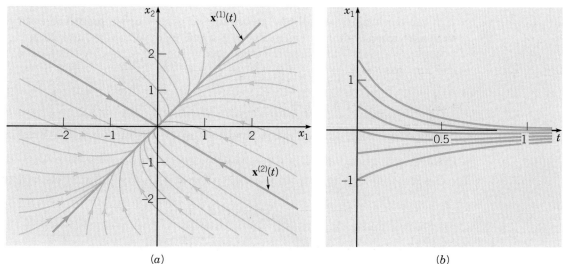

(a) (b)

FIGURE 7.5.4 (a) Trajectories of the system (14); the origin is a node. (b) Plots of x_1 versus t for the system (14).

which the eigenvalues are real, different, and of the same sign. The origin is called a **node** for such a system. If the eigenvalues were positive rather than negative, then the trajectories would be similar but traversed in the outward direction. Nodes are asymptotically stable if the eigenvalues are negative and unstable if the eigenvalues are positive.

Although Figure 7.5.4a was computer-generated, a qualitatively correct sketch of the trajectories can be drawn quickly by hand on the basis of a knowledge of the eigenvalues and eigenvectors.

Some typical plots of x_1 versus t are shown in Figure 7.5.4b. Observe that each of the graphs approaches the t-axis asymptotically as t increases, corresponding to a trajectory that approaches the origin in Figure 7.5.2a. The behavior of x_2 as a function of t is similar.

The two preceding examples illustrate the two main cases for 2×2 systems having eigenvalues that are real and different: The eigenvalues have either opposite signs (Example 1) or the same sign (Example 2). The other possibility is that zero is an eigenvalue, but in this case it follows that $\det \mathbf{A} = 0$, which violates the assumption made at the beginning of this section.

Returning to the general system (1), we proceed as in the examples. To find solutions of the differential equation (1) we must find the eigenvalues and eigenvectors of \mathbf{A} from the associated algebraic system (4). The eigenvalues r_1, \ldots, r_n (which need not all be different) are roots of the nth degree polynomial equation

$$\det(\mathbf{A} - r\mathbf{I}) = 0. \tag{22}$$

The nature of the eigenvalues and the corresponding eigenvectors determines the nature of the general solution of the system (1). If we assume that \mathbf{A} is a real-valued matrix, there are three possibilities for the eigenvalues of \mathbf{A}:

1. All eigenvalues are real and different from each other.
2. Some eigenvalues occur in complex conjugate pairs.
3. Some eigenvalues are repeated.

If the eigenvalues are all real and different, as in the two preceding examples, then associated with each eigenvalue r_i is a real eigenvector $\boldsymbol{\xi}^{(i)}$, and the n eigenvectors $\boldsymbol{\xi}^{(1)}, \ldots, \boldsymbol{\xi}^{(n)}$ are linearly independent. The corresponding solutions of the differential system (1) are

$$\mathbf{x}^{(1)}(t) = \boldsymbol{\xi}^{(1)} e^{r_1 t}, \quad \ldots, \quad \mathbf{x}^{(n)}(t) = \boldsymbol{\xi}^{(n)} e^{r_n t}. \tag{23}$$

To show that these solutions form a fundamental set, we evaluate their Wronskian:

$$W[\mathbf{x}^{(1)}, \ldots, \mathbf{x}^{(n)}](t) = \begin{vmatrix} \xi_1^{(1)} e^{r_1 t} & \cdots & \xi_1^{(n)} e^{r_n t} \\ \vdots & & \vdots \\ \xi_n^{(1)} e^{r_1 t} & \cdots & \xi_n^{(n)} e^{r_n t} \end{vmatrix}$$

$$= e^{(r_1 + \cdots + r_n)t} \begin{vmatrix} \xi_1^{(1)} & \cdots & \xi_1^{(n)} \\ \vdots & & \vdots \\ \xi_n^{(1)} & \cdots & \xi_n^{(n)} \end{vmatrix}. \tag{24}$$

First, we observe that the exponential function is never zero. Next, since the eigen-vectors $\boldsymbol{\xi}^{(1)}, \ldots, \boldsymbol{\xi}^{(n)}$ are linearly independent, the determinant in the last term of Eq. (24) is nonzero. As a consequence, the Wronskian $W[\mathbf{x}^{(1)}, \ldots, \mathbf{x}^{(n)}](t)$ is never zero; hence $\mathbf{x}^{(1)}, \ldots, \mathbf{x}^{(n)}$ form a fundamental set of solutions. Thus the general solution of Eq. (1) is

$$\mathbf{x} = c_1 \boldsymbol{\xi}^{(1)} e^{r_1 t} + \cdots + c_n \boldsymbol{\xi}^{(n)} e^{r_n t}. \tag{25}$$

If \mathbf{A} is real and symmetric (a special case of Hermitian matrices), recall from Section 7.3 that all the eigenvalues r_1, \ldots, r_n must be real. Further, even if some of the eigenvalues are repeated, there is always a full set of n eigenvectors $\boldsymbol{\xi}^{(1)}, \ldots, \boldsymbol{\xi}^{(n)}$ that are linearly independent (in fact, orthogonal). Hence the corresponding solutions of the differential system (1) given by Eq. (23) again form a fundamental set of solutions, and the general solution is again given by Eq. (25). The following example illustrates this case.

**EXAMPLE
3**

Find the general solution of

$$\mathbf{x}' = \begin{pmatrix} 0 & 1 & 1 \\ 1 & 0 & 1 \\ 1 & 1 & 0 \end{pmatrix} \mathbf{x}. \tag{26}$$

Observe that the coefficient matrix is real and symmetric. The eigenvalues and eigenvectors of this matrix were found in Example 5 of Section 7.3:

$$r_1 = 2, \qquad \boldsymbol{\xi}^{(1)} = \begin{pmatrix} 1 \\ 1 \\ 1 \end{pmatrix}; \tag{27}$$

$$r_2 = -1, \qquad r_3 = -1; \qquad \boldsymbol{\xi}^{(2)} = \begin{pmatrix} 1 \\ 0 \\ -1 \end{pmatrix}, \qquad \boldsymbol{\xi}^{(3)} = \begin{pmatrix} 0 \\ 1 \\ -1 \end{pmatrix}. \tag{28}$$

Hence a fundamental set of solutions of Eq. (26) is

$$\mathbf{x}^{(1)}(t) = \begin{pmatrix} 1 \\ 1 \\ 1 \end{pmatrix} e^{2t}, \qquad \mathbf{x}^{(2)}(t) = \begin{pmatrix} 1 \\ 0 \\ -1 \end{pmatrix} e^{-t}, \qquad \mathbf{x}^{(3)}(t) = \begin{pmatrix} 0 \\ 1 \\ -1 \end{pmatrix} e^{-t}, \tag{29}$$

and the general solution is

$$\mathbf{x} = c_1 \begin{pmatrix} 1 \\ 1 \\ 1 \end{pmatrix} e^{2t} + c_2 \begin{pmatrix} 1 \\ 0 \\ -1 \end{pmatrix} e^{-t} + c_3 \begin{pmatrix} 0 \\ 1 \\ -1 \end{pmatrix} e^{-t}. \tag{30}$$

This example illustrates the fact that even though an eigenvalue ($r = -1$) has algebraic mul-tiplicity 2, it may still be possible to find two linearly independent eigenvectors $\boldsymbol{\xi}^{(2)}$ and $\boldsymbol{\xi}^{(3)}$ and, as a consequence, to construct the general solution (30).

The behavior of the solution (30) depends critically on the initial conditions. For large t the first term on the right side of Eq. (30) is the dominant one; therefore, if $c_1 \neq 0$, all components of \mathbf{x} become unbounded as $t \to \infty$. On the other hand, for certain initial points c_1 will be zero. In this case, the solution involves only the negative exponential terms, and $\mathbf{x} \to \mathbf{0}$ as $t \to \infty$. The initial points that cause c_1 to be zero are precisely those that lie in the plane

determined by the eigenvectors $\boldsymbol{\xi}^{(2)}$ and $\boldsymbol{\xi}^{(3)}$ corresponding to the two negative eigenvalues. Thus solutions that start in this plane approach the origin as $t \to \infty$, while all other solutions become unbounded.

If some of the eigenvalues occur in complex conjugate pairs, then there are still n linearly independent solutions of the form (23), provided that all the eigenvalues are different. Of course, the solutions arising from complex eigenvalues are complex-valued. However, as in Section 3.4, it is possible to obtain a full set of real-valued solutions. This is discussed in Section 7.6.

More serious difficulties can occur if an eigenvalue is repeated. In this event the number of corresponding linearly independent eigenvectors may be smaller than the algebraic multiplicity of the eigenvalue. If so, the number of linearly independent solutions of the form $\boldsymbol{\xi} e^{rt}$ will be smaller than n. To construct a fundamental set of solutions, it is then necessary to seek additional solutions of another form. The situation is somewhat analogous to that for an nth order linear equation with constant coefficients; a repeated root of the characteristic equation gave rise to solutions of the form $e^{rt}, te^{rt}, t^2 e^{rt}, \ldots$. The case of repeated eigenvalues is treated in Section 7.8.

Finally, if \mathbf{A} is complex, then complex eigenvalues need not occur in conjugate pairs, and the eigenvectors are normally complex-valued even though the associated eigenvalue may be real. The solutions of the differential equation (1) are still of the form (23), provided that the eigenvalues are distinct, but in general all the solutions are complex-valued.

PROBLEMS

In each of Problems 1 through 6 find the general solution of the given system of equations and describe the behavior of the solution as $t \to \infty$. Also draw a direction field and plot a few trajectories of the system.

1. $\mathbf{x}' = \begin{pmatrix} 3 & -2 \\ 2 & -2 \end{pmatrix} \mathbf{x}$

2. $\mathbf{x}' = \begin{pmatrix} 1 & -2 \\ 3 & -4 \end{pmatrix} \mathbf{x}$

3. $\mathbf{x}' = \begin{pmatrix} 2 & -1 \\ 3 & -2 \end{pmatrix} \mathbf{x}$

4. $\mathbf{x}' = \begin{pmatrix} 1 & 1 \\ 4 & -2 \end{pmatrix} \mathbf{x}$

5. $\mathbf{x}' = \begin{pmatrix} -2 & 1 \\ 1 & -2 \end{pmatrix} \mathbf{x}$

6. $\mathbf{x}' = \begin{pmatrix} \frac{5}{4} & \frac{3}{4} \\ \frac{3}{4} & \frac{5}{4} \end{pmatrix} \mathbf{x}$

In each of Problems 7 and 8 find the general solution of the given system of equations. Also draw a direction field and a few of the trajectories. In each of these problems the coefficient matrix has a zero eigenvalue. As a result, the pattern of trajectories is different from those in the examples in the text.

7. $\mathbf{x}' = \begin{pmatrix} 4 & -3 \\ 8 & -6 \end{pmatrix} \mathbf{x}$

8. $\mathbf{x}' = \begin{pmatrix} 3 & 6 \\ -1 & -2 \end{pmatrix} \mathbf{x}$

In each of Problems 9 through 14 find the general solution of the given system of equations.

9. $\mathbf{x}' = \begin{pmatrix} 1 & i \\ -i & 1 \end{pmatrix} \mathbf{x}$

10. $\mathbf{x}' = \begin{pmatrix} 2 & 2+i \\ -1 & -1-i \end{pmatrix} \mathbf{x}$

11. $\mathbf{x}' = \begin{pmatrix} 1 & 1 & 2 \\ 1 & 2 & 1 \\ 2 & 1 & 1 \end{pmatrix} \mathbf{x}$

12. $\mathbf{x}' = \begin{pmatrix} 3 & 2 & 4 \\ 2 & 0 & 2 \\ 4 & 2 & 3 \end{pmatrix} \mathbf{x}$

13. $\mathbf{x}' = \begin{pmatrix} 1 & 1 & 1 \\ 2 & 1 & -1 \\ -8 & -5 & -3 \end{pmatrix} \mathbf{x}$

14. $\mathbf{x}' = \begin{pmatrix} 1 & -1 & 4 \\ 3 & 2 & -1 \\ 2 & 1 & -1 \end{pmatrix} \mathbf{x}$

In each of Problems 15 through 18 solve the given initial value problem. Describe the behavior of the solution as $t \to \infty$.

15. $\mathbf{x}' = \begin{pmatrix} 5 & -1 \\ 3 & 1 \end{pmatrix} \mathbf{x}, \qquad \mathbf{x}(0) = \begin{pmatrix} 2 \\ -1 \end{pmatrix}$

16. $\mathbf{x}' = \begin{pmatrix} -2 & 1 \\ -5 & 4 \end{pmatrix} \mathbf{x}, \qquad \mathbf{x}(0) = \begin{pmatrix} 1 \\ 3 \end{pmatrix}$

17. $\mathbf{x}' = \begin{pmatrix} 1 & 1 & 2 \\ 0 & 2 & 2 \\ -1 & 1 & 3 \end{pmatrix} \mathbf{x}, \qquad \mathbf{x}(0) = \begin{pmatrix} 2 \\ 0 \\ 1 \end{pmatrix}$

18. $\mathbf{x}' = \begin{pmatrix} 0 & 0 & -1 \\ 2 & 0 & 0 \\ -1 & 2 & 4 \end{pmatrix} \mathbf{x}, \qquad \mathbf{x}(0) = \begin{pmatrix} 7 \\ 5 \\ 5 \end{pmatrix}$

19. The system $t\mathbf{x}' = \mathbf{A}\mathbf{x}$ is analogous to the second order Euler equation (Section 5.5). Assuming that $\mathbf{x} = \boldsymbol{\xi} t^r$, where $\boldsymbol{\xi}$ is a constant vector, show that $\boldsymbol{\xi}$ and r must satisfy $(\mathbf{A} - r\mathbf{I})\boldsymbol{\xi} = \mathbf{0}$ in order to obtain nontrivial solutions of the given differential equation.

Referring to Problem 19, solve the given system of equations in each of Problems 20 through 23. Assume that $t > 0$.

20. $t\mathbf{x}' = \begin{pmatrix} 2 & -1 \\ 3 & -2 \end{pmatrix} \mathbf{x}$

21. $t\mathbf{x}' = \begin{pmatrix} 5 & -1 \\ 3 & 1 \end{pmatrix} \mathbf{x}$

22. $t\mathbf{x}' = \begin{pmatrix} 4 & -3 \\ 8 & -6 \end{pmatrix} \mathbf{x}$

23. $t\mathbf{x}' = \begin{pmatrix} 3 & -2 \\ 2 & -2 \end{pmatrix} \mathbf{x}$

In each of Problems 24 through 27 the eigenvalues and eigenvectors of a matrix \mathbf{A} are given. Consider the corresponding system $\mathbf{x}' = \mathbf{A}\mathbf{x}$.

(a) Sketch a phase portrait of the system.
(b) Sketch the trajectory passing through the initial point $(2, 3)$.
(c) For the trajectory in part (b) sketch the graphs of x_1 versus t and of x_2 versus t on the same set of axes.

24. $r_1 = -1, \quad \boldsymbol{\xi}^{(1)} = \begin{pmatrix} -1 \\ 2 \end{pmatrix}; \qquad r_2 = -2, \quad \boldsymbol{\xi}^{(2)} = \begin{pmatrix} 1 \\ 2 \end{pmatrix}$

25. $r_1 = 1, \quad \boldsymbol{\xi}^{(1)} = \begin{pmatrix} -1 \\ 2 \end{pmatrix}; \qquad r_2 = -2, \quad \boldsymbol{\xi}^{(2)} = \begin{pmatrix} 1 \\ 2 \end{pmatrix}$

26. $r_1 = -1, \quad \boldsymbol{\xi}^{(1)} = \begin{pmatrix} -1 \\ 2 \end{pmatrix}; \qquad r_2 = 2, \quad \boldsymbol{\xi}^{(2)} = \begin{pmatrix} 1 \\ 2 \end{pmatrix}$

27. $r_1 = 1, \quad \boldsymbol{\xi}^{(1)} = \begin{pmatrix} 1 \\ 2 \end{pmatrix}; \qquad r_2 = 2, \quad \boldsymbol{\xi}^{(2)} = \begin{pmatrix} 1 \\ -2 \end{pmatrix}$

28. Consider a 2×2 system $\mathbf{x}' = \mathbf{A}\mathbf{x}$. If we assume that $r_1 \neq r_2$, the general solution is $\mathbf{x} = c_1 \boldsymbol{\xi}^{(1)} e^{r_1 t} + c_2 \boldsymbol{\xi}^{(2)} e^{r_2 t}$, provided that $\boldsymbol{\xi}^{(1)}$ and $\boldsymbol{\xi}^{(2)}$ are linearly independent. In this problem we establish the linear independence of $\boldsymbol{\xi}^{(1)}$ and $\boldsymbol{\xi}^{(2)}$ by assuming that they are linearly dependent, and then showing that this leads to a contradiction.

(a) Note that $\boldsymbol{\xi}^{(1)}$ satisfies the matrix equation $(\mathbf{A} - r_1\mathbf{I})\boldsymbol{\xi}^{(1)} = \mathbf{0}$; similarly, note that $(\mathbf{A} - r_2\mathbf{I})\boldsymbol{\xi}^{(2)} = \mathbf{0}$.

(b) Show that $(\mathbf{A} - r_2\mathbf{I})\boldsymbol{\xi}^{(1)} = (r_1 - r_2)\boldsymbol{\xi}^{(1)}$.

(c) Suppose that $\boldsymbol{\xi}^{(1)}$ and $\boldsymbol{\xi}^{(2)}$ are linearly dependent. Then $c_1\boldsymbol{\xi}^{(1)} + c_2\boldsymbol{\xi}^{(2)} = \mathbf{0}$ and at least one of c_1 and c_2 is not zero; suppose that $c_1 \neq 0$. Show that $(\mathbf{A} - r_2\mathbf{I})(c_1\boldsymbol{\xi}^{(1)} + c_2\boldsymbol{\xi}^{(2)}) = \mathbf{0}$, and also show that $(\mathbf{A} - r_2\mathbf{I})(c_1\boldsymbol{\xi}^{(1)} + c_2\boldsymbol{\xi}^{(2)}) = c_1(r_1 - r_2)\boldsymbol{\xi}^{(1)}$. Hence $c_1 = 0$, which is a contradiction. Therefore $\boldsymbol{\xi}^{(1)}$ and $\boldsymbol{\xi}^{(2)}$ are linearly independent.

(d) Modify the argument of part (c) if we assume that $c_2 \neq 0$.

(e) Carry out a similar argument for the case in which the order n is equal to 3; note that the procedure can be extended to an arbitrary value of n.

29. Consider the equation

$$ay'' + by' + cy = 0, \tag{i}$$

where a, b, and c are constants with $a \neq 0$. In Chapter 3 it was shown that the general solution depended on the roots of the characteristic equation

$$ar^2 + br + c = 0. \tag{ii}$$

(a) Transform Eq. (i) into a system of first order equations by letting $x_1 = y$, $x_2 = y'$. Find the system of equations $\mathbf{x}' = \mathbf{A}\mathbf{x}$ satisfied by $\mathbf{x} = \begin{pmatrix} x_1 \\ x_2 \end{pmatrix}$.

(b) Find the equation that determines the eigenvalues of the coefficient matrix \mathbf{A} in part (a). Note that this equation is just the characteristic equation (ii) of Eq. (i).

30. The two-tank system of Problem 22 in Section 7.1 leads to the initial value problem

$$\mathbf{x}' = \begin{pmatrix} -\frac{1}{10} & \frac{3}{40} \\ \frac{1}{10} & -\frac{1}{5} \end{pmatrix} \mathbf{x}, \qquad \mathbf{x}(0) = \begin{pmatrix} -17 \\ -21 \end{pmatrix},$$

where x_1 and x_2 are the deviations of the salt levels Q_1 and Q_2 from their respective equilibria.

(a) Find the solution of the given initial value problem.

(b) Plot x_1 versus t and x_2 versus t on the same set of axes.

(c) Find the time T such that $|x_1(t)| \leq 0.5$ and $|x_2(t)| \leq 0.5$ for all $t \geq T$.

31. Consider the system

$$\mathbf{x}' = \begin{pmatrix} -1 & -1 \\ -\alpha & -1 \end{pmatrix} \mathbf{x}.$$

(a) Solve the system for $\alpha = 0.5$. What are the eigenvalues of the coefficient matrix? Classify the equilibrium point at the origin as to type.

(b) Solve the system for $\alpha = 2$. What are the eigenvalues of the coefficient matrix? Classify the equilibrium point at the origin as to type.

(c) In parts (a) and (b) solutions of the system exhibit two quite different types of behavior. Find the eigenvalues of the coefficient matrix in terms of α and determine the value of α between 0.5 and 2 where the transition from one type of behavior to the other occurs.

Electric Circuits. Problems 32 and 33 are concerned with the electric circuit described by the system of differential equations in Problem 21 of Section 7.1:

$$\frac{d}{dt}\begin{pmatrix} I \\ V \end{pmatrix} = \begin{pmatrix} -\dfrac{R_1}{L} & -\dfrac{1}{L} \\ \dfrac{1}{C} & -\dfrac{1}{CR_2} \end{pmatrix}\begin{pmatrix} I \\ V \end{pmatrix}. \tag{i}$$

32. (a) Find the general solution of Eq. (i) if $R_1 = 1$ ohm, $R_2 = \frac{3}{5}$ ohm, $L = 2$ henrys, and $C = \frac{2}{3}$ farad.

 (b) Show that $I(t) \to 0$ and $V(t) \to 0$ as $t \to \infty$ regardless of the initial values $I(0)$ and $V(0)$.

33. Consider the preceding system of differential equations (i).

 (a) Find a condition on R_1, R_2, C, and L that must be satisfied if the eigenvalues of the coefficient matrix are to be real and different.

 (b) If the condition found in part (a) is satisfied, show that both eigenvalues are negative. Then show that $I(t) \to 0$ and $V(t) \to 0$ as $t \to \infty$ regardless of the initial conditions.

 (c) If the condition found in part (a) is not satisfied, then the eigenvalues are either complex or repeated. Do you think that $I(t) \to 0$ and $V(t) \to 0$ as $t \to \infty$ in these cases as well?

 Hint: In part (c) one approach is to change the system (i) into a single second order equation. We also discuss complex and repeated eigenvalues in Sections 7.6 and 7.8.

7.6 Complex Eigenvalues

In this section we consider again a system of n linear homogeneous equations with constant coefficients

$$\mathbf{x}' = \mathbf{Ax}, \tag{1}$$

where the coefficient matrix \mathbf{A} is real-valued. If we seek solutions of the form $\mathbf{x} = \boldsymbol{\xi}e^{rt}$, then it follows, as in Section 7.5, that r must be an eigenvalue and $\boldsymbol{\xi}$ a corresponding eigenvector of the coefficient matrix \mathbf{A}. Recall that the eigenvalues r_1, \ldots, r_n of \mathbf{A} are the roots of the equation

$$\det(\mathbf{A} - r\mathbf{I}) = 0, \tag{2}$$

and that the corresponding eigenvectors satisfy

$$(\mathbf{A} - r\mathbf{I})\boldsymbol{\xi} = \mathbf{0}. \tag{3}$$

If \mathbf{A} is real, then the coefficients in the polynomial equation (2) for r are real, and any complex eigenvalues must occur in conjugate pairs. For example, if $r_1 = \lambda + i\mu$, where λ and μ are real, is an eigenvalue of \mathbf{A}, then so is $r_2 = \lambda - i\mu$. Further, the corresponding eigenvectors $\boldsymbol{\xi}^{(1)}$ and $\boldsymbol{\xi}^{(2)}$ are also complex conjugates. To see that this is so, suppose that r_1 and $\boldsymbol{\xi}^{(1)}$ satisfy

$$(\mathbf{A} - r_1\mathbf{I})\boldsymbol{\xi}^{(1)} = \mathbf{0}. \tag{4}$$

On taking the complex conjugate of this equation and noting that \mathbf{A} and \mathbf{I} are real-valued, we obtain

$$(\mathbf{A} - \bar{r}_1\mathbf{I})\bar{\boldsymbol{\xi}}^{(1)} = \mathbf{0}, \tag{5}$$

where \bar{r}_1 and $\bar{\boldsymbol{\xi}}^{(1)}$ are the complex conjugates of r_1 and $\boldsymbol{\xi}^{(1)}$, respectively. In other words, $r_2 = \bar{r}_1$ is also an eigenvalue, and $\boldsymbol{\xi}^{(2)} = \bar{\boldsymbol{\xi}}^{(1)}$ is a corresponding eigenvector. The corresponding solutions

$$\mathbf{x}^{(1)}(t) = \boldsymbol{\xi}^{(1)}e^{r_1 t}, \qquad \mathbf{x}^{(2)}(t) = \bar{\boldsymbol{\xi}}^{(1)}e^{\bar{r}_1 t} \tag{6}$$

of the differential equation (1) are then complex conjugates of each other. Therefore, as in Section 3.4, we can find two real-valued solutions of Eq. (1) corresponding to the eigenvalues r_1 and r_2 by taking the real and imaginary parts of $\mathbf{x}^{(1)}(t)$ or $\mathbf{x}^{(2)}(t)$ given by Eq. (6).

Let us write $\boldsymbol{\xi}^{(1)} = \mathbf{a} + i\mathbf{b}$, where \mathbf{a} and \mathbf{b} are real; then we have

$$\mathbf{x}^{(1)}(t) = (\mathbf{a} + i\mathbf{b})e^{(\lambda + i\mu)t}$$

$$= (\mathbf{a} + i\mathbf{b})e^{\lambda t}(\cos \mu t + i \sin \mu t). \tag{7}$$

Upon separating $\mathbf{x}^{(1)}(t)$ into its real and imaginary parts, we obtain

$$\mathbf{x}^{(1)}(t) = e^{\lambda t}(\mathbf{a} \cos \mu t - \mathbf{b} \sin \mu t) + ie^{\lambda t}(\mathbf{a} \sin \mu t + \mathbf{b} \cos \mu t). \tag{8}$$

If we write $\mathbf{x}^{(1)}(t) = \mathbf{u}(t) + i\mathbf{v}(t)$, then the vectors

$$\begin{aligned}\mathbf{u}(t) &= e^{\lambda t}(\mathbf{a} \cos \mu t - \mathbf{b} \sin \mu t), \\ \mathbf{v}(t) &= e^{\lambda t}(\mathbf{a} \sin \mu t + \mathbf{b} \cos \mu t)\end{aligned} \tag{9}$$

are real-valued solutions of Eq. (1). It is possible to show that \mathbf{u} and \mathbf{v} are linearly independent solutions (see Problem 27).

For example, suppose that $r_1 = \lambda + i\mu$, $r_2 = \lambda - i\mu$, and that r_3, \ldots, r_n are all real and distinct. Let the corresponding eigenvectors be $\boldsymbol{\xi}^{(1)} = \mathbf{a} + i\mathbf{b}$, $\boldsymbol{\xi}^{(2)} = \mathbf{a} - i\mathbf{b}$, $\boldsymbol{\xi}^{(3)}, \ldots, \boldsymbol{\xi}^{(n)}$. Then the general solution of Eq. (1) is

$$\mathbf{x} = c_1\mathbf{u}(t) + c_2\mathbf{v}(t) + c_3\boldsymbol{\xi}^{(3)}e^{r_3 t} + \cdots + c_n\boldsymbol{\xi}^{(n)}e^{r_n t}, \tag{10}$$

where $\mathbf{u}(t)$ and $\mathbf{v}(t)$ are given by Eqs. (9). We emphasize that this analysis applies only if the coefficient matrix \mathbf{A} in Eq. (1) is real, for it is only then that complex eigenvalues and eigenvectors must occur in conjugate pairs.

The following example illustrates the case $n = 2$, both to simplify the algebraic calculations and to permit easy visualization of the solutions in the phase plane.

EXAMPLE 1

Find a fundamental set of real-valued solutions of the system

$$\mathbf{x}' = \begin{pmatrix} -\frac{1}{2} & 1 \\ -1 & -\frac{1}{2} \end{pmatrix}\mathbf{x} \tag{11}$$

and display them graphically.

A direction field for the system (11) is shown in Figure 7.6.1. This plot suggests that the trajectories in the phase plane spiral clockwise toward the origin.

To find a fundamental set of solutions we assume that

$$\mathbf{x} = \boldsymbol{\xi}e^{rt} \tag{12}$$

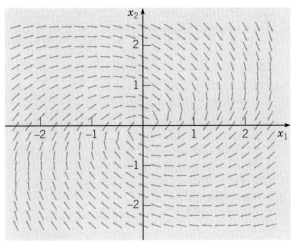

FIGURE 7.6.1 A direction field for the system (11).

and obtain the set of linear algebraic equations

$$\begin{pmatrix} -\frac{1}{2} - r & 1 \\ -1 & -\frac{1}{2} - r \end{pmatrix} \begin{pmatrix} \xi_1 \\ \xi_2 \end{pmatrix} = \begin{pmatrix} 0 \\ 0 \end{pmatrix} \tag{13}$$

for the eigenvalues and eigenvectors of **A**. The characteristic equation is

$$\begin{vmatrix} -\frac{1}{2} - r & 1 \\ -1 & -\frac{1}{2} - r \end{vmatrix} = r^2 + r + \frac{5}{4} = 0; \tag{14}$$

therefore the eigenvalues are $r_1 = -\frac{1}{2} + i$ and $r_2 = -\frac{1}{2} - i$. From Eq. (13) a straightforward calculation shows that the corresponding eigenvectors are

$$\boldsymbol{\xi}^{(1)} = \begin{pmatrix} 1 \\ i \end{pmatrix}, \qquad \boldsymbol{\xi}^{(2)} = \begin{pmatrix} 1 \\ -i \end{pmatrix}. \tag{15}$$

Hence a fundamental set of solutions of the system (11) is

$$\mathbf{x}^{(1)}(t) = \begin{pmatrix} 1 \\ i \end{pmatrix} e^{(-1/2+i)t}, \qquad \mathbf{x}^{(2)}(t) = \begin{pmatrix} 1 \\ -i \end{pmatrix} e^{(-1/2-i)t}. \tag{16}$$

To obtain a set of real-valued solutions, we must find the real and imaginary parts of either $\mathbf{x}^{(1)}$ or $\mathbf{x}^{(2)}$. In fact,

$$\mathbf{x}^{(1)}(t) = \begin{pmatrix} 1 \\ i \end{pmatrix} e^{-t/2}(\cos t + i \sin t) = \begin{pmatrix} e^{-t/2} \cos t \\ -e^{-t/2} \sin t \end{pmatrix} + i \begin{pmatrix} e^{-t/2} \sin t \\ e^{-t/2} \cos t \end{pmatrix}. \tag{17}$$

Hence

$$\mathbf{u}(t) = e^{-t/2} \begin{pmatrix} \cos t \\ -\sin t \end{pmatrix}, \qquad \mathbf{v}(t) = e^{-t/2} \begin{pmatrix} \sin t \\ \cos t \end{pmatrix} \tag{18}$$

is a set of real-valued solutions. To verify that $\mathbf{u}(t)$ and $\mathbf{v}(t)$ are linearly independent, we compute their Wronskian:

$$W(\mathbf{u}, \mathbf{v})(t) = \begin{vmatrix} e^{-t/2} \cos t & e^{-t/2} \sin t \\ -e^{-t/2} \sin t & e^{-t/2} \cos t \end{vmatrix}$$

$$= e^{-t}(\cos^2 t + \sin^2 t) = e^{-t}.$$

Since the Wronskian is never zero, it follows that $\mathbf{u}(t)$ and $\mathbf{v}(t)$ constitute a fundamental set of (real-valued) solutions of the system (11).

The graphs of the solutions $\mathbf{u}(t)$ and $\mathbf{v}(t)$ are shown in Figure 7.6.2a. Since

$$\mathbf{u}(0) = \begin{pmatrix} 1 \\ 0 \end{pmatrix}, \qquad \mathbf{v}(0) = \begin{pmatrix} 0 \\ 1 \end{pmatrix},$$

the graphs of $\mathbf{u}(t)$ and $\mathbf{v}(t)$ pass through the points $(1, 0)$ and $(0, 1)$, respectively. Other solutions of the system (11) are linear combinations of $\mathbf{u}(t)$ and $\mathbf{v}(t)$, and graphs of a few of these solutions are also shown in Figure 7.6.2a. Each trajectory approaches the origin along a spiral path as $t \to \infty$, making infinitely many circuits about the origin; this is due to the fact that the solutions (18) are products of decaying exponential and sine or cosine factors. Some typical graphs of x_1 versus t are shown in Figure 7.6.2b; each one represents a decaying oscillation in time.

Figure 7.6.2a is typical of all second order systems $\mathbf{x}' = \mathbf{A}\mathbf{x}$ whose eigenvalues are complex with negative real part. The origin is called a **spiral point** and is asymptotically stable because all trajectories approach it as t increases. For a system whose eigenvalues have a positive real part, the trajectories are similar to those in Figure 7.6.2a, but the direction of motion is away from the origin and the trajectories become unbounded. In this case, the origin is unstable. If the real part of the eigenvalues is zero, then the trajectories neither approach the origin nor become unbounded but instead traverse repeatedly a closed curve about the origin. Examples of this behavior can be seen in Figures 7.6.3b and 7.6.4b below. In this case the origin is called a **center** and is said to be stable, but not asymptotically stable. In all three cases the direction of motion may be either clockwise, as in this example, or counterclockwise, depending on the elements of the coefficient matrix \mathbf{A}.

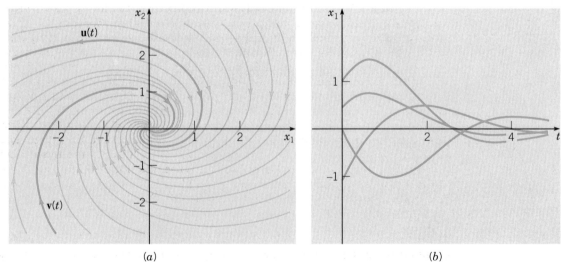

(a) (b)

FIGURE 7.6.2 (a) Trajectories of the system (11); the origin is a spiral point. (b) Plots of x_1 versus t for the system (11).

For second order systems with real coefficients, we have now completed our description of the three main cases that can occur.

1. Eigenvalues are real and have opposite signs; $\mathbf{x} = \mathbf{0}$ is a saddle point.
2. Eigenvalues are real and have the same sign but are unequal; $\mathbf{x} = \mathbf{0}$ is a node.
3. Eigenvalues are complex with nonzero real part; $\mathbf{x} = \mathbf{0}$ is a spiral point.

Other possibilities are of less importance and occur as transitions between two of the cases just listed. For example, a zero eigenvalue occurs during the transition between a saddle point and a node. Purely imaginary eigenvalues occur during a transition between asymptotically stable and unstable spiral points. Finally, real and equal eigenvalues appear during the transition between nodes and spiral points.

EXAMPLE 2

The system

$$\mathbf{x}' = \begin{pmatrix} \alpha & 2 \\ -2 & 0 \end{pmatrix} \mathbf{x} \tag{19}$$

contains a parameter α. Describe how the solutions depend qualitatively on α; in particular, find the critical values of α at which the qualitative behavior of the trajectories in the phase plane changes markedly.

The behavior of the trajectories is controlled by the eigenvalues of the coefficient matrix. The characteristic equation is

$$r^2 - \alpha r + 4 = 0, \tag{20}$$

so the eigenvalues are

$$r = \frac{\alpha \pm \sqrt{\alpha^2 - 16}}{2}. \tag{21}$$

From Eq. (21) it follows that the eigenvalues are complex conjugates for $-4 < \alpha < 4$ and are real otherwise. Thus two critical values are $\alpha = -4$ and $\alpha = 4$, where the eigenvalues change from real to complex, or vice versa. For $\alpha < -4$ both eigenvalues are negative, so all trajectories approach the origin, which is an asymptotically stable node. For $\alpha > 4$ both eigenvalues are positive, so the origin is again a node, this time unstable; all trajectories (except $\mathbf{x} = \mathbf{0}$) become unbounded. In the intermediate range, $-4 < \alpha < 4$, the eigenvalues are complex and the trajectories are spirals. However, for $-4 < \alpha < 0$ the real part of the eigenvalues is negative, the spirals are directed inward, and the origin is asymptotically stable, whereas for $0 < \alpha < 4$ the real part of the eigenvalues is positive and the origin is unstable. Thus $\alpha = 0$ is also a critical value where the direction of the spirals changes from inward to outward. For this value of α, the origin is a center and the trajectories are closed curves about the origin, corresponding to solutions that are periodic in time. The other critical values, $\alpha = \pm 4$, yield eigenvalues that are real and equal. In this case the origin is again a node, but the phase portrait differs somewhat from those in Section 7.5. We take up this case in Section 7.8.

A Multiple Spring–Mass System. Consider the system of two masses and three springs shown in Figure 7.1.1, whose equations of motion are given by Eqs. (1) in Section 7.1. If we assume that there are no external forces, then $F_1(t) = 0, F_2(t) = 0$, and the

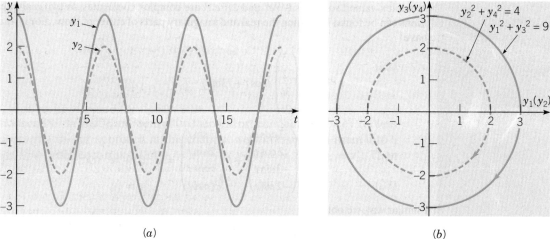

(a) (b)

FIGURE 7.6.3 (*a*) A plot of y_1 versus t and y_2 versus t for the solution $\mathbf{u}^{(1)}(t)$. (*b*) Superposition of projections of trajectories in the y_1y_3- and y_2y_4-planes for the solution $\mathbf{u}^{(1)}(t)$.

y_1y_3- and y_2y_4-planes. Similar graphs (with an appropriate shift in time) are obtained from $\mathbf{v}^{(1)}$ or from a linear combination of $\mathbf{u}^{(1)}$ and $\mathbf{v}^{(1)}$.

The remaining terms on the right side of Eq. (31) describe motions with frequency 2 and period π. Observe that in this case $y_2 = -(4/3)y_1$ and $y_4 = -(4/3)y_3$. This means that the two masses are always moving in opposite directions and that the second mass moves four-thirds as far as the first mass. If we look only at $\mathbf{u}^{(2)}(t)$ and plot y_1 versus t and y_2 versus t on the same axes, we obtain Figure 7.6.4*a*. There is a phase difference of π, and the amplitude of y_2 is four-thirds that of y_1, confirming the preceding statements about the motions of the masses. Figure 7.6.4*b* shows a superposition of the trajectories for the two masses in their respective phase planes. Both graphs are ellipses, the inner one corresponding to the first mass and the outer one to the second. The trajectory on the inner ellipse starts at $(3, 0)$, and the trajectory

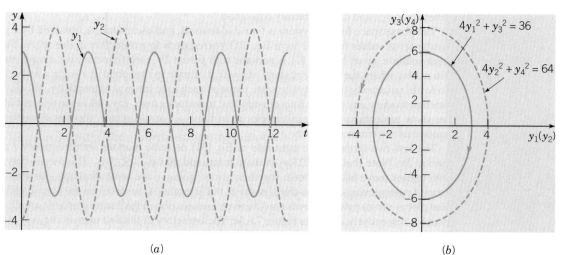

(a) (b)

FIGURE 7.6.4 (*a*) A plot of y_1 versus t and y_2 versus t for the solution $\mathbf{u}^{(2)}(t)$. (*b*) Superposition of projections of trajectories in the y_1y_3- and y_2y_4-planes for the solution $\mathbf{u}^{(2)}(t)$.

EXAMP
3

on the outer ellipse starts at $(-4, 0)$. Both are traversed clockwise, and a circuit is completed in time π. The origin is a center in the respective y_1y_3- and y_2y_4-planes. Once again, similar graphs are obtained from $\mathbf{v}^{(2)}$ or from a linear combination of $\mathbf{u}^{(2)}$ and $\mathbf{v}^{(2)}$.

The types of motion described in the two preceding paragraphs are called **fundamental modes** of vibration for the two-mass system. Each of them results from fairly special initial conditions. For example, to obtain the fundamental mode of frequency 1, both of the constants c_3 and c_4 in Eq. (31) must be zero. This occurs only for initial conditions in which $3y_2(0) = 2y_1(0)$ and $3y_4(0) = 2y_3(0)$. Similarly, the mode of frequency 2 is obtained only when both of the constants c_1 and c_2 in Eq. (31) are zero—that is, when the initial conditions are such that $3y_2(0) = -4y_1(0)$ and $3y_4(0) = -4y_3(0)$.

For more general initial conditions the solution is a combination of the two fundamental modes. A plot of y_1 versus t for a typical case is shown in Figure 7.6.5a, and the projection of the corresponding trajectory in the y_1y_3-plane is shown in Figure 7.6.5b. Observe that this latter figure may be a bit misleading in that it shows the projection of the trajectory crossing itself. This cannot be the case for the actual trajectory in four dimensions, because it would violate the general uniqueness theorem: There cannot be two different solutions issuing from the same initial point.

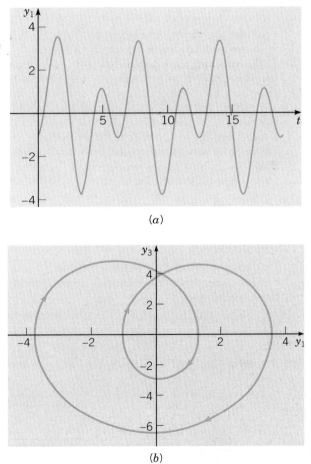

(a)

(b)

FIGURE 7.6.5 A solution of the system (22) satisfying the initial condition $\mathbf{y}(0) = (-1, 4, 1, 1)$. (a) A plot of y_1 versus t. (b) The projection of the trajectory in the y_1y_3-plane.

PROBLEMS In each of Problems 1 through 8 express the general solution of the given system of equations in terms of real-valued functions. In each of Problems 1 through 6 also draw a direction field, sketch a few of the trajectories, and describe the behavior of the solutions as $t \to \infty$.

1. $\mathbf{x}' = \begin{pmatrix} 3 & -2 \\ 4 & -1 \end{pmatrix} \mathbf{x}$

2. $\mathbf{x}' = \begin{pmatrix} -1 & -4 \\ 1 & -1 \end{pmatrix} \mathbf{x}$

3. $\mathbf{x}' = \begin{pmatrix} 2 & -5 \\ 1 & -2 \end{pmatrix} \mathbf{x}$

4. $\mathbf{x}' = \begin{pmatrix} 2 & -\frac{5}{2} \\ \frac{9}{5} & -1 \end{pmatrix} \mathbf{x}$

5. $\mathbf{x}' = \begin{pmatrix} 1 & -1 \\ 5 & -3 \end{pmatrix} \mathbf{x}$

6. $\mathbf{x}' = \begin{pmatrix} 1 & 2 \\ -5 & -1 \end{pmatrix} \mathbf{x}$

7. $\mathbf{x}' = \begin{pmatrix} 1 & 0 & 0 \\ 2 & 1 & -2 \\ 3 & 2 & 1 \end{pmatrix} \mathbf{x}$

8. $\mathbf{x}' = \begin{pmatrix} -3 & 0 & 2 \\ 1 & -1 & 0 \\ -2 & -1 & 0 \end{pmatrix} \mathbf{x}$

In each of Problems 9 and 10 find the solution of the given initial value problem. Describe the behavior of the solution as $t \to \infty$.

9. $\mathbf{x}' = \begin{pmatrix} 1 & -5 \\ 1 & -3 \end{pmatrix} \mathbf{x}, \qquad \mathbf{x}(0) = \begin{pmatrix} 1 \\ 1 \end{pmatrix}$

10. $\mathbf{x}' = \begin{pmatrix} -3 & 2 \\ -1 & -1 \end{pmatrix} \mathbf{x}, \qquad \mathbf{x}(0) = \begin{pmatrix} 1 \\ -2 \end{pmatrix}$

In each of Problems 11 and 12:
(a) Find the eigenvalues of the given system.
(b) Choose an initial point (other than the origin) and draw the corresponding trajectory in the $x_1 x_2$-plane.
(c) For your trajectory in part (b) draw the graphs of x_1 versus t and of x_2 versus t.
(d) For your trajectory in part (b) draw the corresponding graph in three-dimensional $t x_1 x_2$-space.

11. $\mathbf{x}' = \begin{pmatrix} \frac{3}{4} & -2 \\ 1 & -\frac{5}{4} \end{pmatrix} \mathbf{x}$

12. $\mathbf{x}' = \begin{pmatrix} -\frac{4}{5} & 2 \\ -1 & \frac{6}{5} \end{pmatrix} \mathbf{x}$

In each of Problems 13 through 20 the coefficient matrix contains a parameter α. In each of these problems:
(a) Determine the eigenvalues in terms of α.
(b) Find the critical value or values of α where the qualitative nature of the phase portrait for the system changes.
(c) Draw a phase portrait for a value of α slightly below, and for another value slightly above, each critical value.

13. $\mathbf{x}' = \begin{pmatrix} \alpha & 1 \\ -1 & \alpha \end{pmatrix} \mathbf{x}$

14. $\mathbf{x}' = \begin{pmatrix} 0 & -5 \\ 1 & \alpha \end{pmatrix} \mathbf{x}$

15. $\mathbf{x}' = \begin{pmatrix} 2 & -5 \\ \alpha & -2 \end{pmatrix} \mathbf{x}$

16. $\mathbf{x}' = \begin{pmatrix} \frac{5}{4} & \frac{3}{4} \\ \alpha & \frac{5}{4} \end{pmatrix} \mathbf{x}$

17. $\mathbf{x}' = \begin{pmatrix} -1 & \alpha \\ -1 & -1 \end{pmatrix} \mathbf{x}$

18. $\mathbf{x}' = \begin{pmatrix} 3 & \alpha \\ -6 & -4 \end{pmatrix} \mathbf{x}$

 19. $\mathbf{x}' = \begin{pmatrix} \alpha & 10 \\ -1 & -4 \end{pmatrix} \mathbf{x}$ 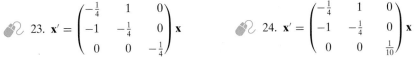 20. $\mathbf{x}' = \begin{pmatrix} 4 & \alpha \\ 8 & -6 \end{pmatrix} \mathbf{x}$

In each of Problems 21 and 22 solve the given system of equations by the method of Problem 19 of Section 7.5. Assume that $t > 0$.

21. $t\mathbf{x}' = \begin{pmatrix} -1 & -1 \\ 2 & -1 \end{pmatrix} \mathbf{x}$ 22. $t\mathbf{x}' = \begin{pmatrix} 2 & -5 \\ 1 & -2 \end{pmatrix} \mathbf{x}$

In each of Problems 23 and 24:

(a) Find the eigenvalues of the given system.

(b) Choose an initial point (other than the origin) and draw the corresponding trajectory in the x_1x_2-plane. Also draw the trajectories in the x_1x_3- and x_2x_3-planes.

(c) For the initial point in part (b) draw the corresponding trajectory in $x_1x_2x_3$-space.

23. $\mathbf{x}' = \begin{pmatrix} -\frac{1}{4} & 1 & 0 \\ -1 & -\frac{1}{4} & 0 \\ 0 & 0 & -\frac{1}{4} \end{pmatrix} \mathbf{x}$ 24. $\mathbf{x}' = \begin{pmatrix} -\frac{1}{4} & 1 & 0 \\ -1 & -\frac{1}{4} & 0 \\ 0 & 0 & \frac{1}{10} \end{pmatrix} \mathbf{x}$

25. Consider the electric circuit shown in Figure 7.6.6. Suppose that $R_1 = R_2 = 4$ ohms, $C = \frac{1}{2}$ farad, and $L = 8$ henrys.

(a) Show that this circuit is described by the system of differential equations

$$\frac{d}{dt}\begin{pmatrix} I \\ V \end{pmatrix} = \begin{pmatrix} -\frac{1}{2} & -\frac{1}{8} \\ 2 & -\frac{1}{2} \end{pmatrix}\begin{pmatrix} I \\ V \end{pmatrix}, \tag{i}$$

where I is the current through the inductor and V is the voltage drop across the capacitor. *Hint:* See Problem 19 of Section 7.1.

(b) Find the general solution of Eqs. (i) in terms of real-valued functions.

(c) Find $I(t)$ and $V(t)$ if $I(0) = 2$ amperes and $V(0) = 3$ volts.

(d) Determine the limiting values of $I(t)$ and $V(t)$ as $t \to \infty$. Do these limiting values depend on the initial conditions?

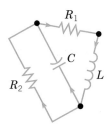

FIGURE 7.6.6 The circuit in Problem 25.

26. The electric circuit shown in Figure 7.6.7 is described by the system of differential equations

$$\frac{d}{dt}\begin{pmatrix} I \\ V \end{pmatrix} = \begin{pmatrix} 0 & \dfrac{1}{L} \\ -\dfrac{1}{C} & -\dfrac{1}{RC} \end{pmatrix}\begin{pmatrix} I \\ V \end{pmatrix}, \tag{i}$$

where I is the current through the inductor and V is the voltage drop across the capacitor. These differential equations were derived in Problem 19 of Section 7.1.

(a) Show that the eigenvalues of the coefficient matrix are real and different if $L > 4R^2C$; show that they are complex conjugates if $L < 4R^2C$.

(b) Suppose that $R = 1$ ohm, $C = \frac{1}{2}$ farad, and $L = 1$ henry. Find the general solution of the system (i) in this case.

(c) Find $I(t)$ and $V(t)$ if $I(0) = 2$ amperes and $V(0) = 1$ volt.

(d) For the circuit of part (b) determine the limiting values of $I(t)$ and $V(t)$ as $t \to \infty$. Do these limiting values depend on the initial conditions?

FIGURE 7.6.7 The circuit in Problem 26.

27. In this problem we indicate how to show that $\mathbf{u}(t)$ and $\mathbf{v}(t)$, as given by Eqs. (9), are linearly independent. Let $r_1 = \lambda + i\mu$ and $\bar{r}_1 = \lambda - i\mu$ be a pair of conjugate eigenvalues of the coefficient matrix \mathbf{A} of Eq. (1); let $\boldsymbol{\xi}^{(1)} = \mathbf{a} + i\mathbf{b}$ and $\bar{\boldsymbol{\xi}}^{(1)} = \mathbf{a} - i\mathbf{b}$ be the corresponding eigenvectors. Recall that it was stated in Section 7.3 that if $r_1 \neq \bar{r}_1$, then $\boldsymbol{\xi}^{(1)}$ and $\bar{\boldsymbol{\xi}}^{(1)}$ are linearly independent.

(a) First we show that \mathbf{a} and \mathbf{b} are linearly independent. Consider the equation $c_1\mathbf{a} + c_2\mathbf{b} = \mathbf{0}$. Express \mathbf{a} and \mathbf{b} in terms of $\boldsymbol{\xi}^{(1)}$ and $\bar{\boldsymbol{\xi}}^{(1)}$, and then show that $(c_1 - ic_2)\boldsymbol{\xi}^{(1)} + (c_1 + ic_2)\bar{\boldsymbol{\xi}}^{(1)} = \mathbf{0}$.

(b) Show that $c_1 - ic_2 = 0$ and $c_1 + ic_2 = 0$ and then that $c_1 = 0$ and $c_2 = 0$. Consequently, \mathbf{a} and \mathbf{b} are linearly independent.

(c) To show that $\mathbf{u}(t)$ and $\mathbf{v}(t)$ are linearly independent, consider the equation $c_1\mathbf{u}(t_0) + c_2\mathbf{v}(t_0) = \mathbf{0}$, where t_0 is an arbitrary point. Rewrite this equation in terms of \mathbf{a} and \mathbf{b}, and then proceed as in part (b) to show that $c_1 = 0$ and $c_2 = 0$. Hence $\mathbf{u}(t)$ and $\mathbf{v}(t)$ are linearly independent at the arbitrary point t_0. Therefore they are linearly independent at every point and on every interval.

28. A mass m on a spring with constant k satisfies the differential equation (see Section 3.8)

$$mu'' + ku = 0,$$

where $u(t)$ is the displacement at time t of the mass from its equilibrium position.

(a) Let $x_1 = u, x_2 = u'$, and show that the resulting system is

$$\mathbf{x}' = \begin{pmatrix} 0 & 1 \\ -k/m & 0 \end{pmatrix} \mathbf{x}.$$

(b) Find the eigenvalues of the matrix for the system in part (a).

(c) Sketch several trajectories of the system. Choose one of your trajectories, and sketch the corresponding graphs of x_1 versus t and of x_2 versus t. Sketch both graphs on one set of axes.

(d) What is the relation between the eigenvalues of the coefficient matrix and the natural frequency of the spring–mass system?

29. Consider the two-mass, three-spring system of Example 3 in the text. Instead of converting the problem into a system of four first order equations, we indicate here how to proceed directly from Eqs. (22).

 (a) Show that Eqs. (22) can be written in the form

 $$\mathbf{x}'' = \begin{pmatrix} -2 & 3/2 \\ 4/3 & -3 \end{pmatrix} \mathbf{x} = \mathbf{A}\mathbf{x}. \tag{i}$$

 (b) Assume that $\mathbf{x} = \boldsymbol{\xi}\, e^{rt}$ and show that

 $$(\mathbf{A} - r^2\mathbf{I})\boldsymbol{\xi} = \mathbf{0}.$$

 Note that r^2 (rather than r) is an eigenvalue of \mathbf{A} corresponding to an eigenvector $\boldsymbol{\xi}$.

 (c) Find the eigenvalues and eigenvectors of \mathbf{A}.

 (d) Write down expressions for x_1 and x_2. There should be four arbitrary constants in these expressions.

 (e) By differentiating the results from part (d), write down expressions for x_1' and x_2'. Your results from parts (d) and (e) should agree with Eq. (31) in the text.

 30. Consider the two-mass, three-spring system whose equations of motion are Eqs. (22) in the text. Let $m_1 = 1, m_2 = 4/3, k_1 = 1, k_2 = 3$, and $k_3 = 4/3$.

 (a) As in the text, convert the system to four first order equations of the form $\mathbf{y}' = \mathbf{A}\mathbf{y}$. Determine the coefficient matrix \mathbf{A}.

 (b) Find the eigenvalues and eigenvectors of \mathbf{A}.

 (c) Write down the general solution of the system.

 (d) Describe the fundamental modes of vibration. For each fundamental mode draw graphs of y_1 versus t and y_2 versus t. Also draw the corresponding trajectories in the $y_1 y_3$- and $y_2 y_4$-planes.

 (e) Consider the initial conditions $\mathbf{y}(0) = (2, 1, 0, 0)^T$. Evaluate the arbitrary constants in the general solution in part (c). What is the period of the motion in this case? Plot graphs of y_1 versus t and y_2 versus t. Also plot the corresponding trajectories in the $y_1 y_3$- and $y_2 y_4$-planes. Be sure you understand how the trajectories are traversed for a full period.

 (f) Consider other initial conditions of your own choice, and plot graphs similar to those requested in part (e).

 31. Consider the two-mass, three-spring system whose equations of motion are Eqs. (22) in the text. Let $m_1 = m_2 = 1$ and $k_1 = k_2 = k_3 = 1$.

 (a) As in the text, convert the system to four first order equations of the form $\mathbf{y}' = \mathbf{A}\mathbf{y}$. Determine the coefficient matrix \mathbf{A}.

 (b) Find the eigenvalues and eigenvectors of \mathbf{A}.

 (c) Write down the general solution of the system.

 (d) Describe the fundamental modes of vibration. For each fundamental mode draw graphs of y_1 versus t and y_2 versus t. Also draw the corresponding trajectories in the $y_1 y_3$- and $y_2 y_4$-planes.

 (e) Consider the initial conditions $\mathbf{y}(0) = (-1, 3, 0, 0)^T$. Evaluate the arbitrary constants in the general solution in part (c). Plot y_1 versus t and y_2 versus t. Do you think the solution is periodic? Also draw the trajectories in the $y_1 y_3$- and $y_2 y_4$-planes.

 (f) Consider other initial conditions of your own choice, and plot graphs similar to those requested in part (e).

7.7 Fundamental Matrices

The structure of the solutions of systems of linear differential equations can be further illuminated by introducing the idea of a fundamental matrix. Suppose that $\mathbf{x}^{(1)}(t), \ldots, \mathbf{x}^{(n)}(t)$ form a fundamental set of solutions for the equation

$$\mathbf{x}' = \mathbf{P}(t)\mathbf{x} \tag{1}$$

on some interval $\alpha < t < \beta$. Then the matrix

$$\mathbf{\Psi}(t) = \begin{pmatrix} x_1^{(1)}(t) & \cdots & x_1^{(n)}(t) \\ \vdots & & \vdots \\ x_n^{(1)}(t) & \cdots & x_n^{(n)}(t) \end{pmatrix}, \tag{2}$$

whose columns are the vectors $\mathbf{x}^{(1)}(t), \ldots, \mathbf{x}^{(n)}(t)$, is said to be a **fundamental matrix** for the system (1). Note that a fundamental matrix is nonsingular since its columns are linearly independent vectors.

EXAMPLE 1

Find a fundamental matrix for the system

$$\mathbf{x}' = \begin{pmatrix} 1 & 1 \\ 4 & 1 \end{pmatrix} \mathbf{x}. \tag{3}$$

In Example 1 of Section 7.5 we found that

$$\mathbf{x}^{(1)}(t) = \begin{pmatrix} e^{3t} \\ 2e^{3t} \end{pmatrix}, \qquad \mathbf{x}^{(2)}(t) = \begin{pmatrix} e^{-t} \\ -2e^{-t} \end{pmatrix}$$

are linearly independent solutions of Eq. (3). Thus a fundamental matrix for the system (3) is

$$\mathbf{\Psi}(t) = \begin{pmatrix} e^{3t} & e^{-t} \\ 2e^{3t} & -2e^{-t} \end{pmatrix}. \tag{4}$$

The solution of an initial value problem can be written very compactly in terms of a fundamental matrix. The general solution of Eq. (1) is

$$\mathbf{x} = c_1 \mathbf{x}^{(1)}(t) + \cdots + c_n \mathbf{x}^{(n)}(t) \tag{5}$$

or, in terms of $\mathbf{\Psi}(t)$,

$$\mathbf{x} = \mathbf{\Psi}(t)\mathbf{c}, \tag{6}$$

where \mathbf{c} is a constant vector with arbitrary components c_1, \ldots, c_n. For an initial value problem consisting of the differential equation (1) and the initial condition

$$\mathbf{x}(t_0) = \mathbf{x}^0, \tag{7}$$

where t_0 is a given point in $\alpha < t < \beta$ and \mathbf{x}^0 is a given initial vector, it is only necessary to choose the vector \mathbf{c} in Eq. (6) so as to satisfy the initial condition (7). Hence \mathbf{c}

must satisfy

$$\boldsymbol{\Psi}(t_0)\mathbf{c} = \mathbf{x}^0. \tag{8}$$

Therefore, since $\boldsymbol{\Psi}(t_0)$ is nonsingular,

$$\mathbf{c} = \boldsymbol{\Psi}^{-1}(t_0)\mathbf{x}^0 \tag{9}$$

and

$$\mathbf{x} = \boldsymbol{\Psi}(t)\boldsymbol{\Psi}^{-1}(t_0)\mathbf{x}^0 \tag{10}$$

is the solution of the initial value problem (1), (7). We emphasize, however, that to solve a given initial value problem, one would ordinarily solve Eq. (8) by row reduction and then substitute for \mathbf{c} in Eq. (6), rather than compute $\boldsymbol{\Psi}^{-1}(t_0)$ and use Eq. (10).

Recall that each column of the fundamental matrix $\boldsymbol{\Psi}$ is a solution of Eq. (1). It follows that $\boldsymbol{\Psi}$ satisfies the matrix differential equation

$$\boldsymbol{\Psi}' = \mathbf{P}(t)\boldsymbol{\Psi}. \tag{11}$$

This relation is readily confirmed by comparing the two sides of Eq. (11) column by column.

Sometimes it is convenient to make use of the special fundamental matrix, denoted by $\boldsymbol{\Phi}(t)$, whose columns are the vectors $\mathbf{x}^{(1)}(t), \ldots, \mathbf{x}^{(n)}(t)$ designated in Theorem 7.4.4. Besides the differential equation (1), these vectors satisfy the initial conditions

$$\mathbf{x}^{(j)}(t_0) = \mathbf{e}^{(j)}, \tag{12}$$

where $\mathbf{e}^{(j)}$ is the unit vector, defined in Theorem 7.4.4, with a one in the jth position and zeros elsewhere. Thus $\boldsymbol{\Phi}(t)$ has the property that

$$\boldsymbol{\Phi}(t_0) = \begin{pmatrix} 1 & 0 & \cdots & 0 \\ 0 & 1 & \cdots & 0 \\ \vdots & \vdots & & \vdots \\ 0 & 0 & \cdots & 1 \end{pmatrix} = \mathbf{I}. \tag{13}$$

We will always reserve the symbol $\boldsymbol{\Phi}$ to denote the fundamental matrix satisfying the initial condition (13) and use $\boldsymbol{\Psi}$ when an arbitrary fundamental matrix is intended. In terms of $\boldsymbol{\Phi}(t)$ the solution of the initial value problem (1), (7) is even simpler in appearance; since $\boldsymbol{\Phi}^{-1}(t_0) = \mathbf{I}$, it follows from Eq. (10) that

$$\mathbf{x} = \boldsymbol{\Phi}(t)\mathbf{x}^0. \tag{14}$$

Although the fundamental matrix $\boldsymbol{\Phi}(t)$ is often more complicated than $\boldsymbol{\Psi}(t)$, it is especially helpful if the same system of differential equations is to be solved repeatedly subject to many different initial conditions. This corresponds to a given physical system that can be started from many different initial states. If the fundamental matrix $\boldsymbol{\Phi}(t)$ has been determined, then the solution for each set of initial conditions can be found simply by matrix multiplication, as indicated by Eq. (14). The matrix $\boldsymbol{\Phi}(t)$ thus represents a transformation of the initial conditions \mathbf{x}^0 into the solution $\mathbf{x}(t)$ at an arbitrary time t. Comparing Eqs. (10) and (14) makes it clear that $\boldsymbol{\Phi}(t) = \boldsymbol{\Psi}(t)\boldsymbol{\Psi}^{-1}(t_0)$.

EXAMPLE 2

For the system (3),

$$\mathbf{x}' = \begin{pmatrix} 1 & 1 \\ 4 & 1 \end{pmatrix} \mathbf{x},$$

in Example 1, find the fundamental matrix $\boldsymbol{\Phi}$ such that $\boldsymbol{\Phi}(0) = \mathbf{I}$.

The columns of $\boldsymbol{\Phi}$ are solutions of Eq. (3) that satisfy the initial conditions

$$\mathbf{x}^{(1)}(0) = \begin{pmatrix} 1 \\ 0 \end{pmatrix}, \qquad \mathbf{x}^{(2)}(0) = \begin{pmatrix} 0 \\ 1 \end{pmatrix}. \tag{15}$$

Since the general solution of Eq. (3) is

$$\mathbf{x} = c_1 \begin{pmatrix} 1 \\ 2 \end{pmatrix} e^{3t} + c_2 \begin{pmatrix} 1 \\ -2 \end{pmatrix} e^{-t},$$

we can find the solution satisfying the first set of these initial conditions by choosing $c_1 = c_2 = \frac{1}{2}$; similarly, we obtain the solution satisfying the second set of initial conditions by choosing $c_1 = \frac{1}{4}$ and $c_2 = -\frac{1}{4}$. Hence

$$\boldsymbol{\Phi}(t) = \begin{pmatrix} \frac{1}{2}e^{3t} + \frac{1}{2}e^{-t} & \frac{1}{4}e^{3t} - \frac{1}{4}e^{-t} \\ e^{3t} - e^{-t} & \frac{1}{2}e^{3t} + \frac{1}{2}e^{-t} \end{pmatrix}. \tag{16}$$

Note that the elements of $\boldsymbol{\Phi}(t)$ are more complicated than those of the fundamental matrix $\boldsymbol{\Psi}(t)$ given by Eq. (4); however, it is now easy to determine the solution corresponding to any set of initial conditions.

The Matrix exp(At). Recall that the solution of the scalar initial value problem

$$x' = ax, \qquad x(0) = x_0, \tag{17}$$

where a is a constant, is

$$x = x_0 \exp(at). \tag{18}$$

Now consider the corresponding initial value problem for an $n \times n$ system, namely,

$$\mathbf{x}' = \mathbf{A}\mathbf{x}, \qquad \mathbf{x}(0) = \mathbf{x}^0, \tag{19}$$

where \mathbf{A} is a constant matrix. Applying the results of this section to the problem (19), we can write its solution as

$$\mathbf{x} = \boldsymbol{\Phi}(t)\mathbf{x}^0, \tag{20}$$

where $\boldsymbol{\Phi}(0) = \mathbf{I}$. Comparing the problems (17) and (19), and their solutions, suggests that the matrix $\boldsymbol{\Phi}(t)$ might have an exponential character. We now explore this possibility.

The scalar exponential function $\exp(at)$ can be represented by the power series

$$\exp(at) = 1 + \sum_{n=1}^{\infty} \frac{a^n t^n}{n!}, \tag{21}$$

which converges for all t. Let us now replace the scalar a by the $n \times n$ constant matrix \mathbf{A} and consider the corresponding series

$$\mathbf{I} + \sum_{n=1}^{\infty} \frac{\mathbf{A}^n t^n}{n!} = \mathbf{I} + \mathbf{A}t + \frac{\mathbf{A}^2 t^2}{2!} + \cdots + \frac{\mathbf{A}^n t^n}{n!} + \cdots. \tag{22}$$

Each term in the series (22) is an $n \times n$ matrix. It is possible to show that each element of this matrix sum converges for all t as $n \to \infty$. Thus the series (22) defines as its sum a new matrix, which we denote by $\exp(\mathbf{A}t)$; that is,

$$\exp(\mathbf{A}t) = \mathbf{I} + \sum_{n=1}^{\infty} \frac{\mathbf{A}^n t^n}{n!}, \tag{23}$$

analogous to the expansion (21) of the scalar function $\exp(at)$.

By differentiating the series (23) term by term we obtain

$$\frac{d}{dt}[\exp(\mathbf{A}t)] = \sum_{n=1}^{\infty} \frac{\mathbf{A}^n t^{n-1}}{(n-1)!} = \mathbf{A} \left[\mathbf{I} + \sum_{n=1}^{\infty} \frac{\mathbf{A}^n t^n}{n!} \right] = \mathbf{A} \exp(\mathbf{A}t). \tag{24}$$

Thus $\exp(\mathbf{A}t)$ satisfies the differential equation

$$\frac{d}{dt} \exp(\mathbf{A}t) = \mathbf{A} \exp(\mathbf{A}t). \tag{25}$$

Further, when $t = 0$, $\exp(\mathbf{A}t)$ satisfies the initial condition

$$\exp(\mathbf{A}t)\Big|_{t=0} = \mathbf{I}. \tag{26}$$

The fundamental matrix $\mathbf{\Phi}$ satisfies the same initial value problem as $\exp(\mathbf{A}t)$, namely,

$$\mathbf{\Phi}' = \mathbf{A}\mathbf{\Phi}, \qquad \mathbf{\Phi}(0) = \mathbf{I}. \tag{27}$$

Thus, by the uniqueness part of Theorem 7.1.2 (extended to matrix differential equations), we conclude that $\exp(\mathbf{A}t)$ and the fundamental matrix $\mathbf{\Phi}(t)$ are the same. Thus we can write the solution of the initial value problem (19) in the form

$$\mathbf{x} = \exp(\mathbf{A}t)\mathbf{x}^0, \tag{28}$$

which is analogous to the solution (18) of the initial value problem (17).

In order to justify more conclusively the use of $\exp(\mathbf{A}t)$ for the sum of the series (22), we should demonstrate that this matrix function does indeed have the properties we associate with the exponential function. One way to do this is outlined in Problem 15.

Diagonalizable Matrices. The basic reason why a system of linear (algebraic or differential) equations presents some difficulty is that the equations are usually *coupled*. In other words, some or all of the equations involve more than one—typically all—of the unknown variables. Hence the equations in the system must be solved *simultaneously*. In contrast, if each equation involves only a single variable, then each equation can be solved independently of all the others, which is a much easier task. This observation suggests that one way to solve a system of equations might be by transforming it into an equivalent *uncoupled* system in which each equation contains only one unknown variable. This corresponds to transforming the coefficient matrix \mathbf{A} into a *diagonal* matrix.

Eigenvectors are useful in accomplishing such a transformation. Suppose that the $n \times n$ matrix \mathbf{A} has a full set of n linearly independent eigenvectors. Recall that this will certainly be the case if the eigenvalues of \mathbf{A} are all different, or if \mathbf{A} is Hermitian. Letting $\boldsymbol{\xi}^{(1)}, \ldots, \boldsymbol{\xi}^{(n)}$ denote these eigenvectors and $\lambda_1, \ldots, \lambda_n$ the

corresponding eigenvalues, form the matrix \mathbf{T} whose columns are the eigenvectors—that is,

$$\mathbf{T} = \begin{pmatrix} \xi_1^{(1)} & \cdots & \xi_1^{(n)} \\ \vdots & & \vdots \\ \xi_n^{(1)} & \cdots & \xi_n^{(n)} \end{pmatrix}. \tag{29}$$

Since the columns of \mathbf{T} are linearly independent vectors, $\det \mathbf{T} \neq 0$; hence \mathbf{T} is non-singular and \mathbf{T}^{-1} exists. A straightforward calculation shows that the columns of the matrix \mathbf{AT} are just the vectors $\mathbf{A}\xi^{(1)}, \ldots, \mathbf{A}\xi^{(n)}$. Since $\mathbf{A}\xi^{(k)} = \lambda_k \xi^{(k)}$, it follows that

$$\mathbf{AT} = \begin{pmatrix} \lambda_1 \xi_1^{(1)} & \cdots & \lambda_n \xi_1^{(n)} \\ \vdots & & \vdots \\ \lambda_1 \xi_n^{(1)} & \cdots & \lambda_n \xi_n^{(n)} \end{pmatrix} = \mathbf{TD}, \tag{30}$$

where

$$\mathbf{D} = \begin{pmatrix} \lambda_1 & 0 & \cdots & 0 \\ 0 & \lambda_2 & \cdots & 0 \\ \vdots & \vdots & & \vdots \\ 0 & 0 & \cdots & \lambda_n \end{pmatrix} \tag{31}$$

is a diagonal matrix whose diagonal elements are the eigenvalues of \mathbf{A}. From Eq. (30) it follows that

$$\mathbf{T}^{-1}\mathbf{AT} = \mathbf{D}. \tag{32}$$

Thus, if the eigenvalues and eigenvectors of \mathbf{A} are known, \mathbf{A} can be transformed into a diagonal matrix by the process shown in Eq. (32). This process is known as a **similarity transformation**, and Eq. (32) is summed up in words by saying that \mathbf{A} is **similar** to the diagonal matrix \mathbf{D}. Alternatively, we may say that \mathbf{A} is **diagonalizable**. Observe that a similarity transformation leaves the eigenvalues of \mathbf{A} unchanged and transforms its eigenvectors into the coordinate vectors $\mathbf{e}^{(1)}, \ldots, \mathbf{e}^{(n)}$.

If \mathbf{A} is Hermitian, then the determination of \mathbf{T}^{-1} is very simple. The eigenvectors $\xi^{(1)}, \ldots, \xi^{(n)}$ of \mathbf{A} are known to be mutually orthogonal, so let us choose them so that they are also normalized by $(\xi^{(i)}, \xi^{(i)}) = 1$ for each i. Then it is easy to verify that $\mathbf{T}^{-1} = \mathbf{T}^*$; in other words, the inverse of \mathbf{T} is the same as its adjoint (the transpose of its complex conjugate).

Finally, we note that if \mathbf{A} has fewer than n linearly independent eigenvectors, then there is no matrix \mathbf{T} such that $\mathbf{T}^{-1}\mathbf{AT} = \mathbf{D}$. In this case, \mathbf{A} is not similar to a diagonal matrix and is not diagonalizable.

EXAMPLE 3

Consider the matrix

$$\mathbf{A} = \begin{pmatrix} 1 & 1 \\ 4 & 1 \end{pmatrix}. \tag{33}$$

Find the similarity transformation matrix \mathbf{T} and show that \mathbf{A} can be diagonalized.

In Example 1 of Section 7.5 we found that the eigenvalues and eigenvectors of \mathbf{A} are

$$r_1 = 3, \quad \xi^{(1)} = \begin{pmatrix} 1 \\ 2 \end{pmatrix}; \qquad r_2 = -1, \quad \xi^{(2)} = \begin{pmatrix} 1 \\ -2 \end{pmatrix}. \tag{34}$$

Thus the transformation matrix \mathbf{T} and its inverse \mathbf{T}^{-1} are

$$\mathbf{T} = \begin{pmatrix} 1 & 1 \\ 2 & -2 \end{pmatrix}; \qquad \mathbf{T}^{-1} = \begin{pmatrix} \frac{1}{2} & \frac{1}{4} \\ \frac{1}{2} & -\frac{1}{4} \end{pmatrix}. \tag{35}$$

Consequently, you can check that

$$\mathbf{T}^{-1}\mathbf{A}\mathbf{T} = \begin{pmatrix} 3 & 0 \\ 0 & -1 \end{pmatrix} = \mathbf{D}. \tag{36}$$

Now let us turn again to the system

$$\mathbf{x}' = \mathbf{A}\mathbf{x}, \tag{37}$$

where \mathbf{A} is a constant matrix. In Sections 7.5 and 7.6 we described how to solve such a system by starting from the assumption that $\mathbf{x} = \boldsymbol{\xi}e^{rt}$. Now we provide another viewpoint, one based on diagonalizing the coefficient matrix \mathbf{A}.

According to the results stated just above, it is possible to diagonalize \mathbf{A} whenever \mathbf{A} has a full set of n linearly independent eigenvectors. Let $\boldsymbol{\xi}^{(1)}, \ldots, \boldsymbol{\xi}^{(n)}$ be eigenvectors of \mathbf{A} corresponding to the eigenvalues r_1, \ldots, r_n and form the transformation matrix \mathbf{T} whose columns are $\boldsymbol{\xi}^{(1)}, \ldots, \boldsymbol{\xi}^{(n)}$. Then, defining a new dependent variable \mathbf{y} by the relation

$$\mathbf{x} = \mathbf{T}\mathbf{y}, \tag{38}$$

we have from Eq. (37) that

$$\mathbf{T}\mathbf{y}' = \mathbf{A}\mathbf{T}\mathbf{y}. \tag{39}$$

Multiplying by \mathbf{T}^{-1}, we then obtain

$$\mathbf{y}' = (\mathbf{T}^{-1}\mathbf{A}\mathbf{T})\mathbf{y}, \tag{40}$$

or, using Eq. (32),

$$\mathbf{y}' = \mathbf{D}\mathbf{y}. \tag{41}$$

Recall that \mathbf{D} is the diagonal matrix with the eigenvalues r_1, \ldots, r_n of \mathbf{A} along the diagonal. A fundamental matrix for the system (41) is the diagonal matrix (see Problem 16)

$$\mathbf{Q}(t) = \exp(\mathbf{D}t) = \begin{pmatrix} e^{r_1 t} & 0 & \cdots & 0 \\ 0 & e^{r_2 t} & \cdots & 0 \\ \vdots & \vdots & & \vdots \\ 0 & 0 & \cdots & e^{r_n t} \end{pmatrix}. \tag{42}$$

A fundamental matrix $\boldsymbol{\Psi}$ for the system (37) is then found from \mathbf{Q} by the transformation (38),

$$\boldsymbol{\Psi} = \mathbf{T}\mathbf{Q}; \tag{43}$$

that is,

$$\boldsymbol{\Psi}(t) = \begin{pmatrix} \xi_1^{(1)}e^{r_1 t} & \cdots & \xi_1^{(n)}e^{r_n t} \\ \vdots & & \vdots \\ \xi_n^{(1)}e^{r_1 t} & \cdots & \xi_n^{(n)}e^{r_n t} \end{pmatrix}. \tag{44}$$

Equation (44) is the same result that was obtained in Section 7.5. This diagonalization procedure does not offer any computational advantage over the method of Section 7.5, since in either case it is necessary to calculate the eigenvalues and eigenvectors of the coefficient matrix in the system of differential equations. Nevertheless, it is noteworthy that the problem of uncoupling a system of differential equations and the problem of diagonalizing a matrix are mathematically the same.

EXAMPLE 4

Consider again the system of differential equations

$$\mathbf{x}' = \mathbf{A}\mathbf{x}, \tag{45}$$

where \mathbf{A} is given by Eq. (33). Using the transformation $\mathbf{x} = \mathbf{T}\mathbf{y}$, where \mathbf{T} is given by Eq. (35), you can reduce the system (45) to the diagonal system

$$\mathbf{y}' = \begin{pmatrix} 3 & 0 \\ 0 & -1 \end{pmatrix} \mathbf{y} = \mathbf{D}\mathbf{y}. \tag{46}$$

Obtain a fundamental matrix for the system (46) and then transform it to obtain a fundamental matrix for the original system (45).

By multiplying \mathbf{D} repeatedly with itself, we find that

$$\mathbf{D}^2 = \begin{pmatrix} 9 & 0 \\ 0 & 1 \end{pmatrix}, \quad \mathbf{D}^3 = \begin{pmatrix} 27 & 0 \\ 0 & -1 \end{pmatrix}, \quad \dots \tag{47}$$

Therefore it follows from Eq. (23) that $\exp(\mathbf{D}t)$ is a diagonal matrix with the entries e^{3t} and e^{-t} on the diagonal; that is,

$$e^{\mathbf{D}t} = \begin{pmatrix} e^{3t} & 0 \\ 0 & e^{-t} \end{pmatrix}. \tag{48}$$

Finally, we obtain the required fundamental matrix $\mathbf{\Psi}(t)$ by multiplying \mathbf{T} and $\exp(\mathbf{D}t)$:

$$\mathbf{\Psi}(t) = \begin{pmatrix} 1 & 1 \\ 2 & -2 \end{pmatrix} \begin{pmatrix} e^{3t} & 0 \\ 0 & e^{-t} \end{pmatrix} = \begin{pmatrix} e^{3t} & e^{-t} \\ 2e^{3t} & -2e^{-t} \end{pmatrix}. \tag{49}$$

Observe that this fundamental matrix is the same as the one found in Example 1.

PROBLEMS

In each of Problems 1 through 10 find a fundamental matrix for the given system of equations. In each case also find the fundamental matrix $\mathbf{\Phi}(t)$ satisfying $\mathbf{\Phi}(0) = \mathbf{I}$.

1. $\mathbf{x}' = \begin{pmatrix} 3 & -2 \\ 2 & -2 \end{pmatrix} \mathbf{x}$

2. $\mathbf{x}' = \begin{pmatrix} -\frac{3}{4} & \frac{1}{2} \\ \frac{1}{8} & -\frac{3}{4} \end{pmatrix} \mathbf{x}$

3. $\mathbf{x}' = \begin{pmatrix} 2 & -1 \\ 3 & -2 \end{pmatrix} \mathbf{x}$

4. $\mathbf{x}' = \begin{pmatrix} 1 & 1 \\ 4 & -2 \end{pmatrix} \mathbf{x}$

5. $\mathbf{x}' = \begin{pmatrix} 2 & -5 \\ 1 & -2 \end{pmatrix} \mathbf{x}$

6. $\mathbf{x}' = \begin{pmatrix} -1 & -4 \\ 1 & -1 \end{pmatrix} \mathbf{x}$

7. $\mathbf{x}' = \begin{pmatrix} 5 & -1 \\ 3 & 1 \end{pmatrix} \mathbf{x}$

8. $\mathbf{x}' = \begin{pmatrix} 1 & -1 \\ 5 & -3 \end{pmatrix} \mathbf{x}$

9. $\mathbf{x}' = \begin{pmatrix} 1 & 1 & 1 \\ 2 & 1 & -1 \\ -8 & -5 & -3 \end{pmatrix} \mathbf{x}$ 10. $\mathbf{x}' = \begin{pmatrix} 1 & -1 & 4 \\ 3 & 2 & -1 \\ 2 & 1 & -1 \end{pmatrix} \mathbf{x}$

11. Solve the initial value problem

$$\mathbf{x}' = \begin{pmatrix} 2 & -1 \\ 3 & -2 \end{pmatrix} \mathbf{x}, \qquad \mathbf{x}(0) = \begin{pmatrix} 2 \\ -1 \end{pmatrix}$$

by using the fundamental matrix $\mathbf{\Phi}(t)$ found in Problem 3.

12. Solve the initial value problem

$$\mathbf{x}' = \begin{pmatrix} -1 & -4 \\ 1 & -1 \end{pmatrix} \mathbf{x}, \qquad \mathbf{x}(0) = \begin{pmatrix} 3 \\ 1 \end{pmatrix}$$

by using the fundamental matrix $\mathbf{\Phi}(t)$ found in Problem 6.

13. Show that $\mathbf{\Phi}(t) = \mathbf{\Psi}(t)\mathbf{\Psi}^{-1}(t_0)$, where $\mathbf{\Phi}(t)$ and $\mathbf{\Psi}(t)$ are as defined in this section.

14. The fundamental matrix $\mathbf{\Phi}(t)$ for the system (3) was found in Example 2. Show that $\mathbf{\Phi}(t)\mathbf{\Phi}(s) = \mathbf{\Phi}(t + s)$ by multiplying $\mathbf{\Phi}(t)$ and $\mathbf{\Phi}(s)$.

15. Let $\mathbf{\Phi}(t)$ denote the fundamental matrix satisfying $\mathbf{\Phi}' = \mathbf{A}\mathbf{\Phi}, \mathbf{\Phi}(0) = \mathbf{I}$. In the text we also denoted this matrix by $\exp(\mathbf{A}t)$. In this problem we show that $\mathbf{\Phi}$ does indeed have the principal algebraic properties associated with the exponential function.

(a) Show that $\mathbf{\Phi}(t)\mathbf{\Phi}(s) = \mathbf{\Phi}(t + s)$; that is, show that $\exp(\mathbf{A}t)\exp(\mathbf{A}s) = \exp[\mathbf{A}(t + s)]$.
Hint: Show that if s is fixed and t is variable, then both $\mathbf{\Phi}(t)\mathbf{\Phi}(s)$ and $\mathbf{\Phi}(t + s)$ satisfy the initial value problem $\mathbf{Z}' = \mathbf{A}\mathbf{Z}, \mathbf{Z}(0) = \mathbf{\Phi}(s)$.

(b) Show that $\mathbf{\Phi}(t)\mathbf{\Phi}(-t) = \mathbf{I}$; that is, $\exp(\mathbf{A}t)\exp[\mathbf{A}(-t)] = \mathbf{I}$. Then show that $\mathbf{\Phi}(-t) = \mathbf{\Phi}^{-1}(t)$.

(c) Show that $\mathbf{\Phi}(t - s) = \mathbf{\Phi}(t)\mathbf{\Phi}^{-1}(s)$.

16. Show that if \mathbf{A} is a diagonal matrix with diagonal elements a_1, a_2, \ldots, a_n, then $\exp(\mathbf{A}t)$ is also a diagonal matrix with diagonal elements $\exp(a_1 t), \exp(a_2 t), \ldots, \exp(a_n t)$.

17. Consider an oscillator satisfying the initial value problem

$$u'' + \omega^2 u = 0, \qquad u(0) = u_0, \quad u'(0) = v_0. \tag{i}$$

(a) Let $x_1 = u, x_2 = u'$, and transform Eqs. (i) into the form

$$\mathbf{x}' = \mathbf{A}\mathbf{x}, \qquad \mathbf{x}(0) = \mathbf{x}^0. \tag{ii}$$

(b) By using the series (23), show that

$$\exp \mathbf{A}t = \mathbf{I}\cos \omega t + \mathbf{A}\frac{\sin \omega t}{\omega}. \tag{iii}$$

(c) Find the solution of the initial value problem (ii).

18. The method of successive approximations (see Section 2.8) can also be applied to systems of equations. For example, consider the initial value problem

$$\mathbf{x}' = \mathbf{A}\mathbf{x}, \qquad \mathbf{x}(0) = \mathbf{x}^0, \tag{i}$$

where \mathbf{A} is a constant matrix and \mathbf{x}^0 is a prescribed vector.

(a) Assuming that a solution $\mathbf{x} = \boldsymbol{\phi}(t)$ exists, show that it must satisfy the integral equation

$$\boldsymbol{\phi}(t) = \mathbf{x}^0 + \int_0^t \mathbf{A}\boldsymbol{\phi}(s)\, ds. \tag{ii}$$

(b) Start with the initial approximation $\boldsymbol{\phi}^{(0)}(t) = \mathbf{x}^0$. Substitute this expression for $\boldsymbol{\phi}(s)$ in the right side of Eq. (ii) and obtain a new approximation $\boldsymbol{\phi}^{(1)}(t)$. Show that

$$\boldsymbol{\phi}^{(1)}(t) = (\mathbf{I} + \mathbf{A}t)\mathbf{x}^0. \tag{iii}$$

(c) Repeat this process and thereby obtain a sequence of approximations $\boldsymbol{\phi}^{(0)}$, $\boldsymbol{\phi}^{(1)}$, $\boldsymbol{\phi}^{(2)}, \ldots, \boldsymbol{\phi}^{(n)}, \ldots$. Use an inductive argument to show that

$$\boldsymbol{\phi}^{(n)}(t) = \left(\mathbf{I} + \mathbf{A}t + \mathbf{A}^2\frac{t^2}{2!} + \cdots + \mathbf{A}^n\frac{t^n}{n!}\right)\mathbf{x}^0. \tag{iv}$$

(d) Let $n \to \infty$ and show that the solution of the initial value problem (i) is

$$\boldsymbol{\phi}(t) = \exp(\mathbf{A}t)\mathbf{x}^0. \tag{v}$$

7.8 Repeated Eigenvalues

We conclude our consideration of the linear homogeneous system with constant coefficients

$$\mathbf{x}' = \mathbf{A}\mathbf{x} \tag{1}$$

with a discussion of the case in which the matrix \mathbf{A} has a repeated eigenvalue. Recall that in Section 7.3 we stated that a repeated eigenvalue with algebraic multiplicity $k \geq 2$ may have a geometric multiplicity less than k. In other words, there may be fewer than k linearly independent eigenvectors associated with this eigenvalue. The following example illustrates this possibility.

EXAMPLE 1

Find the eigenvalues and eigenvectors of the matrix

$$\mathbf{A} = \begin{pmatrix} 1 & -1 \\ 1 & 3 \end{pmatrix}. \tag{2}$$

The eigenvalues r and eigenvectors $\boldsymbol{\xi}$ satisfy the equation $(\mathbf{A} - r\mathbf{I})\boldsymbol{\xi} = \mathbf{0}$, or

$$\begin{pmatrix} 1-r & -1 \\ 1 & 3-r \end{pmatrix}\begin{pmatrix} \xi_1 \\ \xi_2 \end{pmatrix} = \begin{pmatrix} 0 \\ 0 \end{pmatrix}. \tag{3}$$

The eigenvalues are the roots of the equation

$$\det(\mathbf{A} - r\mathbf{I}) = \begin{vmatrix} 1-r & -1 \\ 1 & 3-r \end{vmatrix} = r^2 - 4r + 4 = 0. \tag{4}$$

Thus the two eigenvalues are $r_1 = 2$ and $r_2 = 2$; that is, the eigenvalue 2 has algebraic multiplicity 2.

To determine the eigenvectors, we must return to Eq. (3) and use for r the value 2. This gives

$$\begin{pmatrix} -1 & -1 \\ 1 & 1 \end{pmatrix}\begin{pmatrix} \xi_1 \\ \xi_2 \end{pmatrix} = \begin{pmatrix} 0 \\ 0 \end{pmatrix}. \tag{5}$$

Hence we obtain the single condition $\xi_1 + \xi_2 = 0$, which determines ξ_2 in terms of ξ_1, or vice versa. Thus the eigenvector corresponding to the eigenvalue $r = 2$ is

$$\boldsymbol{\xi}^{(1)} = \begin{pmatrix} 1 \\ -1 \end{pmatrix}, \tag{6}$$

or any nonzero multiple of this vector. Observe that there is only one linearly independent eigenvector associated with the double eigenvalue.

Returning to the system (1), suppose that $r = \rho$ is a k-fold root of the determinantal equation

$$\det(\mathbf{A} - r\mathbf{I}) = 0. \tag{7}$$

Then ρ is an eigenvalue of algebraic multiplicity k of the matrix \mathbf{A}. In this event, there are two possibilities: Either there are k linearly independent eigenvectors corresponding to the eigenvalue ρ, or else there are fewer than k such eigenvectors.

In the first case, let $\boldsymbol{\xi}^{(1)}, \ldots, \boldsymbol{\xi}^{(k)}$ be k linearly independent eigenvectors associated with the eigenvalue ρ of algebraic multiplicity k. Then $\mathbf{x}^{(1)}(t) = \boldsymbol{\xi}^{(1)} e^{\rho t}, \ldots,$ $\mathbf{x}^{(k)}(t) = \boldsymbol{\xi}^{(k)} e^{\rho t}$ are k linearly independent solutions of Eq. (1). Thus in this case it makes no difference that the eigenvalue $r = \rho$ is repeated; there is still a fundamental set of solutions of Eq. (1) of the form $\boldsymbol{\xi} e^{rt}$. This case always occurs if the coefficient matrix \mathbf{A} is Hermitian.

However, if the coefficient matrix is not Hermitian, then there may be fewer than k independent eigenvectors corresponding to an eigenvalue ρ of algebraic multiplicity k, and if so, there will be fewer than k solutions of Eq. (1) of the form $\boldsymbol{\xi} e^{\rho t}$ associated with this eigenvalue. Therefore, to construct the general solution of Eq. (1), it is necessary to find other solutions of a different form. By analogy with previous results for linear equations of order n, it is natural to seek additional solutions involving products of polynomials and exponential functions. We first consider an example.

EXAMPLE 2

Find a fundamental set of solutions of

$$\mathbf{x}' = \mathbf{A}\mathbf{x} = \begin{pmatrix} 1 & -1 \\ 1 & 3 \end{pmatrix} \mathbf{x} \tag{8}$$

and draw a phase portrait for this system.

A direction field for the system (8) is shown in Figure 7.8.1. From this figure it appears that all nonzero solutions depart from the origin.

To solve the system, observe that the coefficient matrix \mathbf{A} is the same as the matrix in Example 1. Thus we know that $r = 2$ is a double eigenvalue and that it has only a single corresponding eigenvector, which we may take as $\boldsymbol{\xi}^T = (1, -1)$. Thus one solution of the system (8) is

$$\mathbf{x}^{(1)}(t) = \begin{pmatrix} 1 \\ -1 \end{pmatrix} e^{2t}, \tag{9}$$

but there is no second solution of the form $\mathbf{x} = \boldsymbol{\xi} e^{rt}$.

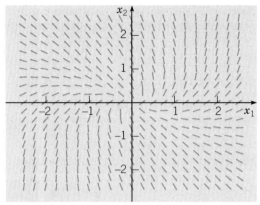

FIGURE 7.8.1 A direction field for the system (8).

Based on the procedure used for second order linear equations in Section 3.5, it may be natural to attempt to find a second solution of the system (8) of the form

$$\mathbf{x} = \boldsymbol{\xi} t e^{2t}, \tag{10}$$

where $\boldsymbol{\xi}$ is a constant vector to be determined. Substituting for \mathbf{x} in Eq. (8) gives

$$2\boldsymbol{\xi} t e^{2t} + \boldsymbol{\xi} e^{2t} - \mathbf{A}\boldsymbol{\xi} t e^{2t} = \mathbf{0}. \tag{11}$$

For Eq. (11) to be satisfied for all t, it is necessary for the coefficients of $t e^{2t}$ and e^{2t} both to be zero. From the term in e^{2t} we find that

$$\boldsymbol{\xi} = \mathbf{0}. \tag{12}$$

Hence there is no nonzero solution of the system (8) of the form (10).

Since Eq. (11) contains terms in both $t e^{2t}$ and e^{2t}, it appears that in addition to $\boldsymbol{\xi} t e^{2t}$, the second solution must contain a term of the form $\boldsymbol{\eta} e^{2t}$; in other words, we need to assume that

$$\mathbf{x} = \boldsymbol{\xi} t e^{2t} + \boldsymbol{\eta} e^{2t}, \tag{13}$$

where $\boldsymbol{\xi}$ and $\boldsymbol{\eta}$ are constant vectors. Upon substituting this expression for \mathbf{x} in Eq. (8), we obtain

$$2\boldsymbol{\xi} t e^{2t} + (\boldsymbol{\xi} + 2\boldsymbol{\eta}) e^{2t} = \mathbf{A}(\boldsymbol{\xi} t e^{2t} + \boldsymbol{\eta} e^{2t}). \tag{14}$$

Equating coefficients of $t e^{2t}$ and e^{2t} on each side of Eq. (14) gives the conditions

$$(\mathbf{A} - 2\mathbf{I})\boldsymbol{\xi} = \mathbf{0} \tag{15}$$

and

$$(\mathbf{A} - 2\mathbf{I})\boldsymbol{\eta} = \boldsymbol{\xi} \tag{16}$$

for the determination of $\boldsymbol{\xi}$ and $\boldsymbol{\eta}$. Equation (15) is satisfied if $\boldsymbol{\xi}$ is an eigenvector of \mathbf{A} corresponding to the eigenvalue $r = 2$, that is, $\boldsymbol{\xi}^T = (1, -1)$. Since $\det(\mathbf{A} - 2\mathbf{I})$ is zero, we might expect that Eq. (16) cannot be solved. However, this is not necessarily true, since for some vectors $\boldsymbol{\xi}$ it is possible to solve Eq. (16). In fact, the augmented matrix for Eq. (16) is

$$\begin{pmatrix} -1 & -1 & | & 1 \\ 1 & 1 & | & -1 \end{pmatrix}.$$

The second row of this matrix is proportional to the first, so the system is solvable. We have

$$-\eta_1 - \eta_2 = 1,$$

so if $\eta_1 = k$, where k is arbitrary, then $\eta_2 = -k - 1$. If we write

$$\boldsymbol{\eta} = \begin{pmatrix} k \\ -1 - k \end{pmatrix} = \begin{pmatrix} 0 \\ -1 \end{pmatrix} + k \begin{pmatrix} 1 \\ -1 \end{pmatrix}, \tag{17}$$

then by substituting for $\boldsymbol{\xi}$ and $\boldsymbol{\eta}$ in Eq. (13), we obtain

$$\mathbf{x} = \begin{pmatrix} 1 \\ -1 \end{pmatrix} te^{2t} + \begin{pmatrix} 0 \\ -1 \end{pmatrix} e^{2t} + k \begin{pmatrix} 1 \\ -1 \end{pmatrix} e^{2t}. \tag{18}$$

The last term in Eq. (18) is merely a multiple of the first solution $\mathbf{x}^{(1)}(t)$ and may be ignored, but the first two terms constitute a new solution:

$$\mathbf{x}^{(2)}(t) = \begin{pmatrix} 1 \\ -1 \end{pmatrix} te^{2t} + \begin{pmatrix} 0 \\ -1 \end{pmatrix} e^{2t}. \tag{19}$$

An elementary calculation shows that $W[\mathbf{x}^{(1)}, \mathbf{x}^{(2)}](t) = -e^{4t}$, and therefore $\mathbf{x}^{(1)}$ and $\mathbf{x}^{(2)}$ form a fundamental set of solutions of the system (8). The general solution is

$$\mathbf{x} = c_1 \mathbf{x}^{(1)}(t) + c_2 \mathbf{x}^{(2)}(t)$$

$$= c_1 \begin{pmatrix} 1 \\ -1 \end{pmatrix} e^{2t} + c_2 \left[\begin{pmatrix} 1 \\ -1 \end{pmatrix} te^{2t} + \begin{pmatrix} 0 \\ -1 \end{pmatrix} e^{2t} \right]. \tag{20}$$

The graph of the solution (20) is a little more difficult to analyze than in some of the previous examples. It is clear that \mathbf{x} becomes unbounded as $t \to \infty$ and that $\mathbf{x} \to \mathbf{0}$ as $t \to -\infty$. It is possible to show that as $t \to -\infty$, all solutions approach the origin tangent to the line $x_2 = -x_1$ determined by the eigenvector. Similarly, as $t \to \infty$, each trajectory is asymptotic to a line of slope -1. The trajectories of the system (8) are shown in Figure 7.8.2a, and some typical plots of x_1 versus t are shown in Figure 7.8.2b. The pattern of trajectories in this figure is typical of second order systems $\mathbf{x}' = \mathbf{A}\mathbf{x}$ with equal eigenvalues and only one independent eigenvector. The origin is called an **improper node** in this case. If the eigenvalues are negative, then the trajectories are similar but are traversed in the inward direction. An improper node is asymptotically stable or unstable, depending on whether the eigenvalues are negative or positive.

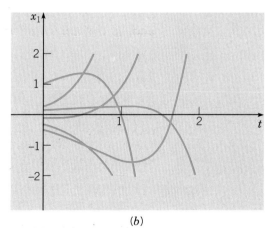

(a) (b)

FIGURE 7.8.2 (a) Trajectories of the system (8); the origin is an improper node. (b) Plots of x_1 versus t for the system (8).

One difference between a system of two first order equations and a single second order equation is evident from the preceding example. Recall that for a second order linear equation with a repeated root r_1 of the characteristic equation, a term $ce^{r_1 t}$ in the second solution is not required since it is a multiple of the first solution. On the other hand, for a system of two first order equations, the term $\boldsymbol{\eta} e^{r_1 t}$ of Eq. (13) with $r_1 = 2$ is not a multiple of the first solution $\boldsymbol{\xi} e^{r_1 t}$, so the term $\boldsymbol{\eta} e^{r_1 t}$ must be retained.

Example 2 is entirely typical of the general case when there is a double eigenvalue and a single associated eigenvector. Consider again the system (1), and suppose that $r = \rho$ is a double eigenvalue of \mathbf{A}, but that there is only one corresponding eigenvector $\boldsymbol{\xi}$. Then one solution [similar to Eq. (9)] is

$$\mathbf{x}^{(1)}(t) = \boldsymbol{\xi} e^{\rho t}, \tag{21}$$

where $\boldsymbol{\xi}$ satisfies

$$(\mathbf{A} - \rho \mathbf{I})\boldsymbol{\xi} = \mathbf{0}. \tag{22}$$

By proceeding as in Example 2, we find that a second solution [similar to Eq. (19)] is

$$\mathbf{x}^{(2)}(t) = \boldsymbol{\xi} t e^{\rho t} + \boldsymbol{\eta} e^{\rho t}, \tag{23}$$

where $\boldsymbol{\xi}$ satisfies Eq. (22) and $\boldsymbol{\eta}$ is determined from

$$(\mathbf{A} - \rho \mathbf{I})\boldsymbol{\eta} = \boldsymbol{\xi}. \tag{24}$$

Even though $\det(\mathbf{A} - \rho \mathbf{I}) = 0$, it can be shown that it is always possible to solve Eq. (24) for $\boldsymbol{\eta}$. The vector $\boldsymbol{\eta}$ is called a **generalized eigenvector** corresponding to the eigenvalue ρ.

Fundamental Matrices. As explained in Section 7.7, fundamental matrices are formed by arranging linearly independent solutions in columns. Thus, for example, a fundamental matrix for the system (8) can be formed from the solutions $\mathbf{x}^{(1)}(t)$ and $\mathbf{x}^{(2)}(t)$ from Eqs. (9) and (19), respectively:

$$\boldsymbol{\Psi}(t) = \begin{pmatrix} e^{2t} & te^{2t} \\ -e^{2t} & -te^{2t} - e^{2t} \end{pmatrix} = e^{2t} \begin{pmatrix} 1 & t \\ -1 & -1-t \end{pmatrix}. \tag{25}$$

The matrix $\boldsymbol{\Phi}$ that satisfies $\boldsymbol{\Phi}(0) = \mathbf{I}$ can also be readily found from the relation $\boldsymbol{\Phi}(t) = \boldsymbol{\Psi}(t)\boldsymbol{\Psi}^{-1}(0)$. For Eq. (8) we have

$$\boldsymbol{\Psi}(0) = \begin{pmatrix} 1 & 0 \\ -1 & -1 \end{pmatrix}, \qquad \boldsymbol{\Psi}^{-1}(0) = \begin{pmatrix} 1 & 0 \\ -1 & -1 \end{pmatrix}, \tag{26}$$

and then

$$\boldsymbol{\Phi}(t) = \boldsymbol{\Psi}(t)\boldsymbol{\Psi}^{-1}(0) = e^{2t} \begin{pmatrix} 1 & t \\ -1 & -1-t \end{pmatrix} \begin{pmatrix} 1 & 0 \\ -1 & -1 \end{pmatrix}$$

$$= e^{2t} \begin{pmatrix} 1-t & -t \\ t & 1+t \end{pmatrix}. \tag{27}$$

The latter matrix is also the exponential matrix $\exp(\mathbf{A}t)$.

Jordan Forms. An $n \times n$ matrix \mathbf{A} can be diagonalized as discussed in Section 7.7 only if it has a full complement of n linearly independent eigenvectors. If there is a shortage

of eigenvectors (because of repeated eigenvalues), then \mathbf{A} can always be transformed into a nearly diagonal matrix called its Jordan[6] form, which has the eigenvalues of \mathbf{A} on the main diagonal, ones in certain positions on the diagonal above the main diagonal, and zeros elsewhere.

Consider again the matrix \mathbf{A} given by Eq. (2). Form the transformation matrix \mathbf{T} with the single eigenvector $\boldsymbol{\xi}$ from Eq. (6) in its first column and the generalized eigenvector $\boldsymbol{\eta}$ from Eq. (17) with $k = 0$ in the second column. Then \mathbf{T} and its inverse are given by

$$\mathbf{T} = \begin{pmatrix} 1 & 0 \\ -1 & -1 \end{pmatrix}, \qquad \mathbf{T}^{-1} = \begin{pmatrix} 1 & 0 \\ -1 & -1 \end{pmatrix}. \tag{28}$$

As you can verify, it follows that

$$\mathbf{T}^{-1}\mathbf{A}\mathbf{T} = \begin{pmatrix} 2 & 1 \\ 0 & 2 \end{pmatrix} = \mathbf{J}. \tag{29}$$

The matrix \mathbf{J} in Eq. (29) is the Jordan form of \mathbf{A}. It is typical of all Jordan forms in that it has a 1 above the main diagonal in the column corresponding to the eigenvector that is lacking (and is replaced in \mathbf{T} by the generalized eigenvector).

If we start again from Eq. (1),

$$\mathbf{x}' = \mathbf{A}\mathbf{x},$$

the transformation $\mathbf{x} = \mathbf{T}\mathbf{y}$, where \mathbf{T} is given by Eq. (28), produces the system

$$\mathbf{y}' = \mathbf{J}\mathbf{y}, \tag{30}$$

where \mathbf{J} is given by Eq. (29). In scalar form the system (30) is

$$y_1' = 2y_1 + y_2, \qquad y_2' = 2y_2. \tag{31}$$

These equations can be solved readily in reverse order. In this way we obtain

$$y_2 = c_1 e^{2t}, \qquad y_1 = c_1 t e^{2t} + c_2 e^{2t}. \tag{32}$$

Thus two independent solutions of the system (30) are

$$\mathbf{y}^{(1)}(t) = \begin{pmatrix} 1 \\ 0 \end{pmatrix} e^{2t}, \qquad \mathbf{y}^{(2)}(t) = \begin{pmatrix} t \\ 1 \end{pmatrix} e^{2t}, \tag{33}$$

and the corresponding fundamental matrix is

$$\hat{\boldsymbol{\Psi}}(t) = \begin{pmatrix} e^{2t} & t e^{2t} \\ 0 & e^{2t} \end{pmatrix}. \tag{34}$$

Since $\hat{\boldsymbol{\Psi}}(0) = \mathbf{I}$, we can also identify the matrix in Eq. (34) as $\exp(\mathbf{J}t)$. The same result can be reached by calculating powers of \mathbf{J} and substituting them into the exponential

[6]Camille Jordan (1838–1921), professor at the École Polytechnique and the Collège de France, made important contributions to analysis, topology, and especially to algebra. The Jordan form of a matrix appeared in his influential book *Traité des substitutions et des équations algébriques*, published in 1870.

series (see Problems 19 through 21). To obtain a fundamental matrix for the original system we now form the product

$$\Psi(t) = \mathbf{T}\exp(\mathbf{J}t) = \begin{pmatrix} e^{2t} & te^{2t} \\ -e^{2t} & -e^{2t} - te^{2t} \end{pmatrix}, \tag{35}$$

which is the same as the fundamental matrix given in Eq. (25).

PROBLEMS

In each of Problems 1 through 6 find the general solution of the given system of equations. In each of Problems 1 through 4 also draw a direction field, sketch a few trajectories, and describe how the solutions behave as $t \to \infty$.

1. $\mathbf{x}' = \begin{pmatrix} 3 & -4 \\ 1 & -1 \end{pmatrix} \mathbf{x}$

2. $\mathbf{x}' = \begin{pmatrix} 4 & -2 \\ 8 & -4 \end{pmatrix} \mathbf{x}$

3. $\mathbf{x}' = \begin{pmatrix} -\frac{3}{2} & 1 \\ -\frac{1}{4} & -\frac{1}{2} \end{pmatrix} \mathbf{x}$

4. $\mathbf{x}' = \begin{pmatrix} -3 & \frac{5}{2} \\ -\frac{5}{2} & 2 \end{pmatrix} \mathbf{x}$

5. $\mathbf{x}' = \begin{pmatrix} 1 & 1 & 1 \\ 2 & 1 & -1 \\ 0 & -1 & 1 \end{pmatrix} \mathbf{x}$

6. $\mathbf{x}' = \begin{pmatrix} 0 & 1 & 1 \\ 1 & 0 & 1 \\ 1 & 1 & 0 \end{pmatrix} \mathbf{x}$

In each of Problems 7 through 10 find the solution of the given initial value problem. Draw the trajectory of the solution in the x_1x_2-plane and also draw the graph of x_1 versus t.

7. $\mathbf{x}' = \begin{pmatrix} 1 & -4 \\ 4 & -7 \end{pmatrix} \mathbf{x}, \quad \mathbf{x}(0) = \begin{pmatrix} 3 \\ 2 \end{pmatrix}$

8. $\mathbf{x}' = \begin{pmatrix} -\frac{5}{2} & \frac{3}{2} \\ -\frac{3}{2} & \frac{1}{2} \end{pmatrix} \mathbf{x}, \quad \mathbf{x}(0) = \begin{pmatrix} 3 \\ -1 \end{pmatrix}$

9. $\mathbf{x}' = \begin{pmatrix} 2 & \frac{3}{2} \\ -\frac{3}{2} & -1 \end{pmatrix} \mathbf{x}, \quad \mathbf{x}(0) = \begin{pmatrix} 3 \\ -2 \end{pmatrix}$

10. $\mathbf{x}' = \begin{pmatrix} 3 & 9 \\ -1 & -3 \end{pmatrix} \mathbf{x}, \quad \mathbf{x}(0) = \begin{pmatrix} 2 \\ 4 \end{pmatrix}$

In each of Problems 11 and 12 find the solution of the given initial value problem. Draw the corresponding trajectory in $x_1x_2x_3$-space and also draw the graph of x_1 versus t.

11. $\mathbf{x}' = \begin{pmatrix} 1 & 0 & 0 \\ -4 & 1 & 0 \\ 3 & 6 & 2 \end{pmatrix} \mathbf{x}, \quad \mathbf{x}(0) = \begin{pmatrix} -1 \\ 2 \\ -30 \end{pmatrix}$

12. $\mathbf{x}' = \begin{pmatrix} -\frac{5}{2} & 1 & 1 \\ 1 & -\frac{5}{2} & 1 \\ 1 & 1 & -\frac{5}{2} \end{pmatrix} \mathbf{x}, \quad \mathbf{x}(0) = \begin{pmatrix} 2 \\ 3 \\ -1 \end{pmatrix}$

In each of Problems 13 and 14 solve the given system of equations by the method of Problem 19 of Section 7.5. Assume that $t > 0$.

13. $t\mathbf{x}' = \begin{pmatrix} 3 & -4 \\ 1 & -1 \end{pmatrix} \mathbf{x}$

14. $t\mathbf{x}' = \begin{pmatrix} 1 & -4 \\ 4 & -7 \end{pmatrix} \mathbf{x}$

15. Show that all solutions of the system

$$\mathbf{x}' = \begin{pmatrix} a & b \\ c & d \end{pmatrix} \mathbf{x}$$

approach zero as $t \to \infty$ if and only if $a + d < 0$ and $ad - bc > 0$. Compare this result with that of Problem 38 in Section 3.5.

16. Consider again the electric circuit in Problem 26 of Section 7.6. This circuit is described by the system of differential equations

$$\frac{d}{dt} \begin{pmatrix} I \\ V \end{pmatrix} = \begin{pmatrix} 0 & \dfrac{1}{L} \\ -\dfrac{1}{C} & -\dfrac{1}{RC} \end{pmatrix} \begin{pmatrix} I \\ V \end{pmatrix}.$$

 (a) Show that the eigenvalues are real and equal if $L = 4R^2C$.

 (b) Suppose that $R = 1$ ohm, $C = 1$ farad, and $L = 4$ henrys. Suppose also that $I(0) = 1$ ampere and $V(0) = 2$ volts. Find $I(t)$ and $V(t)$.

Eigenvalues of Multiplicity 3. If the matrix \mathbf{A} has an eigenvalue of algebraic multiplicity 3, then there may be either one, two, or three corresponding linearly independent eigenvectors. The general solution of the system $\mathbf{x}' = \mathbf{A}\mathbf{x}$ is different, depending on the number of eigenvectors associated with the triple eigenvalue. As noted in the text, there is no difficulty if there are three eigenvectors, since then there are three independent solutions of the form $\mathbf{x} = \boldsymbol{\xi} e^{rt}$. The following two problems illustrate the solution procedure for a triple eigenvalue with one or two eigenvectors, respectively.

17. Consider the system

$$\mathbf{x}' = \mathbf{A}\mathbf{x} = \begin{pmatrix} 1 & 1 & 1 \\ 2 & 1 & -1 \\ -3 & 2 & 4 \end{pmatrix} \mathbf{x}. \qquad (i)$$

 (a) Show that $r = 2$ is an eigenvalue of algebraic multiplicity 3 of the coefficient matrix \mathbf{A} and that there is only one corresponding eigenvector, namely

$$\boldsymbol{\xi}^{(1)} = \begin{pmatrix} 0 \\ 1 \\ -1 \end{pmatrix}.$$

 (b) Using the information in part (a), write down one solution $\mathbf{x}^{(1)}(t)$ of the system (i). There is no other solution of the purely exponential form $\mathbf{x} = \boldsymbol{\xi} e^{rt}$.

 (c) To find a second solution assume that $\mathbf{x} = \boldsymbol{\xi} t e^{2t} + \boldsymbol{\eta} e^{2t}$. Show that $\boldsymbol{\xi}$ and $\boldsymbol{\eta}$ satisfy the equations

$$(\mathbf{A} - 2\mathbf{I})\boldsymbol{\xi} = \mathbf{0}, \qquad (\mathbf{A} - 2\mathbf{I})\boldsymbol{\eta} = \boldsymbol{\xi}.$$

Since $\boldsymbol{\xi}$ has already been found in part (a), solve the second equation for $\boldsymbol{\eta}$. Neglect the multiple of $\boldsymbol{\xi}^{(1)}$ that appears in $\boldsymbol{\eta}$, since it leads only to a multiple of the first solution $\mathbf{x}^{(1)}$. Then write down a second solution $\mathbf{x}^{(2)}(t)$ of the system (i).

 (d) To find a third solution assume that $\mathbf{x} = \boldsymbol{\xi}(t^2/2)e^{2t} + \boldsymbol{\eta} t e^{2t} + \boldsymbol{\zeta} e^{2t}$. Show that $\boldsymbol{\xi}, \boldsymbol{\eta}$, and $\boldsymbol{\zeta}$ satisfy the equations

$$(\mathbf{A} - 2\mathbf{I})\boldsymbol{\xi} = \mathbf{0}, \qquad (\mathbf{A} - 2\mathbf{I})\boldsymbol{\eta} = \boldsymbol{\xi}, \qquad (\mathbf{A} - 2\mathbf{I})\boldsymbol{\zeta} = \boldsymbol{\eta}.$$

The first two equations are the same as in part (c), so solve the third equation for ζ, again neglecting the multiple of $\xi^{(1)}$ that appears. Then write down a third solution $\mathbf{x}^{(3)}(t)$ of the system (i).

(e) Write down a fundamental matrix $\Psi(t)$ for the system (i).

(f) Form a matrix \mathbf{T} with the eigenvector $\xi^{(1)}$ in the first column and the generalized eigenvectors η and ζ in the second and third columns. Then find \mathbf{T}^{-1} and form the product $\mathbf{J} = \mathbf{T}^{-1}\mathbf{AT}$. The matrix \mathbf{J} is the Jordan form of \mathbf{A}.

18. Consider the system

$$\mathbf{x}' = \mathbf{Ax} = \begin{pmatrix} 5 & -3 & -2 \\ 8 & -5 & -4 \\ -4 & 3 & 3 \end{pmatrix} \mathbf{x}. \tag{i}$$

(a) Show that $r = 1$ is a triple eigenvalue of the coefficient matrix \mathbf{A} and that there are only two linearly independent eigenvectors, which we may take as

$$\xi^{(1)} = \begin{pmatrix} 1 \\ 0 \\ 2 \end{pmatrix}, \qquad \xi^{(2)} = \begin{pmatrix} 0 \\ 2 \\ -3 \end{pmatrix}. \tag{ii}$$

Find two linearly independent solutions $\mathbf{x}^{(1)}(t)$ and $\mathbf{x}^{(2)}(t)$ of Eq. (i).

(b) To find a third solution assume that $\mathbf{x} = \xi t e^t + \eta e^t$; then show that ξ and η must satisfy

$$(\mathbf{A} - \mathbf{I})\xi = \mathbf{0}, \tag{iii}$$

$$(\mathbf{A} - \mathbf{I})\eta = \xi. \tag{iv}$$

(c) Show that $\xi = c_1\xi^{(1)} + c_2\xi^{(2)}$, where c_1 and c_2 are arbitrary constants, is the most general solution of Eq. (iii). Show that in order for us to solve Eq. (iv), it is necessary that $c_1 = c_2$.

(d) It is convenient to choose $c_1 = c_2 = 2$. For this choice, show that

$$\xi = \begin{pmatrix} 2 \\ 4 \\ -2 \end{pmatrix}, \qquad \eta = \begin{pmatrix} 0 \\ 0 \\ -1 \end{pmatrix}, \tag{v}$$

where we have dropped the multiples of $\xi^{(1)}$ and $\xi^{(2)}$ that appear in η. Use the results given in Eqs. (v) to find a third linearly independent solution $\mathbf{x}^{(3)}(t)$ of Eq. (i).

(e) Write down a fundamental matrix $\Psi(t)$ for the system (i).

(f) Form a matrix \mathbf{T} with the eigenvector $\xi^{(1)}$ in the first column and with the eigenvector ξ and the generalized eigenvector η from Eqs. (v) in the other two columns. Find \mathbf{T}^{-1} and form the product $\mathbf{J} = \mathbf{T}^{-1}\mathbf{AT}$. The matrix \mathbf{J} is the Jordan form of \mathbf{A}.

19. Let $\mathbf{J} = \begin{pmatrix} \lambda & 1 \\ 0 & \lambda \end{pmatrix}$, where λ is an arbitrary real number.

(a) Find $\mathbf{J}^2, \mathbf{J}^3$, and \mathbf{J}^4.

(b) Use an inductive argument to show that $\mathbf{J}^n = \begin{pmatrix} \lambda^n & n\lambda^{n-1} \\ 0 & \lambda^n \end{pmatrix}$.

(c) Determine $\exp(\mathbf{J}t)$.

(d) Use $\exp(\mathbf{J}t)$ to solve the initial value problem $\mathbf{x}' = \mathbf{Jx}, \mathbf{x}(0) = \mathbf{x}^0$.

20. Let

$$\mathbf{J} = \begin{pmatrix} \lambda & 0 & 0 \\ 0 & \lambda & 1 \\ 0 & 0 & \lambda \end{pmatrix},$$

where λ is an arbitrary real number.
(a) Find $\mathbf{J}^2, \mathbf{J}^3$, and \mathbf{J}^4.
(b) Use an inductive argument to show that

$$\mathbf{J}^n = \begin{pmatrix} \lambda^n & 0 & 0 \\ 0 & \lambda^n & n\lambda^{n-1} \\ 0 & 0 & \lambda^n \end{pmatrix}.$$

(c) Determine $\exp(\mathbf{J}t)$.
(d) Observe that if you choose $\lambda = 1$, then the matrix \mathbf{J} in this problem is the same as the matrix \mathbf{J} in Problem 18(f). Using the matrix \mathbf{T} from Problem 18(f), form the product $\mathbf{T}\exp(\mathbf{J}t)$ with $\lambda = 1$. Is the resulting matrix the same as the fundamental matrix $\boldsymbol{\Psi}(t)$ in Problem 18(e)? If not, explain the discrepancy.

21. Let

$$\mathbf{J} = \begin{pmatrix} \lambda & 1 & 0 \\ 0 & \lambda & 1 \\ 0 & 0 & \lambda \end{pmatrix},$$

where λ is an arbitrary real number.
(a) Find $\mathbf{J}^2, \mathbf{J}^3$, and \mathbf{J}^4.
(b) Use an inductive argument to show that

$$\mathbf{J}^n = \begin{pmatrix} \lambda^n & n\lambda^{n-1} & [n(n-1)/2]\lambda^{n-2} \\ 0 & \lambda^n & n\lambda^{n-1} \\ 0 & 0 & \lambda^n \end{pmatrix}.$$

(c) Determine $\exp(\mathbf{J}t)$.
(d) Observe that if you choose $\lambda = 2$, then the matrix \mathbf{J} in this problem is the same as the matrix \mathbf{J} in Problem 17(f). Using the matrix \mathbf{T} from Problem 17(f), form the product $\mathbf{T}\exp(\mathbf{J}t)$ with $\lambda = 2$. Observe that the resulting matrix is the same as the fundamental matrix $\boldsymbol{\Psi}(t)$ in Problem 17(e).

7.9 Nonhomogeneous Linear Systems

In this section we turn to the nonhomogeneous system

$$\mathbf{x}' = \mathbf{P}(t)\mathbf{x} + \mathbf{g}(t), \tag{1}$$

where the $n \times n$ matrix $\mathbf{P}(t)$ and $n \times 1$ vector $\mathbf{g}(t)$ are continuous for $\alpha < t < \beta$. By the same argument as in Section 3.6 (see also Problem 16 in this section), the general solution of Eq. (1) can be expressed as

$$\mathbf{x} = c_1\mathbf{x}^{(1)}(t) + \cdots + c_n\mathbf{x}^{(n)}(t) + \mathbf{v}(t), \tag{2}$$

where $c_1\mathbf{x}^{(1)}(t) + \cdots + c_n\mathbf{x}^{(n)}(t)$ is the general solution of the homogeneous system $\mathbf{x}' = \mathbf{P}(t)\mathbf{x}$, and $\mathbf{v}(t)$ is a particular solution of the nonhomogeneous system (1). We will briefly describe several methods for determining $\mathbf{v}(t)$.

Diagonalization. We begin with systems of the form

$$\mathbf{x}' = \mathbf{A}\mathbf{x} + \mathbf{g}(t), \tag{3}$$

where \mathbf{A} is an $n \times n$ diagonalizable constant matrix. By diagonalizing the coefficient matrix \mathbf{A}, as indicated in Section 7.7, we can transform Eq. (3) into a system of equations that is readily solvable.

Let \mathbf{T} be the matrix whose columns are the eigenvectors $\boldsymbol{\xi}^{(1)}, \ldots, \boldsymbol{\xi}^{(n)}$ of \mathbf{A}, and define a new dependent variable \mathbf{y} by

$$\mathbf{x} = \mathbf{T}\mathbf{y}. \tag{4}$$

Then, substituting for \mathbf{x} in Eq. (3), we obtain

$$\mathbf{T}\mathbf{y}' = \mathbf{A}\mathbf{T}\mathbf{y} + \mathbf{g}(t).$$

When we multiply by \mathbf{T}^{-1}, it follows that

$$\mathbf{y}' = (\mathbf{T}^{-1}\mathbf{A}\mathbf{T})\mathbf{y} + \mathbf{T}^{-1}\mathbf{g}(t) = \mathbf{D}\mathbf{y} + \mathbf{h}(t), \tag{5}$$

where $\mathbf{h}(t) = \mathbf{T}^{-1}\mathbf{g}(t)$ and where \mathbf{D} is the diagonal matrix whose diagonal entries are the eigenvalues r_1, \ldots, r_n of \mathbf{A}, arranged in the same order as the corresponding eigenvectors $\boldsymbol{\xi}^{(1)}, \ldots, \boldsymbol{\xi}^{(n)}$ that appear as columns of \mathbf{T}. Equation (5) is a system of n *uncoupled* equations for $y_1(t), \ldots, y_n(t)$; as a consequence, the equations can be solved separately. In scalar form Eq. (5) has the form

$$y_j'(t) = r_j y_j(t) + h_j(t), \qquad j = 1, \ldots, n, \tag{6}$$

where $h_j(t)$ is a certain linear combination of $g_1(t), \ldots, g_n(t)$. Equation (6) is a first order linear equation and can be solved by the methods of Section 2.1. In fact, we have

$$y_j(t) = e^{r_j t} \int_{t_0}^{t} e^{-r_j s} h_j(s)\, ds + c_j e^{r_j t}, \qquad j = 1, \ldots, n, \tag{7}$$

where the c_j are arbitrary constants. Finally, the solution \mathbf{x} of Eq. (3) is obtained from Eq. (4). When multiplied by the transformation matrix \mathbf{T}, the second term on the right side of Eq. (7) produces the general solution of the homogeneous equation $\mathbf{x}' = \mathbf{A}\mathbf{x}$, while the first term on the right side of Eq. (7) yields a particular solution of the nonhomogeneous system (3).

EXAMPLE 1

Find the general solution of the system

$$\mathbf{x}' = \begin{pmatrix} -2 & 1 \\ 1 & -2 \end{pmatrix} \mathbf{x} + \begin{pmatrix} 2e^{-t} \\ 3t \end{pmatrix} = \mathbf{A}\mathbf{x} + \mathbf{g}(t). \tag{8}$$

Proceeding as in Section 7.5, we find that the eigenvalues of the coefficient matrix are $r_1 = -3$ and $r_2 = -1$ and that the corresponding eigenvectors are

$$\boldsymbol{\xi}^{(1)} = \begin{pmatrix} 1 \\ -1 \end{pmatrix}, \qquad \boldsymbol{\xi}^{(2)} = \begin{pmatrix} 1 \\ 1 \end{pmatrix}. \tag{9}$$

Thus the general solution of the homogeneous system is

$$\mathbf{x} = c_1 \begin{pmatrix} 1 \\ -1 \end{pmatrix} e^{-3t} + c_2 \begin{pmatrix} 1 \\ 1 \end{pmatrix} e^{-t}. \tag{10}$$

Before writing down the matrix \mathbf{T} of eigenvectors, we recall that eventually we must find \mathbf{T}^{-1}. The coefficient matrix \mathbf{A} is real and symmetric, so we can use the result stated at the end of Section 7.3: \mathbf{T}^{-1} is simply the adjoint or (since \mathbf{T} is real) the transpose of \mathbf{T}, provided that the eigenvectors of \mathbf{A} are normalized so that $(\boldsymbol{\xi}, \boldsymbol{\xi}) = 1$. Hence, upon normalizing $\boldsymbol{\xi}^{(1)}$ and $\boldsymbol{\xi}^{(2)}$, we have

$$\mathbf{T} = \frac{1}{\sqrt{2}} \begin{pmatrix} 1 & 1 \\ -1 & 1 \end{pmatrix}, \qquad \mathbf{T}^{-1} = \frac{1}{\sqrt{2}} \begin{pmatrix} 1 & -1 \\ 1 & 1 \end{pmatrix}. \tag{11}$$

Letting $\mathbf{x} = \mathbf{T}\mathbf{y}$ and substituting for \mathbf{x} in Eq. (8), we obtain the following system of equations for the new dependent variable \mathbf{y}:

$$\mathbf{y}' = \mathbf{D}\mathbf{y} + \mathbf{T}^{-1}\mathbf{g}(t) = \begin{pmatrix} -3 & 0 \\ 0 & -1 \end{pmatrix} \mathbf{y} + \frac{1}{\sqrt{2}} \begin{pmatrix} 2e^{-t} - 3t \\ 2e^{-t} + 3t \end{pmatrix}. \tag{12}$$

Thus

$$y_1' + 3y_1 = \sqrt{2}\, e^{-t} - \frac{3}{\sqrt{2}} t,$$

$$y_2' + y_2 = \sqrt{2}\, e^{-t} + \frac{3}{\sqrt{2}} t. \tag{13}$$

Each of Eqs. (13) is a first order linear equation and so can be solved by the methods of Section 2.1. In this way we obtain

$$y_1 = \frac{\sqrt{2}}{2} e^{-t} - \frac{3}{\sqrt{2}} \left[\left(\frac{t}{3} \right) - \frac{1}{9} \right] + c_1 e^{-3t},$$

$$y_2 = \sqrt{2}\, t e^{-t} + \frac{3}{\sqrt{2}} (t - 1) + c_2 e^{-t}. \tag{14}$$

Finally, we write the solution in terms of the original variables:

$$\begin{aligned}
\mathbf{x} = \mathbf{T}\mathbf{y} &= \frac{1}{\sqrt{2}} \begin{pmatrix} y_1 + y_2 \\ -y_1 + y_2 \end{pmatrix} \\[2mm]
&= \begin{pmatrix} (c_1/\sqrt{2})e^{-3t} + \left[(c_2/\sqrt{2}) + \frac{1}{2} \right] e^{-t} + t - \frac{4}{3} + te^{-t} \\[2mm] -(c_1/\sqrt{2})e^{-3t} + \left[(c_2/\sqrt{2}) - \frac{1}{2} \right] e^{-t} + 2t - \frac{5}{3} + te^{-t} \end{pmatrix} \\[2mm]
&= k_1 \begin{pmatrix} 1 \\ -1 \end{pmatrix} e^{-3t} + k_2 \begin{pmatrix} 1 \\ 1 \end{pmatrix} e^{-t} + \frac{1}{2} \begin{pmatrix} 1 \\ -1 \end{pmatrix} e^{-t} + \begin{pmatrix} 1 \\ 1 \end{pmatrix} te^{-t} + \begin{pmatrix} 1 \\ 2 \end{pmatrix} t - \frac{1}{3} \begin{pmatrix} 4 \\ 5 \end{pmatrix},
\end{aligned} \tag{15}$$

where $k_1 = c_1/\sqrt{2}$ and $k_2 = c_2/\sqrt{2}$. The first two terms on the right side of Eq. (15) form the general solution of the homogeneous system corresponding to Eq. (8). The remaining terms are a particular solution of the nonhomogeneous system.

If the coefficient matrix \mathbf{A} in Eq. (3) is not diagonalizable (because of repeated eigenvalues and a shortage of eigenvectors), it can nevertheless be reduced to a Jordan form \mathbf{J} by a suitable transformation matrix \mathbf{T} involving both eigenvectors and generalized eigenvectors. In this case the differential equations for y_1, \ldots, y_n are not totally uncoupled since some rows of \mathbf{J} have two nonzero elements: an eigenvalue in the diagonal position and a 1 in the adjacent position to the right. However, the equations for y_1, \ldots, y_n can still be solved consecutively, starting with y_n. Then the solution of the original system (3) can be found by the relation $\mathbf{x} = \mathbf{T}\mathbf{y}$.

Undetermined Coefficients. A second way to find a particular solution of the nonhomogeneous system (1) is the method of undetermined coefficients. To make use of this method, one assumes the form of the solution with some or all of the coefficients unspecified, and then seeks to determine these coefficients so as to satisfy the differential equation. As a practical matter, this method is applicable only if the coefficient matrix \mathbf{P} is a constant matrix, and if the components of \mathbf{g} are polynomial, exponential, or sinusoidal functions, or sums or products of these. In these cases the correct form of the solution can be predicted in a simple and systematic manner. The procedure for choosing the form of the solution is substantially the same as that given in Section 3.6 for linear second order equations. The main difference is illustrated by the case of a nonhomogeneous term of the form $\mathbf{u}e^{\lambda t}$, where λ is a simple root of the characteristic equation. In this situation, rather than assuming a solution of the form $\mathbf{a}te^{\lambda t}$, it is necessary to use $\mathbf{a}te^{\lambda t} + \mathbf{b}e^{\lambda t}$, where \mathbf{a} and \mathbf{b} are determined by substituting into the differential equation.

EXAMPLE 2

Use the method of undetermined coefficients to find a particular solution of

$$\mathbf{x}' = \begin{pmatrix} -2 & 1 \\ 1 & -2 \end{pmatrix} \mathbf{x} + \begin{pmatrix} 2e^{-t} \\ 3t \end{pmatrix} = \mathbf{A}\mathbf{x} + \mathbf{g}(t). \tag{16}$$

This is the same system of equations as in Example 1. To use the method of undetermined coefficients, we write $\mathbf{g}(t)$ in the form

$$\mathbf{g}(t) = \begin{pmatrix} 2 \\ 0 \end{pmatrix} e^{-t} + \begin{pmatrix} 0 \\ 3 \end{pmatrix} t. \tag{17}$$

Then we assume that

$$\mathbf{x} = \mathbf{v}(t) = \mathbf{a}te^{-t} + \mathbf{b}e^{-t} + \mathbf{c}t + \mathbf{d}, \tag{18}$$

where $\mathbf{a}, \mathbf{b}, \mathbf{c}$, and \mathbf{d} are vectors to be determined. Observe that $r = -1$ is an eigenvalue of the coefficient matrix, and therefore we must include both $\mathbf{a}te^{-t}$ and $\mathbf{b}e^{-t}$ in the assumed solution. By substituting Eq. (18) into Eq. (16) and collecting terms, we obtain the following algebraic equations for $\mathbf{a}, \mathbf{b}, \mathbf{c}$, and \mathbf{d}:

$$\mathbf{A}\mathbf{a} = -\mathbf{a},$$

$$\mathbf{A}\mathbf{b} = \mathbf{a} - \mathbf{b} - \begin{pmatrix} 2 \\ 0 \end{pmatrix},$$

$$\mathbf{A}\mathbf{c} = -\begin{pmatrix} 0 \\ 3 \end{pmatrix}, \tag{19}$$

$$\mathbf{A}\mathbf{d} = \mathbf{c}.$$

From the first of Eqs. (19) we see that **a** is an eigenvector of **A** corresponding to the eigenvalue $r = -1$. Thus $\mathbf{a}^T = (\alpha, \alpha)$, where α is any nonzero constant. Then we find that the second of Eqs. (19) can be solved only if $\alpha = 1$ and that in this case

$$\mathbf{b} = k \begin{pmatrix} 1 \\ 1 \end{pmatrix} - \begin{pmatrix} 0 \\ 1 \end{pmatrix} \tag{20}$$

for any constant k. The simplest choice is $k = 0$, from which $\mathbf{b}^T = (0, -1)$. Then the third and fourth of Eqs. (19) yield $\mathbf{c}^T = (1, 2)$ and $\mathbf{d}^T = (-\frac{4}{3}, -\frac{5}{3})$, respectively. Finally, from Eq. (18) we obtain the particular solution

$$\mathbf{v}(t) = \begin{pmatrix} 1 \\ 1 \end{pmatrix} te^{-t} - \begin{pmatrix} 0 \\ 1 \end{pmatrix} e^{-t} + \begin{pmatrix} 1 \\ 2 \end{pmatrix} t - \frac{1}{3} \begin{pmatrix} 4 \\ 5 \end{pmatrix}. \tag{21}$$

The particular solution (21) is not identical to the one contained in Eq. (15) of Example 1 because the term in e^{-t} is different. However, if we choose $k = \frac{1}{2}$ in Eq. (20), then $\mathbf{b}^T = (\frac{1}{2}, -\frac{1}{2})$ and the two particular solutions agree.

Variation of Parameters. Now let us turn to more general problems in which the coefficient matrix is not constant or not diagonalizable. Let

$$\mathbf{x}' = \mathbf{P}(t)\mathbf{x} + \mathbf{g}(t), \tag{22}$$

where $\mathbf{P}(t)$ and $\mathbf{g}(t)$ are continuous on $\alpha < t < \beta$. Assume that a fundamental matrix $\mathbf{\Psi}(t)$ for the corresponding homogeneous system

$$\mathbf{x}' = \mathbf{P}(t)\mathbf{x} \tag{23}$$

has been found. We use the method of variation of parameters to construct a particular solution, and hence the general solution, of the nonhomogeneous system (22).

Since the general solution of the homogeneous system (23) is $\mathbf{\Psi}(t)\mathbf{c}$, it is natural to proceed as in Section 3.7 and to seek a solution of the nonhomogeneous system (22) by replacing the constant vector \mathbf{c} by a vector function $\mathbf{u}(t)$. Thus we assume that

$$\mathbf{x} = \mathbf{\Psi}(t)\mathbf{u}(t), \tag{24}$$

where $\mathbf{u}(t)$ is a vector to be found. Upon differentiating \mathbf{x} as given by Eq. (24) and requiring that Eq. (22) be satisfied, we obtain

$$\mathbf{\Psi}'(t)\mathbf{u}(t) + \mathbf{\Psi}(t)\mathbf{u}'(t) = \mathbf{P}(t)\mathbf{\Psi}(t)\mathbf{u}(t) + \mathbf{g}(t). \tag{25}$$

Since $\mathbf{\Psi}(t)$ is a fundamental matrix, $\mathbf{\Psi}'(t) = \mathbf{P}(t)\mathbf{\Psi}(t)$; hence Eq. (25) reduces to

$$\mathbf{\Psi}(t)\mathbf{u}'(t) = \mathbf{g}(t). \tag{26}$$

Recall that $\mathbf{\Psi}(t)$ is nonsingular on any interval where \mathbf{P} is continuous. Hence $\mathbf{\Psi}^{-1}(t)$ exists, and therefore

$$\mathbf{u}'(t) = \mathbf{\Psi}^{-1}(t)\mathbf{g}(t). \tag{27}$$

Thus for $\mathbf{u}(t)$ we can select any vector from the class of vectors that satisfy Eq. (27); these vectors are determined only up to an arbitrary additive constant vector; therefore we denote $\mathbf{u}(t)$ by

$$\mathbf{u}(t) = \int \mathbf{\Psi}^{-1}(t)\mathbf{g}(t)\, dt + \mathbf{c}, \tag{28}$$

where the constant vector **c** is arbitrary. If the integrals in Eq. (28) can be evaluated, then the general solution of the system (22) is found by substituting for $\mathbf{u}(t)$ from Eq. (28) in Eq. (24). However, even if the integrals cannot be evaluated, we can still write the general solution of Eq. (22) in the form

$$\mathbf{x} = \mathbf{\Psi}(t)\mathbf{c} + \mathbf{\Psi}(t) \int_{t_1}^{t} \mathbf{\Psi}^{-1}(s)\mathbf{g}(s)\, ds, \tag{29}$$

where t_1 is any point in the interval (α, β). Observe that the first term on the right side of Eq. (29) is the general solution of the corresponding homogeneous system (23), and the second term is a particular solution of Eq. (22).

Now let us consider the initial value problem consisting of the differential equation (22) and the initial condition

$$\mathbf{x}(t_0) = \mathbf{x}^0. \tag{30}$$

We can find the solution of this problem most conveniently if we choose the lower limit of integration in Eq. (29) to be the initial point t_0. Then the general solution of the differential equation is

$$\mathbf{x} = \mathbf{\Psi}(t)\mathbf{c} + \mathbf{\Psi}(t) \int_{t_0}^{t} \mathbf{\Psi}^{-1}(s)\mathbf{g}(s)\, ds. \tag{31}$$

For $t = t_0$ the integral in Eq. (31) is zero, so the initial condition (30) is also satisfied if we choose

$$\mathbf{c} = \mathbf{\Psi}^{-1}(t_0)\mathbf{x}^0. \tag{32}$$

Therefore

$$\mathbf{x} = \mathbf{\Psi}(t)\mathbf{\Psi}^{-1}(t_0)\mathbf{x}^0 + \mathbf{\Psi}(t) \int_{t_0}^{t} \mathbf{\Psi}^{-1}(s)\mathbf{g}(s)\, ds \tag{33}$$

is the solution of the given initial value problem. Again, although it is helpful to use $\mathbf{\Psi}^{-1}$ to write the solutions (29) and (33), it is usually better in particular cases to solve the necessary equations by row reduction than to calculate $\mathbf{\Psi}^{-1}$ and substitute into Eqs. (29) and (33).

The solution (33) takes a slightly simpler form if we use the fundamental matrix $\mathbf{\Phi}(t)$ satisfying $\mathbf{\Phi}(t_0) = \mathbf{I}$. In this case we have

$$\mathbf{x} = \mathbf{\Phi}(t)\mathbf{x}^0 + \mathbf{\Phi}(t) \int_{t_0}^{t} \mathbf{\Phi}^{-1}(s)\mathbf{g}(s)\, ds. \tag{34}$$

Equation (34) can be simplified further if the coefficient matrix $\mathbf{P}(t)$ is a constant matrix (see Problem 17).

EXAMPLE 3

Use the method of variation of parameters to find the general solution of the system

$$\mathbf{x}' = \begin{pmatrix} -2 & 1 \\ 1 & -2 \end{pmatrix} \mathbf{x} + \begin{pmatrix} 2e^{-t} \\ 3t \end{pmatrix} = \mathbf{A}\mathbf{x} + \mathbf{g}(t). \tag{35}$$

This is the same system of equations as in Examples 1 and 2.

The general solution of the corresponding homogeneous system was given in Eq. (10). Thus

$$\mathbf{\Psi}(t) = \begin{pmatrix} e^{-3t} & e^{-t} \\ -e^{-3t} & e^{-t} \end{pmatrix} \tag{36}$$

is a fundamental matrix. Then the solution \mathbf{x} of Eq. (35) is given by $\mathbf{x} = \mathbf{\Psi}(t)\mathbf{u}(t)$, where $\mathbf{u}(t)$ satisfies $\mathbf{\Psi}(t)\mathbf{u}'(t) = \mathbf{g}(t)$, or

$$\begin{pmatrix} e^{-3t} & e^{-t} \\ -e^{-3t} & e^{-t} \end{pmatrix} \begin{pmatrix} u_1' \\ u_2' \end{pmatrix} = \begin{pmatrix} 2e^{-t} \\ 3t \end{pmatrix}. \tag{37}$$

Solving Eq. (37) by row reduction, we obtain

$$u_1' = e^{2t} - \tfrac{3}{2}te^{3t},$$

$$u_2' = 1 + \tfrac{3}{2}te^t.$$

Hence

$$u_1(t) = \tfrac{1}{2}e^{2t} - \tfrac{1}{2}te^{3t} + \tfrac{1}{6}e^{3t} + c_1,$$

$$u_2(t) = t + \tfrac{3}{2}te^t - \tfrac{3}{2}e^t + c_2,$$

and

$$\mathbf{x} = \mathbf{\Psi}(t)\mathbf{u}(t)$$

$$= c_1 \begin{pmatrix} 1 \\ -1 \end{pmatrix} e^{-3t} + c_2 \begin{pmatrix} 1 \\ 1 \end{pmatrix} e^{-t} + \begin{pmatrix} 1 \\ 1 \end{pmatrix} te^{-t} + \frac{1}{2}\begin{pmatrix} 1 \\ -1 \end{pmatrix} e^{-t} + \begin{pmatrix} 1 \\ 2 \end{pmatrix} t - \frac{1}{3}\begin{pmatrix} 4 \\ 5 \end{pmatrix}, \tag{38}$$

which is the same as the solution obtained previously.

Laplace Transforms. We used the Laplace transform in Chapter 6 to solve linear equations of arbitrary order. It can also be used in very much the same way to solve systems of equations. Since the transform is an integral, the transform of a vector is computed component by component. Thus $\mathcal{L}\{\mathbf{x}(t)\}$ is the vector whose components are the transforms of the respective components of $\mathbf{x}(t)$, and similarly for $\mathcal{L}\{\mathbf{x}'(t)\}$. We will denote $\mathcal{L}\{\mathbf{x}(t)\}$ by $\mathbf{X}(s)$. Then, by an extension of Theorem 6.2.1 to vectors, we also have

$$\mathcal{L}\{\mathbf{x}'(t)\} = s\mathbf{X}(s) - \mathbf{x}(0). \tag{39}$$

EXAMPLE 4

Use the method of Laplace transforms to solve the system

$$\mathbf{x}' = \begin{pmatrix} -2 & 1 \\ 1 & -2 \end{pmatrix} \mathbf{x} + \begin{pmatrix} 2e^{-t} \\ 3t \end{pmatrix} = \mathbf{A}\mathbf{x} + \mathbf{g}(t). \tag{40}$$

This is the same system of equations as in Examples 1 through 3.

We take the Laplace transform of each term in Eq. (40), obtaining

$$s\mathbf{X}(s) - \mathbf{x}(0) = \mathbf{A}\mathbf{X}(s) + \mathbf{G}(s), \tag{41}$$

where $\mathbf{G}(s)$ is the transform of $\mathbf{g}(t)$. The transform $\mathbf{G}(s)$ is given by

$$\mathbf{G}(s) = \begin{pmatrix} 2/(s+1) \\ 3/s^2 \end{pmatrix}. \tag{42}$$

We will simplify the remaining calculations by assuming that $\mathbf{x}(t)$ satisfies the initial condition $\mathbf{x}(0) = \mathbf{0}$. Then Eq. (41) becomes

$$(s\mathbf{I} - \mathbf{A})\mathbf{X}(s) = \mathbf{G}(s), \tag{43}$$

where, as usual, \mathbf{I} is the identity matrix. Consequently, $\mathbf{X}(s)$ is given by

$$\mathbf{X}(s) = (s\mathbf{I} - \mathbf{A})^{-1}\mathbf{G}(s). \tag{44}$$

The matrix $(s\mathbf{I} - \mathbf{A})^{-1}$ is called the **transfer matrix** because multiplying it by the transform of the input vector $\mathbf{g}(t)$ yields the transform of the output vector $\mathbf{x}(t)$. In this example we have

$$s\mathbf{I} - \mathbf{A} = \begin{pmatrix} s+2 & -1 \\ -1 & s+2 \end{pmatrix}, \tag{45}$$

and by a straightforward calculation we obtain

$$(s\mathbf{I} - \mathbf{A})^{-1} = \frac{1}{(s+1)(s+3)} \begin{pmatrix} s+2 & 1 \\ 1 & s+2 \end{pmatrix}. \tag{46}$$

Then, substituting from Eqs. (42) and (46) in Eq. (44) and carrying out the indicated multiplication, we find that

$$\mathbf{X}(s) = \begin{pmatrix} \dfrac{2(s+2)}{(s+1)^2(s+3)} + \dfrac{3}{s^2(s+1)(s+3)} \\[3mm] \dfrac{2}{(s+1)^2(s+3)} + \dfrac{3(s+2)}{s^2(s+1)(s+3)} \end{pmatrix}. \tag{47}$$

Finally, we need to obtain the solution $\mathbf{x}(t)$ from its transform $\mathbf{X}(s)$. This can be done by expanding the expressions in Eq. (47) in partial fractions and using Table 6.2.1, or (more efficiently) by using appropriate computer software. In any case, after some simplification the result is

$$\mathbf{x}(t) = \begin{pmatrix} 2 \\ 1 \end{pmatrix} e^{-t} - \frac{2}{3} \begin{pmatrix} 1 \\ -1 \end{pmatrix} e^{-3t} + \begin{pmatrix} 1 \\ 1 \end{pmatrix} te^{-t} + \begin{pmatrix} 1 \\ 2 \end{pmatrix} t - \frac{1}{3} \begin{pmatrix} 4 \\ 5 \end{pmatrix}. \tag{48}$$

Equation (48) gives the particular solution of the system (40) that satisfies the initial condition $\mathbf{x}(0) = \mathbf{0}$. As a result, it differs slightly from the particular solutions obtained in the preceding three examples. To obtain the general solution of Eq. (40) you must add to the expression in Eq. (48) the general solution (10) of the homogeneous system corresponding to Eq. (40).

Each of the methods for solving nonhomogeneous equations has some advantages and disadvantages. The method of undetermined coefficients requires no integration, but it is limited in scope and may entail the solution of several sets of algebraic equations. The method of diagonalization requires finding the inverse of the transformation matrix and the solution of a set of uncoupled first order linear equations, followed by a matrix multiplication. Its main advantage is that for Hermitian coefficient matrices, the inverse of the transformation matrix can be written down without calculation—a feature that is more important for large systems. The method of Laplace transforms involves a matrix inversion to find the transfer matrix, followed by a multiplication, and finally by the determination of the inverse transform of each term in the resulting expression. It is particularly useful in problems with forcing functions that involve discontinuous or impulsive terms. Variation of parameters is the most general method. On the other hand, it involves the solution of a set of

linear algebraic equations with variable coefficients, followed by an integration and a matrix multiplication, so it may also be the most complicated from a computational viewpoint. For many small systems with constant coefficients, such as the one in the examples in this section, all of these methods work well, and there may be little reason to select one over another.

PROBLEMS In each of Problems 1 through 12 find the general solution of the given system of equations.

1. $\mathbf{x}' = \begin{pmatrix} 2 & -1 \\ 3 & -2 \end{pmatrix} \mathbf{x} + \begin{pmatrix} e^t \\ t \end{pmatrix}$

2. $\mathbf{x}' = \begin{pmatrix} 1 & \sqrt{3} \\ \sqrt{3} & -1 \end{pmatrix} \mathbf{x} + \begin{pmatrix} e^t \\ \sqrt{3}e^{-t} \end{pmatrix}$

3. $\mathbf{x}' = \begin{pmatrix} 2 & -5 \\ 1 & -2 \end{pmatrix} \mathbf{x} + \begin{pmatrix} -\cos t \\ \sin t \end{pmatrix}$

4. $\mathbf{x}' = \begin{pmatrix} 1 & 1 \\ 4 & -2 \end{pmatrix} \mathbf{x} + \begin{pmatrix} e^{-2t} \\ -2e^t \end{pmatrix}$

5. $\mathbf{x}' = \begin{pmatrix} 4 & -2 \\ 8 & -4 \end{pmatrix} \mathbf{x} + \begin{pmatrix} t^{-3} \\ -t^{-2} \end{pmatrix}, \quad t > 0$

6. $\mathbf{x}' = \begin{pmatrix} -4 & 2 \\ 2 & -1 \end{pmatrix} \mathbf{x} + \begin{pmatrix} t^{-1} \\ 2t^{-1} + 4 \end{pmatrix}, \quad t > 0$

7. $\mathbf{x}' = \begin{pmatrix} 1 & 1 \\ 4 & 1 \end{pmatrix} \mathbf{x} + \begin{pmatrix} 2 \\ -1 \end{pmatrix} e^t$

8. $\mathbf{x}' = \begin{pmatrix} 2 & -1 \\ 3 & -2 \end{pmatrix} \mathbf{x} + \begin{pmatrix} 1 \\ -1 \end{pmatrix} e^t$

9. $\mathbf{x}' = \begin{pmatrix} -\frac{5}{4} & \frac{3}{4} \\ \frac{3}{4} & -\frac{5}{4} \end{pmatrix} \mathbf{x} + \begin{pmatrix} 2t \\ e^t \end{pmatrix}$

10. $\mathbf{x}' = \begin{pmatrix} -3 & \sqrt{2} \\ \sqrt{2} & -2 \end{pmatrix} \mathbf{x} + \begin{pmatrix} 1 \\ -1 \end{pmatrix} e^{-t}$

11. $\mathbf{x}' = \begin{pmatrix} 2 & -5 \\ 1 & -2 \end{pmatrix} \mathbf{x} + \begin{pmatrix} 0 \\ \cos t \end{pmatrix}, \quad 0 < t < \pi$

12. $\mathbf{x}' = \begin{pmatrix} 2 & -5 \\ 1 & -2 \end{pmatrix} \mathbf{x} + \begin{pmatrix} \csc t \\ \sec t \end{pmatrix}, \quad \frac{\pi}{2} < t < \pi$

13. The electric circuit shown in Figure 7.9.1 is described by the system of differential equations

$$\frac{d\mathbf{x}}{dt} = \begin{pmatrix} -\frac{1}{2} & -\frac{1}{8} \\ 2 & -\frac{1}{2} \end{pmatrix} \mathbf{x} + \begin{pmatrix} \frac{1}{2} \\ 0 \end{pmatrix} I(t), \tag{i}$$

where x_1 is the current through the inductor, x_2 is the voltage drop across the capacitor, and $I(t)$ is the current supplied by the external source.

FIGURE 7.9.1 The circuit in Problem 13.

(a) Determine a fundamental matrix $\Psi(t)$ for the homogeneous system corresponding to Eq. (i). Refer to Problem 25 of Section 7.6.

(b) If $I(t) = e^{-t/2}$, determine the solution of the system (i) that also satisfies the initial conditions $\mathbf{x}(0) = \mathbf{0}$.

In each of Problems 14 and 15 verify that the given vector is the general solution of the corresponding homogeneous system, and then solve the nonhomogeneous system. Assume that $t > 0$.

14. $t\mathbf{x}' = \begin{pmatrix} 2 & -1 \\ 3 & -2 \end{pmatrix} \mathbf{x} + \begin{pmatrix} 1 - t^2 \\ 2t \end{pmatrix}, \qquad \mathbf{x}^{(c)} = c_1 \begin{pmatrix} 1 \\ 1 \end{pmatrix} t + c_2 \begin{pmatrix} 1 \\ 3 \end{pmatrix} t^{-1}$

15. $t\mathbf{x}' = \begin{pmatrix} 3 & -2 \\ 2 & -2 \end{pmatrix} \mathbf{x} + \begin{pmatrix} -2t \\ t^4 - 1 \end{pmatrix}, \qquad \mathbf{x}^{(c)} = c_1 \begin{pmatrix} 1 \\ 2 \end{pmatrix} t^{-1} + c_2 \begin{pmatrix} 2 \\ 1 \end{pmatrix} t^2$

16. Let $\mathbf{x} = \boldsymbol{\phi}(t)$ be the general solution of $\mathbf{x}' = \mathbf{P}(t)\mathbf{x} + \mathbf{g}(t)$, and let $\mathbf{x} = \mathbf{v}(t)$ be some particular solution of the same system. By considering the difference $\boldsymbol{\phi}(t) - \mathbf{v}(t)$, show that $\boldsymbol{\phi}(t) = \mathbf{u}(t) + \mathbf{v}(t)$, where $\mathbf{u}(t)$ is the general solution of the homogeneous system $\mathbf{x}' = \mathbf{P}(t)\mathbf{x}$.

17. Consider the initial value problem

$$\mathbf{x}' = \mathbf{A}\mathbf{x} + \mathbf{g}(t), \qquad \mathbf{x}(0) = \mathbf{x}^0.$$

(a) By referring to Problem 15(c) in Section 7.7, show that

$$\mathbf{x} = \boldsymbol{\Phi}(t)\mathbf{x}^0 + \int_0^t \boldsymbol{\Phi}(t - s)\mathbf{g}(s)\, ds.$$

(b) Show also that

$$\mathbf{x} = \exp(\mathbf{A}t)\mathbf{x}^0 + \int_0^t \exp[\mathbf{A}(t - s)]\mathbf{g}(s)\, ds.$$

Compare these results with those of Problem 27 in Section 3.7.

REFERENCES

Further information on matrices and linear algebra is available in any introductory book on the subject. The following is a representative sample:

Anton, H. and Rorres, C., *Elementary Linear Algebra* (8th ed.) (New York: Wiley, 2000).

Johnson, L. W., Riess, R. D., and Arnold, J. T., *Introduction to Linear Algebra* (5th ed.) (Reading, MA: Addison-Wesley, 2001).

Kolman, B., *Elementary Linear Algebra* (8th ed.) (Upper Saddle River, NJ: Prentice-Hall, 2003).

Leon, S. J., *Linear Algebra with Applications* (6th ed.) (New York: Macmillan, 2002).

Strang, G., *Linear Algebra and Its Applications* (4th ed.)(New York: Academic Press, 1998).

Numerical Methods

Up to this point we have discussed methods for solving differential equations by using analytical techniques such as integration or series expansions. Usually, the emphasis was on finding an exact expression for the solution. Unfortunately, there are many important problems in engineering and science, especially nonlinear ones, to which these methods either do not apply or are very complicated to use. In this chapter we discuss an alternative approach, the use of numerical approximation methods to obtain an accurate approximation to the solution of an initial value problem. We present these methods in the simplest possible context, namely, a single scalar first order equation. However, they can readily be extended to systems of first order equations, and this is outlined briefly in Section 8.6. The procedures described here can be executed easily on personal computers as well as on some pocket calculators.

8.1 The Euler or Tangent Line Method

To discuss the development and use of numerical procedures, we will concentrate mainly on the first order initial value problem consisting of the differential equation

$$\frac{dy}{dt} = f(t, y) \tag{1}$$

and the initial condition

$$y(t_0) = y_0. \tag{2}$$

We assume that the functions f and f_y are continuous on some rectangle in the ty-plane containing the point (t_0, y_0). Then, by Theorem 2.4.2, there exists a unique solution $y = \phi(t)$ of the given problem in some interval about t_0. If Eq. (1) is nonlinear, then

the interval of existence of the solution may be difficult to determine and may have no simple relationship to the function f. However, in all our discussions we assume that there is a unique solution of the initial value problem (1), (2) in the interval of interest.

In Section 2.7 we described the oldest and simplest numerical approximation method, namely, the Euler or tangent line method. This method is expressed by the equation

$$y_{n+1} = y_n + f(t_n, y_n)(t_{n+1} - t_n), \qquad n = 0, 1, 2, \ldots. \tag{3}$$

If the step size has a uniform value h and if we denote $f(t_n, y_n)$ by f_n, then Eq. (3) simplifies to

$$y_{n+1} = y_n + f_n h, \qquad n = 0, 1, 2, \ldots. \tag{4}$$

Euler's method consists of repeatedly evaluating Eq. (3) or (4), using the result of each step to execute the next step. In this way you obtain a sequence of values $y_0, y_1, y_2, \ldots, y_n, \ldots$ that approximate the values of the solution $\phi(t)$ at the points $t_0, t_1, t_2, \ldots, t_n, \ldots$.

In Section 2.7 we pointed out that a computer program for Euler's method has the structure given below. The specific instructions can be written in any convenient programming language.

The Euler Method

Step 1. **define** $f(t, y)$
Step 2. **input** initial values $t0$ and $y0$
Step 3. **input** step size h and number of steps n
Step 4. **output** $t0$ and $y0$
Step 5. **for** j from 1 to n **do**
Step 6. $k1 = f(t, y)$
 $y = y + h * k1$
 $t = t + h$
Step 7. **output** t and y
Step 8. **end**

Some examples of Euler's method appear in Section 2.7. As another example, consider the initial value problem

$$y' = 1 - t + 4y, \tag{5}$$

$$y(0) = 1. \tag{6}$$

Equation (5) is a first order linear equation, and it is easily verified that the solution satisfying the initial condition (6) is

$$y = \phi(t) = \tfrac{1}{4}t - \tfrac{3}{16} + \tfrac{19}{16}e^{4t}. \tag{7}$$

Since the exact solution is known, we do not need numerical methods to solve the initial value problem (5), (6). On the other hand, the availability of the exact solution makes it easy to determine the accuracy of any numerical procedure that we use on this problem. We will use this problem throughout the chapter to illustrate and compare different numerical methods. The solutions of Eq. (5) diverge rather rapidly from each other, so we should expect that it will be fairly difficult to approximate

the solution (7) well over any considerable interval. Indeed, this is the reason for choosing this particular problem; it will be relatively easy to observe the benefits of using more accurate methods.

EXAMPLE 1

Using the Euler formula (4) and step sizes $h = 0.05, 0.025, 0.01$, and 0.001, determine approximate values of the solution $y = \phi(t)$ of the problem (5), (6) on the interval $0 \leq t \leq 2$.

The indicated calculations were carried out on a computer, and some of the results are shown in Table 8.1.1. Their accuracy is not particularly impressive. For $h = 0.01$ the percentage error is 3.85% at $t = 0.5$, 7.49% at $t = 1.0$, and 14.4% at $t = 2.0$. The corresponding percentage errors for $h = 0.001$ are 0.40%, 0.79%, and 1.58%, respectively. Observe that if $h = 0.001$, then it requires 2000 steps to traverse the interval from $t = 0$ to $t = 2$. Thus considerable computation is needed to obtain even reasonably good accuracy for this problem using the Euler method. When we discuss other numerical approximation methods later in this chapter, we will find that it is possible to obtain comparable or better accuracy with much larger step sizes and many fewer computational steps.

TABLE 8.1.1 A Comparison of Results for the Numerical Solution of $y' = 1 - t + 4y$, $y(0) = 1$ Using the Euler Method for Different Step Sizes h

t	$h = 0.05$	$h = 0.025$	$h = 0.01$	$h = 0.001$	Exact
0.0	1.0000000	1.0000000	1.0000000	1.0000000	1.0000000
0.1	1.5475000	1.5761188	1.5952901	1.6076289	1.6090418
0.2	2.3249000	2.4080117	2.4644587	2.5011159	2.5053299
0.3	3.4333560	3.6143837	3.7390345	3.8207130	3.8301388
0.4	5.0185326	5.3690304	5.6137120	5.7754845	5.7942260
0.5	7.2901870	7.9264062	8.3766865	8.6770692	8.7120041
1.0	45.588400	53.807866	60.037126	64.382558	64.897803
1.5	282.07187	361.75945	426.40818	473.55979	479.25919
2.0	1745.6662	2432.7878	3029.3279	3484.1608	3540.2001

To begin to investigate the errors in using numerical approximations, and also to suggest ways to construct more accurate algorithms, it is helpful to mention some alternative ways to look at the Euler method.

First, let us write the differential equation (1) at the point $t = t_n$ in the form

$$\frac{d\phi}{dt}(t_n) = f[t_n, \phi(t_n)]. \tag{8}$$

Then we approximate the derivative in Eq. (8) by the corresponding (forward) difference quotient, obtaining

$$\frac{\phi(t_{n+1}) - \phi(t_n)}{t_{n+1} - t_n} \cong f[t_n, \phi(t_n)]. \tag{9}$$

Finally, if we replace $\phi(t_{n+1})$ and $\phi(t_n)$ by their approximate values y_{n+1} and y_n, respectively, and solve for y_{n+1}, we obtain the Euler formula (3).

Another way to proceed is to write the problem as an integral equation. Since $y = \phi(t)$ is a solution of the initial value problem (1), (2), by integrating from t_n to t_{n+1} we obtain

$$\int_{t_n}^{t_{n+1}} \phi'(t)\, dt = \int_{t_n}^{t_{n+1}} f[t, \phi(t)]\, dt,$$

or

$$\phi(t_{n+1}) = \phi(t_n) + \int_{t_n}^{t_{n+1}} f[t, \phi(t)]\, dt. \tag{10}$$

The integral in Eq. (10) is represented geometrically as the area under the curve in Figure 8.1.1 between $t = t_n$ and $t = t_{n+1}$. If we approximate the integral by replacing $f[t, \phi(t)]$ by its value $f[t_n, \phi(t_n)]$ at $t = t_n$, then we are approximating the actual area by the area of the shaded rectangle. In this way we obtain

$$\phi(t_{n+1}) \cong \phi(t_n) + f[t_n, \phi(t_n)](t_{n+1} - t_n)$$
$$= \phi(t_n) + hf[t_n, \phi(t_n)]. \tag{11}$$

Finally, to obtain an approximation y_{n+1} for $\phi(t_{n+1})$, we make a second approximation by replacing $\phi(t_n)$ by its approximate value y_n in Eq. (11). This gives the Euler formula $y_{n+1} = y_n + hf(t_n, y_n)$. A more accurate algorithm can be obtained by approximating the integral more accurately. This is discussed in Section 8.2.

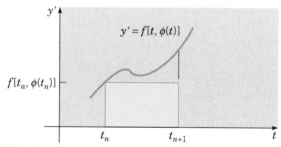

FIGURE 8.1.1 Integral derivation of the Euler method.

A third approach is to assume that the solution $y = \phi(t)$ has a Taylor series about the point t_n. Then

$$\phi(t_n + h) = \phi(t_n) + \phi'(t_n)h + \phi''(t_n)\frac{h^2}{2!} + \cdots,$$

or

$$\phi(t_{n+1}) = \phi(t_n) + f[t_n, \phi(t_n)]h + \phi''(t_n)\frac{h^2}{2!} + \cdots. \tag{12}$$

If the series is terminated after the first two terms, and $\phi(t_{n+1})$ and $\phi(t_n)$ are replaced by their approximate values y_{n+1} and y_n, we again obtain the Euler formula (4). If more terms in the series are retained, a more accurate formula is obtained. Further,

by using a Taylor series with a remainder, it is possible to estimate the magnitude of the error in the formula. This is discussed later in this section.

The Backward Euler Formula. A variation on the Euler formula can be obtained by approximating the derivative in Eq. (8) by the backward difference quotient $[\phi(t_n) - \phi(t_{n-1})]/h$ instead of the forward difference quotient used in Eq. (9). In this way we obtain

$$\phi(t_n) - \phi(t_{n-1}) \cong hf(t_n, y_n),$$

or

$$y_n = y_{n-1} + hf(t_n, y_n).$$

Stepping the index up from n to $n + 1$, we obtain the backward Euler formula

$$y_{n+1} = y_n + hf(t_{n+1}, y_{n+1}). \tag{13}$$

Assuming that y_n is known and y_{n+1} is to be calculated, observe that Eq. (13) does not provide an explicit formula for y_{n+1}. Rather, it is an equation that implicitly defines y_{n+1} and must be solved to determine the value of y_{n+1}. How difficult this is depends entirely on the nature of the function f.

**EXAMPLE
2**

Use the backward Euler formula (13) and step sizes $h = 0.05, 0.025, 0.01$, and 0.001 to find approximate values of the solution of the initial value problem (5), (6) on the interval $0 \le t \le 2$.
 For this problem the backward Euler formula (13) becomes

$$y_{n+1} = y_n + h(1 - t_{n+1} + 4y_{n+1}).$$

We will show the first two steps in detail so that it will be clear how the method works. At the first step we have

$$y_1 = y_0 + h(1 - t_1 + 4y_1) = 1 + (0.05)(1 - 0.05 + 4y_1).$$

Solving this equation for y_1, we obtain

$$y_1 = 1.0475/0.8 = 1.309375.$$

Observe that because the differential equation is linear, the implicit equation for y_1 is also linear and therefore easy to solve. Next,

$$y_2 = y_1 + h(1 - t_2 + 4y_2) = 1.309375 + (0.05)(1 - 0.1 + 4y_2),$$

which leads to

$$y_2 = 1.354375/0.8 = 1.69296875.$$

Continuing the computations on a computer, we obtain the results shown in Table 8.1.2. The values given by the backward Euler method are uniformly too large for this problem, whereas the values obtained from the Euler method were too small. In this problem the errors are somewhat larger for the backward Euler method than for the Euler method, although for small values of h the differences are insignificant. Since the backward Euler method appears to be no more accurate than the Euler method, and is somewhat more complicated, a natural question is why it should even be mentioned. The answer is that it is the simplest example of a class of methods known as backward differentiation formulas that are very useful for certain types of differential equations. We will return to this issue later in this chapter.

TABLE 8.1.2 A Comparison of Results for the Numerical Solution of $y' = 1 - t + 4y$, $y(0) = 1$ Using the Backward Euler Method for Different Step Sizes h

t	$h = 0.05$	$h = 0.025$	$h = 0.01$	$h = 0.001$	Exact
0	1.0000000	1.0000000	1.0000000	1.0000000	1.0000000
0.1	1.6929688	1.6474375	1.6236638	1.6104634	1.6090418
0.2	2.7616699	2.6211306	2.5491368	2.5095731	2.5053299
0.3	4.4174530	4.0920886	3.9285724	3.8396379	3.8301388
0.4	6.9905516	6.3209569	5.9908303	5.8131282	5.7942260
0.5	10.996956	9.7050002	9.0801473	8.7472667	8.7120041
1.0	103.06171	80.402761	70.452395	65.419964	64.897803
1.5	959.44236	661.00731	542.12432	485.05825	479.25919
2.0	8934.0696	5435.7294	4172.7228	3597.4478	3540.2001

Errors in Numerical Approximations. The use of a numerical procedure, such as the Euler formula, to solve an initial value problem raises a number of questions that must be answered before the approximate numerical solution can be accepted as satisfactory. One of these is the question of **convergence**. That is, as the step size h tends to zero, do the values of the numerical solution $y_1, y_2, \ldots, y_n, \ldots$ approach the corresponding values of the actual solution? If we assume that the answer is affirmative, there remains the important practical question of how rapidly the numerical approximation converges to the solution. In other words, how small a step size is needed in order to guarantee a given level of accuracy? We want to use a step size that is small enough to ensure the required accuracy, but not too small. An unnecessarily small step size slows down the calculations, makes them more expensive, and in some cases may even cause a loss of accuracy.

There are two fundamental sources of error in solving an initial value problem numerically. Let us first assume that our computer is such that we can carry out all computations with complete accuracy; that is, we can retain an infinite number of decimal places. The difference E_n between the solution $y = \phi(t)$ of the initial value problem (1), (2) and its numerical approximation is given by

$$E_n = \phi(t_n) - y_n \tag{14}$$

and is known as the **global truncation error**. It arises from two causes: First, at each step we use an approximate formula to determine y_{n+1}; second, the input data at each step are only approximately correct since in general $\phi(t_n)$ is not equal to y_n. If we assume that $y_n = \phi(t_n)$, then the only error in going one step is due to the use of an approximate formula. This error is known as the **local truncation error** e_n.

The second fundamental source of error is that we carry out the computations in arithmetic with only a finite number of digits. This leads to a **round-off error** R_n defined by

$$R_n = y_n - Y_n, \tag{15}$$

where Y_n is the value *actually computed* from the given numerical method.

The absolute value of the total error in computing $\phi(t_n)$ is given by

$$|\phi(t_n) - Y_n| = |\phi(t_n) - y_n + y_n - Y_n|. \tag{16}$$

Making use of the triangle inequality, $|a + b| \leq |a| + |b|$, we obtain, from Eq. (16),

$$|\phi(t_n) - Y_n| \leq |\phi(t_n) - y_n| + |y_n - Y_n|$$

$$\leq |E_n| + |R_n|. \tag{17}$$

Thus the total error is bounded by the sum of the absolute values of the truncation and round-off errors. For the numerical procedures discussed in this book, it is possible to obtain useful estimates of the truncation error. However, we limit our discussion primarily to the local truncation error, which is somewhat simpler. The round-off error is clearly more random in nature. It depends on the type of computer used, the sequence in which the computations are carried out, the method of rounding off, and so forth. An analysis of round-off error is beyond the scope of this book, but it is possible to say more about it than one might at first expect (see, for example, Henrici). Some of the dangers from round-off error are discussed in Problems 25 through 27 and in Section 8.5.

Local Truncation Error for the Euler Method. Let us assume that the solution $y = \phi(t)$ of the initial value problem (1), (2) has a continuous second derivative in the interval of interest. To ensure this, we can assume that f, f_t, and f_y are continuous. Observe that if f has these properties and if ϕ is a solution of the initial value problem (1), (2), then

$$\phi'(t) = f[t, \phi(t)],$$

and, by the chain rule,

$$\phi''(t) = f_t[t, \phi(t)] + f_y[t, \phi(t)]\phi'(t)$$

$$= f_t[t, \phi(t)] + f_y[t, \phi(t)]f[t, \phi(t)]. \tag{18}$$

Since the right side of this equation is continuous, ϕ'' is also continuous.

Then, making use of a Taylor polynomial with a remainder to expand ϕ about t_n, we obtain

$$\phi(t_n + h) = \phi(t_n) + \phi'(t_n)h + \tfrac{1}{2}\phi''(\bar{t}_n)h^2, \tag{19}$$

where \bar{t}_n is some point in the interval $t_n < \bar{t}_n < t_n + h$. Subtracting Eq. (4) from Eq. (19), and noting that $\phi(t_n + h) = \phi(t_{n+1})$ and $\phi'(t_n) = f[t_n, \phi(t_n)]$, we find that

$$\phi(t_{n+1}) - y_{n+1} = [\phi(t_n) - y_n] + h\{f[t_n, \phi(t_n)] - f(t_n, y_n)\} + \tfrac{1}{2}\phi''(\bar{t}_n)h^2. \tag{20}$$

To compute the local truncation error we apply Eq. (20) to the true solution $\phi(t)$; that is, we take y_n to be $\phi(t_n)$. Then we immediately see from Eq. (20) that the local truncation error e_{n+1} is

$$e_{n+1} = \phi(t_{n+1}) - y_{n+1} = \tfrac{1}{2}\phi''(\bar{t}_n)h^2. \tag{21}$$

Thus the local truncation error for the Euler method is proportional to the square of the step size h, and the proportionality factor depends on the second derivative of the solution ϕ. The expression given by Eq. (21) depends on n and, in general, is different for each step. A uniform bound, valid on an interval $[a, b]$, is given by

$$|e_n| \leq Mh^2/2, \tag{22}$$

where M is the maximum of $|\phi''(t)|$ on the interval $[a, b]$. Since Eq. (22) is based on a consideration of the worst possible case—that is, the largest possible value of

$|\phi''(t)|$—it may well be a considerable overestimate of the actual local truncation error in some parts of the interval $[a, b]$. One use of Eq. (22) is to choose a step size that will result in a local truncation error no greater than some given tolerance level. For example, if the local truncation error must be no greater than ϵ, then from Eq. (22) we have

$$h \leq \sqrt{2\epsilon/M}. \tag{23}$$

The primary difficulty in using Eq. (21), (22), or (23) lies in estimating $|\phi''(t)|$ or M. However, the central fact expressed by these equations is that the local truncation error is proportional to h^2. Thus, if h is reduced by a factor of $\frac{1}{2}$, then the error is reduced by $\frac{1}{4}$, and so on.

More important than the local truncation error is the global truncation error E_n. The analysis for estimating E_n is more difficult than that for e_n. However, knowing the local truncation error, we can make an *intuitive* estimate of the global truncation error at a fixed $T > t_0$ as follows. Suppose that we take n steps in going from t_0 to $T = t_0 + nh$. In each step the error is at most $Mh^2/2$; thus the error in n steps is at most $nMh^2/2$. Noting that $n = (T - t_0)/h$, we find that the global truncation error for the Euler method in going from t_0 to T is bounded by

$$n\frac{Mh^2}{2} = (T - t_0)\frac{Mh}{2}. \tag{24}$$

This argument is not complete since it does not take into account the effect that an error at one step will have in succeeding steps. Nevertheless, it can be shown that the global truncation error in using the Euler method on a finite interval is no greater than a constant times h; see Problem 23 for more details. The Euler method is called a first order method because its global truncation error is proportional to the first power of the step size.

Because it is more accessible, we will hereafter use the local truncation error as our principal measure of the accuracy of a numerical method and for comparing different methods. If we have *a priori* information about the solution of the given initial value problem, we can use the result (21) to obtain more precise information about how the local truncation error varies with t. As an example, consider the illustrative problem

$$y' = 1 - t + 4y, \qquad y(0) = 1 \tag{25}$$

on the interval $0 \leq t \leq 2$. Let $y = \phi(t)$ be the solution of the initial value problem (25). Then, as noted previously,

$$\phi(t) = (4t - 3 + 19e^{4t})/16$$

and therefore

$$\phi''(t) = 19e^{4t}.$$

Equation (21) then states that

$$e_{n+1} = \frac{19e^{4\bar{t}_n}h^2}{2}, \qquad t_n < \bar{t}_n < t_n + h. \tag{26}$$

The appearance of the factor 19 and the rapid growth of e^{4t} explain why the results in this section with $h = 0.05$ were not very accurate.

For instance, the error in the first step is

$$e_1 = \phi(t_1) - y_1 = \frac{19e^{4\bar{t}_0}(0.0025)}{2}, \qquad 0 < \bar{t}_0 < 0.05.$$

It is clear that e_1 is positive, and since $e^{4\bar{t}_0} < e^{0.2}$, we have

$$e_1 \leq \frac{19e^{0.2}(0.0025)}{2} \cong 0.02901. \tag{27}$$

Note also that $e^{4\bar{t}_0} > 1$; hence $e_1 > 19(0.0025)/2 = 0.02375$. The actual error is 0.02542. It follows from Eq. (26) that the error becomes progressively worse with increasing t; this is also clearly shown by the results in Table 8.1.1. Similar computations for bounds for the local truncation error give

$$1.0617 \cong \frac{19e^{3.8}(0.0025)}{2} \leq e_{20} \leq \frac{19e^{4}(0.0025)}{2} \cong 1.2967 \tag{28}$$

in going from 0.95 to 1.0 and

$$57.96 \cong \frac{19e^{7.8}(0.0025)}{2} \leq e_{40} \leq \frac{19e^{8}(0.0025)}{2} \cong 70.80 \tag{29}$$

in going from 1.95 to 2.0.

These results indicate that, for this problem, the local truncation error is about 2500 times larger near $t = 2$ than near $t = 0$. Thus, to reduce the local truncation error to an acceptable level throughout $0 \leq t \leq 2$, one must choose a step size h based on an analysis near $t = 2$. Of course, this step size will be much smaller than necessary near $t = 0$. For example, to achieve a local truncation error of 0.01 for this problem, we need a step size of about 0.00059 near $t = 2$ and a step size of about 0.032 near $t = 0$. The use of a uniform step size that is smaller than necessary over much of the interval results in more calculations than necessary, more time consumed, and possibly more danger of unacceptable round-off errors.

Another approach is to keep the local truncation error approximately constant throughout the interval by gradually reducing the step size as t increases. In the example problem we would need to reduce h by a factor of about 50 in going from $t = 0$ to $t = 2$. A method that provides for variations in the step size is called **adaptive**. All modern computer codes for solving differential equations have the capability of adjusting the step size as needed. We will return to this question in the next section.

PROBLEMS In each of Problems 1 through 6 find approximate values of the solution of the given initial value problem at $t = 0.1, 0.2, 0.3$, and 0.4.

(a) Use the Euler method with $h = 0.05$.
(b) Use the Euler method with $h = 0.025$.
(c) Use the backward Euler method with $h = 0.05$.
(d) Use the backward Euler method with $h = 0.025$.

1. $y' = 3 + t - y, \qquad y(0) = 1$ 2. $y' = 5t - 3\sqrt{y}, \qquad y(0) = 2$

3. $y' = 2y - 3t, \qquad y(0) = 1$ 4. $y' = 2t + e^{-ty}, \qquad y(0) = 1$

5. $y' = \dfrac{y^2 + 2ty}{3 + t^2}, \qquad y(0) = 0.5$ 6. $y' = (t^2 - y^2)\sin y, \qquad y(0) = -1$

In each of Problems 7 through 12 find approximate values of the solution of the given initial value problem at $t = 0.5, 1.0, 1.5,$ and 2.0.

(a) Use the Euler method with $h = 0.025$.

(b) Use the Euler method with $h = 0.0125$.

(c) Use the backward Euler method with $h = 0.025$.

(d) Use the backward Euler method with $h = 0.0125$.

7. $y' = 0.5 - t + 2y, \qquad y(0) = 1$ 8. $y' = 5t - 3\sqrt{y}, \qquad y(0) = 2$

9. $y' = \sqrt{t + y}, \qquad y(0) = 3$ 10. $y' = 2t + e^{-ty}, \qquad y(0) = 1$

11. $y' = (4 - ty)/(1 + y^2), \qquad y(0) = -2$

12. $y' = (y^2 + 2ty)/(3 + t^2), \qquad y(0) = 0.5$

13. Complete the calculations leading to the entries in columns three and four of Table 8.1.1.

14. Complete the calculations leading to the entries in columns three and four of Table 8.1.2.

15. Using three terms in the Taylor series given in Eq. (12) and taking $h = 0.1$, determine approximate values of the solution of the illustrative example $y' = 1 - t + 4y, y(0) = 1$ at $t = 0.1$ and 0.2. Compare the results with those using the Euler method and with the exact values.

Hint: If $y' = f(t, y)$, what is y''?

In each of Problems 16 and 17 estimate the local truncation error for the Euler method in terms of the solution $y = \phi(t)$. Obtain a bound for e_{n+1} in terms of t and $\phi(t)$ that is valid on the interval $0 \le t \le 1$. By using a formula for the solution, obtain a more accurate error bound for e_{n+1}. For $h = 0.1$ compute a bound for e_1 and compare it with the actual error at $t = 0.1$. Also compute a bound for the error e_4 in the fourth step.

16. $y' = 2y - 1, \qquad y(0) = 1$ 17. $y' = \frac{1}{2} - t + 2y, \qquad y(0) = 1$

In each of Problems 18 through 21 obtain a formula for the local truncation error for the Euler method in terms of t and the solution ϕ.

18. $y' = t^2 + y^2, \qquad y(0) = 1$ 19. $y' = 5t - 3\sqrt{y}, \qquad y(0) = 2$

20. $y' = \sqrt{t + y}, \qquad y(1) = 3$ 21. $y' = 2t + e^{-ty}, \qquad y(0) = 1$

22. Consider the initial value problem

$$y' = \cos 5\pi t, \qquad y(0) = 1.$$

(a) Determine the solution $y = \phi(t)$ and draw a graph of $y = \phi(t)$ for $0 \le t \le 1$.

(b) Determine approximate values of $\phi(t)$ at $t = 0.2, 0.4,$ and 0.6 using the Euler method with $h = 0.2$. Draw a broken-line graph for the approximate solution and compare it with the graph of the exact solution.

(c) Repeat the computation of part (b) for $0 \le t \le 0.4$, but take $h = 0.1$.

(d) Show by computing the local truncation error that neither of these step sizes is sufficiently small. Determine a value of h to ensure that the local truncation error is less than 0.05 throughout the interval $0 \le t \le 1$. That such a small value of h is required results from the fact that max $|\phi''(t)|$ is large.

23. In this problem we discuss the global truncation error associated with the Euler method for the initial value problem $y' = f(t, y), y(t_0) = y_0$. Assuming that the functions f and f_y are continuous in a closed, bounded region R of the ty-plane that includes the point (t_0, y_0), it can be shown that there exists a constant L such that $|f(t, y) - f(t, \tilde{y})| < L|y - \tilde{y}|$, where (t, y) and (t, \tilde{y}) are any two points in R with the same t coordinate (see Problem 15 of Section 2.8). Further, we assume that f_t is continuous, so the solution ϕ has a continuous second derivative.

(a) Using Eq. (20), show that

$$|E_{n+1}| \leq |E_n| + h|f[t_n, \phi(t_n)] - f(t_n, y_n)| + \tfrac{1}{2}h^2|\phi''(\bar{t}_n)| \leq \alpha|E_n| + \beta h^2, \qquad \text{(i)}$$

where $\alpha = 1 + hL$ and $\beta = \max |\phi''(t)|/2$ on $t_0 \leq t \leq t_n$.

(b) Accepting without proof that if $E_0 = 0$, and if $|E_n|$ satisfies Eq. (i), then $|E_n| \leq \beta h^2(\alpha^n - 1)/(\alpha - 1)$ for $\alpha \neq 1$, show that

$$|E_n| \leq \frac{(1 + hL)^n - 1}{L}\beta h. \qquad \text{(ii)}$$

Equation (ii) gives a bound for $|E_n|$ in terms of h, L, n, and β. Notice that for a fixed h, this error bound increases with increasing n; that is, the error bound increases with distance from the starting point t_0.

(c) Show that $(1 + hL)^n \leq e^{nhL}$; hence

$$|E_n| \leq \frac{e^{nhL} - 1}{L}\beta h = \frac{e^{(t_n - t_0)L} - 1}{L}\beta h.$$

For a fixed point $T = t_0 + nh$ [that is, nh is constant and $h = (T - t_0)/n$], this error bound is of the form of a constant times h and approaches zero as $h \to 0$. Also note that for $nhL = (T - t_0)L$ small, the right side of the preceding equation is approximately $nh^2\beta = (T - t_0)\beta h$, which was obtained in Eq. (24) by an intuitive argument.

24. Derive an expression analogous to Eq. (21) for the local truncation error for the backward Euler formula.

 Hint: Construct a suitable Taylor approximation to $\phi(t)$ about $t = t_{n+1}$.

25. Using a step size $h = 0.05$ and the Euler method, but retaining only three digits throughout the computations, determine approximate values of the solution at $t = 0.1, 0.2, 0.3$, and 0.4 for each of the following initial value problems.

 (a) $y' = 1 - t + 4y, \qquad y(0) = 1$

 (b) $y' = 3 + t - y, \qquad y(0) = 1$

 (c) $y' = 2y - 3t, \qquad y(0) = 1$

 Compare the results with those obtained in Example 1 and in Problems 1 and 3. The small differences between some of those results rounded to three digits and the present results are due to round-off error. The round-off error would become important if the computation required many steps.

26. The following problem illustrates a danger that occurs because of round-off error when nearly equal numbers are subtracted and the difference is then multiplied by a large number. Evaluate the quantity

$$1000 \cdot \begin{vmatrix} 6.010 & 18.04 \\ 2.004 & 6.000 \end{vmatrix}$$

in the following ways:

 (a) First round each entry in the determinant to two digits.

 (b) First round each entry in the determinant to three digits.

 (c) Retain all four digits. Compare this value with the results in parts (a) and (b).

27. The distributive law $a(b - c) = ab - ac$ does not hold, in general, if the products are rounded off to a smaller number of digits. To show this in a specific case, take $a = 0.22$, $b = 3.19$, and $c = 2.17$. After each multiplication, round off the last digit.

8.2 Improvements on the Euler Method

Since for many problems the Euler method requires a very small step size to produce sufficiently accurate results, much effort has been devoted to the development of more efficient methods. In the next three sections we will discuss some of these methods. Consider the initial value problem

$$y' = f(t, y), \qquad y(t_0) = y_0 \qquad (1)$$

and let $y = \phi(t)$ denote its solution. Recall from Eq. (10) of Section 8.1 that by integrating the given differential equation from t_n to t_{n+1}, we obtain

$$\phi(t_{n+1}) = \phi(t_n) + \int_{t_n}^{t_{n+1}} f[t, \phi(t)]\,dt. \qquad (2)$$

The Euler formula

$$y_{n+1} = y_n + hf(t_n, y_n) \qquad (3)$$

is obtained by replacing $f[t, \phi(t)]$ in Eq. (2) by its approximate value $f(t_n, y_n)$ at the left endpoint of the interval of integration.

Improved Euler Formula. A better approximate formula can be obtained if the integrand in Eq. (2) is approximated more accurately. One way to do this is to replace the integrand by the average of its values at the two endpoints, namely, $\{f[t_n, \phi(t_n)] + f[t_{n+1}, \phi(t_{n+1})]\}/2$. This is equivalent to approximating the area under the curve in Figure 8.2.1 between $t = t_n$ and $t = t_{n+1}$ by the area of the shaded trapezoid. Further, we replace $\phi(t_n)$ and $\phi(t_{n+1})$ by their respective approximate values y_n and y_{n+1}. In this way we obtain, from Eq. (2),

$$y_{n+1} = y_n + \frac{f(t_n, y_n) + f(t_{n+1}, y_{n+1})}{2} h. \qquad (4)$$

Since the unknown y_{n+1} appears as one of the arguments of f on the right side of Eq. (4), this equation defines y_{n+1} implicitly rather than explicitly. Depending on the nature of the function f, it may be fairly difficult to solve Eq. (4) for y_{n+1}. This

FIGURE 8.2.1 Derivation of the improved Euler method.

difficulty can be overcome by replacing y_{n+1} on the right side of Eq. (4) by the value obtained using the Euler formula (3). Thus

$$y_{n+1} = y_n + \frac{f(t_n, y_n) + f[t_n + h, y_n + hf(t_n, y_n)]}{2}h$$

$$= y_n + \frac{f_n + f(t_n + h, y_n + hf_n)}{2}h, \tag{5}$$

where t_{n+1} has been replaced by $t_n + h$.

Equation (5) gives an explicit formula for computing y_{n+1}, the approximate value of $\phi(t_{n+1})$, in terms of the data at t_n. This formula is known as the **improved Euler formula** or the **Heun formula**. The improved Euler formula is an example of a two-stage method; that is, we first calculate $y_n + hf_n$ from the Euler formula and then use this result to calculate y_{n+1} from Eq. (5). The improved Euler formula (5) does represent an improvement over the Euler formula (3) because the local truncation error in using Eq. (5) is proportional to h^3, while for the Euler method it is proportional to h^2. This error estimate for the improved Euler formula is established in Problem 14. It can also be shown that for a finite interval the global truncation error for the improved Euler formula is bounded by a constant times h^2, so this method is a second order method. Note that this greater accuracy is achieved at the expense of more computational work, since it is now necessary to evaluate $f(t, y)$ twice in order to go from t_n to t_{n+1}.

If $f(t, y)$ depends only on t and not on y, then solving the differential equation $y' = f(t, y)$ reduces to integrating $f(t)$. In this case the improved Euler formula (5) becomes

$$y_{n+1} - y_n = \frac{h}{2}[f(t_n) + f(t_n + h)], \tag{6}$$

which is just the trapezoid rule for numerical integration.

EXAMPLE 1

Use the improved Euler formula (5) to calculate approximate values of the solution of the initial value problem

$$y' = 1 - t + 4y, \qquad y(0) = 1. \tag{7}$$

To make clear exactly what computations are required, we show a couple of steps in detail. For this problem $f(t, y) = 1 - t + 4y$; hence

$$f_n = 1 - t_n + 4y_n$$

and

$$f(t_n + h, y_n + hf_n) = 1 - (t_n + h) + 4(y_n + hf_n).$$

Further, $t_0 = 0, y_0 = 1$, and $f_0 = 1 - t_0 + 4y_0 = 5$. If $h = 0.025$, then

$$f(t_0 + h, y_0 + hf_0) = 1 - 0.025 + 4[1 + (0.025)(5)] = 5.475.$$

Then, from Eq. (5),

$$y_1 = 1 + (0.5)(5 + 5.475)(0.025) = 1.1309375. \tag{8}$$

At the second step we must calculate

$$f_1 = 1 - 0.025 + 4(1.1309375) = 5.49875,$$

$$y_1 + hf_1 = 1.1309375 + (0.025)(5.49875) = 1.26840625,$$

and

$$f(t_2, y_1 + hf_1) = 1 - 0.05 + 4(1.26840625) = 6.023625.$$

Then, from Eq. (5),

$$y_2 = 1.1309375 + (0.5)(5.49875 + 6.023625)(0.025) = 1.2749671875. \tag{9}$$

Further results for $0 \leq t \leq 2$ obtained by using the improved Euler method with $h = 0.025$ and $h = 0.01$ are given in Table 8.2.1. To compare the results of the improved Euler method with those of the Euler method, note that the improved Euler method requires two evaluations of f at each step, while the Euler method requires only one. This is significant because typically most of the computing time in each step is spent in evaluating f, so counting these evaluations is a reasonable way to estimate the total computing effort. Thus, for a given step size h, the improved Euler method requires twice as many evaluations of f as the Euler method. Alternatively, the improved Euler method for step size h requires the same number of evaluations of f as the Euler method with step size $h/2$.

TABLE 8.2.1 A Comparison of Results Using the Euler and Improved Euler Methods for the Initial Value Problem $y' = 1 - t + 4y, \ y(0) = 1$

	Euler		Improved Euler		
t	$h = 0.01$	$h = 0.001$	$h = 0.025$	$h = 0.01$	Exact
0	1.0000000	1.0000000	1.0000000	1.0000000	1.0000000
0.1	1.5952901	1.6076289	1.6079462	1.6088585	1.6090418
0.2	2.4644587	2.5011159	2.5020618	2.5047827	2.5053299
0.3	3.7390345	3.8207130	3.8228282	3.8289146	3.8301388
0.4	5.6137120	5.7754845	5.7796888	5.7917911	5.7942260
0.5	8.3766865	8.6770692	8.6849039	8.7074637	8.7120041
1.0	60.037126	64.382558	64.497931	64.830722	64.897803
1.5	426.40818	473.55979	474.83402	478.51588	479.25919
2.0	3029.3279	3484.1608	3496.6702	3532.8789	3540.2001

By referring to Table 8.2.1, you can see that the improved Euler method with $h = 0.025$ gives much better results than the Euler method with $h = 0.01$. Note that to reach $t = 2$ with these step sizes, the improved Euler method requires 160 evaluations of f, while the Euler method requires 200. More noteworthy is that the improved Euler method with $h = 0.025$ is also slightly more accurate than the Euler method with $h = 0.001$ (2000 evaluations of f). In other words, with something like one-twelfth of the computing effort, the improved Euler method yields results for this problem that are comparable to, or a bit better than, those generated by the Euler method. This illustrates that, compared to the Euler method, the improved Euler method is clearly more efficient, yielding substantially better results or requiring much less total computing effort, or both.

The percentage errors at $t = 2$ for the improved Euler method are 1.23% for $h = 0.025$ and 0.21% for $h = 0.01$.

A computer program for the Euler method can be readily modified to implement the improved Euler method instead. All that is required is to replace Step 6 in the algorithm in Section 8.1 by the following:

Step 6. $k1 = f(t, y)$
$k2 = f(t + h, y + h * k1)$
$y = y + (h/2) * (k1 + k2)$
$t = t + h$

Variation of Step Size. In Section 8.1 we mentioned the possibility of adjusting the step size as a calculation proceeds so as to maintain the local truncation error at a more or less constant level. The goal is to use no more steps than necessary and, at the same time, to keep some control over the accuracy of the approximation. Here we will describe how this can be done. Suppose that after n steps we have reached the point (t_n, y_n). We choose a step size h and calculate y_{n+1}. Next we need to estimate the error we have made in calculating y_{n+1}. Not knowing the actual solution, the best that we can do is to use a more accurate method and repeat the calculation starting from (t_n, y_n). For example, if we used the Euler method for the original calculation, we might repeat it with the improved Euler method. Then the difference between the two calculated values is an estimate e_{n+1}^{est} of the error in using the original method. If the estimated error is different from the error tolerance ϵ, then we adjust the step size and repeat the calculation. The key to making this adjustment efficiently is knowing how the local truncation error e_{n+1} depends on the step size h. For the Euler method the local truncation error is proportional to h^2, so to bring the estimated error down (or up) to the tolerance level ϵ, we must multiply the original step size by the factor $\sqrt{\epsilon/e_{n+1}^{est}}$.

To illustrate this procedure, consider the example problem (7),

$$y' = 1 - t + 4y, \qquad y(0) = 1.$$

You can verify that after one step with $h = 0.1$ we obtain the values 1.5 and 1.595 from the Euler method and the improved Euler method, respectively. Thus the estimated error in using the Euler method is 0.095. If we have chosen an error tolerance of 0.05, for instance, then we need to adjust the step size downward by the factor $\sqrt{0.05/0.095} \cong 0.73$. Rounding downward to be conservative, let us choose the adjusted step size $h = 0.07$. Then, from the Euler formula, we obtain

$$y_1 = 1 + (0.07)f(0, 1) = 1.35 \cong \phi(0.07).$$

Using the improved Euler method, we obtain $y_1 = 1.39655$, so the estimated error in using the Euler formula is 0.04655, which is slightly less than the specified tolerance. The actual error, based on a comparison with the solution itself, is somewhat greater, namely, 0.05122.

We can follow the same procedure at each step of the calculation, thereby keeping the local truncation error approximately constant throughout the entire numerical process. Modern adaptive codes for solving differential equations adjust the step size as they proceed in very much this way, although they usually use more accurate formulas than the Euler and improved Euler formulas. Consequently, they are able to achieve both efficiency and accuracy by using very small steps only where they are really needed.

PROBLEMS

In each of Problems 1 through 6 find approximate values of the solution of the given initial value problem at $t = 0.1, 0.2, 0.3$, and 0.4. Compare the results with those obtained by the Euler method and the backward Euler method in Section 8.1 and with the exact solution (if available).

(a) Use the improved Euler method with $h = 0.05$.

(b) Use the improved Euler method with $h = 0.025$.

(c) Use the improved Euler method with $h = 0.0125$.

1. $y' = 3 + t - y$, $y(0) = 1$ 2. $y' = 5t - 3\sqrt{y}$, $y(0) = 2$

3. $y' = 2y - 3t$, $y(0) = 1$ 4. $y' = 2t + e^{-ty}$, $y(0) = 1$

5. $y' = \dfrac{y^2 + 2ty}{3 + t^2}$, $y(0) = 0.5$ 6. $y' = (t^2 - y^2)\sin y$, $y(0) = -1$

In each of Problems 7 through 12 find approximate values of the solution of the given initial value problem at $t = 0.5, 1.0, 1.5$, and 2.0.

(a) Use the improved Euler method with $h = 0.025$.

(b) Use the improved Euler method with $h = 0.0125$.

7. $y' = 0.5 - t + 2y$, $y(0) = 1$ 8. $y' = 5t - 3\sqrt{y}$, $y(0) = 2$

9. $y' = \sqrt{t + y}$, $y(0) = 3$ 10. $y' = 2t + e^{-ty}$, $y(0) = 1$

11. $y' = (4 - ty)/(1 + y^2)$, $y(0) = -2$

12. $y' = (y^2 + 2ty)/(3 + t^2)$, $y(0) = 0.5$

13. Complete the calculations leading to the entries in columns four and five of Table 8.2.1.

14. In this problem we establish that the local truncation error for the improved Euler formula is proportional to h^3. If we assume that the solution ϕ of the initial value problem $y' = f(t, y)$, $y(t_0) = y_0$ has derivatives that are continuous through the third order (f has continuous second partial derivatives), then it follows that

$$\phi(t_n + h) = \phi(t_n) + \phi'(t_n)h + \frac{\phi''(t_n)}{2!}h^2 + \frac{\phi'''(\bar{t}_n)}{3!}h^3,$$

where $t_n < \bar{t}_n < t_n + h$. Assume that $y_n = \phi(t_n)$.

(a) Show that, for y_{n+1} as given by Eq. (5),

$$e_{n+1} = \phi(t_{n+1}) - y_{n+1}$$

$$= \frac{\phi''(t_n)h - \{f[t_n + h, y_n + hf(t_n, y_n)] - f(t_n, y_n)\}}{2!}h + \frac{\phi'''(\bar{t}_n)h^3}{3!}. \tag{i}$$

(b) Making use of the facts that $\phi''(t) = f_t[t, \phi(t)] + f_y[t, \phi(t)]\phi'(t)$ and that the Taylor approximation with a remainder for a function $F(t, y)$ of two variables is

$$F(a + h, b + k) = F(a, b) + F_t(a, b)h + F_y(a, b)k$$

$$+ \left. \frac{1}{2!}(h^2 F_{tt} + 2hkF_{ty} + k^2 F_{yy}) \right|_{x=\xi, y=\eta}$$

where ξ lies between a and $a + h$ and η lies between b and $b + k$, show that the first term on the right side of Eq. (i) is proportional to h^3 plus higher order terms. This is the desired result.

(c) Show that if $f(t, y)$ is linear in t and y, then $e_{n+1} = \phi'''(\bar{t}_n)h^3/6$, where $t_n < \bar{t}_n < t_{n+1}$. *Hint:* What are f_{tt}, f_{ty}, and f_{yy}?

15. Consider the improved Euler method for solving the illustrative initial value problem $y' = 1 - t + 4y, y(0) = 1$. Using the result of Problem 14(c) and the exact solution of the initial value problem, determine e_{n+1} and a bound for the error at any step on $0 \le t \le 2$. Compare this error with the one obtained in Eq. (26) of Section 8.1 using the Euler method. Also obtain a bound for e_1 for $h = 0.05$, and compare it with Eq. (27) of Section 8.1.

In each of Problems 16 and 17 use the actual solution $\phi(t)$ to determine e_{n+1} and a bound for e_{n+1} at any step on $0 \le t \le 1$ for the improved Euler method for the given initial value problem. Also obtain a bound for e_1 for $h = 0.1$, and compare it with the similar estimate for the Euler method and with the actual error using the improved Euler method.

16. $y' = 2y - 1, \qquad y(0) = 1$ 17. $y' = 0.5 - t + 2y, \qquad y(0) = 1$

In each of Problems 18 through 21 carry out one step of the Euler method and of the improved Euler method, using the step size $h = 0.1$. Suppose that a local truncation error no greater than 0.0025 is required. Estimate the step size that is needed for the Euler method to satisfy this requirement at the first step.

18. $y' = 0.5 - t + 2y, \qquad y(0) = 1$ 19. $y' = 5t - 3\sqrt{y}, \qquad y(0) = 2$

20. $y' = \sqrt{t + y}, \qquad y(0) = 3$ 21. $y' = (y^2 + 2ty)/(3 + t^2), \qquad y(0) = 0.5$

22. The **modified Euler formula** for the initial value problem $y' = f(t, y), y(t_0) = y_0$ is given by

$$y_{n+1} = y_n + hf[t_n + \tfrac{1}{2}h, y_n + \tfrac{1}{2}hf(t_n, y_n)].$$

Following the procedure outlined in Problem 14, show that the local truncation error in the modified Euler formula is proportional to h^3.

In each of Problems 23 through 26 use the modified Euler formula of Problem 22 with $h = 0.05$ to compute approximate values of the solution of the given initial value problem at $t = 0.1$, 0.2, 0.3, and 0.4. Compare the results with those obtained in Problems 1 through 4.

23. $y' = 3 + t - y, \qquad y(0) = 1$ 24. $y' = 5t - 3\sqrt{y}, \qquad y(0) = 2$

25. $y' = 2y - 3t, \qquad y(0) = 1$ 26. $y' = 2t + e^{-ty}, \qquad y(0) = 1$

27. Show that the modified Euler formula of Problem 22 is identical to the improved Euler formula of Eq. (5) for $y' = f(t, y)$ if f is linear in both t and y.

8.3 The Runge–Kutta Method

In preceding sections we have introduced the Euler formula, the backward Euler formula, and the improved Euler formula as ways to solve the initial value problem

$$y' = f(t, y), \qquad y(t_0) = y_0 \tag{1}$$

numerically. The local truncation errors for these methods are proportional to h^2, h^2, and h^3, respectively. The Euler and improved Euler methods belong to what is now called the Runge–Kutta[1] class of methods.

In this section we discuss the method originally developed by Runge and Kutta. This method is now called the classic fourth order four-stage Runge–Kutta method, but it is often referred to simply as *the* Runge–Kutta method, and we will follow this practice for brevity. This method has a local truncation error that is proportional to h^5. Thus it is two orders of magnitude more accurate than the improved Euler method and three orders of magnitude better than the Euler method. It is relatively simple to use and is sufficiently accurate to handle many problems efficiently. This is especially true of adaptive Runge–Kutta methods, in which provision is made to vary the step size as needed. We return to this issue at the end of the section.

The Runge–Kutta formula involves a weighted average of values of $f(t, y)$ at different points in the interval $t_n \leq t \leq t_{n+1}$. It is given by

$$y_{n+1} = y_n + h \left(\frac{k_{n1} + 2k_{n2} + 2k_{n3} + k_{n4}}{6} \right), \tag{2}$$

where

$$
\begin{aligned}
k_{n1} &= f(t_n, y_n) \\
k_{n2} &= f(t_n + \tfrac{1}{2}h, y_n + \tfrac{1}{2}hk_{n1}), \\
k_{n3} &= f(t_n + \tfrac{1}{2}h, y_n + \tfrac{1}{2}hk_{n2}), \\
k_{n4} &= f(t_n + h, y_n + hk_{n3}).
\end{aligned}
\tag{3}
$$

The sum $(k_{n1} + 2k_{n2} + 2k_{n3} + k_{n4})/6$ can be interpreted as an average slope. Note that k_{n1} is the slope at the left end of the interval, k_{n2} is the slope at the midpoint using the Euler formula to go from t_n to $t_n + h/2$, k_{n3} is a second approximation to the slope at the midpoint, and k_{n4} is the slope at $t_n + h$ using the Euler formula and the slope k_{n3} to go from t_n to $t_n + h$.

Although in principle it is not difficult to show that Eq. (2) differs from the Taylor expansion of the solution ϕ by terms that are proportional to h^5, the algebra is rather lengthy.[2] Thus we will simply accept the fact that the local truncation error in using Eq. (2) is proportional to h^5 and that for a finite interval the global truncation error is at most a constant times h^4. The earlier description of this method as a fourth order four-stage method reflects the facts that the global truncation error is of fourth order in the step size h and that there are four intermediate stages in the calculation (the calculation of k_{n1}, \ldots, k_{n4}).

Clearly the Runge–Kutta formula, Eqs. (2) and (3), is more complicated than any of the formulas discussed previously. This is of relatively little significance, however,

[1]Carl David Runge (1856–1927), German mathematician and physicist, worked for many years in spectroscopy. The analysis of data led him to consider problems in numerical computation, and the Runge–Kutta method originated in his paper on the numerical solution of differential equations in 1895. The method was extended to systems of equations in 1901 by M. Wilhelm Kutta (1867–1944). Kutta was a German mathematician and aerodynamicist who is also well known for his important contributions to classical airfoil theory.

[2]See, for example, Chapter 3 of the book by Henrici listed in the references.

since it is not hard to write a computer program to implement this method. Such a program has the same structure as the algorithm for the Euler method outlined in Section 8.1. To be specific, the lines in Step 6 in the Euler algorithm must be replaced by the following:

Step 6.
$$k1 = f(t, y)$$
$$k2 = f(t + 0.5 * h, y + 0.5 * h * k1)$$
$$k3 = f(t + 0.5 * h, y + 0.5 * h * k2)$$
$$k4 = f(t + h, y + h * k3)$$
$$y = y + (h/6) * (k1 + 2 * k2 + 2 * k3 + k4)$$
$$t = t + h$$

Note that if f does not depend on y, then

$$k_{n1} = f(t_n), \qquad k_{n2} = k_{n3} = f(t_n + h/2), \qquad k_{n4} = f(t_n + h), \qquad (4)$$

and Eq. (2) reduces to

$$y_{n+1} - y_n = \frac{h}{6}[f(t_n) + 4f(t_n + h/2) + f(t_n + h)]. \qquad (5)$$

Equation (5) can be identified as Simpson's[3] rule for the approximate evaluation of the integral of $y' = f(t)$. The fact that Simpson's rule has an error proportional to h^5 is consistent with the local truncation error in the Runge–Kutta formula.

Use the Runge–Kutta method to calculate approximate values of the solution $y = \phi(t)$ of the initial value problem

EXAMPLE 1

$$y' = 1 - t + 4y, \qquad y(0) = 1. \qquad (6)$$

Taking $h = 0.2$, we have

$$k_{01} = f(0, 1) = 5; \qquad hk_{01} = 1.0,$$
$$k_{02} = f(0 + 0.1, 1 + 0.5) = 6.9; \qquad hk_{02} = 1.38,$$
$$k_{03} = f(0 + 0.1, 1 + 0.69) = 7.66; \qquad hk_{03} = 1.532,$$
$$k_{04} = f(0 + 0.2, 1 + 1.532) = 10.928.$$

Thus

$$y_1 = 1 + \frac{0.2}{6}[5 + 2(6.9) + 2(7.66) + 10.928]$$
$$= 1 + 1.5016 = 2.5016.$$

Further results using the Runge–Kutta method with $h = 0.2, h = 0.1,$ and $h = 0.05$ are given in Table 8.3.1. Note that the Runge–Kutta method yields a value at $t = 2$ that differs from the exact solution by only 0.122% if the step size is $h = 0.1$, and by only 0.00903% if $h = 0.05$. In the latter case, the error is less than one part in ten thousand, and the calculated value at $t = 2$ is correct to four digits.

For comparison, note that both the Runge–Kutta method with $h = 0.05$ and the improved Euler method with $h = 0.025$ require 160 evaluations of f to reach $t = 2$. The improved Euler

[3]Simpson's rule is named for Thomas Simpson (1710–1761), an English mathematician and textbook author, who published it in 1743.

method yields a result at $t = 2$ that is in error by 1.23%. Although this error may be acceptable for some purposes, it is more than 135 times the error yielded by the Runge–Kutta method with comparable computing effort. Note also that the Runge–Kutta method with $h = 0.2$, or 40 evaluations of f, produces a value at $t = 2$ with an error of 1.40%, which is only slightly greater than the error in the improved Euler method with $h = 0.025$, or 160 evaluations of f. Thus we see again that a more accurate algorithm is more efficient; it produces better results with similar effort, or similar results with less effort.

TABLE 8.3.1 A Comparison of Results for the Numerical Solution of the Initial Value Problem $y' = 1 - t + 4y$, $y(0) = 1$

	Improved Euler	Runge–Kutta			Exact
t	$h = 0.025$	$h = 0.2$	$h = 0.1$	$h = 0.05$	
0	1.0000000	1.0000000	1.0000000	1.0000000	1.0000000
0.1	1.6079462		1.6089333	1.6090338	1.6090418
0.2	2.5020619	2.5016000	2.5050062	2.5053060	2.5053299
0.3	3.8228282		3.8294145	3.8300854	3.8301388
0.4	5.7796888	5.7776358	5.7927853	5.7941197	5.7942260
0.5	8.6849039		8.7093175	8.7118060	8.7120041
1.0	64.497931	64.441579	64.858107	64.894875	64.897803
1.5	474.83402		478.81928	479.22674	479.25919
2.0	3496.6702	3490.5574	3535.8667	3539.8804	3540.2001

The classic Runge–Kutta method suffers from the same shortcoming as other methods with a fixed step size for problems in which the local truncation error varies widely over the interval of interest. That is, a step size that is small enough to achieve satisfactory accuracy in some parts of the interval may be much smaller than necessary in other parts of the interval. This has stimulated the development of adaptive Runge–Kutta methods that provide for modifying the step size automatically as the computation proceeds, so as to maintain the local truncation error near or below a specified tolerance level. As explained in Section 8.2, this requires the estimation of the local truncation error at each step. One way to do this is to repeat the computation with a fifth order method—which has a local truncation error proportional to h^6—and then to use the difference between the two results as an estimate of the error. If this is done in a straightforward (unsophisticated) manner, then the use of the fifth order method requires at least five more evaluations of f at each step, in addition to those required originally by the fourth order method. However, if we make an appropriate choice of the intermediate points and the weighting coefficients in the expressions for k_{n1}, \ldots in a certain fourth order Runge–Kutta method, then these expressions can be used again, together with one additional stage, in a corresponding fifth order method. This results in a substantial gain in efficiency. It turns out that this can be done in more than one way.[4] The resulting adaptive Runge–Kutta methods

[4]The first widely used fourth and fifth order Runge–Kutta pair was developed by Erwin Fehlberg in the late 1960's. Its popularity was considerably enhanced by the appearance in 1977 of its Fortran implementation RKF45 by Lawrence F. Shampine and H. A. Watts.

are very powerful and efficient means of approximating numerically the solutions of an enormous class of initial value problems. Specific implementations of one or more of them are widely available in commercial software packages.

PROBLEMS

In each of Problems 1 through 6 find approximate values of the solution of the given initial value problem at $t = 0.1, 0.2, 0.3$, and 0.4. Compare the results with those obtained by using other methods and with the exact solution (if available).

(a) Use the Runge-Kutta method with $h = 0.1$.

(b) Use the Runge-Kutta method with $h = 0.05$.

1. $y' = 3 + t - y, \qquad y(0) = 1$
2. $y' = 5t - 3\sqrt{y}, \qquad y(0) = 2$
3. $y' = 2y - 3t, \qquad y(0) = 1$
4. $y' = 2t + e^{-ty}, \qquad y(0) = 1$
5. $y' = \dfrac{y^2 + 2ty}{3 + t^2}, \qquad y(0) = 0.5$
6. $y' = (t^2 - y^2) \sin y, \qquad y(0) = -1$

In each of Problems 7 through 12 find approximate values of the solution of the given initial value problem at $t = 0.5, 1.0, 1.5$, and 2.0. Compare the results with those obtained by other methods.

(a) Use the Runge–Kutta method with $h = 0.1$.

(b) Use the Runge–Kutta method with $h = 0.05$.

7. $y' = 0.5 - t + 2y, \qquad y(0) = 1$
8. $y' = 5t - 3\sqrt{y}, \qquad y(0) = 2$
9. $y' = \sqrt{t + y}, \qquad y(0) = 3$
10. $y' = 2t + e^{-ty}, \qquad y(0) = 1$
11. $y' = (4 - ty)/(1 + y^2), \qquad y(0) = -2$
12. $y' = (y^2 + 2ty)/(3 + t^2), \qquad y(0) = 0.5$

13. Confirm the results in Table 8.3.1 by executing the indicated computations.

14. Consider the initial value problem

$$y' = t^2 + y^2, \qquad y(0) = 1.$$

(a) Draw a direction field for this equation.

(b) Use the Runge–Kutta method or another method to find approximate values of the solution at $t = 0.8, 0.9$, and 0.95. Choose a small enough step size so that you believe your results are accurate to at least four digits.

(c) Try to extend the calculations in part (b) to obtain an accurate approximation to the solution at $t = 1$. If you encounter difficulties in doing this, explain why you think this happens. The direction field in part (a) may be helpful.

15. Consider the initial value problem

$$y' = 3t^2/(3y^2 - 4), \qquad y(0) = 0.$$

(a) Draw a direction field for this equation.

(b) Estimate how far the solution can be extended to the right. Let t_M be the right endpoint of the interval of existence of this solution. What happens at t_M to prevent the solution from continuing farther?

(c) Use the Runge–Kutta method with various step sizes to determine an approximate value of t_M.

See parts (d) and (e) on the next page.

(d) If you continue the computation beyond t_M, you can continue to generate values of y. What significance, if any, do these values have?

(e) Suppose that the initial condition is changed to $y(0) = 1$. Repeat parts (b) and (c) for this problem.

8.4 Multistep Methods

In previous sections we have discussed numerical procedures for solving the initial value problem

$$y' = f(t, y), \qquad y(t_0) = y_0, \tag{1}$$

in which data at the point $t = t_n$ are used to calculate an approximate value of the solution $\phi(t_{n+1})$ at the next mesh point $t = t_{n+1}$. In other words, the calculated value of ϕ at any mesh point depends only on the data at the preceding mesh point. Such methods are called **one-step methods**. However, once approximate values of the solution $y = \phi(t)$ have been obtained at a few points beyond t_0, it is natural to ask whether we can make use of some of this information, rather than just the value at the last point, to calculate the value of $\phi(t)$ at the next point. Specifically, if y_1 at t_1, y_2 at t_2, \ldots, y_n at t_n are known, how can we use this information to determine y_{n+1} at t_{n+1}? Methods that use information at more than the last mesh point are referred to as **multistep methods**. In this section we will describe two types of multistep methods: Adams[5] methods and backward differentiation formulas. Within each type, one can achieve various levels of accuracy, depending on the number of preceding data points that are used. For simplicity, we will assume throughout our discussion that the step size h is constant.

Adams Methods. Recall that

$$\phi(t_{n+1}) - \phi(t_n) = \int_{t_n}^{t_{n+1}} \phi'(t) \, dt, \tag{2}$$

where $\phi(t)$ is the solution of the initial value problem (1). The basic idea of an Adams method is to approximate $\phi'(t)$ by a polynomial $P_k(t)$ of degree k and to use the polynomial to evaluate the integral on the right side of Eq. (2). The coefficients in $P_k(t)$ are determined by using $k + 1$ previously calculated data points. For example, suppose that we wish to use a first degree polynomial $P_1(t) = At + B$. Then we need only the two data points (t_n, y_n) and (t_{n-1}, y_{n-1}). Since P_1 is to be an approximation

[5]John Couch Adams (1819–1892), an English astronomer, is most famous as codiscoverer, with Joseph Leverrier, of the planet Neptune in 1846. Adams was also extremely skilled at computation; his procedure for numerical integration of differential equations appeared in 1883 in a book that he wrote with Francis Bashforth on capillary action.

to ϕ', we require that $P_1(t_n) = f(t_n, y_n)$ and that $P_1(t_{n-1}) = f(t_{n-1}, y_{n-1})$. Recall that we denote $f(t_j, y_j)$ by f_j for an integer j. Then A and B must satisfy the equations

$$At_n + B = f_n,$$
$$At_{n-1} + B = f_{n-1}. \tag{3}$$

Solving for A and B, we obtain

$$A = \frac{f_n - f_{n-1}}{h}, \qquad B = \frac{f_{n-1}t_n - f_n t_{n-1}}{h}. \tag{4}$$

Replacing $\phi'(t)$ by $P_1(t)$ and evaluating the integral in Eq. (2), we find that

$$\phi(t_{n+1}) - \phi(t_n) = \frac{A}{2}(t_{n+1}^2 - t_n^2) + B(t_{n+1} - t_n).$$

Finally, we replace $\phi(t_{n+1})$ and $\phi(t_n)$ by y_{n+1} and y_n, respectively, and carry out some algebraic simplification. For a constant step size h we obtain

$$y_{n+1} = y_n + \tfrac{3}{2}hf_n - \tfrac{1}{2}hf_{n-1}. \tag{5}$$

Equation (5) is the second order Adams–Bashforth formula. It is an explicit formula for y_{n+1} in terms of y_n and y_{n-1} and has a local truncation error proportional to h^3.

We note in passing that the first order Adams–Bashforth formula, based on the polynomial $P_0(t) = f_n$ of degree zero, is just the original Euler formula.

More accurate Adams formulas can be obtained by following the procedure outlined above, but using a higher degree polynomial and correspondingly more data points. For example, suppose that a polynomial $P_3(t)$ of degree three is used. The coefficients are determined from the four points (t_n, y_n), (t_{n-1}, y_{n-1}), (t_{n-2}, y_{n-2}), and (t_{n-3}, y_{n-3}). Substituting this polynomial for $\phi'(t)$ in Eq. (2), evaluating the integral, and simplifying the result, we eventually obtain the fourth order Adams–Bashforth formula,

$$y_{n+1} = y_n + (h/24)(55f_n - 59f_{n-1} + 37f_{n-2} - 9f_{n-3}). \tag{6}$$

The local truncation error of this fourth order formula is proportional to h^5.

A variation on the derivation of the Adams–Bashforth formulas gives another set of formulas called the Adams–Moulton[6] formulas. To see the difference, let us again consider the second order case. Again we use a first degree polynomial $Q_1(t) = \alpha t + \beta$, but we determine the coefficients by using the points (t_n, y_n) and (t_{n+1}, y_{n+1}). Thus α and β must satisfy

$$\alpha t_n + \beta = f_n,$$
$$\alpha t_{n+1} + \beta = f_{n+1}, \tag{7}$$

and it follows that

$$\alpha = \frac{f_{n+1} - f_n}{h}, \qquad \beta = \frac{f_n t_{n+1} - f_{n+1} t_n}{h}. \tag{8}$$

[6]Forest Ray Moulton (1872–1952) was an American astronomer and administrator of science. While calculating ballistics trajectories during World War I, he devised substantial improvements in the Adams formula.

Substituting $Q_1(t)$ for $\phi'(t)$ in Eq. (2) and simplifying, we obtain

$$y_{n+1} = y_n + \tfrac{1}{2}hf_n + \tfrac{1}{2}hf(t_{n+1}, y_{n+1}), \tag{9}$$

which is the second order Adams–Moulton formula. We have written $f(t_{n+1}, y_{n+1})$ in the last term to emphasize that the Adams–Moulton formula is implicit, rather than explicit, since the unknown y_{n+1} appears on both sides of the equation. The local truncation error for the second order Adams–Moulton formula is proportional to h^3.

The first order Adams–Moulton formula is just the backward Euler formula, as you might anticipate by analogy with the first order Adams–Bashforth formula.

More accurate higher order formulas can be obtained by using an approximating polynomial of higher degree. The fourth order Adams–Moulton formula, with a local truncation error proportional to h^5, is

$$y_{n+1} = y_n + (h/24)(9f_{n+1} + 19f_n - 5f_{n-1} + f_{n-2}). \tag{10}$$

Observe that this is also an implicit formula because y_{n+1} appears in f_{n+1}.

Although both the Adams–Bashforth and Adams–Moulton formulas of the same order have local truncation errors proportional to the same power of h, the Adams–Moulton formulas of moderate order are in fact considerably more accurate. For example, for the fourth order formulas (6) and (10), the proportionality constant for the Adams–Moulton formula is less than $1/10$ of the proportionality constant for the Adams–Bashforth formula. Thus the question arises: Should one use the explicit (and faster) Adams–Bashforth formula or the more accurate but implicit (and slower) Adams–Moulton formula? The answer depends on whether, by using the more accurate formula, one can increase the step size, and thereby reduce the number of steps enough to compensate for the additional computations required at each step.

In fact, numerical analysts have attempted to achieve both simplicity and accuracy by combining the two formulas in what is called a **predictor–corrector method**. Once $y_{n-3}, y_{n-2}, y_{n-1}$, and y_n are known, we can compute $f_{n-3}, f_{n-2}, f_{n-1}$, and f_n, and then use the Adams–Bashforth (predictor) formula (6) to obtain a first value for y_{n+1}. Then we compute f_{n+1} and use the Adams–Moulton (corrector) formula (10), which is no longer implicit, to obtain an improved value of y_{n+1}. We can, of course, continue to use the corrector formula (10) if the change in y_{n+1} is too large. However, if it is necessary to use the corrector formula more than once or perhaps twice, it means that the step size h is too large and should be reduced.

In order to use any of the multistep methods, it is necessary first to calculate a few y_j by some other method. For example, the fourth order Adams–Moulton method requires values for y_1 and y_2, while the fourth order Adams–Bashforth method also requires a value for y_3. One way to proceed is to use a one-step method of comparable accuracy to calculate the necessary starting values. Thus, for a fourth order multistep method, one might use the fourth order Runge–Kutta method to calculate the starting values. This is the method used in the next example.

Another approach is to use a low order method with a very small h to calculate y_1, and then to increase gradually both the order and the step size until enough starting values have been determined.

**EXAMPLE
1**

Consider again the initial value problem

$$y' = 1 - t + 4y, \qquad y(0) = 1. \tag{11}$$

With a step size of $h = 0.1$, determine an approximate value of the solution $y = \phi(t)$ at $t = 0.4$ using the fourth order Adams–Bashforth formula, the fourth order Adams–Moulton formula, and the predictor–corrector method.

For starting data, we use the values of y_1, y_2, and y_3 found by the Runge–Kutta method. These are tabulated in Table 8.3.1. Next, calculating the corresponding values of $f(t, y)$, we obtain

$$y_0 = 1, \qquad\qquad f_0 = 5,$$
$$y_1 = 1.6089333, \qquad f_1 = 7.3357332,$$
$$y_2 = 2.5050062, \qquad f_2 = 10.820025,$$
$$y_3 = 3.8294145, \qquad f_3 = 16.017658.$$

Then, from the Adams–Bashforth formula, Eq. (6), we find that $y_4 = 5.7836305$. The exact value of the solution at $t = 0.4$, correct through eight digits, is 5.7942260, so the error is -0.010595.

The Adams–Moulton formula, Eq. (10), leads to the equation

$$y_4 = 4.9251275 + 0.15y_4,$$

from which it follows that $y_4 = 5.7942676$ with an error of only 0.0000416.

Finally, using the result from the Adams–Bashforth formula as a predicted value of $\phi(0.4)$, we can then use Eq. (10) as a corrector. Corresponding to the predicted value of y_4, we find that $f_4 = 23.734522$. Hence, from Eq. (10), the corrected value of y_4 is 5.7926721. This result is in error by -0.0015539.

Observe that the Adams–Bashforth method is the simplest and fastest of these methods, since it involves only the evaluation of a single explicit formula. It is also the least accurate. Using the Adams–Moulton formula as a corrector increases the amount of calculation that is required, but the method is still explicit. In this problem the error in the corrected value of y_4 is reduced by approximately a factor of 7 when compared to the error in the predicted value. The Adams–Moulton method alone yields by far the best result, with an error that is about $1/40$ as large as the error from the predictor–corrector method. Remember, however, that the Adams–Moulton method is implicit, which means that an equation must be solved at each step. In the problem considered here this equation is linear, so the solution is quickly found, but in other problems this part of the procedure may be much more time-consuming.

The Runge–Kutta method with $h = 0.1$ gives $y_4 = 5.7927853$ with an error of -0.0014407; see Table 8.3.1. Thus, for this problem, the Runge–Kutta method is comparable in accuracy to the predictor–corrector method.

Backward Differentiation Formulas. Another type of multistep method uses a polynomial $P_k(t)$ to approximate the solution $\phi(t)$ of the initial value problem (1) rather than its derivative $\phi'(t)$, as in the Adams methods. We then differentiate $P_k(t)$ and set $P_k'(t_{n+1})$ equal to $f(t_{n+1}, y_{n+1})$ to obtain an implicit formula for y_{n+1}. These are called **backward differentiation formulas**.

The simplest case uses a first degree polynomial $P_1(t) = At + B$. The coefficients are chosen to match the computed values of the solution y_n and y_{n+1}. Thus A and B must satisfy

$$At_n + B = y_n,$$
$$At_{n+1} + B = y_{n+1}. \tag{12}$$

Since $P_1'(t) = A$, the requirement that

$$P_1'(t_{n+1}) = f(t_{n+1}, y_{n+1})$$

is just

$$A = f(t_{n+1}, y_{n+1}). \tag{13}$$

Another expression for A comes from subtracting the first of Eqs. (12) from the second, which gives

$$A = (y_{n+1} - y_n)/h.$$

Substituting this value of A into Eq. (13) and rearranging terms, we obtain the first order backward differentiation formula

$$y_{n+1} = y_n + hf(t_{n+1}, y_{n+1}). \tag{14}$$

Note that Eq. (14) is just the backward Euler formula that we first saw in Section 8.1.

By using higher order polynomials and correspondingly more data points, one can obtain backward differentiation formulas of any order. The second order formula is

$$y_{n+1} = \tfrac{1}{3}[4y_n - y_{n-1} + 2hf(t_{n+1}, y_{n+1})], \tag{15}$$

and the fourth order formula is

$$y_{n+1} = \tfrac{1}{25}[48y_n - 36y_{n-1} + 16y_{n-2} - 3y_{n-3} + 12hf(t_{n+1}, y_{n+1})]. \tag{16}$$

These formulas have local truncation errors proportional to h^3 and h^5, respectively.

EXAMPLE 2

Use the fourth order backward differentiation formula with $h = 0.1$ and the data given in Example 1 to determine an approximate value of the solution $y = \phi(t)$ at $t = 0.4$ for the initial value problem (11).

Using Eq. (16) with $n = 3$, $h = 0.1$, and with y_0, \ldots, y_3 given in Example 1, we obtain the equation

$$y_4 = 4.6837842 + 0.192y_4.$$

Thus

$$y_4 = 5.7967626.$$

Comparing the calculated value with the exact value $\phi(0.4) = 5.7942260$, we find that the error is 0.0025366. This is somewhat better than the result using the Adams–Bashforth method, but not as good as the result using the predictor–corrector method, and not nearly as good as the result using the Adams–Moulton method.

A comparison between one-step and multistep methods must take several factors into consideration. The fourth order Runge–Kutta method requires four evaluations of f at each step, while the fourth order Adams–Bashforth method (once past the starting values) requires only one, and the predictor–corrector method only two. Thus, for a given step size h, the latter two methods may well be considerably faster than Runge–Kutta. However, if Runge–Kutta is more accurate and therefore can use fewer steps, then the difference in speed will be reduced and perhaps eliminated. The Adams–Moulton and backward differentiation formulas also require that the difficulty in solving the implicit equation at each step be taken into account. All multistep methods have the possible disadvantage that errors in earlier steps can

feed back into later calculations with unfavorable consequences. On the other hand, the underlying polynomial approximations in multistep methods make it easy to approximate the solution at points between the mesh points, should this be desirable. Multistep methods have become popular largely because it is relatively easy to estimate the error at each step and to adjust the order or the step size to control it. For a further discussion of such questions as these, see the books listed at the end of this chapter; in particular, Shampine (1994) is an authoritative source.

PROBLEMS

In each of Problems 1 through 6 determine an approximate value of the solution at $t = 0.4$ and $t = 0.5$ using the specified method. For starting values use the values given by the Runge–Kutta method; see Problems 1 through 6 of Section 8.3. Compare the results of the various methods with each other and with the actual solution (if available).

(a) Use the fourth order predictor–corrector method with $h = 0.1$. Use the corrector formula once at each step.

(b) Use the fourth order Adams–Moulton method with $h = 0.1$.

(c) Use the fourth order backward differentiation method with $h = 0.1$.

1. $y' = 3 + t - y$, $\quad y(0) = 1$
2. $y' = 5t - 3\sqrt{y}$, $\quad y(0) = 2$
3. $y' = 2y - 3t$, $\quad y(0) = 1$
4. $y' = 2t + e^{-ty}$, $\quad y(0) = 1$
5. $y' = \dfrac{y^2 + 2ty}{3 + t^2}$, $\quad y(0) = 0.5$
6. $y' = (t^2 - y^2)\sin y$, $\quad y(0) = -1$

In each of Problems 7 through 12 find approximate values of the solution of the given initial value problem at $t = 0.5$, 1.0, 1.5, and 2.0, using the specified method. For starting values use the values given by the Runge–Kutta method; see Problems 7 through 12 in Section 8.3. Compare the results of the various methods with each other and with the actual solution (if available).

(a) Use the fourth order predictor–corrector method with $h = 0.05$. Use the corrector formula once at each step.

(b) Use the fourth order Adams–Moulton method with $h = 0.05$.

(c) Use the fourth order backward differentiation method with $h = 0.05$.

7. $y' = 0.5 - t + 2y$, $\quad y(0) = 1$
8. $y' = 5t - 3\sqrt{y}$, $\quad y(0) = 2$
9. $y' = \sqrt{t + y}$, $\quad y(0) = 3$
10. $y' = 2t + e^{-ty}$, $\quad y(0) = 1$
11. $y' = (4 - ty)/(1 + y^2)$, $\quad y(0) = -2$
12. $y' = (y^2 + 2ty)/(3 + t^2)$, $\quad y(0) = 0.5$

13. Show that the first order Adams–Bashforth method is the Euler method and that the first order Adams–Moulton method is the backward Euler method.

14. Show that the third order Adams–Bashforth formula is

$$y_{n+1} = y_n + (h/12)(23f_n - 16f_{n-1} + 5f_{n-2}).$$

15. Show that the third order Adams–Moulton formula is

$$y_{n+1} = y_n + (h/12)(5f_{n+1} + 8f_n - f_{n-1}).$$

16. Derive the second order backward differentiation formula given by Eq. (15) in this section.

8.5 More on Errors; Stability

In Section 8.1 we discussed some ideas related to the errors that can occur in a numerical approximation of the solution of the initial value problem

$$y' = f(t, y), \qquad y(t_0) = y_0. \tag{1}$$

In this section we continue that discussion and also point out some other difficulties that can arise. Some of the points that we wish to make are fairly difficult to treat in detail, so we will illustrate them by means of examples.

Truncation and Round-off Errors. Recall that for the Euler method we showed that the local truncation error is proportional to h^2 and that for a finite interval the global truncation error is at most a constant times h. In general, for a method of order p, the local truncation error is proportional to h^{p+1} and the global truncation error on a finite interval is bounded by a constant times h^p. To achieve high accuracy we normally use a numerical procedure for which p is fairly large, perhaps 4 or higher. As p increases, the formula used in computing y_{n+1} normally becomes more complicated, and hence more calculations are required at each step; however, this is usually not a serious problem unless $f(t, y)$ is very complicated or the calculation must be repeated very many times. If the step size h is decreased, the global truncation error is decreased by the same factor raised to the power p. However, as we mentioned in Section 8.1, if h is very small, a great many steps will be required to cover a fixed interval, and the global round-off error may be larger than the global truncation error. The situation is shown schematically in Figure 8.5.1. We assume that the round-off error R_n is proportional to the number of computations performed and therefore is inversely proportional to the step size h. On the other hand, the truncation error E_n is proportional to a positive power of h. From Eq. (17) of Section 8.1 we know that the total error is bounded by $|E_n| + |R_n|$; hence we wish to choose h so as to minimize this quantity. The optimum value of h occurs when the rate of increase of the truncation error (as h increases) is balanced by the rate of decrease of the round-off error, as indicated in Figure 8.5.1.

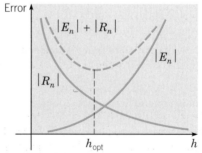

FIGURE 8.5.1 The dependence of truncation and round-off errors on the step size h.

Consider the example problem

EXAMPLE
1

$$y' = 1 - t + 4y, \qquad y(0) = 1. \tag{2}$$

Using the Euler method with various step sizes, calculate approximate values for the solution $\phi(t)$ at $t = 0.5$ and $t = 1$. Try to determine the optimum step size.

TABLE 8.5.1 Approximations to the Solution of the Initial Value
Problem $y' = 1 - t + 4y$, $y(0) = 1$ Using the Euler Method with
Different Step Sizes

h	N	$y_{N/2}$	Error	y_N	Error
0.01	100	8.390	0.322	60.12	4.78
0.005	200	8.551	0.161	62.51	2.39
0.002	500	8.633	0.079	63.75	1.15
0.001	1000	8.656	0.056	63.94	0.96
0.0008	1250	8.636	0.076	63.78	1.12
0.000625	1600	8.616	0.096	64.35	0.55
0.0005	2000	8.772	0.060	64.00	0.90
0.0004	2500	8.507	0.205	63.40	1.50
0.00025	4000	8.231	0.481	56.77	8.13

Keeping only four digits in order to shorten the calculations, we obtain the data shown in Table 8.5.1. The first two columns are the step size h and the number of steps N required to traverse the interval $0 \le t \le 1$. Then $y_{N/2}$ and y_N are approximations to $\phi(0.5) = 8.712$ and $\phi(1) = 64.90$, respectively. These quantities appear in the third and fifth columns. The fourth and sixth columns display the differences between the calculated values and the actual value of the solution.

For relatively large step sizes the round-off error is much less than the global truncation error. Consequently, the total error is approximately the same as the global truncation error, which for the Euler method is bounded by a constant times h. Thus, as the step size is reduced, the error is reduced proportionally. The first three lines in Table 8.5.1 show this type of behavior. For $h = 0.001$ the error has been further reduced, but much less than proportionally; this indicates that round-off error is becoming important. As h is reduced still more, the error begins to fluctuate, and further improvements in accuracy become problematical. For values of h less than 0.0005 the error is clearly increasing, which indicates that round-off error is now the dominant part of the total error.

These results can also be expressed in terms of the number of steps N. For N less than about 1000 accuracy is improved by taking more steps, while for N greater than about 2000 using more steps has an adverse effect. Thus for this problem it is best to use an N somewhere between 1000 and 2000. For the calculations shown in Table 8.5.1 the best result at $t = 0.5$ occurs for $N = 1000$, while at $t = 1.0$ the best result is for $N = 1600$.

One should be careful not to read too much into the results shown in Example 1. The optimum ranges for h and N depend on the differential equation, the numerical method that is used, and the number of digits that are retained in the calculation. Nevertheless, it is generally true that if too many steps are required in a calculation,

then eventually round-off error is likely to accumulate to the point where it seriously degrades the accuracy of the procedure. For many problems this is not a concern: For them, any of the fourth order methods we have discussed in Sections 8.3 and 8.4 will produce good results with a number of steps far less than the level at which round-off error becomes important. For some problems, however, round-off error does become vitally important. For such problems, the choice of method may be crucial. This is also one reason why modern codes provide a means of adjusting the step size as they go along, using a larger step size wherever possible and a very small step size only where necessary.

Vertical Asymptotes. As a second example, consider the problem of determining the solution $y = \phi(t)$ of

$$y' = t^2 + y^2, \qquad y(0) = 1. \tag{3}$$

Since the differential equation is nonlinear, the existence and uniqueness theorem (Theorem 2.4.2) guarantees only that there is a solution in *some* interval about $t = 0$. Suppose that we try to compute a solution of the initial value problem on the interval $0 \le t \le 1$ using different numerical procedures.

If we use the Euler method with $h = 0.1$, 0.05, and 0.01, we find the following approximate values at $t = 1$: 7.189548, 12.32093, and 90.75551, respectively. The large differences among the computed values are convincing evidence that we should use a more accurate numerical procedure—the Runge–Kutta method, for example. Using the Runge–Kutta method with $h = 0.1$, we find the approximate value 735.0991 at $t = 1$, which is quite different from those obtained using the Euler method. Repeating the calculations using step sizes of $h = 0.05$ and $h = 0.01$, we obtain the interesting information shown in Table 8.5.2.

TABLE 8.5.2 Calculation of the Solution of
the Initial Value Problem $y' = t^2 + y^2$,
$y(0) = 1$ Using the Runge–Kutta Method

h	$t = 0.90$	$t = 1.0$
0.1	14.02182	735.0991
0.05	14.27117	1.75863×10^5
0.01	14.30478	2.0913×10^{2893}
0.001	14.30486	

The values at $t = 0.90$ are reasonable and we might well believe that the solution has a value of about 14.305 at $t = 0.90$. However, it is not clear what is happening between $t = 0.9$ and $t = 1.0$. To help clarify this, let us turn to some analytical approximations to the solution of the initial value problem (3). Note that, on $0 \le t \le 1$,

$$y^2 \le t^2 + y^2 \le 1 + y^2. \tag{4}$$

This suggests that the solution $y = \phi_1(t)$ of

$$y' = 1 + y^2, \qquad y(0) = 1 \tag{5}$$

and the solution $y = \phi_2(t)$ of

$$y' = y^2, \qquad y(0) = 1 \tag{6}$$

are upper and lower bounds, respectively, for the solution $y = \phi(t)$ of the original problem, since all these solutions pass through the same initial point. Indeed, it can be shown (for example, by the iteration method of Section 2.8) that $\phi_2(t) \leq \phi(t) \leq \phi_1(t)$ as long as these functions exist. The important thing to note is that we *can solve* Eqs. (5) and (6) for ϕ_1 and ϕ_2 by separation of variables. We find that

$$\phi_1(t) = \tan\left(t + \frac{\pi}{4}\right), \qquad \phi_2(t) = \frac{1}{1 - t}. \tag{7}$$

Thus $\phi_2(t) \to \infty$ as $t \to 1$, and $\phi_1(t) \to \infty$ as $t \to \pi/4 \cong 0.785$. These calculations show that the solution of the original initial value problem exists at least for $0 \leq t < \pi/4$ and at most for $0 \leq t < 1$. The solution of the problem (3) has a vertical asymptote for some t in $\pi/4 \leq t \leq 1$ and thus does not exist on the entire interval $0 \leq t \leq 1$.

Our numerical calculations, however, suggest that we can go beyond $t = \pi/4$, and probably beyond $t = 0.9$. Assuming that the solution of the initial value problem exists at $t = 0.9$ and has the value 14.305, we can obtain a more accurate appraisal of what happens for larger t by considering the initial value problems (5) and (6) with $y(0) = 1$ replaced by $y(0.9) = 14.305$. Then we obtain

$$\phi_1(t) = \tan(t + 0.60100), \qquad \phi_2(t) = 1/(0.96991 - t), \tag{8}$$

where only five decimal places have been kept. Thus $\phi_1(t) \to \infty$ as $t \to \pi/2 - 0.60100 \cong 0.96980$ and $\phi_2(t) \to \infty$ as $t \to 0.96991$. We conclude that the asymptote of the solution of the initial value problem (3) lies between these two values. This example illustrates the sort of information that can be obtained by a judicious combination of analytical and numerical work.

Stability. The concept of stability is associated with the possibility that small errors that are introduced in the course of a mathematical procedure may die out as the procedure continues. Conversely, instability occurs if small errors tend to increase, perhaps without bound. For example, in Section 2.5 we identified equilibrium solutions of a differential equation as (asymptotically) stable or unstable, depending on whether solutions that were initially near the equilibrium solution tended to approach it or to depart from it as t increased. Somewhat more generally, the solution of an initial value problem is asymptotically stable if initially nearby solutions tend to approach the given solution, and unstable if they tend to depart from it. Visually, in an asymptotically stable problem the graphs of solutions will come together, while in an unstable problem they will separate.

If we are solving an initial value problem numerically, the best that we can hope for is that the numerical approximation will mimic the behavior of the actual solution. We cannot make an unstable problem into a stable one merely by solving it numerically. However, it may well happen that a numerical procedure will introduce instabilities that were not part of the original problem, and this can cause trouble in approximating the solution. Avoidance of such instabilities may require us to place restrictions on the step size h.

To illustrate what can happen in the simplest possible context, consider the differential equation

$$dy/dt = ry, \tag{9}$$

where r is a constant. Suppose that in solving this equation we have reached the point (t_n, y_n). Let us compare the exact solution of Eq. (9) that passes through this point, namely,

$$y = y_n \exp[r(t - t_n)], \tag{10}$$

with numerical approximations obtained from the Euler formula

$$y_{n+1} = y_n + hf(t_n, y_n) \tag{11}$$

and from the backward Euler formula

$$y_{n+1} = y_n + hf(t_{n+1}, y_{n+1}). \tag{12}$$

From the Euler formula (11) we obtain

$$y_{n+1} = y_n + hry_n = y_n(1 + rh). \tag{13}$$

Similarly, from the backward Euler formula (12) we have

$$y_{n+1} = y_n + hry_{n+1},$$

or

$$y_{n+1} = \frac{y_n}{1 - rh} = y_n[1 + rh + (rh)^2 + \cdots]. \tag{14}$$

Finally, evaluating the solution (10) at $t_n + h$, we find that

$$y_{n+1} = y_n \exp(rh) = y_n\left[1 + rh + \frac{(rh)^2}{2} + \cdots\right]. \tag{15}$$

Comparing Eqs. (13), (14), and (15), we see that the errors in both the Euler formula and the backward Euler formula are of order h^2, as the theory predicts.

Now suppose that we change the value y_n to $y_n + \delta$. Think, if you wish, of δ as the error that has accumulated by the time we reach $t = t_n$. The question is whether this error increases or decreases in going one more step to t_{n+1}.

For the exact solution (15), the change in y_{n+1} due to the change δ in y_n is just $\delta \exp(rh)$. This quantity is less than δ if $\exp(rh) < 1$, that is, if $r < 0$. This confirms our conclusion in Chapter 2 that Eq. (9) is asymptotically stable if $r < 0$, and is unstable if $r > 0$.

For the backward Euler method, the change in y_{n+1} in Eq. (14) due to δ is $\delta/(1 - rh)$. For $r \leq 0$ the quantity $1/(1 - rh)$ is always nonnegative and never greater than 1. Thus, if the differential equation is stable, then so is the backward Euler method for an arbitrary step size h.

On the other hand, for the Euler method, the change in y_{n+1} in Eq. (13) due to δ is $\delta(1 + rh)$. If we recall that $r \leq 0$ and require that $|1 + rh| \leq 1$, then we find that h must satisfy $h \leq 2/|r|$. Thus the Euler method is not stable for this problem unless h is sufficiently small.

The restriction on the step size h in using the Euler method in the preceding example is rather mild unless $|r|$ is quite large. Nonetheless, the example illustrates that it may be necessary to restrict h in order to achieve stability in the numerical method, even though the initial value problem itself is stable for all values of h. Problems for which a much smaller step size is needed for stability than for accuracy are called **stiff**. The backward differentiation formulas described in Section 8.4 (of which the backward Euler formula is the lowest order example) are the most popular formulas for solving stiff problems. The following example illustrates the kind of instability that can occur when we try to approximate the solution of a stiff problem.

EXAMPLE

2

A Stiff
Problem

Consider the initial value problem

$$y' = -100y + 100t + 1, \qquad y(0) = 1. \tag{16}$$

Find numerical approximations to the solution for $0 \le t \le 1$ using the Euler, backward Euler, and Runge–Kutta methods. Compare the numerical results with the exact solution.

Since the differential equation is linear, it is easy to solve, and the solution of the initial value problem (16) is

$$y = \phi(t) = e^{-100t} + t. \tag{17}$$

Some values of the solution $\phi(t)$, correct to six decimal places, are given in the second column of Table 8.5.3, and a graph of the solution is shown in Figure 8.5.2. There is a thin layer (sometimes called a boundary layer) to the right of $t = 0$ in which the exponential term is significant and the solution varies rapidly. Once past this layer, however, $\phi(t) \cong t$ and the graph of the solution is essentially a straight line. The width of the boundary layer is somewhat arbitrary, but it is certainly small. At $t = 0.1$, for example, $\exp(-100t) \cong 0.000045$.

FIGURE 8.5.2 The solution of the initial value problem (16).

If we plan to approximate the solution (17) numerically, we might intuitively expect that a small step size will be needed only in the boundary layer. To make this expectation a bit more precise, recall from Section 8.1 that the local truncation errors for the Euler and backward Euler methods are proportional to $\phi''(t)$. For this problem $\phi''(t) = 10^4 e^{-100t}$, which varies from a value of 10^4 at $t = 0$ to nearly zero for $t > 0.2$. Thus a very small step size is needed for accuracy near $t = 0$, but a much larger step size is adequate once t is a little larger.

On the other hand, the stability analysis in Eqs. (9) through (15) also applies to this problem. Since $r = -100$ for Eq. (16), it follows that for stability we need $h < 0.02$ for the Euler method, but there is no corresponding restriction for the backward Euler method.

Some results obtained from the Euler method are shown in columns 3 and 4 of Table 8.5.3. The values for $h = 0.025$ are worthless because of instability, while those for $h = 0.01666\ldots$ are reasonably accurate for $t \geq 0.2$. However, comparable accuracy for this range of t is obtained for $h = 0.1$ by using the backward Euler method, as shown by the results in column 7 of the table.

The situation is not improved by using, instead of the Euler method, a more accurate one, such as Runge–Kutta. For this problem the Runge–Kutta method is unstable for $h = 0.033\ldots$ but stable for $h = 0.025$, as shown by the results in columns 5 and 6 of Table 8.5.3.

The results given in the table for $t = 0.05$ and for $t = 0.1$ show that in the boundary layer a smaller step size is needed to obtain an accurate approximation. You are invited to explore this matter further in Problem 3.

TABLE 8.5.3 Numerical Approximations to the Solution of the Initial Value Problem $y' = -100y + 100t + 1,\ y(0) = 1$

t	Exact	Euler 0.025	Euler 0.0166...	Runge–Kutta 0.0333...	Runge–Kutta 0.025	Backward Euler 0.1
0.0	1.000000	1.000000	1.000000	1.000000	1.000000	1.000000
0.05	0.056738	2.300000	−0.246296		0.470471	
0.1	0.100045	5.162500	0.187792	10.6527	0.276796	0.190909
0.2	0.200000	25.8289	0.207707	111.559	0.231257	0.208264
0.4	0.400000	657.241	0.400059	1.24×10^4	0.400977	0.400068
0.6	0.600000	1.68×10^4	0.600000	1.38×10^6	0.600031	0.600001
0.8	0.800000	4.31×10^5	0.800000	1.54×10^8	0.800001	0.800000
1.0	1.000000	1.11×10^7	1.000000	1.71×10^{10}	1.000000	1.000000

As a final example, consider the problem of determining two linearly independent solutions of the second order linear equation

$$y'' - 10\pi^2 y = 0 \tag{18}$$

for $t > 0$. The generalization of numerical techniques for the first order equations to higher order equations or to systems of equations is discussed in Section 8.6, but that is not needed for the present discussion. Two linearly independent solutions of Eq. (18) are $\phi_1(t) = \cosh\sqrt{10}\pi t$ and $\phi_2(t) = \sinh\sqrt{10}\pi t$. The first solution, $\phi_1(t) = \cosh\sqrt{10}\pi t$, is generated by the initial conditions $\phi_1(0) = 1$, $\phi_1'(0) = 0$; the second solution, $\phi_2(t) = \sinh\sqrt{10}\pi t$, is generated by the initial conditions $\phi_2(0) = 0$, $\phi_2'(0) = \sqrt{10}\pi$. Although analytically we can tell the difference between $\cosh\sqrt{10}\pi t$ and $\sinh\sqrt{10}\pi t$, for large t we have $\cosh\sqrt{10}\pi t \sim e^{\sqrt{10}\pi t}/2$ and $\sinh\sqrt{10}\pi t \sim e^{\sqrt{10}\pi t}/2$; numerically these two functions look exactly the same if only a fixed number of digits are retained. For example, correct to eight significant figures, we find that for $t = 1$,

$$\sinh\sqrt{10}\pi = \cosh\sqrt{10}\pi = 10{,}315.894.$$

If the calculations are performed on a machine that carries only eight digits, the two solutions ϕ_1 and ϕ_2 are identical at $t = 1$ and indeed for all $t > 1$. Thus, even though the solutions are linearly independent, their numerical tabulation would show that they are the same because we can retain only a finite number of digits. This phenomenon is called **numerical dependence**.

For the present problem we can partially circumvent this difficulty by computing, instead of $\sinh \sqrt{10}\pi t$ and $\cosh \sqrt{10}\pi t$, the linearly independent solutions $\phi_3(t) = e^{\sqrt{10}\pi t}$ and $\phi_4(t) = e^{-\sqrt{10}\pi t}$ corresponding to the initial conditions $\phi_3(0) = 1$, $\phi_3'(0) = \sqrt{10}\pi$ and $\phi_4(0) = 1$, $\phi_4'(0) = -\sqrt{10}\pi$, respectively. The solution ϕ_3 grows exponentially while ϕ_4 decays exponentially. Even so, we encounter difficulty in calculating ϕ_4 correctly on a large interval. The reason is that at each step of the calculation for ϕ_4 we introduce truncation and round-off errors. Thus, at any point t_n, the data to be used in going to the next point are not precisely the values of $\phi_4(t_n)$ and $\phi_4'(t_n)$. The solution of the initial value problem with these data at t_n involves not only $e^{-\sqrt{10}\pi t}$ but also $e^{\sqrt{10}\pi t}$. Because the error in the data at t_n is small, the latter function appears with a very small coefficient. Nevertheless, since $e^{-\sqrt{10}\pi t}$ tends to zero and $e^{\sqrt{10}\pi t}$ grows very rapidly, the latter eventually dominates, and the calculated solution is simply a multiple of $e^{\sqrt{10}\pi t} = \phi_3(t)$.

To be specific, suppose that we use the Runge–Kutta method to calculate the solution $y = \phi_4(t) = e^{-\sqrt{10}\pi t}$ of the initial value problem

$$y'' - 10\pi^2 y = 0, \qquad y(0) = 1, \qquad y'(0) = -\sqrt{10}\pi.$$

(The Runge–Kutta method for second order systems is described in Section 8.6.) Using single-precision (eight-digit) arithmetic with a step size $h = 0.01$, we obtain the results in Table 8.5.4. It is clear from these results that the numerical approximation

TABLE 8.5.4 Exact Solution of $y'' - 10\pi^2 y = 0$, $y(0) = 1$, $y'(0) = -\sqrt{10}\pi$ and Numerical Approximation Using the Runge–Kutta Method with $h = 0.01$

	y	
t	Numerical	Exact
0.0	1.0	1.0
0.25	8.3439×10^{-2}	8.3438×10^{-2}
0.5	6.9623×10^{-3}	6.9620×10^{-3}
0.75	5.8409×10^{-4}	5.8089×10^{-4}
1.0	8.6688×10^{-5}	4.8469×10^{-5}
1.5	5.4900×10^{-3}	3.3744×10^{-7}
2.0	7.8852×10^{-1}	2.3492×10^{-9}
2.5	1.1326×10^{2}	1.6355×10^{-11}
3.0	1.6268×10^{4}	1.1386×10^{-13}
3.5	2.3368×10^{6}	7.9272×10^{-16}
4.0	3.3565×10^{8}	5.5189×10^{-18}
4.5	4.8211×10^{10}	3.8422×10^{-20}
5.0	6.9249×10^{12}	2.6749×10^{-22}

begins to deviate significantly from the exact solution for $t > 0.5$, and soon differs from it by many orders of magnitude. The reason is the presence, in the numerical approximation, of a small component of the exponentially growing solution $\phi_3(t) = e^{\sqrt{10}\pi t}$. With eight-digit arithmetic we can expect a round-off error of the order of 10^{-8} at each step. Since $e^{\sqrt{10}\pi t}$ grows by a factor of 3.7×10^{21} from $t = 0$ to $t = 5$, an error of order 10^{-8} near $t = 0$ can produce an error of order 10^{13} at $t = 5$ even if no further errors are introduced in the intervening calculations. The results given in Table 8.5.4 demonstrate that this is exactly what happens.

You should bear in mind that the numerical values of the entries in the second column of Table 8.5.4 are extremely sensitive to slight variations in how the calculations are executed. Regardless of such details, however, the exponential growth of the approximation will be clearly evident.

Equation (18) is highly unstable, and the behavior shown in this example is typical of unstable problems. One can track a solution accurately for a while, and the interval can be extended by using smaller step sizes or more accurate methods, but eventually the instability in the problem itself takes over and leads to large errors.

Some Comments on Numerical Methods. In this chapter we have introduced several numerical methods for approximating the solution of an initial value problem. We have tried to emphasize some important ideas while maintaining a reasonable level of complexity. For one thing, we have always used a uniform step size, whereas production codes that are currently in use provide for varying the step size as the calculation proceeds.

There are several considerations that must be taken into account in choosing step sizes. Of course, one is accuracy; too large a step size leads to an inaccurate result. Normally, an error tolerance is prescribed in advance, and the step size at each step must be consistent with this requirement. As we have seen, the step size must also be chosen so that the method is stable. Otherwise, small errors will grow and soon render the results worthless. Finally, for implicit methods an equation must be solved at each step, and the method used to solve the equation may impose additional restrictions on the step size.

In choosing a method, one must also balance the considerations of accuracy and stability against the amount of time required to execute each step. An implicit method, such as the Adams–Moulton method, requires more calculations for each step, but if its accuracy and stability permit a larger step size (and consequently fewer steps), then this may more than compensate for the additional calculations. The backward differentiation formulas of moderate order, say, four, are highly stable and are therefore indicated for stiff problems, for which stability is the controlling factor.

Some current production codes also permit the order of the method to be varied, as well as the step size, as the calculation proceeds. The error is estimated at each step, and the order and step size are chosen to satisfy the prescribed error tolerance. In practice, Adams methods up to order twelve and backward differentiation formulas up to order five are in use. Higher order backward differentiation formulas are unsuitable because of a lack of stability.

Finally, we note that the smoothness of the function f—that is, the number of continuous derivatives that it possesses—is a factor in choosing the order of the method to be used. High order methods lose some of their accuracy if f is not smooth to a corresponding order.

PROBLEMS

1. To obtain some idea of the possible dangers of small errors in the initial conditions, such as those due to round-off, consider the initial value problem

$$y' = t + y - 3, \qquad y(0) = 2.$$

(a) Show that the solution is $y = \phi_1(t) = 2 - t$.

(b) Suppose that in the initial condition a mistake is made, and 2.001 is used instead of 2. Determine the solution $y = \phi_2(t)$ in this case, and compare the difference $\phi_2(t) - \phi_1(t)$ at $t = 1$ and as $t \to \infty$.

2. Consider the initial value problem

$$y' = t^2 + e^y, \qquad y(0) = 0. \tag{i}$$

Using the Runge–Kutta method with step size h, we obtain the results in Table 8.5.5. These results suggest that the solution has a vertical asymptote between $t = 0.9$ and $t = 1.0$.

(a) Let $y = \phi(t)$ be the solution of problem (i). Further, let $y = \phi_1(t)$ be the solution of

$$y' = 1 + e^y, \qquad y(0) = 0, \tag{ii}$$

and let $y = \phi_2(t)$ be the solution of

$$y' = e^y, \qquad y(0) = 0. \tag{iii}$$

Show that

$$\phi_2(t) \leq \phi(t) \leq \phi_1(t) \tag{iv}$$

on some interval, contained in $0 \leq t \leq 1$, where all three solutions exists.

(b) Determine $\phi_1(t)$ and $\phi_2(t)$. Then show that $\phi(t) \to \infty$ for some t between $t = \ln 2 \cong 0.69315$ and $t = 1$.

(c) Solve the differential equations $y' = e^y$ and $y' = 1 + e^y$, respectively, with the initial condition $y(0.9) = 3.4298$. Use the results to show that $\phi(t) \to \infty$ when $t \cong 0.932$.

TABLE 8.5.5 Calculation of the Solution of the Initial Value Problem $y' = t^2 + e^y,\ y(0) = 0$ Using the Runge–Kutta Method

h	$t = 0.90$	$t = 1.0$
0.02	3.42985	$> 10^{38}$
0.01	3.42982	$> 10^{38}$

 3. Consider again the initial value problem (16) from Example 2. Investigate how small a step size h must be chosen to ensure that the error at $t = 0.05$ and at $t = 0.1$ is less than 0.0005.

(a) Use the Euler method.

(b) Use the backward Euler method.

(c) Use the Runge–Kutta method.

 4. Consider the initial value problem

$$y' = -10y + 2.5t^2 + 0.5t, \qquad y(0) = 4.$$

(a) Find the solution $y = \phi(t)$ and draw its graph for $0 \le t \le 5$.

(b) The stability analysis in the text suggests that, for this problem, the Euler method is stable only for $h < 0.2$. Confirm that this is true by applying the Euler method to this problem for $0 \le t \le 5$ with step sizes near 0.2.

(c) Apply the Runge–Kutta method to this problem for $0 \le t \le 5$ with various step sizes. What can you conclude about the stability of this method?

(d) Apply the backward Euler method to this problem for $0 \le t \le 5$ with various step sizes. What step size is needed to ensure that the error at $t = 5$ is less than 0.01?

In each of Problems 5 and 6:

(a) Find a formula for the solution of the initial value problem, and note that it is independent of λ.

(b) Use the Runge–Kutta method with $h = 0.01$ to compute approximate values of the solution for $0 \le t \le 1$ for various values of λ such as $\lambda = 1, 10, 20$, and 50.

(c) Explain the differences, if any, between the exact solution and the numerical approximations.

 5. $y' - \lambda y = 1 - \lambda t, \qquad y(0) = 0$ 6. $y' - \lambda y = 2t - \lambda t^2, \qquad y(0) = 0$

8.6 Systems of First Order Equations

In the preceding sections we discussed numerical methods for solving initial value problems associated with a first order differential equation. These methods can also be applied to a system of first order equations. Since a higher order equation can always be reduced to a system of first order equations, it is sufficient to deal with systems of first order equations alone. For simplicity, we consider a system of two first order equations

$$x' = f(t, x, y), \qquad y' = g(t, x, y), \tag{1}$$

with the initial conditions

$$x(t_0) = x_0, \qquad y(t_0) = y_0. \tag{2}$$

The functions f and g are assumed to satisfy the conditions of Theorem 7.1.1 so that the initial value problem (1), (2) has a unique solution in some interval of the t-axis containing the point t_0. We wish to determine approximate values $x_1, x_2, \ldots, x_n, \ldots$ and $y_1, y_2, \ldots, y_n, \ldots$ of the solution $x = \phi(t), y = \psi(t)$ at the points $t_n = t_0 + nh$ with $n = 1, 2, \ldots$.

In vector notation the initial value problem (1), (2) can be written as

$$\mathbf{x}' = \mathbf{f}(t, \mathbf{x}), \qquad \mathbf{x}(t_0) = \mathbf{x}_0, \tag{3}$$

where \mathbf{x} is the vector with components x and y, \mathbf{f} is the vector function with components f and g, and \mathbf{x}_0 is the vector with components x_0 and y_0. The methods of the previous sections can be readily generalized to handle systems of two (or more)

equations. All that is needed (formally) is to replace the scalar variable x by the vector \mathbf{x} and the scalar function f by the vector function \mathbf{f} in the appropriate equations. For example, the Euler formula becomes

$$\mathbf{x}_{n+1} = \mathbf{x}_n + h\mathbf{f}_n, \tag{4}$$

or, in component form,

$$\begin{pmatrix} x_{n+1} \\ y_{n+1} \end{pmatrix} = \begin{pmatrix} x_n \\ y_n \end{pmatrix} + h \begin{pmatrix} f(t_n, x_n, y_n) \\ g(t_n, x_n, y_n) \end{pmatrix}. \tag{5}$$

The initial conditions are used to determine \mathbf{f}_0, which is the vector tangent to the graph of the solution $\mathbf{x} = \boldsymbol{\phi}(t)$ in the xy-plane. We move in the direction of this tangent vector for a time step h in order to find the next point \mathbf{x}_1. Then we calculate a new tangent vector \mathbf{f}_1, move along it for a time step h to find \mathbf{x}_2, and so forth.

In a similar way, the Runge–Kutta method can be extended to a system. For the step from t_n to t_{n+1} we have

$$\mathbf{x}_{n+1} = \mathbf{x}_n + (h/6)(\mathbf{k}_{n1} + 2\mathbf{k}_{n2} + 2\mathbf{k}_{n3} + \mathbf{k}_{n4}), \tag{6}$$

where

$$\begin{aligned}
\mathbf{k}_{n1} &= \mathbf{f}(t_n, \mathbf{x}_n), \\
\mathbf{k}_{n2} &= \mathbf{f}[t_n + (h/2), \mathbf{x}_n + (h/2)\mathbf{k}_{n1}], \\
\mathbf{k}_{n3} &= \mathbf{f}[t_n + (h/2), \mathbf{x}_n + (h/2)\mathbf{k}_{n2}], \\
\mathbf{k}_{n4} &= \mathbf{f}(t_n + h, \mathbf{x}_n + h\mathbf{k}_{n3}).
\end{aligned} \tag{7}$$

The formulas for the Adams–Moulton predictor–corrector method as it applies to the initial value problem (1), (2) are given in Problem 9.

The vector equations (3), (4), (6), and (7) are, in fact, valid in any number of dimensions. All that is needed is to interpret the vectors as having n components rather than two.

EXAMPLE 1

Determine approximate values of the solution $x = \phi(t), y = \psi(t)$ of the initial value problem

$$x' = x - 4y, \qquad y' = -x + y, \tag{8}$$
$$x(0) = 1, \qquad y(0) = 0, \tag{9}$$

at the point $t = 0.2$. Use the Euler method with $h = 0.1$ and the Runge–Kutta method with $h = 0.2$. Compare the results with the values of the exact solution:

$$\phi(t) = \frac{e^{-t} + e^{3t}}{2}, \qquad \psi(t) = \frac{e^{-t} - e^{3t}}{4}. \tag{10}$$

Let us first use the Euler method. For this problem $f_n = x_n - 4y_n$ and $g_n = -x_n + y_n$; hence

$$f_0 = 1 - (4)(0) = 1, \qquad g_0 = -1 + 0 = -1.$$

Then, from the Euler formulas (4) and (5), we obtain

$$x_1 = 1 + (0.1)(1) = 1.1, \qquad y_1 = 0 + (0.1)(-1) = -0.1.$$

At the next step

$$f_1 = 1.1 - (4)(-0.1) = 1.5, \qquad g_1 = -1.1 + (-0.1) = -1.2.$$

Consequently,

$$x_2 = 1.1 + (0.1)(1.5) = 1.25, \qquad y_2 = -0.1 + (0.1)(-1.2) = -0.22.$$

The values of the exact solution, correct to eight digits, are $\phi(0.2) = 1.3204248$ and $\psi(0.2) = -0.25084701$. Thus the values calculated from the Euler method are in error by about 0.0704 and 0.0308, respectively, corresponding to percentage errors of about 5.3% and 12.3%.

Now let us use the Runge–Kutta method to approximate $\phi(0.2)$ and $\psi(0.2)$. With $h = 0.2$ we obtain the following values from Eqs. (7):

$$\mathbf{k}_{01} = \begin{pmatrix} f(1,0) \\ g(1,0) \end{pmatrix} = \begin{pmatrix} 1 \\ -1 \end{pmatrix};$$

$$\mathbf{k}_{02} = \begin{pmatrix} f(1.1,-0.1) \\ g(1.1,-0.1) \end{pmatrix} = \begin{pmatrix} 1.5 \\ -1.2 \end{pmatrix};$$

$$\mathbf{k}_{03} = \begin{pmatrix} f(1.15,-0.12) \\ g(1.15,-0.12) \end{pmatrix} = \begin{pmatrix} 1.63 \\ -1.27 \end{pmatrix};$$

$$\mathbf{k}_{04} = \begin{pmatrix} f(1.326,-0.254) \\ g(1.326,-0.254) \end{pmatrix} = \begin{pmatrix} 2.342 \\ -1.580 \end{pmatrix}.$$

Then, substituting these values in Eq. (6), we obtain

$$\mathbf{x}_1 = \begin{pmatrix} 1 \\ 0 \end{pmatrix} + \frac{0.2}{6} \begin{pmatrix} 9.602 \\ -7.52 \end{pmatrix} = \begin{pmatrix} 1.3200667 \\ -0.25066667 \end{pmatrix}.$$

These values of x_1 and y_1 are in error by about 0.000358 and 0.000180, respectively, with percentage errors much less than one-tenth of 1%.

This example again illustrates the great gains in accuracy that are obtainable by using a more accurate approximation method, such as the Runge–Kutta method. In the calculations we have just outlined, the Runge–Kutta method requires only twice as many function evaluations as the Euler method, but the error in the Runge–Kutta method is about 200 times less than in the Euler method.

PROBLEMS

In each of Problems 1 through 6 determine approximate values of the solution $x = \phi(t)$, $y = \psi(t)$ of the given initial value problem at $t = 0.2, 0.4, 0.6, 0.8,$ and 1.0. Compare the results obtained by different methods and different step sizes.

(a) Use the Euler method with $h = 0.1$.

(b) Use the Runge–Kutta method with $h = 0.2$.

(c) Use the Runge–Kutta method with $h = 0.1$.

1. $x' = x + y + t, \quad y' = 4x - 2y; \qquad x(0) = 1, \quad y(0) = 0$

2. $x' = 2x + ty, \quad y' = xy; \qquad x(0) = 1, \quad y(0) = 1$

3. $x' = -tx - y - 1, \quad y' = x; \qquad x(0) = 1, \quad y(0) = 1$

4. $x' = x - y + xy, \quad y' = 3x - 2y - xy; \qquad x(0) = 0, \quad y(0) = 1$

5. $x' = x(1 - 0.5x - 0.5y), \quad y' = y(-0.25 + 0.5x); \qquad x(0) = 4, \quad y(0) = 1$

6. $x' = \exp(-x + y) - \cos x, \quad y' = \sin(x - 3y); \qquad x(0) = 1, \quad y(0) = 2$

7. Consider the example problem $x' = x - 4y$, $y' = -x + y$ with the initial conditions $x(0) = 1$ and $y(0) = 0$. Use the Runge–Kutta method to find approximate values of the solution of this problem on the interval $0 \leq t \leq 1$. Start with $h = 0.2$ and then repeat the calculation with step sizes $h = 0.1, 0.05, \ldots$, each half as long as in the preceding case. Continue the process until the first five digits of the solution at $t = 1$ are unchanged for successive step sizes. Determine whether these digits are accurate by comparing them with the exact solution given in Eqs. (10) in the text.

8. Consider the initial value problem

$$x'' + t^2 x' + 3x = t, \qquad x(0) = 1, \quad x'(0) = 2.$$

Convert this problem to a system of two first order equations, and determine approximate values of the solution at $t = 0.5$ and $t = 1.0$ using the Runge–Kutta method with $h = 0.1$.

9. Consider the initial value problem $x' = f(t, x, y)$ and $y' = g(t, x, y)$ with $x(t_0) = x_0$ and $y(t_0) = y_0$. The generalization of the Adams–Moulton predictor–corrector method of Section 8.4 is

$$x_{n+1} = x_n + \tfrac{1}{24} h(55 f_n - 59 f_{n-1} + 37 f_{n-2} - 9 f_{n-3}),$$

$$y_{n+1} = y_n + \tfrac{1}{24} h(55 g_n - 59 g_{n-1} + 37 g_{n-2} - 9 g_{n-3})$$

and

$$x_{n+1} = x_n + \tfrac{1}{24} h(9 f_{n+1} + 19 f_n - 5 f_{n-1} + f_{n-2}),$$

$$y_{n+1} = y_n + \tfrac{1}{24} h(9 g_{n+1} + 19 g_n - 5 g_{n-1} + g_{n-2}).$$

Determine an approximate value of the solution at $t = 0.4$ for the example initial value problem $x' = x - 4y$, $y' = -x + y$ with $x(0) = 1$, $y(0) = 0$. Take $h = 0.1$. Correct the predicted value once. For the values of x_1, \ldots, y_3 use the values of the exact solution rounded to six digits: $x_1 = 1.12735$, $x_2 = 1.32042$, $x_3 = 1.60021$, $y_1 = -0.111255$, $y_2 = -0.250847$, and $y_3 = -0.429696$.

REFERENCES There are many books of varying degrees of sophistication that deal with numerical analysis in general and the numerical solution of ordinary differential equations in particular. Among these are

Ascher, Uri M., and Petzold, Linda R., *Computer Methods for Ordinary Differential Equations and Differential-Algebraic Equations* (Philadelphia: Society for Industrial and Applied Mathematics, 1998).

Gear, C. William, *Numerical Initial Value Problems in Ordinary Differential Equations* (Englewood Cliffs, NJ: Prentice-Hall, 1971).

Henrici, Peter, *Discrete Variable Methods in Ordinary Differential Equations* (New York: Wiley, 1962).

Shampine, Lawrence F., *Numerical Solution of Ordinary Differential Equations* (New York: Chapman and Hall, 1994).

A detailed exposition of Adams predictor–corrector methods, including practical guidelines for implementation, may be found in

Shampine, L. F., and Gordon, M. K., *Computer Solution of Ordinary Differential Equations: The Initial Value Problem* (San Francisco: Freeman, 1975).

Many books on numerical analysis have chapters on differential equations. For example, at an elementary level, see

Burden, R. L., and Faires, J. D., *Numerical Analysis* (7th ed.) (Pacific Grove, CA: Brooks/Cole, 2001).

Nonlinear Differential Equations and Stability

There are many differential equations, especially nonlinear ones, that are not susceptible to analytical solution in any reasonably convenient manner. Numerical methods, such as those discussed in the preceding chapter, provide one means of dealing with these equations. Another approach, presented in this chapter, is geometrical in character and leads to a qualitative understanding of the behavior of solutions rather than to detailed quantitative information.

9.1 The Phase Plane: Linear Systems

Since many differential equations cannot be solved conveniently by analytical methods, it is important to consider what qualitative[1] information can be obtained about their solutions without actually solving the equations. The questions that we consider in this chapter are associated with the idea of stability of a solution, and the methods that we employ are basically geometrical. Both the concept of stability and the use

[1]The qualitative theory of differential equations was created by Henri Poincaré (1854–1912) in several major papers between 1880 and 1886. Poincaré was professor at the University of Paris and is generally considered the leading mathematician of his time. He made fundamental discoveries in several different areas of mathematics, including complex function theory, partial differential equations, and celestial mechanics. In a series of papers beginning in 1894 he initiated the use of modern methods in topology. In differential equations he was a pioneer in the use of asymptotic series, one of the most powerful tools of contemporary applied mathematics. Among other things, he used asymptotic expansions to obtain solutions about irregular singular points, thereby extending the work of Fuchs and Frobenius discussed in Chapter 5.

483

of geometrical analysis were introduced in Chapter 1 and used in Section 2.5 for first order autonomous equations

$$dy/dt = f(y). \tag{1}$$

In this chapter we refine the ideas and extend the discussion to systems of equations.

We start with a consideration of the simplest system, namely, a second order linear homogeneous system with constant coefficients. Such a system has the form

$$d\mathbf{x}/dt = \mathbf{A}\mathbf{x}, \tag{2}$$

where \mathbf{A} is a 2×2 constant matrix and \mathbf{x} is a 2×1 vector. Systems of this kind were solved in Sections 7.5 through 7.8. Recall that if we seek solutions of the form $\mathbf{x} = \boldsymbol{\xi}e^{rt}$, then by substitution for \mathbf{x} in Eq. (2) we find that

$$(\mathbf{A} - r\mathbf{I})\boldsymbol{\xi} = \mathbf{0}. \tag{3}$$

Thus r must be an eigenvalue and $\boldsymbol{\xi}$ a corresponding eigenvector of the coefficient matrix \mathbf{A}. The eigenvalues are the roots of the polynomial equation

$$\det(\mathbf{A} - r\mathbf{I}) = 0, \tag{4}$$

and the eigenvectors are determined from Eq. (3) up to an arbitrary multiplicative constant.

In Section 2.5 we found that points where the right side of Eq. (1) is zero are of special importance. Such points correspond to constant solutions, or **equilibrium solutions**, of Eq. (1) and are often called **critical points**. Similarly, for the system (2), points where $\mathbf{A}\mathbf{x} = \mathbf{0}$ correspond to equilibrium (constant) solutions, and again they are called critical points. We will assume that \mathbf{A} is nonsingular, or that $\det \mathbf{A} \neq 0$. It follows that $\mathbf{x} = \mathbf{0}$ is the only critical point of the system (2).

Recall that a solution of Eq. (2) is a vector function $\mathbf{x} = \boldsymbol{\phi}(t)$ that satisfies the differential equation. Such a function can be viewed as a parametric representation for a curve in the x_1x_2-plane. It is often useful to regard this curve as the path, or **trajectory**, traversed by a moving particle whose velocity $d\mathbf{x}/dt$ is specified by the differential equation. The x_1x_2-plane itself is called the **phase plane**, and a representative set of trajectories is referred to as a **phase portrait**.

In analyzing the system (2) we must consider several different cases, depending on the nature of the eigenvalues of \mathbf{A}. These cases also occurred in Sections 7.5 through 7.8, where we were primarily interested in finding a convenient formula for the general solution. Now our main goal is to characterize the differential equation according to the geometric pattern formed by its trajectories. In each case we discuss the behavior of the trajectories in general and illustrate it with an example. It is important that you become familiar with the types of behavior that the trajectories have for each case, because these are the basic ingredients of the qualitative theory of differential equations.

CASE 1 **Real Unequal Eigenvalues of the Same Sign.** The general solution of Eq. (2) is

$$\mathbf{x} = c_1\boldsymbol{\xi}^{(1)}e^{r_1t} + c_2\boldsymbol{\xi}^{(2)}e^{r_2t}, \tag{5}$$

where r_1 and r_2 are either both positive or both negative. Suppose first that $r_1 < r_2 < 0$, and that the eigenvectors $\boldsymbol{\xi}^{(1)}$ and $\boldsymbol{\xi}^{(2)}$ are as shown in Figure 9.1.1a. It follows from Eq. (5) that $\mathbf{x} \to \mathbf{0}$ as $t \to \infty$ regardless of the values of c_1 and c_2; in

other words, all solutions approach the critical point at the origin as $t \to \infty$. If the solution starts at an initial point on the line through $\boldsymbol{\xi}^{(1)}$, then $c_2 = 0$. Consequently, the solution remains on the line through $\boldsymbol{\xi}^{(1)}$ for all t and approaches the origin as $t \to \infty$. Similarly, if the initial point is on the line through $\boldsymbol{\xi}^{(2)}$, then the solution approaches the origin along that line. In the general situation, it is helpful to rewrite Eq. (5) in the form

$$\mathbf{x} = e^{r_2 t}[c_1 \boldsymbol{\xi}^{(1)} e^{(r_1 - r_2)t} + c_2 \boldsymbol{\xi}^{(2)}]. \tag{6}$$

Observe that $r_1 - r_2 < 0$. Therefore, as long as $c_2 \neq 0$, the term $c_1 \boldsymbol{\xi}^{(1)} \exp[(r_1 - r_2)t]$ is negligible compared to $c_2 \boldsymbol{\xi}^{(2)}$ for t sufficiently large. Thus, as $t \to \infty$, the trajectory not only approaches the origin but also tends toward the line through $\boldsymbol{\xi}^{(2)}$. Hence all solutions are tangent to $\boldsymbol{\xi}^{(2)}$ at the critical point except for those solutions that start exactly on the line through $\boldsymbol{\xi}^{(1)}$. Several trajectories are sketched in Figure 9.1.1a. Some typical graphs of x_1 versus t are shown in Figure 9.1.1b, illustrating that all solutions exhibit exponential decay in time. The behavior of x_2 versus t is similar. This type of critical point is called a **node** or a **nodal sink**.

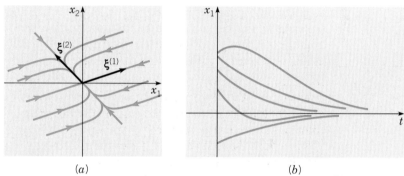

(a) $\qquad\qquad\qquad\qquad\qquad\qquad\qquad$ (b)

FIGURE 9.1.1 A node; $r_1 < r_2 < 0$. (a) The phase plane. (b) x_1 versus t.

If r_1 and r_2 are both positive, and $0 < r_2 < r_1$, then the trajectories have the same pattern as in Figure 9.1.1a, but the direction of motion is away from, rather than toward, the critical point at the origin. In this case x_1 and x_2 grow exponentially as functions of t. Again the critical point is called a node or a **nodal source**.

An example of a node occurs in Example 2 of Section 7.5, and its trajectories are shown in Figure 7.5.4.

CASE 2 **Real Eigenvalues of Opposite Sign.** The general solution of Eq. (2) is

$$\mathbf{x} = c_1 \boldsymbol{\xi}^{(1)} e^{r_1 t} + c_2 \boldsymbol{\xi}^{(2)} e^{r_2 t}, \tag{7}$$

where $r_1 > 0$ and $r_2 < 0$. Suppose that the eigenvectors $\boldsymbol{\xi}^{(1)}$ and $\boldsymbol{\xi}^{(2)}$ are as shown in Figure 9.1.2a. If the solution starts at an initial point on the line through $\boldsymbol{\xi}^{(1)}$, then it follows that $c_2 = 0$. Consequently, the solution remains on the line through $\boldsymbol{\xi}^{(1)}$ for all t, and since $r_1 > 0$, $\|\mathbf{x}\| \to \infty$ as $t \to \infty$. If the solution starts at an initial point on the line through $\boldsymbol{\xi}^{(2)}$, then the situation is similar except that $\|\mathbf{x}\| \to 0$ as $t \to \infty$ because $r_2 < 0$. Solutions starting at other initial points follow trajectories such as those shown in Figure 9.1.2a. The positive exponential is the dominant term

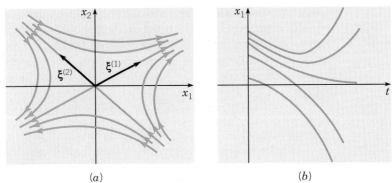

FIGURE 9.1.2 A saddle point; $r_1 > 0, r_2 < 0$. (*a*) The phase plane. (*b*) x_1 versus t.

in Eq. (7) for large t, so eventually all these solutions approach infinity asymptotic to the line determined by the eigenvector $\boldsymbol{\xi}^{(1)}$ corresponding to the positive eigenvalue r_1. The only solutions that approach the critical point at the origin are those that start precisely on the line determined by $\boldsymbol{\xi}^{(2)}$. Figure 9.1.2*b* shows some typical graphs of x_1 versus t. For certain initial conditions the positive exponential term is absent from the solution, so $x_1 \to 0$ as $t \to \infty$. For all other initial conditions the positive exponential term eventually dominates and causes x_1 to become unbounded. The behavior of x_2 is similar. The origin is called a **saddle point** in this case.

An example of a saddle point occurs in Example 1 of Section 7.5, and its trajectories are shown in Figure 7.5.2.

CASE 3 **Equal Eigenvalues.** We now suppose that $r_1 = r_2 = r$. We consider the case in which the eigenvalues are negative; if they are positive, the trajectories are similar but the direction of motion is reversed. There are two subcases, depending on whether the repeated eigenvalue has two independent eigenvectors or only one.

(a) **Two independent eigenvectors.** The general solution of Eq. (2) is

$$\mathbf{x} = c_1 \boldsymbol{\xi}^{(1)} e^{rt} + c_2 \boldsymbol{\xi}^{(2)} e^{rt}, \tag{8}$$

where $\boldsymbol{\xi}^{(1)}$ and $\boldsymbol{\xi}^{(2)}$ are the independent eigenvectors. The ratio x_2/x_1 is independent of t, but depends on the components of $\boldsymbol{\xi}^{(1)}$ and $\boldsymbol{\xi}^{(2)}$ and on the arbitrary constants c_1 and c_2. Thus every trajectory lies on a straight line through the origin, as shown in Figure 9.1.3*a*. Typical graphs of x_1 or x_2 versus t are shown in Figure 9.1.3*b*. The critical point is called a **proper node**, or sometimes a **star point**.

(b) **One independent eigenvector.** As shown in Section 7.8, the general solution of Eq. (2) in this case is

$$\mathbf{x} = c_1 \boldsymbol{\xi} e^{rt} + c_2 (\boldsymbol{\xi} t e^{rt} + \boldsymbol{\eta} e^{rt}), \tag{9}$$

where $\boldsymbol{\xi}$ is the eigenvector and $\boldsymbol{\eta}$ is the generalized eigenvector associated with the repeated eigenvalue. For large t the dominant term in Eq. (9) is $c_2 \boldsymbol{\xi} t e^{rt}$. Thus, as $t \to \infty$, every trajectory approaches the origin tangent to the line through the eigenvector. This is true even if $c_2 = 0$, for then the solution $\mathbf{x} = c_1 \boldsymbol{\xi} e^{rt}$ lies on this line. Similarly, for large negative t the term $c_2 \boldsymbol{\xi} t e^{rt}$ is again the dominant one, so as $t \to -\infty$, each trajectory is asymptotic to a line parallel to $\boldsymbol{\xi}$.

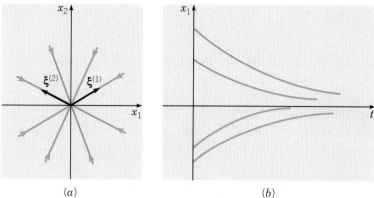

(a) (b)

FIGURE 9.1.3 A proper node, two independent eigenvectors; $r_1 = r_2 < 0$. (a) The phase plane. (b) x_1 versus t.

The orientation of the trajectories depends on the relative positions of ξ and η. One possible situation is shown in Figure 9.1.4a. To locate the trajectories, it is helpful to write the solution (9) in the form

$$\mathbf{x} = [(c_1\xi + c_2\eta) + c_2\xi t]e^{rt} = \mathbf{y}e^{rt}, \tag{10}$$

where $\mathbf{y} = (c_1\xi + c_2\eta) + c_2\xi t$. Observe that the vector \mathbf{y} determines the direction of \mathbf{x}, whereas the scalar quantity e^{rt} affects only the magnitude of \mathbf{x}. Also note that, for fixed values of c_1 and c_2, the expression for \mathbf{y} is a vector equation of the straight line through the point $c_1\xi + c_2\eta$ and parallel to ξ.

To sketch the trajectory corresponding to a given pair of values of c_1 and c_2, you can proceed in the following way. First, draw the line given by $(c_1\xi + c_2\eta) + c_2\xi t$ and note the direction of increasing t on this line. Two such lines are shown in Figure 9.1.4a, one for $c_2 > 0$ and the other for $c_2 < 0$. Next, note that the given trajectory passes through the point $c_1\xi + c_2\eta$ when $t = 0$. Further, as t increases, the direction of the vector \mathbf{x} given by Eq. (10) follows the direction of increasing t on the line, but the magnitude of \mathbf{x} rapidly decreases and approaches zero because of the decaying exponential factor e^{rt}. Finally, as t decreases toward $-\infty$, the direction of \mathbf{x} is determined by points on the corresponding part of the line, and the magnitude of \mathbf{x} approaches infinity. In this way we obtain the heavy trajectories in Figure 9.1.4a. A few other trajectories are lightly sketched as well to help complete the diagram. Typical graphs of x_1 versus t are shown in Figure 9.1.4b.

The other possible situation is shown in Figure 9.1.4c, where the relative orientation of ξ and η is reversed. As indicated in the figure, this results in a reversal in the orientation of the trajectories.

If $r_1 = r_2 > 0$, you can sketch the trajectories by following the same procedure. In this event the trajectories are traversed in the outward direction, and the orientation of the trajectories with respect to that of ξ and η is also reversed.

When a double eigenvalue has only a single independent eigenvector, the critical point is called an **improper** or **degenerate node**. A specific example of this case is Example 2 in Section 7.8; the trajectories are shown in Figure 7.8.2.

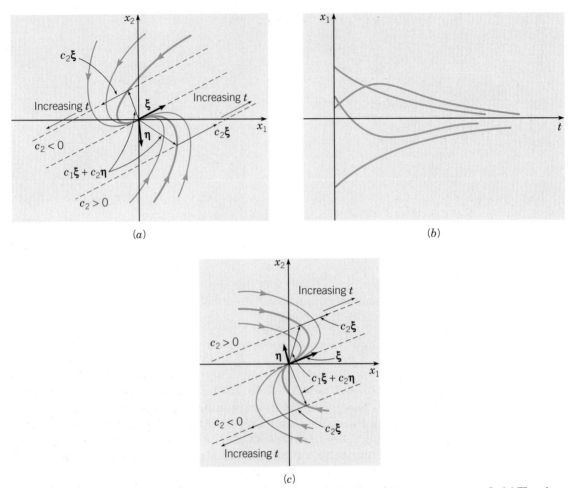

FIGURE 9.1.4 An improper node, one independent eigenvector; $r_1 = r_2 < 0$. (*a*) The phase plane. (*b*) x_1 versus t. (*c*) The phase plane.

CASE 4 **Complex Eigenvalues.** Suppose that the eigenvalues are $\lambda \pm i\mu$, where λ and μ are real, $\lambda \neq 0$, and $\mu > 0$. It is possible to write down the general solution in terms of the eigenvalues and eigenvectors, as shown in Section 7.6. However, we proceed in a different way.

Systems having the eigenvalues $\lambda \pm i\mu$ are typified by

$$\mathbf{x}' = \begin{pmatrix} \lambda & \mu \\ -\mu & \lambda \end{pmatrix} \mathbf{x} \tag{11}$$

or, in scalar form,

$$x_1' = \lambda x_1 + \mu x_2, \qquad x_2' = -\mu x_1 + \lambda x_2. \tag{12}$$

We introduce the polar coordinates r, θ given by

$$r^2 = x_1^2 + x_2^2, \qquad \tan\theta = x_2/x_1.$$

By differentiating these equations we obtain

$$rr' = x_1 x_1' + x_2 x_2', \qquad (\sec^2 \theta)\theta' = (x_1 x_2' - x_2 x_1')/x_1^2. \tag{13}$$

Substituting from Eqs. (12) in the first of Eqs. (13), we find that

$$r' = \lambda r, \tag{14}$$

and hence

$$r = ce^{\lambda t}, \tag{15}$$

where c is a constant. Similarly, substituting from Eqs. (12) in the second of Eqs. (13), and using the fact that $\sec^2 \theta = r^2/x_1^2$, we have

$$\theta' = -\mu. \tag{16}$$

Hence

$$\theta = -\mu t + \theta_0, \tag{17}$$

where θ_0 is the value of θ when $t = 0$.

Equations (15) and (17) are parametric equations in polar coordinates of the trajectories of the system (11). Since $\mu > 0$, it follows from Eq. (17) that θ decreases as t increases, so the direction of motion on a trajectory is clockwise. As $t \to \infty$, we see from Eq. (15) that $r \to 0$ if $\lambda < 0$ and $r \to \infty$ if $\lambda > 0$. Thus the trajectories are spirals, which approach or recede from the origin depending on the sign of λ. Both possibilities are shown in Figure 9.1.5, along with some typical graphs of x_1 versus t. The critical point is called a **spiral point** in this case. Frequently, the terms **spiral sink** and **spiral source**, respectively, are used to refer to spiral points whose trajectories approach, or depart from, the critical point.

More generally, it is possible to show that for any system with complex eigenvalues $\lambda \pm i\mu$, where $\lambda \neq 0$, the trajectories are always spirals. They are directed inward or outward, respectively, depending on whether λ is negative or positive. They may be elongated and skewed with respect to the coordinate axes, and the direction of motion may be either clockwise or counterclockwise. While a detailed analysis is moderately difficult, it is easy to obtain a general idea of the orientation of the trajectories directly from the differential equations. Suppose that

$$\begin{pmatrix} dx/dt \\ dy/dt \end{pmatrix} = \begin{pmatrix} a & b \\ c & d \end{pmatrix} \begin{pmatrix} x \\ y \end{pmatrix} \tag{18}$$

has complex eigenvalues $\lambda \pm i\mu$, and look at the point $(0, 1)$ on the positive y-axis. At this point it follows from Eqs. (18) that $dx/dt = b$ and $dy/dt = d$. Depending on the signs of b and d, one can infer the direction of motion and the approximate orientation of the trajectories. For instance, if both b and d are negative, then the trajectories cross the positive y-axis so as to move down and into the second quadrant. If $\lambda < 0$ also, then the trajectories must be inward-pointing spirals resembling the one in Figure 9.1.6. Another case was given in Example 1 of Section 7.6, whose trajectories are shown in Figure 7.6.2.

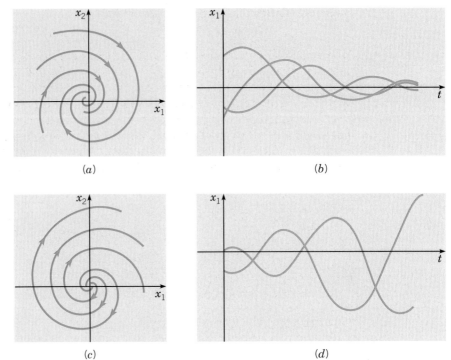

FIGURE 9.1.5 A spiral point; $r_1 = \lambda + i\mu, r_2 = \lambda - i\mu$. (*a*) $\lambda < 0$, the phase plane. (*b*) $\lambda < 0$, x_1 versus t. (*c*) $\lambda > 0$, the phase plane. (*d*) $\lambda > 0, x_1$ versus t.

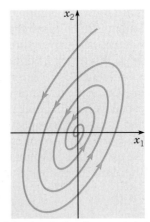

FIGURE 9.1.6 A spiral point; $r = \lambda \pm i\mu$ with $\lambda < 0$.

CASE 5 **Pure Imaginary Eigenvalues.** In this case $\lambda = 0$ and the system (11) reduces to

$$\mathbf{x}' = \begin{pmatrix} 0 & \mu \\ -\mu & 0 \end{pmatrix} \mathbf{x} \tag{19}$$

with eigenvalues $\pm i\mu$. Using the same argument as in Case 4, we find that

$$r' = 0, \qquad \theta' = -\mu, \tag{20}$$

and consequently,

$$r = c, \qquad \theta = -\mu t + \theta_0, \qquad (21)$$

where c and θ_0 are constants. Thus the trajectories are circles, with center at the origin, that are traversed clockwise if $\mu > 0$ and counterclockwise if $\mu < 0$. A complete circuit about the origin is made in a time interval of length $2\pi/\mu$, so all solutions are periodic with period $2\pi/\mu$. The critical point is called a **center**.

In general, when the eigenvalues are pure imaginary, it is possible to show (see Problem 19) that the trajectories are ellipses centered at the origin. A typical situation is shown in Figure 9.1.7, which also includes some typical graphs of x_1 versus t. See also Example 3 in Section 7.6, especially Figures 7.6.3 and 7.6.4.

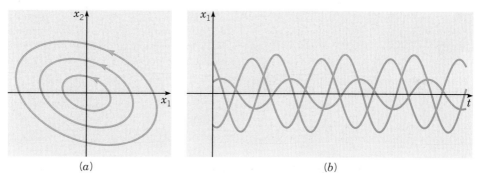

(a) (b)

FIGURE 9.1.7 A center; $r_1 = i\mu, r_2 = -i\mu$. (a) The phase plane. (b) x_1 versus t.

By reflecting on these five cases and by examining the corresponding figures, we can make several observations:

1. After a long time, each individual trajectory exhibits one of only three types of behavior. As $t \to \infty$, each trajectory approaches infinity, approaches the critical point $\mathbf{x} = \mathbf{0}$, or repeatedly traverses a closed curve, corresponding to a periodic solution, that surrounds the critical point.

2. Viewed as a whole, the pattern of trajectories in each case is relatively simple. To be more specific, through each point (x_0, y_0) in the phase plane there is only one trajectory; thus the trajectories do not cross each other. Do not be misled by the figures, in which it sometimes appears that many trajectories pass through the critical point $\mathbf{x} = \mathbf{0}$. In fact, the only solution passing through the origin is the equilibrium solution $\mathbf{x} = \mathbf{0}$. The other solutions that appear to pass through the origin actually only approach this point as $t \to \infty$ or $t \to -\infty$.

3. In each case the set of all trajectories is such that one of three situations occurs.

 (a) All trajectories approach the critical point $\mathbf{x} = \mathbf{0}$ as $t \to \infty$. This is the case if the eigenvalues are real and negative or complex with negative real part. The origin is either a nodal or a spiral sink.

 (b) All trajectories remain bounded but do not approach the origin as $t \to \infty$. This is the case if the eigenvalues are pure imaginary. The origin is a center.

 (c) Some trajectories, and possibly all trajectories except $\mathbf{x} = \mathbf{0}$, tend to infinity as $t \to \infty$. This is the case if at least one of the eigenvalues is positive or if the eigenvalues have positive real part. The origin is a nodal source, a spiral source, or a saddle point.

The situations described in 3(a), (b), and (c) above illustrate the concepts of asymptotic stability, stability, and instability, respectively, of the equilibrium solution $\mathbf{x} = \mathbf{0}$ of the system (2). The precise definitions of these terms are given in Section 9.2, but their basic meaning should be clear from the geometrical discussion in this section. The information that we have obtained about the system (2) is summarized in Table 9.1.1. Also see Problems 20 and 21.

TABLE 9.1.1 Stability Properties of Linear Systems $\mathbf{x}' = \mathbf{A}\mathbf{x}$ with $\det(\mathbf{A} - r\mathbf{I}) = 0$ and $\det \mathbf{A} \neq 0$

Eigenvalues	Type of Critical Point	Stability
$r_1 > r_2 > 0$	Node	Unstable
$r_1 < r_2 < 0$	Node	Asymptotically stable
$r_2 < 0 < r_1$	Saddle point	Unstable
$r_1 = r_2 > 0$	Proper or improper node	Unstable
$r_1 = r_2 < 0$	Proper or improper node	Asymptotically stable
$r_1, r_2 = \lambda \pm i\mu$	Spiral point	
$\quad \lambda > 0$		Unstable
$\quad \lambda < 0$		Asymptotically stable
$r_1 = i\mu, r_2 = -i\mu$	Center	Stable

The analysis in this section applies only to second order systems $\mathbf{x}' = \mathbf{A}\mathbf{x}$ whose solutions are represented geometrically as curves in the phase plane. A similar, though more complicated, analysis can be carried out for an nth order system, with an $n \times n$ coefficient matrix \mathbf{A}, whose solutions are curves in an n-dimensional phase space. The cases that can occur in higher order systems are essentially combinations of those we have seen in two dimensions. For instance, in a third order system with a three-dimensional phase space, one possibility is that solutions in a certain plane may spiral toward the origin, while other solutions may tend to infinity along a line transverse to this plane. This would be the case if the coefficient matrix has two complex eigenvalues with negative real part and one positive real eigenvalue. However, because of their complexity, we will not discuss systems of higher than second order.

PROBLEMS

For each of the systems in Problems 1 through 12:

(a) Find the eigenvalues and eigenvectors.

(b) Classify the critical point $(0, 0)$ as to type and determine whether it is stable, asymptotically stable, or unstable.

(c) Sketch several trajectories in the phase plane and also sketch some typical graphs of x_1 versus t.

(d) Use a computer to plot accurately the curves requested in part (c).

1. $\dfrac{d\mathbf{x}}{dt} = \begin{pmatrix} 3 & -2 \\ 2 & -2 \end{pmatrix} \mathbf{x}$

2. $\dfrac{d\mathbf{x}}{dt} = \begin{pmatrix} 5 & -1 \\ 3 & 1 \end{pmatrix} \mathbf{x}$

3. $\dfrac{d\mathbf{x}}{dt} = \begin{pmatrix} 2 & -1 \\ 3 & -2 \end{pmatrix} \mathbf{x}$

4. $\dfrac{d\mathbf{x}}{dt} = \begin{pmatrix} 1 & -4 \\ 4 & -7 \end{pmatrix} \mathbf{x}$

5. $\dfrac{d\mathbf{x}}{dt} = \begin{pmatrix} 1 & -5 \\ 1 & -3 \end{pmatrix} \mathbf{x}$

6. $\dfrac{d\mathbf{x}}{dt} = \begin{pmatrix} 2 & -5 \\ 1 & -2 \end{pmatrix} \mathbf{x}$

7. $\dfrac{d\mathbf{x}}{dt} = \begin{pmatrix} 3 & -2 \\ 4 & -1 \end{pmatrix} \mathbf{x}$

8. $\dfrac{d\mathbf{x}}{dt} = \begin{pmatrix} -1 & -1 \\ 0 & -0.25 \end{pmatrix} \mathbf{x}$

9. $\dfrac{d\mathbf{x}}{dt} = \begin{pmatrix} 3 & -4 \\ 1 & -1 \end{pmatrix} \mathbf{x}$

10. $\dfrac{d\mathbf{x}}{dt} = \begin{pmatrix} 1 & 2 \\ -5 & -1 \end{pmatrix} \mathbf{x}$

11. $\dfrac{d\mathbf{x}}{dt} = \begin{pmatrix} -1 & 0 \\ 0 & -1 \end{pmatrix} \mathbf{x}$

12. $\dfrac{d\mathbf{x}}{dt} = \begin{pmatrix} 2 & -\frac{5}{2} \\ \frac{9}{5} & -1 \end{pmatrix} \mathbf{x}$

In each of Problems 13 through 16 determine the critical point $\mathbf{x} = \mathbf{x}^0$, and then classify its type and examine its stability by making the transformation $\mathbf{x} = \mathbf{x}^0 + \mathbf{u}$.

13. $\dfrac{d\mathbf{x}}{dt} = \begin{pmatrix} 1 & 1 \\ 1 & -1 \end{pmatrix} \mathbf{x} - \begin{pmatrix} 2 \\ 0 \end{pmatrix}$

14. $\dfrac{d\mathbf{x}}{dt} = \begin{pmatrix} -2 & 1 \\ 1 & -2 \end{pmatrix} \mathbf{x} + \begin{pmatrix} -2 \\ 1 \end{pmatrix}$

15. $\dfrac{d\mathbf{x}}{dt} = \begin{pmatrix} -1 & -1 \\ 2 & -1 \end{pmatrix} \mathbf{x} + \begin{pmatrix} -1 \\ 5 \end{pmatrix}$

16. $\dfrac{d\mathbf{x}}{dt} = \begin{pmatrix} 0 & -\beta \\ \delta & 0 \end{pmatrix} \mathbf{x} + \begin{pmatrix} \alpha \\ -\gamma \end{pmatrix}; \qquad \alpha, \beta, \gamma, \delta > 0$

17. The equation of motion of a spring–mass system with damping (see Section 3.8) is

$$m \frac{d^2 u}{dt^2} + c \frac{du}{dt} + ku = 0,$$

where m, c, and k are positive. Write this second order equation as a system of two first order equations for $x = u$, $y = du/dt$. Show that $x = 0$, $y = 0$ is a critical point, and analyze the nature and stability of the critical point as a function of the parameters m, c, and k. A similar analysis can be applied to the electric circuit equation (see Section 3.8)

$$L \frac{d^2 I}{dt^2} + R \frac{dI}{dt} + \frac{1}{C} I = 0.$$

18. Consider the system $\mathbf{x}' = \mathbf{A}\mathbf{x}$, and suppose that \mathbf{A} has one zero eigenvalue.

 (a) Show that $\mathbf{x} = \mathbf{0}$ is a critical point and that, in addition, every point on a certain straight line through the origin is also a critical point.

 (b) Let $r_1 = 0$ and $r_2 \neq 0$, and let $\boldsymbol{\xi}^{(1)}$ and $\boldsymbol{\xi}^{(2)}$ be corresponding eigenvectors. Show that the trajectories are as indicated in Figure 9.1.8. What is the direction of motion on the trajectories?

19. In this problem we indicate how to show that the trajectories are ellipses when the eigenvalues are pure imaginary. Consider the system

$$\begin{pmatrix} x \\ y \end{pmatrix}' = \begin{pmatrix} a_{11} & a_{12} \\ a_{21} & a_{22} \end{pmatrix} \begin{pmatrix} x \\ y \end{pmatrix}. \tag{i}$$

 (a) Show that the eigenvalues of the coefficient matrix are pure imaginary if and only if

$$a_{11} + a_{22} = 0, \qquad a_{11}a_{22} - a_{12}a_{21} > 0. \tag{ii}$$

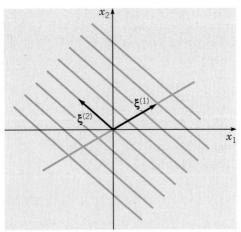

FIGURE 9.1.8 Nonisolated critical points; $r_1 = 0, r_2 \neq 0$. Every point on the line through $\xi^{(1)}$ is a critical point.

(b) The trajectories of the system (i) can be found by converting Eqs. (i) into the single equation

$$\frac{dy}{dx} = \frac{dy/dt}{dx/dt} = \frac{a_{21}x + a_{22}y}{a_{11}x + a_{12}y}. \tag{iii}$$

Use the first of Eqs. (ii) to show that Eq. (iii) is exact.

(c) By integrating Eq. (iii) show that

$$a_{21}x^2 + 2a_{22}xy - a_{12}y^2 = k, \tag{iv}$$

where k is a constant. Use Eqs. (ii) to conclude that the graph of Eq. (iv) is always an ellipse.

Hint: What is the discriminant of the quadratic form in Eq. (iv)?

20. Consider the linear system

$$dx/dt = a_{11}x + a_{12}y, \qquad dy/dt = a_{21}x + a_{22}y,$$

where a_{11}, \ldots, a_{22} are real constants. Let $p = a_{11} + a_{22}, q = a_{11}a_{22} - a_{12}a_{21},$ and $\Delta = p^2 - 4q$. Observe that p and q are the trace and determinant, respectively, of the coefficient matrix of the given system. Show that the critical point $(0,0)$ is a

(a) Node if $q > 0$ and $\Delta \geq 0$; (b) Saddle point if $q < 0$;

(c) Spiral point if $p \neq 0$ and $\Delta < 0$; (d) Center if $p = 0$ and $q > 0$.

Hint: These conclusions can be obtained by studying the eigenvalues r_1 and r_2. It may also be helpful to establish, and then to use, the relations $r_1 r_2 = q$ and $r_1 + r_2 = p$.

21. Continuing Problem 20, show that the critical point $(0,0)$ is

(a) Asymptotically stable if $q > 0$ and $p < 0$;

(b) Stable if $q > 0$ and $p = 0$;

(c) Unstable if $q < 0$ or $p > 0$.

The results of Problems 20 and 21 are summarized visually in Figure 9.1.9.

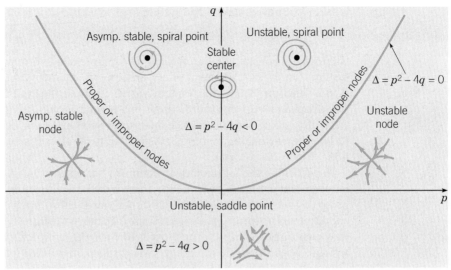

FIGURE 9.1.9 Stability diagram.

9.2 Autonomous Systems and Stability

In this section we begin to draw together, and to expand on, the geometrical ideas introduced in Section 2.5 for certain first order equations and in Section 9.1 for second order linear homogeneous systems with constant coefficients. These ideas concern the qualitative study of differential equations and the concept of stability, an idea that will be defined precisely later in this section.

Autonomous Systems. We are concerned with systems of two simultaneous differential equations of the form

$$dx/dt = F(x, y), \qquad dy/dt = G(x, y). \tag{1}$$

We assume that the functions F and G are continuous and have continuous partial derivatives in some domain D of the xy-plane. If (x_0, y_0) is a point in this domain, then by Theorem 7.1.1 there exists a unique solution $x = \phi(t), y = \psi(t)$ of the system (1) satisfying the initial conditions

$$x(t_0) = x_0, \qquad y(t_0) = y_0. \tag{2}$$

The solution is defined in some time interval I that contains the point t_0.

Frequently, we will write the initial value problem (1), (2) in the vector form

$$d\mathbf{x}/dt = \mathbf{f}(\mathbf{x}), \qquad \mathbf{x}(t_0) = \mathbf{x}^0, \tag{3}$$

where $\mathbf{x} = x\mathbf{i} + y\mathbf{j}$, $\mathbf{f}(\mathbf{x}) = F(x, y)\mathbf{i} + G(x, y)\mathbf{j}$, and $\mathbf{x}^0 = x_0\mathbf{i} + y_0\mathbf{j}$. In this case the solution is expressed as $\mathbf{x} = \boldsymbol{\phi}(t)$, where $\boldsymbol{\phi}(t) = \phi(t)\mathbf{i} + \psi(t)\mathbf{j}$. As usual, we interpret a solution $\mathbf{x} = \boldsymbol{\phi}(t)$ as a curve traced by a moving point in the xy-plane, the phase plane.

Observe that the functions F and G in Eqs. (1) do not depend on the independent variable t, but only on the dependent variables x and y. A system with this property

is said to be **autonomous**. The system

$$\mathbf{x}' = \mathbf{A}\mathbf{x}, \tag{4}$$

where \mathbf{A} is a constant matrix, is a simple example of a two-dimensional autonomous system. On the other hand, if one or more of the elements of the coefficient matrix \mathbf{A} is a function of the independent variable t, then the system is nonautonomous. The distinction between autonomous and nonautonomous systems is important because the geometrical qualitative analysis in Section 9.1 can be effectively extended to two-dimensional autonomous systems in general, but is not nearly as useful for nonautonomous systems.

In particular, the autonomous system (1) has an associated direction field that is independent of time. Consequently, there is only one trajectory passing through each point (x_0, y_0) in the phase plane. In other words, all solutions that satisfy an initial condition of the form (2) lie on the same trajectory, regardless of the time t_0 at which they pass through (x_0, y_0). Thus, just as for the constant coefficient linear system (4), a single phase portrait simultaneously displays important qualitative information about all solutions of the system (1). We will see this fact confirmed repeatedly in this chapter.

Autonomous systems occur frequently in applications. Physically, an autonomous system is one whose configuration, including physical parameters and external forces or effects, is independent of time. The response of the system to given initial conditions is then independent of the time at which the conditions are imposed.

Stability and Instability. The concepts of stability, asymptotic stability, and instability have already been mentioned several times in this book. It is now time to give a precise mathematical definition of these concepts, at least for autonomous systems of the form

$$\mathbf{x}' = \mathbf{f}(\mathbf{x}). \tag{5}$$

In the following definitions, and elsewhere, we use the notation $\|\mathbf{x}\|$ to designate the length, or magnitude, of the vector \mathbf{x}.

The points, if any, where $\mathbf{f}(\mathbf{x}) = \mathbf{0}$ are called **critical points** of the autonomous system (5). At such points $\mathbf{x}' = \mathbf{0}$ also, so critical points correspond to constant, or equilibrium, solutions of the system of differential equations. A critical point \mathbf{x}^0 of the system (5) is said to be **stable** if, given any $\epsilon > 0$, there is a $\delta > 0$ such that every solution $\mathbf{x} = \boldsymbol{\phi}(t)$ of the system (1), which at $t = 0$ satisfies

$$\|\boldsymbol{\phi}(0) - \mathbf{x}^0\| < \delta, \tag{6}$$

exists for all positive t and satisfies

$$\|\boldsymbol{\phi}(t) - \mathbf{x}^0\| < \epsilon \tag{7}$$

for all $t \geq 0$. This is illustrated geometrically in Figures 9.2.1*a* and 9.2.1*b*. These mathematical statements say that all solutions that start "sufficiently close" (that is, within the distance δ) to \mathbf{x}^0 stay "close" (within the distance ϵ) to \mathbf{x}^0. Note that in Figure 9.2.1*a* the trajectory is within the circle $\|\mathbf{x} - \mathbf{x}^0\| = \delta$ at $t = 0$ and, although it soon passes outside of this circle, it remains within the circle $\|\mathbf{x} - \mathbf{x}^0\| = \epsilon$ for all $t \geq 0$. However, the trajectory of the solution does not have to approach the critical point \mathbf{x}^0 as $t \to \infty$, as illustrated in Figure 9.2.1*b*. A critical point that is not stable is said to be **unstable**.

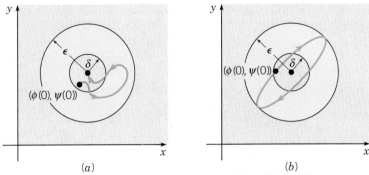

FIGURE 9.2.1 (*a*) Asymptotic stability. (*b*) Stability.

A critical point \mathbf{x}^0 is said to be **asymptotically stable** if it is stable and if there exists a δ_0 ($\delta_0 > 0$) such that if a solution $\mathbf{x} = \boldsymbol{\phi}(t)$ satisfies

$$\|\boldsymbol{\phi}(0) - \mathbf{x}^0\| < \delta_0, \tag{8}$$

then

$$\lim_{t \to \infty} \boldsymbol{\phi}(t) = \mathbf{x}^0. \tag{9}$$

Thus trajectories that start "sufficiently close" to \mathbf{x}^0 not only must stay "close" but must eventually approach \mathbf{x}^0 as $t \to \infty$. This is the case for the trajectory in Figure 9.2.1*a* but not for the one in Figure 9.2.1*b*. Note that asymptotic stability is a stronger property than stability, since a critical point must be stable before we can even talk about whether it might be asymptotically stable. On the other hand, the limit condition (9), which is an essential feature of asymptotic stability, does not by itself imply even ordinary stability. Indeed, examples can be constructed in which all the trajectories approach \mathbf{x}^0 as $t \to \infty$, but for which \mathbf{x}^0 is not a stable critical point. Geometrically, all that is needed is a family of trajectories having members that start arbitrarily close to \mathbf{x}^0 and then depart an arbitrarily large distance before eventually approaching \mathbf{x}^0 as $t \to \infty$.

In this chapter we are concentrating on systems of second order, but the definitions just given are independent of the order of the system. If you interpret the vectors in Eqs. (5) through (9) as *n*-dimensional, then the definitions of stability, asymptotic stability, and instability apply also to *n*th order systems. These definitions can be made more concrete by interpreting them in terms of a specific physical problem.

The Oscillating Pendulum. The concepts of asymptotic stability, stability, and instability can be easily visualized in terms of an oscillating pendulum. Consider the configuration shown in Figure 9.2.2, in which a mass *m* is attached to one end of a rigid, but weightless, rod of length *L*. The other end of the rod is supported at the origin *O*, and the rod is free to rotate in the plane of the paper. The position of the pendulum is described by the angle θ between the rod and the downward vertical direction, with the counterclockwise direction taken as positive. The gravitational force *mg* acts downward, while the damping force $c|d\theta/dt|$, where *c* is positive, is always opposite to the direction of motion. We assume that θ and $d\theta/dt$ are both positive. The equation of motion can be quickly derived from the principle of angular momentum, which states that the time rate of change of angular momentum about any point is

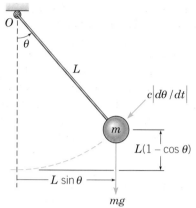

FIGURE 9.2.2 An oscillating pendulum.

equal to the moment of the resultant force about that point. The angular momentum about the origin is $mL^2(d\theta/dt)$, so the governing equation is

$$mL^2\frac{d^2\theta}{dt^2} = -cL\frac{d\theta}{dt} - mgL\sin\theta. \tag{10}$$

The factors L and $L\sin\theta$ on the right side of Eq. (10) are the moment arms of the resistive force and of the gravitational force, respectively; the minus signs are due to the fact that the two forces tend to make the pendulum rotate in the clockwise (negative) direction. You should verify, as an exercise, that the same equation is obtained for the other three possible sign combinations of θ and $d\theta/dt$.

By straightforward algebraic operations, we can write Eq. (10) in the standard form

$$\frac{d^2\theta}{dt^2} + \frac{c}{mL}\frac{d\theta}{dt} + \frac{g}{L}\sin\theta = 0, \tag{11}$$

or

$$\frac{d^2\theta}{dt^2} + \gamma\frac{d\theta}{dt} + \omega^2\sin\theta = 0, \tag{12}$$

where $\gamma = c/mL$ and $\omega^2 = g/L$. To convert Eq. (12) to a system of two first order equations, we let $x = \theta$ and $y = d\theta/dt$; then

$$\frac{dx}{dt} = y, \qquad \frac{dy}{dt} = -\omega^2\sin x - \gamma y. \tag{13}$$

Since γ and ω^2 are constants, the system (13) is an autonomous system of the form (1).

The critical points of Eqs. (13) are found by solving the equations

$$y = 0, \qquad -\omega^2\sin x - \gamma y = 0.$$

We obtain $y = 0$ and $x = \pm n\pi$, where n is an integer. These points correspond to two physical equilibrium positions, one with the mass directly below the point of support ($\theta = 0$) and the other with the mass directly above the point of support ($\theta = \pi$). Our intuition suggests that the first is stable and the second is unstable.

More precisely, if the mass is slightly displaced from the lower equilibrium position, it will oscillate back and forth with gradually decreasing amplitude, eventually approaching the equilibrium position as the initial potential energy is dissipated by the damping force. This type of motion illustrates *asymptotic stability.*

On the other hand, if the mass is slightly displaced from the upper equilibrium position, it will rapidly fall, under the influence of gravity, and will ultimately approach the lower equilibrium position in this case also. This type of motion illustrates *instability.* In practice, it is impossible to maintain the pendulum in its upward equilibrium position for any length of time without an external constraint mechanism, since the slightest perturbation will cause the mass to fall.

Finally, consider the ideal situation in which the damping coefficient c (or γ) is zero. In this case, if the mass is displaced slightly from its lower equilibrium position, it will oscillate indefinitely with constant amplitude about the equilibrium position. Since there is no dissipation in the system, the mass will remain near the equilibrium position but will not approach it asymptotically. This type of motion is *stable* but not asymptotically stable. In general, this motion is impossible to achieve experimentally, because the slightest degree of air resistance or friction at the point of support will eventually cause the pendulum to approach its rest position.

These three types of motion are illustrated schematically in Figure 9.2.3. Solutions of the pendulum equations are discussed in more detail in the next section.

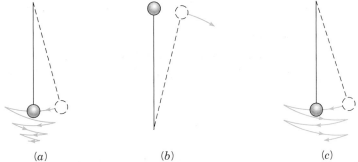

(a) (b) (c)

FIGURE 9.2.3 Qualitative motion of a pendulum. (*a*) With air resistance. (*b*) With or without air resistance. (*c*) Without air resistance.

Determination of Trajectories. The trajectories of a two-dimensional autonomous system can sometimes be found by solving a related first order differential equation. From Eqs. (1) we have

$$\frac{dy}{dx} = \frac{dy/dt}{dx/dt} = \frac{G(x,y)}{F(x,y)}, \tag{14}$$

which is a first order equation in the variables x and y. Observe that such a reduction is not usually possible if F and G depend also on t. If Eq. (14) can be solved by any of the methods of Chapter 2, and if we write solutions (implicitly) in the form

$$H(x,y) = c, \tag{15}$$

then Eq. (15) is an equation for the trajectories of the system (14). In other words, the trajectories lie on the level curves of $H(x,y)$. Keep in mind that there is no general

way of solving Eq. (14) to obtain the function H, so this approach is applicable only in special cases.

Find the trajectories of the system

$$dx/dt = y, \qquad dy/dt = x. \tag{16}$$

In this case, Eq. (14) becomes

$$\frac{dy}{dx} = \frac{x}{y}. \tag{17}$$

This equation is separable since it can be written as

$$y\,dy = x\,dx,$$

and its solutions are given by

$$H(x, y) = y^2 - x^2 = c, \tag{18}$$

where c is arbitrary. Therefore the trajectories of the system (16) are the hyperbolas shown in Figure 9.2.4. The direction of motion on the trajectories can be inferred from the fact that both dx/dt and $dy\,dt$ are positive in the first quadrant. The only critical point is the saddle point at the origin.

Another way to obtain the trajectories is to solve the system (16) by the methods of Section 7.5. We omit the details, but the result is

$$x = c_1 e^t + c_2 e^{-t}, \qquad y = c_1 e^t - c_2 e^{-t}.$$

Eliminating t between these two equations again leads to Eq. (18).

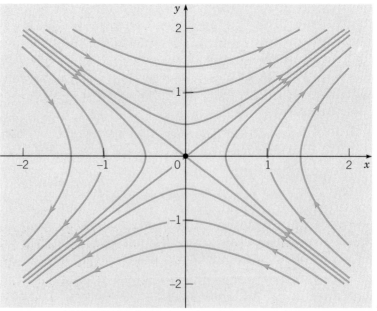

FIGURE 9.2.4 Trajectories of the system (16).

EXAMPLE
2

Find the trajectories of the system

$$\frac{dx}{dt} = 4 - 2y, \qquad \frac{dy}{dt} = 12 - 3x^2. \tag{19}$$

From the equations

$$4 - 2y = 0, \qquad 12 - 3x^2 = 0$$

we find that the critical points of the system (19) are the points $(-2, 2)$ and $(2, 2)$. To determine the trajectories, note that for this system Eq. (14) becomes

$$\frac{dy}{dx} = \frac{12 - 3x^2}{4 - 2y}. \tag{20}$$

Separating the variables in Eq. (20) and integrating, we find that solutions satisfy

$$H(x, y) = 4y - y^2 - 12x + x^3 = c, \tag{21}$$

where c is an arbitrary constant. A computer plotting routine is helpful in displaying the level curves of $H(x, y)$, some of which are shown in Figure 9.2.5. The direction of motion on the trajectories can be determined by drawing a direction field for the system (19), or by evaluating dx/dt and dy/dt at one or two selected points. From Figure 9.2.5 you can see that the critical point $(2, 2)$ is a saddle point and the point $(-2, 2)$ is a center. Observe that one trajectory leaves the saddle point (at $t = -\infty$), loops around the center, and returns to the saddle point (at $t = +\infty$).

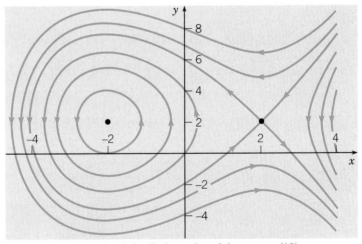

FIGURE 9.2.5 Trajectories of the system (19).

PROBLEMS

In each of Problems 1 through 4 sketch the trajectory corresponding to the solution satisfying the specified initial conditions, and indicate the direction of motion for increasing t.

1. $dx/dt = -x, \qquad dy/dt = -2y; \qquad x(0) = 4, \quad y(0) = 2$
2. $dx/dt = -x, \qquad dy/dt = 2y; \qquad x(0) = 4, \quad y(0) = 2 \quad$ and $\quad x(0) = 4, \quad y(0) = 0$
3. $dx/dt = -y, \qquad dy/dt = x; \qquad x(0) = 4, \quad y(0) = 0 \quad$ and $\quad x(0) = 0, \quad y(0) = 4$
4. $dx/dt = ay, \qquad dy/dt = -bx, \qquad a > 0, \quad b > 0; \qquad x(0) = \sqrt{a}, \quad y(0) = 0$

For each of the systems in Problems 5 through 14:

(a) Find all the critical points (equilibrium solutions).

(b) Use a computer to draw a direction field and phase portrait for the system.

(c) From the plot(s) in part (b), determine whether each critical point is asymptotically stable, stable, or unstable, and classify it as to type.

5. $dx/dt = x - xy$, $dy/dt = y + 2xy$

6. $dx/dt = 1 + 2y$, $dy/dt = 1 - 3x^2$

7. $dx/dt = x - x^2 - xy$, $dy/dt = \frac{1}{2}y - \frac{1}{4}y^2 - \frac{3}{4}xy$

8. $dx/dt = -(x - y)(1 - x - y)$, $dy/dt = x(2 + y)$

9. $dx/dt = y(2 - x - y)$, $dy/dt = -x - y - 2xy$

10. $dx/dt = (2 + x)(y - x)$, $dy/dt = y(2 + x - x^2)$

11. $dx/dt = -x + 2xy$, $dy/dt = y - x^2 - y^2$

12. $dx/dt = y$, $dy/dt = x - \frac{1}{6}x^3 - \frac{1}{5}y$

13. $dx/dt = (2 + x)(y - x)$, $dy/dt = (4 - x)(y + x)$

14. The van der Pol equation: $dx/dt = y$, $dy/dt = (1 - x^2)y - x$

In each of Problems 15 through 22:

(a) Find an equation of the form $H(x, y) = c$ satisfied by the trajectories.

(b) Plot several level curves of the function H. These are trajectories of the given system. Indicate the direction of motion on each trajectory.

15. $dx/dt = 2y$, $dy/dt = 8x$

16. $dx/dt = 2y$, $dy/dt = -8x$

17. $dx/dt = y$, $dy/dt = 2x + y$

18. $dx/dt = -x + y$, $dy/dt = -x - y$

19. $dx/dt = -x + y + x^2$, $dy/dt = y - 2xy$

20. $dx/dt = 2x^2y - 3x^2 - 4y$, $dy/dt = -2xy^2 + 6xy$

21. Undamped pendulum: $dx/dt = y$, $dy/dt = -\sin x$

22. Duffing's equation: $dx/dt = y$, $dy/dt = -x + (x^3/6)$

23. Given that $x = \phi(t), y = \psi(t)$ is a solution of the autonomous system

$$dx/dt = F(x, y), dy/dt = G(x, y)$$

for $\alpha < t < \beta$, show that $x = \Phi(t) = \phi(t - s)$, $y = \Psi(t) = \psi(t - s)$ is a solution for $\alpha + s < t < \beta + s$ for any real number s.

24. Prove that, for the system

$$dx/dt = F(x, y), dy/dt = G(x, y),$$

there is at most one trajectory passing through a given point (x_0, y_0).
Hint: Let C_0 be the trajectory generated by the solution $x = \phi_0(t)$, $y = \psi_0(t)$, with $\phi_0(t_0) = x_0, \psi_0(t_0) = y_0$, and let C_1 be the trajectory generated by the solution $x = \phi_1(t)$, $y = \psi_1(t)$, with $\phi_1(t_1) = x_0, \psi_1(t_1) = y_0$. Use the fact that the system is autonomous, and also the existence and uniqueness theorem, to show that C_0 and C_1 are the same.

25. Prove that if a trajectory starts at a noncritical point of the system

$$dx/dt = F(x, y), dy/dt = G(x, y),$$

then it cannot reach a critical point (x_0, y_0) in a finite length of time.
Hint: Assume the contrary; that is, assume that the solution $x = \phi(t), y = \psi(t)$ satisfies $\phi(a) = x_0, \psi(a) = y_0$. Then use the fact that $x = x_0, y = y_0$ is a solution of the given system satisfying the initial condition $x = x_0, y = y_0$ at $t = a$.

26. Assuming that the trajectory corresponding to a solution $x = \phi(t), y = \psi(t), -\infty < t < \infty$, of an autonomous system is closed, show that the solution is periodic.

 Hint: Since the trajectory is closed, there exists at least one point (x_0, y_0) such that $\phi(t_0) = x_0, \psi(t_0) = y_0$ and a number $T > 0$ such that $\phi(t_0 + T) = x_0, \psi(t_0 + T) = y_0$. Show that $x = \Phi(t) = \phi(t + T)$ and $y = \Psi(t) = \psi(t + T)$ is a solution, and then use the existence and uniqueness theorem to show that $\Phi(t) = \phi(t)$ and $\Psi(t) = \psi(t)$ for all t.

9.3 Almost Linear Systems

In Section 9.1 we gave an informal description of the stability properties of the equilibrium solution $\mathbf{x} = \mathbf{0}$ of the two-dimensional linear system

$$\mathbf{x}' = \mathbf{A}\mathbf{x}. \tag{1}$$

The results are summarized in Table 9.1.1. Recall that we required that $\det \mathbf{A} \neq 0$, so $\mathbf{x} = \mathbf{0}$ is the only critical point of the system (1). Now that we have defined the concepts of asymptotic stability, stability, and instability more precisely, we can restate these results in the following theorem.

Theorem 9.3.1

The critical point $\mathbf{x} = \mathbf{0}$ of the linear system (1) is asymptotically stable if the eigenvalues r_1, r_2 are real and negative or have negative real part; stable, but not asymptotically stable, if r_1 and r_2 are pure imaginary; unstable if r_1 and r_2 are real and either is positive, or if they have positive real part.

It is apparent from this theorem or from Table 9.1.1 that the eigenvalues r_1, r_2 of the coefficient matrix \mathbf{A} determine the type of critical point at $\mathbf{x} = \mathbf{0}$ and its stability characteristics. In turn, the values of r_1 and r_2 depend on the coefficients in the system (1). When such a system arises in some applied field, the coefficients usually result from the measurements of certain physical quantities. Such measurements are often subject to small uncertainties, so it is of interest to investigate whether small changes (perturbations) in the coefficients can affect the stability or instability of a critical point and/or significantly alter the pattern of trajectories.

Recall that the eigenvalues r_1, r_2 are the roots of the polynomial equation

$$\det(\mathbf{A} - r\mathbf{I}) = 0. \tag{2}$$

It is possible to show that *small* perturbations in some or all the coefficients are reflected in *small* perturbations in the eigenvalues. The most sensitive situation occurs when $r_1 = i\mu$ and $r_2 = -i\mu$, that is, when the critical point is a center and the trajectories are closed curves surrounding it. If a slight change is made in the coefficients, then the eigenvalues r_1 and r_2 will take on new values $r_1' = \lambda' + i\mu'$ and $r_2' = \lambda' - i\mu'$, where λ' is small in magnitude and $\mu' \cong \mu$ (see Figure 9.3.1). If $\lambda' \neq 0$, which almost always occurs, then the trajectories of the perturbed system are spirals rather than closed curves. The system is asymptotically stable if $\lambda' < 0$ but unstable if $\lambda' > 0$. Thus, in the case of a center, small perturbations in the coefficients may well change a stable system into an unstable one, and in any case may be expected to alter radically the pattern of trajectories in the phase plane (see Problem 25).

FIGURE 9.3.1 Schematic perturbation of $r_1 = i\mu, r_2 = -i\mu$.

Another slightly less sensitive case occurs if the eigenvalues r_1 and r_2 are equal; in this case the critical point is a node. Small perturbations in the coefficients will normally cause the two equal roots to separate (bifurcate). If the separated roots are real, then the critical point of the perturbed system remains a node, but if the separated roots are complex conjugates, then the critical point becomes a spiral point. These two possibilities are shown schematically in Figure 9.3.2. In this case the stability or instability of the system is not affected by small perturbations in the coefficients, but the trajectories may be altered considerably (see Problem 26).

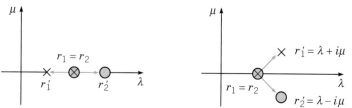

FIGURE 9.3.2 Schematic perturbation of $r_1 = r_2$.

In all other cases the stability or instability of the system is not changed, nor is the type of critical point altered, by sufficiently small perturbations in the coefficients of the system. For example, if r_1 and r_2 are real, negative, and unequal, then a *small* change in the coefficients will neither change the sign of r_1 and r_2 nor allow them to coalesce. Thus the critical point remains an asymptotically stable node.

Now let us consider a nonlinear two-dimensional autonomous system

$$\mathbf{x}' = \mathbf{f}(\mathbf{x}). \tag{3}$$

Our main object is to investigate the behavior of trajectories of the system (3) near a critical point \mathbf{x}^0. We will seek to do this by approximating the nonlinear system (3) by an appropriate linear system, whose trajectories are easy to describe. The crucial question is whether the trajectories of the linear system are good approximations to those of the nonlinear system. Of course, we also need to know how to find the approximating linear system.

It is convenient to choose the critical point to be the origin. This involves no loss of generality, since if $\mathbf{x}^0 \neq \mathbf{0}$, it is always possible to make the substitution $\mathbf{u} = \mathbf{x} - \mathbf{x}^0$ in Eq. (3). Then \mathbf{u} will satisfy an autonomous system with a critical point at the origin.

First, let us consider what it means for a nonlinear system (3) to be "close" to a linear system (1). Accordingly, suppose that

$$\mathbf{x}' = \mathbf{A}\mathbf{x} + \mathbf{g}(\mathbf{x}) \qquad (4)$$

and that $\mathbf{x} = \mathbf{0}$ is an **isolated** critical point of the system (4). This means that there is some circle about the origin within which there are no other critical points. In addition, we assume that $\det \mathbf{A} \neq 0$, so that $\mathbf{x} = \mathbf{0}$ is also an isolated critical point of the linear system $\mathbf{x}' = \mathbf{A}\mathbf{x}$. For the nonlinear system (4) to be close to the linear system $\mathbf{x}' = \mathbf{A}\mathbf{x}$, we must assume that $\mathbf{g}(\mathbf{x})$ is small. More precisely, we assume that the components of \mathbf{g} have continuous first partial derivatives and satisfy the limit condition

$$\|\mathbf{g}(\mathbf{x})\|/\|\mathbf{x}\| \to 0 \qquad \text{as} \quad \mathbf{x} \to \mathbf{0}; \qquad (5)$$

that is, $\|\mathbf{g}\|$ is small in comparison to $\|\mathbf{x}\|$ itself near the origin. Such a system is called an **almost linear system** in the neighborhood of the critical point $\mathbf{x} = \mathbf{0}$.

It may be helpful to express the condition (5) in scalar form. If we let $\mathbf{x}^T = (x, y)$, then $\|\mathbf{x}\| = (x^2 + y^2)^{1/2} = r$. Similarly, if $\mathbf{g}^T(\mathbf{x}) = (g_1(x, y), g_2(x, y))$, then $\|\mathbf{g}(\mathbf{x})\| = [g_1^2(x, y) + g_2^2(x, y)]^{1/2}$. Then it follows that condition (5) is satisfied if and only if

$$g_1(x, y)/r \to 0, \qquad g_2(x, y)/r \to 0 \qquad \text{as} \quad r \to 0. \qquad (6)$$

Determine whether the system

EXAMPLE 1

$$\begin{pmatrix} x \\ y \end{pmatrix}' = \begin{pmatrix} 1 & 0 \\ 0 & 0.5 \end{pmatrix} \begin{pmatrix} x \\ y \end{pmatrix} + \begin{pmatrix} -x^2 - xy \\ -0.75xy - 0.25y^2 \end{pmatrix} \qquad (7)$$

is almost linear in the neighborhood of the origin.

Observe that the system (7) is of the form (4), that $(0,0)$ is a critical point, and that $\det \mathbf{A} \neq 0$. It is not hard to show that the other critical points of Eqs. (7) are $(0, 2), (1, 0)$, and $(0.5, 0.5)$; consequently, the origin is an isolated critical point. In checking the condition (6), it is convenient to introduce polar coordinates by letting $x = r\cos\theta, y = r\sin\theta$. Then

$$\frac{g_1(x, y)}{r} = \frac{-x^2 - xy}{r} = \frac{-r^2\cos^2\theta - r^2\sin\theta\cos\theta}{r}$$

$$= -r(\cos^2\theta + \sin\theta\cos\theta) \to 0$$

as $r \to 0$. In a similar way you can show that $g_2(x, y)/r \to 0$ as $r \to 0$. Hence the system (7) is almost linear near the origin.

The motion of a pendulum is described by the system [see Eq. (13) of Section 9.2]

EXAMPLE 2

$$\frac{dx}{dt} = y, \qquad \frac{dy}{dt} = -\omega^2\sin x - \gamma y. \qquad (8)$$

The critical points are $(0, 0), (\pm\pi, 0), (\pm 2\pi, 0), \ldots$, so the origin is an isolated critical point of this system. Show that the system is almost linear near the origin.

To compare Eqs. (8) with Eq. (4), we must rewrite the former so that the linear and nonlinear terms are clearly identified. If we write $\sin x = x + (\sin x - x)$ and substitute this expression in the second of Eqs. (8), we obtain the equivalent system

$$\begin{pmatrix} x \\ y \end{pmatrix}' = \begin{pmatrix} 0 & 1 \\ -\omega^2 & -\gamma \end{pmatrix} \begin{pmatrix} x \\ y \end{pmatrix} - \omega^2 \begin{pmatrix} 0 \\ \sin x - x \end{pmatrix}. \tag{9}$$

On comparing Eq. (9) with Eq. (4), we see that $g_1(x, y) = 0$ and $g_2(x, y) = -\omega^2(\sin x - x)$. From the Taylor series for $\sin x$, we know that $\sin x - x$ behaves like $-x^3/3! = -(r^3 \cos^3 \theta)/3!$ when x is small. Consequently, $(\sin x - x)/r \to 0$ as $r \to 0$. Thus the conditions (6) are satisfied and the system (9) is almost linear near the origin.

Let us now return to the general nonlinear system (3), which we write in the scalar form

$$x' = F(x, y), \qquad y' = G(x, y). \tag{10}$$

The system (10) is almost linear in the neighborhood of a critical point (x_0, y_0) whenever the functions F and G have continuous partial derivatives up to order 2. To show this, we use Taylor expansions about the point (x_0, y_0) to write $F(x, y)$ and $G(x, y)$ in the form

$$F(x, y) = F(x_0, y_0) + F_x(x_0, y_0)(x - x_0) + F_y(x_0, y_0)(y - y_0) + \eta_1(x, y),$$

$$G(x, y) = G(x_0, y_0) + G_x(x_0, y_0)(x - x_0) + G_y(x_0, y_0)(y - y_0) + \eta_2(x, y),$$

where $\eta_1(x, y)/[(x - x_0)^2 + (y - y_0)^2]^{1/2} \to 0$ as $(x, y) \to (x_0, y_0)$, and similarly for η_2. Note that $F(x_0, y_0) = G(x_0, y_0) = 0$, and that $dx/dt = d(x - x_0)/dt$ and $dy/dt = d(y - y_0)/dt$. Then the system (10) reduces to

$$\frac{d}{dt} \begin{pmatrix} x - x_0 \\ y - y_0 \end{pmatrix} = \begin{pmatrix} F_x(x_0, y_0) & F_y(x_0, y_0) \\ G_x(x_0, y_0) & G_y(x_0, y_0) \end{pmatrix} \begin{pmatrix} x - x_0 \\ y - y_0 \end{pmatrix} + \begin{pmatrix} \eta_1(x, y) \\ \eta_2(x, y) \end{pmatrix}, \tag{11}$$

or, in vector notation,

$$\frac{d\mathbf{u}}{dt} = \frac{d\mathbf{f}}{d\mathbf{x}}(\mathbf{x}^0)\mathbf{u} + \boldsymbol{\eta}(\mathbf{x}), \tag{12}$$

where $\mathbf{u} = (x - x_0, y - y_0)^T$ and $\boldsymbol{\eta} = (\eta_1, \eta_2)^T$.

The significance of this result is twofold. First, if the functions F and G are twice differentiable, then the system (10) is almost linear, and it is unnecessary to resort to the limiting process used in Examples 1 and 2. Second, the linear system that approximates the nonlinear system (10) near (x_0, y_0) is given by the linear part of Eqs. (11) or (12):

$$\frac{d}{dt} \begin{pmatrix} u_1 \\ u_2 \end{pmatrix} = \begin{pmatrix} F_x(x_0, y_0) & F_y(x_0, y_0) \\ G_x(x_0, y_0) & G_y(x_0, y_0) \end{pmatrix} \begin{pmatrix} u_1 \\ u_2 \end{pmatrix}, \tag{13}$$

where $u_1 = x - x_0$ and $u_2 = y - y_0$. Equation (13) provides a simple and general method for finding the linear system corresponding to an almost linear system near a given critical point.

**EXAMPLE
3**

Use Eq. (13) to find the linear system corresponding to the pendulum equations (8) near the origin; near the critical point $(\pi, 0)$.

In this case we have, from Eq. (8),

$$F(x, y) = y, \qquad G(x, y) = -\omega^2 \sin x - \gamma y; \tag{14}$$

since these functions are differentiable as many times as necessary, the system (8) is almost linear near each critical point. The derivatives of F and G are

$$F_x = 0, \qquad F_y = 1, \qquad G_x = -\omega^2 \cos x, \qquad G_y = -\gamma. \tag{15}$$

Thus, at the origin the corresponding linear system is

$$\frac{d}{dt} \begin{pmatrix} x \\ y \end{pmatrix} = \begin{pmatrix} 0 & 1 \\ -\omega^2 & -\gamma \end{pmatrix} \begin{pmatrix} x \\ y \end{pmatrix}, \tag{16}$$

which agrees with Eq. (9).

Similarly, evaluating the partial derivatives in Eq. (15) at $(\pi, 0)$, we obtain

$$\frac{d}{dt} \begin{pmatrix} u \\ v \end{pmatrix} = \begin{pmatrix} 0 & 1 \\ \omega^2 & -\gamma \end{pmatrix} \begin{pmatrix} u \\ v \end{pmatrix}, \tag{17}$$

where $u = x - \pi$, $v = y$. This is the linear system corresponding to Eqs. (8) near the point $(\pi, 0)$.

We now return to the almost linear system (4). Since the nonlinear term $\mathbf{g}(\mathbf{x})$ is small compared to the linear term \mathbf{Ax} when \mathbf{x} is small, it is reasonable to hope that the trajectories of the linear system (1) are good approximations to those of the nonlinear system (4), at least near the origin. This turns out to be true in many (but not all) cases, as the following theorem states.

Theorem 9.3.2 Let r_1 and r_2 be the eigenvalues of the linear system (1) corresponding to the almost linear system (4). Then the type and stability of the critical point $(0, 0)$ of the linear system (1) and the almost linear system (4) are as shown in Table 9.3.1.

At this stage, the proof of Theorem 9.3.2 is too difficult to give, so we will accept the results without proof. The statements for asymptotic stability and instability follow as a consequence of a result discussed in Section 9.6, and a proof is sketched in Problems 10 to 12 of that section. Essentially, Theorem 9.3.2 says that for small \mathbf{x} (or $\mathbf{x} - \mathbf{x}^0$) the nonlinear terms are also small and do not affect the stability and type of critical point as determined by the linear terms except in two sensitive cases: r_1 and r_2 pure imaginary, and r_1 and r_2 real and equal. Recall that earlier in this section we stated that small perturbations in the coefficients of the linear system (1), and hence in the eigenvalues r_1 and r_2, can alter the type and stability of the critical point only in these two sensitive cases. It is reasonable to expect that the small nonlinear term in Eq. (4) might have a similar substantial effect, at least in these two sensitive cases. This is so, but the main significance of Theorem 9.3.2 is that in *all other cases* the small nonlinear term does not alter the type or stability of the critical point. Thus, except in the two sensitive cases, the type and stability of the critical point of the nonlinear system (4) can be determined from a study of the much simpler linear system (1).

TABLE 9.3.1 Stability and Instability Properties of Linear and Almost Linear Systems

| r_1, r_2 | Linear System | | Almost Linear System | |
	Type	Stability	Type	Stability
$r_1 > r_2 > 0$	N	Unstable	N	Unstable
$r_1 < r_2 < 0$	N	Asymptotically stable	N	Asymptotically stable
$r_2 < 0 < r_1$	SP	Unstable	SP	Unstable
$r_1 = r_2 > 0$	PN or IN	Unstable	N or SpP	Unstable
$r_1 = r_2 < 0$	PN or IN	Asymptotically stable	N or SpP	Asymptotically stable
$r_1, r_2 = \lambda \pm i\mu$				
$\quad \lambda > 0$	SpP	Unstable	SpP	Unstable
$\quad \lambda < 0$	SpP	Asymptotically stable	SpP	Asymptotically stable
$r_1 = i\mu, r_2 = -i\mu$	C	Stable	C or SpP	Indeterminate

Note: N, node; IN, improper node; PN, proper node; SP, saddle point; SpP, spiral point; C, center.

Even if the critical point is of the same type as that of the linear system, the trajectories of the almost linear system may be considerably different in appearance from those of the corresponding linear system, except very near the critical point. However, it can be shown that the slopes at which trajectories "enter" or "leave" the critical point are given correctly by the linear equation.

Damped Pendulum. We continue the discussion of the damped pendulum begun in Examples 2 and 3. Near the origin the nonlinear equations (8) are approximated by the linear system (16), whose eigenvalues are

$$r_1, r_2 = \frac{-\gamma \pm \sqrt{\gamma^2 - 4\omega^2}}{2}. \tag{18}$$

The nature of the solutions of Eqs. (8) and (16) depends on the sign of $\gamma^2 - 4\omega^2$ as follows:

1. If $\gamma^2 - 4\omega^2 > 0$, then the eigenvalues are real, unequal, and negative. The critical point $(0,0)$ is an asymptotically stable node of the linear system (16) and of the almost linear system (8).
2. If $\gamma^2 - 4\omega^2 = 0$, then the eigenvalues are real, equal, and negative. The critical point $(0,0)$ is an asymptotically stable (proper or improper) node of the linear system (16). It may be either an asymptotically stable node or spiral point of the almost linear system (8).
3. If $\gamma^2 - 4\omega^2 < 0$, then the eigenvalues are complex with negative real part. The critical point $(0,0)$ is an asymptotically stable spiral point of the linear system (16) and of the almost linear system (8).

Thus the critical point $(0,0)$ is a spiral point of the system (8) if the damping is small and a node if the damping is large enough. In either case, the origin is asymptotically stable.

Let us now consider the case $\gamma^2 - 4\omega^2 < 0$, corresponding to small damping, in more detail. The direction of motion on the spirals near $(0,0)$ can be obtained directly from Eqs. (8). Consider the point at which a spiral intersects the positive y-axis ($x = 0$ and $y > 0$). At such a point it follows from Eqs. (8) that $dx/dt > 0$. Thus the point (x, y) on the trajectory is moving to the right, so the direction of motion on the spirals is clockwise.

The behavior of the pendulum near the critical points $(\pm n\pi, 0)$, with n even, is the same as its behavior near the origin. We expect this on physical grounds since all these critical points correspond to the downward equilibrium position of the pendulum. The conclusion can be confirmed by repeating the analysis carried out above for the origin. Figure 9.3.3 shows the clockwise spirals at a few of these critical points.

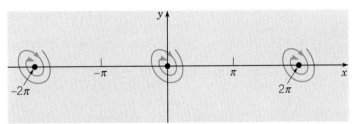

FIGURE 9.3.3 Asymptotically stable spiral points for the damped pendulum.

Now let us consider the critical point $(\pi, 0)$. Here the nonlinear equations (8) are approximated by the linear system (17), whose eigenvalues are

$$r_1, r_2 = \frac{-\gamma \pm \sqrt{\gamma^2 + 4\omega^2}}{2}. \tag{19}$$

One eigenvalue (r_1) is positive and the other (r_2) is negative. Therefore, regardless of the amount of damping, the critical point $x = \pi, y = 0$ is an unstable saddle point both of the linear system (17) and of the almost linear system (8).

To examine the behavior of trajectories near the saddle point $(\pi, 0)$ in more detail, we write down the general solution of Eqs. (17), namely,

$$\begin{pmatrix} u \\ v \end{pmatrix} = C_1 \begin{pmatrix} 1 \\ r_1 \end{pmatrix} e^{r_1 t} + C_2 \begin{pmatrix} 1 \\ r_2 \end{pmatrix} e^{r_2 t}, \tag{20}$$

where C_1 and C_2 are arbitrary constants. Since $r_1 > 0$ and $r_2 < 0$, it follows that the solution that approaches zero as $t \to \infty$ corresponds to $C_1 = 0$. For this solution $v/u = r_2$, so the slope of the entering trajectories is negative; one lies in the second quadrant ($C_2 < 0$), and the other lies in the fourth quadrant ($C_2 > 0$). For $C_2 = 0$ we obtain the pair of trajectories "exiting" from the saddle point. These trajectories have slope $r_1 > 0$; one lies in the first quadrant ($C_1 > 0$), and the other lies in the third quadrant ($C_1 < 0$).

The situation is the same at other critical points $(n\pi, 0)$ with n odd. These all correspond to the upward equilibrium position of the pendulum, so we expect them to be unstable. The analysis at $(\pi, 0)$ can be repeated to show that they are saddle points oriented in the same way as the one at $(\pi, 0)$. Diagrams of the trajectories in the neighborhood of two saddle points are shown in Figure 9.3.4.

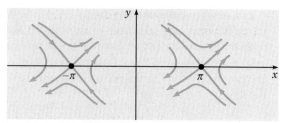

FIGURE 9.3.4 Unstable saddle points for the damped pendulum.

EXAMPLE

4

The equations of motion of a certain pendulum are

$$dx/dt = y, \qquad dy/dt = -9\sin x - \tfrac{1}{5}y, \tag{21}$$

where $x = \theta$ and $y = d\theta/dt$. Draw a phase portrait for this system and explain how it shows the possible motions of the pendulum.

By plotting the trajectories starting at various initial points in the phase plane, we obtain the phase portrait shown in Figure 9.3.5. As we have seen, the critical points (equilibrium solutions) are the points $(n\pi, 0)$, where $n = 0, \pm 1, \pm 2, \ldots$. Even values of n, including zero, correspond to the downward position of the pendulum, while odd values of n correspond to the upward position. Near each of the asymptotically stable critical points, the trajectories are clockwise spirals that represent a decaying oscillation about the downward equilibrium position. The wavy horizontal portions of the trajectories that occur for larger values of $|y|$ represent whirling motions of the pendulum. Note that a whirling motion cannot continue indefinitely, no matter how large $|y|$ is; eventually the angular velocity is so much reduced by the damping term that the pendulum can no longer go over the top, and instead begins to oscillate about its downward position.

FIGURE 9.3.5 Phase portrait for the damped pendulum of Example 4.

The trajectories that enter the saddle points separate the phase plane into regions. Such a trajectory is called a **separatrix**. Each region contains exactly one of the asymptotically stable spiral points. The initial conditions on θ and $d\theta/dt$ determine the position of an initial point (x, y) in the phase plane. The subsequent motion of the pendulum is represented by

the trajectory passing through the initial point as it spirals toward the asymptotically stable critical point in that region. The set of all initial points from which trajectories approach a given asymptotically stable critical point is called the **basin of attraction** or the **region of asymptotic stability** for that critical point. Each asymptotically stable critical point has its own basin of attraction, which is bounded by the separatrices through the neighboring unstable saddle points. The basin of attraction for the origin is shown in blue in Figure 9.3.5. Note that it is mathematically possible (but physically unrealizable) to choose initial conditions on a separatrix so that the resulting motion leads to a balanced pendulum in a vertically upward position of unstable equilibrium.

An important difference between nonlinear autonomous systems and the linear systems discussed in Section 9.1 is illustrated by the pendulum equations. Recall that the linear system (1) has only the single critical point $\mathbf{x} = \mathbf{0}$ if $\det \mathbf{A} \neq 0$. Thus, if the origin is asymptotically stable, then not only do trajectories that start close to the origin approach it, but, in fact, every trajectory approaches the origin. In this case the critical point $\mathbf{x} = \mathbf{0}$ is said to be **globally asymptotically stable**. This property of linear systems is not, in general, true for nonlinear systems. For nonlinear systems it is important to determine (or to estimate) the basin of attraction for each asymptotically stable critical point.

PROBLEMS

In each of Problems 1 through 4 verify that $(0,0)$ is a critical point, show that the system is almost linear, and discuss the type and stability of the critical point $(0,0)$ by examining the corresponding linear system.

1. $dx/dt = x - y^2, \qquad dy/dt = x - 2y + x^2$

2. $dx/dt = -x + y + 2xy, \qquad dy/dt = -4x - y + x^2 - y^2$

3. $dx/dt = (1 + x)\sin y, \qquad dy/dt = 1 - x - \cos y$

4. $dx/dt = x + y^2, \qquad dy/dt = x + y$

In each of Problems 5 through 16:

(a) Determine all critical points of the given system of equations.

(b) Find the corresponding linear system near each critical point.

(c) Find the eigenvalues of each linear system. What conclusions can you then draw about the nonlinear system?

(d) Draw a phase portrait of the nonlinear system to confirm your conclusions, or to extend them in those cases where the linear system does not provide definite information about the nonlinear system.

5. $dx/dt = (2 + x)(y - x), \qquad dy/dt = (4 - x)(y + x)$

6. $dx/dt = x - x^2 - xy, \qquad dy/dt = 3y - xy - 2y^2$

7. $dx/dt = 1 - y, \qquad dy/dt = x^2 - y^2$

8. $dx/dt = x - x^2 - xy, \qquad dy/dt = \frac{1}{2}y - \frac{1}{4}y^2 - \frac{3}{4}xy$

9. $dx/dt = -(x - y)(1 - x - y), \qquad dy/dt = x(2 + y)$

10. $dx/dt = x + x^2 + y^2, \qquad dy/dt = y - xy$

11. $dx/dt = 2x + y + xy^3, \qquad dy/dt = x - 2y - xy$

12. $dx/dt = (1 + x)\sin y, \qquad dy/dt = 1 - x - \cos y$

13. $dx/dt = x - y^2, \qquad dy/dt = y - x^2$

14. $dx/dt = 1 - xy, \qquad dy/dt = x - y^3$

15. $dx/dt = -2x - y - x(x^2 + y^2), \qquad dy/dt = x - y + y(x^2 + y^2)$

16. $dx/dt = y + x(1 - x^2 - y^2), \qquad dy/dt = -x + y(1 - x^2 - y^2)$

17. Consider the autonomous system

$$dx/dt = y, \qquad dy/dt = x + 2x^3.$$

(a) Show that the critical point $(0,0)$ is a saddle point.

(b) Sketch the trajectories for the corresponding linear system by integrating the equation for dy/dx. Show from the parametric form of the solution that the only trajectory on which $x \to 0, y \to 0$ as $t \to \infty$ is $y = -x$.

(c) Determine the trajectories for the nonlinear system by integrating the equation for dy/dx. Sketch the trajectories for the nonlinear system that correspond to $y = -x$ and $y = x$ for the linear system.

18. Consider the autonomous system

$$dx/dt = x, \qquad dy/dt = -2y + x^3.$$

(a) Show that the critical point $(0,0)$ is a saddle point.

(b) Sketch the trajectories for the corresponding linear system, and show that the trajectory for which $x \to 0, y \to 0$ as $t \to \infty$ is given by $x = 0$.

(c) Determine the trajectories for the nonlinear system for $x \neq 0$ by integrating the equation for dy/dx. Show that the trajectory corresponding to $x = 0$ for the linear system is unaltered, but that the one corresponding to $y = 0$ is $y = x^3/5$. Sketch several of the trajectories for the nonlinear system.

19. The equation of motion of an undamped pendulum is $d^2\theta/dt^2 + \omega^2 \sin\theta = 0$, where $\omega^2 = g/L$. Let $x = \theta, y = d\theta/dt$ to obtain the system of equations

$$dx/dt = y, \qquad dy/dt = -\omega^2 \sin x.$$

(a) Show that the critical points are $(\pm n\pi, 0), n = 0, 1, 2, \ldots$, and that the system is almost linear in the neighborhood of each critical point.

(b) Show that the critical point $(0,0)$ is a (stable) center of the corresponding linear system. Using Theorem 9.3.2, what can you say about the nonlinear system? The situation is similar at the critical points $(\pm 2n\pi, 0), n = 1, 2, 3, \ldots$. What is the physical interpretation of these critical points?

(c) Show that the critical point $(\pi, 0)$ is an (unstable) saddle point of the corresponding linear system. What conclusion can you draw about the nonlinear system? The situation is similar at the critical points $[\pm(2n-1)\pi, 0], n = 1, 2, 3, \ldots$. What is the physical interpretation of these critical points?

(d) Choose a value for ω^2 and plot a few trajectories of the nonlinear system in the neighborhood of the origin. Can you now draw any further conclusion about the nature of the critical point at $(0,0)$ for the nonlinear system?

(e) Using the value of ω^2 from part (d), draw a phase portrait for the pendulum. Compare your plot with Figure 9.3.5 for the damped pendulum.

20. (a) By solving the equation for dy/dx, show that the equation of the trajectories of the undamped pendulum of Problem 19 can be written as

$$\tfrac{1}{2}y^2 + \omega^2(1 - \cos x) = c, \tag{i}$$

where c is a constant of integration.

(b) Multiply Eq. (i) by mL^2. Then express the result in terms of θ to obtain

$$\frac{1}{2}mL^2\left(\frac{d\theta}{dt}\right)^2 + mgL(1 - \cos\theta) = E, \tag{ii}$$

where $E = mL^2 c$.

(c) Show that the first term in Eq. (ii) is the kinetic energy of the pendulum and that the second term is the potential energy due to gravity. Thus the total energy E of the pendulum is constant along any trajectory; its value is determined by the initial conditions.

21. The motion of a certain undamped pendulum is described by the equations

$$dx/dt = y, \qquad dy/dt = -4\sin x.$$

If the pendulum is set in motion with an angular displacement A and no initial velocity, then the initial conditions are $x(0) = A, y(0) = 0$.

(a) Let $A = 0.25$ and plot x versus t. From the graph, estimate the amplitude R and period T of the resulting motion of the pendulum.

(b) Repeat part (a) for $A = 0.5, 1.0, 1.5$, and 2.0.

(c) How do the amplitude and period of the pendulum's motion depend on the initial position A? Draw a graph to show each of these relationships. Can you say anything about the limiting value of the period as $A \to 0$?

(d) Let $A = 4$ and plot x versus t. Explain why this graph differs from those in parts (a) and (b). For what value of A does the transition take place?

22. Consider again the pendulum equations (see Problem 21)

$$dx/dt = y, \qquad dy/dt = -4\sin x.$$

If the pendulum is set in motion from its downward equilibrium position with angular velocity v, then the initial conditions are $x(0) = 0, y(0) = v$.

(a) Plot x versus t for $v = 2$ and also for $v = 5$. Explain the differing motions of the pendulum that these two graphs represent.

(b) There is a critical value of v, which we denote by v_c, such that one type of motion occurs for $v < v_c$ and the other for $v > v_c$. Estimate the value of v_c.

23. This problem extends Problem 22 to a damped pendulum. The equations of motion are

$$dx/dt = y, \qquad dy/dt = -4\sin x - \gamma y,$$

where γ is the damping coefficient, with the initial conditions $x(0) = 0, y(0) = v$.

(a) For $\gamma = 1/4$ plot x versus t for $v = 2$ and for $v = 5$. Explain these plots in terms of the motions of the pendulum that they represent. Also explain how they relate to the corresponding graphs in Problem 22(a).

(b) Estimate the critical value v_c of the initial velocity where the transition from one type of motion to the other occurs.

(c) Repeat part (b) for other values of γ and determine how v_c depends on γ.

24. Theorem 9.3.2 provides no information about the stability of a critical point of an almost linear system if that point is a center of the corresponding linear system. That this must be the case is illustrated by the systems

$$\begin{aligned} dx/dt &= y + x(x^2 + y^2), \\ dy/dt &= -x + y(x^2 + y^2) \end{aligned} \qquad \text{(i)}$$

and

$$\begin{aligned} dx/dt &= y - x(x^2 + y^2), \\ dy/dt &= -x - y(x^2 + y^2). \end{aligned} \qquad \text{(ii)}$$

(a) Show that $(0,0)$ is a critical point of each system and, furthermore, is a center of the corresponding linear system.

(b) Show that each system is almost linear.

(c) Let $r^2 = x^2 + y^2$, and note that $x\,dx/dt + y\,dy/dt = r\,dr/dt$. For system (ii) show that $dr/dt < 0$ and that $r \to 0$ as $t \to \infty$; hence the critical point is asymptotically stable. For system (i) show that the solution of the initial value problem for r with $r = r_0$ at $t = 0$ becomes unbounded as $t \to 1/2r_0^2$, and hence the critical point is unstable.

25. In this problem we show how small changes in the coefficients of a system of linear equations can affect a critical point that is a center. Consider the system

$$\mathbf{x}' = \begin{pmatrix} 0 & 1 \\ -1 & 0 \end{pmatrix} \mathbf{x}.$$

Show that the eigenvalues are $\pm i$ so that $(0,0)$ is a center. Now consider the system

$$\mathbf{x}' = \begin{pmatrix} \epsilon & 1 \\ -1 & \epsilon \end{pmatrix} \mathbf{x},$$

where $|\epsilon|$ is arbitrarily small. Show that the eigenvalues are $\epsilon \pm i$. Thus no matter how small $|\epsilon| \neq 0$ is, the center becomes a spiral point. If $\epsilon < 0$, the spiral point is asymptotically stable; if $\epsilon > 0$, the spiral point is unstable.

26. In this problem we show how small changes in the coefficients of a system of linear equations can affect the nature of a critical point when the eigenvalues are equal. Consider the system

$$\mathbf{x}' = \begin{pmatrix} -1 & 1 \\ 0 & -1 \end{pmatrix} \mathbf{x}.$$

Show that the eigenvalues are $r_1 = -1, r_2 = -1$ so that the critical point $(0,0)$ is an asymptotically stable node. Now consider the system

$$\mathbf{x}' = \begin{pmatrix} -1 & 1 \\ -\epsilon & -1 \end{pmatrix} \mathbf{x},$$

where $|\epsilon|$ is arbitrarily small. Show that if $\epsilon > 0$, then the eigenvalues are $-1 \pm i\sqrt{\epsilon}$, so that the asymptotically stable node becomes an asymptotically stable spiral point. If $\epsilon < 0$, then the roots are $-1 \pm \sqrt{|\epsilon|}$, and the critical point remains an asymptotically stable node.

27. In this problem we derive a formula for the natural period of an undamped nonlinear pendulum [$c = 0$ in Eq. (10) of Section 9.2]. Suppose that the bob is pulled through a positive angle α and then released with zero velocity.

(a) We usually think of θ and $d\theta/dt$ as functions of t. However, if we reverse the roles of t and θ, we can regard t as a function of θ and, consequently, can also think of $d\theta/dt$ as a function of θ. Then derive the following sequence of equations:

$$\frac{1}{2}mL^2 \frac{d}{d\theta}\left[\left(\frac{d\theta}{dt}\right)^2\right] = -mgL\sin\theta,$$

$$\frac{1}{2}m\left(L\frac{d\theta}{dt}\right)^2 = mgL(\cos\theta - \cos\alpha),$$

$$dt = -\sqrt{\frac{L}{2g}}\frac{d\theta}{\sqrt{\cos\theta - \cos\alpha}}.$$

Why was the negative square root chosen in the last equation?

(b) If T is the natural period of oscillation, derive the formula

$$\frac{T}{4} = -\sqrt{\frac{L}{2g}} \int_{\alpha}^{0} \frac{d\theta}{\sqrt{\cos\theta - \cos\alpha}}.$$

(c) By using the identities $\cos\theta = 1 - 2\sin^2(\theta/2)$ and $\cos\alpha = 1 - 2\sin^2(\alpha/2)$, followed by the change of variable $\sin(\theta/2) = k\sin\phi$ with $k = \sin(\alpha/2)$, show that

$$T = 4\sqrt{\frac{L}{g}} \int_{0}^{\pi/2} \frac{d\phi}{\sqrt{1 - k^2\sin^2\phi}}.$$

The integral is called the **elliptic integral** of the first kind. Note that the period depends on the ratio L/g and also on the initial displacement α through $k = \sin(\alpha/2)$.

(d) By evaluating the integral in the expression for T, obtain values for T that you can compare with the graphical estimates you obtained in Problem 21.

28. A generalization of the damped pendulum equation discussed in the text, or a damped spring–mass system, is the Liénard[2] equation

$$\frac{d^2x}{dt^2} + c(x)\frac{dx}{dt} + g(x) = 0.$$

If $c(x)$ is a constant and $g(x) = kx$, then this equation has the form of the linear pendulum equation [replace $\sin\theta$ with θ in Eq. (12) of Section 9.2]; otherwise, the damping force $c(x)\,dx/dt$ and the restoring force $g(x)$ are nonlinear. Assume that c is continuously differentiable, g is twice continuously differentiable, and $g(0) = 0$.

(a) Write the Liénard equation as a system of two first order equations by introducing the variable $y = dx/dt$.

(b) Show that $(0, 0)$ is a critical point and that the system is almost linear in the neighborhood of $(0, 0)$.

(c) Show that if $c(0) > 0$ and $g'(0) > 0$, then the critical point is asymptotically stable, and that if $c(0) < 0$ or $g'(0) < 0$, then the critical point is unstable.

Hint: Use Taylor series to approximate c and g in the neighborhood of $x = 0$.

9.4 Competing Species

In this section and the next, we explore the application of phase plane analysis to some problems in population dynamics. These problems involve two interacting populations and are extensions of those discussed in Section 2.5, which dealt with a single population. Although the equations discussed here are extremely simple compared to the very complex relationships that exist in nature, it is still possible to acquire some insight into ecological principles from a study of these model problems.

Suppose that in some closed environment there are two similar species competing for a limited food supply—for example, two species of fish in a pond that do not prey

[2]Alfred-Marie Liénard (1869–1958), professor at l'École des Mines in Paris, worked in electricity, mechanics, and applied mathematics. His investigation of this differential equation was published in 1928.

on each other but do compete for the available food. Let x and y be the populations of the two species at time t. As discussed in Section 2.5, we assume that the population of each of the species, in the absence of the other, is governed by a logistic equation. Thus

$$dx/dt = x(\epsilon_1 - \sigma_1 x), \tag{1a}$$

$$dy/dt = y(\epsilon_2 - \sigma_2 y), \tag{1b}$$

respectively, where ϵ_1 and ϵ_2 are the growth rates of the two populations, and ϵ_1/σ_1 and ϵ_2/σ_2 are their saturation levels. However, when both species are present, each will impinge on the available food supply for the other. In effect, they reduce each other's growth rates and saturation populations. The simplest expression for reducing the growth rate of species x due to the presence of species y is to replace the growth rate factor $\epsilon_1 - \sigma_1 x$ in Eq. (1a) by $\epsilon_1 - \sigma_1 x - \alpha_1 y$, where α_1 is a measure of the degree to which species y interferes with species x. Similarly, in Eq. (1b) we replace $\epsilon_2 - \sigma_2 y$ by $\epsilon_2 - \sigma_2 y - \alpha_2 x$. Thus we have the system of equations

$$\begin{aligned} dx/dt &= x(\epsilon_1 - \sigma_1 x - \alpha_1 y), \\ dy/dt &= y(\epsilon_2 - \sigma_2 y - \alpha_2 x). \end{aligned} \tag{2}$$

The values of the positive constants $\epsilon_1, \sigma_1, \alpha_1, \epsilon_2, \sigma_2$, and α_2 depend on the particular species under consideration and in general must be determined from observations. We are interested in solutions of Eqs. (2) for which x and y are nonnegative. In the following two examples we discuss two typical problems in some detail. At the end of the section we return to the general equations (2).

EXAMPLE 1

Discuss the qualitative behavior of solutions of the system

$$\begin{aligned} dx/dt &= x(1 - x - y), \\ dy/dt &= y(0.75 - y - 0.5x). \end{aligned} \tag{3}$$

We find the critical points by solving the system of algebraic equations

$$x(1 - x - y) = 0, \qquad y(0.75 - y - 0.5x) = 0. \tag{4}$$

There are four points that satisfy Eqs. (4), namely $(0,0)$, $(0,0.75)$, $(1,0)$, and $(0.5,0.5)$; they correspond to equilibrium solutions of the system (3). The first three of these points involve the extinction of one or both species; only the last corresponds to the long-term survival of both species. Other solutions are represented as curves or trajectories in the xy-plane that describe the evolution of the populations in time. To begin to discover their qualitative behavior we can proceed in the following way.

A direction field for the system (3) in the positive quadrant is shown in Figure 9.4.1; the heavy dots in this figure are the critical points or equilibrium solutions. Based on the direction field, it appears that the point $(0.5, 0.5)$ attracts other solutions and is therefore asymptotically stable, while the other three critical points are unstable. To confirm these conclusions, we can look at the linear approximations near each critical point.

The system (3) is almost linear in the neighborhood of each critical point. There are two ways to obtain the linear system near a critical point (X, Y). First, we can use the substitution

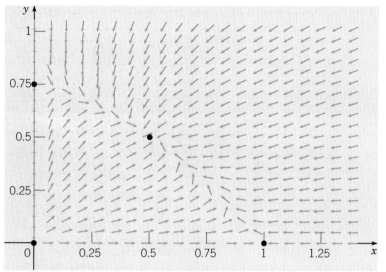

FIGURE 9.4.1 Critical points and direction field for the system (3).

$x = X + u, y = Y + v$ in Eqs. (3), retaining only the terms that are linear in u and v. Alternatively, we can use Eq. (13) of Section 9.3—that is,

$$\frac{d}{dt}\begin{pmatrix} u \\ v \end{pmatrix} = \begin{pmatrix} F_x(X,Y) & F_y(X,Y) \\ G_x(X,Y) & G_y(X,Y) \end{pmatrix} \begin{pmatrix} u \\ v \end{pmatrix}, \tag{5}$$

where, for the system (3),

$$F(x,y) = x(1 - x - y), \qquad G(x,y) = y(0.75 - y - 0.5x). \tag{6}$$

Thus Eq. (5) becomes

$$\frac{d}{dt}\begin{pmatrix} u \\ v \end{pmatrix} = \begin{pmatrix} 1 - 2X - Y & -X \\ -0.5Y & 0.75 - 2Y - 0.5X \end{pmatrix} \begin{pmatrix} u \\ v \end{pmatrix}. \tag{7}$$

$x = 0, y = 0$. This critical point corresponds to a state in which both species die as a result of their competition. By rewriting the system (3) in the form

$$\frac{d}{dt}\begin{pmatrix} x \\ y \end{pmatrix} = \begin{pmatrix} 1 & 0 \\ 0 & 0.75 \end{pmatrix} \begin{pmatrix} x \\ y \end{pmatrix} - \begin{pmatrix} x^2 + xy \\ 0.5xy + y^2 \end{pmatrix}, \tag{8}$$

or by setting $X = Y = 0$ in Eq. (7), we see that near the origin the corresponding linear system is

$$\frac{d}{dt}\begin{pmatrix} x \\ y \end{pmatrix} = \begin{pmatrix} 1 & 0 \\ 0 & 0.75 \end{pmatrix} \begin{pmatrix} x \\ y \end{pmatrix}. \tag{9}$$

The eigenvalues and eigenvectors of the system (9) are

$$r_1 = 1, \quad \xi^{(1)} = \begin{pmatrix} 1 \\ 0 \end{pmatrix}; \quad r_2 = 0.75, \quad \xi^{(2)} = \begin{pmatrix} 0 \\ 1 \end{pmatrix}, \tag{10}$$

so the general solution of the system is

$$\begin{pmatrix} x \\ y \end{pmatrix} = c_1 \begin{pmatrix} 1 \\ 0 \end{pmatrix} e^t + c_2 \begin{pmatrix} 0 \\ 1 \end{pmatrix} e^{0.75t}. \tag{11}$$

Thus the origin is an unstable node of both the linear system (9) and the nonlinear system (8) or (3). In the neighborhood of the origin, all trajectories are tangent to the y-axis except for one trajectory that lies along the x-axis.

$x = 1, y = 0.$ This corresponds to a state in which species x survives the competition but species y does not. The corresponding linear system is

$$\frac{d}{dt} \begin{pmatrix} u \\ v \end{pmatrix} = \begin{pmatrix} -1 & -1 \\ 0 & 0.25 \end{pmatrix} \begin{pmatrix} u \\ v \end{pmatrix}. \tag{12}$$

Its eigenvalues and eigenvectors are

$$r_1 = -1, \quad \boldsymbol{\xi}^{(1)} = \begin{pmatrix} 1 \\ 0 \end{pmatrix}; \qquad r_2 = 0.25, \quad \boldsymbol{\xi}^{(2)} = \begin{pmatrix} 4 \\ -5 \end{pmatrix}, \tag{13}$$

and its general solution is

$$\begin{pmatrix} u \\ v \end{pmatrix} = c_1 \begin{pmatrix} 1 \\ 0 \end{pmatrix} e^{-t} + c_2 \begin{pmatrix} 4 \\ -5 \end{pmatrix} e^{0.25t}. \tag{14}$$

Since the eigenvalues have opposite signs, the point $(1, 0)$ is a saddle point, and hence is an unstable equilibrium point of the linear system (12) and of the nonlinear system (3). The behavior of the trajectories near $(1, 0)$ can be seen from Eq. (14). If $c_2 = 0$, then there is one pair of trajectories that approaches the critical point along the x-axis. All other trajectories depart from the neighborhood of $(1, 0)$.

$x = 0, y = 0.75.$ In this case species y survives but x does not. The analysis is similar to that for the point $(1, 0)$. The corresponding linear system is

$$\frac{d}{dt} \begin{pmatrix} u \\ v \end{pmatrix} = \begin{pmatrix} 0.25 & 0 \\ -0.375 & -0.75 \end{pmatrix} \begin{pmatrix} u \\ v \end{pmatrix}. \tag{15}$$

The eigenvalues and eigenvectors are

$$r_1 = 0.25, \quad \boldsymbol{\xi}^{(1)} = \begin{pmatrix} 8 \\ -3 \end{pmatrix}; \qquad r_2 = -0.75, \quad \boldsymbol{\xi}^{(2)} = \begin{pmatrix} 0 \\ 1 \end{pmatrix}, \tag{16}$$

so the general solution of Eq. (15) is

$$\begin{pmatrix} u \\ v \end{pmatrix} = c_1 \begin{pmatrix} 8 \\ -3 \end{pmatrix} e^{0.25t} + c_2 \begin{pmatrix} 0 \\ 1 \end{pmatrix} e^{-0.75t}. \tag{17}$$

Thus the point $(0, 0.75)$ is also a saddle point. All trajectories leave the neighborhood of this point except one pair that approaches along the y-axis.

$x = 0.5, y = 0.5$. This critical point corresponds to a mixed equilibrium state, or coexistence, in the competition between the two species. The eigenvalues and eigenvectors of the corresponding linear system

$$\frac{d}{dt} \begin{pmatrix} u \\ v \end{pmatrix} = \begin{pmatrix} -0.5 & -0.5 \\ -0.25 & -0.5 \end{pmatrix} \begin{pmatrix} u \\ v \end{pmatrix} \tag{18}$$

are

$$r_1 = (-2 + \sqrt{2})/4 \cong -0.146, \qquad \boldsymbol{\xi}^{(1)} = \begin{pmatrix} \sqrt{2} \\ -1 \end{pmatrix};$$

$$r_2 = (-2 - \sqrt{2})/4 \cong -0.854, \qquad \boldsymbol{\xi}^{(2)} = \begin{pmatrix} \sqrt{2} \\ 1 \end{pmatrix}. \tag{19}$$

Therefore the general solution of Eq. (18) is

$$\begin{pmatrix} u \\ v \end{pmatrix} = c_1 \begin{pmatrix} \sqrt{2} \\ -1 \end{pmatrix} e^{-0.146t} + c_2 \begin{pmatrix} \sqrt{2} \\ 1 \end{pmatrix} e^{-0.854t}. \tag{20}$$

Since both eigenvalues are negative, the critical point $(0.5, 0.5)$ is an asymptotically stable node of the linear system (18) and of the nonlinear system (3). All trajectories approach the critical point as $t \to \infty$. One pair of trajectories approaches the critical point along the line with slope $\sqrt{2}/2$ determined from the eigenvector $\boldsymbol{\xi}^{(2)}$. All other trajectories approach the critical point tangent to the line with slope $-\sqrt{2}/2$ determined from the eigenvector $\boldsymbol{\xi}^{(1)}$.

A phase portrait for the system (3) is shown in Figure 9.4.2. By looking closely at the trajectories near each critical point, you can see that they behave in the manner predicted by the linear system near that point. In addition, note that the quadratic terms on the right side of Eqs. (3) are all negative. Since for x and y large and positive these terms are the dominant ones, it follows that far from the origin in the first quadrant both x' and y' are negative; that is, the trajectories are directed inward. Thus all trajectories that start at a point (x_0, y_0) with $x_0 > 0$ and $y_0 > 0$ eventually approach the point $(0.5, 0.5)$.

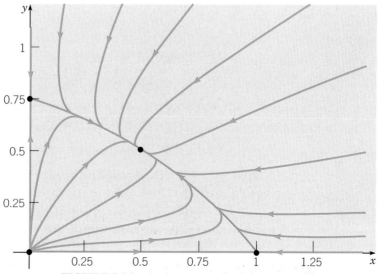

FIGURE 9.4.2 A phase portrait of the system (3).

EXAMPLE

2

Discuss the qualitative behavior of the solutions of the system

$$dx/dt = x(1 - x - y),$$
$$dy/dt = y(0.5 - 0.25y - 0.75x),$$

(21)

when x and y are nonnegative. Observe that this system is also a special case of the system (2) for two competing species.

Once again, there are four critical points, namely $(0,0)$, $(1,0)$, $(0,2)$, and $(0.5, 0.5)$, corresponding to equilibrium solutions of the system (21). Figure 9.4.3 shows a direction field for the system (21), together with the four critical points. From the direction field it appears that the mixed equilibrium solution $(0.5, 0.5)$ is a saddle point, and therefore unstable, while the points $(1,0)$ and $(0,2)$ are asymptotically stable. Thus, for competition described by Eqs. (21), one species will eventually overwhelm the other and drive it to extinction. The surviving species is determined by the initial state of the system. To confirm these conclusions, we can look at the linear approximations near each critical point.

FIGURE 9.4.3 Critical points and direction field for the system (21).

$x = 0, y = 0$. Neglecting the nonlinear terms in Eqs. (21), we obtain the linear system

$$\frac{d}{dt}\begin{pmatrix} x \\ y \end{pmatrix} = \begin{pmatrix} 1 & 0 \\ 0 & 0.5 \end{pmatrix}\begin{pmatrix} x \\ y \end{pmatrix},$$

(22)

which is valid near the origin. The eigenvalues and eigenvectors of the system (22) are

$$r_1 = 1, \quad \boldsymbol{\xi}^{(1)} = \begin{pmatrix} 1 \\ 0 \end{pmatrix}; \qquad r_2 = 0.5, \quad \boldsymbol{\xi}^{(2)} = \begin{pmatrix} 0 \\ 1 \end{pmatrix}, \tag{23}$$

so the general solution is

$$\begin{pmatrix} x \\ y \end{pmatrix} = c_1 \begin{pmatrix} 1 \\ 0 \end{pmatrix} e^t + c_2 \begin{pmatrix} 0 \\ 1 \end{pmatrix} e^{0.5t}. \tag{24}$$

Therefore the origin is an unstable node of the linear system (22) and also of the nonlinear system (21). All trajectories leave the origin tangent to the y-axis except for one trajectory that lies along the x-axis.

$x = 1, y = 0.$ The corresponding linear system is

$$\frac{d}{dt} \begin{pmatrix} u \\ v \end{pmatrix} = \begin{pmatrix} -1 & -1 \\ 0 & -0.25 \end{pmatrix} \begin{pmatrix} u \\ v \end{pmatrix}. \tag{25}$$

Its eigenvalues and eigenvectors are

$$r_1 = -1, \quad \boldsymbol{\xi}^{(1)} = \begin{pmatrix} 1 \\ 0 \end{pmatrix}; \qquad r_2 = -0.25, \quad \boldsymbol{\xi}^{(2)} = \begin{pmatrix} 4 \\ -3 \end{pmatrix}, \tag{26}$$

and its general solution is

$$\begin{pmatrix} u \\ v \end{pmatrix} = c_1 \begin{pmatrix} 1 \\ 0 \end{pmatrix} e^{-t} + c_2 \begin{pmatrix} 4 \\ -3 \end{pmatrix} e^{-0.25t}. \tag{27}$$

The point $(1, 0)$ is an asymptotically stable node of the linear system (25) and of the nonlinear system (21). If the initial values of x and y are sufficiently close to $(1, 0)$, then the interaction process will lead ultimately to that state; that is, to the survival of species x and the extinction of species y. There is one pair of trajectories that approaches the critical point along the x-axis. All other trajectories approach $(1, 0)$ tangent to the line with slope $-3/4$ that is determined by the eigenvector $\boldsymbol{\xi}^{(2)}$.

$x = 0, y = 2.$ The analysis in this case is similar to that for the point $(1, 0)$. The appropriate linear system is

$$\frac{d}{dt} \begin{pmatrix} u \\ v \end{pmatrix} = \begin{pmatrix} -1 & 0 \\ -1.5 & -0.5 \end{pmatrix} \begin{pmatrix} u \\ v \end{pmatrix}. \tag{28}$$

The eigenvalues and eigenvectors of this system are

$$r_1 = -1, \quad \boldsymbol{\xi}^{(1)} = \begin{pmatrix} 1 \\ 3 \end{pmatrix}; \qquad r_2 = -0.5, \quad \boldsymbol{\xi}^{(2)} = \begin{pmatrix} 0 \\ 1 \end{pmatrix}, \tag{29}$$

and its general solution is

$$\begin{pmatrix} u \\ v \end{pmatrix} = c_1 \begin{pmatrix} 1 \\ 3 \end{pmatrix} e^{-t} + c_2 \begin{pmatrix} 0 \\ 1 \end{pmatrix} e^{-0.5t}. \tag{30}$$

Thus the critical point $(0, 2)$ is an asymptotically stable node of both the linear system (28) and the nonlinear system (21). All trajectories approach the critical point tangent to the y-axis except for one trajectory that approaches along the line with slope 3.

$x = 0.5, y = 0.5.$ The corresponding linear system is

$$\frac{d}{dt}\begin{pmatrix} u \\ v \end{pmatrix} = \begin{pmatrix} -0.5 & -0.5 \\ -0.375 & -0.125 \end{pmatrix}\begin{pmatrix} u \\ v \end{pmatrix}. \tag{31}$$

The eigenvalues and eigenvectors are

$$r_1 = \frac{-5 + \sqrt{57}}{16} \cong 0.1594, \qquad \xi^{(1)} = \begin{pmatrix} 1 \\ (-3 - \sqrt{57})/8 \end{pmatrix} \cong \begin{pmatrix} 1 \\ -1.3187 \end{pmatrix},$$

$$r_2 = \frac{-5 - \sqrt{57}}{16} \cong -0.7844, \qquad \xi^{(2)} = \begin{pmatrix} 1 \\ (-3 + \sqrt{57})/8 \end{pmatrix} \cong \begin{pmatrix} 1 \\ 0.5687 \end{pmatrix}, \tag{32}$$

so the general solution is

$$\begin{pmatrix} u \\ v \end{pmatrix} = c_1 \begin{pmatrix} 1 \\ -1.3187 \end{pmatrix} e^{0.1594t} + c_2 \begin{pmatrix} 1 \\ 0.5687 \end{pmatrix} e^{-0.7844t}. \tag{33}$$

Since the eigenvalues are of opposite sign, the critical point $(0.5, 0.5)$ is a saddle point and therefore is unstable, as we had surmised earlier. One pair of trajectories approaches the critical point as $t \to \infty$; the others depart from it. As they approach the critical point, the entering trajectories are tangent to the line with slope $(\sqrt{57} - 3)/8 \cong 0.5687$ determined from the eigenvector $\xi^{(2)}$.

A phase portrait for the system (21) is shown in Figure 9.4.4. Near each of the critical points the trajectories of the nonlinear system behave as predicted by the corresponding

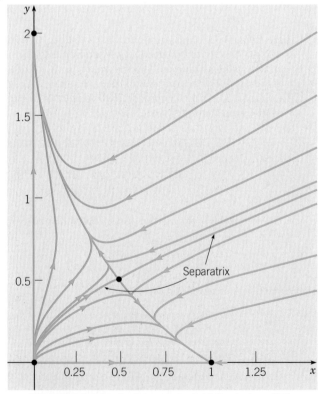

FIGURE 9.4.4 A phase portrait of the system (21).

linear approximation. Of particular interest is the pair of trajectories that enter the saddle point. These trajectories form a separatrix that divides the first quadrant into two basins of attraction. Trajectories starting above the separatrix ultimately approach the node at $(0,2)$, while trajectories starting below the separatrix approach the node at $(1,0)$. If the initial state lies precisely on the separatrix, then the solution (x, y) will approach the saddle point as $t \to \infty$. However, the slightest perturbation as one follows this trajectory will dislodge the point (x, y) from the separatrix and cause it to approach one of the nodes instead. Thus, in practice, one species will survive the competition and the other will not.

Examples 1 and 2 show that in some cases the competition between two species leads to an equilibrium state of coexistence, while in other cases the competition results in the eventual extinction of one of the species. To understand more clearly how and why this happens, and to learn how to predict which situation will occur, it is useful to look again at the general system (2). There are four cases to be considered, depending on the relative orientation of the lines

$$\epsilon_1 - \sigma_1 x - \alpha_1 y = 0 \quad \text{and} \quad \epsilon_2 - \sigma_2 y - \alpha_2 x = 0, \tag{34}$$

as shown in Figure 9.4.5. These lines are called the x and y **nullclines**, respectively, because x' is zero on the first and y' is zero on the second. Let (X, Y) denote any critical point in any one of the four cases. As in Examples 1 and 2, the system (2) is almost linear in the neighborhood of this point because the right side of each differential equation is a quadratic polynomial. To study the system (2) in the neighborhood of this critical point, we can look at the corresponding linear system obtained from

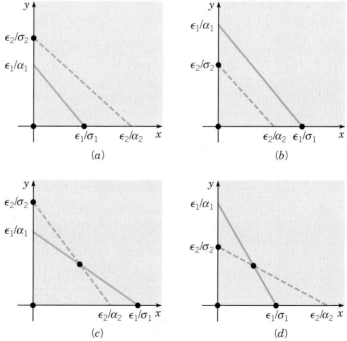

FIGURE 9.4.5 The various cases for the competing-species system (2).

Eq. (13) of Section 9.3,

$$\frac{d}{dt}\begin{pmatrix} u \\ v \end{pmatrix} = \begin{pmatrix} \epsilon_1 - 2\sigma_1 X - \alpha_1 Y & -\alpha_1 X \\ -\alpha_2 Y & \epsilon_2 - 2\sigma_2 Y - \alpha_2 X \end{pmatrix}\begin{pmatrix} u \\ v \end{pmatrix}. \tag{35}$$

We now use Eq. (35) to determine the conditions under which the model described by Eqs. (2) permits the coexistence of the two species x and y. Of the four possible cases shown in Figure 9.4.5, coexistence is possible only in cases (c) and (d). In these cases the nonzero values of X and Y are readily obtained by solving the algebraic equations (34); the result is

$$X = \frac{\epsilon_1 \sigma_2 - \epsilon_2 \alpha_1}{\sigma_1 \sigma_2 - \alpha_1 \alpha_2}, \qquad Y = \frac{\epsilon_2 \sigma_1 - \epsilon_1 \alpha_2}{\sigma_1 \sigma_2 - \alpha_1 \alpha_2}. \tag{36}$$

Further, since $\epsilon_1 - \sigma_1 X - \alpha_1 Y = 0$ and $\epsilon_2 - \sigma_2 Y - \alpha_2 X = 0$, Eq. (35) immediately reduces to

$$\frac{d}{dt}\begin{pmatrix} u \\ v \end{pmatrix} = \begin{pmatrix} -\sigma_1 X & -\alpha_1 X \\ -\alpha_2 Y & -\sigma_2 Y \end{pmatrix}\begin{pmatrix} u \\ v \end{pmatrix}. \tag{37}$$

The eigenvalues of the system (37) are found from the equation

$$r^2 + (\sigma_1 X + \sigma_2 Y)r + (\sigma_1 \sigma_2 - \alpha_1 \alpha_2)XY = 0. \tag{38}$$

Thus

$$r_{1,2} = \frac{-(\sigma_1 X + \sigma_2 Y) \pm \sqrt{(\sigma_1 X + \sigma_2 Y)^2 - 4(\sigma_1 \sigma_2 - \alpha_1 \alpha_2)XY}}{2}. \tag{39}$$

If $\sigma_1 \sigma_2 - \alpha_1 \alpha_2 < 0$, then the radicand of Eq. (39) is positive and greater than $(\sigma_1 X + \sigma_2 Y)^2$. Thus the eigenvalues are real and of opposite sign. Consequently, the critical point (X, Y) is an (unstable) saddle point, and coexistence is not possible. This is the case in Example 2, where $\sigma_1 = 1$, $\alpha_1 = 1$, $\sigma_2 = 0.25$, $\alpha_2 = 0.75$, and $\sigma_1 \sigma_2 - \alpha_1 \alpha_2 = -0.5$.

On the other hand, if $\sigma_1 \sigma_2 - \alpha_1 \alpha_2 > 0$, then the radicand of Eq. (39) is less than $(\sigma_1 X + \sigma_2 Y)^2$. Thus the eigenvalues are real, negative, and unequal, or complex with negative real part. A straightforward analysis of the radicand of Eq. (39) shows that the eigenvalues cannot be complex (see Problem 7). Thus the critical point is an asymptotically stable node, and sustained coexistence is possible. This is illustrated by Example 1, where $\sigma_1 = 1$, $\alpha_1 = 1$, $\sigma_2 = 1$, $\alpha_2 = 0.5$, and $\sigma_1 \sigma_2 - \alpha_1 \alpha_2 = 0.5$.

Let us relate this result to Figures 9.4.5c and 9.4.5d. In Figure 9.4.5c we have

$$\frac{\epsilon_1}{\sigma_1} > \frac{\epsilon_2}{\alpha_2} \quad \text{or} \quad \epsilon_1 \alpha_2 > \epsilon_2 \sigma_1 \quad \text{and} \quad \frac{\epsilon_2}{\sigma_2} > \frac{\epsilon_1}{\alpha_1} \quad \text{or} \quad \epsilon_2 \alpha_1 > \epsilon_1 \sigma_2. \tag{40}$$

These inequalities, coupled with the condition that X and Y given by Eqs. (36) be positive, yield the inequality $\sigma_1 \sigma_2 < \alpha_1 \alpha_2$. Hence in this case the critical point is a saddle point. On the other hand, in Figure 9.4.5d we have

$$\frac{\epsilon_1}{\sigma_1} < \frac{\epsilon_2}{\alpha_2} \quad \text{or} \quad \epsilon_1 \alpha_2 < \epsilon_2 \sigma_1 \quad \text{and} \quad \frac{\epsilon_2}{\sigma_2} < \frac{\epsilon_1}{\alpha_1} \quad \text{or} \quad \epsilon_2 \alpha_1 < \epsilon_1 \sigma_2. \tag{41}$$

Now the condition that X and Y be positive yields $\sigma_1 \sigma_2 > \alpha_1 \alpha_2$. Hence the critical point is asymptotically stable. For this case we can also show that the other critical points $(0, 0)$, $(\epsilon_1/\sigma_1, 0)$, and $(0, \epsilon_2/\sigma_2)$ are unstable. Thus for any positive initial values

of x and y, the two populations approach the equilibrium state of coexistence given by Eqs. (36).

Equations (2) provide the biological interpretation of the result that coexistence occurs or not depending on whether $\sigma_1\sigma_2 - \alpha_1\alpha_2$ is positive or negative. The σ's are a measure of the inhibitory effect that the growth of each population has on itself, while the α's are a measure of the inhibiting effect that the growth of each population has on the other species. Thus, when $\sigma_1\sigma_2 > \alpha_1\alpha_2$, interaction (competition) is "weak" and the species can coexist; when $\sigma_1\sigma_2 < \alpha_1\alpha_2$, interaction (competition) is "strong" and the species cannot coexist—one must die out.

PROBLEMS

Each of Problems 1 through 6 can be interpreted as describing the interaction of two species with populations x and y. In each of these problems carry out the following steps.

(a) Draw a direction field and describe how solutions seem to behave.

(b) Find the critical points.

(c) For each critical point find the corresponding linear system. Find the eigenvalues and eigenvectors of the linear system; classify each critical point as to type, and determine whether it is asymptotically stable, stable, or unstable.

(d) Sketch the trajectories in the neighborhood of each critical point.

(e) Compute and plot enough trajectories of the given system to show clearly the behavior of the solutions.

(f) Determine the limiting behavior of x and y as $t \to \infty$, and interpret the results in terms of the populations of the two species.

1. $dx/dt = x(1.5 - x - 0.5y)$
 $dy/dt = y(2 - y - 0.75x)$

2. $dx/dt = x(1.5 - x - 0.5y)$
 $dy/dt = y(2 - 0.5y - 1.5x)$

3. $dx/dt = x(1.5 - 0.5x - y)$
 $dy/dt = y(2 - y - 1.125x)$

4. $dx/dt = x(1.5 - 0.5x - y)$
 $dy/dt = y(0.75 - y - 0.125x)$

5. $dx/dt = x(1 - x - y)$
 $dy/dt = y(1.5 - y - x)$

6. $dx/dt = x(1 - x + 0.5y)$
 $dy/dt = y(2.5 - 1.5y + 0.25x)$

7. Show that

$$(\sigma_1 X + \sigma_2 Y)^2 - 4(\sigma_1\sigma_2 - \alpha_1\alpha_2)XY = (\sigma_1 X - \sigma_2 Y)^2 + 4\alpha_1\alpha_2 XY.$$

Hence conclude that the eigenvalues given by Eq. (39) can never be complex.

8. Two species of fish that compete with each other for food, but do not prey on each other, are bluegill and redear. Suppose that a pond is stocked with bluegill and redear, and let x and y be the populations of bluegill and redear, respectively, at time t. Suppose further that the competition is modeled by the equations

$$dx/dt = x(\epsilon_1 - \sigma_1 x - \alpha_1 y),$$
$$dy/dt = y(\epsilon_2 - \sigma_2 y - \alpha_2 x).$$

(a) If $\epsilon_2/\alpha_2 > \epsilon_1/\sigma_1$ and $\epsilon_2/\sigma_2 > \epsilon_1/\alpha_1$, show that the only equilibrium populations in the pond are no fish, no redear, or no bluegill. What will happen?

(b) If $\epsilon_1/\sigma_1 > \epsilon_2/\alpha_2$ and $\epsilon_1/\alpha_1 > \epsilon_2/\sigma_2$, show that the only equilibrium populations in the pond are no fish, no redear, or no bluegill. What will happen?

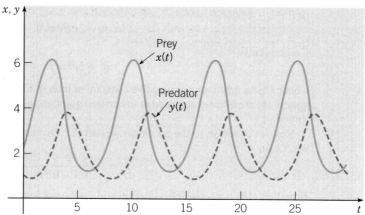

FIGURE 9.5.3 Variations of the prey and predator populations with time for the system (2).

there is little predation. Then the predators, with abundant food, increase in population also. This causes heavier predation, and the prey tend to decrease. Finally, with a diminished food supply, the predator population also decreases, and the system returns to the original state.

The general system (1) can be analyzed in exactly the same way as in the example. The critical points of the system (1) are the solutions of

$$x(a - \alpha y) = 0, \qquad y(-c + \gamma x) = 0,$$

that is, the points $(0, 0)$ and $(c/\gamma, a/\alpha)$. We first examine the solutions of the corresponding linear system near each critical point.

In the neighborhood of the origin the corresponding linear system is

$$\frac{d}{dt} \begin{pmatrix} x \\ y \end{pmatrix} = \begin{pmatrix} a & 0 \\ 0 & -c \end{pmatrix} \begin{pmatrix} x \\ y \end{pmatrix}. \tag{14}$$

The eigenvalues and eigenvectors are

$$r_1 = a, \quad \boldsymbol{\xi}^{(1)} = \begin{pmatrix} 1 \\ 0 \end{pmatrix}; \qquad r_2 = -c, \quad \boldsymbol{\xi}^{(2)} = \begin{pmatrix} 0 \\ 1 \end{pmatrix}, \tag{15}$$

so the general solution is

$$\begin{pmatrix} x \\ y \end{pmatrix} = c_1 \begin{pmatrix} 1 \\ 0 \end{pmatrix} e^{at} + c_2 \begin{pmatrix} 0 \\ 1 \end{pmatrix} e^{-ct}. \tag{16}$$

Thus the origin is a saddle point and hence unstable. Entrance to the saddle point is along the y-axis; all other trajectories depart from the neighborhood of the critical point.

Next consider the critical point $(c/\gamma, a/\alpha)$. If $x = (c/\gamma) + u$ and $y = (a/\alpha) + v$, then the corresponding linear system is

$$\frac{d}{dt} \begin{pmatrix} u \\ v \end{pmatrix} = \begin{pmatrix} 0 & -\alpha c/\gamma \\ \gamma a/\alpha & 0 \end{pmatrix} \begin{pmatrix} u \\ v \end{pmatrix}. \tag{17}$$

The eigenvalues of the system (17) are $r = \pm i\sqrt{ac}$, so the critical point is a (stable) center of the linear system. To find the trajectories of the system (17), we can divide the second equation by the first to obtain

$$\frac{dv}{du} = \frac{dv/dt}{du/dt} = -\frac{(\gamma a/\alpha)u}{(\alpha c/\gamma)v}, \tag{18}$$

or

$$\gamma^2 au\,du + \alpha^2 cv\,dv = 0. \tag{19}$$

Consequently,

$$\gamma^2 au^2 + \alpha^2 cv^2 = k, \tag{20}$$

where k is a nonnegative constant of integration. Thus the trajectories of the linear system (17) are ellipses, just as in the example.

Returning briefly to the nonlinear system (1), observe that it can be reduced to the single equation

$$\frac{dy}{dx} = \frac{dy/dt}{dx/dt} = \frac{y(-c + \gamma x)}{x(a - \alpha y)}. \tag{21}$$

Equation (21) is separable and has the solution

$$a \ln y - \alpha y + c \ln x - \gamma x = C, \tag{22}$$

where C is a constant of integration. Again it is possible to show that, for fixed C, the graph of Eq. (22) is a closed curve surrounding the critical point $(c/\gamma, a/\alpha)$. Thus this critical point is also a center for the general nonlinear system (1).

The cyclic variation of the predator and prey populations can be analyzed in more detail when the deviations from the point $(c/\gamma, a/\alpha)$ are small and the linear system (17) can be used. The solution of the system (17) can be written in the form

$$u = \frac{c}{\gamma} K \cos(\sqrt{ac}\,t + \phi), \qquad v = \frac{a}{\alpha}\sqrt{\frac{c}{a}} K \sin(\sqrt{ac}\,t + \phi), \tag{23}$$

where the constants K and ϕ are determined by the initial conditions. Thus

$$
\begin{aligned}
x &= \frac{c}{\gamma} + \frac{c}{\gamma} K \cos(\sqrt{ac}\,t + \phi), \\[4pt]
y &= \frac{a}{\alpha} + \frac{a}{\alpha}\sqrt{\frac{c}{a}} K \sin(\sqrt{ac}\,t + \phi).
\end{aligned}
\tag{24}
$$

These equations are good approximations for the nearly elliptical trajectories close to the critical point $(c/\gamma, a/\alpha)$. We can use them to draw several conclusions about the cyclic variation of the predator and prey on such trajectories.

1. The sizes of the predator and prey populations vary sinusoidally with period $2\pi/\sqrt{ac}$. This period of oscillation is independent of the initial conditions.

2. The predator and prey populations are out of phase by one-quarter of a cycle. The prey leads and the predator lags, as explained in the example.

3. The amplitudes of the oscillations are Kc/γ for the prey and $a\sqrt{c}K/\alpha\sqrt{a}$ for the predator and hence depend on the initial conditions as well as on the parameters of the problem.

4. The average populations of predator and prey over one complete cycle are c/γ and a/α, respectively. These are the same as the equilibrium populations; see Problem 10.

Cyclic variations of predator and prey as predicted by Eqs. (1) have been observed in nature. One striking example is described by Odum (pp. 191–192); based on the records of the Hudson Bay Company of Canada, the abundance of lynx and snowshoe hare, as indicated by the number of pelts turned in over the period 1845–1935, shows a distinct periodic variation with period of 9 to 10 years. The peaks of abundance are followed by very rapid declines, and the peaks of abundance of the lynx and hare are out of phase, with that of the hare preceding that of the lynx by a year or more.

The Lotka–Volterra model of the predator–prey problem has revealed a cyclic variation that perhaps could have been anticipated. On the other hand, applying the Lotka–Volterra model in other situations can lead to conclusions that are not intuitively obvious. An example that suggests a possible danger in using insecticides is given in Problem 12.

One criticism of the Lotka–Volterra equations is that in the absence of the predator, the prey will grow without bound. This can be corrected by allowing for the natural inhibiting effect that an increasing population has on the growth rate of the population. For example, the first of Eqs. (1) can be modified so that when $y = 0$, it reduces to a logistic equation for x (see Problem 13). The most important consequence of this modification is that the critical point at $(c/\gamma, a/\alpha)$ moves to $(c/\gamma, a/\alpha - \sigma c/\alpha\gamma)$ and becomes an asymptotically stable point. It is either a node or a spiral point, depending on the values of the parameters in the differential equations. In either case, other trajectories are no longer closed curves but approach the critical point as $t \to \infty$.

PROBLEMS

Each of Problems 1 through 5 can be interpreted as describing the interaction of two species with population densities x and y. In each of these problems carry out the following steps.

(a) Draw a direction field and describe how solutions seem to behave.

(b) Find the critical points.

(c) For each critical point find the corresponding linear system. Find the eigenvalues and eigenvectors of the linear system; classify each critical point as to type, and determine whether it is asymptotically stable, stable, or unstable.

(d) Sketch the trajectories in the neighborhood of each critical point.

(e) Draw a phase portrait for the system.

(f) Determine the limiting behavior of x and y as $t \to \infty$ and interpret the results in terms of the populations of the two species.

1. $dx/dt = x(1.5 - 0.5y)$
 $dy/dt = y(-0.5 + x)$

2. $dx/dt = x(1 - 0.5y)$
 $dy/dt = y(-0.25 + 0.5x)$

3. $dx/dt = x(1 - 0.5x - 0.5y)$
 $dy/dt = y(-0.25 + 0.5x)$

4. $dx/dt = x(1.125 - x - 0.5y)$
 $dy/dt = y(-1 + x)$

5. $dx/dt = x(-1 + 2.5x - 0.3y - x^2)$
 $dy/dt = y(-1.5 + x)$

6. In this problem we examine the phase difference between the cyclic variations of the predator and prey populations as given by Eqs. (24) of this section. Suppose we assume that $K > 0$ and that t is measured from the time that the prey population (x) is a maximum; then $\phi = 0$. Show that the predator population (y) is a maximum at $t = \pi/2\sqrt{ac} = T/4$,

where T is the period of the oscillation. When is the prey population increasing most rapidly? decreasing most rapidly? a minimum? Answer the same questions for the predator population. Draw a typical elliptic trajectory enclosing the point $(c/\gamma, a/\alpha)$, and mark these points on it.

7. (a) Find the ratio of the amplitudes of the oscillations of the prey and predator populations about the critical point $(c/\gamma, a/\alpha)$, using the approximation (24), which is valid for small oscillations. Observe that the ratio is independent of the initial conditions.

(b) Evaluate the ratio found in part (a) for the system (2).

(c) Estimate the amplitude ratio for the solution of the nonlinear system (2) shown in Figure 9.5.3. Does the result agree with that obtained from the linear approximation?

(d) Determine the prey–predator amplitude ratio for other solutions of the system (2), that is, for solutions satisfying other initial conditions. Is the ratio independent of the initial conditions?

8. (a) Find the period of the oscillations of the prey and predator populations, using the approximation (24), which is valid for small oscillations. Note that the period is independent of the amplitude of the oscillations.

(b) For the solution of the nonlinear system (2) shown in Figure 9.5.3, estimate the period as well as possible. Is the result the same as for the linear approximation?

(c) Calculate other solutions of the system (2), that is, solutions satisfying other initial conditions, and determine their periods. Is the period the same for all initial conditions?

9. Consider the system

$$dx/dt = ax[1 - (y/2)], \qquad dy/dt = by[-1 + (x/3)],$$

where a and b are positive constants. Observe that this system is the same as in the example in the text if $a = 1$ and $b = 0.75$. Suppose the initial conditions are $x(0) = 5$ and $y(0) = 2$.

(a) Let $a = 1$ and $b = 1$. Plot the trajectory in the phase plane and determine (or estimate) the period of the oscillation.

(b) Repeat part (a) for $a = 3$ and $a = 1/3$, with $b = 1$.

(c) Repeat part (a) for $b = 3$ and $b = 1/3$, with $a = 1$.

(d) Describe how the period and the shape of the trajectory depend on a and b.

10. The average sizes of the prey and predator populations are defined as

$$\bar{x} = \frac{1}{T} \int_{A}^{A+T} x(t)\, dt, \qquad \bar{y} = \frac{1}{T} \int_{A}^{A+T} y(t)\, dt,$$

respectively, where T is the period of a full cycle, and A is any nonnegative constant.

(a) Using the approximation (24), which is valid near the critical point, show that $\bar{x} = c/\gamma$ and $\bar{y} = a/\alpha$.

(b) For the solution of the nonlinear system (2) shown in Figure 9.5.3, estimate \bar{x} and \bar{y} as well as you can. Try to determine whether \bar{x} and \bar{y} are given by c/γ and a/α, respectively, in this case.

Hint: Consider how you might estimate the value of an integral even though you do not have a formula for the integrand.

(c) Calculate other solutions of the system (2), that is, solutions satisfying other initial conditions, and determine \bar{x} and \bar{y} for these solutions. Are the values of \bar{x} and \bar{y} the same for all solutions?

11. Suppose that the predator–prey equations (1) of the text describe foxes (y) and rabbits (x) in a forest. A trapping company is engaged in trapping foxes and rabbits for their pelts. Explain why it is reasonable for the company to conduct its operation so as to move the population of each species closer to the center $(c/\gamma, a/\alpha)$. When is it best to trap foxes? Rabbits? Rabbits and foxes? Neither?

 Hint: See Problem 6. An intuitive argument is all that is required.

12. Suppose that an insect population x is controlled by a natural predator population y according to the model (1), so that there are small cyclic variations of the populations about the critical point $(c/\gamma, a/\alpha)$. Suppose that an insecticide is employed with the goal of reducing the population of the insects, and suppose that this insecticide is also toxic to the predators. Indeed, suppose that the insecticide kills both prey and predator at rates proportional to their respective populations. Write down the modified differential equations, determine the new equilibrium point, and compare it to the original equilibrium point.

 To ban insecticides on the basis of this simple model and counterintuitive result would certainly be ill-advised. On the other hand, it is also rash to ignore the possible genuine existence of a phenomenon suggested by such a model.

13. As mentioned in the text, one improvement in the predator–prey model is to modify the equation for the prey so that it has the form of a logistic equation in the absence of the predator. Thus, in place of Eqs. (1), we consider the system

$$dx/dt = x(a - \sigma x - \alpha y), \qquad dy/dt = y(-c + \gamma x),$$

where a, σ, α, c, and γ are positive constants. Note that Problems 3 and 4 are of this form. Determine all critical points and discuss their nature and stability characteristics. Assume that $a/\sigma \gg c/\gamma$. What happens for initial data $x \neq 0, y \neq 0$?

9.6 Liapunov's Second Method

In Section 9.3 we showed how the stability of a critical point of an almost linear system can usually be determined from a study of the corresponding linear system. However, no conclusion can be drawn when the critical point is a center of the corresponding linear system. Examples of this situation are the undamped pendulum, Equations (1) and (2) below, and the predator–prey problem discussed in Section 9.5. Also, for an asymptotically stable critical point, it may be important to investigate the basin of attraction—that is, the domain such that all solutions starting within that domain approach the critical point. Since the theory of almost linear systems is a local theory, it provides no information about this question.

In this section we discuss another approach, known as **Liapunov's**[6] **second method** or **direct method**. The method is referred to as a direct method because no knowledge

[6]Alexandr M. Liapunov (1857–1918), a student of Chebyshev at St. Petersburg, taught at the University of Kharkov from 1885 to 1901, when he became academician in applied mathematics at the St. Petersburg Academy of Sciences. In 1917 he moved to Odessa because of his wife's frail health. His research in stability encompassed both theoretical analysis and applications to various physical problems. His second method formed part of his most influential work, *General Problem of Stability of Motion*, published in 1892.

of the solution of the system of differential equations is required. Rather, conclusions about the stability or instability of a critical point are obtained by constructing a suitable auxiliary function. The technique is a very powerful one that provides a more global type of information, for example, an estimate of the extent of the basin of attraction of a critical point. Liapunov's second method can also be used to study systems of equations that are not almost linear; however, we will not discuss such problems.

Basically, Liapunov's second method is a generalization of two physical principles for conservative systems, namely, (i) a rest position is stable if the potential energy is a local minimum, otherwise it is unstable, and (ii) the total energy is a constant during any motion. To illustrate these concepts, again consider the undamped pendulum (a conservative mechanical system), which is governed by the equation

$$\frac{d^2\theta}{dt^2} + \frac{g}{L}\sin\theta = 0. \tag{1}$$

The corresponding system of first order equations is

$$\frac{dx}{dt} = y, \qquad \frac{dy}{dt} = -\frac{g}{L}\sin x, \tag{2}$$

where $x = \theta$ and $y = d\theta/dt$. If we omit an arbitrary constant, the potential energy U is the work done in lifting the pendulum above its lowest position, namely,

$$U(x, y) = mgL(1 - \cos x); \tag{3}$$

see Figure 9.2.2. The critical points of the system (2) are $x = \pm n\pi, y = 0, n = 0, 1, 2, 3, \ldots$, corresponding to $\theta = \pm n\pi, d\theta/dt = 0$. Physically, we expect the points $x = 0, y = 0$; $x = \pm 2\pi, y = 0$; \ldots, corresponding to $\theta = 0, \pm 2\pi, \ldots$, to be stable, since for them the pendulum bob is vertical with the weight down; further, we expect the points $x = \pm\pi, y = 0$; $x = \pm 3\pi, y = 0$; \ldots, corresponding to $\theta = \pm\pi, \pm 3\pi, \ldots$, to be unstable, since for them the pendulum bob is vertical with the weight up. This agrees with statement (i), for at the former points U is a minimum equal to zero, and at the latter points U is a maximum equal to $2mgL$.

Next consider the total energy V, which is the sum of the potential energy U and the kinetic energy $\frac{1}{2}mL^2(d\theta/dt)^2$. In terms of x and y,

$$V(x, y) = mgL(1 - \cos x) + \tfrac{1}{2}mL^2y^2. \tag{4}$$

On a trajectory corresponding to a solution $x = \phi(t), y = \psi(t)$ of Eqs. (2), V can be considered a function of t. The derivative of $V[\phi(t), \psi(t)]$ with respect to t is called the rate of change of V following the trajectory. By the chain rule,

$$\frac{dV[\phi(t), \psi(t)]}{dt} = V_x[\phi(t), \psi(t)]\frac{d\phi(t)}{dt} + V_y[\phi(t), \psi(t)]\frac{d\psi(t)}{dt}$$

$$= (mgL\sin x)\frac{dx}{dt} + mL^2y\frac{dy}{dt}, \tag{5}$$

where it is understood that $x = \phi(t), y = \psi(t)$. Finally, substituting in Eq. (5) for dx/dt and dy/dt from Eqs. (2), we find that $dV/dt = 0$. Hence V is a constant along any trajectory of the system (2), which is statement (ii).

It is important to note that, at any point (x, y), the rate of change of V along the trajectory through that point was computed *without actually solving* the system (2). It is precisely this fact that enables us to use Liapunov's second method for systems whose solution we do not know, which is the main reason for its importance.

At the stable critical points, $x = \pm 2n\pi$, $y = 0$, $n = 0, 1, 2, \ldots$, the energy V is zero. If the initial state, say, (x_1, y_1), of the pendulum is sufficiently near a stable critical point, then the energy $V(x_1, y_1)$ is small, and the motion (trajectory) associated with this energy stays close to the critical point. It can be shown that if $V(x_1, y_1)$ is sufficiently small, then the trajectory is closed and contains the critical point. For example, suppose that (x_1, y_1) is near $(0, 0)$ and that $V(x_1, y_1)$ is very small. The equation of the trajectory with energy $V(x_1, y_1)$ is

$$V(x, y) = mgL(1 - \cos x) + \tfrac{1}{2}mL^2 y^2 = V(x_1, y_1).$$

For x small we have $1 - \cos x = 1 - (1 - x^2/2! + \cdots) \cong x^2/2$. Thus the equation of the trajectory is approximately

$$\tfrac{1}{2}mgLx^2 + \tfrac{1}{2}mL^2 y^2 = V(x_1, y_1),$$

or

$$\frac{x^2}{2V(x_1, y_1)/mgL} + \frac{y^2}{2V(x_1, y_1)/mL^2} = 1.$$

This is an ellipse enclosing the critical point $(0, 0)$; the smaller $V(x_1, y_1)$ is, the smaller are the major and minor axes of the ellipse. Physically, the closed trajectory corresponds to a solution that is periodic in time—the motion is a small oscillation about the equilibrium point.

If damping is present, however, it is natural to expect that the amplitude of the motion decays in time and that the stable critical point (center) becomes an asymptotically stable critical point (spiral point). See the phase portrait for the damped pendulum in Figure 9.3.5. This can almost be argued from a consideration of dV/dt. For the damped pendulum, the total energy is still given by Eq. (4), but now, from Eqs. (8) of Section 9.3, $dx/dt = y$ and $dy/dt = -(g/L)\sin x - (c/Lm)y$. Substituting for dx/dt and dy/dt in Eq. (5) gives $dV/dt = -cLy^2 \le 0$. Thus the energy is nonincreasing along any trajectory, and except for the line $y = 0$, the motion is such that the energy decreases. Hence each trajectory must approach a point of minimum energy—a stable equilibrium point. If $dV/dt < 0$ instead of $dV/dt \le 0$, it is reasonable to expect that this would be true for all trajectories that start sufficiently close to the origin.

To pursue these ideas further, consider the autonomous system

$$dx/dt = F(x, y), \qquad dy/dt = G(x, y), \tag{6}$$

and suppose that the point $x = 0$, $y = 0$ is an asymptotically stable critical point. Then there exists some domain D containing $(0, 0)$ such that every trajectory that starts in D must approach the origin as $t \to \infty$. Suppose that there exists an "energy" function V such that $V \ge 0$ for (x, y) in D with $V = 0$ only at the origin. Since each trajectory in D approaches the origin as $t \to \infty$, then following any particular trajectory, V decreases to zero as t approaches infinity. The type of result we want to prove is essentially the converse: If, on every trajectory, V decreases to zero as t increases,

then the trajectories must approach the origin as $t \to \infty$, and hence the origin is asymptotically stable. First, however, it is necessary to make several definitions.

Let V be defined on some domain D containing the origin. Then V is said to be **positive definite** on D if $V(0,0) = 0$ and $V(x, y) > 0$ for all other points in D. Similarly, V is said to be **negative definite** on D if $V(0,0) = 0$ and $V(x, y) < 0$ for all other points in D. If the inequalities $>$ and $<$ are replaced by \geq and \leq, then V is said to be **positive semidefinite** and **negative semidefinite**, respectively. We emphasize that when we speak of a positive definite (negative definite, ...) function on a domain D containing the origin, the function must be zero at the origin in addition to satisfying the proper inequality at all other points in D.

EXAMPLE

1

The function
$$V(x, y) = \sin(x^2 + y^2)$$
is positive definite on $x^2 + y^2 < \pi/2$ since $V(0,0) = 0$ and $V(x, y) > 0$ for $0 < x^2 + y^2 < \pi/2$. However, the function
$$V(x, y) = (x + y)^2$$
is only positive semidefinite since $V(x, y) = 0$ on the line $y = -x$.

We also want to consider the function
$$\dot{V}(x, y) = V_x(x, y)F(x, y) + V_y(x, y)G(x, y), \tag{7}$$

where F and G are the same functions as in Eqs. (6). We choose this notation because $\dot{V}(x, y)$ can be identified as the rate of change of V along the trajectory of the system (6) that passes through the point (x, y). That is, if $x = \phi(t)$, $y = \psi(t)$ is a solution of the system (6), then

$$\frac{dV[\phi(t), \psi(t)]}{dt} = V_x[\phi(t), \psi(t)]\frac{d\phi(t)}{dt} + V_y[\phi(t), \psi(t)]\frac{d\psi(t)}{dt}$$

$$= V_x(x, y)F(x, y) + V_y(x, y)G(x, y)$$

$$= \dot{V}(x, y). \tag{8}$$

The function \dot{V} is sometimes referred to as the derivative of V with respect to the system (6).

We now state two Liapunov theorems, the first dealing with stability, the second with instability.

Theorem 9.6.1

Suppose that the autonomous system (6) has an isolated critical point at the origin. If there exists a function V that is continuous and has continuous first partial derivatives, is positive definite, and for which the function \dot{V} given by Eq. (7) is negative definite on some domain D in the xy-plane containing $(0,0)$, then the origin is an asymptotically stable critical point. If \dot{V} is negative semidefinite, then the origin is a stable critical point.

Theorem 9.6.2 Let the origin be an isolated critical point of the autonomous system (6). Let V be a function that is continuous and has continuous first partial derivatives. Suppose that $V(0,0) = 0$ and that in every neighborhood of the origin there is at least one point at which V is positive (negative). If there exists a domain D containing the origin such that the function \dot{V} given by Eq. (7) is positive definite (negative definite) on D, then the origin is an unstable critical point.

The function V is called a **Liapunov function**. Before sketching geometrical arguments for Theorems 9.6.1 and 9.6.2, we note that the difficulty in using these theorems is that they tell us nothing about how to construct a Liapunov function, assuming that one exists. In cases where the autonomous system (6) represents a physical problem, it is natural to consider first the actual total energy function of the system as a possible Liapunov function. However, Theorems 9.6.1 and 9.6.2 are applicable in cases where the concept of physical energy is not pertinent. In such cases a judicious trial-and-error approach may be necessary.

Now consider the second part of Theorem 9.6.1, that is, the case $\dot{V} \leq 0$. Let $c \geq 0$ be a constant and consider the curve in the xy-plane given by $V(x, y) = c$. For $c = 0$ the curve reduces to the single point $x = 0$, $y = 0$. We assume that if $0 < c_1 < c_2$, then the curve $V(x, y) = c_1$ contains the origin and lies within the curve $V(x, y) = c_2$, as illustrated in Figure 9.6.1a. We show that a trajectory starting inside a closed curve $V(x, y) = c$ cannot cross to the outside. Thus, given a circle of radius ϵ about the origin, by taking c sufficiently small we can ensure that every trajectory starting inside the closed curve $V(x, y) = c$ stays within the circle of radius ϵ; indeed, it stays within the closed curve $V(x, y) = c$ itself. Thus the origin is a stable critical point.

To show this, recall from calculus that the vector

$$\nabla V(x, y) = V_x(x, y)\mathbf{i} + V_y(x, y)\mathbf{j}, \tag{9}$$

known as the gradient of V, is normal to the curve $V(x, y) = c$ and points in the direction of increasing V. In the present case, V increases outward from the origin, so ∇V points away from the origin, as indicated in Figure 9.6.1b. Next, consider a trajectory $x = \phi(t)$, $y = \psi(t)$ of the system (6), and recall that the vector $\mathbf{T}(t) = \phi'(t)\mathbf{i} + \psi'(t)\mathbf{j}$ is tangent to the trajectory at each point; see Figure 9.6.1b. Let $x_1 = \phi(t_1)$, $y_1 = \psi(t_1)$ be a point of intersection of the trajectory and a closed curve $V(x, y) = c$. At this

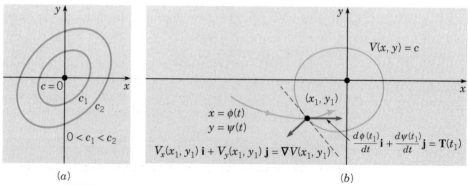

(a) (b)

FIGURE 9.6.1 Geometrical interpretation of Liapunov's method.

point $\phi'(t_1) = F(x_1, y_1), \psi'(t_1) = G(x_1, y_1)$, so from Eq. (7) we obtain

$$\begin{aligned}
\dot{V}(x_1, y_1) &= V_x(x_1, y_1)\phi'(t_1) + V_y(x_1, y_1)\psi'(t_1) \\
&= [V_x(x_1, y_1)\mathbf{i} + V_y(x_1, y_1)\mathbf{j}] \cdot [\phi'(t_1)\mathbf{i} + \psi'(t_1)\mathbf{j}] \\
&= \nabla V(x_1, y_1) \cdot \mathbf{T}(t_1).
\end{aligned} \tag{10}$$

Thus $\dot{V}(x_1, y_1)$ is the scalar product of the vector $\nabla V(x_1, y_1)$ and the vector $\mathbf{T}(t_1)$. Since $\dot{V}(x_1, y_1) \leq 0$, it follows that the cosine of the angle between $\nabla V(x_1, y_1)$ and $\mathbf{T}(t_1)$ is also less than or equal to zero; hence the angle itself is in the range $[\pi/2, 3\pi/2]$. Thus the direction of motion on the trajectory is inward with respect to $V(x_1, y_1) = c$ or, at worst, tangent to this curve. Trajectories starting inside a closed curve $V(x_1, y_1) = c$ (no matter how small c is) cannot escape, so the origin is a stable point. If $\dot{V}(x_1, y_1) < 0$, then the trajectories passing through points on the curve are actually pointed inward. As a consequence, it can be shown that trajectories starting sufficiently close to the origin must approach the origin; hence the origin is asymptotically stable.

A geometric argument for Theorem 9.6.2 follows in a somewhat similar manner. Briefly, suppose that \dot{V} is positive definite, and suppose that given any circle about the origin, there is an interior point (x_1, y_1) at which $V(x_1, y_1) > 0$. Consider a trajectory that starts at (x_1, y_1). Along this trajectory it follows from Eq. (8) that V must increase since $\dot{V}(x_1, y_1) > 0$; furthermore, since $V(x_1, y_1) > 0$, the trajectory cannot approach the origin because $V(0, 0) = 0$. This shows that the origin cannot be asymptotically stable. By further exploiting the fact that $\dot{V}(x, y) > 0$, it is possible to show that the origin is an unstable point; however, we will not pursue this argument.

EXAMPLE 2

Use Theorem 9.6.1 to show that $(0, 0)$ is a stable critical point for the undamped pendulum equations (2). Also use Theorem 9.6.2 to show that $(\pi, 0)$ is an unstable critical point.

Let V be the total energy given by Eq. (4):

$$V(x, y) = mgL(1 - \cos x) + \tfrac{1}{2}mL^2 y^2. \tag{4}$$

If we take D to be the domain $-\pi/2 < x < \pi/2, -\infty < y < \infty$, then V is positive there except at the origin, where it is zero. Thus V is positive definite on D. Further, as we have already seen,

$$\dot{V} = (mgL \sin x)(y) + (mL^2 y)(-g \sin x)/L = 0$$

for all x and y. Thus \dot{V} is negative semidefinite on D. Consequently, by the last statement in Theorem 9.6.1, the origin is a stable critical point for the undamped pendulum. Observe that this conclusion cannot be obtained from Theorem 9.3.2 because $(0, 0)$ is a center for the corresponding linear system.

Now consider the critical point $(\pi, 0)$. The Liapunov function given by Eq. (4) is no longer suitable because Theorem 9.6.2 calls for a function V for which \dot{V} is either positive or negative definite. To analyze the point $(\pi, 0)$ it is convenient to move this point to the origin by the change of variables $x = \pi + u, y = v$. Then the differential equations (2) become

$$du/dt = v, \quad dv/dt = \frac{g}{L} \sin u, \tag{11}$$

and the critical point is $(0, 0)$ in the uv-plane. Consider the function

$$V(u, v) = v \sin u \tag{12}$$

and let D be the domain $-\pi/4 < u < \pi/4, -\infty < v < \infty$. Then

$$\dot{V} = (v\cos u)(v) + (\sin u)[(g/L)\sin u] = v^2 \cos u + (g/L)\sin^2 u \tag{13}$$

is positive definite in D. The only remaining question is whether there are points in every neighborhood of the origin where V itself is positive. From Eq. (12) we see that $V(u,v) > 0$ in the first quadrant (where both $\sin u$ and v are positive) and in the third quadrant (where both are negative). Thus the conditions of Theorem 9.6.2 are satisfied, and the point $(0,0)$ in the uv-plane, or the point $(\pi, 0)$ in the xy-plane, is unstable.

The damped pendulum equations are discussed in Problem 7.

From a practical point of view, we are often more interested in the basin of attraction. The following theorem provides some information on this subject.

Theorem 9.6.3 Let the origin be an isolated critical point of the autonomous system (6). Let the function V be continuous and have continuous first partial derivatives. If there is a bounded domain D_K containing the origin where $V(x,y) < K$, V is positive definite, and \dot{V} is negative definite, then every solution of Eqs. (6) that starts at a point in D_K approaches the origin as t approaches infinity.

In other words, Theorem 9.6.3 says that if $x = \phi(t), y = \psi(t)$ is the solution of Eqs. (6) for initial data lying in D_K, then (x,y) approaches the critical point $(0,0)$ as $t \to \infty$. Thus D_K gives a region of asymptotic stability; of course, it may not be the entire basin of attraction. This theorem is proved by showing that (i) there are no periodic solutions of the system (6) in D_K, and (ii) there are no other critical points in D_K. It then follows that trajectories starting in D_K cannot escape and therefore must tend to the origin as t tends to infinity.

Theorems 9.6.1 and 9.6.2 give sufficient conditions for stability and instability, respectively. However, these conditions are not necessary, nor does our failure to determine a suitable Liapunov function mean that there is not one. Unfortunately, there are no general methods for the construction of Liapunov functions; however, there has been extensive work on the construction of Liapunov functions for special classes of equations. An elementary algebraic result that is often useful in constructing positive definite or negative definite functions is stated without proof in the following theorem.

Theorem 9.6.4 The function

$$V(x,y) = ax^2 + bxy + cy^2 \tag{14}$$

is positive definite if, and only if,

$$a > 0 \quad \text{and} \quad 4ac - b^2 > 0, \tag{15}$$

and is negative definite if, and only if,

$$a < 0 \quad \text{and} \quad 4ac - b^2 > 0. \tag{16}$$

The use of Theorem 9.6.4 is illustrated in the following example.

EXAMPLE 3

Show that the critical point $(0,0)$ of the autonomous system

$$dx/dt = -x - xy^2, \qquad dy/dt = -y - x^2y \tag{17}$$

is asymptotically stable.

We try to construct a Liapunov function of the form (14). Then $V_x(x,y) = 2ax + by$, $V_y(x,y) = bx + 2cy$, so

$$\dot{V}(x,y) = (2ax + by)(-x - xy^2) + (bx + 2cy)(-y - x^2y)$$
$$= -\left[2a(x^2 + x^2y^2) + b(2xy + xy^3 + x^3y) + 2c(y^2 + x^2y^2)\right].$$

If we choose $b = 0$, and a and c to be any positive numbers, then \dot{V} is negative definite and V is positive definite by Theorem 9.6.4. Thus, by Theorem 9.6.1, the origin is an asymptotically stable critical point.

EXAMPLE 4

Consider the system

$$dx/dt = x(1 - x - y),$$
$$dy/dt = y(0.75 - y - 0.5x). \tag{18}$$

In Example 1 of Section 9.4 we found that this system models a certain pair of competing species, and that the critical point $(0.5, 0.5)$ is asymptotically stable. Confirm this conclusion by finding a suitable Liapunov function.

It is helpful to transform the point $(0.5, 0.5)$ to the origin. To this end, let

$$x = 0.5 + u, \qquad y = 0.5 + v. \tag{19}$$

Then, substituting for x and y in Eqs. (18), we obtain the new system

$$du/dt = -0.5u - 0.5v - u^2 - uv,$$
$$dv/dt = -0.25u - 0.5v - 0.5uv - v^2. \tag{20}$$

To keep the calculations relatively simple, consider the function $V(u,v) = u^2 + v^2$ as a possible Liapunov function. This function is clearly positive definite, so we need only to determine whether there is a region containing the origin in the uv-plane where the derivative \dot{V} with respect to the system (20) is negative definite. We compute $\dot{V}(u,v)$ and find that

$$\dot{V}(u,v) = V_u \frac{du}{dt} + V_v \frac{dv}{dt}$$
$$= 2u(-0.5u - 0.5v - u^2 - uv) + 2v(-0.25u - 0.5v - 0.5uv - v^2),$$

or

$$\dot{V}(u,v) = -\left[(u^2 + 1.5uv + v^2) + (2u^3 + 2u^2v + uv^2 + 2v^3)\right], \tag{21}$$

where we have collected together the quadratic and cubic terms. We want to show that the expression in square brackets in Eq. (21) is positive definite, at least for u and v sufficiently small. Observe that the quadratic terms can be written as

$$u^2 + 1.5uv + v^2 = 0.25(u^2 + v^2) + 0.75(u + v)^2, \tag{22}$$

so these terms are positive definite. On the other hand, the cubic terms in Eq. (21) may be of either sign. Thus we must show that, in some neighborhood of $u = 0$, $v = 0$, the cubic terms are smaller in magnitude than the quadratic terms; that is,

$$\left|2u^3 + 2u^2v + uv^2 + 2v^3\right| < 0.25(u^2 + v^2) + 0.75(u + v)^2. \tag{23}$$

To estimate the left side of Eq. (23) we introduce polar coordinates $u = r\cos\theta$, $v = r\sin\theta$. Then

$$\left|2u^3 + 2u^2v + uv^2 + 2v^3\right| = r^3 \left|2\cos^3\theta + 2\cos^2\theta\sin\theta + \cos\theta\sin^2\theta + 2\sin^3\theta\right|$$
$$\leq r^3 \left[2|\cos^3\theta| + 2\cos^2\theta|\sin\theta| + |\cos\theta|\sin^2\theta + 2|\sin^3\theta|\right]$$
$$\leq 7r^3,$$

since $|\sin\theta|, |\cos\theta| \leq 1$. To satisfy Eq. (23) it is now certainly sufficient to satisfy the more stringent requirement

$$7r^3 < 0.25(u^2 + v^2) = 0.25r^2,$$

which yields $r < 1/28$. Thus, at least in this disk, the hypotheses of Theorem 9.6.1 are satisfied, so the origin is an asymptotically stable critical point of the system (20). The same is then true of the critical point $(0.5, 0.5)$ of the original system (18).

If we refer to Theorem 9.6.3, the preceding argument also shows that the disk with center $(0.5, 0.5)$ and radius $1/28$ is a region of asymptotic stability for the system (18). This is a severe underestimate of the full basin of attraction, as the discussion in Section 9.4 shows. To obtain a better estimate of the actual basin of attraction from Theorem 9.6.3, one would have to estimate the terms in Eq. (23) more accurately, use a better (and presumably more complicated) Liapunov function, or both.

PROBLEMS

In each of Problems 1 through 4 construct a suitable Liapunov function of the form $ax^2 + cy^2$, where a and c are to be determined. Then show that the critical point at the origin is of the indicated type.

1. $dx/dt = -x^3 + xy^2$, $dy/dt = -2x^2y - y^3$; asymptotically stable
2. $dx/dt = -\frac{1}{2}x^3 + 2xy^2$, $dy/dt = -y^3$; asymptotically stable
3. $dx/dt = -x^3 + 2y^3$, $dy/dt = -2xy^2$; stable (at least)
4. $dx/dt = x^3 - y^3$, $dy/dt = 2xy^2 + 4x^2y + 2y^3$; unstable
5. Consider the system of equations

$$dx/dt = y - xf(x, y), \qquad dy/dt = -x - yf(x, y),$$

where f is continuous and has continuous first partial derivatives. Show that if $f(x, y) > 0$ in some neighborhood of the origin, then the origin is an asymptotically stable critical point, and if $f(x, y) < 0$ in some neighborhood of the origin, then the origin is an unstable critical point.

Hint: Construct a Liapunov function of the form $c(x^2 + y^2)$.

6. A generalization of the undamped pendulum equation is

$$d^2u/dt^2 + g(u) = 0, \tag{i}$$

where $g(0) = 0$, $g(u) > 0$ for $0 < u < k$, and $g(u) < 0$ for $-k < u < 0$; that is, $ug(u) > 0$ for $u \neq 0$, $-k < u < k$. Notice that $g(u) = \sin u$ has this property on $(-\pi/2, \pi/2)$.

(a) Letting $x = u$, $y = du/dt$, write Eq. (i) as a system of two equations, and show that $x = 0, y = 0$ is a critical point.

(b) Show that

$$V(x,y) = \tfrac{1}{2}y^2 + \int_0^x g(s)\, ds, \qquad -k < x < k \qquad \text{(ii)}$$

is positive definite, and use this result to show that the critical point $(0,0)$ is stable. Note that the Liapunov function V given by Eq. (ii) corresponds to the energy function $V(x,y) = \tfrac{1}{2}y^2 + (1 - \cos x)$ for the case $g(u) = \sin u$.

7. By introducing suitable dimensionless variables, we can write the system of nonlinear equations for the damped pendulum [Eqs. (8) of Section 9.3] as

$$dx/dt = y, \qquad dy/dt = -y - \sin x.$$

(a) Show that the origin is a critical point.

(b) Show that while $V(x,y) = x^2 + y^2$ is positive definite, $\dot{V}(x,y)$ takes on both positive and negative values in any domain containing the origin, so that V is not a Liapunov function.
Hint: $x - \sin x > 0$ for $x > 0$ and $x - \sin x < 0$ for $x < 0$. Consider these cases with y positive but y so small that y^2 can be ignored compared to y.

(c) Using the energy function $V(x,y) = \tfrac{1}{2}y^2 + (1 - \cos x)$ mentioned in Problem 6(b), show that the origin is a stable critical point. Since there is damping in the system, we can expect that the origin is asymptotically stable. However, it is not possible to draw this conclusion using this Liapunov function.

(d) To show asymptotic stability it is necessary to construct a better Liapunov function than the one used in part (c). Show that $V(x,y) = \tfrac{1}{2}(x + y)^2 + x^2 + \tfrac{1}{2}y^2$ is such a Liapunov function, and conclude that the origin is an asymptotically stable critical point.
Hint: From Taylor's formula with a remainder it follows that $\sin x = x - \alpha x^3/3!$, where α depends on x but $0 < \alpha < 1$ for $-\pi/2 < x < \pi/2$. Then, letting $x = r\cos\theta$, $y = r\sin\theta$, show that $\dot{V}(r\cos\theta, r\sin\theta) = -r^2[1 + h(r,\theta)]$, where $|h(r,\theta)| < 1$ if r is sufficiently small.

8. The Liénard equation (Problem 28 of Section 9.3) is

$$\frac{d^2u}{dt^2} + c(u)\frac{du}{dt} + g(u) = 0,$$

where g satisfies the conditions of Problem 6 and $c(u) \geq 0$. Show that the point $u = 0$, $du/dt = 0$ is a stable critical point.

9. (a) A special case of the Liénard equation of Problem 8 is

$$\frac{d^2u}{dt^2} + \frac{du}{dt} + g(u) = 0,$$

where g satisfies the conditions of Problem 6. Letting $x = u$, $y = du/dt$, show that the origin is a critical point of the resulting system. This equation can be interpreted as describing the motion of a spring–mass system with damping proportional to the velocity and a nonlinear restoring force. Using the Liapunov function of Problem 6, show that the origin is a stable critical point, but note that even with damping, we cannot conclude asymptotic stability using this Liapunov function.

(b) Asymptotic stability of the critical point $(0,0)$ can be shown by constructing a better Liapunov function as was done in part (d) of Problem 7. However, the analysis for a general function g is somewhat sophisticated, and we mention only that an appropriate

form for V is

$$V(x, y) = \tfrac{1}{2}y^2 + Ayg(x) + \int_0^x g(s)\, ds,$$

where A is a positive constant to be chosen so that V is positive definite and \dot{V} is negative definite. For the pendulum problem $[g(x) = \sin x]$, use V as given by the preceding equation with $A = \tfrac{1}{2}$ to show that the origin is asymptotically stable.

Hint: Use $\sin x = x - \alpha x^3/3!$ and $\cos x = 1 - \beta x^2/2!$, where α and β depend on x, but $0 < \alpha < 1$ and $0 < \beta < 1$ for $-\pi/2 < x < \pi/2$; let $x = r\cos\theta$, $y = r\sin\theta$, and show that $\dot{V}(r\cos\theta, r\sin\theta) = -\tfrac{1}{2}r^2[1 + \tfrac{1}{2}\sin 2\theta + h(r, \theta)]$, where $|h(r, \theta)| < \tfrac{1}{2}$ if r is sufficiently small. To show that V is positive definite, use $\cos x = 1 - x^2/2 + \gamma x^4/4!$, where γ depends on x, but $0 < \gamma < 1$ for $-\pi/2 < x < \pi/2$.

In Problems 10 and 11 we will prove part of Theorem 9.3.2: If the critical point $(0, 0)$ of the almost linear system

$$dx/dt = a_{11}x + a_{12}y + F_1(x, y), \qquad dy/dt = a_{21}x + a_{22}y + G_1(x, y) \tag{i}$$

is an asymptotically stable critical point of the corresponding linear system

$$dx/dt = a_{11}x + a_{12}y, \qquad dy/dt = a_{21}x + a_{22}y, \tag{ii}$$

then it is an asymptotically stable critical point of the almost linear system (i). Problem 12 deals with the corresponding result for instability.

10. Consider the linear system (ii).

 (a) Since $(0, 0)$ is an asymptotically stable critical point, show that $a_{11} + a_{22} < 0$ and $a_{11}a_{22} - a_{12}a_{21} > 0$. (See Problem 21 of Section 9.1.)

 (b) Construct a Liapunov function $V(x, y) = Ax^2 + Bxy + Cy^2$ such that V is positive definite and \dot{V} is negative definite. One way to ensure that \dot{V} is negative definite is to choose A, B, and C so that $\dot{V}(x, y) = -x^2 - y^2$. Show that this leads to the result

$$A = -\frac{a_{21}^2 + a_{22}^2 + (a_{11}a_{22} - a_{12}a_{21})}{2\Delta}, \qquad B = \frac{a_{12}a_{22} + a_{11}a_{21}}{\Delta},$$

$$C = -\frac{a_{11}^2 + a_{12}^2 + (a_{11}a_{22} - a_{12}a_{21})}{2\Delta},$$

 where $\Delta = (a_{11} + a_{22})(a_{11}a_{22} - a_{12}a_{21})$.

 (c) Using the result of part (a), show that $A > 0$ and then show (several steps of algebra are required) that

$$4AC - B^2 = \frac{(a_{11}^2 + a_{12}^2 + a_{21}^2 + a_{22}^2)(a_{11}a_{22} - a_{12}a_{21}) + 2(a_{11}a_{22} - a_{12}a_{21})^2}{\Delta^2} > 0.$$

 Thus, by Theorem 9.6.4, V is positive definite.

11. In this problem we show that the Liapunov function constructed in the preceding problem is also a Liapunov function for the almost linear system (i). We must show that there is some region containing the origin for which \dot{V} is negative definite.

 (a) Show that

$$\dot{V}(x, y) = -(x^2 + y^2) + (2Ax + By)F_1(x, y) + (Bx + 2Cy)G_1(x, y).$$

(b) Recall that $F_1(x, y)/r \to 0$ and $G_1(x, y)/r \to 0$ as $r = (x^2 + y^2)^{1/2} \to 0$. This means that, given any $\epsilon > 0$, there exists a circle $r = R$ about the origin such that for $0 < r < R$, $|F_1(x, y)| < \epsilon r$ and $|G_1(x, y)| < \epsilon r$. Letting M be the maximum of $|2A|$, $|B|$, and $|2C|$, show by introducing polar coordinates that R can be chosen so that $\dot{V}(x, y) < 0$ for $r < R$. *Hint:* Choose ϵ sufficiently small in terms of M.

12. In this problem we prove a part of Theorem 9.3.2 related to instability.

(a) Show that if $a_{11} + a_{22} > 0$ and $a_{11}a_{22} - a_{12}a_{21} > 0$, then the critical point $(0,0)$ of the linear system (ii) is unstable.

(b) The same result holds for the almost linear system (i). As in Problems 10 and 11, construct a positive definite function V such that $\dot{V}(x, y) = x^2 + y^2$ and hence is positive definite, and then invoke Theorem 9.6.2.

9.7 Periodic Solutions and Limit Cycles

In this section we discuss further the possible existence of periodic solutions of second order autonomous systems

$$\mathbf{x}' = \mathbf{f}(\mathbf{x}). \tag{1}$$

Such solutions satisfy the relation

$$\mathbf{x}(t + T) = \mathbf{x}(t) \tag{2}$$

for all t and for some nonnegative constant T called the period. The corresponding trajectories are *closed curves* in the phase plane. Periodic solutions often play an important role in physical problems because they represent phenomena that occur repeatedly. In many situations a periodic solution represents a "final state" toward which all "neighboring" solutions tend as the transients due to the initial conditions die out.

A special case of a periodic solution is a constant solution $\mathbf{x} = \mathbf{x}^0$, which corresponds to a critical point of the autonomous system. Such a solution is clearly periodic with any period. In this section, when we speak of a periodic solution, we mean a nonconstant periodic solution. In this case the period T is positive and is usually chosen as the smallest positive number for which Eq. (2) is valid.

Recall that the solutions of the linear autonomous system

$$\mathbf{x}' = \mathbf{A}\mathbf{x} \tag{3}$$

are periodic if and only if the eigenvalues of \mathbf{A} are pure imaginary. In this case the critical point at the origin is a center, as discussed in Section 9.1. We emphasize that if the eigenvalues of \mathbf{A} are pure imaginary, then every solution of the linear system (3) is periodic, while if the eigenvalues are not pure imaginary, then there are no (nonconstant) periodic solutions. The predator–prey equations discussed in Section 9.5, although nonlinear, behave similarly: All solutions in the first quadrant are periodic. The following example illustrates a different way in which periodic solutions of nonlinear autonomous systems can occur.

EXAMPLE 1

Discuss the solutions of the system

$$\begin{pmatrix} x \\ y \end{pmatrix}' = \begin{pmatrix} x + y - x(x^2 + y^2) \\ -x + y - y(x^2 + y^2) \end{pmatrix}. \tag{4}$$

It is not difficult to show that $(0,0)$ is the only critical point of the system (4), and also that the system is almost linear in the neighborhood of the origin. The corresponding linear system

$$\begin{pmatrix} x \\ y \end{pmatrix}' = \begin{pmatrix} 1 & 1 \\ -1 & 1 \end{pmatrix} \begin{pmatrix} x \\ y \end{pmatrix} \tag{5}$$

has eigenvalues $1 \pm i$. Therefore the origin is an unstable spiral point for both the linear system (5) and the nonlinear system (4). Thus any solution that starts near the origin in the phase plane will spiral away from the origin. Since there are no other critical points, we might think that all solutions of Eqs. (4) correspond to trajectories that spiral out to infinity. However, we now show that this is incorrect, because far away from the origin the trajectories are directed inward.

It is convenient to introduce polar coordinates r and θ, where

$$x = r\cos\theta, \qquad y = r\sin\theta, \tag{6}$$

and $r \geq 0$. If we multiply the first of Eqs. (4) by x, the second by y, and add, we then obtain

$$x\frac{dx}{dt} + y\frac{dy}{dt} = (x^2 + y^2) - (x^2 + y^2)^2. \tag{7}$$

Since $r^2 = x^2 + y^2$ and $r(dr/dt) = x(dx/dt) + y(dy/dt)$, it follows from Eq. (7) that

$$r\frac{dr}{dt} = r^2(1 - r^2). \tag{8}$$

This is similar to the equations discussed in Section 2.5. The critical points (for $r \geq 0$) are the origin and the point $r = 1$, which corresponds to the unit circle in the phase plane. From Eq. (8) it follows that $dr/dt > 0$ if $r < 1$ and $dr/dt < 0$ if $r > 1$. Thus, inside the unit circle the trajectories are directed outward, while outside the unit circle they are directed inward. Apparently, the circle $r = 1$ is a limiting trajectory for this system.

To determine an equation for θ, we multiply the first of Eqs. (4) by y, the second by x, and subtract, obtaining

$$y\frac{dx}{dt} - x\frac{dy}{dt} = x^2 + y^2. \tag{9}$$

Upon calculating dx/dt and dy/dt from Eqs. (6), we find that the left side of Eq. (9) is $-r^2(d\theta/dt)$, so Eq. (9) reduces to

$$\frac{d\theta}{dt} = -1. \tag{10}$$

The system of equations (8), (10) for r and θ is equivalent to the original system (4). One solution of the system (8), (10) is

$$r = 1, \qquad \theta = -t + t_0, \tag{11}$$

where t_0 is an arbitrary constant. As t increases, a point satisfying Eqs. (11) moves clockwise around the unit circle. Thus the autonomous system (4) has a periodic solution. Other solutions can be obtained by solving Eq. (8) by separation of variables; if $r \neq 0$ and $r \neq 1$, then

$$\frac{dr}{r(1 - r^2)} = dt. \tag{12}$$

Equation (12) can be solved by using partial fractions to rewrite the left side and then integrating. By performing these calculations, we find that the solution of Eqs. (10) and (12) is

$$r = \frac{1}{\sqrt{1 + c_0 e^{-2t}}}, \qquad \theta = -t + t_0, \qquad (13)$$

where c_0 and t_0 are arbitrary constants. The solution (13) also contains the solution (11), which is obtained by setting $c_0 = 0$ in the first of Eqs. (13).

The solution satisfying the initial conditions $r = \rho, \theta = \alpha$ at $t = 0$ is given by

$$r = \frac{1}{\sqrt{1 + [(1/\rho^2) - 1]e^{-2t}}}, \qquad \theta = -(t - \alpha). \qquad (14)$$

If $\rho < 1$, then $r \to 1$ from the inside as $t \to \infty$; if $\rho > 1$, then $r \to 1$ from the outside as $t \to \infty$. Thus in all cases the trajectories spiral toward the circle $r = 1$ as $t \to \infty$. Several trajectories are shown in Figure 9.7.1.

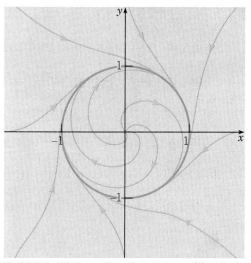

FIGURE 9.7.1 Trajectories of the system (4); a limit cycle.

In this example, the circle $r = 1$ not only corresponds to periodic solutions of the system (4), but it also attracts other nonclosed trajectories that spiral toward it as $t \to \infty$. In general, a closed trajectory in the phase plane such that other nonclosed trajectories spiral toward it, either from the inside or the outside, as $t \to \infty$, is called a **limit cycle**. Thus the circle $r = 1$ is a limit cycle for the system (4). If all trajectories that start near a closed trajectory (both inside and outside) spiral toward the closed trajectory as $t \to \infty$, then the limit cycle is **asymptotically stable**. Since the limiting trajectory is itself a periodic orbit rather than an equilibrium point, this type of stability is often called **orbital stability**. If the trajectories on one side spiral toward the closed trajectory, while those on the other side spiral away as $t \to \infty$, then the limit cycle is said to be **semistable**. If the trajectories on both sides of the closed trajectory spiral away as $t \to \infty$, then the closed trajectory is **unstable**. It is also possible to

have closed trajectories that other trajectories neither approach nor depart from—for example, the periodic solutions of the predator–prey equations in Section 9.5. In this case the closed trajectory is **stable**.

In Example 1 the existence of an asymptotically stable limit cycle was established by solving the equations explicitly. Unfortunately, this is usually not possible, so it is worthwhile to know general theorems concerning the existence or nonexistence of limit cycles of nonlinear autonomous systems. In discussing these theorems it is convenient to rewrite the system (1) in the scalar form

$$dx/dt = F(x, y), \qquad dy/dt = G(x, y). \tag{15}$$

Theorem 9.7.1 Let the functions F and G have continuous first partial derivatives in a domain D of the xy-plane. A closed trajectory of the system (15) must necessarily enclose at least one critical (equilibrium) point. If it encloses only one critical point, the critical point cannot be a saddle point.

Although we omit the proof of this theorem, it is easy to show examples of it. One is given by Example 1 and Figure 9.7.1 in which the closed trajectory encloses the critical point $(0, 0)$, a spiral point. Another example is the system of predator–prey equations in Section 9.5; see Figure 9.5.2. Each closed trajectory surrounds the critical point $(3, 2)$; in this case the critical point is a center.

Theorem 9.7.1 is also useful in a negative sense. If a given region contains no critical points, then there can be no closed trajectory lying entirely in the region. The same conclusion is true if the region contains only one critical point, and this point is a saddle point. For instance, in Example 2 of Section 9.4, an example of competing species, the only critical point in the interior of the first quadrant is the saddle point $(0.5, 0.5)$. Therefore this system has no closed trajectory lying in the first quadrant.

A second result about the nonexistence of closed trajectories is given in the following theorem.

Theorem 9.7.2 Let the functions F and G have continuous first partial derivatives in a simply connected domain D of the xy-plane. If $F_x + G_y$ has the same sign throughout D, then there is no closed trajectory of the system (15) lying entirely in D.

A simply connected two-dimensional domain is one with no holes. Theorem 9.7.2 is a straightforward consequence of Green's theorem in the plane; see Problem 13. Note that if $F_x + G_y$ changes sign in the domain, then no conclusion can be drawn; there may or may not be closed trajectories in D.

To illustrate Theorem 9.7.2 consider the system (4). A routine calculation shows that

$$F_x(x, y) + G_y(x, y) = 2 - 4(x^2 + y^2) = 2(1 - 2r^2), \tag{16}$$

where, as usual, $r^2 = x^2 + y^2$. Hence $F_x + G_y$ is positive for $0 \leq r < 1/\sqrt{2}$, so there is no closed trajectory in this circular disk. Of course, we showed in Example 1 that there is no closed trajectory in the larger region $r < 1$. This illustrates that the information given by Theorem 9.7.2 may not be the best possible result. Again referring to Eq. (16), note that $F_x + G_y < 0$ for $r > 1/\sqrt{2}$. However, the theorem is not applicable in this case because this annular region is not simply connected. Indeed, as shown in Example 1, it does contain a limit cycle.

The following theorem gives conditions that guarantee the existence of a closed trajectory.

Theorem 9.7.3

(Poincaré–Bendixson[7] Theorem) Let the functions F and G have continuous first partial derivatives in a domain D of the xy-plane. Let D_1 be a bounded subdomain in D, and let R be the region that consists of D_1 plus its boundary (all points of R are in D). Suppose that R contains no critical point of the system (15). If there exists a constant t_0 such that $x = \phi(t), y = \psi(t)$ is a solution of the system (15) that exists and stays in R for all $t \geq t_0$, then either $x = \phi(t), y = \psi(t)$ is a periodic solution (closed trajectory), or $x = \phi(t), y = \psi(t)$ spirals toward a closed trajectory as $t \to \infty$. In either case, the system (15) has a periodic solution in R.

Note that if R does contain a closed trajectory, then necessarily, by Theorem 9.7.1, this trajectory must enclose a critical point. However, this critical point cannot be in R. Thus R cannot be simply connected; it must have a hole.

As an application of the Poincaré–Bendixson theorem, consider again the system (4). Since the origin is a critical point, it must be excluded. For instance, we can consider the region R defined by $0.5 \leq r \leq 2$. Next, we must show that there is a solution whose trajectory stays in R for all t greater than or equal to some t_0. This follows immediately from Eq. (8). For $r = 0.5$, $dr/dt > 0$, so r increases, while for $r = 2$, $dr/dt < 0$, so r decreases. Thus any trajectory that crosses the boundary of R is entering R. Consequently, any solution of Eqs. (4) that starts in R at $t = t_0$ cannot leave but must stay in R for $t > t_0$. Of course, other numbers could be used instead of 0.5 and 2; all that is important is that $r = 1$ is included.

One should not infer from this discussion of the preceding theorems that it is easy to determine whether a given nonlinear autonomous system has periodic solutions; often it is not a simple matter at all. Theorems 9.7.1 and 9.7.2 are frequently inconclusive, and for Theorem 9.7.3 it is often difficult to determine a region R and a solution that always remains within it.

We close this section with another example of a nonlinear system that has a limit cycle.

[7]Ivar Otto Bendixson (1861–1935) a Swedish mathematician, received his doctorate from Uppsala University and was professor and for many years rector of Stockholm University. This theorem improved on an earlier result of Poincaré and was published in a paper by Bendixson in *Acta Mathematica* in 1901.

EXAMPLE

2

The van der Pol[8] equation

$$u'' - \mu(1 - u^2)u' + u = 0, \tag{17}$$

where μ is a nonnegative constant, describes the current u in a triode oscillator. Discuss the solutions of this equation.

If $\mu = 0$, Eq. (17) reduces to $u'' + u = 0$, whose solutions are sine or cosine waves of period 2π. For $\mu > 0$, the second term on the left side of Eq. (17) must also be considered. This is the resistance term, proportional to u', with a coefficient $-\mu(1 - u^2)$ that depends on u. For large u, this term is positive and acts as usual to reduce the amplitude of the response. However, for small u, the resistance term is negative and so causes the response to grow. This suggests that perhaps there is a solution of intermediate size that other solutions approach as t increases.

To analyze Eq. (17) more carefully, we write it as a second order system by introducing the variables $x = u, y = u'$. Then it follows that

$$x' = y, \qquad y' = -x + \mu(1 - x^2)y. \tag{18}$$

The only critical point of the system (18) is the origin. Near the origin the corresponding linear system is

$$\begin{pmatrix} x \\ y \end{pmatrix}' = \begin{pmatrix} 0 & 1 \\ -1 & \mu \end{pmatrix} \begin{pmatrix} x \\ y \end{pmatrix}, \tag{19}$$

whose eigenvalues are $(\mu \pm \sqrt{\mu^2 - 4})/2$. Thus the origin is an unstable spiral point for $0 < \mu < 2$ and an unstable node for $\mu \geq 2$. In all cases, a solution that starts near the origin grows as t increases.

With regard to periodic solutions, Theorems 9.7.1 and 9.7.2 provide only partial information. From Theorem 9.7.1 we conclude that if there are closed trajectories, they must enclose the origin. Next we calculate $F_x(x, y) + G_y(x, y)$, with the result that

$$F_x(x, y) + G_y(x, y) = \mu(1 - x^2). \tag{20}$$

Then it follows from Theorem 9.7.2 that closed trajectories, if there are any, are not contained in the strip $|x| < 1$ where $F_x + G_y > 0$.

The application of the Poincaré–Bendixson theorem to this problem is not nearly as simple as for Example 1. If we introduce polar coordinates, we find that the equation for the radial variable r is

$$r' = \mu(1 - r^2 \cos^2 \theta)r \sin^2 \theta. \tag{21}$$

Again, consider an annular region R given by $r_1 \leq r \leq r_2$, where r_1 is small and r_2 is large. When $r = r_1$, the linear term on the right side of Eq. (21) dominates, and $r' > 0$ except on the x-axis, where $\sin \theta = 0$ and consequently $r' = 0$ also. Thus trajectories are entering R at every point on the circle $r = r_1$, except possibly for those on the x-axis, where the trajectories are tangent to the circle. When $r = r_2$, the cubic term on the right side of Eq. (21) is the dominant one. Thus $r' < 0$, except for points on the x-axis where $r' = 0$ and for points near the y-axis where $r^2 \cos^2 \theta < 1$ and the linear term makes $r' > 0$. Thus, no matter how large a circle is chosen, there will be points on it (namely, the points on or near the y-axis) where trajectories are leaving R. Therefore, the Poincaré–Bendixson theorem is not applicable unless we consider more complicated regions.

[8]Balthasar van der Pol (1889–1959) was a Dutch physicist and electrical engineer who worked at the Philips Research Laboratory in Eindhoven. He was a pioneer in the experimental study of nonlinear phenomena and investigated the equation that bears his name in a paper published in 1926.

It is possible to show, by a more intricate analysis, that the van der Pol equation does have a unique limit cycle. However, we will not follow this line of argument further. We turn instead to a different approach in which we plot numerically computed solutions. Experimental observations indicate that the van der Pol equation has an asymptotically stable periodic solution whose period and amplitude depend on the parameter μ. By looking at graphs of trajectories in the phase plane and of u versus t, we can gain some understanding of this periodic behavior.

Figure 9.7.2 shows two trajectories of the van der Pol equation in the phase plane for $\mu = 0.2$. The trajectory starting near the origin spirals outward in the clockwise direction; this is consistent with the behavior of the linear approximation near the origin. The other trajectory passes through $(-3, 2)$ and spirals inward, again in the clockwise direction. Both trajectories approach a closed curve that corresponds to a stable periodic solution. In Figure 9.7.3 we show the plots of u versus t for the solutions corresponding to the trajectories in Figure 9.7.2. The solution that is initially smaller gradually increases in amplitude, while the larger solution gradually decays. Both solutions approach a stable periodic motion that corresponds to the limit cycle. Figure 9.7.3 also shows that there is a phase difference between the two solutions as they approach the limit cycle. The plots of u versus t are nearly sinusoidal in shape, consistent with the nearly circular limit cycle in this case.

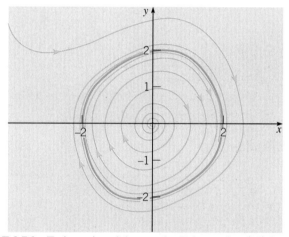

FIGURE 9.7.2 Trajectories of the van der Pol equation (17) for $\mu = 0.2$.

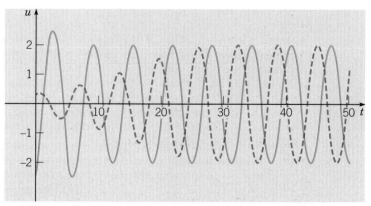

FIGURE 9.7.3 Plots of u versus t for the trajectories in Figure 9.7.2.

Figures 9.7.4 and 9.7.5 show similar plots for the case $\mu = 1$. Trajectories again move clockwise in the phase plane, but the limit cycle is considerably different from a circle. The plots of u versus t tend more rapidly to the limiting oscillation, and again show a phase difference. The oscillations are somewhat less symmetric in this case, rising somewhat more steeply than they fall.

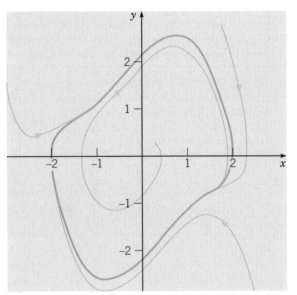

FIGURE 9.7.4 Trajectories of the van der Pol equation (17) for $\mu = 1$.

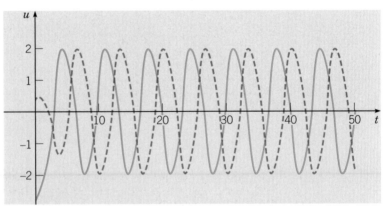

FIGURE 9.7.5 Plots of u versus t for the trajectories in Figure 9.7.4.

Figure 9.7.6 shows the phase plane for $\mu = 5$. The motion remains clockwise, and the limit cycle is even more elongated, especially in the y direction. In Figure 9.7.7 is a plot of u versus t. Although the solution starts far from the limit cycle, the limiting oscillation is virtually reached in a fraction of a period. Starting from one of its extreme values on the x-axis in the phase plane, the solution moves toward the other extreme position slowly at first, but once a certain

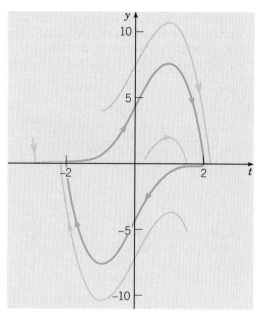

FIGURE 9.7.6 Trajectories of the van der Pol equation (17) for $\mu = 5$.

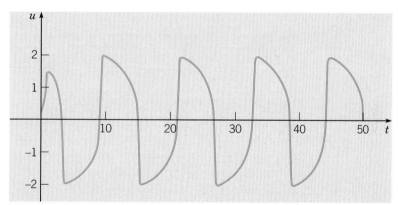

FIGURE 9.7.7 Plot of u versus t for the outward spiralling trajectory in Figure 9.7.6.

point on the trajectory is reached, the remainder of the transition is completed very swiftly. The process is then repeated in the opposite direction. The waveform of the limit cycle, as shown in Figure 9.7.7, is quite different from a sine wave.

These graphs clearly show that, in the absence of external excitation, the van der Pol oscillator has a certain characteristic mode of vibration for each value of μ. The graphs of u versus t show that the amplitude of this oscillation changes very little with μ, but the period increases as μ increases. At the same time, the waveform changes from one that is very nearly sinusoidal to one that is much less smooth.

The presence of a single periodic motion that attracts all (nearby) solutions, that is, an asymptotically stable limit cycle, is one of the characteristic phenomena associated with nonlinear differential equations.

PROBLEMS

In each of Problems 1 through 6 an autonomous system is expressed in polar coordinates. Determine all periodic solutions, all limit cycles, and determine their stability characteristics.

1. $dr/dt = r^2(1 - r^2)$, $d\theta/dt = 1$ 2. $dr/dt = r(1 - r)^2$, $d\theta/dt = -1$

3. $dr/dt = r(r - 1)(r - 3)$, $d\theta/dt = 1$ 4. $dr/dt = r(1 - r)(r - 2)$, $d\theta/dt = -1$

5. $dr/dt = \sin \pi r$, $d\theta/dt = 1$

6. $dr/dt = r|r - 2|(r - 3)$, $d\theta/dt = -1$

7. If $x = r \cos \theta, y = r \sin \theta$, show that $y(dx/dt) - x(dy/dt) = -r^2(d\theta/dt)$.

8. (a) Show that the system

$$dx/dt = -y + xf(r)/r, \qquad dy/dt = x + yf(r)/r$$

has periodic solutions corresponding to the zeros of $f(r)$. What is the direction of motion on the closed trajectories in the phase plane?

(b) Let $f(r) = r(r - 2)^2(r^2 - 4r + 3)$. Determine all periodic solutions and determine their stability characteristics.

9. Determine the periodic solutions, if any, of the system

$$\frac{dx}{dt} = y + \frac{x}{\sqrt{x^2 + y^2}}(x^2 + y^2 - 2), \qquad \frac{dy}{dt} = -x + \frac{y}{\sqrt{x^2 + y^2}}(x^2 + y^2 - 2).$$

10. Using Theorem 9.7.2, show that the linear autonomous system

$$dx/dt = a_{11}x + a_{12}y, \qquad dy/dt = a_{21}x + a_{22}y$$

does not have a periodic solution (other than $x = 0, y = 0$) if $a_{11} + a_{22} \neq 0$.

In each of Problems 11 and 12 show that the given system has no periodic solutions other than constant solutions.

11. $dx/dt = x + y + x^3 - y^2$, $dy/dt = -x + 2y + x^2y + y^3/3$

12. $dx/dt = -2x - 3y - xy^2$, $dy/dt = y + x^3 - x^2y$

13. Prove Theorem 9.7.2 by completing the following argument. According to Green's theorem in the plane, if C is a sufficiently smooth simple closed curve, and if F and G are continuous and have continuous first partial derivatives, then

$$\int_C \left[F(x, y)\, dy - G(x, y)\, dx \right] = \iint_R \left[F_x(x, y) + G_y(x, y) \right] dA,$$

where C is traversed counterclockwise and R is the region enclosed by C. Assume that $x = \phi(t), y = \psi(t)$ is a solution of the system (15) that is periodic with period T. Let C be the closed curve given by $x = \phi(t), y = \psi(t)$ for $0 \leq t \leq T$. Show that for this curve the line integral is zero. Then show that the conclusion of Theorem 9.7.2 must follow.

14. (a) By examining the graphs of u versus t in Figures 9.7.4, 9.7.6, and 9.7.8, estimate the period T of the van der Pol oscillator in these cases.

(b) Calculate and plot the graphs of solutions of the van der Pol equation for other values of the parameter μ. Estimate the period T in these cases also.

(c) Plot the estimated values of T versus μ. Describe how T depends on μ.

15. The equation

$$u'' - \mu(1 - \tfrac{1}{3}u'^2)u' + u = 0$$

is often called the Rayleigh[9] equation.

(a) Write the Rayleigh equation as a system of two first order equations.

(b) Show that the origin is the only critical point of this system. Determine its type and whether it is asymptotically stable, stable, or unstable.

(c) Let $\mu = 1$. Choose initial conditions and compute the corresponding solution of the system on an interval such as $0 \le t \le 20$ or longer. Plot u versus t and also plot the trajectory in the phase plane. Observe that the trajectory approaches a closed curve (limit cycle). Estimate the amplitude A and the period T of the limit cycle.

(d) Repeat part (c) for other values of μ, such as $\mu = 0.2, 0.5, 2$, and 5. In each case estimate the amplitude A and the period T.

(e) Describe how the limit cycle changes as μ increases. For example, make a table of values and/or plot A and T as functions of μ.

16. Consider the system of equations

$$x' = \mu x + y - x(x^2 + y^2), \qquad y' = -x + \mu y - y(x^2 + y^2), \qquad \text{(i)}$$

where μ is a parameter. Observe that this system is the same as the one in Example 1, except for the introduction of μ.

(a) Show that the origin is the only critical point.

(b) Find the linear system that approximates Eqs. (i) near the origin and find its eigenvalues. Determine the type and stability of the critical point at the origin. How does this classification depend on μ?

(c) Referring to Example 1 if necessary, rewrite Eqs. (i) in polar coordinates.

(d) Show that when $\mu > 0$ there is a periodic solution $r = \sqrt{\mu}$. By solving the system found in part (c), or by plotting numerically computed solutions, conclude that this periodic solution attracts all other nonzero solutions.

Note: As the parameter μ increases through the value zero, the previously asymptotically stable critical point at the origin loses its stability, and simultaneously a new asymptotically stable solution (the limit cycle) emerges. Thus the point $\mu = 0$ is a bifurcation point; this type of bifurcation is called a **Hopf**[10] **bifurcation**.

17. There are certain chemical reactions in which the constituent concentrations oscillate periodically over time. The system

$$x' = 1 - (b + 1)x + x^2 y/4, \qquad y' = bx - x^2 y/4$$

is a special case of a model, known as the Brusselator, of this kind of reaction. Assume that b is a positive parameter, and consider solutions in the first quadrant of the xy-plane.

(a) Show that the only critical point is $(1, 4b)$.

(b) Find the eigenvalues of the approximate linear system at the critical point.

[9]John William Strutt (1842–1919), third Lord Rayleigh, made notable contributions in several areas of mathematical physics. Apart from five years as Cavendish professor of physics at Cambridge, he worked primarily in his private laboratory at home. He was awarded the Nobel Prize in 1904 for the discovery of argon.

[10]Eberhard Hopf (1902–1983) was born in Austria and educated at the University of Berlin but spent much of his life in the United States, mainly at Indiana University. Hopf bifurcations are named for him in honor of his rigorous treatment of them in a 1942 paper.

(c) Classify the critical point as to type and stability. How does the classification depend on b?

(d) As b increases through a certain value b_0, the critical point changes from asymptotically stable to unstable. What is that value b_0?

(e) Plot trajectories in the phase plane for values of b slightly less than and slightly greater than b_0. Observe the limit cycle when $b > b_0$; the Brusselator has a Hopf bifurcation point at b_0.

(f) Plot trajectories for several values of $b > b_0$ and observe how the limit cycle deforms as b increases.

18. The system
$$x' = 3(x + y - \tfrac{1}{3}x^3 - k), \qquad y' = -\tfrac{1}{3}(x + 0.8y - 0.7)$$

is a special case of the Fitzhugh–Nagumo[11] equations, which model the transmission of neural impulses along an axon. The parameter k is the external stimulus.

(a) Show that the system has one critical point regardless of the value of k.

(b) Find the critical point for $k = 0$ and show that it is an asymptotically stable spiral point. Repeat the analysis for $k = 0.5$ and show that the critical point is now an unstable spiral point. Draw a phase portrait for the system in each case.

(c) Find the value k_0 where the critical point changes from asymptotically stable to unstable. Find the critical point and draw a phase portrait for the system for $k = k_0$.

(d) For $k > k_0$ the system exhibits an asymptotically stable limit cycle; the system has a Hopf bifurcation point at k_0. Draw a phase portrait for $k = 0.4, 0.5$, and 0.6; observe that the limit cycle is not small when k is near k_0. Also plot x versus t and estimate the period T in each case.

(e) As k increases further, there is a value k_1 at which the critical point again becomes asymptotically stable and the limit cycle vanishes. Find k_1.

9.8 Chaos and Strange Attractors: The Lorenz Equations

In principle, the methods described in this chapter for second order autonomous systems can be applied to higher order systems as well. In practice, several difficulties arise when we try to do this. One problem is that there is simply a greater number of possible cases that can occur, and the number increases with the order of the system (and the dimension of the phase space). Another problem is the difficulty of graphing trajectories accurately in a phase space of higher than two dimensions; even in three dimensions it may not be easy to construct a clear and understandable plot of the trajectories, and it becomes more difficult as the number of variables increases. Finally, and this has been clearly realized only in the last few years, there are different and very complex phenomena that can occur, and frequently do occur, in systems of third and higher order that are not present in second order systems. Our goal in this section is to provide a brief introduction to some of these phenomena by discussing

[11]Richard Fitzhugh (1922–) of the United States Public Health Service and Jin-Ichi Nagumo (1926–1999) of the University of Tokyo independently proposed a simplification of the Hodgkin–Huxley model of neural transmission about 1961.

one particular third order autonomous system that has been intensively studied in recent years. In some respects, the presentation here is similar to the treatment of the logistic difference equation in Section 2.9.

An important problem in meteorology, and in other applications of fluid dynamics, concerns the motion of a layer of fluid, such as the earth's atmosphere, that is warmer at the bottom than at the top; see Figure 9.8.1. If the vertical temperature difference ΔT is small, then there is a linear variation of temperature with altitude but no significant motion of the fluid layer. However, if ΔT is large enough, then the warmer air rises, displacing the cooler air above it, and a steady convective motion results. If the temperature difference increases further, then eventually the steady convective flow breaks up and a more complex and turbulent motion ensues.

FIGURE 9.8.1 A layer of fluid heated from below.

While investigating this phenomenon, Edward N. Lorenz[12] was led (by a process too involved to describe here) to the nonlinear autonomous third order system

$$
\begin{aligned}
dx/dt &= \sigma(-x + y), \\
dy/dt &= rx - y - xz, \\
dz/dt &= -bz + xy.
\end{aligned}
\tag{1}
$$

Equations (1) are now commonly referred to as the Lorenz equations.[13] Observe that the second and third equations involve quadratic nonlinearities. However, except for being a third order system, superficially the Lorenz equations appear no more complicated than the competing species or predator–prey equations discussed in Sections 9.4 and 9.5. The variable x in Eqs. (1) is related to the intensity of the fluid motion, while the variables y and z are related to the temperature variations in the horizontal and vertical directions. The Lorenz equations also involve three parameters σ, r, and b, all of which are real and positive. The parameters σ and b depend on the material and geometrical properties of the fluid layer. For the earth's atmosphere, reasonable values of these parameters are $\sigma = 10$ and $b = 8/3$; they will be assigned these values in much of what follows in this section. The parameter r, on

[12]Edward N. Lorenz (1917–), American meteorologist, received his Ph.D. from the Massachusetts Institute of Technology in 1948 and has been associated with that institution throughout his scientific career. His first studies of the system (1) appeared in a famous 1963 paper dealing with the stability of fluid flows in the atmosphere.

[13]A very thorough treatment of the Lorenz equations appears in the book by Sparrow listed in the references at the end of this chapter.

the other hand, is proportional to the temperature difference ΔT, and our purpose is to investigate how the nature of the solutions of Eqs. (1) changes with r.

The first step in analyzing the Lorenz equations is to locate the critical points by solving the algebraic system

$$\sigma x - \sigma y = 0,$$
$$rx - y - xz = 0, \tag{2}$$
$$-bz + xy = 0.$$

From the first equation we have $y = x$. Then, eliminating y from the second and third equations, we obtain

$$x(r - 1 - z) = 0, \tag{3}$$
$$-bz + x^2 = 0. \tag{4}$$

One way to satisfy Eq. (3) is to choose $x = 0$. Then it follows that $y = 0$ and, from Eq. (4), $z = 0$. Alternatively, we can satisfy Eq. (3) by choosing $z = r - 1$. Then Eq. (4) requires that $x = \pm\sqrt{b(r-1)}$ and then $y = \pm\sqrt{b(r-1)}$ also. Observe that these expressions for x and y are real only when $r \geq 1$. Thus $(0, 0, 0)$, which we will denote by P_1, is a critical point for all values of r, and it is the only critical point for $r < 1$. However, when $r > 1$, there are also two other critical points, namely $(\sqrt{b(r-1)}, \sqrt{b(r-1)}, r-1)$ and $(-\sqrt{b(r-1)}, -\sqrt{b(r-1)}, r-1)$. We will denote the latter two points by P_2 and P_3, respectively. Note that all three critical points coincide when $r = 1$. As r increases through the value 1, the critical point P_1 at the origin *bifurcates*, and the critical points P_2 and P_3 come into existence.

Next we will determine the local behavior of solutions in the neighborhood of each critical point. Although much of the following analysis can be carried out for arbitrary values of σ and b, we will simplify our work by using the values $\sigma = 10$ and $b = 8/3$. Near the origin (the critical point P_1) the approximating linear system is

$$\begin{pmatrix} x \\ y \\ z \end{pmatrix}' = \begin{pmatrix} -10 & 10 & 0 \\ r & -1 & 0 \\ 0 & 0 & -8/3 \end{pmatrix} \begin{pmatrix} x \\ y \\ z \end{pmatrix}. \tag{5}$$

The eigenvalues are determined from the equation

$$\begin{vmatrix} -10 - \lambda & 10 & 0 \\ r & -1 - \lambda & 0 \\ 0 & 0 & -8/3 - \lambda \end{vmatrix} = -(8/3 + \lambda)[\lambda^2 + 11\lambda - 10(r - 1)] = 0. \tag{6}$$

Therefore

$$\lambda_1 = -\frac{8}{3}, \qquad \lambda_2 = \frac{-11 - \sqrt{81 + 40r}}{2}, \qquad \lambda_3 = \frac{-11 + \sqrt{81 + 40r}}{2}. \tag{7}$$

Note that all three eigenvalues are negative for $r < 1$; for example, when $r = 1/2$, the eigenvalues are $\lambda_1 = -8/3, \lambda_2 = -10.52494, \lambda_3 = -0.47506$. Hence the origin is asymptotically stable for this range of r both for the linear approximation (5) and for the original system (1). However, λ_3 changes sign when $r = 1$ and is positive for $r > 1$. The value $r = 1$ corresponds to the initiation of convective flow in the physical problem described earlier. The origin is unstable for $r > 1$; all solutions starting near the origin tend to grow, except for those lying precisely in the plane determined by the

eigenvectors associated with λ_1 and λ_2 [or, for the nonlinear system (1), in a certain surface tangent to this plane at the origin].

Next let us consider the neighborhood of the critical point $P_2(\sqrt{8(r-1)/3},$ $\sqrt{8(r-1)/3}, r-1)$ for $r > 1$. If u, v, and w are the perturbations from the critical point in the x, y, and z directions, respectively, then the approximating linear system is

$$\begin{pmatrix} u \\ v \\ w \end{pmatrix}' = \begin{pmatrix} -10 & 10 & 0 \\ 1 & -1 & -\sqrt{8(r-1)/3} \\ \sqrt{8(r-1)/3} & \sqrt{8(r-1)/3} & -8/3 \end{pmatrix} \begin{pmatrix} u \\ v \\ w \end{pmatrix}. \qquad (8)$$

The eigenvalues of the coefficient matrix of Eq. (8) are determined from the equation

$$3\lambda^3 + 41\lambda^2 + 8(r+10)\lambda + 160(r-1) = 0, \qquad (9)$$

which is obtained by straightforward algebraic steps that are omitted here. The solutions of Eq. (9) depend on r in the following way:

For $1 < r < r_1 \cong 1.3456$ there are three negative real eigenvalues.

For $r_1 < r < r_2 \cong 24.737$ there are one negative real eigenvalue and two complex eigenvalues with negative real part.

For $r_2 < r$ there are one negative real eigenvalue and two complex eigenvalues with positive real part.

The same results are obtained for the critical point P_3. Thus there are several different situations.

For $0 < r < 1$ the only critical point is P_1 and it is asymptotically stable. All solutions approach this point (the origin) as $t \to \infty$.

For $1 < r < r_1$ the critical points P_2 and P_3 are asymptotically stable and P_1 is unstable. All nearby solutions approach one or the other of the points P_2 and P_3 exponentially.

For $r_1 < r < r_2$ the critical points P_2 and P_3 are asymptotically stable and P_1 is unstable. All nearby solutions approach one or the other of the points P_2 and P_3; most of them spiral inward to the critical point.

For $r_2 < r$ all three critical points are unstable. Most solutions near P_2 or P_3 spiral away from the critical point.

However, this is by no means the end of the story. Let us consider solutions for r somewhat greater than r_2. In this case P_1 has one positive eigenvalue and each of P_2 and P_3 has a pair of complex eigenvalues with positive real part. A trajectory can approach any one of the critical points only on certain highly restricted paths. The slightest deviation from these paths causes the trajectory to depart from the critical point. Since none of the critical points is stable, one might expect that most trajectories would approach infinity for large t. However, it can be shown that all solutions remain bounded as $t \to \infty$; see Problem 5. In fact, it can be shown that all solutions ultimately approach a certain limiting set of points that has zero volume. Indeed, this is true not only for $r > r_2$ but for all positive values of r.

A plot of computed values of x versus t for a typical solution with $r > r_2$ is shown in Figure 9.8.2. Note that the solution oscillates back and forth between positive and negative values in a rather erratic manner. Indeed, the graph of x versus t resembles a random vibration, although the Lorenz equations are entirely deterministic and

the solution is completely determined by the initial conditions. Nevertheless, the solution also exhibits a certain *regularity* in that the frequency and amplitude of the oscillations are essentially constant in time.

The solutions of the Lorenz equations are also extremely sensitive to perturbations in the initial conditions. Figure 9.8.3 shows the graphs of computed values of x versus t for the two solutions whose initial points are $(5, 5, 5)$ and $(5.01, 5, 5)$. The dashed graph is the same as the one in Figure 9.8.2, while the solid graph starts at a nearby point. The two solutions remain close until t is near 10, after which they are quite different and, indeed, seem to have no relation to each other. It was this property that particularly attracted the attention of Lorenz in his original study of these equations, and caused him to conclude that detailed long-range weather predictions are probably not possible.

The attracting set in this case, although of zero volume, has a rather complicated structure and is called a **strange attractor**. The term **chaotic** has come into general use to describe solutions such as those shown in Figures 9.8.2 and 9.8.3.

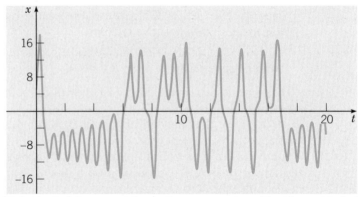

FIGURE 9.8.2 A plot of x versus t for the Lorenz equations (1) with $r = 28$; the initial point is $(5, 5, 5)$.

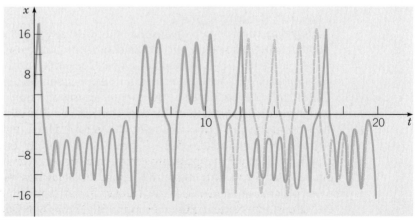

FIGURE 9.8.3 Plots of x versus t for two initially nearby solutions of Lorenz equations with $r = 28$; the initial point is $(5, 5, 5)$ for the dashed curve and is $(5.01, 5, 5)$ for the solid curve.

To determine how and when the strange attractor is created, it is illuminating to investigate solutions for smaller values of r. For $r = 21$, solutions starting at three different initial points are shown in Figure 9.8.4. For the initial point $(3, 8, 0)$ the solution begins to converge to the point P_3 almost at once; see Figure 9.8.4a. For the second initial point $(5, 5, 5)$ there is a fairly short interval of transient behavior, after which the solution converges to P_2; see Figure 9.8.4b. However, as shown in Figure 9.8.4c, for the third initial point $(5, 5, 10)$ there is a much longer interval of transient chaotic behavior before the solution eventually converges to P_2. As r increases, the duration of the chaotic transient behavior also increases. When $r = r_3 \cong 24.06$, the chaotic transients appear to continue indefinitely, and the strange attractor comes into being.

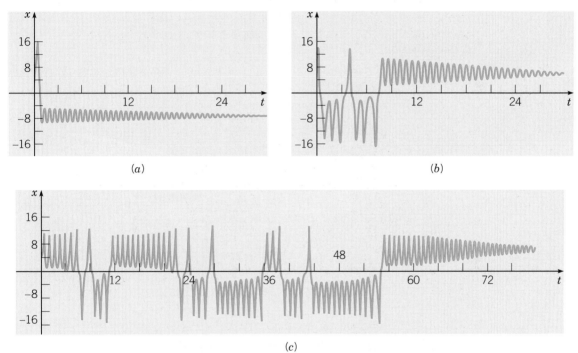

(a)

(b)

(c)

FIGURE 9.8.4 Plots of x versus t for three solutions of Lorenz equations with $r = 21$. (a) Initial point is $(3, 8, 0)$. (b) Initial point is $(5, 5, 5)$. (c) Initial point is $(5, 5, 10)$.

One can also show the trajectories of the Lorenz equations in the three-dimensional phase space, or at least projections of them in various planes. Figures 9.8.5 and 9.8.6 show projections in the xy- and xz-planes, respectively, of the trajectory starting at $(5, 5, 5)$. Observe that the graphs in these figures appear to cross over themselves repeatedly, but this cannot be true for the actual trajectories in three-dimensional space because of the general uniqueness theorem. The apparent crossings are due wholly to the two-dimensional character of the figures.

The sensitivity of solutions to perturbations of the initial data also has implications for numerical computations, such as those reported here. Different step sizes, different numerical algorithms, or even the execution of the same algorithm on different

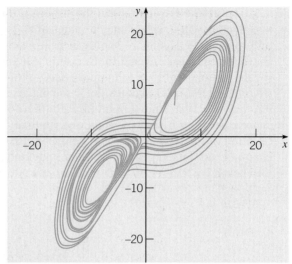

FIGURE 9.8.5 Projections of a trajectory of the Lorenz equations (with $r = 28$) in the xy-plane.

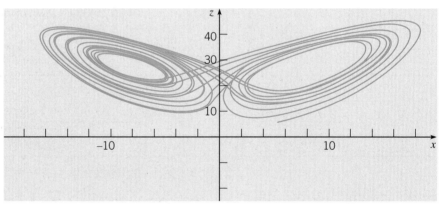

FIGURE 9.8.6 Projections of a trajectory of the Lorenz equations (with $r = 28$) in the xz-plane.

machines will introduce small differences in the computed solution, which eventually lead to large deviations. For example, the exact sequence of positive and negative loops in the calculated solution depends strongly on the precise numerical algorithm and its implementation, as well as on the initial conditions. However, the general appearance of the solution and the structure of the attracting set are independent of all these factors.

Solutions of the Lorenz equations for other parameter ranges exhibit other interesting types of behavior. For example, for certain values of r greater than r_2, intermittent bursts of chaotic behavior separate long intervals of apparently steady periodic oscillation. For other ranges of r, solutions show the period-doubling property that we saw in Section 2.9 for the logistic difference equation. Some of these features are taken up in the problems.

Since about 1975 the Lorenz equations and other higher order autonomous systems have been studied intensively, and this is one of the most active areas of current mathematical research. Chaotic behavior of solutions appears to be much more common than was suspected at first, and many questions remain unanswered. Some of these are mathematical in nature, while others relate to the physical applications or interpretations of solutions.

PROBLEMS

Problems 1 through 3 ask you to fill in some of the details of the analysis of the Lorenz equations in this section.

1. (a) Show that the eigenvalues of the linear system (5), valid near the origin, are given by Eq. (7).

 (b) Determine the corresponding eigenvectors.

 (c) Determine the eigenvalues and eigenvectors of the system (5) in the case where $r = 28$.

 2. (a) Show that the linear approximation valid near the critical point P_2 is given by Eq. (8).

 (b) Show that the eigenvalues of the system (8) satisfy Eq. (9).

 (c) For $r = 28$, solve Eq. (9) and thereby determine the eigenvalues of the system (8).

3. (a) By solving Eq. (9) numerically, show that the real part of the complex roots changes sign when $r \cong 24.737$.

 (b) Show that a cubic polynomial $x^3 + Ax^2 + Bx + C$ has one real zero and two pure imaginary zeros only if $AB = C$.

 (c) By applying the result of part (b) to Eq. (9), show that the real part of the complex roots changes sign when $r = 470/19$.

4. Use the Liapunov function $V(x, y, z) = x^2 + \sigma y^2 + \sigma z^2$ to show that the origin is a globally asymptotically stable critical point for the Lorenz equations (1) if $r < 1$.

5. Consider the ellipsoid

$$V(x, y, z) = rx^2 + \sigma y^2 + \sigma(z - 2r)^2 = c > 0.$$

 (a) Calculate dV/dt along trajectories of the Lorenz equations (1).

 (b) Determine a sufficient condition on c so that every trajectory crossing $V(x, y, z) = c$ is directed inward.

 (c) Evaluate the condition found in part (b) for the case $\sigma = 10, b = 8/3, r = 28$.

In each of Problems 6 through 10 carry out the indicated investigations of the Lorenz equations.

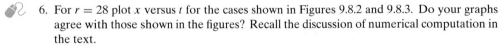 6. For $r = 28$ plot x versus t for the cases shown in Figures 9.8.2 and 9.8.3. Do your graphs agree with those shown in the figures? Recall the discussion of numerical computation in the text.

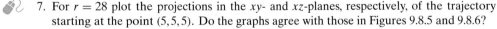 7. For $r = 28$ plot the projections in the xy- and xz-planes, respectively, of the trajectory starting at the point $(5, 5, 5)$. Do the graphs agree with those in Figures 9.8.5 and 9.8.6?

 8. (a) For $r = 21$ plot x versus t for the solutions starting at the initial points $(3, 8, 0), (5, 5, 5)$, and $(5, 5, 10)$. Use a t interval of at least $0 \le t \le 30$. Compare your graphs with those in Figure 9.8.4.

 (b) Repeat the calculation in part (a) for $r = 22, r = 23$, and $r = 24$. Increase the t interval as necessary so that you can determine when each solution begins to converge to one of

In this section we suppose that a function f is given. If this function is periodic with period $2L$ and integrable on the interval $[-L, L]$, then a set of coefficients a_m and b_m can be computed from Eqs. (2) and (3), and a series of the form (1) can be formally constructed. The question is whether this series converges for each value of x and, if so, whether its sum is $f(x)$. Examples have been discovered showing that the Fourier series corresponding to a function f may not converge to $f(x)$ or may even diverge. Functions whose Fourier series do not converge to the value of the function at isolated points are easily constructed, and examples will be presented later in this section. Functions whose Fourier series diverge at one or more points are more pathological, and we will not consider them in this book.

To guarantee convergence of a Fourier series to the function from which its coefficients were computed, it is essential to place additional conditions on the function. From a practical point of view, such conditions should be broad enough to cover all situations of interest, yet simple enough to be easily checked for particular functions. Through the years several sets of conditions have been devised to serve this purpose.

Before stating a convergence theorem for Fourier series, we define a term that appears in the theorem. A function f is said to be **piecewise continuous** on an interval $a \leq x \leq b$ if the interval can be partitioned by a finite number of points $a = x_0 < x_1 < \cdots < x_n = b$ so that

1. f is continuous on each open subinterval $x_{i-1} < x < x_i$.
2. f approaches a finite limit as the endpoints of each subinterval are approached from within the subinterval.

The graph of a piecewise continuous function is shown in Figure 10.3.1.

FIGURE 10.3.1 A piecewise continuous function.

The notation $f(c+)$ is used to denote the limit of $f(x)$ as $x \to c$ from the right; similarly, $f(c-)$ denotes the limit of $f(x)$ as x approaches c from the left.

Note that it is not essential that the function even be defined at the partition points x_i. For example, in the following theorem we assume that f' is piecewise continuous; but certainly f' does not exist at those points where f itself is discontinuous. It is also not essential that the interval be closed; it may also be open, or open at one end and closed at the other.

Theorem 10.3.1 Suppose that f and f' are piecewise continuous on the interval $-L \leq x < L$. Further, suppose that f is defined outside the interval $-L \leq x < L$ so that it is periodic with period $2L$. Then f has a Fourier series

$$f(x) = \frac{a_0}{2} + \sum_{m=1}^{\infty} \left(a_m \cos \frac{m\pi x}{L} + b_m \sin \frac{m\pi x}{L} \right), \tag{4}$$

whose coefficients are given by Eqs. (2) and (3). The Fourier series converges to $f(x)$ at all points where f is continuous, and to $[f(x+) + f(x-)]/2$ at all points where f is discontinuous.

Note that $[f(x+) + f(x-)]/2$ is the mean value of the right- and left-hand limits at the point x. At any point where f is continuous, $f(x+) = f(x-) = f(x)$. Thus it is correct to say that the Fourier series converges to $[f(x+) + f(x-)]/2$ at all points. Whenever we say that a Fourier series converges to a function f, we always mean that it converges in this sense.

It should be emphasized that the conditions given in this theorem are only sufficient for the convergence of a Fourier series; they are by no means necessary. Nor are they the most general sufficient conditions that have been discovered. In spite of this, the proof of the theorem is fairly difficult and we do not discuss it here.[4] Under more restrictive conditions a much simpler convergence proof is possible; see Problem 18.

To obtain a better understanding of the content of the theorem, it is helpful to consider some classes of functions that fail to satisfy the assumed conditions. Functions that are not included in the theorem are primarily those with infinite discontinuities in the interval $[-L, L]$, such as $1/x^2$ as $x \to 0$, or $\ln|x - L|$ as $x \to L$. Functions having an infinite number of jump discontinuities in this interval are also excluded; however, such functions are rarely encountered.

It is noteworthy that a Fourier series may converge to a sum that is not differentiable, or even continuous, in spite of the fact that each term in the series (4) is continuous, and even differentiable infinitely many times. The example below is an illustration of this, as is Example 2 in Section 10.2.

EXAMPLE 1

Let

$$f(x) = \begin{cases} 0, & -L < x < 0, \\ L, & 0 < x < L, \end{cases} \tag{5}$$

and let f be defined outside this interval so that $f(x + 2L) = f(x)$ for all x. We will temporarily leave open the definition of f at the points $x = 0, \pm L$, except to say that its value must be finite. Find the Fourier series for this function and determine where it converges.

The equation $y = f(x)$ has the graph shown in Figure 10.3.2, extended to infinity in both directions. It can be thought of as representing a square wave. The interval $[-L, L]$ can be partitioned to give the two open subintervals $(-L, 0)$ and $(0, L)$. In $(0, L)$, $f(x) = L$ and

[4]Proofs of the convergence of a Fourier series can be found in most books on advanced calculus. See, for example, Kaplan (Chapter 7) or Buck (Chapter 6).

FIGURE 10.3.2 Square wave.

$f'(x) = 0$. Clearly, both f and f' are continuous and furthermore have limits as $x \to 0$ from the right and as $x \to L$ from the left. The situation in $(-L, 0)$ is similar. Consequently, both f and f' are piecewise continuous on $[-L, L)$, so f satisfies the conditions of Theorem 10.3.1. If the coefficients a_m and b_m are computed from Eqs. (2) and (3), the convergence of the resulting Fourier series to $f(x)$ is ensured at all points where f is continuous. Note that the values of a_m and b_m are the same regardless of the definition of f at its points of discontinuity. This is true because the value of an integral is unaffected by changing the value of the integrand at a finite number of points. From Eq. (2),

$$a_0 = \frac{1}{L} \int_{-L}^{L} f(x)\,dx = \int_{0}^{L} dx = L;$$

$$a_m = \frac{1}{L} \int_{-L}^{L} f(x) \cos \frac{m\pi x}{L}\,dx = \int_{0}^{L} \cos \frac{m\pi x}{L}\,dx$$

$$= 0, \qquad m \neq 0.$$

Similarly, from Eq. (3),

$$b_m = \frac{1}{L} \int_{-L}^{L} f(x) \sin \frac{m\pi x}{L}\,dx = \int_{0}^{L} \sin \frac{m\pi x}{L}\,dx$$

$$= \frac{L}{m\pi}(1 - \cos m\pi)$$

$$= \begin{cases} 0, & m \text{ even;} \\ 2L/m\pi, & m \text{ odd.} \end{cases}$$

Hence

$$f(x) = \frac{L}{2} + \frac{2L}{\pi}\left(\sin \frac{\pi x}{L} + \frac{1}{3}\sin \frac{3\pi x}{L} + \frac{1}{5}\sin \frac{5\pi x}{L} + \cdots\right)$$

$$= \frac{L}{2} + \frac{2L}{\pi} \sum_{m=1,3,5,\ldots}^{\infty} \frac{\sin(m\pi x/L)}{m}$$

$$= \frac{L}{2} + \frac{2L}{\pi} \sum_{n=1}^{\infty} \frac{\sin(2n - 1)\pi x/L}{2n - 1}. \tag{6}$$

At the points $x = 0, \pm nL$, where the function f in the example is not continuous, all terms in the series after the first vanish and the sum is $L/2$. This is the mean value of the limits from the right and left, as it should be. Thus we might as well define f at these points to have the value $L/2$. If we choose to define it otherwise, the series still gives the value $L/2$ at these points, since none of the preceding calculations is altered in any detail; it simply does not converge to the function at those points unless f is defined to have this value. This illustrates

the possibility that the Fourier series corresponding to a function may not converge to it at points of discontinuity unless the function is suitably defined at such points.

The manner in which the partial sums

$$s_n(x) = \frac{L}{2} + \frac{2L}{\pi}\left(\sin\frac{\pi x}{L} + \cdots + \frac{1}{2n-1}\sin\frac{(2n-1)\pi x}{L}\right), \qquad n = 1, 2, \ldots$$

of the Fourier series (6) converge to $f(x)$ is indicated in Figure 10.3.3, where L has been chosen to be 1 and the graph of $s_8(x)$ is plotted. The figure suggests that at points where f is continuous the partial sums do approach $f(x)$ as n increases. However, in the neighborhood of points of discontinuity, such as $x = 0$ and $x = L$, the partial sums do not converge smoothly to the mean value. Instead they tend to overshoot the mark at each end of the jump, as though they cannot quite accommodate themselves to the sharp turn required at this point. This behavior is typical of Fourier series at points of discontinuity and is known as the Gibbs[5] phenomenon.

Additional insight is attained by considering the error $e_n(x) = f(x) - s_n(x)$. Figure 10.3.4 shows a plot of $|e_n(x)|$ versus x for $n = 8$ and for $L = 1$. The least upper bound of $|e_8(x)|$

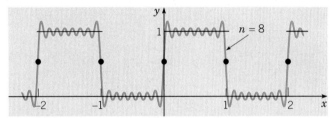

FIGURE 10.3.3 The partial sum $s_8(x)$ in the Fourier series, Eq. (6), for the square wave.

FIGURE 10.3.4 A plot of the error $|e_8(x)|$ versus x for the square wave.

[5]The Gibbs phenomenon is named after Josiah Willard Gibbs (1839–1903), who is better known for his work on vector analysis and statistical mechanics. Gibbs was professor of mathematical physics at Yale and one of the first American scientists to achieve an international reputation. The Gibbs phenomenon is discussed in more detail by Carslaw (Chapter 9).

is 0.5 and is approached as $x \to 0$ and as $x \to 1$. As n increases, the error decreases in the interior of the interval [where $f(x)$ is continuous], but the least upper bound does not diminish with increasing n. Thus one cannot uniformly reduce the error throughout the interval by increasing the number of terms.

Figures 10.3.3 and 10.3.4 also show that the series in this example converges more slowly than the one in Example 1 in Section 10.2. This is due to the fact that the coefficients in the series (6) are proportional only to $1/(2n-1)$.

PROBLEMS

In each of Problems 1 through 6 assume that the given function is periodically extended outside the original interval.

(a) Find the Fourier series for the extended function.

(b) Sketch the graph of the function to which the series converges for three periods.

1. $f(x) = \begin{cases} -1, & -1 \le x < 0, \\ 1, & 0 \le x < 1 \end{cases}$

2. $f(x) = \begin{cases} 0, & -\pi \le x < 0, \\ x, & 0 \le x < \pi \end{cases}$

3. $f(x) = \begin{cases} L + x, & -L \le x < 0, \\ L - x, & 0 \le x < L \end{cases}$

4. $f(x) = 1 - x^2, \quad -1 \le x < 1$

5. $f(x) = \begin{cases} 0, & -\pi \le x < -\pi/2, \\ 1, & -\pi/2 \le x < \pi/2, \\ 0, & \pi/2 \le x < \pi \end{cases}$

6. $f(x) = \begin{cases} 0, & -1 \le x < 0, \\ x^2, & 0 \le x < 1 \end{cases}$

In each of Problems 7 through 12 assume that the given function is periodically extended outside the original interval.

(a) Find the Fourier series for the given function.

(b) Let $e_n(x) = f(x) - s_n(x)$. Find the least upper bound or the maximum value (if it exists) of $|e_n(x)|$ for $n = 10, 20$, and 40.

(c) If possible, find the smallest n for which $|e_n(x)| \le 0.01$ for all x.

 7. $f(x) = \begin{cases} x, & -\pi \le x < 0, \\ 0, & 0 \le x < \pi; \end{cases}$ $\quad f(x + 2\pi) = f(x)$ (see Section 10.2, Problem 15)

8. $f(x) = \begin{cases} x + 1, & -1 \le x < 0, \\ 1 - x, & 0 \le x < 1; \end{cases}$ $\quad f(x + 2) = f(x)$ (see Section 10.2, Problem 16)

9. $f(x) = x, \quad -1 \le x < 1; \quad f(x + 2) = f(x)$ (see Section 10.2, Problem 20)

10. $f(x) = \begin{cases} x + 2, & -2 \le x < 0, \\ 2 - 2x, & 0 \le x < 2; \end{cases}$ $\quad f(x + 4) = f(x)$ (see Section 10.2, Problem 22)

11. $f(x) = \begin{cases} 0, & -1 \le x < 0, \\ x^2, & 0 \le x < 1; \end{cases}$ $\quad f(x + 2) = f(x)$ (see Problem 6)

12. $f(x) = x - x^3, \quad -1 \le x < 1; \quad f(x + 2) = f(x)$

Periodic Forcing Terms. In this chapter we are concerned mainly with the use of Fourier series to solve boundary value problems for certain partial differential equations. However, Fourier series are also useful in many other situations where periodic phenomena occur. Problems 13 through 16 indicate how they can be employed to solve initial value problems with periodic forcing terms.

13. Find the solution of the initial value problem

$$y'' + \omega^2 y = \sin nt, \qquad y(0) = 0, \quad y'(0) = 0,$$

where n is a positive integer and $\omega^2 \neq n^2$. What happens if $\omega^2 = n^2$?

14. Find the formal solution of the initial value problem

$$y'' + \omega^2 y = \sum_{n=1}^{\infty} b_n \sin nt, \qquad y(0) = 0, \quad y'(0) = 0,$$

where $\omega > 0$ is not equal to a positive integer. How is the solution altered if $\omega = m$, where m is a positive integer?

15. Find the formal solution of the initial value problem

$$y'' + \omega^2 y = f(t), \qquad y(0) = 0, \quad y'(0) = 0,$$

where f is periodic with period 2π and

$$f(t) = \begin{cases} 1, & 0 < t < \pi; \\ 0, & t = 0, \pi, 2\pi; \\ -1, & \pi < t < 2\pi. \end{cases}$$

See Problem 1.

16. Find the formal solution of the initial value problem

$$y'' + \omega^2 y = f(t), \qquad y(0) = 1, \quad y'(0) = 0,$$

where f is periodic with period 2 and

$$f(t) = \begin{cases} 1 - t, & 0 \leq t < 1; \\ -1 + t, & 1 \leq t < 2. \end{cases}$$

See Problem 8.

17. Assuming that

$$f(x) = \frac{a_0}{2} + \sum_{n=1}^{\infty} \left(a_n \cos \frac{n\pi x}{L} + b_n \sin \frac{n\pi x}{L} \right), \qquad \text{(i)}$$

show formally that

$$\frac{1}{L} \int_{-L}^{L} [f(x)]^2 \, dx = \frac{a_0^2}{2} + \sum_{n=1}^{\infty} (a_n^2 + b_n^2).$$

This relation between a function f and its Fourier coefficients is known as Parseval's[6] equation. This relation is very important in the theory of Fourier series; see Problem 9 in Section 11.6.

Hint: Multiply Eq. (i) by $f(x)$, integrate from $-L$ to L, and use the Euler–Fourier formulas.

[6]Marc-Antoine Parseval (1755–1836) was a relatively obscure French mathematician for whom an important result has been named. He presented a forerunner of this result in 1799, though not in the context of Fourier series.

18. This problem indicates a proof of convergence of a Fourier series under conditions more restrictive than those in Theorem 10.3.1.

(a) If f and f' are piecewise continuous on $-L \leq x < L$, and if f is periodic with period $2L$, show that na_n and nb_n are bounded as $n \to \infty$.
Hint: Use integration by parts.

(b) If f is continuous on $-L \leq x \leq L$ and periodic with period $2L$, and if f' and f'' are piecewise continuous on $-L \leq x < L$, show that $n^2 a_n$ and $n^2 b_n$ are bounded as $n \to \infty$. If f is continuous on the *closed* interval, then it is continuous for all x. Why is this important?
Hint: Again, use integration by parts.

(c) Using the result of part (b), show that $\sum_{n=1}^{\infty} |a_n|$ and $\sum_{n=1}^{\infty} |b_n|$ converge.

(d) From the result in part (c), show that the Fourier series (4) converges absolutely[7] for all x.

Acceleration of Convergence. In the next problem we show how it is sometimes possible to improve the speed of convergence of a Fourier series.

19. Suppose that we wish to calculate values of the function g, where

$$g(x) = \sum_{n=1}^{\infty} \frac{(2n-1)}{1 + (2n-1)^2} \sin(2n-1)\pi x. \tag{i}$$

It is possible to show that this series converges, albeit rather slowly. However, observe that for large n the terms in the series (i) are approximately equal to $[\sin(2n-1)\pi x]/(2n-1)$ and that the latter terms are similar to those in the example in the text, Eq. (6).

(a) Show that

$$\sum_{n=1}^{\infty} [\sin(2n-1)\pi x]/(2n-1) = (\pi/2)\left[f(x) - \tfrac{1}{2}\right], \tag{ii}$$

where f is the square wave in the example with $L = 1$.

(b) Subtract Eq. (ii) from Eq. (i) and show that

$$g(x) = \frac{\pi}{2}\left[f(x) - \tfrac{1}{2}\right] - \sum_{n=1}^{\infty} \frac{\sin(2n-1)\pi x}{(2n-1)[1 + (2n-1)^2]}. \tag{iii}$$

The series (iii) converges much faster than the series (i) and thus provides a better way to calculate values of $g(x)$.

10.4 Even and Odd Functions

Before looking at further examples of Fourier series it is useful to distinguish two classes of functions for which the Euler–Fourier formulas can be simplified. These are even and odd functions, which are characterized geometrically by the property of symmetry with respect to the y-axis and the origin, respectively (see Figure 10.4.1).

[7]It also converges uniformly; for an explanation of what this means, see a book on advanced calculus or analysis.

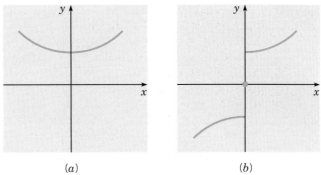

(a) (b)

FIGURE 10.4.1 (a) An even function. (b) An odd function.

Analytically, f is an **even function** if its domain contains the point $-x$ whenever it contains the point x, and if

$$f(-x) = f(x) \tag{1}$$

for each x in the domain of f. Similarly, f is an **odd function** if its domain contains $-x$ whenever it contains x, and if

$$f(-x) = -f(x) \tag{2}$$

for each x in the domain of f. Examples of even functions are 1, x^2, $\cos nx$, $|x|$, and x^{2n}. The functions x, x^3, $\sin nx$, and x^{2n+1} are examples of odd functions. Note that according to Eq. (2), $f(0)$ must be zero if f is an odd function whose domain contains the origin. Most functions are neither even nor odd, for instance, e^x. Only one function, f identically zero, is both even and odd.

Elementary properties of even and odd functions include the following:

1. The sum (difference) and product (quotient) of two even functions are even.
2. The sum (difference) of two odd functions is odd; the product (quotient) of two odd functions is even.
3. The sum (difference) of an odd function and an even function is neither even nor odd; the product (quotient) of two such functions is odd.[8]

The proofs of all these assertions are simple and follow directly from the definitions. For example, if both f_1 and f_2 are odd, and if $g(x) = f_1(x) + f_2(x)$, then

$$g(-x) = f_1(-x) + f_2(-x) = -f_1(x) - f_2(x)$$
$$= -[f_1(x) + f_2(x)] = -g(x), \tag{3}$$

so $f_1 + f_2$ is an odd function also. Similarly, if $h(x) = f_1(x)f_2(x)$, then

$$h(-x) = f_1(-x)f_2(-x) = [-f_1(x)][-f_2(x)] = f_1(x)f_2(x) = h(x), \tag{4}$$

so that $f_1 f_2$ is even.

Also of importance are the following two integral properties of even and odd functions:

[8]These statements may need to be modified if either function vanishes identically.

4. If f is an even function, then

$$\int_{-L}^{L} f(x)\,dx = 2\int_{0}^{L} f(x)\,dx. \tag{5}$$

5. If f is an odd function, then

$$\int_{-L}^{L} f(x)\,dx = 0. \tag{6}$$

These properties are intuitively clear from the interpretation of an integral in terms of area under a curve, and they also follow immediately from the definitions. For example, if f is even, then

$$\int_{-L}^{L} f(x)\,dx = \int_{-L}^{0} f(x)\,dx + \int_{0}^{L} f(x)\,dx.$$

Letting $x = -s$ in the first term on the right side and using Eq. (1), we obtain

$$\int_{-L}^{L} f(x)\,dx = -\int_{L}^{0} f(s)\,ds + \int_{0}^{L} f(x)\,dx = 2\int_{0}^{L} f(x)\,dx.$$

The proof of the corresponding property for odd functions is similar.

Even and odd functions are particularly important in applications of Fourier series since their Fourier series have special forms, which occur frequently in physical problems.

Cosine Series. Suppose that f and f' are piecewise continuous on $-L \le x < L$ and that f is an even periodic function with period $2L$. Then it follows from properties 1 and 3 that $f(x)\cos(n\pi x/L)$ is even and $f(x)\sin(n\pi x/L)$ is odd. As a consequence of Eqs. (5) and (6), the Fourier coefficients of f are then given by

$$a_n = \frac{2}{L}\int_{0}^{L} f(x)\cos\frac{n\pi x}{L}\,dx, \qquad n = 0, 1, 2, \ldots;$$
$$b_n = 0, \qquad n = 1, 2, \ldots. \tag{7}$$

Thus f has the Fourier series

$$f(x) = \frac{a_0}{2} + \sum_{n=1}^{\infty} a_n \cos\frac{n\pi x}{L}.$$

In other words, the Fourier series of any even function consists only of the even trigonometric functions $\cos(n\pi x/L)$ and the constant term; it is natural to call such a series a **Fourier cosine series**. From a computational point of view, observe that only the coefficients a_n, for $n = 0, 1, 2, \ldots$, need to be calculated from the integral formula (7). Each of the b_n, for $n = 1, 2, \ldots$, is automatically zero for any even function and so does not need to be calculated by integration.

Sine Series. Suppose that f and f' are piecewise continuous on $-L \le x < L$ and that f is an odd periodic function of period $2L$. Then it follows from properties 2 and 3

that $f(x)\cos(n\pi x/L)$ is odd and $f(x)\sin(n\pi x/L)$ is even. In this case the Fourier coefficients of f are

$$a_n = 0, \qquad n = 0,1,2,\ldots,$$

$$b_n = \frac{2}{L}\int_0^L f(x)\sin\frac{n\pi x}{L}\,dx, \qquad n = 1,2,\ldots, \tag{8}$$

and the Fourier series for f is of the form

$$f(x) = \sum_{n=1}^{\infty} b_n \sin\frac{n\pi x}{L}.$$

Thus the Fourier series for any odd function consists only of the odd trigonometric functions $\sin(n\pi x/L)$; such a series is called a **Fourier sine series**. Again observe that only half of the coefficients need to be calculated by integration, since each a_n, for $n = 0,1,2,\ldots$, is zero for any odd function.

EXAMPLE 1

Let $f(x) = x$, $-L < x < L$, and let $f(-L) = f(L) = 0$. Let f be defined elsewhere so that it is periodic of period $2L$ (see Figure 10.4.2). The function defined in this manner is known as a sawtooth wave. Find the Fourier series for this function.

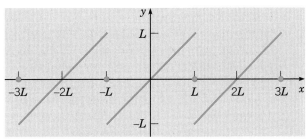

FIGURE 10.4.2 Sawtooth wave.

Since f is an odd function, its Fourier coefficients are, according to Eq. (8),

$$a_n = 0, \qquad n = 0,1,2,\ldots;$$

$$b_n = \frac{2}{L}\int_0^L x\sin\frac{n\pi x}{L}\,dx$$

$$= \frac{2}{L}\left(\frac{L}{n\pi}\right)^2\left\{\sin\frac{n\pi x}{L} - \frac{n\pi x}{L}\cos\frac{n\pi x}{L}\right\}\Bigg|_0^L$$

$$= \frac{2L}{n\pi}(-1)^{n+1}, \qquad n = 1,2,\ldots.$$

Hence the Fourier series for f, the sawtooth wave, is

$$f(x) = \frac{2L}{\pi} \sum_{n=1}^{\infty} \frac{(-1)^{n+1}}{n} \sin \frac{n\pi x}{L}. \tag{9}$$

Observe that the periodic function f is discontinuous at the points $\pm L, \pm 3L, \ldots$, as shown in Figure 10.4.2. At these points the series (9) converges to the mean value of the left and right limits, namely zero. The partial sum of the series (9) for $n = 9$ is shown in Figure 10.4.3. The Gibbs phenomenon (mentioned in Section 10.3) again occurs near the points of discontinuity.

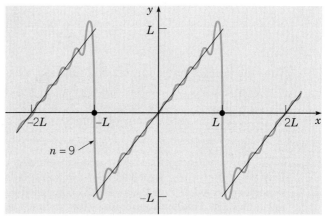

FIGURE 10.4.3 A partial sum in the Fourier series, Eq. (9), for the sawtooth wave.

Note that in this example $f(-L) = f(L) = 0$, as well as $f(0) = 0$. This is required if the function f is to be both odd and periodic with period $2L$. When we speak of constructing a sine series for a function defined on $0 \le x \le L$, it is understood that, if necessary, we must first redefine the function to be zero at $x = 0$ and $x = L$.

It is worthwhile to observe that the triangular wave function (Example 1 of Section 10.2) and the sawtooth wave function just considered are identical on the interval $0 \le x < L$. Therefore, their Fourier series converge to the same function, $f(x) = x$, on this interval. Thus, if it is required to represent the function $f(x) = x$ on $0 \le x < L$ by a Fourier series, it is possible to do this by *either a cosine series or a sine series*. In the former case f is extended as an *even* function into the interval $-L < x < 0$ and elsewhere periodically (the triangular wave). In the latter case f is extended into $-L < x < 0$ as an *odd* function and elsewhere periodically (the sawtooth wave). If f is extended in any other way, the resulting Fourier series will still converge to x in $0 \le x < L$ but will involve both sine and cosine terms.

In solving problems in differential equations it is often useful to expand in a Fourier series of period $2L$ a function f originally defined only on the interval $[0, L]$. As indicated previously for the function $f(x) = x$, several alternatives are available. Explicitly, we can

1. Define a function g of period $2L$ so that

$$g(x) = \begin{cases} f(x), & 0 \le x \le L, \\ f(-x), & -L < x < 0. \end{cases} \tag{10}$$

The function g is thus the even periodic extension of f. Its Fourier series, which is a cosine series, represents f on $[0, L]$.

2. Define a function h of period $2L$ so that

$$h(x) = \begin{cases} f(x), & 0 < x < L, \\ 0, & x = 0, L, \\ -f(-x), & -L < x < 0. \end{cases} \tag{11}$$

The function h is thus the odd periodic extension of f. Its Fourier series, which is a sine series, also represents f on $(0, L)$.

3. Define a function k of period $2L$ so that

$$k(x) = f(x), \qquad 0 \le x \le L, \tag{12}$$

and let $k(x)$ be defined for $(-L, 0)$ in any way consistent with the conditions of Theorem 10.3.1. Sometimes it is convenient to define $k(x)$ to be zero for $-L < x < 0$. The Fourier series for k, which involves both sine and cosine terms, also represents f on $[0, L]$, regardless of the manner in which $k(x)$ is defined in $(-L, 0)$. Thus there are infinitely many such series, all of which converge to $f(x)$ in the original interval.

Usually, the form of the expansion to be used will be dictated (or at least suggested) by the purpose for which it is needed. However, if there is a choice as to the kind of Fourier series to be used, the selection can sometimes be based on the rapidity of convergence. For example, the cosine series for the triangular wave [Eq. (20) of Section 10.2] converges more rapidly than the sine series for the sawtooth wave [Eq. (9) in this section], although both converge to the same function for $0 \le x < L$. This is because the triangular wave is a smoother function than the sawtooth wave and is therefore easier approximate. In general, the more continuous derivatives possessed by a function over the entire interval $-\infty < x < \infty$, the faster its Fourier series will converge. See Problem 18 of Section 10.3.

EXAMPLE 2

Suppose that

$$f(x) = \begin{cases} 1 - x, & 0 < x \le 1, \\ 0, & 1 < x \le 2. \end{cases} \tag{13}$$

As indicated previously, we can represent f either by a cosine series or by a sine series. Sketch the graph of the sum of each of these series for $-6 \le x \le 6$.

In this example $L = 2$, so the cosine series for f converges to the even periodic extension of f of period 4, whose graph is sketched in Figure 10.4.4.

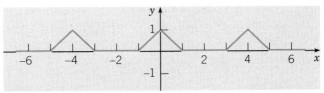

FIGURE 10.4.4 Even periodic extension of $f(x)$ given by Eq. (13).

Similarly, the sine series for f converges to the odd periodic extension of f of period 4. The graph of this function is shown in Figure 10.4.5.

FIGURE 10.4.5 Odd periodic extension of $f(x)$ given by Eq. (13).

PROBLEMS

In each of Problems 1 through 6 determine whether the given function is even, odd, or neither.

1. $x^3 - 2x$

2. $x^3 - 2x + 1$

3. $\tan 2x$

4. $\sec x$

5. $|x|^3$

6. e^{-x}

In each of Problems 7 through 12 a function f is given on an interval of length L. In each case sketch the graphs of the even and odd extensions of f of period $2L$.

7. $f(x) = \begin{cases} x, & 0 \le x < 2, \\ 1, & 2 \le x < 3 \end{cases}$

8. $f(x) = \begin{cases} 0, & 0 \le x < 1, \\ x - 1, & 1 \le x < 2 \end{cases}$

9. $f(x) = 2 - x, \quad 0 < x < 2$

10. $f(x) = x - 3, \quad 0 < x < 4$

11. $f(x) = \begin{cases} 0, & 0 \le x < 1, \\ 1, & 1 \le x < 2 \end{cases}$

12. $f(x) = 4 - x^2, \quad 0 < x < 1$

13. Prove that any function can be expressed as the sum of two other functions, one of which is even and the other odd. That is, for any function f, whose domain contains $-x$ whenever it contains x, show that there are an even function g and an odd function h such that $f(x) = g(x) + h(x)$.
 Hint: What can you say about $f(x) + f(-x)$?

14. Find the coefficients in the cosine and sine series described in Example 2.

In each of Problems 15 through 22 find the required Fourier series for the given function, and sketch the graph of the function to which the series converges over three periods.

15. $f(x) = \begin{cases} 1, & 0 < x < 1, \\ 0, & 1 < x < 2; \end{cases}$ cosine series, period 4
 Compare with Example 1 and Problem 5 of Section 10.3.

16. $f(x) = \begin{cases} x, & 0 \le x < 1, \\ 1, & 1 \le x < 2; \end{cases}$ sine series, period 4

17. $f(x) = 1, \quad 0 \le x \le \pi;$ cosine series, period 2π

18. $f(x) = 1, \quad 0 < x < \pi;$ sine series, period 2π

19. $f(x) = \begin{cases} 0, & 0 < x < \pi, \\ 1, & \pi < x < 2\pi, \\ 2, & 2\pi < x < 3\pi; \end{cases}$ sine series, period 6π

20. $f(x) = x$, $0 \le x < 1$; series of period 1

21. $f(x) = L - x$, $0 \le x \le L$; cosine series, period $2L$
 Compare with Example 1 of Section 10.2.

22. $f(x) = L - x$, $0 < x < L$; sine series, period $2L$

In each of Problems 23 through 26:

(a) Find the required Fourier series for the given function.

(b) Sketch the graph of the function to which the series converges for three periods.

(c) Plot one or more partial sums of the series.

23. $f(x) = \begin{cases} x, & 0 < x < \pi, \\ 0, & \pi < x < 2\pi; \end{cases}$ cosine series, period 4π

24. $f(x) = -x$, $-\pi < x < 0$; sine series, period 2π

25. $f(x) = 2 - x^2$, $0 < x < 2$; sine series, period 4

26. $f(x) = x^2 - 2x$, $0 < x < 4$; cosine series, period 8

In each of Problems 27 through 30 a function is given on an interval $0 < x < L$.

(a) Sketch the graphs of the even extension $g(x)$ and the odd extension $h(x)$ of the given function of period $2L$ over three periods.

(b) Find the Fourier cosine and sine series for the given function.

(c) Plot a few partial sums of each series.

(d) For each series investigate the dependence on n of the maximum error on $[0, L]$.

27. $f(x) = 3 - x$, $0 < x < 3$

28. $f(x) = \begin{cases} x, & 0 < x < 1, \\ 0, & 1 < x < 2 \end{cases}$

29. $f(x) = (4x^2 - 4x - 3)/4$, $0 < x < 2$

30. $f(x) = x^3 - 5x^2 + 5x + 1$, $0 < x < 3$

31. Prove that if f is an odd function, then

$$\int_{-L}^{L} f(x)\, dx = 0.$$

32. Prove properties 2 and 3 of even and odd functions, as stated in the text.

33. Prove that the derivative of an even function is odd and that the derivative of an odd function is even.

34. Let $F(x) = \int_{0}^{x} f(t)\, dt$. Show that if f is even, then F is odd, and that if f is odd, then F is even.

35. From the Fourier series for the square wave in Example 1 of Section 10.3, show that

$$\frac{\pi}{4} = 1 - \frac{1}{3} + \frac{1}{5} - \frac{1}{7} + \cdots = \sum_{n=0}^{\infty} \frac{(-1)^n}{2n+1}.$$

This relation between π and the odd positive integers was discovered by Leibniz in 1674.

36. From the Fourier series for the triangular wave (Example 1 of Section 10.2), show that

$$\frac{\pi^2}{8} = 1 + \frac{1}{3^2} + \frac{1}{5^2} + \cdots = \sum_{n=0}^{\infty} \frac{1}{(2n+1)^2}.$$

37. Assume that f has a Fourier sine series

$$f(x) = \sum_{n=1}^{\infty} b_n \sin(n\pi x/L), \qquad 0 \leq x \leq L.$$

(a) Show formally that

$$\frac{2}{L} \int_0^L [f(x)]^2 \, dx = \sum_{n=1}^{\infty} b_n^2.$$

Compare this result (Parseval's equation) with that of Problem 17 in Section 10.3. What is the corresponding result if f has a cosine series?

(b) Apply the result of part (a) to the series for the sawtooth wave given in Eq. (9), and thereby show that

$$\frac{\pi^2}{6} = 1 + \frac{1}{2^2} + \frac{1}{3^2} + \cdots = \sum_{n=1}^{\infty} \frac{1}{n^2}.$$

This relation was discovered by Euler about 1735.

More Specialized Fourier Series. Let f be a function originally defined on $0 \leq x \leq L$ and satisfying there the continuity conditions of Theorem 10.3.1. In this section we have shown that it is possible to represent f by either a sine series or a cosine series by constructing odd or even periodic extensions of f, respectively. Problems 38 through 40 concern some other, more specialized Fourier series that converge to the given function f on $(0, L)$.

38. Let f be extended into $(L, 2L]$ in an arbitrary manner. Then extend the resulting function into $(-2L, 0)$ as an odd function and elsewhere as a periodic function of period $4L$ (see Figure 10.4.6). Show that this function has a Fourier sine series in terms of the functions $\sin(n\pi x/2L), n = 1, 2, 3, \ldots$; that is,

$$f(x) = \sum_{n=1}^{\infty} b_n \sin(n\pi x/2L),$$

where

$$b_n = \frac{1}{L} \int_0^{2L} f(x) \sin(n\pi x/2L) \, dx.$$

This series converges to the original function on $(0, L)$.

39. Let f first be extended into $(L, 2L)$ so that it is symmetric about $x = L$; that is, so as to satisfy $f(2L - x) = f(x)$ for $0 \leq x < L$. Let the resulting function be extended into $(-2L, 0)$ as an odd function and elsewhere (see Figure 10.4.7) as a periodic function of period $4L$. Show that this function has a Fourier series in terms of the functions $\sin(\pi x/2L), \sin(3\pi x/2L), \sin(5\pi x/2L), \ldots$; that is,

$$f(x) = \sum_{n=1}^{\infty} b_n \sin \frac{(2n-1)\pi x}{2L},$$

where

$$b_n = \frac{2}{L} \int_0^L f(x) \sin \frac{(2n-1)\pi x}{2L} \, dx.$$

This series converges to the original function on $(0, L]$.

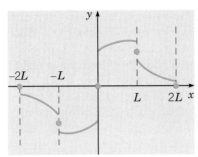

FIGURE 10.4.6 Graph of the function in Problem 38.

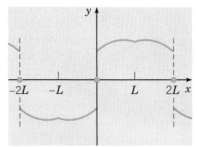

FIGURE 10.4.7 Graph of the function in Problem 39.

40. How should f, originally defined on $[0, L]$, be extended so as to obtain a Fourier series involving only the functions $\cos(\pi x/2L), \cos(3\pi x/2L), \cos(5\pi x/2L), \ldots$? Refer to Problems 38 and 39. If $f(x) = x$ for $0 \le x \le L$, sketch the function to which the Fourier series converges for $-4L \le x \le 4L$.

10.5 Separation of Variables; Heat Conduction in a Rod

The basic partial differential equations of heat conduction, wave propagation, and potential theory that we discuss in this chapter are associated with three distinct types of physical phenomena: diffusive processes, oscillatory processes, and time-independent or steady processes. Consequently, they are of fundamental importance in many branches of physics. They are also of considerable significance from a mathematical point of view. The partial differential equations whose theory is best developed and whose applications are most significant and varied are the linear equations of second order. All such equations can be classified into one of three categories: The heat conduction equation, the wave equation, and the potential equation, respectively, are prototypes of each of these categories. Thus a study of these three equations yields much information about more general second order linear partial differential equations.

During the last two centuries several methods have been developed for solving partial differential equations. The method of separation of variables is the oldest systematic method, having been used by D'Alembert, Daniel Bernoulli, and Euler about 1750 in their investigations of waves and vibrations. It has been considerably refined and generalized in the meantime, and it remains a method of great importance and frequent use today. To show how the method of separation of variables works we consider first a basic problem of heat conduction in a solid body. The mathematical study of heat conduction originated[9] about 1800, and it continues to command the

[9]The first important investigation of heat conduction was carried out by Joseph Fourier (1768–1830) while he was serving as prefect of the department of Isère (Grenoble) from 1801 to 1815. He presented basic papers on the subject to the Academy of Sciences of Paris in 1807 and 1811. However, these papers were criticized by the referees (principally Lagrange) for lack of rigor and so were not published. Fourier continued to develop his ideas and eventually wrote one of the classics of applied mathematics, *Théorie analytique de la chaleur*, published in 1822.

attention of modern scientists. For example, analysis of the dissipation and transfer of heat away from its sources in high-speed machinery is frequently an important technological problem.

Let us now consider a heat conduction problem for a straight bar of uniform cross section and homogeneous material. Let the x-axis be chosen to lie along the axis of the bar, and let $x = 0$ and $x = L$ denote the ends of the bar (see Figure 10.5.1). Suppose further that the sides of the bar are perfectly insulated so that no heat passes through them. We also assume that the cross-sectional dimensions are so small that the temperature u can be considered constant on any given cross section. Then u is a function only of the axial coordinate x and the time t.

$u(x, t)$

$x = 0$ $x = L$

FIGURE 10.5.1 A heat-conducting solid bar.

The variation of temperature in the bar is governed by a partial differential equation whose derivation appears in Appendix A at the end of this chapter. The equation is called the **heat conduction equation** and has the form

$$\alpha^2 u_{xx} = u_t, \qquad 0 < x < L, \qquad t > 0, \tag{1}$$

where α^2 is a constant known as the **thermal diffusivity**. The parameter α^2 depends only on the material from which the bar is made and is defined by

$$\alpha^2 = \kappa/\rho s, \tag{2}$$

where κ is the thermal conductivity, ρ is the density, and s is the specific heat of the material in the bar. The units of α^2 are (length)2/time. Typical values of α^2 are given in Table 10.5.1.

TABLE 10.5.1 Values of the Thermal Diffusivity for Some Common Materials

Material	α^2 (cm^2/sec)
Silver	1.71
Copper	1.14
Aluminum	0.86
Cast iron	0.12
Granite	0.011
Brick	0.0038
Water	0.00144

In addition, we assume that the initial temperature distribution in the bar is given; thus

$$u(x, 0) = f(x), \qquad 0 \le x \le L, \tag{3}$$

where f is a given function. Finally, we assume that the ends of the bar are held at fixed temperatures: the temperature T_1 at $x = 0$ and the temperature T_2 at $x = L$. However, it turns out that we need only consider the case where $T_1 = T_2 = 0$. We show in Section 10.6 how to reduce the more general problem to this special case. Thus in this section we will assume that u is always zero when $x = 0$ or $x = L$:

$$u(0, t) = 0, \qquad u(L, t) = 0, \qquad t > 0. \tag{4}$$

The fundamental problem of heat conduction is to find $u(x, t)$ that satisfies the differential equation (1) for $0 < x < L$ and for $t > 0$, the initial condition (3) when $t = 0$, and the boundary conditions (4) at $x = 0$ and $x = L$.

The problem described by Eqs. (1), (3), and (4) is an initial value problem in the time variable t; an initial condition is given and the differential equation governs what happens later. However, with respect to the space variable x, the problem is a boundary value problem; boundary conditions are imposed at each end of the bar and the differential equation describes the evolution of the temperature in the interval between them. Alternatively, we can consider the problem as a boundary value problem in the xt-plane (see Figure 10.5.2). The solution $u(x, t)$ of Eq. (1) is sought in the semi-infinite strip $0 < x < L, t > 0$, subject to the requirement that $u(x, t)$ must assume a prescribed value at each point on the boundary of this strip.

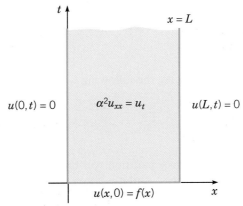

FIGURE 10.5.2 Boundary value problem for the heat conduction equation.

The heat conduction problem (1), (3), (4) is *linear* since u appears only to the first power throughout. The differential equation and boundary conditions are also *homogeneous*. This suggests that we might approach the problem by seeking solutions of the differential equation and boundary conditions, and then superposing them to satisfy the initial condition. The remainder of this section describes how this plan can be implemented.

One solution of the differential equation (1) that satisfies the boundary conditions (4) is the function $u(x,t) = 0$, but this solution does not satisfy the initial condition (3) except in the trivial case in which $f(x)$ is also zero. Thus our goal is to find other, nonzero solutions of the differential equation and boundary conditions. To find the needed solutions we start by making a basic assumption about the form of the solutions that has far-reaching, and perhaps unforeseen, consequences. The assumption is that $u(x,t)$ is a product of two other functions, one depending only on x and the other depending only on t; thus

$$u(x,t) = X(x)T(t). \tag{5}$$

Substituting from Eq. (5) for u in the differential equation (1) yields

$$\alpha^2 X''T = XT', \tag{6}$$

where primes refer to ordinary differentiation with respect to the independent variable, whether x or t. Equation (6) is equivalent to

$$\frac{X''}{X} = \frac{1}{\alpha^2}\frac{T'}{T}, \tag{7}$$

in which the variables are separated; that is, the left side depends only on x and the right side only on t. For Eq. (7) to be valid for $0 < x < L, t > 0$, it is necessary that both sides of Eq. (7) be equal to the same constant. Otherwise, if one independent variable (say x) were kept fixed and the other were allowed to vary, one side (the left in this case) of Eq. (7) would remain unchanged while the other varied, thus violating the equality. If we call this separation constant $-\lambda$, then Eq. (7) becomes

$$\frac{X''}{X} = \frac{1}{\alpha^2}\frac{T'}{T} = -\lambda. \tag{8}$$

Hence we obtain the following two ordinary differential equations for $X(x)$ and $T(t)$:

$$X'' + \lambda X = 0, \tag{9}$$

$$T' + \alpha^2\lambda T = 0. \tag{10}$$

We denote the separation constant by $-\lambda$ (rather than λ) because it turns out that it must be negative, and it is convenient to exhibit the minus sign explicitly.

The assumption (5) has led to the replacement of the partial differential equation (1) by the two ordinary differential equations (9) and (10). Each of these equations can be readily solved for *any* value of λ. The product of two solutions of Eq. (9) and (10), respectively, provides a solution of the partial differential equation (1). However, we are interested only in those solutions of Eq. (1) that also satisfy the boundary conditions (4). As we now show, this severely restricts the possible values of λ.

Substituting for $u(x,t)$ from Eq. (5) in the boundary condition at $x = 0$, we obtain

$$u(0,t) = X(0)T(t) = 0. \tag{11}$$

If Eq. (11) is satisfied by choosing $T(t)$ to be zero for all t, then $u(x, t)$ is zero for all x and t, and we have already rejected this possibility. Therefore Eq. (11) must be satisfied by requiring that

$$X(0) = 0. \tag{12}$$

Similarly, the boundary condition at $x = L$ requires that

$$X(L) = 0. \tag{13}$$

We now want to consider Eq. (9) subject to the boundary conditions (12) and (13). This is an eigenvalue problem and, in fact, is the same problem that we discussed in detail at the end of Section 10.1; see especially the paragraph following Eq. (29) in that section. The only difference is that the dependent variable there was called y rather than X. If we refer to the results obtained earlier [Eq. (31) of Section 10.1], the only nontrivial solutions of Eqs. (9), (12), and (13) are the eigenfunctions

$$X_n(x) = \sin(n\pi x/L), \qquad n = 1, 2, 3, \ldots \tag{14}$$

associated with the eigenvalues

$$\lambda_n = n^2\pi^2/L^2, \qquad n = 1, 2, 3, \ldots. \tag{15}$$

Turning now to Eq. (10) for $T(t)$ and substituting $n^2\pi^2/L^2$ for λ, we have

$$T' + (n^2\pi^2\alpha^2/L^2)T = 0. \tag{16}$$

Thus $T(t)$ is proportional to $\exp(-n^2\pi^2\alpha^2 t/L^2)$. Hence, multiplying solutions of Eqs. (9) and (10) together, and neglecting arbitrary constants of proportionality, we conclude that the functions

$$u_n(x, t) = e^{-n^2\pi^2\alpha^2 t/L^2} \sin(n\pi x/L), \qquad n = 1, 2, 3, \ldots \tag{17}$$

satisfy the partial differential equation (1) and the boundary conditions (4) for each positive integer value of n. The functions u_n are sometimes called fundamental solutions of the heat conduction problem (1), (3), and (4).

It remains only to satisfy the initial condition (3),

$$u(x, 0) = f(x), \qquad 0 \le x \le L. \tag{18}$$

Recall that we have often solved initial value problems by forming linear combinations of a set of fundamental solutions and then choosing the coefficients to satisfy the initial conditions. The analogous step in the present problem is to form a linear combination of the functions (17) and then to choose the coefficients to satisfy Eq. (18). The main difference from earlier problems is that there are infinitely many functions (17), so a general linear combination of them is an infinite series. Thus we assume that

$$u(x, t) = \sum_{n=1}^{\infty} c_n u_n(x, t) = \sum_{n=1}^{\infty} c_n e^{-n^2\pi^2\alpha^2 t/L^2} \sin\frac{n\pi x}{L}, \tag{19}$$

where the coefficients c_n are as yet undetermined. The individual terms in the series (19) satisfy the differential equation (1) and boundary conditions (4). We will assume that the infinite series of Eq. (19) converges and also satisfies Eqs. (1) and (4). To satisfy the initial condition (3) we must have

$$u(x,0) = \sum_{n=1}^{\infty} c_n \sin \frac{n\pi x}{L} = f(x). \tag{20}$$

In other words, we need to choose the coefficients c_n so that the series of sine functions in Eq. (20) converges to the initial temperature distribution $f(x)$ for $0 \le x \le L$. The series in Eq. (20) is just the Fourier sine series for f; according to Eq. (8) of Section 10.4 its coefficients are given by

$$c_n = \frac{2}{L} \int_0^L f(x) \sin \frac{n\pi x}{L} \, dx. \tag{21}$$

Hence the solution of the heat conduction problem of Eqs. (1), (3), and (4) is given by the series in Eq. (19) with the coefficients computed from Eq. (21).

EXAMPLE 1

Find the temperature $u(x,t)$ at any time in a metal rod 50 cm long, insulated on the sides, which initially has a uniform temperature of 20°C throughout and whose ends are maintained at 0°C for all $t > 0$.

The temperature in the rod satisfies the heat conduction problem (1), (3), (4) with $L = 50$ and $f(x) = 20$ for $0 < x < 50$. Thus, from Eq. (19), the solution is

$$u(x,t) = \sum_{n=1}^{\infty} c_n e^{-n^2 \pi^2 \alpha^2 t / 2500} \sin \frac{n\pi x}{50}, \tag{22}$$

where, from Eq. (21),

$$c_n = \frac{4}{5} \int_0^{50} \sin \frac{n\pi x}{50} \, dx$$

$$= \frac{40}{n\pi} (1 - \cos n\pi) = \begin{cases} 80/n\pi, & n \text{ odd}; \\ 0, & n \text{ even}. \end{cases} \tag{23}$$

Finally, by substituting for c_n in Eq. (22), we obtain

$$u(x,t) = \frac{80}{\pi} \sum_{n=1,3,5,\dots}^{\infty} \frac{1}{n} e^{-n^2 \pi^2 \alpha^2 t / 2500} \sin \frac{n\pi x}{50}. \tag{24}$$

The expression (24) for the temperature is moderately complicated, but the negative exponential factor in each term of the series causes the series to converge quite rapidly, except for small values of t or α^2. Therefore accurate results can usually be obtained by using only a few terms of the series.

In order to display quantitative results, let us measure t in seconds; then α^2 has the units of cm²/sec. If we choose $\alpha^2 = 1$ for convenience, this corresponds to a rod of a material

whose thermal properties are somewhere between copper and aluminum. The behavior of the solution can be seen from the graphs in Figures 10.5.3 through 10.5.5. In Figure 10.5.3 we show the temperature distribution in the bar at several different times. Observe that the temperature diminishes steadily as heat in the bar is lost through the end points. The way in which the temperature decays at a given point in the bar is indicated in Figure 10.5.4, where temperature is plotted against time for a few selected points in the bar. Finally, Figure 10.5.5 is a three-dimensional plot of u versus both x and t. Observe that we obtain the graphs in Figures 10.5.3 and 10.5.4 by intersecting the surface in Figure 10.5.5 by planes on which either t or x is constant. The slight waviness in Figure 10.5.5 at $t = 0$ results from using only a finite number of terms in the series for $u(x, t)$ and from the slow convergence of the series for $t = 0$.

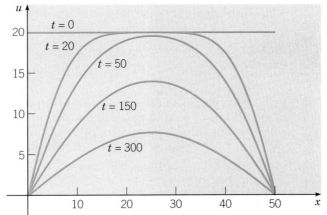

FIGURE 10.5.3 Temperature distributions at several times for the heat conduction problem of Example 1.

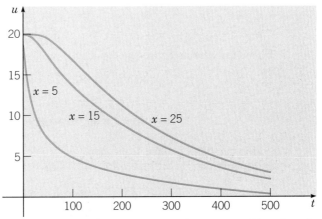

FIGURE 10.5.4 Dependence of temperature on time at several locations for the heat conduction problem of Example 1.

A problem with possible practical implications is to determine the time τ at which the entire bar has cooled to a specified temperature. For example, when is the temperature in the entire bar no greater than $1°C$? Because of the symmetry of the initial temperature distribution and

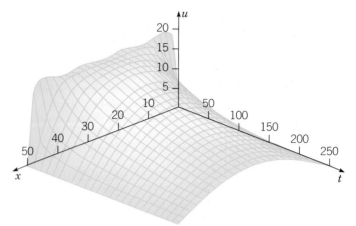

FIGURE 10.5.5 Plot of temperature u versus x and t for the heat conduction problem of Example 1.

the boundary conditions, the warmest point in the bar is always the center. Thus τ is found by solving $u(25, t) = 1$ for t. Using one term in the series expansion (24), we obtain

$$\tau = \frac{2500}{\pi^2} \ln(80/\pi) \cong 820 \text{ sec.}$$

PROBLEMS

In each of Problems 1 through 6 determine whether the method of separation of variables can be used to replace the given partial differential equation by a pair of ordinary differential equations. If so, find the equations.

1. $xu_{xx} + u_t = 0$

2. $tu_{xx} + xu_t = 0$

3. $u_{xx} + u_{xt} + u_t = 0$

4. $[p(x)u_x]_x - r(x)u_{tt} = 0$

5. $u_{xx} + (x + y)u_{yy} = 0$

6. $u_{xx} + u_{yy} + xu = 0$

7. Find the solution of the heat conduction problem

$$100u_{xx} = u_t, \qquad 0 < x < 1, \quad t > 0;$$

$$u(0, t) = 0, \qquad u(1, t) = 0, \quad t > 0;$$

$$u(x, 0) = \sin 2\pi x - \sin 5\pi x, \qquad 0 \le x \le 1.$$

8. Find the solution of the heat conduction problem

$$u_{xx} = 4u_t, \qquad 0 < x < 2, \quad t > 0;$$

$$u(0, t) = 0, \qquad u(2, t) = 0, \quad t > 0;$$

$$u(x, 0) = 2 \sin(\pi x/2) - \sin \pi x + 4 \sin 2\pi x, \qquad 0 \le x \le 2.$$

Consider the conduction of heat in a rod 40 cm in length whose ends are maintained at $0°C$ for all $t > 0$. In each of Problems 9 through 12 find an expression for the temperature $u(x, t)$ if the initial temperature distribution in the rod is the given function. Suppose that $\alpha^2 = 1$.

9. $u(x, 0) = 50, \qquad 0 < x < 40$

10. $u(x,0) = \begin{cases} x, & 0 \le x < 20, \\ 40 - x, & 20 \le x \le 40 \end{cases}$

11. $u(x,0) = \begin{cases} 0, & 0 \le x < 10, \\ 50, & 10 \le x \le 30, \\ 0, & 30 < x \le 40 \end{cases}$

12. $u(x,0) = x, \qquad 0 < x < 40$

13. Consider again the rod in Problem 9. For $t = 5$ and $x = 20$ determine how many terms are needed to find the solution correct to three decimal places. A reasonable way to do this is to find n so that including one more term does not change the first three decimal places of $u(20,5)$. Repeat for $t = 20$ and $t = 80$. Form a conclusion about the speed of convergence of the series for $u(x,t)$.

14. For the rod in Problem 9:

(a) Plot u versus x for $t = 5, 10, 20, 40, 100,$ and 200. Put all of the graphs on the same set of axes and thereby obtain a picture of the way in which the temperature distribution changes with time.

(b) Plot u versus t for $x = 5, 10, 15,$ and 20.

(c) Draw a three-dimensional plot of u versus x and t.

(d) How long does it take for the entire rod to cool off to a temperature of no more than $1°C$?

15. Follow the instructions in Problem 14 for the rod in Problem 10.

16. Follow the instructions in Problem 14 for the rod in Problem 11.

17. For the rod in Problem 12:

(a) Plot u versus x for $t = 5, 10, 20, 40, 100,$ and 200.

(b) For each value of t used in part (a) estimate the value of x for which the temperature is greatest. Plot these values versus t to see how the location of the warmest point in the rod changes with time.

(c) Plot u versus t for $x = 10, 20,$ and 30.

(d) Draw a three-dimensional plot of u versus x and t.

(e) How long does it take for the entire rod to cool off to a temperature of no more than $1°C$?

18. Let a metallic rod 20 cm long be heated to a uniform temperature of $100°C$. Suppose that at $t = 0$ the ends of the bar are plunged into an ice bath at $0°C$, and thereafter maintained at this temperature, but that no heat is allowed to escape through the lateral surface. Find an expression for the temperature at any point in the bar at any later time. Determine the temperature at the center of the bar at time $t = 30$ sec if the bar is made of (a) silver, (b) aluminum, or (c) cast iron.

19. For the rod of Problem 18 find the time that will elapse before the center of the bar cools to a temperature of $5°C$ if the bar is made of (a) silver, (b) aluminum, or (c) cast iron.

20. In solving differential equations the computations can almost always be simplified by the use of **dimensionless variables**. Show that if the dimensionless variable $\xi = x/L$ is introduced, the heat conduction equation becomes

$$\frac{\partial^2 u}{\partial \xi^2} = \frac{L^2}{\alpha^2} \frac{\partial u}{\partial t}, \qquad 0 < \xi < 1, \quad t > 0.$$

Bar with Insulated Ends. A slightly different problem occurs if the ends of the bar are insulated so that there is no passage of heat through them. According to Eq. (2) in Appendix A, the rate of flow of heat across a cross section is proportional to the rate of change of temperature in the x direction. Thus, in the case of no heat flow, the boundary conditions are

$$u_x(0,t) = 0, \qquad u_x(L,t) = 0, \qquad t > 0. \tag{24}$$

The problem posed by Eqs. (1), (3), and (24) can also be solved by the method of separation of variables. If we let

$$u(x,t) = X(x)T(t), \tag{25}$$

and substitute for u in Eq. (1), then it follows, as in Section 10.5, that

$$\frac{X''}{X} = \frac{1}{\alpha^2}\frac{T'}{T} = -\lambda, \tag{26}$$

where λ is a constant. Thus we obtain again the two ordinary differential equations

$$X'' + \lambda X = 0, \tag{27}$$
$$T' + \alpha^2 \lambda T = 0. \tag{28}$$

For any value of λ a product of solutions of Eqs. (27) and (28) is a solution of the partial differential equation (1). However, we are interested only in those solutions that also satisfy the boundary conditions (24).

If we substitute for $u(x,t)$ from Eq. (25) in the boundary condition at $x = 0$, we obtain $X'(0)T(t) = 0$. We cannot permit $T(t)$ to be zero for all t, since then $u(x,t)$ would also be zero for all t. Hence we must have

$$X'(0) = 0. \tag{29}$$

Proceeding in the same way with the boundary condition at $x = L$, we find that

$$X'(L) = 0. \tag{30}$$

Thus we wish to solve Eq. (27) subject to the boundary conditions (29) and (30). It is possible to show that nontrivial solutions of this problem can exist only if λ is real. One way to show this is indicated in Problem 18; alternatively, we can appeal to a more general theory to be discussed in Section 11.2. We will assume that λ is real and will consider in turn the three cases $\lambda < 0, \lambda = 0,$ and $\lambda > 0$.

If $\lambda < 0$, it is convenient to let $\lambda = -\mu^2$, where μ is real and positive. Then Eq. (27) becomes $X'' - \mu^2 X = 0$, and its general solution is

$$X(x) = k_1 \sinh \mu x + k_2 \cosh \mu x. \tag{31}$$

In this case the boundary conditions can be satisfied only by choosing $k_1 = k_2 = 0$. Since this is unacceptable, it follows that λ cannot be negative; in other words, the problem (27), (29), (30) has no negative eigenvalues.

If $\lambda = 0$, then Eq. (27) is $X'' = 0$, and therefore

$$X(x) = k_1 x + k_2. \tag{32}$$

The boundary conditions (29) and (30) require that $k_1 = 0$ but do not determine k_2. Thus $\lambda = 0$ is an eigenvalue, corresponding to the eigenfunction $X(x) = 1$. For $\lambda = 0$ it follows from Eq. (28) that $T(t)$ is also a constant, which can be combined with k_2. Hence, for $\lambda = 0$, we obtain the constant solution $u(x, t) = k_2$.

Finally, if $\lambda > 0$, let $\lambda = \mu^2$, where μ is real and positive. Then Eq. (27) becomes $X'' + \mu^2 X = 0$, and consequently

$$X(x) = k_1 \sin \mu x + k_2 \cos \mu x. \tag{33}$$

The boundary condition (29) requires that $k_1 = 0$, and the boundary condition (30) requires that $\mu = n\pi/L$ for $n = 1, 2, 3, \ldots$ but leaves k_2 arbitrary. Thus the problem (27), (29), (30) has an infinite sequence of positive eigenvalues $\lambda = n^2 \pi^2/L^2$ with the corresponding eigenfunctions $X(x) = \cos(n\pi x/L)$. For these values of λ the solutions $T(t)$ of Eq. (28) are proportional to $\exp(-n^2 \pi^2 \alpha^2 t/L^2)$.

Combining all these results, we have the following fundamental solutions for the problem (1), (3), and (24):

$$u_0(x, t) = 1,$$
$$u_n(x, t) = e^{-n^2 \pi^2 \alpha^2 t/L^2} \cos \frac{n\pi x}{L}, \qquad n = 1, 2, \ldots, \tag{34}$$

where arbitrary constants of proportionality have been dropped. Each of these functions satisfies the differential equation (1) and the boundary conditions (24). Because both the differential equation and the boundary conditions are linear and homogeneous, any finite linear combination of the fundamental solutions satisfies them. We will assume that this is true for convergent infinite linear combinations of fundamental solutions as well. Thus, to satisfy the initial condition (3), we assume that $u(x, t)$ has the form

$$u(x, t) = \frac{c_0}{2} u_0(x, t) + \sum_{n=1}^{\infty} c_n u_n(x, t)$$

$$= \frac{c_0}{2} + \sum_{n=1}^{\infty} c_n e^{-n^2 \pi^2 \alpha^2 t/L^2} \cos \frac{n\pi x}{L}. \tag{35}$$

The coefficients c_n are determined by the requirement that

$$u(x, 0) = \frac{c_0}{2} + \sum_{n=1}^{\infty} c_n \cos \frac{n\pi x}{L} = f(x). \tag{36}$$

Thus the unknown coefficients in Eq. (35) must be the coefficients in the Fourier cosine series of period $2L$ for f. Hence

$$c_n = \frac{2}{L} \int_0^L f(x) \cos \frac{n\pi x}{L} \, dx, \qquad n = 0, 1, 2, \ldots. \tag{37}$$

With this choice of the coefficients c_0, c_1, c_2, \ldots, the series (35) provides the solution to the heat conduction problem for a rod with insulated ends, Eqs. (1), (3), and (24).

It is worth observing that the solution (35) can also be thought of as the sum of a steady-state temperature distribution (given by the constant $c_0/2$), which is independent of time t, and a transient distribution (given by the rest of the infinite series) that vanishes in the limit as t approaches infinity. That the steady state is a constant is consistent with the expectation that the process of heat conduction will gradually smooth out the temperature distribution in the bar as long as no heat is allowed to escape to the outside. The physical interpretation of the term

$$\frac{c_0}{2} = \frac{1}{L} \int_0^L f(x)\, dx \tag{38}$$

is that it is the mean value of the original temperature distribution.

EXAMPLE 2

Find the temperature $u(x, t)$ in a metal rod of length 25 cm that is insulated on the ends as well as on the sides and whose initial temperature distribution is $u(x, 0) = x$ for $0 < x < 25$.

The temperature in the rod satisfies the heat conduction problem (1), (3), (24) with $L = 25$. Thus, from Eq. (35), the solution is

$$u(x, t) = \frac{c_0}{2} + \sum_{n=1}^{\infty} c_n e^{-n^2 \pi^2 \alpha^2 t/625} \cos \frac{n\pi x}{25}, \tag{39}$$

where the coefficients are determined from Eq. (37). We have

$$c_0 = \frac{2}{25} \int_0^{25} x\, dx = 25 \tag{40}$$

and, for $n \geq 1$,

$$c_n = \frac{2}{25} \int_0^{25} x \cos \frac{n\pi x}{25}\, dx$$

$$= 50(\cos n\pi - 1)/(n\pi)^2 = \begin{cases} -100/(n\pi)^2, & n \text{ odd}; \\ 0, & n \text{ even}. \end{cases} \tag{41}$$

Thus

$$u(x, t) = \frac{25}{2} - \frac{100}{\pi^2} \sum_{n=1,3,5,\ldots}^{\infty} \frac{1}{n^2} e^{-n^2 \pi^2 \alpha^2 t/625} \cos(n\pi x/25) \tag{42}$$

is the solution of the given problem.

For $\alpha^2 = 1$, Figure 10.6.2 shows plots of the temperature distribution in the bar at several times. Again the convergence of the series is rapid so that only a relatively few terms are needed to generate the graphs.

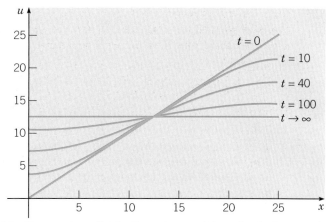

FIGURE 10.6.2 Temperature distributions at several times for the heat conduction problem of Example 2.

More General Problems. The method of separation of variables can also be used to solve heat conduction problems with other boundary conditions than those given by Eqs. (8) and Eqs. (24). For example, the left end of the bar might be held at a fixed temperature T while the other end is insulated. In this case the boundary conditions are

$$u(0,t) = T, \qquad u_x(L,t) = 0, \quad t > 0. \tag{43}$$

The first step in solving this problem is to reduce the given boundary conditions to homogeneous ones by subtracting the steady-state solution. The resulting problem is solved by essentially the same procedure as in the problems previously considered. However, the extension of the initial function f outside of the interval $[0, L]$ is somewhat different from that in any case considered so far (see Problem 15).

A more general type of boundary condition occurs when the rate of heat flow through the end of the bar is proportional to the temperature. It is shown in Appendix A that the boundary conditions in this case are of the form

$$u_x(0,t) - h_1 u(0,t) = 0, \qquad u_x(L,t) + h_2 u(L,t) = 0, \quad t > 0, \tag{44}$$

where h_1 and h_2 are nonnegative constants. If we apply the method of separation of variables to the problem consisting of Eqs. (1), (3), and (44), we find that $X(x)$ must be a solution of

$$X'' + \lambda X = 0, \qquad X'(0) - h_1 X(0) = 0, \qquad X'(L) + h_2 X(L) = 0, \tag{45}$$

where λ is the separation constant. Once again it is possible to show that nontrivial solutions can exist only for certain nonnegative real values of λ, the eigenvalues, but these values are not given by a simple formula (see Problem 20). It is also possible to show that the corresponding solutions of Eqs. (45), the eigenfunctions, satisfy an orthogonality relation and that one can satisfy the initial condition (3) by superposing solutions of Eqs. (45). However, the resulting series is not included in the discussion of this chapter. There is more general theory that covers such problems, and it is outlined in Chapter 11.

PROBLEMS In each of Problems 1 through 8 find the steady-state solution of the heat conduction equation
 $\alpha^2 u_{xx} = u_t$ that satisfies the given set of boundary conditions.

1. $u(0,t) = 10$, $u(50,t) = 40$ 2. $u(0,t) = 30$, $u(40,t) = -20$

3. $u_x(0,t) = 0$, $u(L,t) = 0$ 4. $u_x(0,t) = 0$, $u(L,t) = T$

5. $u(0,t) = 0$, $u_x(L,t) = 0$ 6. $u(0,t) = T$, $u_x(L,t) = 0$

7. $u_x(0,t) - u(0,t) = 0$, $u(L,t) = T$ 8. $u(0,t) = T$, $u_x(L,t) + u(L,t) = 0$

 9. Let an aluminum rod of length 20 cm be initially at the uniform temperature of 25°C.
Suppose that at time $t = 0$ the end $x = 0$ is cooled to 0°C while the end $x = 20$ is heated
to 60°C, and both are thereafter maintained at those temperatures.

(a) Find the temperature distribution in the rod at any time t.

(b) Plot the initial temperature distribution, the final (steady-state) temperature distri-
bution, and the temperature distributions at two representative intermediate times on the
same set of axes.

(c) Plot u versus t for $x = 5, 10$, and 15.

(d) Determine how much time must elapse before the temperature at $x = 5$ cm comes
(and remains) within 1% of its steady-state value.

 10. (a) Let the ends of a copper rod 100 cm long be maintained at 0°C. Suppose that the
center of the bar is heated to 100°C by an external heat source and that this situation is
maintained until a steady state results. Find this steady-state temperature distribution.

(b) At a time $t = 0$ [after the steady state of part (a) has been reached], let the heat source
be removed. At the same instant let the end $x = 0$ be placed in thermal contact with a
reservoir at 20°C, while the other end remains at 0°C. Find the temperature as a function
of position and time.

(c) Plot u versus x for several values of t. Also plot u versus t for several values of x.

(d) What limiting value does the temperature at the center of the rod approach after a
long time? How much time must elapse before the center of the rod cools to within 1° of
its limiting value?

 11. Consider a rod of length 30 for which $\alpha^2 = 1$. Suppose the initial temperature distribution
is given by $u(x,0) = x(60 - x)/30$ and that the boundary conditions are $u(0,t) = 30$ and
$u(30,t) = 0$.

(a) Find the temperature in the rod as a function of position and time.

(b) Plot u versus x for several values of t. Also plot u versus t for several values of x.

(c) Plot u versus t for $x = 12$. Observe that u initially decreases, then increases for a
while, and finally decreases to approach its steady-state value. Explain physically why this
behavior occurs at this point.

 12. Consider a uniform rod of length L with an initial temperature given by $u(x,0) =
\sin(\pi x/L), 0 \le x \le L$. Assume that both ends of the bar are insulated.

(a) Find the temperature $u(x,t)$.

(b) What is the steady-state temperature as $t \to \infty$?

(c) Let $\alpha^2 = 1$ and $L = 40$. Plot u versus x for several values of t. Also plot u versus t for
several values of x.

(d) Describe briefly how the temperature in the rod changes as time progresses.

13. Consider a bar of length 40 cm whose initial temperature is given by $u(x,0) = x(60 - x)/30$.
Suppose that $\alpha^2 = 1/4$ cm^2/sec and that both ends of the bar are insulated.

(a) Find the temperature $u(x, t)$.

(b) Plot u versus x for several values of t. Also plot u versus t for several values of x.

(c) Determine the steady-state temperature in the bar.

(d) Determine how much time must elapse before the temperature at $x = 40$ comes within $1°$ of its steady-state value.

14. Consider a bar 30 cm long that is made of a material for which $\alpha^2 = 1$ and whose ends are insulated. Suppose that the initial temperature is zero except for the interval $5 < x < 10$, where the initial temperature is $25°$.

(a) Find the temperature $u(x, t)$.

(b) Plot u versus x for several values of t. Also plot u versus t for several values of x.

(c) Plot $u(4, t)$ and $u(11, t)$ versus t. Observe that the points $x = 4$ and $x = 11$ are symmetrically located with respect to the initial temperature pulse, yet their temperature plots are significantly different. Explain physically why this is so.

15. Consider a uniform bar of length L having an initial temperature distribution given by $f(x), 0 \leq x \leq L$. Assume that the temperature at the end $x = 0$ is held at $0°C$, while the end $x = L$ is insulated so that no heat passes through it.

(a) Show that the fundamental solutions of the partial differential equation and boundary conditions are

$$u_n(x, t) = e^{-(2n-1)^2 \pi^2 \alpha^2 t / 4L^2} \sin[(2n - 1)\pi x / 2L], \qquad n = 1, 2, 3, \dots.$$

(b) Find a formal series expansion for the temperature $u(x, t)$,

$$u(x, t) = \sum_{n=1}^{\infty} c_n u_n(x, t),$$

that also satisfies the initial condition $u(x, 0) = f(x)$.
Hint: Even though the fundamental solutions involve only the odd sines, it is still possible to represent f by a Fourier series involving only these functions. See Problem 39 of Section 10.4.

16. In the bar of Problem 15 suppose that $L = 30, \alpha^2 = 1$, and the initial temperature distribution is $f(x) = 30 - x$ for $0 < x < 30$.

(a) Find the temperature $u(x, t)$.

(b) Plot u versus x for several values of t. Also plot u versus t for several values of x.

(c) How does the location x_m of the warmest point in the bar change as t increases? Draw a graph of x_m versus t.

(d) Plot the maximum temperature in the bar versus t.

17. Suppose that the conditions are as in Problems 15 and 16 except that the boundary condition at $x = 0$ is $u(0, t) = 40$.

(a) Find the temperature $u(x, t)$.

(b) Plot u versus x for several values of t. Also plot u versus t for several values of x.

(c) Compare the plots you obtained in this problem with those from Problem 16. Explain how the change in the boundary condition at $x = 0$ causes the observed differences in the behavior of the temperature in the bar.

18. Consider the problem

$$X'' + \lambda X = 0, \qquad X'(0) = 0, \qquad X'(L) = 0. \tag{i}$$

Let $\lambda = \mu^2$, where $\mu = \nu + i\sigma$ with ν and σ real. Show that if $\sigma \neq 0$, then the only solution of Eqs. (i) is the trivial solution $X(x) = 0$.

Hint: Use an argument similar to that in Problem 23 of Section 10.1.

19. The right end of a bar of length a with thermal conductivity κ_1 and cross-sectional area A_1 is joined to the left end of a bar of thermal conductivity κ_2 and cross-sectional area A_2. The composite bar has a total length L. Suppose that the end $x = 0$ is held at temperature zero, while the end $x = L$ is held at temperature T. Find the steady-state temperature in the composite bar, assuming that the temperature and rate of heat flow are continuous at $x = a$.

 Hint: See Eq. (2) of Appendix A.

20. Consider the problem

$$\alpha^2 u_{xx} = u_t, \qquad 0 < x < L, \quad t > 0;$$

$$u(0,t) = 0, \qquad u_x(L,t) + \gamma u(L,t) = 0, \qquad t > 0; \tag{i}$$

$$u(x,0) = f(x), \qquad 0 \le x \le L.$$

 (a) Let $u(x,t) = X(x)T(t)$ and show that

$$X'' + \lambda X = 0, \qquad X(0) = 0, \qquad X'(L) + \gamma X(L) = 0, \tag{ii}$$

 and

$$T' + \lambda \alpha^2 T = 0,$$

 where λ is the separation constant.

 (b) Assume that λ is real, and show that problem (ii) has no nontrivial solutions if $\lambda \le 0$.

 (c) If $\lambda > 0$, let $\lambda = \mu^2$ with $\mu > 0$. Show that problem (ii) has nontrivial solutions only if μ is a solution of the equation

$$\mu \cos \mu L + \gamma \sin \mu L = 0. \tag{iii}$$

 (d) Rewrite Eq. (iii) as $\tan \mu L = -\mu/\gamma$. Then, by drawing the graphs of $y = \tan \mu L$ and $y = -\mu L/\gamma L$ for $\mu > 0$ on the same set of axes, show that Eq. (iii) is satisfied by infinitely many positive values of μ; denote these by $\mu_1, \mu_2, \dots, \mu_n, \dots$, ordered in increasing size.

 (e) Determine the set of fundamental solutions $u_n(x,t)$ corresponding to the values μ_n found in part (d).

An External Heat Source. Consider the heat conduction problem in a bar that is in thermal contact with an external heat source or sink. Then the modified heat conduction equation is

$$u_t = \alpha^2 u_{xx} + s(x), \tag{i}$$

where the term $s(x)$ describes the effect of the external agency; $s(x)$ is positive for a source and negative for a sink. Suppose that the boundary conditions are

$$u(0,t) = T_1, \qquad u(L,t) = T_2 \tag{ii}$$

and the initial condition is

$$u(x,0) = f(x). \tag{iii}$$

Problems 21 through 23 deal with this kind of problem.

21. Write $u(x,t) = v(x) + w(x,t)$, where v and w are the steady state and transient parts of the solution, respectively. State the boundary value problems that $v(x)$ and $w(x,t)$, respectively, satisfy. Observe that the problem for w is the fundamental heat conduction problem discussed in Section 10.5, with a modified initial temperature distribution.

22. (a) Suppose that $\alpha^2 = 1$ and $s(x) = k$, a constant, in Eq. (i). Find $v(x)$.

 (b) Assume that $T_1 = 0, T_2 = 0, L = 20, k = 1/5$, and that $f(x) = 0$ for $0 < x < L$. Determine $w(x,t)$. Then plot $u(x,t)$ versus x for several values of t; on the same axes also plot the steady-state part of the solution $v(x)$.

23. (a) Let $\alpha^2 = 1$ and $s(x) = kx/L$, where k is a constant, in Eq. (i). Find $v(x)$.

 (b) Assume that $T_1 = 10, T_2 = 30, L = 20, k = 1/2$, and that $f(x) = 0$ for $0 < x < L$. Determine $w(x,t)$. Then plot $u(x,t)$ versus x for several values of t; on the same axes also plot the steady-state part of the solution $v(x)$.

10.7 The Wave Equation: Vibrations of an Elastic String

A second partial differential equation that occurs frequently in applied mathematics is the wave[10] equation. Some form of this equation, or a generalization of it, almost inevitably arises in any mathematical analysis of phenomena involving the propagation of waves in a continuous medium. For example, the studies of acoustic waves, water waves, electromagnetic waves, and seismic waves are all based on this equation.

 Perhaps the easiest situation to visualize occurs in the investigation of mechanical vibrations. Suppose that an elastic string of length L is tightly stretched between two supports at the same horizontal level, so that the x-axis lies along the string (see Figure 10.7.1). The elastic string may be thought of as a violin string, a guy wire, or possibly an electric power line. Suppose that the string is set in motion (by plucking, for example) so that it vibrates in a vertical plane, and let $u(x,t)$ denote the vertical displacement experienced by the string at the point x at time t. If damping effects, such as air resistance, are neglected, and if the amplitude of the motion is not too large, then $u(x,t)$ satisfies the partial differential equation

$$a^2 u_{xx} = u_{tt} \tag{1}$$

in the domain $0 < x < L, t > 0$. Equation (1) is known as the one-dimensional **wave equation** and is derived in Appendix B at the end of the chapter. The constant coefficient a^2 appearing in Eq. (1) is given by

$$a^2 = T/\rho, \tag{2}$$

where T is the tension (force) in the string, and ρ is the mass per unit length of the string material. It follows that a has the units of length/time—that is, of velocity. In Problem 14 it is shown that a is the velocity of propagation of waves along the string.

 To describe the motion of the string completely it is necessary also to specify suitable initial and boundary conditions for the displacement $u(x,t)$. The ends are

[10]The solution of the wave equation was one of the major mathematical problems of the mid-eighteenth century. The wave equation was first derived and studied by D'Alembert in 1746. It also attracted the attention of Euler (1748), Daniel Bernoulli (1753), and Lagrange (1759). Solutions were obtained in several different forms, and the merits of, and relations among, these solutions were argued, sometimes heatedly, in a series of papers extending over more than 25 years. The major points at issue concerned the nature of a function and the kinds of functions that can be represented by trigonometric series. These questions were not resolved until the nineteenth century.

FIGURE 10.7.1 A vibrating string.

assumed to remain fixed, and therefore the boundary conditions are

$$u(0,t) = 0, \qquad u(L,t) = 0, \qquad t \geq 0. \tag{3}$$

Since the differential equation (1) is of second order with respect to t, it is plausible to prescribe two initial conditions. These are the initial position of the string,

$$u(x,0) = f(x), \qquad 0 \leq x \leq L, \tag{4}$$

and its initial velocity,

$$u_t(x,0) = g(x), \qquad 0 \leq x \leq L, \tag{5}$$

where f and g are given functions. In order for Eqs. (3), (4), and (5) to be consistent it is also necessary to require that

$$f(0) = f(L) = 0, \qquad g(0) = g(L) = 0. \tag{6}$$

 The mathematical problem then is to determine the solution of the wave equation (1) that also satisfies the boundary conditions (3) and the initial conditions (4) and (5). Like the heat conduction problem of Sections 10.5 and 10.6, this problem is an initial value problem in the time variable t and a boundary value problem in the space variable x. Alternatively, it can be considered as a boundary value problem in the semi-infinite strip $0 < x < L, t > 0$ of the xt-plane (see Figure 10.7.2). One condition is imposed at each point on the semi-infinite sides, and two are imposed at each point on the finite base.

 It is important to realize that Eq. (1) governs a large number of other wave problems besides the transverse vibrations of an elastic string. For example, it is only

FIGURE 10.7.2 Boundary value problem for the wave equation.

necessary to interpret the function u and the constant a appropriately to have problems dealing with water waves in an ocean, acoustic or electromagnetic waves in the atmosphere, or elastic waves in a solid body. If more than one space dimension is significant, then Eq. (1) must be slightly generalized. The two-dimensional wave equation is

$$a^2(u_{xx} + u_{yy}) = u_{tt}. \tag{7}$$

This equation would arise, for example, if we considered the motion of a thin elastic sheet, such as a drumhead. Similarly, in three dimensions the wave equation is

$$a^2(u_{xx} + u_{yy} + u_{zz}) = u_{tt}. \tag{8}$$

In connection with the latter two equations, the boundary and initial conditions must also be suitably generalized.

We now solve some typical boundary value problems involving the one-dimensional wave equation.

Elastic String with Nonzero Initial Displacement. First suppose that the string is disturbed from its equilibrium position and then released at time $t = 0$ with zero velocity to vibrate freely. Then the vertical displacement $u(x, t)$ must satisfy the wave equation (1),

$$a^2 u_{xx} = u_{tt}, \qquad 0 < x < L, \quad t > 0;$$

the boundary conditions (3),

$$u(0, t) = 0, \qquad u(L, t) = 0, \qquad t \geq 0;$$

and the initial conditions

$$u(x, 0) = f(x), \qquad u_t(x, 0) = 0, \qquad 0 \leq x \leq L, \tag{9}$$

where f is a given function describing the configuration of the string at $t = 0$.

The method of separation of variables can be used to obtain the solution of Eqs. (1), (3), and (9). Assuming that

$$u(x, t) = X(x)T(t) \tag{10}$$

and substituting for u in Eq. (1), we obtain

$$\frac{X''}{X} = \frac{1}{a^2}\frac{T''}{T} = -\lambda, \tag{11}$$

where λ is a separation constant. Thus we find that $X(x)$ and $T(t)$ satisfy the ordinary differential equations

$$X'' + \lambda X = 0, \tag{12}$$

$$T'' + a^2\lambda T = 0. \tag{13}$$

Further, by substituting from Eq. (10) for $u(x, t)$ in the boundary conditions (3), we find that $X(x)$ must satisfy the boundary conditions

$$X(0) = 0, \qquad X(L) = 0. \tag{14}$$

Finally, by substituting from Eq. (10) into the second of the initial conditions (9), we also find that $T(t)$ must satisfy the initial condition

$$T'(0) = 0. \tag{15}$$

Our next task is to determine $X(x)$, $T(t)$, and λ by solving Eq. (12) subject to the boundary conditions (14) and Eq. (13) subject to the initial condition (15).

The problem of solving the differential equation (12) subject to the boundary conditions (14) is *precisely the same problem* that arose in Section 10.5 in connection with the heat conduction equation. Thus we can use the results obtained there and at the end of Section 10.1: The problem (12), (14) has nontrivial solutions if and only if λ is an eigenvalue,

$$\lambda = n^2\pi^2/L^2, \qquad n = 1, 2, \ldots, \tag{16}$$

and $X(x)$ is proportional to the corresponding eigenfunction $\sin(n\pi x/L)$.

Using the values of λ given by Eq. (16) in Eq. (13), we obtain

$$T'' + \frac{n^2\pi^2 a^2}{L^2}T = 0. \tag{17}$$

Therefore

$$T(t) = k_1 \cos\frac{n\pi at}{L} + k_2 \sin\frac{n\pi at}{L}, \tag{18}$$

where k_1 and k_2 are arbitrary constants. The initial condition (15) requires that $k_2 = 0$, so $T(t)$ must be proportional to $\cos(n\pi at/L)$.

Thus the functions

$$u_n(x, t) = \sin\frac{n\pi x}{L}\cos\frac{n\pi at}{L}, \qquad n = 1, 2, \ldots, \tag{19}$$

satisfy the partial differential equation (1), the boundary conditions (3), and the second initial condition (9). These functions are the fundamental solutions for the given problem.

To satisfy the remaining (nonhomogeneous) initial condition (9) we will consider a superposition of the fundamental solutions (19) with properly chosen coefficients. Thus we assume that $u(x, t)$ has the form

$$u(x, t) = \sum_{n=1}^{\infty} c_n u_n(x, t) = \sum_{n=1}^{\infty} c_n \sin\frac{n\pi x}{L}\cos\frac{n\pi at}{L}, \tag{20}$$

where the constants c_n remain to be chosen. The initial condition $u(x, 0) = f(x)$ requires that

$$u(x, 0) = \sum_{n=1}^{\infty} c_n \sin\frac{n\pi x}{L} = f(x). \tag{21}$$

Consequently, the coefficients c_n must be the coefficients in the Fourier sine series of period $2L$ for f; hence

$$c_n = \frac{2}{L}\int_0^L f(x)\sin\frac{n\pi x}{L}\,dx, \qquad n = 1, 2, \ldots. \tag{22}$$

Thus the formal solution of the problem of Eqs. (1), (3), and (9) is given by Eq. (20) with the coefficients calculated from Eq. (22).

For a fixed value of n the expression $\sin(n\pi x/L)\cos(n\pi at/L)$ in Eq. (19) is periodic in time t with the period $2L/na$; it therefore represents a vibratory motion of the string having this period, or having the frequency $n\pi a/L$. The quantities $\lambda a = n\pi a/L$

for $n = 1, 2, \ldots$ are the **natural frequencies** of the string—that is, the frequencies at which the string will freely vibrate. The factor $\sin(n\pi x/L)$ represents the displacement pattern occurring in the string when it is executing vibrations of the given frequency. Each displacement pattern is called a **natural mode** of vibration and is periodic in the space variable x; the spatial period $2L/n$ is called the **wavelength** of the mode of frequency $n\pi a/L$. Thus the eigenvalues $n^2\pi^2/L^2$ of the problem (12), (14) are proportional to the squares of the natural frequencies, and the eigenfunctions $\sin(n\pi x/L)$ give the natural modes. The first three natural modes are sketched in Figure 10.7.3. The total motion of the string, given by the function $u(x,t)$ of Eq. (20), is thus a combination of the natural modes of vibration and is also a periodic function of time with period $2L/a$.

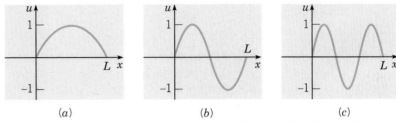

FIGURE 10.7.3 First three fundamental modes of vibration of an elastic string.
(*a*) Frequency $= \pi a/L$, wavelength $= 2L$; (*b*) frequency $= 2\pi a/L$, wavelength $= L$;
(*c*) frequency $= 3\pi a/L$, wavelength $= 2L/3$.

EXAMPLE 1

Consider a vibrating string of length $L = 30$ that satisfies the wave equation

$$4u_{xx} = u_{tt}, \qquad 0 < x < 30, \quad t > 0. \tag{23}$$

Assume that the ends of the string are fixed and that the string is set in motion with no initial velocity from the initial position

$$u(x,0) = f(x) = \begin{cases} x/10, & 0 \le x \le 10, \\ (30 - x)/20, & 10 < x \le 30. \end{cases} \tag{24}$$

Find the displacement $u(x,t)$ of the string and describe its motion through one period.

The solution is given by Eq. (20) with $a = 2$ and $L = 30$; that is,

$$u(x,t) = \sum_{n=1}^{\infty} c_n \sin\frac{n\pi x}{30} \cos\frac{2n\pi t}{30}, \tag{25}$$

where c_n is calculated from Eq. (22). Substituting from Eq. (24) into Eq. (22), we obtain

$$c_n = \frac{2}{30} \int_0^{10} \frac{x}{10} \sin\frac{n\pi x}{30}\, dx + \frac{2}{30} \int_{10}^{30} \frac{30 - x}{20} \sin\frac{n\pi x}{30}\, dx. \tag{26}$$

By evaluating the integrals in Eq. (26), we find that

$$c_n = \frac{9}{n^2\pi^2} \sin\frac{n\pi}{3}, \qquad n = 1, 2, \ldots. \tag{27}$$

The solution (25), (27) gives the displacement of the string at any point x at any time t. The motion is periodic in time with period 30, so it is sufficient to analyze the solution for $0 \leq t \leq 30$.

The best way to visualize the solution is by a computer animation showing the dynamic behavior of the vibrating string. Here we indicate the motion of the string in Figures 10.7.4, 10.7.5, and 10.7.6. Plots of u versus x for $t = 0, 4, 7.5, 11$, and 15 are shown in Figure 10.7.4. Observe that the maximum initial displacement is positive and occurs at $x = 10$, while at $t = 15$, a half-period later, the maximum displacement is negative and occurs at $x = 20$. The string then retraces its motion and returns to its original configuration at $t = 30$. Figure 10.7.5 shows the behavior of the points $x = 10, 15$, and 20 by plots of u versus t for these fixed values of x. The plots confirm that the motion is indeed periodic with period 30. Observe also that each interior point on the string is motionless for one-third of each period. Figure 10.7.6 shows a three-dimensional plot of u versus both x and t, from which the overall nature of the solution is apparent. Of course, the curves in Figures 10.7.4 and 10.7.5 lie on the surface shown in Figure 10.7.6.

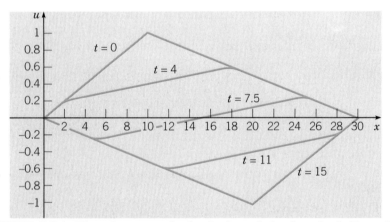

FIGURE 10.7.4 Plots of u versus x for fixed values of t for the string in Example 1.

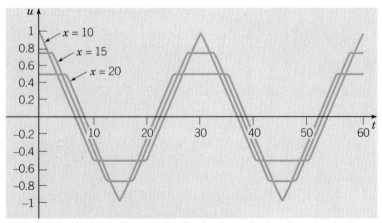

FIGURE 10.7.5 Plots of u versus t for fixed values of x for the string in Example 1.

FIGURE 10.7.6 Plot of u versus x and t for the string in Example 1.

Justification of the Solution. As in the heat conduction problem considered earlier, Eq. (20) with the coefficients c_n given by Eq. (22) is only a *formal* solution of Eqs. (1), (3), and (9). To ascertain whether Eq. (20) *actually* represents the solution of the given problem requires some further investigation. As in the heat conduction problem, it is tempting to try to show this directly by substituting Eq. (20) for $u(x,t)$ in Eqs. (1), (3), and (9). However, upon formally computing u_{xx}, for example, we obtain

$$u_{xx}(x,t) = -\sum_{n=1}^{\infty} c_n \left(\frac{n\pi}{L}\right)^2 \sin \frac{n\pi x}{L} \cos \frac{n\pi at}{L};$$

due to the presence of the n^2 factor in the numerator, this series may not converge. This would not necessarily mean that the series (20) for $u(x,t)$ is incorrect, but only that the series (20) cannot be used to calculate u_{xx} and u_{tt}. A basic difference between solutions of the wave equation and those of the heat conduction equation is that the latter contain negative exponential terms that approach zero very rapidly with increasing n, which ensures the convergence of the series solution and its derivatives. In contrast, series solutions of the wave equation contain only oscillatory terms that do not decay with increasing n.

However, there is an alternative way to establish the validity of Eq. (20) indirectly. At the same time, we will gain additional information about the structure of the solution. First we will show that Eq. (20) is equivalent to

$$u(x,t) = \tfrac{1}{2}\left[h(x - at) + h(x + at)\right], \tag{28}$$

where h is the function obtained by extending the initial data f into $(-L, 0)$ as an odd function, and to other values of x as a periodic function of period $2L$. That is,

$$h(x) = \begin{cases} f(x), & 0 \le x \le L, \\ -f(-x), & -L < x < 0; \end{cases} \tag{29}$$

$$h(x + 2L) = h(x).$$

discussed previously, it is important to note that it can also be solved simply by adding together the two solutions that we obtained above. To show that this is true, let $v(x, t)$ be the solution of the problem (1), (3), and (9), and let $w(x, t)$ be the solution of the problem (1), (3), and (31). Thus $v(x, t)$ is given by Eqs. (20) and (22), and $w(x, t)$ is given by Eqs. (34) and (36). Now let $u(x, t) = v(x, t) + w(x, t)$; what problem does $u(x, t)$ satisfy? First, observe that

$$a^2 u_{xx} - u_{tt} = (a^2 v_{xx} - v_{tt}) + (a^2 w_{xx} - w_{tt}) = 0 + 0 = 0, \tag{38}$$

so $u(x, t)$ satisfies the wave equation (1). Next, we have

$$u(0, t) = v(0, t) + w(0, t) = 0 + 0 = 0, \qquad u(L, t) = v(L, t) + w(L, t) = 0 + 0 = 0, \tag{39}$$

so $u(x, t)$ also satisfies the boundary conditions (3). Finally, we have

$$u(x, 0) = v(x, 0) + w(x, 0) = f(x) + 0 = f(x) \tag{40}$$

and

$$u_t(x, 0) = v_t(x, 0) + w_t(x, 0) = 0 + g(x) = g(x). \tag{41}$$

Thus $u(x, t)$ satisfies the general initial conditions (37).

We can restate the result we have just obtained in the following way. To solve the wave equation with the general initial conditions (37) you can solve instead the somewhat simpler problems with the initial conditions (9) and (31), respectively, and then add together the two solutions. This is another use of the principle of superposition.

PROBLEMS

Consider an elastic string of length L whose ends are held fixed. The string is set in motion with no initial velocity from an initial position $u(x, 0) = f(x)$. In each of Problems 1 through 4 carry out the following steps. Let $L = 10$ and $a = 1$ in parts (b) through (d).

(a) Find the displacement $u(x, t)$ for the given initial position $f(x)$.

(b) Plot $u(x, t)$ versus x for $0 \leq x \leq 10$ and for several values of t between $t = 0$ and $t = 20$.

(c) Plot $u(x, t)$ versus t for $0 \leq t \leq 20$ and for several values of x.

(d) Construct an animation of the solution in time for at least one period.

(e) Describe the motion of the string in a few sentences.

1. $f(x) = \begin{cases} 2x/L, & 0 \leq x \leq L/2, \\ 2(L-x)/L, & L/2 < x \leq L \end{cases}$

2. $f(x) = \begin{cases} 4x/L, & 0 \leq x \leq L/4, \\ 1, & L/4 < x < 3L/4, \\ 4(L-x)/L, & 3L/4 \leq x \leq L \end{cases}$

3. $f(x) = 8x(L-x)^2/L^3$

4. $f(x) = \begin{cases} 1, & L/2 - 1 < x < L/2 + 1 \quad (L > 2), \\ 0, & \text{otherwise} \end{cases}$

Consider an elastic string of length L whose ends are held fixed. The string is set in motion from its equilibrium position with an initial velocity $u_t(x,0) = g(x)$. In each of Problems 5 through 8 carry out the following steps. Let $L = 10$ and $a = 1$ in parts (b) through (d).

(a) Find the displacement $u(x,t)$ for the given $g(x)$.

(b) Plot $u(x,t)$ versus x for $0 \le x \le 10$ and for several values of t between $t = 0$ and $t = 20$.

(c) Plot $u(x,t)$ versus t for $0 \le t \le 20$ and for several values of x.

(d) Construct an animation of the solution in time for at least one period.

(e) Describe the motion of the string in a few sentences.

5. $g(x) = \begin{cases} 2x/L, & 0 \le x \le L/2, \\ 2(L-x)/L, & L/2 < x \le L \end{cases}$

6. $g(x) = \begin{cases} 4x/L, & 0 \le x \le L/4, \\ 1, & L/4 < x < 3L/4, \\ 4(L-x)/L, & 3L/4 \le x \le L \end{cases}$

7. $g(x) = 8x(L-x)^2/L^3$

8. $g(x) = \begin{cases} 1, & L/2 - 1 < x < L/2 + 1 \quad (L > 2), \\ 0, & \text{otherwise} \end{cases}$

9. If an elastic string is free at one end, the boundary condition to be satisfied there is that $u_x = 0$. Find the displacement $u(x,t)$ in an elastic string of length L, fixed at $x = 0$ and free at $x = L$, set in motion with no initial velocity from the initial position $u(x,0) = f(x)$, where f is a given function.

 Hint: Show that the fundamental solutions for this problem, satisfying all conditions except the nonhomogeneous initial condition, are

$$u_n(x,t) = \sin \lambda_n x \cos \lambda_n a t,$$

where $\lambda_n = (2n-1)\pi/2L, n = 1, 2, \ldots$. Compare this problem with Problem 15 of Section 10.6; pay particular attention to the extension of the initial data out of the original interval $[0, L]$.

10. Consider an elastic string of length L. The end $x = 0$ is held fixed, while the end $x = L$ is free; thus the boundary conditions are $u(0,t) = 0$ and $u_x(L,t) = 0$. The string is set in motion with no initial velocity from the initial position $u(x,0) = f(x)$, where

$$f(x) = \begin{cases} 1, & L/2 - 1 < x < L/2 + 1 \quad (L > 2), \\ 0, & \text{otherwise.} \end{cases}$$

(a) Find the displacement $u(x,t)$.

(b) With $L = 10$ and $a = 1$, plot u versus x for $0 \le x \le 10$ and for several values of t. Pay particular attention to values of t between 3 and 7. Observe how the initial disturbance is reflected at each end of the string.

(c) With $L = 10$ and $a = 1$, plot u versus t for several values of x.

(d) Construct an animation of the solution in time for at least one period.

(e) Describe the motion of the string in a few sentences.

11. Suppose that the string in Problem 10 is started instead from the initial position $f(x) = 8x(L-x)^2/L^3$. Follow the instructions in Problem 10 for this new problem.

12. Dimensionless variables can be introduced into the wave equation $a^2 u_{xx} = u_{tt}$ in the following manner. Let $s = x/L$ and show that the wave equation becomes

$$a^2 u_{ss} = L^2 u_{tt}.$$

Then show that L/a has the dimensions of time and thus can be used as the unit on the time scale. Finally, let $\tau = at/L$ and show that the wave equation then reduces to

$$u_{ss} = u_{\tau\tau}.$$

Problems 13 and 14 indicate the form of the general solution of the wave equation and the physical significance of the constant a.

13. Show that the wave equation

$$a^2 u_{xx} = u_{tt}$$

can be reduced to the form $u_{\xi\eta} = 0$ by the change of variables $\xi = x - at, \eta = x + at$. Show that $u(x,t)$ can be written as

$$u(x,t) = \phi(x - at) + \psi(x + at),$$

where ϕ and ψ are arbitrary functions.

14. Plot the value of $\phi(x - at)$ for $t = 0, 1/a, 2/a$, and t_0/a if $\phi(s) = \sin s$. Note that for any $t \neq 0$ the graph of $y = \phi(x - at)$ is the same as that of $y = \phi(x)$ when $t = 0$, but displaced a distance at in the positive x direction. Thus a represents the velocity at which a disturbance moves along the string. What is the interpretation of $\phi(x + at)$?

15. A steel wire 5 ft in length is stretched by a tensile force of 50 lb. The wire has a weight per unit length of 0.026 lb/ft.

 (a) Find the velocity of propagation of transverse waves in the wire.

 (b) Find the natural frequencies of vibration.

 (c) If the tension in the wire is increased, how are the natural frequencies changed? Are the natural modes also changed?

16. Consider the wave equation

$$a^2 u_{xx} = u_{tt}$$

in an infinite one-dimensional medium subject to the initial conditions

$$u(x,0) = f(x), \qquad u_t(x,0) = 0, \qquad -\infty < x < \infty.$$

 (a) Using the form of the solution obtained in Problem 13, show that ϕ and ψ must satisfy

$$\phi(x) + \psi(x) = f(x),$$
$$-\phi'(x) + \psi'(x) = 0.$$

 (b) Solve the equations of part (a) for ϕ and ψ, and thereby show that

$$u(x,t) = \tfrac{1}{2}\left[f(x - at) + f(x + at)\right].$$

This form of the solution was obtained by D'Alembert in 1746.

Hint: Note that the equation $\psi'(x) = \phi'(x)$ is solved by choosing $\psi(x) = \phi(x) + c$.

(c) Let

$$f(x) = \begin{cases} 2, & -1 < x < 1, \\ 0, & \text{otherwise.} \end{cases}$$

Show that

$$f(x - at) = \begin{cases} 2, & -1 + at < x < 1 + at, \\ 0, & \text{otherwise.} \end{cases}$$

Also determine $f(x + at)$.

(d) Sketch the solution found in part (b) at $t = 0$, $t = 1/2a$, $t = 1/a$, and $t = 2/a$, obtaining the results shown in Figure 10.7.7. Observe that an initial displacement produces two waves moving in opposite directions away from the original location; each wave consists of one-half of the initial displacement.

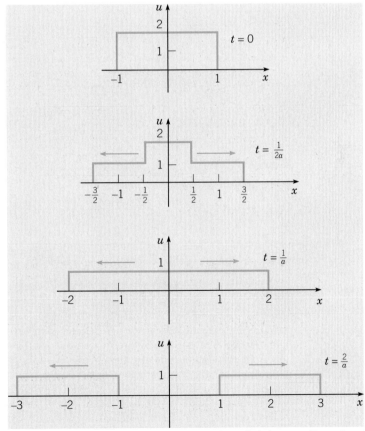

FIGURE 10.7.7 Propagation of initial disturbance in an infinite one-dimensional medium.

17. Consider the wave equation

$$a^2 u_{xx} = u_{tt}$$

in an infinite one-dimensional medium subject to the initial conditions

$$u(x,0) = 0, \qquad u_t(x,0) = g(x), \qquad -\infty < x < \infty.$$

(a) Using the form of the solution obtained in Problem 13, show that

$$\phi(x) + \psi(x) = 0,$$
$$-a\phi'(x) + a\psi'(x) = g(x).$$

(b) Use the first equation of part (a) to show that $\psi'(x) = -\phi'(x)$. Then use the second equation to show that $-2a\phi'(x) = g(x)$ and therefore that

$$\phi(x) = -\frac{1}{2a} \int_{x_0}^{x} g(\xi)\, d\xi + \phi(x_0),$$

where x_0 is arbitrary. Finally, determine $\psi(x)$.

(c) Show that

$$u(x,t) = \frac{1}{2a} \int_{x-at}^{x+at} g(\xi)\, d\xi.$$

18. By combining the results of Problems 16 and 17, show that the solution of the problem

$$a^2 u_{xx} = u_{tt},$$
$$u(x,0) = f(x), \qquad u_t(x,0) = g(x), \qquad -\infty < x < \infty$$

is given by

$$u(x,t) = \frac{1}{2}\left[f(x - at) + f(x + at)\right] + \frac{1}{2a} \int_{x-at}^{x+at} g(\xi)\, d\xi.$$

Problems 19 and 20 indicate how the formal solution (20), (22) of Eqs. (1), (3), and (9) can be shown to constitute the actual solution of that problem.

19. By using the trigonometric identity $\sin A \cos B = \frac{1}{2}\left[\sin(A + B) + \sin(A - B)\right]$ show that the solution (20) of the problem of Eqs. (1), (3), and (9) can be written in the form (28).

20. Let $h(\xi)$ represent the initial displacement in $[0, L]$, extended into $(-L, 0)$ as an odd function and extended elsewhere as a periodic function of period $2L$. Assuming that h, h', and h'' are continuous, show by direct differentiation that $u(x,t)$ as given in Eq. (28) satisfies the wave equation (1) and also the initial conditions (9). Note also that since Eq. (20) clearly satisfies the boundary conditions (3), the same is true of Eq. (28). Comparing Eq. (28) with the solution of the corresponding problem for the infinite string (Problem 16), we see that they have the same form, provided that the initial data for the finite string, defined originally only on the interval $0 \le x \le L$, are extended in the given manner over the entire x-axis. If this is done, the solution for the infinite string is also applicable to the finite one.

21. The motion of a circular elastic membrane, such as a drumhead, is governed by the two-dimensional wave equation in polar coordinates

$$u_{rr} + (1/r)u_r + (1/r^2)u_{\theta\theta} = a^{-2}u_{tt}.$$

Assuming that $u(r,\theta,t) = R(r)\Theta(\theta)T(t)$, find ordinary differential equations satisfied by $R(r)$, $\Theta(\theta)$, and $T(t)$.

22. The total energy $E(t)$ of the vibrating string is given as a function of time by

$$E(t) = \int_0^L \left[\tfrac{1}{2}\rho u_t^2(x,t) + \tfrac{1}{2}Tu_x^2(x,t) \right] dx; \tag{i}$$

the first term is the kinetic energy due to the motion of the string, and the second term is the potential energy created by the displacement of the string away from its equilibrium position.

 For the displacement $u(x,t)$ given by Eq. (20)—that is, for the solution of the string problem with zero initial velocity—show that

$$E(t) = \frac{\pi^2 T}{4L} \sum_{n=1}^{\infty} n^2 c_n^2. \tag{ii}$$

Note that the right side of Eq. (ii) does not depend on t. Thus the total energy E is a constant and therefore is *conserved* during the motion of the string.

Hint: Use Parseval's equation (Problem 37 of Section 10.4 and Problem 17 of Section 10.3), and recall that $a^2 = T/\rho$.

23. **Dispersive Waves.** Consider the modified wave equation

$$a^{-2}u_{tt} + \gamma^2 u = u_{xx}, \qquad 0 < x < L, \quad t > 0 \tag{i}$$

with the boundary conditions

$$u(0,t) = 0, \qquad u(L,t) = 0, \qquad t > 0 \tag{ii}$$

and the initial conditions

$$u(x,0) = f(x), \qquad u_t(x,0) = 0, \qquad 0 < x < L. \tag{iii}$$

(a) Show that the solution can be written as

$$u(x,t) = \sum_{n=1}^{\infty} c_n \cos\left(\sqrt{\frac{n^2\pi^2}{L^2} + \gamma^2}\, at \right) \sin \frac{n\pi x}{L},$$

where

$$c_n = \frac{2}{L} \int_0^L f(x) \sin \frac{n\pi x}{L}\, dx.$$

(b) By using trigonometric identities, rewrite the solution as

$$u(x,t) = \frac{1}{2} \sum_{n=1}^{\infty} c_n \left[\sin \frac{n\pi}{L}(x + a_n t) + \sin \frac{n\pi}{L}(x - a_n t) \right].$$

Determine a_n, the speed of wave propagation.

(c) Observe that a_n, found in part (b), depends on n. This means that components of different wave lengths (or frequencies) are propagated at different speeds, resulting in a distortion of the original wave form over time. This phenomenon is called **dispersion**. Find the condition under which a_n is independent of n—that is, there is no dispersion.

 24. Consider the situation in Problem 23 with $a^2 = 1, L = 10$, and

$$f(x) = \begin{cases} x - 4, & 4 \le x \le 5, \\ 6 - x, & 5 \le x \le 6, \\ 0, & \text{otherwise.} \end{cases}$$

(a) Determine the coefficients c_n in the solution of Problem 23(a). By plotting

$$\sum_{n=1}^{N} c_n \sin \frac{n\pi x}{10}$$ for $0 \le x \le 10$, choose N large enough so that the plot accurately displays the graph of $f(x)$. Use this value of N for the remaining plots called for in this problem.

(b) Let $\gamma = 0$. Plot $u(x, t)$ versus x for $t = 60$.

(c) Let $\gamma = 1/8$. Plot $u(x, t)$ versus x for $t = 20, 40$, and 60.

(d) Let $\gamma = 1/4$. Plot $u(x, t)$ versus x for $t = 20, 40$, and 60.

10.8 Laplace's Equation

One of the most important of all partial differential equations occurring in applied mathematics is that associated with the name of Laplace[11]: in two dimensions

$$u_{xx} + u_{yy} = 0, \tag{1}$$

and in three dimensions

$$u_{xx} + u_{yy} + u_{zz} = 0. \tag{2}$$

For example, in a two-dimensional heat conduction problem, the temperature $u(x, y, t)$ must satisfy the differential equation

$$\alpha^2 (u_{xx} + u_{yy}) = u_t, \tag{3}$$

where α^2 is the thermal diffusivity. If a steady state exists, u is a function of x and y only, and the time derivative vanishes; in this case Eq. (3) reduces to Eq. (1). Similarly, for the steady-state heat conduction problem in three dimensions, the temperature must satisfy the three-dimensional form of Laplace's equation. Equations (1) and (2) also occur in other branches of mathematical physics. In the consideration of electrostatic fields, the electric potential function in a dielectric medium containing no electric charges must satisfy either Eq. (1) or Eq. (2), depending on the number of space dimensions involved. Similarly, the potential function of a particle in free space acted on only by gravitational forces satisfies the same equations. Consequently, Laplace's

[11]Laplace's equation is named for Pierre-Simon de Laplace, who, beginning in 1782, studied its solutions extensively while investigating the gravitational attraction of arbitrary bodies in space. However, the equation first appeared in 1752 in a paper by Euler on hydrodynamics.

equation is often referred to as the **potential equation**. Another example arises in the study of the steady (time-independent), two-dimensional, inviscid, irrotational motion of an incompressible fluid. This study centers on two functions, known as the velocity potential function and the stream function, both of which satisfy Eq. (1). In elasticity, the displacements that occur when a perfectly elastic bar is twisted are described in terms of the so-called warping function, which also satisfies Eq. (1).

Since there is no time dependence in any of the problems mentioned previously, there are no initial conditions to be satisfied by the solutions of Eq. (1) or (2). They must, however, satisfy certain boundary conditions on the bounding curve or surface of the region in which the differential equation is to be solved. Since Laplace's equation is of second order, it might be plausible to expect that two boundary conditions would be required to determine the solution completely. This, however, is not the case. Recall that in the heat conduction problem for the finite bar (Sections 10.5 and 10.6) it was necessary to prescribe one condition at each end of the bar—that is, *one condition at each point of the boundary*. If we generalize this observation to multidimensional problems, it is then natural to prescribe one condition on the function u at each point on the boundary of the region in which a solution of Eq. (1) or (2) is sought. The most common boundary condition occurs when the value of u is specified at each boundary point; in terms of the heat conduction problem this corresponds to prescribing the temperature on the boundary. In some problems the value of the derivative, or rate of change, of u in the direction normal to the boundary is specified instead; the condition on the boundary of a thermally insulated body, for example, is of this type. It is entirely possible for more complicated boundary conditions to occur; for example, u might be prescribed on part of the boundary and its normal derivative specified on the remainder. The problem of finding a solution of Laplace's equation that takes on given boundary values is known as a **Dirichlet problem**, in honor of P. G. L. Dirichlet.[12] In contrast, if the values of the normal derivative are prescribed on the boundary, the problem is said to be a **Neumann problem**, in honor of K. G. Neumann.[13] The Dirichlet and Neumann problems are also known as the first and second boundary value problems of potential theory, respectively.

Physically, it is plausible to expect that the types of boundary conditions just mentioned will be sufficient to determine the solution completely. Indeed, it is possible to establish the existence and uniqueness of the solution of Laplace's equation under the boundary conditions mentioned, provided that the shape of the boundary and the functions appearing in the boundary conditions satisfy certain very mild requirements. However, the proofs of these theorems, and even their accurate statement, are beyond the scope of the present book. Our only concern will be solving some typical problems by means of separation of variables and Fourier series.

[12] Peter Gustav Lejeune Dirichlet (1805–1859) was a professor at Berlin and, after the death of Gauss, at Göttingen. In 1829 he gave the first set of conditions sufficient to guarantee the convergence of a Fourier series. The definition of *function* that is usually used today in elementary calculus is essentially the one given by Dirichlet in 1837. Although he is best known for his work in analysis and differential equations, Dirichlet was also one of the leading number theorists of the nineteenth century.

[13] Karl Gottfried Neumann (1832–1925), professor at Leipzig, made contributions to differential equations, integral equations, and complex variables.

While the problems chosen as examples have interesting physical interpretations (in terms of electrostatic potentials or steady-state temperature distributions, for instance), our purpose here is primarily to point out some of the features that may occur during their mathematical solution. It is also worth noting again that more complicated problems can sometimes be solved by expressing the solution as the sum of solutions of several simpler problems (see Problems 3 and 4).

Dirichlet Problem for a Rectangle. Consider the mathematical problem of finding the function u satisfying Laplace's equation (1),

$$u_{xx} + u_{yy} = 0,$$

in the rectangle $0 < x < a, 0 < y < b$, and also satisfying the boundary conditions

$$u(x,0) = 0, \qquad u(x,b) = 0, \qquad 0 < x < a,$$
$$u(0,y) = 0, \qquad u(a,y) = f(y), \qquad 0 \le y \le b, \tag{4}$$

where f is a given function on $0 \le y \le b$ (see Figure 10.8.1).

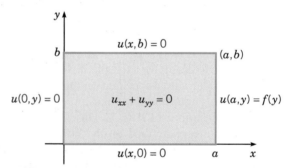

FIGURE 10.8.1 Dirichlet problem for a rectangle.

To solve this problem we wish to construct a fundamental set of solutions satisfying the partial differential equation and the homogeneous boundary conditions; then we will superpose these solutions so as to satisfy the remaining boundary condition. Let us assume that

$$u(x,y) = X(x)Y(y) \tag{5}$$

and substitute for u in Eq. (1). This yields

$$\frac{X''}{X} = -\frac{Y''}{Y} = \lambda,$$

where λ is the separation constant. Thus we obtain the two ordinary differential equations

$$X'' - \lambda X = 0, \tag{6}$$
$$Y'' + \lambda Y = 0. \tag{7}$$

If we now substitute for u from Eq. (5) in each of the homogeneous boundary conditions, we find that

$$X(0) = 0 \tag{8}$$

and

$$Y(0) = 0, \qquad Y(b) = 0. \tag{9}$$

We will first determine the solution of the differential equation (7) subject to the boundary conditions (9). However, this problem is essentially identical to one encountered previously in Sections 10.1, 10.5, and 10.7. We conclude that there are nontrivial solutions if and only if λ is an eigenvalue, namely,

$$\lambda = (n\pi/b)^2, \qquad n = 1, 2, \ldots, \tag{10}$$

and $Y(y)$ is proportional to the corresponding eigenfunction $\sin(n\pi y/b)$. Next, we substitute from Eq. (10) for λ in Eq. (6) and solve this equation subject to the boundary condition (8). It is convenient to write the general solution of Eq. (6) as

$$X(x) = k_1 \cosh(n\pi x/b) + k_2 \sinh(n\pi x/b), \tag{11}$$

and the boundary condition (8) then requires that $k_1 = 0$. Therefore $X(x)$ must be proportional to $\sinh(n\pi x/b)$. Thus we obtain the fundamental solutions

$$u_n(x, y) = \sinh \frac{n\pi x}{b} \sin \frac{n\pi y}{b}, \qquad n = 1, 2, \ldots. \tag{12}$$

These functions satisfy the differential equation (1) and all the homogeneous boundary conditions for each value of n.

To satisfy the remaining nonhomogeneous boundary condition at $x = a$, we assume, as usual, that we can represent the solution $u(x, y)$ in the form

$$u(x, y) = \sum_{n=1}^{\infty} c_n u_n(x, y) = \sum_{n=1}^{\infty} c_n \sinh \frac{n\pi x}{b} \sin \frac{n\pi y}{b}. \tag{13}$$

The coefficients c_n are determined by the boundary condition

$$u(a, y) = \sum_{n=1}^{\infty} c_n \sinh \frac{n\pi a}{b} \sin \frac{n\pi y}{b} = f(y). \tag{14}$$

Therefore the quantities $c_n \sinh(n\pi a/b)$ must be the coefficients in the Fourier sine series of period $2b$ for f and are given by

$$c_n \sinh \frac{n\pi a}{b} = \frac{2}{b} \int_0^b f(y) \sin \frac{n\pi y}{b} \, dy. \tag{15}$$

Thus the solution of the partial differential equation (1) satisfying the boundary conditions (4) is given by Eq. (13) with the coefficients c_n computed from Eq. (15).

From Eqs. (13) and (15) we see that the solution contains the factor $\sinh(n\pi x/b)/\sinh(n\pi a/b)$. To estimate this quantity for large n we can use the approximation $\sinh \xi \cong e^\xi/2$, and thereby obtain

$$\frac{\sinh(n\pi x/b)}{\sinh(n\pi a/b)} \cong \frac{\frac{1}{2}\exp(n\pi x/b)}{\frac{1}{2}\exp(n\pi a/b)} = \exp[-n\pi(a - x)/b].$$

Thus this factor has the character of a negative exponential; consequently, the series (13) converges quite rapidly unless $a - x$ is very small.

EXAMPLE 1

To illustrate these results, let $a = 3, b = 2$, and

$$f(y) = \begin{cases} y, & 0 \le y \le 1, \\ 2 - y, & 1 \le y \le 2. \end{cases} \tag{16}$$

By evaluating c_n from Eq. (15) we find that

$$c_n = \frac{8 \sin(n\pi/2)}{n^2 \pi^2 \sinh(3n\pi/2)}. \tag{17}$$

Then $u(x, y)$ is given by Eq. (13). Keeping 20 terms in the series, we can plot u versus x and y, as shown in Figure 10.8.2. Alternatively, we can construct a contour plot showing level curves of $u(x, y)$; Figure 10.8.3 is such a plot, with an increment of 0.1 between adjacent curves.

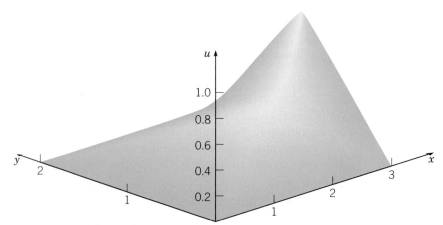

FIGURE 10.8.2 Plot of u versus x and y for Example 1.

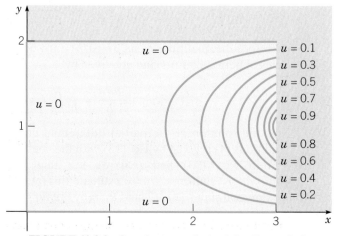

FIGURE 10.8.3 Level curves of $u(x, y)$ for Example 1.

Dirichlet Problem for a Circle. Consider the problem of solving Laplace's equation in a circular region $r < a$ subject to the boundary condition

$$u(a, \theta) = f(\theta), \tag{18}$$

where f is a given function on $0 \leq \theta < 2\pi$ (see Figure 10.8.4). In polar coordinates Laplace's equation takes the form

$$u_{rr} + \frac{1}{r} u_r + \frac{1}{r^2} u_{\theta\theta} = 0. \tag{19}$$

To complete the statement of the problem we note that for $u(r, \theta)$ to be single-valued, it is necessary that u be periodic in θ with period 2π. Moreover, we state explicitly that $u(r, \theta)$ must be bounded for $r \leq a$, since this will become important later.

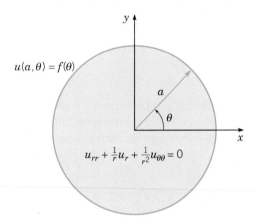

FIGURE 10.8.4 Dirichlet problem for a circle.

To apply the method of separation of variables to this problem we assume that

$$u(r, \theta) = R(r)\Theta(\theta), \tag{20}$$

and substitute for u in the differential equation (19). This yields

$$R''\Theta + \frac{1}{r} R'\Theta + \frac{1}{r^2} R\Theta'' = 0,$$

or

$$r^2 \frac{R''}{R} + r \frac{R'}{R} = -\frac{\Theta''}{\Theta} = \lambda, \tag{21}$$

where λ is the separation constant. Thus we obtain the two ordinary differential equations

$$r^2 R'' + r R' - \lambda R = 0, \tag{22}$$

$$\Theta'' + \lambda\Theta = 0. \tag{23}$$

In this problem there are no homogeneous boundary conditions; recall, however, that solutions must be bounded and also periodic in θ with period 2π. It is possible

to show (Problem 9) that the periodicity condition requires λ to be real. We will consider in turn the cases in which λ is negative, zero, and positive.

If $\lambda < 0$, let $\lambda = -\mu^2$, where $\mu > 0$. Then Eq. (23) becomes $\Theta'' - \mu^2\Theta = 0$, and consequently

$$\Theta(\theta) = c_1 e^{\mu\theta} + c_2 e^{-\mu\theta}. \tag{24}$$

Thus $\Theta(\theta)$ can be periodic only if $c_1 = c_2 = 0$, and we conclude that λ cannot be negative.

If $\lambda = 0$, then Eq. (23) becomes $\Theta'' = 0$, and thus

$$\Theta(\theta) = c_1 + c_2\theta. \tag{25}$$

For $\Theta(\theta)$ to be periodic we must have $c_2 = 0$, so that $\Theta(\theta)$ is a constant. Further, for $\lambda = 0$, Eq. (22) becomes

$$r^2 R'' + rR' = 0. \tag{26}$$

This equation is of the Euler type and has the solution

$$R(r) = k_1 + k_2 \ln r. \tag{27}$$

The logarithmic term cannot be accepted if $u(r, \theta)$ is to remain bounded as $r \to 0$; hence $k_2 = 0$. Thus, corresponding to $\lambda = 0$, we conclude that $u(r, \theta)$ must be a constant—that is, proportional to the solution

$$u_0(r, \theta) = 1. \tag{28}$$

Finally, if $\lambda > 0$, we let $\lambda = \mu^2$, where $\mu > 0$. Then Eqs. (22) and (23) become

$$r^2 R'' + rR' - \mu^2 R = 0 \tag{29}$$

and

$$\Theta'' + \mu^2\Theta = 0, \tag{30}$$

respectively. Equation (29) is an Euler equation and has the solution

$$R(r) = k_1 r^\mu + k_2 r^{-\mu}, \tag{31}$$

while Eq. (30) has the solution

$$\Theta(\theta) = c_1 \sin \mu\theta + c_2 \cos \mu\theta. \tag{32}$$

In order that Θ be periodic with period 2π it is necessary that μ be a positive integer n. With $\mu = n$ it follows that the solution $r^{-\mu}$ in Eq. (31) must be discarded since it becomes unbounded as $r \to 0$. Consequently, $k_2 = 0$ and the appropriate solutions of Eq. (19) are

$$u_n(r, \theta) = r^n \cos n\theta, \qquad v_n(r, \theta) = r^n \sin n\theta, \qquad n = 1, 2, \ldots. \tag{33}$$

These functions, together with $u_0(r, \theta) = 1$, form a set of fundamental solutions for the present problem.

In the usual way we now assume that u can be expressed as a linear combination of the fundamental solutions; that is,

$$u(r,\theta) = \frac{c_0}{2} + \sum_{n=1}^{\infty} r^n(c_n \cos n\theta + k_n \sin n\theta). \tag{34}$$

The boundary condition (18) then requires that

$$u(a,\theta) = \frac{c_0}{2} + \sum_{n=1}^{\infty} a^n(c_n \cos n\theta + k_n \sin n\theta) = f(\theta) \tag{35}$$

for $0 \le \theta < 2\pi$. The function f may be extended outside this interval so that it is periodic with period 2π and therefore has a Fourier series of the form (35). Since the extended function has period 2π, we may compute its Fourier coefficients by integrating over any period of the function. In particular, it is convenient to use the original interval $(0, 2\pi)$; then

$$a^n c_n = \frac{1}{\pi} \int_0^{2\pi} f(\theta) \cos n\theta \, d\theta, \qquad n = 0, 1, 2, \ldots; \tag{36}$$

$$a^n k_n = \frac{1}{\pi} \int_0^{2\pi} f(\theta) \sin n\theta \, d\theta, \qquad n = 1, 2, \ldots. \tag{37}$$

With this choice of the coefficients, Eq. (34) represents the solution of the boundary value problem of Eqs. (18) and (19). Note that in this problem we needed both sine and cosine terms in the solution. This is because the boundary data were given on $0 \le \theta < 2\pi$ and have period 2π. As a consequence, the full Fourier series is required, rather than sine or cosine terms alone.

PROBLEMS

1. (a) Find the solution $u(x, y)$ of Laplace's equation in the rectangle $0 < x < a, 0 < y < b$, that satisfies the boundary conditions

$$u(0, y) = 0, \qquad u(a, y) = 0, \qquad 0 < y < b,$$
$$u(x, 0) = 0, \qquad u(x, b) = g(x), \qquad 0 \le x \le a.$$

(b) Find the solution if

$$g(x) = \begin{cases} x, & 0 \le x \le a/2, \\ a - x, & a/2 \le x \le a. \end{cases}$$

(c) For $a = 3$ and $b = 1$ plot u versus x for several values of y and also plot u versus y for several values of x.

(d) Plot u versus both x and y in three dimensions. Also draw a contour plot showing several level curves of $u(x, y)$ in the xy-plane.

2. Find the solution $u(x, y)$ of Laplace's equation in the rectangle $0 < x < a, 0 < y < b$, that satisfies the boundary conditions

$$u(0, y) = 0, \qquad u(a, y) = 0, \qquad 0 < y < b,$$
$$u(x, 0) = h(x), \qquad u(x, b) = 0, \qquad 0 \le x \le a.$$

 3. (a) Find the solution $u(x, y)$ of Laplace's equation in the rectangle $0 < x < a, 0 < y < b$, that satisfies the boundary conditions

$$u(0, y) = 0, \qquad u(a, y) = f(y), \qquad 0 < y < b,$$
$$u(x, 0) = h(x), \qquad u(x, b) = 0, \qquad 0 \le x \le a.$$

Hint: Consider the possibility of adding the solutions of two problems, one with homogeneous boundary conditions except for $u(a, y) = f(y)$, and the other with homogeneous boundary conditions except for $u(x, 0) = h(x)$.

(b) Find the solution if $h(x) = (x/a)^2$ and $f(y) = 1 - (y/b)$.

(c) Let $a = 2$ and $b = 2$. Plot the solution in several ways: u versus x, u versus y, u versus both x and y, and a contour plot.

4. Show how to find the solution $u(x, y)$ of Laplace's equation in the rectangle $0 < x < a$, $0 < y < b$, that satisfies the boundary conditions

$$u(0, y) = k(y), \qquad u(a, y) = f(y), \qquad 0 < y < b,$$
$$u(x, 0) = h(x), \qquad u(x, b) = g(x), \qquad 0 \le x \le a.$$

Hint: See Problem 3.

5. Find the solution $u(r, \theta)$ of Laplace's equation

$$u_{rr} + (1/r)u_r + (1/r^2)u_{\theta\theta} = 0$$

outside the circle $r = a$, that satisfies the boundary condition

$$u(a, \theta) = f(\theta), \qquad 0 \le \theta < 2\pi,$$

on the circle. Assume that $u(r, \theta)$ is single-valued and bounded for $r > a$.

 6. (a) Find the solution $u(r, \theta)$ of Laplace's equation in the semicircular region $r < a$, $0 < \theta < \pi$, that satisfies the boundary conditions

$$u(r, 0) = 0, \qquad u(r, \pi) = 0, \qquad 0 \le r < a,$$
$$u(a, \theta) = f(\theta), \qquad 0 \le \theta \le \pi.$$

Assume that u is single-valued and bounded in the given region.

(b) Find the solution if $f(\theta) = \theta(\pi - \theta)$.

(c) Let $a = 2$ and plot the solution in several ways: u versus r, u versus θ, u versus both r and θ, and a contour plot.

7. Find the solution $u(r, \theta)$ of Laplace's equation in the circular sector $0 < r < a$, $0 < \theta < \alpha$, that satisfies the boundary conditions

$$u(r, 0) = 0, \qquad u(r, \alpha) = 0, \qquad 0 \le r < a,$$
$$u(a, \theta) = f(\theta), \qquad 0 \le \theta \le \alpha.$$

Assume that u is single-valued and bounded in the sector.

 8. (a) Find the solution $u(x, y)$ of Laplace's equation in the semi-infinite strip $0 < x < a$, $y > 0$, that satisfies the boundary conditions

$$u(0, y) = 0, \qquad u(a, y) = 0, \qquad y > 0,$$
$$u(x, 0) = f(x), \qquad 0 \le x \le a$$

and the additional condition that $u(x, y) \to 0$ as $y \to \infty$.

(b) Find the solution if $f(x) = x(a - x)$.

(c) Let $a = 5$. Find the smallest value of y_0 for which $u(x, y) \leq 0.1$ for all $y \geq y_0$.

9. Show that Eq. (23) has periodic solutions only if λ is real.
 Hint: Let $\lambda = -\mu^2$, where $\mu = \nu + i\sigma$ with ν and σ real.

10. Consider the problem of finding a solution $u(x, y)$ of Laplace's equation in the rectangle $0 < x < a, 0 < y < b$, that satisfies the boundary conditions

$$u_x(0, y) = 0, \quad u_x(a, y) = f(y), \quad 0 < y < b,$$
$$u_y(x, 0) = 0, \quad u_y(x, b) = 0, \quad 0 \leq x \leq a.$$

This is an example of a Neumann problem.

(a) Show that Laplace's equation and the homogeneous boundary conditions determine the fundamental set of solutions

$$u_0(x, y) = c_0,$$
$$u_n(x, y) = c_n \cosh(n\pi x/b) \cos(n\pi y/b), \quad n = 1, 2, 3, \ldots.$$

(b) By superposing the fundamental solutions of part (a), formally determine a function u satisfying the nonhomogeneous boundary condition $u_x(a, y) = f(y)$. Note that when $u_x(a, y)$ is calculated, the constant term in $u(x, y)$ is eliminated, and there is no condition from which to determine c_0. Furthermore, it must be possible to express f by means of a Fourier cosine series of period $2b$, which does not have a constant term. This means that

$$\int_0^b f(y) \, dy = 0$$

is a necessary condition for the given problem to be solvable. Finally, note that c_0 remains arbitrary, and hence the solution is determined only up to this additive constant. This is a property of all Neumann problems.

11. Find a solution $u(r, \theta)$ of Laplace's equation inside the circle $r = a$, that satisfies the boundary condition on the circle

$$u_r(a, \theta) = g(\theta), \quad 0 \leq \theta < 2\pi.$$

Note that this is a Neumann problem and that its solution is determined only up to an arbitrary additive constant. State a necessary condition on $g(\theta)$ for this problem to be solvable by the method of separation of variables (see Problem 10).

12. (a) Find the solution $u(x, y)$ of Laplace's equation in the rectangle $0 < x < a$, $0 < y < b$, that satisfies the boundary conditions

$$u(0, y) = 0, \quad u(a, y) = 0, \quad 0 < y < b,$$
$$u_y(x, 0) = 0, \quad u(x, b) = g(x), \quad 0 \leq x \leq a.$$

Note that this is neither a Dirichlet nor a Neumann problem, but a mixed problem in which u is prescribed on part of the boundary and its normal derivative on the rest.

(b) Find the solution if

$$g(x) = \begin{cases} x, & 0 \leq x \leq a/2, \\ a - x, & a/2 \leq x \leq a. \end{cases}$$

(c) Let $a = 3$ and $b = 1$. By drawing suitable plots compare this solution with the solution of Problem 1.

13. (a) Find the solution $u(x,y)$ of Laplace's equation in the rectangle $0 < x < a$, $0 < y < b$, that satisfies the boundary conditions

$$u(0,y) = 0, \quad u(a,y) = f(y), \quad 0 < y < b,$$
$$u(x,0) = 0, \quad u_y(x,b) = 0, \quad 0 \le x \le a.$$

Hint: Eventually it will be necessary to expand $f(y)$ in a series that makes use of the functions $\sin(\pi y/2b), \sin(3\pi y/2b), \sin(5\pi y/2b), \ldots$ (see Problem 39 of Section 10.4).
(b) Find the solution if $f(y) = y(2b - y)$.
(c) Let $a = 3$ and $b = 2$; plot the solution in several ways.

14. (a) Find the solution $u(x,y)$ of Laplace's equation in the rectangle $0 < x < a$, $0 < y < b$, that satisfies the boundary conditions

$$u_x(0,y) = 0, \quad u_x(a,y) = 0, \quad 0 < y < b,$$
$$u(x,0) = 0, \quad u(x,b) = g(x), \quad 0 \le x \le a.$$

(b) Find the solution if $g(x) = 1 + x^2(x - a)^2$.
(c) Let $a = 3$ and $b = 2$; plot the solution in several ways.

15. By writing Laplace's equation in cylindrical coordinates r, θ, and z and then assuming that the solution is axially symmetric (no dependence on θ), we obtain the equation

$$u_{rr} + (1/r)u_r + u_{zz} = 0.$$

Assuming that $u(r,z) = R(r)Z(z)$, show that R and Z satisfy the equations

$$rR'' + R' + \lambda^2 rR = 0, \quad Z'' - \lambda^2 Z = 0.$$

The equation for R is Bessel's equation of order zero with independent variable λr.

16. **Flow in an Aquifer.** Consider the flow of water in a porous medium, such as sand, in an aquifer. The flow is driven by the hydraulic head, a measure of the potential energy of the water above the aquifer. Let $R : 0 < x < a, 0 < z < b$ be a vertical section of an aquifer. In a uniform, homogeneous medium the hydraulic head $u(x,z)$ satisfies Laplace's equation

$$u_{xx} + u_{zz} = 0 \quad \text{in } R. \tag{i}$$

If water cannot flow through the sides and bottom of R, then the boundary conditions there are

$$u_x(0,z) = 0, \quad u_x(a,z) = 0, \quad 0 \le z \le b \tag{ii}$$

$$u_z(x,0) = 0, \quad 0 \le x \le a \tag{iii}$$

Finally, suppose that the boundary condition at $z = b$ is

$$u(x,b) = b + \alpha x, \quad 0 \le x \le a, \tag{iv}$$

where α is the slope of the water table.
(a) Solve the given boundary value problem for $u(x,z)$.
(b) Let $a = 1000, b = 500$, and $\alpha = 0.1$. Draw a contour plot of the solution in R; that is, plot some level curves of $u(x,z)$.
(c) Water flows along paths in R that are orthogonal to the level curves of $u(x,z)$ in the direction of decreasing u. Plot some of the flow paths.

Derivation of the Heat Conduction Equation. In this section we derive the differential equation that, to a first approximation at least, governs the conduction of heat in solids. It is important to understand that the mathematical analysis of a physical situation or process such as this ultimately rests on a foundation of empirical knowledge of the phenomenon involved. The mathematician must have a place to start, so to speak, and this place is furnished by experience. Consider a uniform rod insulated on the lateral surfaces so that heat can flow only in the axial direction. It has been demonstrated many times that if two parallel cross sections of the same area A and different temperatures T_1 and T_2, respectively, are separated by a small distance d, an amount of heat per unit time will pass from the warmer section to the cooler one. Moreover, this amount of heat is proportional to the area A and to the temperature difference $|T_2 - T_1|$, and is inversely proportional to the separation distance d. Thus

$$\text{Amount of heat per unit time} = \kappa A |T_2 - T_1|/d, \tag{1}$$

where the positive proportionality factor κ is called the thermal conductivity and depends primarily on the material[14] of the rod. The relation (1) is often called Fourier's law of heat conduction. We repeat that Eq. (1) is an empirical, not a theoretical, result and that it can be, and has often been, verified by careful experiment. It is the basis of the mathematical theory of heat conduction.

Now consider a straight rod of uniform cross section and homogeneous material, oriented so that the x-axis lies along the axis of the rod (see Figure 10.A.1). Let $x = 0$ and $x = L$ designate the ends of the bar.

FIGURE 10.A.1 Conduction of heat in an element of a rod.

We will assume that the sides of the bar are perfectly insulated so that there is no passage of heat through them. We will also assume that the temperature u depends only on the axial position x and the time t, and not on the lateral coordinates y and z. In other words, we assume that the temperature remains constant on any cross section of the bar. This assumption is usually satisfactory when the lateral dimensions of the rod are small compared to its length.

The differential equation governing the temperature in the bar is an expression of a fundamental physical balance; the rate at which heat flows into any portion of the bar is equal to the rate at which heat is absorbed in that portion of the bar. The terms in the equation are called the flux (flow) term and the absorption term, respectively.

[14]Actually, κ also depends on the temperature, but if the temperature range is not too great, it is satisfactory to assume that κ is independent of temperature.

We will first calculate the flux term. Consider an element of the bar lying between the cross sections $x = x_0$ and $x = x_0 + \Delta x$, where x_0 is arbitrary and Δx is small. The instantaneous rate of heat transfer $H(x_0, t)$ from left to right across the cross section $x = x_0$ is given by

$$H(x_0, t) = -\lim_{d \to 0} \kappa A \frac{u(x_0 + d/2, t) - u(x_0 - d/2, t)}{d}$$

$$= -\kappa A u_x(x_0, t). \tag{2}$$

The minus sign appears in this equation because there will be a positive flow of heat from left to right only if the temperature is greater to the left of $x = x_0$ than to the right; in this case $u_x(x_0, t)$ is negative. In a similar manner, the rate at which heat passes from left to right through the cross section $x = x_0 + \Delta x$ is given by

$$H(x_0 + \Delta x, t) = -\kappa A u_x(x_0 + \Delta x, t). \tag{3}$$

The net rate at which heat flows into the segment of the bar between $x = x_0$ and $x = x_0 + \Delta x$ is thus given by

$$Q = H(x_0, t) - H(x_0 + \Delta x, t) = \kappa A[u_x(x_0 + \Delta x, t) - u_x(x_0, t)], \tag{4}$$

and the amount of heat entering this bar element in time Δt is

$$Q \Delta t = \kappa A[u_x(x_0 + \Delta x, t) - u_x(x_0, t)] \Delta t. \tag{5}$$

Let us now calculate the absorption term. The average change in temperature Δu, in the time interval Δt, is proportional to the amount of heat $Q \Delta t$ introduced and inversely proportional to the mass Δm of the element. Thus

$$\Delta u = \frac{1}{s} \frac{Q \Delta t}{\Delta m} = \frac{Q \Delta t}{s \rho A \Delta x}, \tag{6}$$

where the constant of proportionality s is known as the specific heat of the material of the bar, and ρ is its density.[15] The average temperature change Δu in the bar element under consideration is the actual temperature change at some intermediate point $x = x_0 + \theta \Delta x$, where $0 < \theta < 1$. Thus Eq. (6) can be written as

$$u(x_0 + \theta \Delta x, t + \Delta t) - u(x_0 + \theta \Delta x, t) = \frac{Q \Delta t}{s \rho A \Delta x}, \tag{7}$$

or as

$$Q \Delta t = [u(x_0 + \theta \Delta x, t + \Delta t) - u(x_0 + \theta \Delta x, t)] s \rho A \Delta x. \tag{8}$$

To balance the flux and absorption terms, we equate the two expressions for $Q \Delta t$:

$$\kappa A[u_x(x_0 + \Delta x, t) - u_x(x_0, t)] \Delta t$$
$$= s \rho A[u(x_0 + \theta \Delta x, t + \Delta t) - u(x_0 + \theta \Delta x, t)] \Delta x. \tag{9}$$

[15]The dependence of the density and specific heat on temperature is relatively small and will be neglected. Thus both ρ and s will be considered as constants.

Upon dividing Eq. (9) by $\Delta x \, \Delta t$ and then letting $\Delta x \to 0$ and $\Delta t \to 0$, we obtain the *heat conduction* or *diffusion* equation

$$\alpha^2 u_{xx} = u_t. \tag{10}$$

The quantity α^2 defined by

$$\alpha^2 = \kappa/\rho s \tag{11}$$

is called the *thermal diffusivity* and is a parameter that depends only on the material of the bar. The units of α^2 are $(\text{length})^2/\text{time}$. Typical values of α^2 are given in Table 10.5.1.

Several relatively simple conditions may be imposed at the ends of the bar. For example, the temperature at an end may be maintained at some constant value T. This might be accomplished by placing the end of the bar in thermal contact with some reservoir of sufficient size so that any heat that flows between the bar and the reservoir does not appreciably alter the temperature of the reservoir. At an end where this is done the boundary condition is

$$u = T. \tag{12}$$

Another simple boundary condition occurs if the end is insulated so that no heat passes through it. Recalling the expression (2) for the amount of heat crossing any cross section of the bar, we conclude that the condition of insulation is that this quantity vanish. Thus

$$u_x = 0 \tag{13}$$

is the boundary condition at an insulated end.

A more general type of boundary condition occurs if the rate of flow of heat through an end of the bar is proportional to the temperature there. Let us consider the end $x = 0$, where the rate of flow of heat from left to right is given by $-\kappa A u_x(0, t)$; see Eq. (2). Hence the rate of heat flow out of the bar (from right to left) at $x = 0$ is $\kappa A u_x(0, t)$. If this quantity is proportional to the temperature $u(0, t)$, then we obtain the boundary condition

$$u_x(0, t) - h_1 u(0, t) = 0, \qquad t > 0, \tag{14}$$

where h_1 is a nonnegative constant of proportionality. Note that $h_1 = 0$ corresponds to an insulated end and that $h_1 \to \infty$ corresponds to an end held at zero temperature.

If heat flow is taking place at the right end of the bar ($x = L$), then in a similar way we obtain the boundary condition

$$u_x(L, t) + h_2 u(L, t) = 0, \qquad t > 0, \tag{15}$$

where again h_2 is a nonnegative constant of proportionality.

Finally, to determine completely the flow of heat in the bar it is necessary to state the temperature distribution at one fixed instant, usually taken as the initial time $t = 0$. This initial condition is of the form

$$u(x, 0) = f(x), \qquad 0 \le x \le L. \tag{16}$$

The problem then is to determine the solution of the differential equation (10) subject to one of the boundary conditions (12) to (15) at each end and to the initial condition (16) at $t = 0$.

Several generalizations of the heat equation (10) also occur in practice. First, the bar material may be nonuniform and the cross section may not be constant along the length of the bar. In this case, the parameters κ, ρ, s, and A may depend on the axial variable x. Going back to Eq. (2), we see that the rate of heat transfer from left to right across the cross section at $x = x_0$ is now given by

$$H(x_0, t) = -\kappa(x_0)A(x_0)u_x(x_0, t) \tag{17}$$

with a similar expression for $H(x_0 + \Delta x, t)$. If we introduce these quantities into Eq. (4) and eventually into Eq. (9), and proceed as before, we obtain the partial differential equation

$$(\kappa A u_x)_x = s\rho A u_t. \tag{18}$$

We will usually write Eq. (18) in the form

$$r(x)u_t = [p(x)u_x]_x, \tag{19}$$

where $p(x) = \kappa(x)A(x)$ and $r(x) = s(x)\rho(x)A(x)$. Note that both of these quantities are intrinsically positive.

A second generalization occurs if there are other ways in which heat enters or leaves the bar. Suppose that there is a *source* that adds heat to the bar at a rate $G(x, t, u)$ per unit time per unit length, where $G(x, t, u) > 0$. In this case we must add the term $G(x, t, u) \, \Delta x \, \Delta t$ to the left side of Eq. (9), and this leads to the differential equation

$$r(x)u_t = [p(x)u_x]_x + G(x, t, u). \tag{20}$$

If $G(x, t, u) < 0$, then we speak of a *sink* that removes heat from the bar at the rate $G(x, t, u)$ per unit time per unit length. To make the problem tractable we must restrict the form of the function G. In particular, we assume that G is linear in u and that the coefficient of u does not depend on t. Thus we write

$$G(x, t, u) = F(x, t) - q(x)u. \tag{21}$$

The minus sign in Eq. (21) has been introduced so that certain equations that appear later will have their customary forms. Substituting from Eq. (21) into Eq. (20), we obtain

$$r(x)u_t = [p(x)u_x]_x - q(x)u + F(x, t). \tag{22}$$

This equation is sometimes called the generalized heat conduction equation. Boundary value problems for Eq. (22) will be discussed to some extent in Chapter 11.

Finally, if instead of a one-dimensional bar, we consider a body with more than one significant space dimension, then the temperature is a function of two or three space coordinates rather than of x alone. Considerations similar to those leading to Eq. (10) can be employed to derive the heat conduction equation in two dimensions,

$$\alpha^2(u_{xx} + u_{yy}) = u_t, \tag{23}$$

or in three dimensions,

$$\alpha^2(u_{xx} + u_{yy} + u_{zz}) = u_t. \tag{24}$$

The boundary conditions corresponding to Eqs. (12) and (13) for multidimensional problems correspond to a prescribed temperature distribution on the boundary, or to an insulated boundary. Similarly, the initial temperature distribution will in general be a function of x and y for Eq. (23) and a function of x, y, and z for Eq. (24).

Derivation of the Wave Equation. In this appendix we derive the wave equation in one space dimension as it applies to the transverse vibrations of an elastic string, or cable; the elastic string may be thought of as a violin string, a guy wire, or possibly an electric power line. The same equation, however, with the variables properly interpreted, occurs in many other wave problems having only one significant space variable.

Consider a perfectly flexible elastic string stretched tightly between supports fixed at the same horizontal level (see Figure 10.B.1a). Let the x-axis lie along the string with the endpoints located at $x = 0$ and $x = L$. If the string is set in motion at some initial time $t = 0$ (by plucking, for example) and is thereafter left undisturbed, it will vibrate freely in a vertical plane, provided that damping effects, such as air resistance, are neglected. To determine the differential equation governing this motion we will consider the forces acting on a small element of the string of length Δx lying between the points x and $x + \Delta x$ (see Figure 10.B.1b). We assume that the motion of the string is small and that, as a consequence, each point on the string moves solely in a vertical line. We denote by $u(x, t)$ the vertical displacement of the point x at the time t. Let the tension in the string, which always acts in the tangential direction, be denoted by $T(x, t)$, and let ρ denote the mass per unit length of the string.

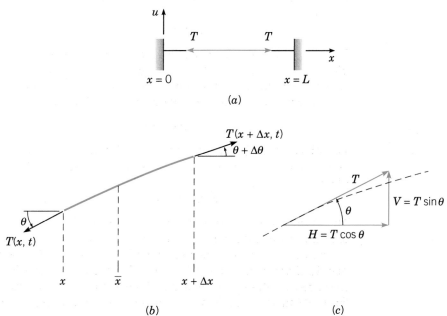

FIGURE 10.B.1 (a) An elastic string under tension. (b) An element of the displaced string. (c) Resolution of the tension T into components.

Newton's law, as it applies to the element Δx of the string, states that the net external force, due to the tension at the ends of the element, must be equal to the product of the mass of the element and the acceleration of its mass center. Since

there is no horizontal acceleration, the horizontal components must satisfy

$$T(x + \Delta x, t) \cos(\theta + \Delta\theta) - T(x, t) \cos\theta = 0. \tag{1}$$

If we denote the horizontal component of the tension (see Figure 10.B.1c) by H, then Eq. (1) states that H is independent of x.

On the other hand, the vertical components satisfy

$$T(x + \Delta x, t) \sin(\theta + \Delta\theta) - T(x, t) \sin\theta = \rho \,\Delta x\, u_{tt}(\overline{x}, t), \tag{2}$$

where \overline{x} is the coordinate of the center of mass of the element of the string under consideration. Clearly, \overline{x} lies in the interval $x < \overline{x} < x + \Delta x$. The weight of the string, which acts vertically downward, is assumed to be negligible and has been neglected in Eq. (2).

If the vertical component of T is denoted by V, then Eq. (2) can be written as

$$\frac{V(x + \Delta x, t) - V(x, t)}{\Delta x} = \rho u_{tt}(\overline{x}, t).$$

Passing to the limit as $\Delta x \to 0$ gives

$$V_x(x, t) = \rho u_{tt}(x, t). \tag{3}$$

To express Eq. (3) entirely in terms of u we note that

$$V(x, t) = H(t) \tan\theta = H(t) u_x(x, t).$$

Hence Eq. (3) becomes

$$(Hu_x)_x = \rho u_{tt},$$

or, since H is independent of x,

$$Hu_{xx} = \rho u_{tt}. \tag{4}$$

For small motions of the string it is permissible to replace $H = T \cos\theta$ by T. Then Eq. (4) takes its customary form

$$a^2 u_{xx} = u_{tt}, \tag{5}$$

where

$$a^2 = T/\rho. \tag{6}$$

We will assume further that a^2 is a constant, although this is not required in our derivation, even for small motions. Equation (5) is called the wave equation for one space dimension. Since T has the dimension of force, and ρ that of mass/length, it follows that the constant a has the dimension of velocity. It is possible to identify a as the velocity with which a small disturbance (wave) moves along the string. According to Eq. (6), the wave velocity a varies directly with the tension in the string, but inversely with the density of the string material. These facts are in agreement with experience.

As in the case of the heat conduction equation, there are various generalizations of the wave equation (5). One important equation is known as the **telegraph equation** and has the form

$$u_{tt} + cu_t + ku = a^2 u_{xx} + F(x, t), \tag{7}$$

where c and k are nonnegative constants. The terms cu_t, ku, and $F(x,t)$ arise from a viscous damping force, an elastic restoring force, and an external force, respectively. Note the similarity of Eq. (7), except for the term $a^2 u_{xx}$, with the equation for the spring–mass system derived in Section 3.8; the additional term $a^2 u_{xx}$ arises from a consideration of internal elastic forces.

The telegraph equation also governs the flow of voltage, or current, in a transmission line (hence its name); in this case the coefficients are related to electrical parameters in the line.

For a vibrating system with more than one significant space coordinate, it may be necessary to consider the wave equation in two dimensions,

$$a^2(u_{xx} + u_{yy}) = u_{tt}, \tag{8}$$

or in three dimensions,

$$a^2(u_{xx} + u_{yy} + u_{zz}) = u_{tt}. \tag{9}$$

REFERENCES

The following books contain additional information on Fourier series:

Buck, R. C., and Buck, E. F., *Advanced Calculus* (3rd ed.) (New York: McGraw-Hill, 1978).

Carslaw, H. S., *Introduction to the Theory of Fourier's Series and Integrals* (3rd ed.) (Cambridge: Cambridge University Press, 1930; New York: Dover, 1952).

Courant, R., and John, F., *Introduction to Calculus and Analysis* (New York: Wiley-Interscience, 1965; reprinted by Springer-Verlag, New York, 1989).

Kaplan, W., *Advanced Calculus* (5th ed.) (Reading, MA: Addison-Wesley, 2003).

A brief biography of Fourier and an annotated copy of his 1807 paper are contained in

Grattan-Guinness, I., *Joseph Fourier 1768–1830* (Cambridge, MA: MIT Press, 1973).

Useful references on partial differential equations and the method of separation of variables include the following:

Churchill, R. V., and Brown, J. W., *Fourier Series and Boundary Value Problems* (6th ed.) (New York: McGraw-Hill, 2000).

Haberman, R., *Elementary Applied Partial Differential Equations* (3rd ed.) (Englewood Cliffs, NJ: Prentice-Hall, 1998).

Pinsky, M. A., *Partial Differential Equations and Boundary Value Problems with Applications* (3rd ed.) (Boston: WCB/McGraw-Hill, 1998).

Powers, D. L., *Boundary Value Problems* (4th ed.) (San Diego: Academic Press, 1999).

Strauss, W. A., *Partial Differential Equations, an Introduction* (New York: Wiley, 1992).

Weinberger, H. F., *A First Course in Partial Differential Equations* (New York: Wiley, 1965; New York: Dover, 1995).

Boundary Value Problems and Sturm–Liouville Theory

As a result of separating variables in a partial differential equation in Chapter 10, we repeatedly encountered the differential equation

$$X'' + \lambda X = 0, \qquad 0 < x < L$$

with the boundary conditions

$$X(0) = 0, \qquad X(L) = 0.$$

This boundary value problem is the prototype of a large class of problems that are important in applied mathematics. These problems are known as Sturm–Liouville boundary value problems. In this chapter we discuss the major properties of Sturm–Liouville problems and their solutions; in the process we are able to generalize somewhat the method of separation of variables for partial differential equations.

11.1 The Occurrence of Two-Point Boundary Value Problems

In Chapter 10 we described the method of separation of variables as a means of solving certain problems involving partial differential equations. The heat conduction problem consisting of the partial differential equation

$$\alpha^2 u_{xx} = u_t, \qquad 0 < x < L, \quad t > 0, \tag{1}$$

subject to the boundary conditions

$$u(0, t) = 0, \qquad u(L, t) = 0, \qquad t > 0 \tag{2}$$

657

and the initial condition

$$u(x,0) = f(x), \qquad 0 \leq x \leq L, \tag{3}$$

is typical of the problems considered there. A crucial part of the process of solving such problems is to find the eigenvalues and eigenfunctions of the differential equation

$$X'' + \lambda X = 0, \qquad 0 < x < L \tag{4}$$

with the boundary conditions

$$X(0) = 0, \qquad X(L) = 0 \tag{5}$$

or perhaps

$$X'(0) = 0, \qquad X'(L) = 0. \tag{6}$$

The sine or cosine functions that result from solving Eq. (4) subject to the boundary conditions (5) or (6) are used to expand the initial temperature distribution $f(x)$ in a Fourier series.

In this chapter we extend and generalize the results of Chapter 10. Our main goal is to show how the method of separation of variables can be used to solve problems somewhat more general than that of Eqs. (1), (2), and (3). We are interested in three types of generalizations.

First, we wish to consider more general partial differential equations—for example, the equation

$$r(x)u_t = [p(x)u_x]_x - q(x)u + F(x,t). \tag{7}$$

This equation can arise, as indicated in Appendix A of Chapter 10, in the study of heat conduction in a bar of variable material properties in the presence of heat sources. If p and r are constants, and if the source terms qu and F are zero, then Eq. (7) reduces to Eq. (1). The partial differential equation (7) also occurs in the investigation of other phenomena of a diffusive character.

A second generalization is to allow more general boundary conditions. In particular, we wish to consider boundary conditions of the form

$$u_x(0,t) - h_1 u(0,t) = 0, \qquad u_x(L,t) + h_2 u(L,t) = 0. \tag{8}$$

Such conditions occur when the rate of heat flow through an end of the bar is proportional to the temperature there. Usually, h_1 and h_2 are nonnegative constants, but in some cases they may be negative or depend on t. The boundary conditions (2) are obtained in the limit as $h_1 \to \infty$ and $h_2 \to \infty$. The other important limiting case, $h_1 = h_2 = 0$, gives the boundary conditions for insulated ends.

The final generalization that we discuss in this chapter concerns the geometry of the region in which the problem is posed. The results of Chapter 10 are adequate only for a rather restricted class of problems, chiefly those in which the region of interest is rectangular or, in a few cases, circular. Later in this chapter we consider certain problems posed in a few other geometrical regions.

Let us consider the equation

$$r(x)u_t = [p(x)u_x]_x - q(x)u \tag{9}$$

obtained by setting the term $F(x,t)$ in Eq. (7) equal to zero. To separate the variables we assume that

$$u(x,t) = X(x)T(t),\tag{10}$$

and substitute for u in Eq. (9). We obtain

$$r(x)XT' = [p(x)X']'T - q(x)XT\tag{11}$$

or, upon dividing by $r(x)XT$,

$$\frac{T'}{T} = \frac{[p(x)X']'}{r(x)X} - \frac{q(x)}{r(x)} = -\lambda.\tag{12}$$

We have denoted the separation constant by $-\lambda$ in anticipation of the fact that usually it will turn out to be real and negative. From Eq. (12) we obtain the following two ordinary differential equations for X and T:

$$[p(x)X']' - q(x)X + \lambda r(x)X = 0,\tag{13}$$

$$T' + \lambda T = 0.\tag{14}$$

If we substitute from Eq. (10) for u in Eqs. (8) and assume that h_1 and h_2 are constants, then we obtain the boundary conditions

$$X'(0) - h_1 X(0) = 0, \qquad X'(L) + h_2 X(L) = 0.\tag{15}$$

To proceed further, we need to solve Eq. (13) subject to the boundary conditions (15). Although this is a more general linear homogeneous two-point boundary value problem than the problem consisting of the differential equation (4) and the boundary conditions (5) or (6), the solutions behave in very much the same way. For every value of λ, the problem (13), (15) has the trivial solution $X(x) = 0$. For certain values of λ, called **eigenvalues**, there are also other, nontrivial solutions, called **eigenfunctions**. These eigenfunctions form the basis for series solutions of a variety of problems in partial differential equations, such as the generalized heat conduction equation (9) subject to the boundary conditions (8) and the initial condition (3).

In this chapter we discuss some of the properties of solutions of two-point boundary value problems for second order linear equations. Sometimes we consider the general linear homogeneous equation

$$P(x)y'' + Q(x)y' + R(x)y = 0,\tag{16}$$

investigated in Chapter 3. However, for most purposes it is better to discuss equations in which the first and second derivative terms are related as in Eq. (13). It is always possible to transform the general equation (16) so that the derivative terms appear as in Eq. (13) (see Problem 11).

Boundary value problems with higher order differential equations can also occur; in them the number of boundary conditions must equal the order of the differential equation. As a rule, the order of the differential equation is even, and half the boundary conditions are given at each end of the interval. It is also possible for a single boundary condition to involve values of the solution and/or its derivatives at both boundary points; for example,

$$y(0) - y(L) = 0.\tag{17}$$

The following example involves one boundary condition of the form (15) and is therefore more complicated than the problems in Section 10.1.

EXAMPLE

1

Find the eigenvalues and the corresponding eigenfunctions of the boundary value problem

$$y'' + \lambda y = 0, \tag{18}$$

$$y(0) = 0, \qquad y'(1) + y(1) = 0. \tag{19}$$

One place where this problem occurs is in the heat conduction problem in a bar of unit length. The boundary condition at $x = 0$ corresponds to a zero temperature there. The boundary condition at $x = 1$ corresponds to a rate of heat flow that is proportional to the temperature there, and units are chosen so that the constant of proportionality is 1 (see Appendix A of Chapter 10).

The solution of the differential equation may have one of several forms, depending on λ, so it is necessary to consider several cases. First, if $\lambda = 0$, the general solution of the differential equation is

$$y = c_1 x + c_2. \tag{20}$$

The two boundary conditions require that

$$c_2 = 0, \qquad 2c_1 + c_2 = 0, \tag{21}$$

respectively. The only solution of Eqs. (21) is $c_1 = c_2 = 0$, so the boundary value problem has no nontrivial solution in this case. Hence $\lambda = 0$ is not an eigenvalue.

If $\lambda > 0$, then the general solution of the differential equation (18) is

$$y = c_1 \sin \sqrt{\lambda}\, x + c_2 \cos \sqrt{\lambda}\, x, \tag{22}$$

where $\sqrt{\lambda} > 0$. The boundary condition at $x = 0$ requires that $c_2 = 0$; from the boundary condition at $x = 1$ we then obtain the equation

$$c_1(\sin \sqrt{\lambda} + \sqrt{\lambda} \cos \sqrt{\lambda}) = 0.$$

For a nontrivial solution y we must have $c_1 \neq 0$, and thus λ must satisfy

$$\sin \sqrt{\lambda} + \sqrt{\lambda} \cos \sqrt{\lambda} = 0. \tag{23}$$

Note that if λ is such that $\cos \sqrt{\lambda} = 0$, then $\sin \sqrt{\lambda} \neq 0$, and Eq. (23) is not satisfied. Hence we may assume that $\cos \sqrt{\lambda} \neq 0$; dividing Eq. (23) by $\cos \sqrt{\lambda}$, we obtain

$$\sqrt{\lambda} = -\tan \sqrt{\lambda}. \tag{24}$$

The solutions of Eq. (24) can be determined numerically. They can also be found approximately by sketching the graphs of $f(\sqrt{\lambda}) = \sqrt{\lambda}$ and $g(\sqrt{\lambda}) = -\tan \sqrt{\lambda}$ for $\sqrt{\lambda} > 0$ on the same set of axes, and identifying the points of intersection of the two curves (see Figure 11.1.1). The point $\sqrt{\lambda} = 0$ is specifically excluded from this argument because the solution (22) is valid only for $\sqrt{\lambda} \neq 0$. Despite the fact that the curves intersect there, $\lambda = 0$ is not an eigenvalue, as we have already shown. The first three positive solutions of Eq. (24) are $\sqrt{\lambda_1} \cong 2.029$, $\sqrt{\lambda_2} \cong 4.913$, and $\sqrt{\lambda_3} \cong 7.979$. As can be seen from Figure 11.1.1, the other roots are given

with reasonable accuracy by $\sqrt{\lambda_n} \cong (2n-1)\pi/2$ for $n = 4, 5, \ldots,$ the precision of this estimate improving as n increases. Hence the eigenvalues are

$$\lambda_1 \cong 4.116, \qquad \lambda_2 \cong 24.14,$$
$$\lambda_3 \cong 63.66, \qquad \lambda_n \cong (2n-1)^2\pi^2/4 \qquad \text{for } n = 4, 5, \ldots. \tag{25}$$

Finally, since $c_2 = 0$, the eigenfunction corresponding to the eigenvalue λ_n is

$$\phi_n(x, \lambda_n) = k_n \sin \sqrt{\lambda_n}\, x; \qquad n = 1, 2, \ldots, \tag{26}$$

where the constant k_n remains arbitrary.

Next consider $\lambda < 0$. In this case it is convenient to let $\lambda = -\mu$ so that $\mu > 0$. Then Eq. (14) becomes

$$y'' - \mu y = 0, \tag{27}$$

and its general solution is

$$y = c_1 \sinh \sqrt{\mu}\, x + c_2 \cosh \sqrt{\mu}\, x, \tag{28}$$

where $\sqrt{\mu} > 0$. Proceeding as in the previous case, we find that μ must satisfy the equation

$$\sqrt{\mu} = -\tanh \sqrt{\mu}. \tag{29}$$

From Figure 11.1.2 it is clear that the graphs of $f(\sqrt{\mu}) = \sqrt{\mu}$ and $g(\sqrt{\mu}) = -\tanh \sqrt{\mu}$ intersect only at the origin. Hence there are no positive values of $\sqrt{\mu}$ that satisfy Eq. (29), and hence the boundary value problem (18), (19) has no negative eigenvalues.

FIGURE 11.1.1 Graphical solution of $\sqrt{\lambda} = -\tan\sqrt{\lambda}$.

FIGURE 11.1.2 Graphical solution of $\sqrt{\mu} = -\tanh \sqrt{\mu}$.

Finally, it is necessary to consider the possibility that λ may be complex. It is possible to show by direct calculation that the problem (18), (19) has no complex eigenvalues. However, in Section 11.2 we consider in more detail a large class of problems that includes this example. One of the things we show there is that every problem in this class has only real eigenvalues. Therefore we omit the discussion of the nonexistence of complex eigenvalues here. Thus we conclude that all the eigenvalues and eigenfunctions of the problem (18), (19) are given by Eqs. (25) and (26).

PROBLEMS

In each of Problems 1 through 6 state whether the given boundary value problem is homogeneous or nonhomogeneous.

1. $y'' + 4y = 0$, $y(-1) = 0$, $y(1) = 0$
2. $[(1+x^2)y']' + 4y = 0$, $y(0) = 0$, $y(1) = 1$
3. $y'' + 4y = \sin x$, $y(0) = 0$, $y(1) = 0$
4. $-y'' + x^2y = \lambda y$, $y'(0) - y(0) = 0$, $y'(1) + y(1) = 0$
5. $-[(1+x^2)y']' = \lambda y + 1$, $y(-1) = 0$, $y(1) = 0$
6. $-y'' = \lambda(1+x^2)y$, $y(0) = 0$, $y'(1) + 3y(1) = 0$

In each of Problems 7 through 10 determine the form of the eigenfunctions and the determinantal equation satisfied by the nonzero eigenvalues. Determine whether $\lambda = 0$ is an eigenvalue, and find approximate values for λ_1 and λ_2, the nonzero eigenvalues of smallest absolute value. Estimate λ_n for large values of n. Assume that all eigenvalues are real.

7. $y'' + \lambda y = 0$, $y(0) = 0$, $y(\pi) + y'(\pi) = 0$
8. $y'' + \lambda y = 0$, $y'(0) = 0$, $y(1) + y'(1) = 0$
9. $y'' + \lambda y = 0$, $y(0) - y'(0) = 0$, $y(1) + y'(1) = 0$
10. $y'' - \lambda y = 0$, $y(0) + y'(0) = 0$, $y(1) = 0$

11. Consider the general linear homogeneous second order equation

$$P(x)y'' + Q(x)y' + R(x)y = 0. \tag{i}$$

We seek an integrating factor $\mu(x)$ such that, upon multiplying Eq. (i) by $\mu(x)$, we can write the resulting equation in the form

$$[\mu(x)P(x)y']' + \mu(x)R(x)y = 0. \tag{ii}$$

(a) By equating coefficients of y' in Eqs. (i) and (ii) show that μ must be a solution of

$$P\mu' = (Q - P')\mu. \qquad \text{(iii)}$$

(b) Solve Eq. (iii) and thereby show that

$$\mu(x) = \frac{1}{P(x)} \exp \int_{x_0}^{x} \frac{Q(s)}{P(s)} \, ds. \qquad \text{(iv)}$$

Compare this result with that of Problem 28 in Section 3.2.

In each of Problems 12 through 15 use the method of Problem 11 to transform the given equation into the form $[p(x)y']' + q(x)y = 0$.

12. $y'' - 2xy' + \lambda y = 0,$ Hermite equation

13. $x^2 y'' + xy' + (x^2 - v^2)y = 0,$ Bessel equation

14. $xy'' + (1 - x)y' + \lambda y = 0,$ Laguerre equation

15. $(1 - x^2)y'' - xy' + \alpha^2 y = 0,$ Chebyshev equation

16. The equation

$$u_{tt} + cu_t + ku = a^2 u_{xx} + F(x, t), \qquad \text{(i)}$$

where $a^2 > 0, c \geq 0,$ and $k \geq 0$ are constants, is known as the **telegraph equation**. It arises in the study of an elastic string under tension (see Appendix B of Chapter 10). Equation (i) also occurs in other applications. Assuming that $F(x, t) = 0$, let $u(x, t) = X(x)T(t)$, separate the variables in Eq. (i), and derive ordinary differential equations for X and T.

17. Consider the boundary value problem

$$y'' - 2y' + (1 + \lambda)y = 0, \qquad y(0) = 0, \quad y(1) = 0.$$

(a) Introduce a new dependent variable u by the relation $y = s(x)u$. Determine $s(x)$ so that the differential equation for u has no u' term.

(b) Solve the boundary value problem for u and thereby determine the eigenvalues and eigenfunctions of the original problem. Assume that all eigenvalues are real.

(c) Also solve the given problem directly (without introducing u).

18. Consider the boundary value problem

$$y'' + 4y' + (4 + 9\lambda)y = 0, \qquad y(0) = 0, \quad y'(L) = 0.$$

(a) Determine, at least approximately, the real eigenvalues and the corresponding eigenfunctions by proceeding as in Problem 17(a, b).

(b) Also solve the given problem directly (without introducing a new variable).
Hint: In part (a) be sure to pay attention to the boundary conditions as well as the differential equation.

The differential equations in Problems 19 and 20 differ from those in previous problems in that the parameter λ multiplies the y' term as well as the y term. In each of these problems determine the real eigenvalues and the corresponding eigenfunctions.

19. $y'' + y' + \lambda(y' + y) = 0,$ $y'(0) = 0,$ $y(1) = 0$

20. $x^2 y'' - \lambda(xy' - y) = 0,$ $y(1) = 0,$ $y(2) - y'(2) = 0$

21. Consider the problem

$$y'' + \lambda y = 0, \qquad 2y(0) + y'(0) = 0, \quad y(1) = 0.$$

(a) Find the determinantal equation satisfied by the positive eigenvalues. Show that there is an infinite sequence of such eigenvalues. Find λ_1 and λ_2. Then show that $\lambda_n \cong [(2n+1)\pi/2]^2$ for large n.

(b) Find the determinantal equation satisfied by the negative eigenvalues. Show that there is exactly one negative eigenvalue and find its value.

22. Consider the problem

$$y'' + \lambda y = 0, \qquad \alpha y(0) + y'(0) = 0, \quad y(1) = 0,$$

where α is a given constant.

(a) Show that for all values of α there is an infinite sequence of positive eigenvalues.

(b) If $\alpha < 1$, show that all (real) eigenvalues are positive. Show that the smallest eigenvalue approaches zero as α approaches 1 from below.

(c) Show that $\lambda = 0$ is an eigenvalue only if $\alpha = 1$.

(d) If $\alpha > 1$, show that there is exactly one negative eigenvalue and that this eigenvalue decreases as α increases.

23. Consider the problem

$$y'' + \lambda y = 0, \qquad y(0) = 0, \quad y'(L) = 0.$$

Show that if ϕ_m and ϕ_n are eigenfunctions corresponding to the eigenvalues λ_m and λ_n, respectively, with $\lambda_m \neq \lambda_n$, then

$$\int_0^L \phi_m(x)\phi_n(x)\,dx = 0.$$

Hint: Note that

$$\phi_m'' + \lambda_m \phi_m = 0, \qquad \phi_n'' + \lambda_n \phi_n = 0.$$

Multiply the first of these equations by ϕ_n, the second by ϕ_m, and integrate from 0 to L, using integration by parts. Finally, subtract one equation from the other.

24. In this problem we consider a higher order eigenvalue problem. In the study of transverse vibrations of a uniform elastic bar one is led to the differential equation

$$y^{(4)} - \lambda y = 0,$$

where y is the transverse displacement and $\lambda = m\omega^2/EI$; m is the mass per unit length of the rod, E is Young's modulus, I is the moment of inertia of the cross section about an axis through the centroid perpendicular to the plane of vibration, and ω is the frequency of vibration. Thus for a bar whose material and geometrical properties are given, the eigenvalues determine the natural frequencies of vibration. Boundary conditions at each end are usually one of the following types:

$$y = y' = 0, \qquad \text{clamped end,}$$

$$y = y'' = 0, \qquad \text{simply supported or hinged end,}$$

$$y'' = y''' = 0, \qquad \text{free end.}$$

For each of the following three cases find the form of the eigenfunctions and the equation satisfied by the eigenvalues of this fourth order boundary value problem. Determine λ_1 and λ_2, the two eigenvalues of smallest magnitude. Assume that the eigenvalues are real and positive.

(a) $y(0) = y''(0) = 0,$ $y(L) = y''(L) = 0$

(b) $y(0) = y''(0) = 0,$ $y(L) = y'(L) = 0$

(c) $y(0) = y'(0) = 0,$ $y''(L) = y'''(L) = 0$ (cantilevered bar)

25. This problem illustrates that the eigenvalue parameter sometimes appears in the boundary conditions as well as in the differential equation. Consider the longitudinal vibrations of a uniform straight elastic bar of length L. It can be shown that the axial displacement $u(x, t)$ satisfies the partial differential equation

$$(E/\rho)u_{xx} = u_{tt}; \qquad 0 < x < L, \quad t > 0, \tag{i}$$

where E is Young's modulus and ρ is the mass per unit volume. If the end $x = 0$ is fixed, then the boundary condition there is

$$u(0, t) = 0, \qquad t > 0. \tag{ii}$$

Suppose that the end $x = L$ is rigidly attached to a mass m but is otherwise unrestrained. We can obtain the boundary condition here by writing Newton's law for the mass. From the theory of elasticity it can be shown that the force exerted by the bar on the mass is given by $-EAu_x(L, t)$. Hence the boundary condition is

$$EAu_x(L, t) + mu_{tt}(L, t) = 0, \qquad t > 0. \tag{iii}$$

(a) Assume that $u(x, t) = X(x)T(t)$, and show that $X(x)$ and $T(t)$ satisfy the differential equations

$$X'' + \lambda X = 0, \tag{iv}$$

$$T'' + \lambda(E/\rho)T = 0. \tag{v}$$

(b) Show that the boundary conditions are

$$X(0) = 0, \qquad X'(L) - \gamma\lambda LX(L) = 0, \tag{vi}$$

where $\gamma = m/\rho AL$ is a dimensionless parameter that gives the ratio of the end mass to the mass of the rod.
Hint: Use the differential equation for $T(t)$ in simplifying the boundary condition at $x = L$.

(c) Determine the form of the eigenfunctions and the equation satisfied by the real eigenvalues of Eqs. (iv) and (vi). Find the first two eigenvalues λ_1 and λ_2 if $\gamma = 0.5$.

11.2 Sturm–Liouville Boundary Value Problems

We now consider two-point boundary value problems of the type obtained in Section 11.1 by separating the variables in a heat conduction problem for a bar of variable material properties and with a source term proportional to the temperature. This kind of problem also occurs in many other applications.

These boundary value problems are commonly associated with the names of Sturm and Liouville.[1] They consist of a differential equation of the form

$$[p(x)y']' - q(x)y + \lambda r(x)y = 0 \tag{1}$$

on the interval $0 < x < 1$, together with the boundary conditions

$$\alpha_1 y(0) + \alpha_2 y'(0) = 0, \qquad \beta_1 y(1) + \beta_2 y'(1) = 0 \tag{2}$$

at the endpoints. It is often convenient to introduce the linear homogeneous differential operator L defined by

$$L[y] = -[p(x)y']' + q(x)y. \tag{3}$$

Then the differential equation (1) can be written as

$$L[y] = \lambda r(x)y. \tag{4}$$

We assume that the functions p, p', q, and r are continuous on the interval $0 \le x \le 1$ and, further, that $p(x) > 0$ and $r(x) > 0$ at all points in $0 \le x \le 1$. These assumptions are necessary to render the theory as simple as possible while retaining considerable generality. It turns out that these conditions are satisfied in many significant problems in mathematical physics. For example, the equation $y'' + \lambda y = 0$, which arose repeatedly in the preceding chapter, is of the form (1) with $p(x) = 1$, $q(x) = 0$, and $r(x) = 1$. The boundary conditions (2) are said to be **separated**; that is, each involves only one of the boundary points. These are the most general separated boundary conditions that are possible for a second order differential equation.

Before proceeding to establish some of the properties of the Sturm–Liouville problem (1), (2), it is necessary to derive an identity, known as **Lagrange's identity**, which is basic to the study of linear boundary value problems. Let u and v be functions having continuous second derivatives on the interval $0 \le x \le 1$. Then[2]

$$\int_0^1 L[u]v \, dx = \int_0^1 [-(pu')'v + quv] \, dx.$$

Integrating the first term on the right side twice by parts, we obtain

$$\int_0^1 L[u]v \, dx = -p(x)u'(x)v(x) \Big|_0^1 + p(x)u(x)v'(x) \Big|_0^1 + \int_0^1 [-u(pv')' + uqv] \, dx$$

$$= -p(x)[u'(x)v(x) - u(x)v'(x)] \Big|_0^1 + \int_0^1 uL[v] \, dx.$$

[1]Charles-François Sturm (1803–1855) and Joseph Liouville (1809–1882), in a series of papers in 1836 and 1837, set forth many properties of the class of boundary value problems associated with their names, including the results stated in Theorems 11.2.1 to 11.2.4. Sturm is also famous for a theorem on the number of real zeros of a polynomial and, in addition, did extensive work in physics and mechanics. Besides his own research in analysis, algebra, and number theory, Liouville was the founder, and for 39 years the editor, of the influential *Journal de mathématiques pures et appliquées*. One of his most important results was the proof (in 1844) of the existence of transcendental numbers.

[2]For brevity we sometimes use the notation $\int_0^1 f \, dx$ rather than $\int_0^1 f(x) \, dx$ in this chapter.

Hence, upon transposing the integral on the right side, we have

$$\int_0^1 \{L[u]v - uL[v]\}\, dx = -p(x)\left[u'(x)v(x) - u(x)v'(x)\right]\Big|_0^1, \tag{5}$$

which is Lagrange's identity.

Now let us suppose that the functions u and v in Eq. (5) also satisfy the boundary conditions (2). Then, if we assume that $\alpha_2 \neq 0$ and $\beta_2 \neq 0$, the right side of Eq. (5) becomes

$$-p(x)\left[u'(x)v(x) - u(x)v'(x)\right]\Big|_0^1$$

$$= -p(1)\left[u'(1)v(1) - u(1)v'(1)\right] + p(0)\left[u'(0)v(0) - u(0)v'(0)\right]$$

$$= -p(1)\left[-\frac{\beta_1}{\beta_2}u(1)v(1) + \frac{\beta_1}{\beta_2}u(1)v(1)\right] + p(0)\left[-\frac{\alpha_1}{\alpha_2}u(0)v(0) + \frac{\alpha_1}{\alpha_2}u(0)v(0)\right]$$

$$= 0.$$

The same result holds if either α_2 or β_2 is zero; the proof in this case is even simpler and is left for you. Thus, if the differential operator L is defined by Eq. (3), and if the functions u and v satisfy the boundary conditions (2), then Lagrange's identity reduces to

$$\int_0^1 \{L[u]v - uL[v]\}\, dx = 0. \tag{6}$$

Let us now write Eq. (6) in a slightly different way. In Eq. (4) of Section 10.2 we introduced the inner product (u, v) of two real-valued functions u and v on a given interval; using the interval $0 \leq x \leq 1$, we have

$$(u, v) = \int_0^1 u(x)v(x)\, dx. \tag{7}$$

In this notation Eq. (6) becomes

$$(L[u], v) - (u, L[v]) = 0. \tag{8}$$

In proving Theorem 11.2.1 below it is necessary to deal with complex-valued functions. By analogy with the definition in Section 7.2 for vectors, we define the inner product of two complex-valued functions on $0 \leq x \leq 1$ as

$$(u, v) = \int_0^1 u(x)\overline{v}(x)\, dx, \tag{9}$$

where \overline{v} is the complex conjugate of v. Clearly, Eq. (9) coincides with Eq. (7) if $u(x)$ and $v(x)$ are real. It is important to know that Eq. (8) remains valid under the stated conditions if u and v are complex-valued functions and if the inner product (9) is used. To see this, one can start with the quantity $\int_0^1 L[u]\overline{v}\, dx$ and retrace the steps leading to Eq. (6), making use of the fact that $p(x), q(x), \alpha_1, \alpha_2, \beta_1$, and β_2 are all real quantities (see Problem 22).

We now consider some of the implications of Eq. (8) for the Sturm–Liouville boundary value problem (1), (2). We assume without proof[3] that this problem actually has eigenvalues and eigenfunctions. In Theorems 11.2.1 to 11.2.4 below, we state several of their important, but relatively elementary, properties. Each of these properties is illustrated by the basic Sturm–Liouville problem

$$y'' + \lambda y = 0, \qquad y(0) = 0, \quad y(1) = 0, \tag{10}$$

whose eigenvalues are $\lambda_n = n^2\pi^2$, with the corresponding eigenfunctions $\phi_n(x) = \sin n\pi x$.

Theorem 11.2.1

All the eigenvalues of the Sturm–Liouville problem (1), (2) are real.

To prove this theorem let us suppose that λ is a (possibly complex) eigenvalue of the problem (1), (2) and that ϕ is a corresponding eigenfunction, also possibly complex-valued. Let us write $\lambda = \mu + i\nu$ and $\phi(x) = U(x) + iV(x)$, where $\mu, \nu, U(x)$, and $V(x)$ are real. Then, if we let $u = \phi$ and also $v = \phi$ in Eq. (8), we have

$$(L[\phi], \phi) = (\phi, L[\phi]). \tag{11}$$

However, we know that $L[\phi] = \lambda r\phi$, so Eq. (11) becomes

$$(\lambda r\phi, \phi) = (\phi, \lambda r\phi). \tag{12}$$

Writing out Eq. (12) in full, using the definition (9) of the inner product, we obtain

$$\int_0^1 \lambda r(x)\phi(x)\overline{\phi}(x)\, dx = \int_0^1 \phi(x)\overline{\lambda}\overline{r}(x)\overline{\phi}(x)\, dx. \tag{13}$$

Since $r(x)$ is real, Eq. (13) reduces to

$$(\lambda - \overline{\lambda}) \int_0^1 r(x)\phi(x)\overline{\phi}(x)\, dx = 0,$$

or

$$(\lambda - \overline{\lambda}) \int_0^1 r(x)[U^2(x) + V^2(x)]\, dx = 0. \tag{14}$$

The integrand in Eq. (14) is nonnegative and not identically zero. Since the integrand is also continuous, it follows that the integral is positive. Therefore, the factor $\lambda - \overline{\lambda} = 2i\nu$ must be zero. Hence $\nu = 0$ and λ is real, so the theorem is proved.

An important consequence of Theorem 11.2.1 is that in finding eigenvalues and eigenfunctions of a Sturm–Liouville boundary value problem, one need look only for real eigenvalues. Recall that this is what we did in Chapter 10. It is also possible to show that the eigenfunctions of the boundary value problem (1), (2) are real. A proof is sketched in Problem 23.

[3]The proof may be found, for example, in the references by Sagan (Chapter 5) or Birkhoff and Rota (Chapter 10).

Theorem 11.2.2

If ϕ_1 and ϕ_2 are two eigenfunctions of the Sturm–Liouville problem (1), (2) corresponding to eigenvalues λ_1 and λ_2, respectively, and if $\lambda_1 \neq \lambda_2$, then

$$\int_0^1 r(x)\phi_1(x)\phi_2(x)\,dx = 0. \tag{15}$$

This theorem expresses the property of **orthogonality** of the eigenfunctions with respect to the weight function r. To prove the theorem we note that ϕ_1 and ϕ_2 satisfy the differential equations

$$L[\phi_1] = \lambda_1 r\phi_1 \tag{16}$$

and

$$L[\phi_2] = \lambda_2 r\phi_2, \tag{17}$$

respectively. If we let $u = \phi_1, v = \phi_2$, and substitute for $L[u]$ and $L[v]$ in Eq. (8), we obtain

$$(\lambda_1 r\phi_1, \phi_2) - (\phi_1, \lambda_2 r\phi_2) = 0,$$

or, using Eq. (9),

$$\lambda_1 \int_0^1 r(x)\phi_1(x)\overline{\phi}_2(x)\,dx - \overline{\lambda}_2 \int_0^1 \phi_1(x)\overline{r}(x)\overline{\phi}_2(x)\,dx = 0.$$

Because $\lambda_2, r(x)$, and $\phi_2(x)$ are real, this equation becomes

$$(\lambda_1 - \lambda_2) \int_0^1 r(x)\phi_1(x)\phi_2(x)\,dx = 0. \tag{18}$$

Since by hypothesis $\lambda_1 \neq \lambda_2$, it follows that ϕ_1 and ϕ_2 must satisfy Eq. (15), and the theorem is proved.

Theorem 11.2.3

The eigenvalues of the Sturm–Liouville problem (1), (2) are all simple; that is, to each eigenvalue there corresponds only one linearly independent eigenfunction. Further, the eigenvalues form an infinite sequence and can be ordered according to increasing magnitude so that

$$\lambda_1 < \lambda_2 < \lambda_3 < \cdots < \lambda_n < \cdots.$$

Moreover, $\lambda_n \to \infty$ as $n \to \infty$.

The proof of this theorem is somewhat more advanced than those of the two previous theorems and will be omitted. However, a proof that the eigenvalues are simple is outlined in Problem 20.

Again we note that all the properties stated in Theorems 11.2.1 to 11.2.3 are exemplified by the eigenvalues $\lambda_n = n^2\pi^2$ and eigenfunctions $\phi_n(x) = \sin n\pi x$ of the example problem (10). Clearly, the eigenvalues are real. The eigenfunctions satisfy the orthogonality relation

$$\int_0^1 \phi_m(x)\phi_n(x)\,dx = \int_0^1 \sin m\pi x \sin n\pi x\,dx = 0, \qquad m \neq n, \tag{19}$$

which was established in Section 10.2 by direct integration. Further, the eigenvalues can be ordered so that $\lambda_1 < \lambda_2 < \cdots$, and $\lambda_n \to \infty$ as $n \to \infty$. Finally, to each eigenvalue there corresponds a single linearly independent eigenfunction.

We will now assume that the eigenvalues of the Sturm–Liouville problem (1), (2) are ordered as indicated in Theorem 11.2.3. Associated with the eigenvalue λ_n is a corresponding eigenfunction ϕ_n, determined up to a multiplicative constant. It is often convenient to choose the arbitrary constant multiplying each eigenfunction so as to satisfy the condition

$$\int_0^1 r(x)\phi_n^2(x)\,dx = 1, \qquad n = 1, 2, \ldots. \tag{20}$$

Equation (20) is called a normalization condition, and eigenfunctions satisfying this condition are said to be **normalized**. Indeed, in this case, the eigenfunctions are said to form an **orthonormal set** (with respect to the weight function r) since they already satisfy the orthogonality relation (15). It is sometimes useful to combine Eqs. (15) and (20) into a single equation. To this end we introduce the symbol δ_{mn}, known as the Kronecker[4] delta and defined by

$$\delta_{mn} = \begin{cases} 0, & \text{if } m \neq n, \\ 1, & \text{if } m = n. \end{cases} \tag{21}$$

Making use of the Kronecker delta, we can write Eqs. (15) and (20) as

$$\int_0^1 r(x)\phi_m(x)\phi_n(x)\,dx = \delta_{mn}. \tag{22}$$

EXAMPLE 1

Determine the normalized eigenfunctions of the problem (10):

$$y'' + \lambda y = 0, \qquad y(0) = 0, \quad y(1) = 0.$$

The eigenvalues of this problem are $\lambda_1 = \pi^2$, $\lambda_2 = 4\pi^2, \ldots, \lambda_n = n^2\pi^2, \ldots$, and the corresponding eigenfunctions are $k_1 \sin \pi x$, $k_2 \sin 2\pi x, \ldots, k_n \sin n\pi x, \ldots$, respectively. In this case the weight function is $r(x) = 1$. To satisfy Eq. (20) we must choose k_n so that

$$\int_0^1 (k_n \sin n\pi x)^2 \, dx = 1 \tag{23}$$

for each value of n. Since

$$k_n^2 \int_0^1 \sin^2 n\pi x \, dx = k_n^2 \int_0^1 \left(\tfrac{1}{2} - \tfrac{1}{2}\cos 2n\pi x\right) dx = \tfrac{1}{2}k_n^2,$$

Eq. (23) is satisfied if k_n is chosen to be $\sqrt{2}$ for each value of n. Hence the normalized eigenfunctions of the given boundary value problem are

$$\phi_n(x) = \sqrt{2}\sin n\pi x, \qquad n = 1, 2, 3, \ldots. \tag{24}$$

[4]Leopold Kronecker (1823–1891), a student of Dirichlet, was associated with the University of Berlin for most of his life, although (since he was independently wealthy) he held a faculty position only from 1883 onward. He worked in number theory, elliptic functions, algebra, and their interconnections.

Determine the normalized eigenfunctions of the problem

$$y'' + \lambda y = 0, \qquad y(0) = 0, \quad y'(1) + y(1) = 0. \tag{25}$$

EXAMPLE 2

In Example 1 of Section 11.1 we found that the eigenvalues λ_n satisfy the equation

$$\sin \sqrt{\lambda_n} + \sqrt{\lambda_n} \cos \sqrt{\lambda_n} = 0 \tag{26}$$

and that the corresponding eigenfunctions are

$$\phi_n(x) = k_n \sin \sqrt{\lambda_n}\, x, \tag{27}$$

where k_n is arbitrary. We can determine k_n from the normalization condition (20). Since $r(x) = 1$ in this problem, we have

$$\int_0^1 \phi_n^2(x)\, dx = k_n^2 \int_0^1 \sin^2 \sqrt{\lambda_n}\, x\, dx$$

$$= k_n^2 \int_0^1 \left(\tfrac{1}{2} - \tfrac{1}{2} \cos 2\sqrt{\lambda_n}\, x \right) dx = k_n^2 \left(\frac{x}{2} - \frac{\sin 2\sqrt{\lambda_n}\, x}{4\sqrt{\lambda_n}} \right) \Bigg|_0^1$$

$$= k_n^2 \frac{2\sqrt{\lambda_n} - \sin 2\sqrt{\lambda_n}}{4\sqrt{\lambda_n}} = k_n^2 \frac{\sqrt{\lambda_n} - \sin \sqrt{\lambda_n} \cos \sqrt{\lambda_n}}{2\sqrt{\lambda_n}}$$

$$= k_n^2 \frac{1 + \cos^2 \sqrt{\lambda_n}}{2},$$

where in the last step we have used Eq. (26). Hence, to normalize the eigenfunctions ϕ_n we must choose

$$k_n = \left(\frac{2}{1 + \cos^2 \sqrt{\lambda_n}} \right)^{1/2}. \tag{28}$$

The normalized eigenfunctions of the given problem are

$$\phi_n(x) = \frac{\sqrt{2} \sin \sqrt{\lambda_n}\, x}{(1 + \cos^2 \sqrt{\lambda_n})^{1/2}}; \qquad n = 1, 2, \ldots. \tag{29}$$

We now turn to the question of expressing a given function f as a series of eigenfunctions of the Sturm–Liouville problem (1), (2). We have already seen examples of such expansions in Sections 10.2 to 10.4. For example, it was shown there that if f is continuous and has a piecewise continuous derivative on $0 \le x \le 1$, and satisfies the boundary conditions $f(0) = f(1) = 0$, then f can be expanded in a Fourier sine series of the form

$$f(x) = \sum_{n=1}^{\infty} b_n \sin n\pi x. \tag{30}$$

The functions $\sin n\pi x, n = 1, 2, \ldots$, are precisely the eigenfunctions of the boundary value problem (10). The coefficients b_n are given by

$$b_n = 2 \int_0^1 f(x) \sin n\pi x\, dx \tag{31}$$

and the series (30) converges for each x in $0 \le x \le 1$. In a similar way f can be expanded in a Fourier cosine series using the eigenfunctions $\cos n\pi x, n = 0, 1, 2, \ldots$, of the boundary value problem $y'' + \lambda y = 0$, $y'(0) = 0$, $y'(1) = 0$.

Now suppose that a given function f, satisfying suitable conditions, can be expanded in an infinite series of eigenfunctions of the more general Sturm–Liouville problem (1), (2). If this can be done, then we have

$$f(x) = \sum_{n=1}^{\infty} c_n \phi_n(x), \tag{32}$$

where the functions $\phi_n(x)$ satisfy Eqs. (1), (2) and also the orthogonality condition (22). To compute the coefficients in the series (32) we multiply Eq. (32) by $r(x)\phi_m(x)$, where m is a fixed positive integer, and integrate from $x = 0$ to $x = 1$. Assuming that the series can be integrated term by term, we obtain

$$\int_0^1 r(x)f(x)\phi_m(x)\, dx = \sum_{n=1}^{\infty} c_n \int_0^1 r(x)\phi_m(x)\phi_n(x)\, dx = \sum_{n=1}^{\infty} c_n \delta_{mn}. \tag{33}$$

Hence, using the definition of δ_{mn}, we have

$$c_m = \int_0^1 r(x)f(x)\phi_m(x)\, dx = (f, r\phi_m), \qquad m = 1, 2, \ldots. \tag{34}$$

The coefficients in the series (32) have thus been formally determined. Equation (34) has the same structure as the Euler–Fourier formulas for the coefficients in a Fourier series, and the eigenfunction series (32) also has convergence properties similar to those of Fourier series. The following theorem is analogous to Theorem 10.3.1.

Theorem 11.2.4 Let $\phi_1, \phi_2, \ldots, \phi_n, \ldots$ be the normalized eigenfunctions of the Sturm–Liouville problem (1), (2):

$$[p(x)y']' - q(x)y + \lambda r(x)y = 0,$$

$$\alpha_1 y(0) + \alpha_2 y'(0) = 0, \qquad \beta_1 y(1) + \beta_2 y'(1) = 0.$$

Let f and f' be piecewise continuous on $0 \le x \le 1$. Then the series (32) whose coefficients c_m are given by Eq. (34) converges to $[f(x+) + f(x-)]/2$ at each point in the open interval $0 < x < 1$.

If f satisfies further conditions, then a stronger conclusion can be established. Suppose that, in addition to the hypotheses of Theorem 11.2.4, the function f is continuous on $0 \le x \le 1$. If $\alpha_2 = 0$ in the first of Eqs. (2) [so that $\phi_n(0) = 0$], then assume that $f(0) = 0$. Similarly, if $\beta_2 = 0$ in the second of Eqs. (2), assume that $f(1) = 0$. Otherwise no boundary conditions need be prescribed for f. Then the series (32) converges to $f(x)$ at each point in the closed interval $0 \le x \le 1$.

Expand the function

$$f(x) = x, \qquad 0 \le x \le 1 \tag{35}$$

in terms of the normalized eigenfunctions $\phi_n(x)$ of the problem (25).

In Example 2 we found the normalized eigenfunctions to be

$$\phi_n(x) = k_n \sin \sqrt{\lambda_n}\, x, \tag{36}$$

where k_n is given by Eq. (28) and λ_n satisfies Eq. (26). To find the expansion for f in terms of the eigenfunctions ϕ_n we write

$$f(x) = \sum_{n=1}^{\infty} c_n \phi_n(x), \tag{37}$$

where the coefficients are given by Eq. (34). Thus

$$c_n = \int_0^1 f(x)\phi_n(x)\, dx = k_n \int_0^1 x \sin \sqrt{\lambda_n}\, x\, dx.$$

Integrating by parts, we obtain

$$c_n = k_n \left(\frac{\sin \sqrt{\lambda_n}}{\lambda_n} - \frac{\cos \sqrt{\lambda_n}}{\sqrt{\lambda_n}} \right) = k_n \frac{2 \sin \sqrt{\lambda_n}}{\lambda_n},$$

where we have used Eq. (26) in the last step. Upon substituting for k_n from Eq. (28) we obtain

$$c_n = \frac{2\sqrt{2} \sin \sqrt{\lambda_n}}{\lambda_n (1 + \cos^2 \sqrt{\lambda_n})^{1/2}}. \tag{38}$$

Thus

$$f(x) = 4 \sum_{n=1}^{\infty} \frac{\sin \sqrt{\lambda_n} \sin \sqrt{\lambda_n}\, x}{\lambda_n (1 + \cos^2 \sqrt{\lambda_n})}. \tag{39}$$

Observe that although the right side of Eq. (39) is a series of sines, it is not included in the discussion of Fourier sine series in Section 10.4.

Self-Adjoint Problems. Sturm–Liouville boundary value problems are of great importance in their own right, but they can also be viewed as belonging to a much more extensive class of problems that have many of the same properties. For example, there are many similarities between Sturm–Liouville problems and the algebraic system

$$\mathbf{A}\mathbf{x} = \lambda \mathbf{x}, \tag{40}$$

where the $n \times n$ matrix \mathbf{A} is real symmetric or Hermitian. Comparing the results mentioned in Section 7.3 with those of this section, we note that in both cases the eigenvalues are real and the eigenfunctions or eigenvectors form an orthogonal set. Further, the eigenfunctions or eigenvectors can be used as the basis for expressing an essentially arbitrary function or vector, respectively, as a sum. The most important difference is that a matrix has only a finite number of eigenvalues and eigenvectors, while a Sturm–Liouville problem has infinitely many. It is interesting and of fundamental importance in mathematics that these seemingly different problems—the

matrix problem (40) and the Sturm–Liouville problem (1), (2)—which arise in different ways, are actually parts of a single underlying theory. This theory is usually referred to as linear operator theory and is part of the subject of functional analysis.

We now point out some ways in which Sturm–Liouville problems can be generalized, while still preserving the main results of Theorems 11.2.1 to 11.2.4—the existence of a sequence of real eigenvalues tending to infinity, the orthogonality of the eigenfunctions, and the possibility of expressing an arbitrary function as a series of eigenfunctions. These generalizations depend on the continued validity of the crucial relation (8).

Let us consider the boundary value problem consisting of the differential equation

$$L[y] = \lambda r(x)y, \qquad 0 < x < 1, \tag{41}$$

where

$$L[y] = P_n(x)\frac{d^n y}{dx^n} + \cdots + P_1(x)\frac{dy}{dx} + P_0(x)y, \tag{42}$$

and n linear homogeneous boundary conditions at the endpoints. If Eq. (8) is valid for every pair of sufficiently differentiable functions that satisfy the boundary conditions, then the given problem is said to be **self-adjoint**. It is important to observe that Eq. (8) involves restrictions on both the differential equation and the boundary conditions. The differential operator L must be such that the same operator appears in both terms of Eq (8). This requires that L be of even order. Further, a second order operator must have the form (3), a fourth order operator must have the form

$$L[y] = [p(x)y'']'' - [q(x)y']' + s(x)y, \tag{43}$$

and higher order operators must have an analogous structure. In addition, the boundary conditions must be such as to eliminate the boundary terms that arise during the integration by parts used in deriving Eq. (8). For example, in a second order problem this is true for the separated boundary conditions (2) and also in certain other cases, one of which is given in Example 4 below.

Let us suppose that we have a self-adjoint boundary value problem for Eq. (41), where $L[y]$ is given now by Eq. (43). We assume that p, q, r, and s are continuous on $0 \le x \le 1$ and that the derivatives of p and q indicated in Eq. (43) are also continuous. If in addition $p(x) > 0$ and $r(x) > 0$ for $0 \le x \le 1$, then there is an infinite sequence of real eigenvalues tending to $+\infty$, the eigenfunctions are orthogonal with respect to the weight function r, and an arbitrary function can be expressed as a series of eigenfunctions. However, the eigenfunctions may not be simple in these more general problems.

We turn now to the relation between Sturm–Liouville problems and Fourier series. We have noted previously that Fourier sine and cosine series can be obtained by using the eigenfunctions of certain Sturm–Liouville problems involving the differential equation $y'' + \lambda y = 0$. This raises the question of whether we can obtain a full Fourier series, including both sine and cosine terms, by choosing a suitable set of boundary conditions. The answer is provided by the following example, which also serves to illustrate the occurrence of nonseparated boundary conditions.

Find the eigenvalues and eigenfunctions of the boundary value problem

EXAMPLE
4

$$y'' + \lambda y = 0, \qquad (44)$$

$$y(-L) - y(L) = 0, \qquad y'(-L) - y'(L) = 0. \qquad (45)$$

This is not a Sturm–Liouville problem because the boundary conditions are not separated. The boundary conditions (45) are called **periodic boundary conditions** since they require that y and y' assume the same values at $x = L$ as at $x = -L$. Nevertheless, it is straightforward to show that the problem (44), (45) is self-adjoint. A simple calculation establishes that $\lambda_0 = 0$ is an eigenvalue and that the corresponding eigenfunction is $\phi_0(x) = 1$. Further, there are additional eigenvalues $\lambda_1 = (\pi/L)^2, \lambda_2 = (2\pi/L)^2, \ldots, \lambda_n = (n\pi/L)^2, \ldots$. To each of these nonzero eigenvalues there correspond *two* linearly independent eigenfunctions; for example, corresponding to λ_n are the two eigenfunctions $\phi_n(x) = \cos(n\pi x/L)$ and $\psi_n(x) = \sin(n\pi x/L)$. This illustrates that the eigenvalues may not be simple when the boundary conditions are not separated. Further, if we seek to expand a given function f of period $2L$ in a series of eigenfunctions of the problem (44), (45), we obtain the series

$$f(x) = \frac{a_0}{2} + \sum_{n=1}^{\infty} \left(a_n \cos\frac{n\pi x}{L} + b_n \sin\frac{n\pi x}{L} \right),$$

which is just the Fourier series for f.

We will not give further consideration to problems that have nonseparated boundary conditions, nor will we deal with problems of higher than second order, except in a few problems. There is, however, one other kind of generalization that we do wish to discuss. That is the case in which the coefficients p, q, and r in Eq. (1) do not quite satisfy the rather strict continuity and positivity requirements laid down at the beginning of this section. Such problems are called singular Sturm–Liouville problems and are the subject of Section 11.4.

PROBLEMS

In each of Problems 1 through 5 determine the normalized eigenfunctions of the given problem.

1. $y'' + \lambda y = 0$, $\quad y(0) = 0$, $\quad y'(1) = 0$
2. $y'' + \lambda y = 0$, $\quad y'(0) = 0$, $\quad y(1) = 0$
3. $y'' + \lambda y = 0$, $\quad y'(0) = 0$, $\quad y'(1) = 0$
4. $y'' + \lambda y = 0$, $\quad y'(0) = 0$, $\quad y'(1) + y(1) = 0$; \quad see Section 11.1, Problem 8.
5. $y'' - 2y' + (1 + \lambda)y = 0$, $\quad y(0) = 0$, $\quad y(1) = 0$; \quad see Section 11.1, Problem 17.

In each of Problems 6 through 9 find the coefficients in the eigenfunction expansion $\sum_{n=1}^{\infty} a_n \phi_n(x)$ of the given function, using the normalized eigenfunctions of Problem 1.

6. $f(x) = 1$, $\quad 0 \le x \le 1$ \qquad 7. $f(x) = x$, $\quad 0 \le x \le 1$

8. $f(x) = \begin{cases} 1, & 0 \le x < \frac{1}{2} \\ 0, & \frac{1}{2} \le x \le 1 \end{cases}$ \qquad 9. $f(x) = \begin{cases} 2x, & 0 \le x < \frac{1}{2} \\ 1, & \frac{1}{2} \le x \le 1 \end{cases}$

In each of Problems 10 through 13 find the coefficients in the eigenfunction expansion $\sum_{n=1}^{\infty} a_n \phi_n(x)$ of the given function, using the normalized eigenfunctions of Problem 4.

10. $f(x) = 1, \qquad 0 \le x \le 1$　　　　　　11. $f(x) = x, \qquad 0 \le x \le 1$

12. $f(x) = 1 - x, \qquad 0 \le x \le 1$　　　13. $f(x) = \begin{cases} 1, & 0 \le x < \frac{1}{2} \\ 0, & \frac{1}{2} \le x \le 1 \end{cases}$

In each of Problems 14 through 18 determine whether the given boundary value problem is self-adjoint.

14. $y'' + y' + 2y = 0, \qquad y(0) = 0, \quad y(1) = 0$

15. $(1 + x^2)y'' + 2xy' + y = 0, \qquad y'(0) = 0, \quad y(1) + 2y'(1) = 0$

16. $y'' + y = \lambda y, \qquad y(0) - y'(1) = 0, \quad y'(0) - y(1) = 0$

17. $(1 + x^2)y'' + 2xy' + y = \lambda(1 + x^2)y, \qquad y(0) - y'(1) = 0, \quad y'(0) + 2y(1) = 0$

18. $y'' + \lambda y = 0, \qquad y(0) = 0, \quad y(\pi) + y'(\pi) = 0$

19. Show that if the functions u and v satisfy Eqs. (2), and either $\alpha_2 = 0$ or $\beta_2 = 0$, or both, then

$$p(x)\left[u'(x)v(x) - u(x)v'(x)\right]\Big|_0^1 = 0.$$

20. In this problem we outline a proof of the first part of Theorem 11.2.3: that the eigenvalues of the Sturm–Liouville problem (1), (2) are simple. For a given λ suppose that ϕ_1 and ϕ_2 are two linearly independent eigenfunctions. Compute the Wronskian $W(\phi_1, \phi_2)(x)$ and use the boundary conditions (2) to show that $W(\phi_1, \phi_2)(0) = 0$. Then use Theorems 3.3.2 and 3.3.3 to conclude that ϕ_1 and ϕ_2 cannot be linearly independent as assumed.

21. Consider the Sturm–Liouville problem

$$-[p(x)y']' + q(x)y = \lambda r(x)y,$$

$$\alpha_1 y(0) + \alpha_2 y'(0) = 0, \qquad \beta_1 y(1) + \beta_2 y'(1) = 0,$$

where p, q, and r satisfy the conditions stated in the text.

(a) Show that if λ is an eigenvalue and ϕ a corresponding eigenfunction, then

$$\lambda \int_0^1 r\phi^2 \, dx = \int_0^1 (p\phi'^2 + q\phi^2) \, dx + \frac{\beta_1}{\beta_2} p(1)\phi^2(1) - \frac{\alpha_1}{\alpha_2} p(0)\phi^2(0),$$

provided that $\alpha_2 \ne 0$ and $\beta_2 \ne 0$. How must this result be modified if $\alpha_2 = 0$ or $\beta_2 = 0$?

(b) Show that if $q(x) \ge 0$ and if β_1/β_2 and $-\alpha_1/\alpha_2$ are nonnegative, then the eigenvalue λ is nonnegative.

(c) Under the conditions of part (b) show that the eigenvalue λ is strictly positive unless $\alpha_1 = \beta_1 = 0$ and $q(x) = 0$ for each x in $0 \le x \le 1$.

22. Derive Eq. (8) using the inner product (9) and assuming that u and v are complex-valued functions.

Hint: Consider the quantity $\int_0^1 L[u]\bar{v} \, dx$, split u and v into real and imaginary parts, and proceed as in the text.

23. In this problem we outline a proof that the eigenfunctions of the Sturm–Liouville problem (1), (2) are real.

 (a) Let λ be an eigenvalue and ϕ a corresponding eigenfunction. Let $\phi(x) = U(x) + iV(x)$, and show that U and V are also eigenfunctions corresponding to λ.

 (b) Using Theorem 11.2.3, or the result of Problem 20, show that U and V are linearly dependent.

 (c) Show that ϕ must be real, apart from an arbitrary multiplicative constant that may be complex.

24. Consider the problem

$$x^2 y'' = \lambda(xy' - y), \qquad y(1) = 0, \quad y(2) = 0.$$

Note that λ appears as a coefficient of y' as well as of y itself. It is possible to extend the definition of self-adjointness to this type of problem and to show that this particular problem is not self-adjoint. Show that the problem has eigenvalues but that none of them is real. This illustrates that in general nonself-adjoint problems may have eigenvalues that are not real.

Buckling of an Elastic Column. In an investigation of the buckling of a uniform elastic column of length L by an axial load P (Figure 11.2.1a), one is led to the differential equation

$$y^{(4)} + \lambda y'' = 0, \qquad 0 < x < L. \tag{i}$$

The parameter λ is equal to P/EI, where E is Young's modulus and I is the moment of inertia of the cross section about an axis through the centroid perpendicular to the xy-plane. The boundary conditions at $x = 0$ and $x = L$ depend on how the ends of the column are supported. Typical boundary conditions are

$$y = y' = 0, \qquad \text{clamped end;}$$

$$y = y'' = 0, \qquad \text{simply supported (hinged) end.}$$

The bar shown in Figure 11.2.1a is simply supported at $x = 0$ and clamped at $x = L$. It is desired to determine the eigenvalues and eigenfunctions of Eq. (i) subject to suitable boundary conditions. In particular, the smallest eigenvalue λ_1 gives the load at which the column buckles, or can assume a curved equilibrium position, as shown in Figure 11.2.1b. The corresponding eigenfunction describes the configuration of the buckled column. Note that the differential equation (i) does not fall within the theory discussed in this section. It is possible to show, however, that in each of the cases given here all the eigenvalues are real and positive. Problems 25 and 26 deal with column buckling problems.

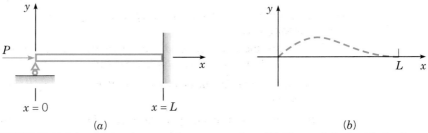

FIGURE 11.2.1 (*a*) A column under compression. (*b*) Shape of the buckled column.

25. For each of the following boundary conditions find the smallest eigenvalue (the buckling load) of $y^{(4)} + \lambda y'' = 0$, and also find the corresponding eigenfunction (the shape of the buckled column).

 (a) $y(0) = y''(0) = 0$, $y(L) = y''(L) = 0$
 (b) $y(0) = y''(0) = 0$, $y(L) = y'(L) = 0$
 (c) $y(0) = y'(0) = 0$, $y(L) = y'(L) = 0$

26. In some buckling problems the eigenvalue parameter appears in the boundary conditions as well as in the differential equation. One such case occurs when one end of the column is clamped and the other end is free. In this case the differential equation $y^{(4)} + \lambda y'' = 0$ must be solved subject to the boundary conditions

 $$y(0) = 0, \qquad y'(0) = 0, \qquad y''(L) = 0, \qquad y'''(L) + \lambda y'(L) = 0.$$

 Find the smallest eigenvalue and the corresponding eigenfunction.

27. Solutes in an aquifer are transported by two separate mechanisms. The process by which a solute is transported by the bulk motion of the flowing groundwater is called **advection**. In addition, the solute is spread by small-scale fluctuations in the groundwater velocity along the tortuous flow paths within individual pores, a process called **mechanical dispersion**. The one-dimensional form of the advection-dispersion equation for a nonreactive dissolved solute in a saturated, homogeneous, isotropic porous medium under steady, uniform flow is

 $$c_t + vc_x = Dc_{xx}, \qquad 0 < x < L, \quad t > 0, \tag{i}$$

 where $c(x,t)$ is the concentration of the solute, v is the average linear groundwater velocity, D is the coefficient of hydrodynamic dispersion, and L is the length of the aquifer. Suppose that the boundary conditions are

 $$c(0,t) = 0, \qquad c_x(L,t) = 0, \qquad t > 0 \tag{ii}$$

 and that the initial condition is

 $$c(x,0) = f(x), \qquad 0 < x < L, \tag{iii}$$

 where $f(x)$ is the given initial concentration of the solute.

 (a) Assume that $c(x,t) = X(x)T(t)$, use the method of separation of variables, and find the equations satisfied by $X(x)$ and $T(t)$, respectively. Show that the problem for $X(x)$ can be written in the Sturm-Liouville form

 $$[p(x)X']' + \lambda r(x)X = 0, \qquad 0 < x < L, \tag{iv}$$

 $$X(0) = 0, \qquad X'(L) = 0, \tag{v}$$

 where $p(x) = r(x) = \exp(-vx/D)$. Hence the eigenvalues are real and the eigenfunctions are orthogonal with respect to the weight function $r(x)$.

 (b) Let $\mu^2 = \lambda - (v^2/4D^2)$. Show that the eigenfunctions are

 $$X_n(x) = e^{vx/2D} \sin \mu_n x, \tag{vi}$$

 where μ_n satisfies the equation

 $$\tan \mu L = -2D\mu/v. \tag{vii}$$

 (c) Show graphically that Eq. (vii) has an infinite sequence of positive roots and that $\mu_n \cong (2n-1)\pi/2L$ for large n.

(d) Show that

$$\int_0^L r(x) X_n^2(x)\, dx = \frac{L}{2} + \frac{v}{4D\mu_n^2} \sin^2 \mu_n L.$$

(e) Find a formal solution of the problem (i), (ii), (iii) in terms of a series of the eigenfunctions $X_n(x)$.

(f) Let $v = 1, D = 0.5, L = 10$, and $f(x) = \delta(x - 3)$, where δ is the Dirac delta[5] function. Using the solution found in part (e), plot $c(x,t)$ versus x for several values of t, such as $t = 0.5, 1, 3, 6,$ and 10. Also plot $c(x,t)$ versus t for several values of x. Note that the number of terms that are needed to obtain an accurate plot depends strongly on the values of t and x.

(g) Describe in a few words how the solution evolves as time advances.

 28. A nonreactive tracer at concentration c_0 is continuously introduced into a steady flow at the upstream end of a column of length L packed with a homogeneous granular medium. Assuming that the tracer concentration in the column is initially zero, the boundary value problem that models this process is

$$c_t + vc_x = Dc_{xx}, \qquad 0 < x < L, \quad t > 0,$$
$$c(0,t) = c_0, \qquad c_x(L,t) = 0, \qquad t > 0,$$
$$c(x,0) = 0, \qquad 0 < x < L,$$

where $c(x,t), v,$ and D are as in Problem 27.

(a) Assume that $c(x,t) = c_0 + u(x,t)$ and find the boundary value problem satisfied by $u(x,t)$.

(b) Proceeding as in Problem 27, find $u(x,t)$ in terms of an eigenfunction expansion.

(c) Let $v = 1, D = 0.5, c_0 = 1,$ and $L = 10$. Plot $c(x,t)$ versus x for several values of t, and also plot $c(x,t)$ versus t for several values of x.

(d) Describe in a few words how the solution evolves with time. For example, about how long does it take for the steady-state solution to be essentially attained?

11.3 Nonhomogeneous Boundary Value Problems

In this section we discuss how to solve nonhomogeneous boundary value problems for both ordinary and partial differential equations. Most of our attention is directed toward problems in which the differential equation alone is nonhomogeneous, while the boundary conditions are homogeneous. We assume that the solution can be expanded in a series of eigenfunctions of a related homogeneous problem, and then we determine the coefficients in this series so that the nonhomogeneous problem is satisfied. We first describe this method as it applies to boundary value problems for second order linear ordinary differential equations. Later we illustrate its use for partial differential equations by solving a heat conduction problem in a bar with variable material properties and in the presence of source terms.

[5] See Section 6.5, especially Eq. (16) of that section.

Nonhomogeneous Sturm–Liouville Problems. Consider the boundary value problem consisting of the nonhomogeneous differential equation

$$L[y] = -[p(x)y']' + q(x)y = \mu r(x)y + f(x), \tag{1}$$

where μ is a given constant and f is a given function on $0 \le x \le 1$, and the boundary conditions

$$\alpha_1 y(0) + \alpha_2 y'(0) = 0, \qquad \beta_1 y(1) + \beta_2 y'(1) = 0. \tag{2}$$

As in Section 11.2, we assume that p, p', q, and r are continuous on $0 \le x \le 1$ and that $p(x) > 0$ and $r(x) > 0$ there. We will solve the problem (1), (2) by making use of the eigenfunctions of the corresponding homogeneous problem consisting of the differential equation

$$L[y] = \lambda r(x)y \tag{3}$$

and the boundary conditions (2). Let $\lambda_1 < \lambda_2 < \cdots < \lambda_n < \cdots$ be the eigenvalues of this problem, and let $\phi_1, \phi_2, \ldots, \phi_n, \ldots$ be the corresponding normalized eigenfunctions.

We now assume that the solution $y = \phi(x)$ of the nonhomogeneous problem (1), (2) can be expressed as a series of the form

$$\phi(x) = \sum_{n=1}^{\infty} b_n \phi_n(x). \tag{4}$$

From Eq. (34) of Section 11.2 we know that

$$b_n = \int_0^1 r(x)\phi(x)\phi_n(x)\,dx, \qquad n = 1, 2, \ldots. \tag{5}$$

However, since we do not know $\phi(x)$, we cannot use Eq. (5) to calculate b_n. Instead, we will try to determine b_n so that the problem (1), (2) is satisfied and then use Eq. (4) to find $\phi(x)$. Note first that ϕ as given by Eq. (4) always satisfies the boundary conditions (2) since each ϕ_n does.

Now consider the differential equation that ϕ must satisfy. This is just Eq. (1) with y replaced by ϕ:

$$L[\phi](x) = \mu r(x)\phi(x) + f(x). \tag{6}$$

We substitute the series (4) into the differential equation (6) and attempt to determine b_n so that the differential equation is satisfied. The term on the left side of Eq. (6) becomes

$$L[\phi](x) = L\left[\sum_{n=1}^{\infty} b_n \phi_n\right](x) = \sum_{n=1}^{\infty} b_n L[\phi_n](x)$$

$$= \sum_{n=1}^{\infty} b_n \lambda_n r(x)\phi_n(x), \tag{7}$$

where we have assumed that we can interchange the operations of summation and differentiation.

Note that the function r appears in Eq. (7) and also in the term $\mu r(x)\phi(x)$ in Eq. (6). This suggests that we rewrite the nonhomogeneous term in Eq. (6) as $r(x)[f(x)/r(x)]$ so that $r(x)$ also appears as a multiplier in this term. If the function f/r satisfies the conditions of Theorem 11.2.4, then

$$\frac{f(x)}{r(x)} = \sum_{n=1}^{\infty} c_n \phi_n(x), \tag{8}$$

where, using Eq. (5) with ϕ replaced by f/r,

$$c_n = \int_0^1 r(x)\frac{f(x)}{r(x)}\phi_n(x)\,dx = \int_0^1 f(x)\phi_n(x)\,dx, \qquad n = 1, 2, \ldots. \tag{9}$$

Upon substituting for $\phi(x)$, $L[\phi](x)$, and $f(x)$ in Eq. (6) from Eqs. (4), (7), and (8), respectively, we find that

$$\sum_{n=1}^{\infty} b_n \lambda_n r(x)\phi_n(x) = \mu r(x)\sum_{n=1}^{\infty} b_n \phi_n(x) + r(x)\sum_{n=1}^{\infty} c_n \phi_n(x).$$

After collecting terms and canceling the common nonzero factor $r(x)$, we have

$$\sum_{n=1}^{\infty}\left[(\lambda_n - \mu)b_n - c_n\right]\phi_n(x) = 0. \tag{10}$$

If Eq. (10) is to hold for each x in the interval $0 \le x \le 1$, then the coefficient of $\phi_n(x)$ must be zero for each n; see Problem 14 for a proof of this fact. Hence

$$(\lambda_n - \mu)b_n - c_n = 0, \qquad n = 1, 2, \ldots. \tag{11}$$

We must now distinguish two main cases, one of which also has two subcases.

First suppose that $\mu \ne \lambda_n$ for $n = 1, 2, 3, \ldots$; that is, μ is not equal to any eigenvalue of the corresponding homogeneous problem. Then

$$b_n = \frac{c_n}{\lambda_n - \mu}, \qquad n = 1, 2, 3, \ldots, \tag{12}$$

and

$$y = \phi(x) = \sum_{n=1}^{\infty} \frac{c_n}{\lambda_n - \mu}\phi_n(x). \tag{13}$$

Equation (13), with c_n given by Eq. (9), is a formal solution of the nonhomogeneous boundary value problem (1), (2). Our argument does not prove that the series (13) converges. However, any solution of the boundary value problem (1), (2) clearly satisfies the conditions of Theorem 11.2.4; indeed, it satisfies the more stringent conditions given in the paragraph following that theorem. Thus it is reasonable to expect that the series (13) does converge at each point, and this fact can be established, provided, for example, that f is continuous.

Now suppose that μ is equal to one of the eigenvalues of the corresponding homogeneous problem, say, $\mu = \lambda_m$; then the situation is quite different. In this event, for $n = m$ Eq. (11) has the form $0 \cdot b_m - c_m = 0$. Again we must consider two cases.

If $\mu = \lambda_m$ and $c_m \neq 0$, then it is impossible to solve Eq. (11) for b_m, and the nonhomogeneous problem (1), (2) has no solution.

If $\mu = \lambda_m$ and $c_m = 0$, then Eq. (11) is satisfied regardless of the value of b_m; in other words, b_m remains arbitrary. In this case the boundary value problem (1), (2) does have a solution, but it is not unique, since it contains an arbitrary multiple of the eigenfunction ϕ_m.

Since c_m is given by Eq. (9), the condition $c_m = 0$ means that

$$\int_0^1 f(x)\phi_m(x)\,dx = 0. \tag{14}$$

Thus, if $\mu = \lambda_m$, the nonhomogeneous boundary value problem (1), (2) can be solved only if f is orthogonal to the eigenfunction corresponding to the eigenvalue λ_m.

The results we have formally obtained are summarized in the following theorem.

Theorem 11.3.1 The nonhomogeneous boundary value problem (1), (2) has a unique solution for each continuous f whenever μ is different from all the eigenvalues of the corresponding homogeneous problem; the solution is given by Eq. (13), and the series converges for each x in $0 \leq x \leq 1$. If μ is equal to an eigenvalue λ_m of the corresponding homogeneous problem, then the nonhomogeneous boundary value problem has no solution unless f is orthogonal to ϕ_m, that is, unless the condition (14) holds. In that case, the solution is not unique and contains an arbitrary multiple of $\phi_m(x)$.

The main part of Theorem 11.3.1 is sometimes stated in the following way:

Theorem 11.3.2 For a given value of μ, either the nonhomogeneous problem (1), (2) has a unique solution for each continuous f (if μ is not equal to any eigenvalue λ_m of the corresponding homogeneous problem), or else the homogeneous problem (3), (2) has a nontrivial solution (the eigenfunction corresponding to λ_m).

This latter form of the theorem is known as the Fredholm[6] alternative theorem. This is one of the basic theorems of mathematical analysis and occurs in many different contexts. You may be familiar with it in connection with sets of linear algebraic equations where the vanishing or nonvanishing of the determinant of coefficients replaces the statements about μ and λ_m. See the discussion in Section 7.3.

[6]The Swedish mathematician Erik Ivar Fredholm (1866–1927), professor at the University of Stockholm, established the modern theory of integral equations in a fundamental paper in 1903. Fredholm's work emphasized the similarities between integral equations and systems of linear algebraic equations. There are also many interrelations between differential and integral equations; for example, see Section 2.8 and Problem 22 of Section 6.6.

EXAMPLE
1

Solve the boundary value problem

$$y'' + 2y = -x, \tag{15}$$

$$y(0) = 0, \qquad y(1) + y'(1) = 0. \tag{16}$$

This particular problem can be solved directly in an elementary way and has the solution

$$y = \frac{\sin \sqrt{2}\, x}{\sin \sqrt{2} + \sqrt{2} \cos \sqrt{2}} - \frac{x}{2}. \tag{17}$$

The method of solution described below illustrates the use of eigenfunction expansions, a method that can be employed in many problems not accessible by elementary procedures. To identify Eq. (15) with Eq. (1) it is helpful to write the former as

$$-y'' = 2y + x. \tag{18}$$

We seek the solution of the given problem as a series of normalized eigenfunctions ϕ_n of the corresponding homogeneous problem

$$y'' + \lambda y = 0, \qquad y(0) = 0, \quad y(1) + y'(1) = 0. \tag{19}$$

These eigenfunctions were found in Example 2 of Section 11.2 and are

$$\phi_n(x) = k_n \sin \sqrt{\lambda_n}\, x, \tag{20}$$

where

$$k_n = \left(\frac{2}{1 + \cos^2 \sqrt{\lambda_n}} \right)^{1/2} \tag{21}$$

and λ_n satisfies

$$\sin \sqrt{\lambda_n} + \sqrt{\lambda_n} \cos \sqrt{\lambda_n} = 0. \tag{22}$$

Recall that in Example 1 of Section 11.1 we found that

$$\lambda_1 \cong 4.116, \qquad \lambda_2 \cong 24.14,$$

$$\lambda_3 \cong 63.66, \qquad \lambda_n \cong (2n-1)^2 \pi^2/4 \qquad \text{for } n = 4, 5, \dots.$$

We assume that y is given by Eq. (4),

$$y = \sum_{n=1}^{\infty} b_n \phi_n(x),$$

and it follows that the coefficients b_n are found from Eq. (12),

$$b_n = \frac{c_n}{\lambda_n - 2},$$

where the c_n are the expansion coefficients of the nonhomogeneous term $f(x) = x$ in Eq. (18) in terms of the eigenfunctions ϕ_n. These coefficients were found in Example 3 of Section 11.2 and are

$$c_n = \frac{2\sqrt{2}\,\sin\sqrt{\lambda_n}}{\lambda_n(1 + \cos^2\sqrt{\lambda_n})^{1/2}}. \tag{23}$$

Putting everything together, we finally obtain the solution

$$y = 4\sum_{n=1}^{\infty} \frac{\sin\sqrt{\lambda_n}}{\lambda_n(\lambda_n - 2)(1 + \cos^2\sqrt{\lambda_n})}\,\sin\sqrt{\lambda_n}\,x. \tag{24}$$

Although Eqs. (17) and (24) are quite different in appearance, they are actually two different expressions for the same function. This follows from the uniqueness part of Theorem 11.3.1 or 11.3.2 since $\lambda = 2$ is not an eigenvalue of the homogeneous problem (19). Alternatively, one can show the equivalence of Eqs. (17) and (24) by expanding the right side of Eq. (17) in terms of the eigenfunctions $\phi_n(x)$. For this problem it is fairly obvious that the formula (17) is more convenient than the series formula (24). However, we emphasize again that in other problems we may not be able to obtain the solution except by series (or numerical) methods.

Nonhomogeneous Heat Conduction Problems. To show how eigenfunction expansions can be used to solve nonhomogeneous problems for partial differential equations, let us consider the generalized heat conduction equation

$$r(x)u_t = [p(x)u_x]_x - q(x)u + F(x, t) \tag{25}$$

with the boundary conditions

$$u_x(0, t) - h_1 u(0, t) = 0, \qquad u_x(1, t) + h_2 u(1, t) = 0 \tag{26}$$

and the initial condition

$$u(x, 0) = f(x). \tag{27}$$

This problem was previously discussed in Appendix A of Chapter 10 and in Section 11.1. In the latter section we let $u(x, t) = X(x)T(t)$ in the homogeneous equation obtained by setting $F(x, t) = 0$, and showed that $X(x)$ must be a solution of the boundary value problem

$$-[p(x)X']' + q(x)X = \lambda r(x)X, \tag{28}$$

$$X'(0) - h_1 X(0) = 0, \qquad X'(1) + h_2 X(1) = 0. \tag{29}$$

If we assume that p, q, and r satisfy the proper continuity requirements and that $p(x)$ and $r(x)$ are always positive, the problem (28), (29) is a Sturm–Liouville problem as discussed in Section 11.2. Thus we obtain a sequence of eigenvalues $\lambda_1 < \lambda_2 < \cdots < \lambda_n < \cdots$ and corresponding normalized eigenfunctions $\phi_1(x), \phi_2(x), \ldots, \phi_n(x), \ldots$.

We will solve the given nonhomogeneous boundary value problem (25) to (27) by assuming that $u(x, t)$ can be expressed as a series of eigenfunctions,

$$u(x, t) = \sum_{n=1}^{\infty} b_n(t)\phi_n(x), \tag{30}$$

and then showing how to determine the coefficients $b_n(t)$. The procedure is basically the same as that used in the problem (1), (2) considered earlier, although it is more complicated in certain respects. For instance, the coefficients b_n must now depend

on t, because otherwise u would be a function of x only. Note that the boundary conditions (26) are automatically satisfied by an expression of the form (30) because each $\phi_n(x)$ satisfies the boundary conditions (29).

Next we substitute from Eq. (30) for u in Eq. (25). From the first two terms on the right side of Eq. (25) we formally obtain

$$[p(x)u_x]_x - q(x)u = \frac{\partial}{\partial x}\left[p(x)\sum_{n=1}^{\infty} b_n(t)\phi_n'(x)\right] - q(x)\sum_{n=1}^{\infty} b_n(t)\phi_n(x)$$

$$= \sum_{n=1}^{\infty} b_n(t)\{[p(x)\phi_n'(x)]' - q(x)\phi_n(x)\}. \tag{31}$$

Since $[p(x)\phi_n'(x)]' - q(x)\phi_n(x) = -\lambda_n r(x)\phi_n(x)$, we obtain finally

$$[p(x)u_x]_x - q(x)u = -r(x)\sum_{n=1}^{\infty} b_n(t)\lambda_n\phi_n(x). \tag{32}$$

Now consider the term on the left side of Eq. (25). We have

$$r(x)u_t = r(x)\frac{\partial}{\partial t}\sum_{n=1}^{\infty} b_n(t)\phi_n(x)$$

$$= r(x)\sum_{n=1}^{\infty} b_n'(t)\phi_n(x). \tag{33}$$

We must also express the nonhomogeneous term in Eq. (25) as a series of eigenfunctions. Once again, it is convenient to look at the ratio $F(x,t)/r(x)$ and to write

$$\frac{F(x,t)}{r(x)} = \sum_{n=1}^{\infty} \gamma_n(t)\phi_n(x), \tag{34}$$

where the coefficients are given by

$$\gamma_n(t) = \int_0^1 r(x)\frac{F(x,t)}{r(x)}\phi_n(x)\,dx$$

$$= \int_0^1 F(x,t)\phi_n(x)\,dx, \qquad n = 1, 2, \ldots. \tag{35}$$

Since $F(x,t)$ is given, we can consider the functions $\gamma_n(t)$ to be known.

Gathering all these results together, we substitute from Eqs. (32), (33), and (34) in Eq. (25), and find that

$$r(x)\sum_{n=1}^{\infty} b_n'(t)\phi_n(x) = -r(x)\sum_{n=1}^{\infty} b_n(t)\lambda_n\phi_n(x) + r(x)\sum_{n=1}^{\infty} \gamma_n(t)\phi_n(x). \tag{36}$$

To simplify Eq. (36) we cancel the common nonzero factor $r(x)$ from all terms and write everything in one summation:

$$\sum_{n=1}^{\infty} \left[b_n'(t) + \lambda_n b_n(t) - \gamma_n(t)\right]\phi_n(x) = 0. \tag{37}$$

Once again, if Eq. (37) is to hold for all x in $0 < x < 1$, it is necessary that the quantity in square brackets be zero for each n (again see Problem 14). Hence $b_n(t)$ is a solution of the first order linear ordinary differential equation

$$b_n'(t) + \lambda_n b_n(t) = \gamma_n(t), \qquad n = 1, 2, \ldots, \tag{38}$$

where $\gamma_n(t)$ is given by Eq. (35). To determine $b_n(t)$ completely we must have an initial condition

$$b_n(0) = B_n, \qquad n = 1, 2, \ldots \tag{39}$$

for Eq. (38). This we obtain from the initial condition (27). Setting $t = 0$ in Eq. (30) and using Eq. (27), we have

$$u(x, 0) = \sum_{n=1}^{\infty} b_n(0)\phi_n(x) = \sum_{n=1}^{\infty} B_n\phi_n(x) = f(x). \tag{40}$$

Thus the initial values B_n are the coefficients in the eigenfunction expansion for $f(x)$. Therefore,

$$B_n = \int_0^1 r(x)f(x)\phi_n(x)\,dx, \qquad n = 1, 2, \ldots. \tag{41}$$

Note that everything on the right side of Eq. (41) is known, so we can consider B_n as known.

The initial value problem (38), (39) is solved by the methods of Section 2.1. The integrating factor is $\mu(t) = \exp(\lambda_n t)$, and it follows that

$$b_n(t) = B_n e^{-\lambda_n t} + \int_0^t e^{-\lambda_n(t-s)}\gamma_n(s)\,ds, \qquad n = 1, 2, \ldots. \tag{42}$$

The details of this calculation are left to you. Note that the first term on the right side of Eq. (42) depends on the function f through the coefficients B_n, while the second depends on the nonhomogeneous term F through the coefficients $\gamma_n(s)$.

Thus an explicit solution of the boundary value problem (25) to (27) is given by Eq. (30),

$$u(x, t) = \sum_{n=1}^{\infty} b_n(t)\phi_n(x),$$

where the coefficients $b_n(t)$ are determined from Eq. (42). The quantities B_n and $\gamma_n(s)$ in Eq. (42) are found in turn from Eqs. (41) and (35), respectively.

Summarizing, to use this method to solve a boundary value problem such as that given by Eqs. (25) to (27), we must

1. Find the eigenvalues λ_n and the normalized eigenfunctions ϕ_n of the homogeneous problem (28), (29).
2. Calculate the coefficients B_n and $\gamma_n(t)$ from Eqs. (41) and (35), respectively.
3. Evaluate the integral in Eq. (42) to determine $b_n(t)$.
4. Sum the infinite series (30).

Since any or all of these steps may be difficult, the entire process can be quite formidable. One redeeming feature is that often the series (30) converges rapidly, in which case only a very few terms may be needed to obtain an adequate approximation to the solution.

Find the solution of the heat conduction problem

EXAMPLE

2

$$u_t = u_{xx} + xe^{-t}, \tag{43}$$

$$u(0,t) = 0, \qquad u_x(1,t) + u(1,t) = 0, \tag{44}$$

$$u(x,0) = 0. \tag{45}$$

Again we use the normalized eigenfunctions ϕ_n of the problem (19) and assume that u is given by Eq. (30),

$$u(x,t) = \sum_{n=1}^{\infty} b_n(t)\phi_n(x).$$

The coefficients b_n are determined from the differential equation

$$b_n' + \lambda_n b_n = \gamma_n(t), \tag{46}$$

where λ_n is the nth eigenvalue of problem (19) and $\gamma_n(t)$ is the nth expansion coefficient of the nonhomogeneous term xe^{-t} in terms of the eigenfunctions ϕ_n. Thus we have

$$\gamma_n(t) = \int_0^1 xe^{-t}\phi_n(x)\,dx = e^{-t}\int_0^1 x\phi_n(x)\,dx$$

$$= c_n e^{-t}, \tag{47}$$

where $c_n = \int_0^1 x\phi_n(x)\,dx$ is given by Eq. (23). The initial condition for Eq. (46) is

$$b_n(0) = 0 \tag{48}$$

since the initial temperature distribution (45) is zero everywhere. The solution of the initial value problem (46), (48) is

$$b_n(t) = e^{-\lambda_n t}\int_0^t e^{\lambda_n s}c_n e^{-s}\,ds = c_n e^{-\lambda_n t}\,\frac{e^{(\lambda_n - 1)t} - 1}{\lambda_n - 1}$$

$$= \frac{c_n}{\lambda_n - 1}(e^{-t} - e^{-\lambda_n t}). \tag{49}$$

Thus the solution of the heat conduction problem (43) to (45) is given by

$$u(x,t) = 4\sum_{n=1}^{\infty}\frac{(\sin\sqrt{\lambda_n})\,(e^{-t} - e^{-\lambda_n t})\sin\sqrt{\lambda_n}\,x}{\lambda_n(\lambda_n - 1)(1 + \cos^2\sqrt{\lambda_n})}. \tag{50}$$

The solution given by Eq. (50) is exact but complicated. To judge whether a satisfactory approximation to the solution can be obtained by using only a few terms in this series, we must estimate its speed of convergence. First we split the right side of Eq. (50) into two parts:

$$u(x,t) = 4e^{-t}\sum_{n=1}^{\infty}\frac{\sin\sqrt{\lambda_n}\,\sin\sqrt{\lambda_n}\,x}{\lambda_n(\lambda_n - 1)(1 + \cos^2\sqrt{\lambda_n})} - 4\sum_{n=1}^{\infty}\frac{e^{-\lambda_n t}\sin\sqrt{\lambda_n}\,\sin\sqrt{\lambda_n}\,x}{\lambda_n(\lambda_n - 1)(1 + \cos^2\sqrt{\lambda_n})}. \tag{51}$$

Recall from Example 1 in Section 11.1 that the eigenvalues λ_n are very nearly proportional to n^2. In the first series on the right side of Eq. (51) the trigonometric factors are all bounded

as $n \to \infty$; therefore, this series converges similarly to the series $\sum_{n=1}^{\infty} \lambda_n^{-2}$ or $\sum_{n=1}^{\infty} n^{-4}$. Hence at most two or three terms are required for us to obtain an excellent approximation to this part of the solution. The second series contains the additional factor $e^{-\lambda_n t}$, so its convergence is even more rapid for $t > 0$; all terms after the first are almost surely negligible.

Further Discussion. Eigenfunction expansions can be used to solve a much greater variety of problems than the preceding discussion and examples may suggest. For example, time-independent nonhomogeneus boundary conditions can be handled much as in Section 10.6. To reduce the problem to one with homogeneous boundary conditions, subtract from u a function v that is chosen to satisfy the given boundary conditions. Then the difference $w = u - v$ satisfies a problem with homogeneous boundary conditions, but with a modified forcing term and initial condition. This problem can be solved by the procedure described in this section.

One potential difficulty in using eigenfunction expansions is that the normalized eigenfunctions of the corresponding homogeneous problem must be found. For a differential equation with variable coefficients this may be difficult, if not impossible. In such a case it is sometimes possible to use other functions, such as eigenfunctions of a simpler problem, that satisfy the same boundary conditions. For instance, if the boundary conditions are

$$u(0, t) = 0, \qquad u(1, t) = 0, \tag{52}$$

then it may be convenient to replace the functions $\phi_n(x)$ in Eq. (30) by $\sin n\pi x$. These functions at least satisfy the correct boundary conditions, although in general they are not solutions of the corresponding homogeneous differential equation. Next we expand the nonhomogeneous term $F(x, t)$ in a series of the form (34), again with $\phi_n(x)$ replaced by $\sin n\pi x$, and then substitute for both u and F in Eq. (25). Upon collecting the coefficients of $\sin n\pi x$ for each n, we have an infinite set of linear first order differential equations from which to determine $b_1(t), b_2(t), \ldots$. The essential difference between this case and the one considered earlier is that now the equations for the functions $b_n(t)$ are *coupled*. Thus they cannot be solved one by one, as before, but must be dealt with simultaneously. In practice, the infinite system is replaced by an approximating finite system, from which approximations to a finite number of coefficients are calculated.

Boundary value problems for equations of higher than second order can also often be solved by eigenfunction expansions. In some cases the procedure parallels almost exactly that for second order problems. However, a variety of complications can also arise.

Finally, we emphasize that the discussion in this section has been purely formal. Separate and sometimes elaborate arguments must be used to establish convergence of eigenfunction expansions or to justify some of the steps used, such as term-by-term differentiation of eigenfunction series.

There are also other, altogether different methods for solving nonhomogeneous boundary value problems. One of these leads to a solution expressed as a definite integral rather than as an infinite series. This approach involves certain functions known as Green's functions and, for ordinary differential equations, is the subject of Problems 28 through 36.

PROBLEMS In each of Problems 1 through 5 solve the given problem by means of an eigenfunction expansion.

1. $y'' + 2y = -x,$ $y(0) = 0,$ $y(1) = 0$
2. $y'' + 2y = -x,$ $y(0) = 0,$ $y'(1) = 0;$ see Section 11.2, Problem 7.
3. $y'' + 2y = -x,$ $y'(0) = 0,$ $y'(1) = 0;$ see Section 11.2, Problem 3.
4. $y'' + 2y = -x,$ $y'(0) = 0,$ $y'(1) + y(1) = 0;$ see Section 11.2, Problem 11.
5. $y'' + 2y = -1 + |1 - 2x|,$ $y(0) = 0,$ $y(1) = 0$

In each of Problems 6 through 9 determine a formal eigenfunction series expansion for the solution of the given problem. Assume that f satisfies the conditions of Theorem 11.3.1. State the values of μ for which the solution exists.

6. $y'' + \mu y = -f(x),$ $y(0) = 0,$ $y'(1) = 0$
7. $y'' + \mu y = -f(x),$ $y'(0) = 0,$ $y(1) = 0$
8. $y'' + \mu y = -f(x),$ $y'(0) = 0,$ $y'(1) = 0$
9. $y'' + \mu y = -f(x),$ $y'(0) = 0,$ $y'(1) + y(1) = 0$

In each of Problems 10 through 13 determine whether there is any value of the constant a for which the problem has a solution. Find the solution for each such value.

10. $y'' + \pi^2 y = a + x,$ $y(0) = 0,$ $y(1) = 0$
11. $y'' + 4\pi^2 y = a + x,$ $y(0) = 0,$ $y(1) = 0$
12. $y'' + \pi^2 y = a,$ $y'(0) = 0,$ $y'(1) = 0$
13. $y'' + \pi^2 y = a - \cos \pi x,$ $y(0) = 0,$ $y(1) = 0$

14. Let $\phi_1, \ldots, \phi_n, \ldots$ be the normalized eigenfunctions of the differential equation (3) subject to the boundary conditions (2). If $\sum\limits_{n=1}^{\infty} c_n \phi_n(x)$ converges to $f(x)$, where $f(x) = 0$ for each x in $0 \le x \le 1$, show that $c_n = 0$ for each n.
 Hint: Multiply by $r(x)\phi_m(x)$, integrate, and use the orthogonality property of the eigenfunctions.

15. Let L be a second order linear differential operator. Show that the solution $y = \phi(x)$ of the problem

$$L[y] = f(x),$$

$$\alpha_1 y(0) + \alpha_2 y'(0) = a, \qquad \beta_1 y(1) + \beta_2 y'(1) = b$$

can be written as $y = u + v,$ where $u = \phi_1(x)$ and $v = \phi_2(x)$ are solutions of the problems

$$L[u] = 0,$$

$$\alpha_1 u(0) + \alpha_2 u'(0) = a, \qquad \beta_1 u(1) + \beta_2 u'(1) = b$$

and

$$L[v] = f(x),$$

$$\alpha_1 v(0) + \alpha_2 v'(0) = 0, \qquad \beta_1 v(1) + \beta_2 v'(1) = 0,$$

respectively.

16. Show that the problem

$$y'' + \pi^2 y = \pi^2 x, \qquad y(0) = 1, \quad y(1) = 0$$

has the solution

$$y = c_1 \sin \pi x + \cos \pi x + x.$$

Also show that this solution cannot be obtained by splitting the problem as suggested in Problem 15, since neither of the two subsidiary problems can be solved in this case.

17. Consider the problem

$$y'' + p(x)y' + q(x)y = 0, \qquad y(0) = a, \quad y(1) = b.$$

Let $y = u + v$, where v is any twice differentiable function satisfying the boundary conditions (but not necessarily the differential equation). Show that u is a solution of the problem

$$u'' + p(x)u' + q(x)u = g(x), \qquad u(0) = 0, \quad u(1) = 0,$$

where $g(x) = -[v'' + p(x)v' + q(x)v]$ and is known once v is chosen. Thus nonhomogeneities can be transferred from the boundary conditions to the differential equation. Find a function v for this problem.

18. Using the method of Problem 17, transform the problem

$$y'' + 2y = 2 - 4x, \qquad y(0) = 1, \quad y(1) + y'(1) = -2$$

into a new problem in which the boundary conditions are homogeneous. Solve the latter problem by reference to Example 1 of the text.

In each of Problems 19 through 22 use eigenfunction expansions to find the solution of the given boundary value problem.

19. $u_t = u_{xx} - x$, $u(0,t) = 0$, $u_x(1,t) = 0$, $u(x,0) = \sin(\pi x/2)$;
 see Problem 2.

20. $u_t = u_{xx} + e^{-t}$, $u_x(0,t) = 0$, $u_x(1,t) + u(1,t) = 0$, $u(x,0) = 1 - x$;
 see Section 11.2, Problems 10 and 12.

21. $u_t = u_{xx} + 1 - |1 - 2x|$, $u(0,t) = 0$, $u(1,t) = 0$, $u(x,0) = 0$;
 see Problem 5.

22. $u_t = u_{xx} + e^{-t}(1 - x)$, $u(0,t) = 0$, $u_x(1,t) = 0$, $u(x,0) = 0$;
 see Section 11.2, Problems 6 and 7.

23. Consider the boundary value problem

$$r(x)u_t = [p(x)u_x]_x - q(x)u + F(x),$$

$$u(0,t) = T_1, \qquad u(1,t) = T_2, \qquad u(x,0) = f(x).$$

(a) Let $v(x)$ be a solution of the problem

$$[p(x)v']' - q(x)v = -F(x), \qquad v(0) = T_1, \quad v(1) = T_2.$$

If $w(x,t) = u(x,t) - v(x)$, find the boundary value problem satisfied by w. Note that this problem can be solved by the method of this section.

(b) Generalize the procedure of part (a) to the case where u satisfies the boundary conditions

$$u_x(0,t) - h_1 u(0,t) = T_1, \qquad u_x(1,t) + h_2 u(1,t) = T_2.$$

In each of Problems 24 and 25 use the method indicated in Problem 23 to solve the given boundary value problem.

24. $u_t = u_{xx} - 2$,
$u(0,t) = 1$, $\quad u(1,t) = 0$,
$u(x,0) = x^2 - 2x + 2$

25. $u_t = u_{xx} - \pi^2 \cos \pi x$,
$u_x(0,t) = 0$, $\quad u(1,t) = 1$,
$u(x,0) = \cos(3\pi x/2) - \cos \pi x$

26. The method of eigenfunction expansions is often useful for nonhomogeneous problems related to the wave equation or its generalizations. Consider the problem

$$r(x)u_{tt} = [p(x)u_x]_x - q(x)u + F(x,t), \tag{i}$$

$$u_x(0,t) - h_1 u(0,t) = 0, \qquad u_x(1,t) + h_2 u(1,t) = 0, \tag{ii}$$

$$u(x,0) = f(x), \qquad u_t(x,0) = g(x). \tag{iii}$$

This problem can arise in connection with generalizations of the telegraph equation (Problem 16 in Section 11.1) or the longitudinal vibrations of an elastic bar (Problem 25 in Section 11.1).

(a) Let $u(x,t) = X(x)T(t)$ in the homogeneous equation corresponding to Eq. (i), and show that $X(x)$ satisfies Eqs. (28) and (29) of the text. Let λ_n and $\phi_n(x)$ denote the eigenvalues and normalized eigenfunctions of this problem.

(b) Assume that $u(x,t) = \sum_{n=1}^{\infty} b_n(t)\phi_n(x)$, and show that $b_n(t)$ must satisfy the initial value problem

$$b_n''(t) + \lambda_n b_n(t) = \gamma_n(t), \qquad b_n(0) = \alpha_n, \quad b_n'(0) = \beta_n,$$

where α_n, β_n, and $\gamma_n(t)$ are the expansion coefficients for $f(x)$, $g(x)$, and $F(x,t)/r(x)$ in terms of the eigenfunctions $\phi_1(x), \ldots, \phi_n(x), \ldots$.

27. In this problem we explore a little further the analogy between Sturm–Liouville boundary value problems and Hermitian matrices. Let \mathbf{A} be an $n \times n$ Hermitian matrix with eigenvalues $\lambda_1, \ldots, \lambda_n$ and corresponding orthogonal eigenvectors $\boldsymbol{\xi}^{(1)}, \ldots, \boldsymbol{\xi}^{(n)}$.

Consider the nonhomogeneous system of equations

$$\mathbf{Ax} - \mu\mathbf{x} = \mathbf{b}, \tag{i}$$

where μ is a given real number and \mathbf{b} is a given vector. We will point out a way of solving Eq. (i) that is analogous to the method presented in the text for solving Eqs. (1) and (2).

(a) Show that $\mathbf{b} = \sum_{i=1}^{n} b_i \boldsymbol{\xi}^{(i)}$, where $b_i = (\mathbf{b}, \boldsymbol{\xi}^{(i)})$.

(b) Assume that $\mathbf{x} = \sum_{i=1}^{n} a_i \boldsymbol{\xi}^{(i)}$ and show that for Eq. (i) to be satisfied, it is necessary that $a_i = b_i/(\lambda_i - \mu)$. Thus

$$\mathbf{x} = \sum_{i=1}^{n} \frac{(\mathbf{b}, \boldsymbol{\xi}^{(i)})}{\lambda_i - \mu} \boldsymbol{\xi}^{(i)}, \tag{ii}$$

provided that μ is not one of the eigenvalues of \mathbf{A}, $\mu \neq \lambda_i$ for $i = 1, \ldots, n$. Compare this result with Eq. (13).

Green's[7] Functions. Consider the nonhomogeneous system of algebraic equations

$$\mathbf{A}\mathbf{x} - \mu\mathbf{x} = \mathbf{b},\tag{i}$$

where \mathbf{A} is an $n \times n$ Hermitian matrix, μ is a given real number, and \mathbf{b} is a given vector. Instead of using an eigenvector expansion as in Problem 27, we can solve Eq. (i) by computing the inverse matrix $(\mathbf{A} - \mu\mathbf{I})^{-1}$, which exists if μ is not an eigenvalue of \mathbf{A}. Then

$$\mathbf{x} = (\mathbf{A} - \mu\mathbf{I})^{-1}\mathbf{b}.\tag{ii}$$

Problems 28 through 36 indicate a way of solving nonhomogeneous boundary value problems that is analogous to using the inverse matrix for a system of linear algebraic equations. The Green's function plays a part similar to the inverse of the matrix of coefficients. This method leads to solutions expressed as definite integrals rather than as infinite series. Except in Problem 35, we will assume that $\mu = 0$ for simplicity.

28. (a) Show by the method of variation of parameters that the general solution of the differential equation

$$-y'' = f(x)$$

can be written in the form

$$y = \phi(x) = c_1 + c_2 x - \int_0^x (x - s) f(s)\, ds,$$

where c_1 and c_2 are arbitrary constants.

(b) Let $y = \phi(x)$ also be required to satisfy the boundary conditions $y(0) = 0$, $y(1) = 0$. Show that in this case

$$c_1 = 0, \qquad c_2 = \int_0^1 (1 - s) f(s)\, ds.$$

(c) Show that, under the conditions of parts (a) and (b), $\phi(x)$ can be written in the form

$$\phi(x) = \int_0^x s(1 - x) f(s)\, ds + \int_x^1 x(1 - s) f(s)\, ds.$$

(d) Defining

$$G(x, s) = \begin{cases} s(1 - x), & 0 \le s \le x, \\ x(1 - s), & x \le s \le 1, \end{cases}$$

show that the solution takes the form

$$\phi(x) = \int_0^1 G(x, s) f(s)\, ds.$$

The function $G(x, s)$ appearing under the integral sign is a Green's function. The usefulness of a Green's function solution rests on the fact that the Green's function is independent of

[7]Green's functions are named after George Green (1793–1841) of England. He was almost entirely self-taught in mathematics and made significant contributions to electricity and magnetism, fluid mechanics, and partial differential equations. His most important work was an essay on electricity and magnetism that was published privately in 1828. In this paper Green was the first to recognize the importance of potential functions. He introduced the functions now known as Green's functions as a means of solving boundary value problems and developed the integral transformation theorems of which Green's theorem in the plane is a particular case. However, these results did not become widely known until Green's essay was republished in the 1850s through the efforts of William Thomson (Lord Kelvin).

the nonhomogeneous term in the differential equation. Thus, once the Green's function is determined, the solution of the boundary value problem for any nonhomogeneous term $f(x)$ is obtained by a single integration. Note further that no determination of arbitrary constants is required, since $\phi(x)$ as given by the Green's function integral formula automatically satisfies the boundary conditions.

29. By a procedure similar to that in Problem 28 show that the solution of the boundary value problem

$$-(y'' + y) = f(x), \qquad y(0) = 0, \quad y(1) = 0$$

is

$$y = \phi(x) = \int_0^1 G(x, s)f(s)\, ds,$$

where

$$G(x, s) = \begin{cases} \dfrac{\sin s\, \sin(1 - x)}{\sin 1}, & 0 \leq s \leq x, \\[2mm] \dfrac{\sin x\, \sin(1 - s)}{\sin 1}, & x \leq s \leq 1. \end{cases}$$

30. It is possible to show that the Sturm–Liouville problem

$$L[y] = -[p(x)y']' + q(x)y = f(x), \tag{i}$$

$$\alpha_1 y(0) + \alpha_2 y'(0) = 0, \qquad \beta_1 y(1) + \beta_2 y'(1) = 0 \tag{ii}$$

has a Green's function solution

$$y = \phi(x) = \int_0^1 G(x, s)f(s)\, ds, \tag{iii}$$

provided that $\lambda = 0$ is not an eigenvalue of $L[y] = \lambda y$ subject to the boundary conditions (ii). Further, $G(x, s)$ is given by

$$G(x, s) = \begin{cases} -y_1(s)y_2(x)/p(x)W(y_1, y_2)(x), & 0 \leq s \leq x, \\[2mm] -y_1(x)y_2(s)/p(x)W(y_1, y_2)(x), & x \leq s \leq 1, \end{cases} \tag{iv}$$

where y_1 is a solution of $L[y] = 0$ satisfying the boundary condition at $x = 0$, y_2 is a solution of $L[y] = 0$ satisfying the boundary condition at $x = 1$, and $W(y_1, y_2)$ is the Wronskian of y_1 and y_2.

(a) Verify that the Green's function obtained in Problem 28 is given by formula (iv).

(b) Verify that the Green's function obtained in Problem 29 is given by formula (iv).

(c) Show that $p(x)W(y_1, y_2)(x)$ is a constant by showing that its derivative is zero.

(d) Using Eq. (iv) and the result of part (c), show that $G(x, s) = G(s, x)$.

(e) Verify that $y = \phi(x)$ from Eq. (iii) with $G(x, s)$ given by Eq. (iv) satisfies the differential equation (i) and the boundary conditions (ii).

In each of Problems 31 through 34 solve the given boundary value problem by determining the appropriate Green's function and expressing the solution as a definite integral. Use Eqs. (i) to (iv) of Problem 30.

31. $-y'' = f(x)$, $\qquad y'(0) = 0$, $\quad y(1) = 0$

32. $-y'' = f(x)$, $\qquad y(0) = 0$, $\quad y(1) + y'(1) = 0$

33. $-(y'' + y) = f(x), \qquad y'(0) = 0, \quad y(1) = 0$

34. $-y'' = f(x), \qquad y(0) = 0, \quad y'(1) = 0$

35. Consider the boundary value problem

$$L[y] = -[p(x)y']' + q(x)y = \mu r(x)y + f(x), \tag{i}$$

$$\alpha_1 y(0) + \alpha_2 y'(0) = 0, \qquad \beta_1 y(1) + \beta_2 y'(1) = 0. \tag{ii}$$

According to the text, the solution $y = \phi(x)$ is given by Eq. (13), where c_n is defined by Eq. (9), provided that μ is not an eigenvalue of the corresponding homogeneous problem. In this case it can also be shown that the solution is given by a Green's function integral of the form

$$y = \phi(x) = \int_0^1 G(x, s, \mu) f(s)\, ds. \tag{iii}$$

Note that in this problem the Green's function also depends on the parameter μ.

(a) Show that if these two expressions for $\phi(x)$ are to be equivalent, then

$$G(x, s, \mu) = \sum_{i=1}^{\infty} \frac{\phi_i(x)\phi_i(s)}{\lambda_i - \mu}, \tag{iv}$$

where λ_i and ϕ_i are the eigenvalues and eigenfunctions, respectively, of Eqs. (3), (2) of the text. Again we see from Eq. (iv) that μ cannot be equal to any eigenvalue λ_i.

(b) Derive Eq. (iv) directly by assuming that $G(x, s, \mu)$ has the eigenfunction expansion

$$G(x, s, \mu) = \sum_{i=1}^{\infty} a_i(x, \mu)\phi_i(s). \tag{v}$$

Determine $a_i(x, \mu)$ by multiplying Eq. (v) by $r(s)\phi_j(s)$ and integrating with respect to s from $s = 0$ to $s = 1$.
Hint: Show first that λ_i and ϕ_i satisfy the equation

$$\phi_i(x) = (\lambda_i - \mu) \int_0^1 G(x, s, \mu) r(s)\phi_i(s)\, ds. \tag{vi}$$

36. Consider the boundary value problem

$$-d^2 y/ds^2 = \delta(s - x), \qquad y(0) = 0, \quad y(1) = 0,$$

where s is the independent variable, $s = x$ is a definite point in the interval $0 < s < 1$, and δ is the Dirac delta function (see Section 6.5). Show that the solution of this problem is the Green's function $G(x, s)$ obtained in Problem 28.

 In solving the given problem, note that $\delta(s - x) = 0$ in the intervals $0 \leq s < x$ and $x < s \leq 1$. Note further that $-dy/ds$ experiences a jump of magnitude 1 as s passes through the value x.

 This problem illustrates a general property, namely, that the Green's function $G(x, s)$ can be identified as the response at the point s to a unit impulse at the point x. A more general nonhomogeneous term f on $0 \leq x \leq 1$ can be regarded as a continuous distribution

of impulses with magnitude $f(x)$ at the point x. The solution of a nonhomogeneous boundary value problem in terms of a Green's function integral can then be interpreted as the result of superposing the responses to the set of impulses represented by the nonhomogeneous term $f(x)$.

11.4 Singular Sturm–Liouville Problems

In the preceding sections of this chapter we considered Sturm–Liouville boundary value problems: the differential equation

$$L[y] = -[p(x)y']' + q(x)y = \lambda r(x)y, \qquad 0 < x < 1, \tag{1}$$

together with boundary conditions of the form

$$\alpha_1 y(0) + \alpha_2 y'(0) = 0, \tag{2}$$

$$\beta_1 y(1) + \beta_2 y'(1) = 0. \tag{3}$$

Until now, we have always assumed that the problem is regular; that is, p is differentiable, q and r are continuous, and $p(x) > 0$ and $r(x) > 0$ at all points in the *closed* interval. However, there are also equations of physical interest in which some of these conditions are not satisfied.

For example, suppose that we wish to study Bessel's equation of order ν on the interval $0 < x < 1$. This equation is sometimes written in the form[8]

$$-(xy')' + \frac{\nu^2}{x}y = \lambda xy \tag{4}$$

so that $p(x) = x$, $q(x) = \nu^2/x$, and $r(x) = x$. Thus $p(0) = 0$, $r(0) = 0$, and $q(x)$ is unbounded and hence discontinuous as $x \to 0$. However, the conditions imposed on regular Sturm–Liouville problems are met elsewhere in the interval.

Similarly, for Legendre's equation we have

$$-[(1 - x^2)y']' = \lambda y, \qquad -1 < x < 1, \tag{5}$$

where $\lambda = \alpha(\alpha + 1)$, $p(x) = 1 - x^2$, $q(x) = 0$, and $r(x) = 1$. Here the required conditions on p, q, and r are satisfied in the interval $0 \le x \le 1$ except at $x = 1$, where p is zero.

We use the term singular Sturm–Liouville problem to refer to a certain class of boundary value problems for the differential equation (1) in which the functions p, q, and r satisfy the conditions stated earlier on the open interval $0 < x < 1$, but at least one of these functions fails to satisfy them at one or both of the boundary points. We also prescribe suitable separated boundary conditions of a kind to be described in more detail later in this section. Singular problems also occur if the interval is

[8]The substitution $t = \sqrt{\lambda}\,x$ reduces Eq. (4) to the standard form $t^2 y'' + t y' + (t^2 - \nu^2)y = 0$.

unbounded, for example, $0 \leq x < \infty$. We do not consider this latter kind of singular problem in this book.

As an example of a singular problem on a finite interval, consider the equation

$$xy'' + y' + \lambda xy = 0, \tag{6}$$

or

$$-(xy')' = \lambda xy, \tag{7}$$

on the interval $0 < x < 1$, and suppose that $\lambda > 0$. This equation arises in the study of free vibrations of a circular elastic membrane and is discussed further in Section 11.5. If we introduce the new independent variable t defined by $t = \sqrt{\lambda}\, x$, then

$$\frac{dy}{dx} = \sqrt{\lambda}\,\frac{dy}{dt}, \qquad \frac{d^2y}{dx^2} = \lambda\,\frac{d^2y}{dt^2}.$$

Hence Eq. (6) becomes

$$\frac{t}{\sqrt{\lambda}}\,\lambda\,\frac{d^2y}{dt^2} + \sqrt{\lambda}\,\frac{dy}{dt} + \lambda\,\frac{t}{\sqrt{\lambda}}y = 0,$$

or, if we cancel the common factor $\sqrt{\lambda}$ in each term,

$$t\frac{d^2y}{dt^2} + \frac{dy}{dt} + ty = 0. \tag{8}$$

Equation (8) is Bessel's equation of order zero (see Section 5.8). The general solution of Eq. (8) for $t > 0$ is

$$y = c_1 J_0(t) + c_2 Y_0(t);$$

hence the general solution of Eq. (7) for $x > 0$ is

$$y = c_1 J_0(\sqrt{\lambda}\, x) + c_2 Y_0(\sqrt{\lambda}\, x), \tag{9}$$

where J_0 and Y_0 denote the Bessel functions of the first and second kinds of order zero. From Eqs. (7) and (13) of Section 5.8 we have

$$J_0(\sqrt{\lambda}\, x) = 1 + \sum_{m=1}^{\infty} \frac{(-1)^m \lambda^m x^{2m}}{2^{2m}(m!)^2}, \qquad x > 0, \tag{10}$$

$$Y_0(\sqrt{\lambda}\, x) = \frac{2}{\pi}\left[\left(\gamma + \ln\frac{\sqrt{\lambda}\, x}{2}\right) J_0(\sqrt{\lambda}\, x) + \sum_{m=1}^{\infty} \frac{(-1)^{m+1} H_m \lambda^m x^{2m}}{2^{2m}(m!)^2}\right], \qquad x > 0, \tag{11}$$

where $H_m = 1 + (1/2) + \cdots + (1/m)$ and $\gamma = \lim_{m\to\infty}(H_m - \ln m)$. The graphs of $y = J_0(x)$ and $y = Y_0(x)$ are given in Figure 5.8.2.

Suppose that we seek a solution of Eq. (7) that also satisfies the boundary conditions

$$y(0) = 0, \tag{12}$$

$$y(1) = 0, \tag{13}$$

which are typical of those we have met in other problems in this chapter. Since $J_0(0) = 1$ and $Y_0(x) \to -\infty$ as $x \to 0$, the condition $y(0) = 0$ can be satisfied only by

choosing $c_1 = c_2 = 0$ in Eq. (9). Thus the boundary value problem (7), (12), (13) has only the trivial solution.

One interpretation of this result is that the boundary condition (12) at $x = 0$ is too restrictive for the differential equation (7). This illustrates the general situation, namely, that at a singular boundary point it is necessary to consider a modified type of boundary condition. In the present problem, suppose that we require only that the solution (9) and its derivative remain bounded. In other words, we take as the boundary condition at $x = 0$ the requirement

$$y, \ y' \text{ bounded as } x \to 0. \tag{14}$$

This condition can be satisfied by choosing $c_2 = 0$ in Eq. (9), so as to eliminate the unbounded solution Y_0. The second boundary condition, $y(1) = 0$, then yields

$$J_0(\sqrt{\lambda}) = 0. \tag{15}$$

It is possible to show[9] that Eq. (15) has an infinite set of discrete positive roots, which yield the eigenvalues $0 < \lambda_1 < \lambda_2 < \cdots < \lambda_n < \cdots$ of the given problem. The corresponding eigenfunctions are

$$\phi_n(x) = J_0\left(\sqrt{\lambda_n}\, x\right), \tag{16}$$

determined only up to a multiplicative constant. The boundary value problem (7), (13), and (14) is an example of a singular Sturm–Liouville problem. This example illustrates that if the boundary conditions are relaxed in an appropriate way, then a singular Sturm–Liouville problem may have an infinite sequence of eigenvalues and eigenfunctions, just as a regular Sturm–Liouville problem does.

Because of their importance in applications, it is worthwhile to investigate singular boundary value problems a little further. There are two main questions that are of concern:

1. Precisely what type of boundary conditions can be allowed in a singular Sturm–Liouville problem?

2. To what extent do the eigenvalues and eigenfunctions of a singular problem share the properties of eigenvalues and eigenfunctions of regular Sturm–Liouville problems? In particular, are the eigenvalues real, are the eigenfunctions orthogonal, and can a given function be expanded as a series of eigenfunctions?

Both these questions can be answered by a study of the identity

$$\int_0^1 \{L[u]v - uL[v]\}\, dx = 0, \tag{17}$$

which played an essential part in the development of the theory of regular Sturm–Liouville problems. We therefore investigate the conditions under which this relation holds for singular problems, where the integral in Eq. (17) may now have to be examined as an improper integral. To be definite, we consider the differential equation (1) and assume that $x = 0$ is a singular boundary point but that $x = 1$ is not. The

[9]The function J_0 is well tabulated; the roots of Eq. (15) can be found in various tables, for example, those in Jahnke and Emde or Abramowitz and Stegun. The first three roots of Eq. (15) are $\sqrt{\lambda} = 2.405, 5.520$, and 8.654, respectively, to four significant figures; $\sqrt{\lambda_n} \cong (n - 1/4)\pi$ for large n.

boundary condition (3) is imposed at the nonsingular boundary point $x = 1$, but we leave unspecified, for the moment, the boundary condition at $x = 0$. Indeed, our principal object is to determine what kinds of boundary conditions are allowable at a singular boundary point if Eq. (17) is to hold.

Since the boundary value problem under investigation is singular at $x = 0$, we choose $\epsilon > 0$ and consider the integral $\int_\epsilon^1 L[u]v\, dx$, instead of $\int_0^1 L[u]v\, dx$, as in Section 11.2. Afterwards we let ϵ approach zero. Assuming that u and v have at least two continuous derivatives on $\epsilon \le x \le 1$, and integrating twice by parts, we find that

$$\int_\epsilon^1 \{L[u]v - uL[v]\}\, dx = -p(x)\left[u'(x)v(x) - u(x)v'(x)\right]\Big|_\epsilon^1. \tag{18}$$

The boundary term at $x = 1$ is again eliminated if both u and v satisfy the boundary condition (3), and thus,

$$\int_\epsilon^1 \{L[u]v - uL[v]\}\, dx = p(\epsilon)\left[u'(\epsilon)v(\epsilon) - u(\epsilon)v'(\epsilon)\right]. \tag{19}$$

Taking the limit as $\epsilon \to 0$ yields

$$\int_0^1 \{L[u]v - uL[v]\}\, dx = \lim_{\epsilon \to 0} p(\epsilon)\left[u'(\epsilon)v(\epsilon) - u(\epsilon)v'(\epsilon)\right]. \tag{20}$$

Hence Eq. (17) holds if and only if, in addition to the assumptions stated previously,

$$\lim_{\epsilon \to 0} p(\epsilon)\left[u'(\epsilon)v(\epsilon) - u(\epsilon)v'(\epsilon)\right] = 0 \tag{21}$$

for every pair of functions u and v in the class under consideration. Equation (21) is therefore the criterion that determines what boundary conditions are allowable at $x = 0$ if that point is a singular boundary point. A similar condition applies at $x = 1$ if that boundary point is singular, namely,

$$\lim_{\epsilon \to 0} p(1 - \epsilon)\left[u'(1 - \epsilon)v(1 - \epsilon) - u(1 - \epsilon)v'(1 - \epsilon)\right] = 0. \tag{22}$$

In summary, as in Section 11.2, a singular boundary value problem for Eq. (1) is said to be **self-adjoint** if Eq. (17) is valid, possibly as an improper integral, for each pair of functions u and v with the following properties: They are twice continuously differentiable on the open interval $0 < x < 1$, they satisfy a boundary condition of the form (2) at each regular boundary point, and they satisfy a boundary condition sufficient to ensure Eq. (21) if $x = 0$ is a singular boundary point, or Eq. (22) if $x = 1$ is a singular boundary point. If at least one boundary point is singular, then the differential equation (1), together with two boundary conditions of the type just described, are said to form a **singular Sturm–Liouville problem**.

For example, for Eq. (7) we have $p(x) = x$. If both u and v satisfy the boundary condition (14) at $x = 0$, it is clear that Eq. (21) will hold. Hence the singular boundary value problem, consisting of the differential equation (7), the boundary condition (14) at $x = 0$, and any boundary condition of the form (3) at $x = 1$, is self-adjoint.

The most striking difference between regular and singular Sturm–Liouville problems is that in a singular problem the eigenvalues may not be discrete. That is, the problem may have nontrivial solutions for every value of λ, or for every value of λ in some interval. In such a case the problem is said to have a **continuous spectrum**. It may happen that a singular problem has a mixture of discrete eigenvalues and also a continuous spectrum. Finally, it is possible that only a discrete set of eigenvalues exists, just as in the regular case discussed in Section 11.2. For example, this is true of the problem consisting of Eqs. (7), (13), and (14). In general, it may be difficult to determine which case actually occurs in a given problem.

A systematic discussion of singular Sturm–Liouville problems is quite sophisticated[10] indeed, requiring a substantial extension of the methods presented in this book. We restrict ourselves to some examples related to physical applications; in each of these examples it is known that there is an infinite set of discrete eigenvalues.

If a singular Sturm–Liouville problem does have only a discrete set of eigenvalues and eigenfunctions, then Eq. (17) can be used, just as in Section 11.2, to prove that the eigenvalues of such a problem are real and that the eigenfunctions are orthogonal with respect to the weight function r. The expansion of a given function in terms of a series of eigenfunctions then follows as in Section 11.2.

Such expansions are useful, as in the regular case, for solving nonhomogeneous boundary value problems. The procedure is very similar to that described in Section 11.3. Some examples for ordinary differential equations are indicated in Problems 1 to 4, and some problems for partial differential equations appear in Section 11.5.

For instance, the eigenfunctions $\phi_n(x) = J_0(\sqrt{\lambda_n}\,x)$ of the singular Sturm–Liouville problem

$$-(xy')' = \lambda xy, \qquad 0 < x < 1,$$

$$y,\ y'\ \text{bounded as } x \to 0, \qquad y(1) = 0$$

satisfy the orthogonality relation

$$\int_0^1 x\phi_m(x)\phi_n(x)\,dx = 0, \qquad m \neq n \tag{23}$$

with respect to the weight function $r(x) = x$. Then, if f is a given function, we assume that

$$f(x) = \sum_{n=1}^{\infty} c_n J_0(\sqrt{\lambda_n}\,x). \tag{24}$$

Multiplying Eq. (24) by $xJ_0(\sqrt{\lambda_m}\,x)$ and integrating term by term from $x = 0$ to $x = 1$ yield

$$\int_0^1 xf(x)J_0(\sqrt{\lambda_m}\,x)\,dx = \sum_{n=1}^{\infty} c_n \int_0^1 xJ_0(\sqrt{\lambda_m}\,x)\,J_0(\sqrt{\lambda_n}\,x)\,dx. \tag{25}$$

[10]See, for example, Chapter 5 of the book by Yosida listed in the references at the end of this chapter.

Because of the orthogonality condition (23), the right side of Eq. (25) collapses to a single term; hence

$$c_m = \frac{\displaystyle\int_0^1 xf(x)J_0(\sqrt{\lambda_m}\,x)\,dx}{\displaystyle\int_0^1 xJ_0^2(\sqrt{\lambda_m}\,x)\,dx}, \tag{26}$$

which determines the coefficients in the series (24).

The convergence of the series (24) is established by an extension of Theorem 11.2.4 to cover this case. This theorem can also be shown to hold for other sets of Bessel functions, which are solutions of appropriate boundary value problems, for Legendre polynomials, and for solutions of a number of other singular Sturm–Liouville problems of considerable interest.

It must be emphasized that the singular problems mentioned here are not necessarily typical. In general, singular boundary value problems are characterized by continuous spectra, rather than by discrete sets of eigenvalues. The corresponding sets of eigenfunctions are therefore not denumerable, and series expansions of the type described in Theorem 11.2.4 do not exist. They are replaced by appropriate integral representations.

PROBLEMS

1. Find a formal solution of the nonhomogeneous boundary value problem

$$-(xy')' = \mu xy + f(x),$$

$$y,\ y'\ \text{bounded as}\ x \to 0, \qquad y(1) = 0,$$

where f is a given continuous function on $0 \le x \le 1$, and μ is not an eigenvalue of the corresponding homogeneous problem.

Hint: Use a series expansion similar to those in Section 11.3.

2. Consider the boundary value problem

$$-(xy')' = \lambda xy,$$

$$y,\ y'\ \text{bounded as}\ x \to 0, \qquad y'(1) = 0.$$

(a) Show that $\lambda_0 = 0$ is an eigenvalue of this problem corresponding to the eigenfunction $\phi_0(x) = 1$. If $\lambda > 0$, show formally that the eigenfunctions are given by $\phi_n(x) = J_0(\sqrt{\lambda_n}\,x)$, where $\sqrt{\lambda_n}$ is the nth positive root (in increasing order) of the equation $J_0'(\sqrt{\lambda}) = 0$. It is possible to show that there is an infinite sequence of such roots.

(b) Show that if $m,\ n = 0, 1, 2, \ldots$, then

$$\int_0^1 x\phi_m(x)\phi_n(x)\,dx = 0, \qquad m \ne n.$$

(c) Find a formal solution to the nonhomogeneous problem

$$-(xy')' = \mu xy + f(x),$$

$$y,\ y'\ \text{bounded as}\ x \to 0, \qquad y'(1) = 0,$$

where f is a given continuous function on $0 \le x \le 1$, and μ is not an eigenvalue of the corresponding homogeneous problem.

3. Consider the problem

$$-(xy')' + (k^2/x)y = \lambda xy,$$

$$y, \ y' \text{ bounded as } x \to 0, \qquad y(1) = 0,$$

where k is a positive integer.

(a) Using the substitution $t = \sqrt{\lambda}\, x$, show that the given differential equation reduces to Bessel's equation of order k (see Problem 9 of Section 5.8). One solution is $J_k(t)$; a second linearly independent solution, denoted by $Y_k(t)$, is unbounded as $t \to 0$.

(b) Show formally that the eigenvalues $\lambda_1, \lambda_2, \ldots$ of the given problem are the squares of the positive zeros of $J_k(\sqrt{\lambda})$ and that the corresponding eigenfunctions are $\phi_n(x) = J_k(\sqrt{\lambda_n}\, x)$. It is possible to show that there is an infinite sequence of such zeros.

(c) Show that the eigenfunctions $\phi_n(x)$ satisfy the orthogonality relation

$$\int_0^1 x\phi_m(x)\phi_n(x)\, dx = 0, \qquad m \neq n.$$

(d) Determine the coefficients in the formal series expansion

$$f(x) = \sum_{n=1}^{\infty} a_n \phi_n(x).$$

(e) Find a formal solution of the nonhomogeneous problem

$$-(xy')' + (k^2/x)y = \mu xy + f(x),$$

$$y, \ y' \text{ bounded as } x \to 0, \qquad y(1) = 0,$$

where f is a given continuous function on $0 \leq x \leq 1$, and μ is not an eigenvalue of the corresponding homogeneous problem.

4. Consider Legendre's equation (see Problems 22 through 24 of Section 5.3)

$$-[(1 - x^2)y']' = \lambda y$$

subject to the boundary conditions

$$y(0) = 0, \qquad y, \ y' \text{ bounded as } x \to 1.$$

The eigenfunctions of this problem are the odd Legendre polynomials $\phi_1(x) = P_1(x) = x$, $\phi_2(x) = P_3(x) = (5x^3 - 3x)/2, \ldots, \phi_n(x) = P_{2n-1}(x), \ldots$ corresponding to the eigenvalues $\lambda_1 = 2, \lambda_2 = 4 \cdot 3, \ldots, \lambda_n = 2n(2n - 1), \ldots$.

(a) Show that

$$\int_0^1 \phi_m(x)\phi_n(x)\, dx = 0, \qquad m \neq n.$$

(b) Find a formal solution of the nonhomogeneous problem

$$-[(1 - x^2)y']' = \mu y + f(x),$$

$$y(0) = 0, \qquad y, \ y' \text{ bounded as } x \to 1,$$

where f is a given continuous function on $0 \leq x \leq 1$, and μ is not an eigenvalue of the corresponding homogeneous problem.

5. The equation

$$(1 - x^2)y'' - xy' + \lambda y = 0 \tag{i}$$

is Chebyshev's equation; see Problem 10 of Section 5.3.

(a) Show that Eq. (i) can be written in the form

$$-[(1 - x^2)^{1/2}y']' = \lambda(1 - x^2)^{-1/2}y, \qquad -1 < x < 1. \tag{ii}$$

(b) Consider the boundary conditions

$$y, \; y' \text{ bounded as } x \to -1, \qquad y, \; y' \text{ bounded as } x \to 1. \tag{iii}$$

Show that the boundary value problem (ii), (iii) is self-adjoint.

(c) It can be shown that the boundary value problem (ii), (iii) has the eigenvalues $\lambda_0 = 0, \lambda_1 = 1, \lambda_2 = 4, \dots, \lambda_n = n^2, \dots$. The corresponding eigenfunctions are the Chebyshev polynomials $T_n(x)$: $T_0(x) = 1, T_1(x) = x, T_2(x) = 1 - 2x^2, \dots$. Show that

$$\int_{-1}^{1} \frac{T_m(x)T_n(x)}{(1 - x^2)^{1/2}} \, dx = 0, \qquad m \neq n. \tag{iv}$$

Note that this is a convergent improper integral.

11.5 Further Remarks on the Method of Separation of Variables: A Bessel Series Expansion

In this chapter we are interested in extending the method of separation of variables developed in Chapter 10 to a larger class of problems—to problems involving more general differential equations, more general boundary conditions, or different geometrical regions. We indicated in Section 11.3 how to deal with a class of more general differential equations or boundary conditions. Here we concentrate on problems posed in various geometrical regions, with emphasis on those leading to singular Sturm–Liouville problems when the variables are separated.

Because of its relative simplicity, as well as the considerable physical significance of many problems to which it is applicable, the method of separation of variables merits its important place in the theory and application of partial differential equations. However, this method does have certain limitations that should not be forgotten. In the first place, the problem must be linear so that the principle of superposition can be invoked to construct additional solutions by forming linear combinations of the fundamental solutions of an appropriate homogeneous problem.

As a practical matter, we must also be able to solve the ordinary differential equations, obtained after separating the variables, in a reasonably convenient manner. In some problems to which the method of separation of variables can be applied in principle, it is of very limited practical value due to a lack of information about the solutions of the ordinary differential equations that appear.

Furthermore, the geometry of the region involved in the problem is subject to rather severe restrictions. On the one hand, a coordinate system must be employed in which the variables can be separated, and the partial differential equation replaced

by a set of ordinary differential equations. For Laplace's equation there are about a dozen such coordinate systems; only rectangular, circular cylindrical, and spherical coordinates are likely to be familiar to most readers of this book. On the other hand, the boundary of the region of interest must consist of coordinate curves or surfaces—that is, curves or surfaces on which one variable remains constant. Thus, at an elementary level, one is limited to regions bounded by straight lines or circular arcs in two dimensions, or by planes, circular cylinders, circular cones, or spheres in three dimensions.

In three-dimensional problems the separation of variables in Laplace's operator $u_{xx} + u_{yy} + u_{zz}$ leads to the equation $X'' + \lambda X = 0$ in rectangular coordinates, to Bessel's equation in cylindrical coordinates, and to Legendre's equation in spherical coordinates. It is this fact that is largely responsible for the intensive study that has been made of these equations and the functions defined by them. It is also noteworthy that two of the three most important situations lead to singular, rather than regular, Sturm–Liouville problems. Thus singular problems are by no means exceptional and may be of even greater interest than regular ones. The remainder of this section is devoted to an example involving an expansion of a given function as a series of Bessel functions.

The Vibrations of a Circular Elastic Membrane. In Section 10.7 [Eq. (7)] we noted that the transverse vibrations of a thin elastic membrane are governed by the two-dimensional wave equation

$$a^2(u_{xx} + u_{yy}) = u_{tt}. \tag{1}$$

To study the motion of a circular membrane it is convenient to write Eq. (1) in polar coordinates:

$$a^2\left(u_{rr} + \frac{1}{r}u_r + \frac{1}{r^2}u_{\theta\theta}\right) = u_{tt}. \tag{2}$$

We will assume that the membrane has unit radius, that it is fixed securely around its circumference, and that initially it occupies a displaced position independent of the angular variable θ, from which it is released at time $t = 0$. Because of the circular symmetry of the initial and boundary conditions, it is natural to assume also that u is independent of θ; that is, u is a function of r and t only. In this event the differential equation (2) becomes

$$a^2\left(u_{rr} + \frac{1}{r}u_r\right) = u_{tt}, \qquad 0 < r < 1, \quad t > 0. \tag{3}$$

The boundary condition at $r = 1$ is

$$u(1, t) = 0, \qquad t \geq 0, \tag{4}$$

and the initial conditions are

$$u(r, 0) = f(r), \qquad 0 \leq r \leq 1, \tag{5}$$

$$u_t(r, 0) = 0, \qquad 0 \leq r \leq 1, \tag{6}$$

where $f(r)$ describes the initial configuration of the membrane. For consistency we also require that $f(1) = 0$. Finally, we state explicitly the requirement that $u(r, t)$ is to be bounded for $0 \leq r \leq 1$.

Assuming that $u(r,t) = R(r)T(t)$, and substituting for $u(r,t)$ in Eq. (3), we obtain

$$\frac{R'' + (1/r)R'}{R} = \frac{1}{a^2}\frac{T''}{T} = -\lambda^2. \tag{7}$$

We have anticipated that the separation constant must be negative by writing it as $-\lambda^2$ with $\lambda > 0$.[11] Then Eq. (7) yields the following two ordinary differential equations:

$$r^2 R'' + r R' + \lambda^2 r^2 R = 0, \tag{8}$$

$$T'' + \lambda^2 a^2 T = 0. \tag{9}$$

Thus, from Eq. (9),

$$T(t) = k_1 \sin \lambda a t + k_2 \cos \lambda a t. \tag{10}$$

Introducing the new independent variable $\xi = \lambda r$ into Eq. (8), we obtain

$$\xi^2 \frac{d^2 R}{d\xi^2} + \xi \frac{dR}{d\xi} + \xi^2 R = 0, \tag{11}$$

which is Bessel's equation of order zero. Thus

$$R = c_1 J_0(\xi) + c_2 Y_0(\xi), \tag{12}$$

where J_0 and Y_0 are Bessel functions of the first and second kinds, respectively, of order zero (see Section 11.4). In terms of r we have

$$R = c_1 J_0(\lambda r) + c_2 Y_0(\lambda r). \tag{13}$$

The boundedness condition on $u(r,t)$ requires that R remain bounded as $r \to 0$. Since $Y_0(\lambda r) \to -\infty$ as $r \to 0$, we must choose $c_2 = 0$. The boundary condition (4) then requires that

$$J_0(\lambda) = 0. \tag{14}$$

Consequently, the allowable values of the separation constant are obtained from the roots of the transcendental equation (14). Recall from Section 11.4 that $J_0(\lambda)$ has an infinite set of discrete positive zeros, which we denote by $\lambda_1, \lambda_2, \lambda_3, \ldots, \lambda_n, \ldots$, ordered in increasing magnitude. Further, the functions $J_0(\lambda_n r)$ are the eigenfunctions of a singular Sturm–Liouville problem and can be used as the basis of a series expansion for the given function f. The fundamental solutions of this problem, satisfying the partial differential equation (3), the boundary condition (4), and boundedness condition, are

$$u_n(r,t) = J_0(\lambda_n r) \sin \lambda_n a t, \qquad n = 1, 2, \ldots, \tag{15}$$

$$v_n(r,t) = J_0(\lambda_n r) \cos \lambda_n a t, \qquad n = 1, 2, \ldots. \tag{16}$$

[11]By denoting the separation constant by $-\lambda^2$, rather than simply by $-\lambda$, we avoid the appearance of numerous radical signs in the following discussion.

Next we assume that $u(r, t)$ can be expressed as an infinite linear combination of the fundamental solutions (15), (16):

$$u(r, t) = \sum_{n=1}^{\infty} \left[k_n u_n(r, t) + c_n v_n(r, t) \right]$$

$$= \sum_{n=1}^{\infty} [k_n J_0(\lambda_n r) \, \sin \lambda_n at + c_n J_0(\lambda_n r) \, \cos \lambda_n at]. \tag{17}$$

The initial conditions require that

$$u(r, 0) = \sum_{n=1}^{\infty} c_n J_0(\lambda_n r) = f(r) \tag{18}$$

and

$$u_t(r, 0) = \sum_{n=1}^{\infty} \lambda_n a k_n J_0(\lambda_n r) = 0. \tag{19}$$

From Eq. (26) of Section 11.4 we obtain

$$k_n = 0, \qquad c_n = \frac{\displaystyle\int_0^1 r f(r) J_0(\lambda_n r) \, dr}{\displaystyle\int_0^1 r [J_0(\lambda_n r)]^2 \, dr}; \qquad n = 1, 2, \ldots. \tag{20}$$

Thus the solution of the partial differential equation (3) satisfying the boundary condition (4) and the initial conditions (5) and (6) is given by

$$u(r, t) = \sum_{n=1}^{\infty} c_n J_0(\lambda_n r) \, \cos \lambda_n at \tag{21}$$

with the coefficients c_n defined by Eq. (20).

PROBLEMS

1. Consider Laplace's equation $u_{xx} + u_{yy} = 0$ in the parallelogram whose vertices are $(0, 0)$, $(2, 0)$, $(3, 2)$, and $(1, 2)$. Suppose that on the side $y = 2$ the boundary condition is $u(x, 2) = f(x)$ for $1 \leq x \leq 3$, and that on the other three sides $u = 0$ (see Figure 11.5.1).

(a) Show that there are no nontrivial solutions of the partial differential equation of the form $u(x, y) = X(x)Y(y)$ that also satisfy the homogeneous boundary conditions.

(b) Let $\xi = x - \frac{1}{2} y, \eta = y$. Show that the given parallelogram in the xy-plane transforms into the square $0 \leq \xi \leq 2, 0 \leq \eta \leq 2$ in the $\xi\eta$-plane. Show that the differential equation transforms into

$$\tfrac{5}{4} u_{\xi\xi} - u_{\xi\eta} + u_{\eta\eta} = 0.$$

How are the boundary conditions transformed?

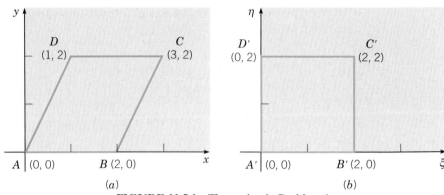

FIGURE 11.5.1 The region in Problem 1.

(c) Show that in the $\xi\eta$-plane the differential equation possesses no solution of the form

$$u(\xi,\eta) = U(\xi)V(\eta).$$

Thus in the xy-plane the shape of the boundary precludes a solution by the method of the separation of variables, while in the $\xi\eta$-plane the region is acceptable but the variables in the differential equation can no longer be separated.

2. Find the displacement $u(r,t)$ in a vibrating circular elastic membrane of radius 1 that satisfies the boundary condition

$$u(1,t) = 0, \qquad t \geq 0,$$

and the initial conditions

$$u(r,0) = 0, \qquad u_t(r,0) = g(r), \qquad 0 \leq r \leq 1,$$

where $g(1) = 0$.
Hint: The differential equation to be satisfied is Eq. (3) of this section.

3. Find the displacement $u(r,t)$ in a vibrating circular elastic membrane of radius 1 that satisfies the boundary condition

$$u(1,t) = 0, \qquad t \geq 0,$$

and the initial conditions

$$u(r,0) = f(r), \qquad u_t(r,0) = g(r), \qquad 0 \leq r \leq 1,$$

where $f(1) = g(1) = 0$.

4. The wave equation in polar coordinates is

$$u_{rr} + (1/r)u_r + (1/r^2)u_{\theta\theta} = a^{-2}u_{tt}.$$

Show that if $u(r,\theta,t) = R(r)\Theta(\theta)T(t)$, then R, Θ, and T satisfy the ordinary differential equations

$$r^2R'' + rR' + (\lambda^2 r^2 - n^2)R = 0,$$

$$\Theta'' + n^2\Theta = 0,$$

$$T'' + \lambda^2 a^2 T = 0.$$

5. In the circular cylindrical coordinates r, θ, z defined by

$$x = r\cos\theta, \qquad y = r\sin\theta, \qquad z = z,$$

Laplace's equation is

$$u_{rr} + (1/r)u_r + (1/r^2)u_{\theta\theta} + u_{zz} = 0.$$

(a) Show that if $u(r, \theta, z) = R(r)\Theta(\theta)Z(z)$, then R, Θ, and Z satisfy the ordinary differential equations

$$r^2 R'' + rR' + (\lambda^2 r^2 - n^2)R = 0,$$
$$\Theta'' + n^2\Theta = 0,$$
$$Z'' - \lambda^2 Z = 0.$$

(b) Show that if $u(r, \theta, z)$ is independent of θ, then the first equation in part (a) becomes

$$r^2 R'' + rR' + \lambda^2 r^2 R = 0,$$

the second is omitted altogether, and the third is unchanged.

6. Find the steady-state temperature in a semi-infinite rod $0 < z < \infty, 0 \leq r < 1$, if the temperature is independent of θ and approaches zero as $z \to \infty$. Assume that the temperature $u(r, z)$ satisfies the boundary conditions

$$u(1, z) = 0, \qquad z > 0,$$
$$u(r, 0) = f(r), \qquad 0 \leq r \leq 1.$$

Hint: Refer to Problem 5.

7. The equation

$$v_{xx} + v_{yy} + k^2 v = 0$$

is a generalization of Laplace's equation and is sometimes called the Helmholtz[12] equation.

(a) In polar coordinates the Helmholtz equation is

$$v_{rr} + (1/r)v_r + (1/r^2)v_{\theta\theta} + k^2 v = 0.$$

If $v(r, \theta) = R(r)\Theta(\theta)$, show that R and Θ satisfy the ordinary differential equations

$$r^2 R'' + rR' + (k^2 r^2 - \lambda^2)R = 0, \qquad \Theta'' + \lambda^2\Theta = 0.$$

(b) Consider the Helmholtz equation in the disk $r < c$. Find the solution that remains bounded at all points in the disk, that is periodic in θ with period 2π, and that satisfies the boundary condition $v(c, \theta) = f(\theta)$, where f is a given function on $0 \leq \theta < 2\pi$.
Hint: The equation for R is a Bessel equation. See Problem 3 of Section 11.4.

[12]Hermann von Helmholtz (1821–1894) was trained in medicine and physiology; early in his career he made important contributions to physiological optics and acoustics, including the invention of the ophthalmoscope in 1851. Later his interests turned to physics, especially fluid mechanics and electrodynamics. During his lifetime he held chairs in physiology or physics at several German universities.

8. Consider the flow of heat in an infinitely long cylinder of radius 1: $0 \le r < 1, 0 \le \theta < 2\pi$, $-\infty < z < \infty$. Let the surface of the cylinder be held at temperature zero, and let the initial temperature distribution be a function of the radial variable r only. Then the temperature u is a function of r and t only, and satisfies the heat conduction equation

$$\alpha^2[u_{rr} + (1/r)u_r] = u_t, \qquad 0 < r < 1, \quad t > 0,$$

and the following initial and boundary conditions:

$$u(r,0) = f(r), \qquad 0 \le r \le 1,$$
$$u(1,t) = 0, \qquad t > 0.$$

Show that

$$u(r,t) = \sum_{n=1}^{\infty} c_n J_0(\lambda_n r) e^{-\alpha^2 \lambda_n^2 t},$$

where $J_0(\lambda_n) = 0$. Find a formula for c_n.

9. In the spherical coordinates ρ, θ, ϕ ($\rho > 0, 0 \le \theta < 2\pi, 0 \le \phi \le \pi$) defined by the equations

$$x = \rho \cos\theta \sin\phi, \qquad y = \rho \sin\theta \sin\phi, \qquad z = \rho \cos\phi,$$

Laplace's equation is

$$\rho^2 u_{\rho\rho} + 2\rho u_\rho + (\csc^2\phi)u_{\theta\theta} + u_{\phi\phi} + (\cot\phi)u_\phi = 0.$$

(a) Show that if $u(\rho,\theta,\phi) = P(\rho)\Theta(\theta)\Phi(\phi)$, then P, Θ, and Φ satisfy ordinary differential equations of the form

$$\rho^2 P'' + 2\rho P' - \mu^2 P = 0,$$
$$\Theta'' + \lambda^2 \Theta = 0,$$
$$(\sin^2\phi)\Phi'' + (\sin\phi\cos\phi)\Phi' + (\mu^2\sin^2\phi - \lambda^2)\Phi = 0.$$

The first of these equations is of the Euler type, while the third is related to Legendre's equation.

(b) Show that if $u(\rho,\theta,\phi)$ is independent of θ, then the first equation in part (a) is unchanged, the second is omitted, and the third becomes

$$(\sin^2\phi)\Phi'' + (\sin\phi\cos\phi)\Phi' + (\mu^2\sin^2\phi)\Phi = 0.$$

(c) Show that if a new independent variable is defined by $s = \cos\phi$, then the equation for Φ in part (b) becomes

$$(1 - s^2)\frac{d^2\Phi}{ds^2} - 2s\frac{d\Phi}{ds} + \mu^2\Phi = 0, \qquad -1 \le s \le 1.$$

Note that this is Legendre's equation.

10. Find the steady-state temperature $u(\rho,\phi)$ in a sphere of unit radius if the temperature is independent of θ and satisfies the boundary condition

$$u(1,\phi) = f(\phi), \qquad 0 \le \phi \le \pi.$$

Hint: Refer to Problem 9 and to Problems 22 through 29 of Section 5.3. Use the fact that the only solutions of Legendre's equation that are finite at both ±1 are the Legendre polynomials.

11.6 Series of Orthogonal Functions: Mean Convergence

In Section 11.2 we stated that, under certain restrictions, a given function f can be expanded in a series of eigenfunctions of a Sturm–Liouville boundary value problem, the series converging to $[f(x+) + f(x-)]/2$ at each point in the open interval. Under somewhat more restrictive conditions the series converges to $f(x)$ at each point in the closed interval. This type of convergence is referred to as pointwise convergence. In this section we describe a different kind of convergence that is especially useful for series of orthogonal functions, such as eigenfunctions.

Suppose that we are given the set of functions $\phi_1, \phi_2, \ldots, \phi_n$, which are continuous on the interval $0 \le x \le 1$ and satisfy the orthonormality condition

$$\int_0^1 r(x)\phi_i(x)\phi_j(x)\, dx = \begin{cases} 0, & i \ne j, \\ 1, & i = j, \end{cases} \tag{1}$$

where r is a nonnegative weight function. Suppose also that we wish to approximate a given function f, defined on $0 \le x \le 1$, by a linear combination of ϕ_1, \ldots, ϕ_n. That is, if

$$S_n(x) = \sum_{i=1}^n a_i \phi_i(x), \tag{2}$$

we wish to choose the coefficients a_1, \ldots, a_n so that the function S_n will best approximate f on $0 \le x \le 1$. The first problem that we must face in doing this is to state precisely what we mean by "best approximate f on $0 \le x \le 1$." There are several reasonable meanings that can be attached to this phrase.

1. We can choose n points x_1, \ldots, x_n in the interval $0 \le x \le 1$ and require that $S_n(x)$ have the same value as $f(x)$ at each of these points. The coefficients a_1, \ldots, a_n are found by solving the set of linear algebraic equations

$$\sum_{i=1}^n a_i \phi_i(x_j) = f(x_j), \qquad j = 1, \ldots, n. \tag{3}$$

This procedure is known as the **method of collocation**. It has the advantage that it is very easy to write down Eqs. (3); one needs only to evaluate the functions involved at the points x_1, \ldots, x_n. If these points are well chosen, and if n is fairly large, then presumably $S_n(x)$ will not only be equal to $f(x)$ at the chosen points but will be reasonably close to it at other points as well. However, collocation has several deficiencies. One is that if one more base function ϕ_{n+1} is added, then one more point x_{n+1} is required, and *all* the coefficients must be recomputed. Thus it is inconvenient to improve the accuracy of a collocation approximation by including additional terms. Further, the coefficients a_i depend on the location of the points x_1, \ldots, x_n, and it is not obvious how best to select these points.

2. Alternatively, we can consider the difference $|f(x) - S_n(x)|$ and try to make it as small as possible. The trouble here is that $|f(x) - S_n(x)|$ is a function of x as well as of the coefficients a_1, \ldots, a_n, and it is not clear how to calculate a_i. The choice of a_i that makes $|f(x) - S_n(x)|$ small at one point may make it large at another. One way to proceed is to consider instead the least upper bound[13] of $|f(x) - S_n(x)|$ for x in $0 \leq x \leq 1$, and then to choose a_1, \ldots, a_n so as to make this quantity as small as possible. That is, if

$$E_n(a_1, \ldots, a_n) = \operatorname*{lub}_{0 \leq x \leq 1} |f(x) - S_n(x)|, \tag{4}$$

then choose a_1, \ldots, a_n so as to minimize E_n. This approach is intuitively appealing and is often used in theoretical calculations. In practice, however, it is usually very hard, if not impossible, to write down an explicit formula for $E_n(a_1, \ldots, a_n)$. Further, this procedure also shares one of the disadvantages of collocation: Upon adding an additional term to $S_n(x)$, one must recompute all the preceding coefficients. Thus it is not often useful in practical problems.

3. Another way to proceed is to consider

$$I_n(a_1, \ldots, a_n) = \int_0^1 r(x)|f(x) - S_n(x)| \, dx. \tag{5}$$

If $r(x) = 1$, then I_n is the area between the graphs of $y = f(x)$ and $y = S_n(x)$ (see Figure 11.6.1). We can then determine the coefficients a_i so as to minimize I_n. To avoid the complications resulting from calculations with absolute values, it is more convenient to consider instead

$$R_n(a_1, \ldots, a_n) = \int_0^1 r(x)[f(x) - S_n(x)]^2 \, dx \tag{6}$$

as our measure of the quality of approximation of the linear combination $S_n(x)$ to $f(x)$. Although R_n is clearly similar in some ways to I_n, it lacks the simple geometric interpretation of the latter. Nevertheless, it is much easier mathematically to deal with R_n than with I_n. The quantity R_n is called the **mean square error** of the approximation S_n to f. If a_1, \ldots, a_n are chosen so as to minimize R_n, then S_n is said to approximate f in the mean square sense.

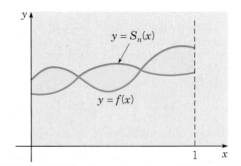

FIGURE 11.6.1 Approximation of $f(x)$ by $S_n(x)$.

[13]The least upper bound (lub) is an upper bound that is smaller than any other upper bound. The lub of a bounded function always exists and is equal to the function's maximum if it has one.

To choose a_1, \ldots, a_n so as to minimize R_n we must satisfy the necessary conditions

$$\partial R_n / \partial a_i = 0, \qquad i = 1, \ldots, n. \tag{7}$$

Writing out Eq. (7), and noting that $\partial S_n(x; a_1, \ldots, a_n) / \partial a_i$ is equal to $\phi_i(x)$, we obtain

$$-\frac{\partial R_n}{\partial a_i} = 2 \int_0^1 r(x)[f(x) - S_n(x)]\phi_i(x)\, dx = 0. \tag{8}$$

Substituting for $S_n(x)$ from Eq. (2) and making use of the orthogonality relation (1) lead to

$$a_i = \int_0^1 r(x) f(x) \phi_i(x)\, dx, \qquad i = 1, \ldots, n. \tag{9}$$

The coefficients defined by Eq. (9) are called the Fourier coefficients of f with respect to the orthonormal set $\phi_1, \phi_2, \ldots, \phi_n$ and the weight function r. Since the conditions (7) are only necessary and not sufficient for R_n to be a minimum, a separate argument is required to show that R_n is actually minimized if the a_i are chosen by Eq. (9). This argument is outlined in Problem 5.

Note that the coefficients (9) are the same as those in the eigenfunction series whose convergence, under certain conditions, was stated in Theorem 11.2.4. Thus $S_n(x)$ is the nth partial sum in this series and constitutes the best mean square approximation to $f(x)$ that is possible with the functions ϕ_1, \ldots, ϕ_n. We will assume hereafter that the coefficients a_i in $S_n(x)$ are given by Eq. (9).

Equation (9) is noteworthy in two other important respects. In the first place, it gives a formula for each a_i *separately*, rather than a set of linear algebraic equations for a_1, \ldots, a_n as in the method of collocation, for example. This is due to the orthogonality of the base functions ϕ_1, \ldots, ϕ_n. Further, the formula for a_i is *independent* of n, the number of terms in $S_n(x)$. The practical significance of this is as follows. Suppose that, to obtain a better approximation to f, we desire to use an approximation with more terms—say, k terms, where $k > n$. It is then unnecessary to recompute the first n coefficients in $S_k(x)$. All that is required is to compute, from Eq. (9), the coefficients a_{n+1}, \ldots, a_k arising from the additional base functions $\phi_{n+1}, \ldots, \phi_k$. Of course, if f, r, and the ϕ_n are complicated functions, it may be necessary to evaluate the integrals numerically.

Now let us suppose that there is an infinite sequence of functions $\phi_1, \ldots, \phi_n, \ldots$, which are continuous and orthonormal on the interval $0 \leq x \leq 1$. Suppose further that as n increases without bound the mean square error R_n approaches zero. In this event the infinite series

$$\sum_{i=1}^{\infty} a_i \phi_i(x)$$

is said to converge in the **mean square sense** (or, more simply, in the **mean**) to $f(x)$. Mean convergence is an essentially different type of convergence from the pointwise convergence considered up to now. A series may converge in the mean without converging at each point. This is plausible geometrically because the area between two curves, which behaves in the same way as the mean square error, may be zero even though the functions are not the same at every point. They may differ on any finite set of points, for example, without affecting the mean square error. It is

less obvious, but also true, that even if an infinite series converges at every point, it may not converge in the mean. Indeed, the mean square error may even become unbounded. An example of this phenomenon is given in Problem 4.

Now suppose that we wish to know what class of functions, defined on $0 \leq x \leq 1$, can be represented as an infinite series of the orthonormal set ϕ_i, $i = 1, 2, \ldots$. The answer depends on what kind of convergence we require. We say that the set $\phi_1, \ldots, \phi_n, \ldots$ is **complete** with respect to mean square convergence for a set of functions \mathcal{F}, if for each function f in \mathcal{F}, the series

$$f(x) = \sum_{i=1}^{\infty} a_i \phi_i(x), \tag{10}$$

with coefficients given by Eq. (9), converges in the mean. There is a similar definition for completeness with respect to pointwise convergence.

Theorems having to do with the convergence of series such as that in Eq. (10) can now be restated in terms of the idea of completeness. For example, Theorem 11.2.4 can be restated as follows: The eigenfunctions of the Sturm–Liouville problem

$$-[p(x)y']' + q(x)y = \lambda r(x)y, \qquad 0 < x < 1, \tag{11}$$

$$\alpha_1 y(0) + \alpha_2 y'(0) = 0, \qquad \beta_1 y(1) + \beta_2 y'(1) = 0 \tag{12}$$

are complete with respect to ordinary pointwise convergence for the set of functions that are continuous on $0 \leq x \leq 1$ and that have piecewise continuous derivatives there.

If pointwise convergence is replaced by mean convergence, Theorem 11.2.4 can be considerably generalized. Before we state such a companion theorem to Theorem 11.2.4, we first define what is meant by a square integrable function. A function f is said to be **square integrable** on the interval $0 \leq x \leq 1$ if both f and f^2 are integrable[14] on that interval. The following theorem is similar to Theorem 11.2.4 except that it involves mean convergence.

Theorem 11.6.1

The eigenfunctions ϕ_i of the Sturm–Liouville problem (11), (12) are complete with respect to mean convergence for the set of functions that are square integrable on $0 \leq x \leq 1$. In other words, given any square integrable function f, the series (10), whose coefficients are given by Eq. (9), converges to $f(x)$ in the mean square sense.

It is significant that the class of functions specified in Theorem 11.6.1 is very large indeed. The class of square integrable functions contains some functions with many

[14]For the Riemann integral used in elementary calculus the hypotheses that f and f^2 are integrable are independent; that is, there are functions such that f is integrable but f^2 is not, and conversely (see Problem 6). A generalized integral, known as the Lebesgue integral, has the property (among others) that if f^2 is integrable, then f is also necessarily integrable. The term *square integrable* came into common use in connection with this type of integration.

discontinuities, including some kinds of infinite discontinuities, as well as some functions that are not differentiable at any point. All these functions have mean convergent expansions in the eigenfunctions of the boundary value problem (11), (12). However, in many cases these series do not converge pointwise, at least not at every point. Thus mean convergence is more naturally associated with series of orthogonal functions, such as eigenfunctions, than ordinary pointwise convergence.

The theory of Fourier series discussed in Chapter 10 is just a special case of the general theory of Sturm–Liouville problems. For instance, the functions

$$\phi_n(x) = \sqrt{2}\sin n\pi x \tag{13}$$

are the normalized eigenfunctions of the Sturm–Liouville problem

$$y'' + \lambda y = 0, \qquad y(0) = 0, \quad y(1) = 0. \tag{14}$$

Thus, if f is a given square integrable function on $0 \le x \le 1$, then according to Theorem 11.6.1, the series

$$f(x) = \sum_{m=1}^{\infty} b_m \phi_m(x) = \sqrt{2}\sum_{m=1}^{\infty} b_m \sin m\pi x, \tag{15}$$

where

$$b_m = \int_0^1 f(x)\phi_m(x)\,dx = \sqrt{2}\int_0^1 f(x)\,\sin m\pi x\,dx, \tag{16}$$

converges in the mean. The series (15) is precisely the Fourier sine series discussed in Section 10.4. If f satisfies the further conditions stated in Theorem 11.2.4, then this series converges pointwise as well as in the mean. Similarly, a Fourier cosine series is associated with the Sturm–Liouville problem

$$y'' + \lambda y = 0, \qquad y'(0) = 0, \quad y'(1) = 0. \tag{17}$$

EXAMPLE 1

Let $f(x) = 1$ for $0 < x < 1$. Expand $f(x)$ using the eigenfunctions (13) and discuss the pointwise and mean square convergence of the resulting series.

The series has the form (15) and its coefficients b_m are given by Eq. (16). Thus

$$b_m = \sqrt{2}\int_0^1 \sin m\pi x\,dx = \frac{\sqrt{2}}{m\pi}(1 - \cos m\pi), \tag{18}$$

and the nth partial sum of the series is

$$S_n(x) = 2\sum_{m=1}^{n} \frac{1 - \cos m\pi}{m\pi}\sin m\pi x. \tag{19}$$

The mean square error is then

$$R_n = \int_0^1 [f(x) - S_n(x)]^2\,dx. \tag{20}$$

By calculating R_n for several values of n and plotting the results, we obtain Figure 11.6.2. This figure indicates that R_n steadily decreases as n increases. Of course, Theorem 11.6.1 asserts that $R_n \to 0$ as $n \to \infty$. Pointwise, we know that $S_n(x) \to f(x) = 1$ as $n \to \infty$ for $0 < x < 1$; further, $S_n(x)$ has the value zero for $x = 0$ or $x = 1$ for every n. Although the series converges

FIGURE 11.6.2 Dependence of the mean square error R_n on n in Example 1.

pointwise for each value of x, the least upper bound of the error does not diminish as n increases. For each n there are points close to $x = 0$ and $x = 1$ where the error is arbitrarily close to 1.

Theorem 11.6.1 can be extended to cover self-adjoint boundary value problems having periodic boundary conditions, such as the problem

$$y'' + \lambda y = 0, \tag{21}$$

$$y(-L) - y(L) = 0, \qquad y'(-L) - y'(L) = 0 \tag{22}$$

considered in Example 4 of Section 11.2. The eigenfunctions of the problem $(21),(22)$ are $\phi_n(x) = \cos(n\pi x/L)$ for $n = 0, 1, 2, \ldots$ and $\psi_n(x) = \sin(n\pi x/L)$ for $n = 1, 2, \ldots$. If f is a given square integrable function on $-L \le x \le L$, then its expansion in terms of the eigenfunctions ϕ_n and ψ_n is of the form

$$f(x) = \frac{a_0}{2} + \sum_{n=1}^{\infty} \left(a_n \cos \frac{n\pi x}{L} + b_n \sin \frac{n\pi x}{L} \right), \tag{23}$$

where

$$a_n = \frac{1}{L} \int_{-L}^{L} f(x) \cos \frac{n\pi x}{L} \, dx, \qquad n = 0, 1, 2, \ldots, \tag{24}$$

$$b_n = \frac{1}{L} \int_{-L}^{L} f(x) \sin \frac{n\pi x}{L} \, dx, \qquad n = 1, 2, \ldots. \tag{25}$$

This expansion is exactly the Fourier series for f discussed in Sections 10.2 and 10.3. According to the generalization of Theorem 11.6.1, the series (23) converges in the mean for any square integrable function f, even though f may not satisfy the conditions of Theorem 10.3.1, which ensure pointwise convergence.

PROBLEMS

1. Extend the results of Example 1 by finding the smallest value of n for which $R_n < 0.02$, where R_n is given by Eq. (20).

2. Let $f(x) = x$ for $0 < x < 1$ and let $\phi_m(x) = \sqrt{2} \sin m\pi x$.

 (a) Find the coefficients b_m in the expansion of $f(x)$ in terms of $\phi_1(x), \phi_2(x), \ldots$.

 (b) Calculate the mean square error R_n for several values of n and plot the results.

 (c) Find the smallest value of n for which $R_n < 0.01$.

3. Follow the instructions for Problem 2 using $f(x) = x(1-x)$ for $0 < x < 1$.

4. In this problem we show that pointwise convergence of a sequence $S_n(x)$ does not imply mean convergence, and conversely.

 (a) Let $S_n(x) = n\sqrt{x}e^{-nx^2/2}$, $0 \leq x \leq 1$. Show that $S_n(x) \to 0$ as $n \to \infty$ for each x in $0 \leq x \leq 1$. Show also that

 $$R_n = \int_0^1 [0 - S_n(x)]^2 \, dx = \frac{n}{2}(1 - e^{-n})$$

 and hence $R_n \to \infty$ as $n \to \infty$. Thus pointwise convergence does not imply mean convergence.

 (b) Let $S_n(x) = x^n$ for $0 \leq x \leq 1$ and let $f(x) = 0$ for $0 \leq x \leq 1$. Show that

 $$R_n = \int_0^1 [f(x) - S_n(x)]^2 \, dx = \frac{1}{2n+1},$$

 and hence $S_n(x)$ converges to $f(x)$ in the mean. Also show that $S_n(x)$ does not converge to $f(x)$ pointwise throughout $0 \leq x \leq 1$. Thus mean convergence does not imply pointwise convergence.

5. Suppose that the functions ϕ_1, \ldots, ϕ_n satisfy the orthonormality relation (1) and that a given function f is to be approximated by $S_n(x) = c_1\phi_1(x) + \cdots + c_n\phi_n(x)$, where the coefficients c_i are not necessarily those of Eq. (9). Show that the mean square error R_n given by Eq. (6) may be written in the form

 $$R_n = \int_0^1 r(x)f^2(x) \, dx - \sum_{i=1}^n a_i^2 + \sum_{i=1}^n (c_i - a_i)^2,$$

 where the a_i are the Fourier coefficients given by Eq. (9). Show that R_n is minimized if $c_i = a_i$ for each i.

6. In this problem we show by examples that the (Riemann) integrability of f and f^2 are independent.

 (a) Let $f(x) = \begin{cases} x^{-1/2}, & 0 < x \leq 1, \\ 0, & x = 0. \end{cases}$

 Show that $\int_0^1 f(x) \, dx$ exists as an improper integral, but $\int_0^1 f^2(x) \, dx$ does not.

 (b) Let $f(x) = \begin{cases} 1, & x \text{ rational}, \\ -1, & x \text{ irrational}. \end{cases}$

 Show that $\int_0^1 f^2(x) \, dx$ exists, but $\int_0^1 f(x) \, dx$ does not.

7. Suppose that it is desired to construct a set of polynomials $f_0(x), f_1(x), f_2(x), \ldots, f_k(x), \ldots$, where $f_k(x)$ is of degree k, that are orthonormal on the interval $0 \leq x \leq 1$. That is, the set of polynomials must satisfy

$$(f_j, f_k) = \int_0^1 f_j(x) f_k(x)\, dx = \delta_{jk}.$$

 (a) Find $f_0(x)$ by choosing the polynomial of degree zero such that $(f_0, f_0) = 1$.

 (b) Find $f_1(x)$ by determining the polynomial of degree one such that $(f_0, f_1) = 0$ and $(f_1, f_1) = 1$.

 (c) Find $f_2(x)$.

 (d) The normalization condition $(f_k, f_k) = 1$ is somewhat awkward to apply. Let $g_0(x)$, $g_1(x), \ldots, g_k(x), \ldots$ be the sequence of polynomials that are orthogonal on $0 \leq x \leq 1$ and that are normalized by the condition $g_k(1) = 1$. Find $g_0(x), g_1(x)$, and $g_2(x)$ and compare them with $f_0(x), f_1(x)$, and $f_2(x)$.

8. Suppose that it is desired to construct a set of polynomials $P_0(x), P_1(x), \ldots, P_k(x), \ldots$, where $P_k(x)$ is of degree k, that are orthogonal on the interval $-1 \leq x \leq 1$; see Problem 7. Suppose further that $P_k(x)$ is normalized by the condition $P_k(1) = 1$. Find $P_0(x), P_1(x)$, $P_2(x)$, and $P_3(x)$. Note that these are the first four Legendre polynomials (see Problem 24 of Section 5.3).

9. This problem develops some further results associated with mean convergence. Let $R_n(a_1, \ldots, a_n), S_n(x)$, and a_i be defined by Eqs. (6), (2), and (9), respectively.

 (a) Show that

$$R_n = \int_0^1 r(x) f^2(x)\, dx - \sum_{i=1}^n a_i^2.$$

 Hint: Substitute for $S_n(x)$ in Eq. (6) and integrate, using the orthogonality relation (1).

 (b) Show that $\sum_{i=1}^n a_i^2 \leq \int_0^1 r(x) f^2(x)\, dx$. This result is known as Bessel's inequality.

 (c) Show that $\sum_{i=1}^\infty a_i^2$ converges.

 (d) Show that $\lim_{n \to \infty} R_n = \int_0^1 r(x) f^2(x)\, dx - \sum_{i=1}^\infty a_i^2$.

 (e) Show that $\sum_{i=1}^\infty a_i \phi_i(x)$ converges to $f(x)$ in the mean if and only if

$$\int_0^1 r(x) f^2(x)\, dx = \sum_{i=1}^\infty a_i^2.$$

 This result is known as Parseval's equation.

 In Problems 10 through 12 let $\phi_1, \phi_2, \ldots, \phi_n, \ldots$ be the normalized eigenfunctions of the Sturm–Liouville problem (11), (12).

10. Show that if a_n is the nth Fourier coefficient of a square integrable function f, then

$$\lim_{n \to \infty} a_n = 0.$$

 Hint: Use Bessel's inequality, Problem 9(b).

11. Show that the series

$$\phi_1(x) + \phi_2(x) + \cdots + \phi_n(x) + \cdots$$

cannot be the eigenfunction series for any square integrable function.
Hint: See Problem 10.

12. Show that the series

$$\phi_1(x) + \frac{\phi_2(x)}{\sqrt{2}} + \cdots + \frac{\phi_n(x)}{\sqrt{n}} + \cdots$$

is not the eigenfunction series for any square integrable function.
Hint: Use Bessel's inequality, Problem 9(b).

13. Show that Parseval's equation in Problem 9(e) is obtained formally by squaring the series (10) corresponding to f, multiplying by the weight function r, and integrating term by term.

REFERENCES

The following books were mentioned in the text in connection with certain theorems about Sturm–Liouville problems:

Birkhoff, G., and Rota, G.-C., *Ordinary Differential Equations* (4th ed.) (New York: Wiley, 1989).

Sagan, H., *Boundary and Eigenvalue Problems in Mathematical Physics* (New York: Wiley, 1961; New York: Dover, 1989).

Weinberger, H., *A First Course in Partial Differential Equations* (New York: Wiley, 1965; New York: Dover, 1995).

Yosida, K., *Lectures on Differential and Integral Equations* (New York: Wiley-Interscience, 1960; New York: Dover, 1991).

The following books are convenient sources of numerical and graphical data about Bessel and Legendre functions:

Abramowitz, M., and Stegun, I. A. (eds.), *Handbook of Mathematical Functions* (New York: Dover, 1965); originally published by the National Bureau of Standards, Washington, DC, 1964.

Jahnke, E., and Emde, F., *Tables of Functions with Formulae and Curves* (Leipzig: Teubner, 1938; New York: Dover, 1945).

The following books also contain much information about Sturm–Liouville problems:

Cole, R. H., *Theory of Ordinary Differential Equations* (New York: Irvington, 1968).

Hochstadt, H., *Differential Equations: A Modern Approach* (New York: Holt, 1964; New York: Dover, 1975).

Miller, R. K., and Michel, A. N., *Ordinary Differential Equations* (New York: Academic Press, 1982).

Tricomi, F. G., *Differential Equations* (New York: Hafner, 1961).

Section 1.2, page 15

1. (a) $y = 5 + (y_0 - 5)e^{-t}$ (b) $y = (5/2) + [y_0 - (5/2)]e^{-2t}$
 (c) $y = 5 + (y_0 - 5)e^{-2t}$
 Equilibrium solution is $y = 5$ in (a) and (c), $y = 5/2$ in (b); solution approaches equilibrium faster in (b) and (c) than in (a).

2. (a) $y = 5 + (y_0 - 5)e^{t}$ (b) $y = (5/2) + [y_0 - (5/2)]e^{2t}$
 (c) $y = 5 + (y_0 - 5)e^{2t}$
 Equilibrium solution is $y = 5$ in (a) and (c), $y = 5/2$ in (b); solution diverges from equilibrium faster in (b) and (c) than in (a).

3. (a) $y = ce^{-at} + (b/a)$
 (c) (i) Equilibrium is lower and is approached more rapidly. (ii) Equilibrium is higher. (iii) Equilibrium remains the same and is approached more rapidly.

4. (a) $y_e = b/a$ (b) $Y' = aY$

5. (a) $y_1(t) = ce^{at}$ (b) $y = ce^{at} + (b/a)$

6. $y = ce^{-at} + (b/a)$

7. (a) $T = 2\ln 18 \cong 5.78$ months (b) $T = 2\ln[900/(900 - p_0)]$ months
 (c) $p_0 = 900(1 - e^{-6}) \cong 897.8$

8. (a) $r = (\ln 2)/30$ day^{-1} (b) $r = (\ln 2)/N$ day^{-1}

9. (a) $T = 5\ln 50 \cong 19.56$ sec (b) 718.34 m

10. (a) $dv/dt = 9.8$, $v(0) = 0$ (b) $T = \sqrt{300/4.9} \cong 7.82$ sec
 (c) $v \cong 76.68$ m/sec

11. (b) $v = 49\tanh(t/5)$ m/sec (e) $x = 245\ln\cosh(t/5)$ m
 (f) $T \cong 9.48$ sec

12. (a) $r \cong 0.02828$ day^{-1} (b) $Q(t) = 100e^{-0.02828t}$ (c) $T \cong 24.5$ days

14. $1620\ln(4/3)/\ln 2 \cong 672.4$ years

15. (a) $u = T + (u_0 - T)e^{-kt}$ (b) $k\tau = \ln 2$

16. 6.69 hr

17. (a) $Q(t) = CV(1 - e^{-t/RC})$ (b) $Q(t) \rightarrow CV = Q_L$
 (c) $Q(t) = CV\exp[-(t - t_1)/RC]$

18. (a) $Q' = 3(1 - 10^{-4}Q)$, $Q(0) = 0$
 (b) $Q(t) = 10^4(1 - e^{-3t/10^4})$, t in hrs; after 1 year $Q \cong 9277.77$ g
 (c) $Q' = -3Q/10^4$, $Q(0) = 9277.77$
 (d) $Q(t) = 9277.77e^{-3t/10^4}$, t in hrs; after 1 year $Q \cong 670.07$ g
 (e) $T \cong 2.60$ years

19. (a) $q' = -q/300$, $q(0) = 5000$ g (b) $q(t) = 5000e^{-t/300}$ (c) no
 (d) $T = 300\ln(25/6) \cong 7.136$ hr (e) $r = 250\ln(25/6) \cong 256.78$ gal/min

Section 1.3, page 24

1. Second order, linear 2. Second order, nonlinear
3. Fourth order, linear 4. First order, nonlinear
5. Second order, nonlinear 6. Third order, linear
15. $r = -2$ 16. $r = \pm 1$
17. $r = 2, -3$ 18. $r = 0, 1, 2$
19. $r = -1, -2$ 20. $r = 1, 4$
21. Second order, linear 22. Second order, nonlinear
23. Fourth order, linear 24. Second order, nonlinear

C H A P T E R 2 **Section 2.1, page 39**

1. (c) $y = ce^{-3t} + (t/3) - (1/9) + e^{-2t}$; y is asymptotic to $t/3 - 1/9$ as $t \to \infty$
2. (c) $y = ce^{2t} + t^3 e^{2t}/3$; $y \to \infty$ as $t \to \infty$
3. (c) $y = ce^{-t} + 1 + t^2 e^{-t}/2$; $y \to 1$ as $t \to \infty$
4. (c) $y = (c/t) + (3\cos 2t)/4t + (3\sin 2t)/2$; y is asymptotic to $(3\sin 2t)/2$ as $t \to \infty$
5. (c) $y = ce^{2t} - 3e^{t}$; $y \to \infty$ or $-\infty$ as $t \to \infty$
6. (c) $y = (c - t\cos t + \sin t)/t^2$; $y \to 0$ as $t \to \infty$
7. (c) $y = t^2 e^{-t^2} + ce^{-t^2}$; $y \to 0$ as $t \to \infty$
8. (c) $y = (\arctan t + c)/(1 + t^2)^2$; $y \to 0$ as $t \to \infty$
9. (c) $y = ce^{-t/2} + 3t - 6$; y is asymptotic to $3t - 6$ as $t \to \infty$
10. (c) $y = -te^{-t} + ct$; $y \to \infty, 0,$ or $-\infty$ as $t \to \infty$
11. (c) $y = ce^{-t} + \sin 2t - 2\cos 2t$; y is asymptotic to $\sin 2t - 2\cos 2t$ as $t \to \infty$
12. (c) $y = ce^{-t/2} + 3t^2 - 12t + 24$; y is asymptotic to $3t^2 - 12t + 24$ as $t \to \infty$
13. $y = 3e^{t} + 2(t - 1)e^{2t}$ 14. $y = (t^2 - 1)e^{-2t}/2$
15. $y = (3t^4 - 4t^3 + 6t^2 + 1)/12t^2$ 16. $y = (\sin t)/t^2$
17. $y = (t + 2)e^{2t}$ 18. $y = t^{-2}[(\pi^2/4) - 1 - t\cos t + \sin t]$
19. $y = -(1 + t)e^{-t}/t^4$, $t \neq 0$ 20. $y = (t - 1 + 2e^{-t})/t$, $t \neq 0$
21. (b) $y = -\frac{4}{5}\cos t + \frac{8}{5}\sin t + (a + \frac{4}{5})e^{t/2}$; $a_0 = -\frac{4}{5}$
 (c) y oscillates for $a = a_0$
22. (b) $y = -3e^{t/3} + (a + 3)e^{t/2}$; $a_0 = -3$
 (c) $y \to -\infty$ for $a = a_0$
23. (b) $y = [2 + a(3\pi + 4)e^{2t/3} - 2e^{-\pi t/2}]/(3\pi + 4)$; $a_0 = -2/(3\pi + 4)$
 (c) $y \to 0$ for $a = a_0$
24. (b) $y = te^{-t} + (ea - 1)e^{-t}/t$; $a_0 = 1/e$ (c) $y \to 0$ as $t \to 0$ for $a = a_0$
25. (b) $y = -\cos t/t^2 + \pi^2 a/4t^2$; $a_0 = 4/\pi^2$
 (c) $y \to \frac{1}{2}$ as $t \to 0$ for $a = a_0$
26. (b) $y = (e^{t} - e + a\sin 1)/\sin t$; $a_0 = (e - 1)/\sin 1$
 (c) $y \to 1$ for $a = a_0$
27. $(t, y) = (1.364312, 0.820082)$ 28. $y_0 = -1.642876$
29. (a) $y = 12 + \frac{8}{65}\cos 2t + \frac{64}{65}\sin 2t - \frac{788}{65}e^{-t/4}$; y oscillates about 12 as $t \to \infty$
 (b) $t = 10.065778$
30. $y_0 = -5/2$
31. $y_0 = -16/3$; $y \to -\infty$ as $t \to \infty$ for $y_0 = -16/3$
39. See Problem 2. 40. See Problem 4.
41. See Problem 6. 42. See Problem 12.

Section 2.2, page 47

1. $3y^2 - 2x^3 = c$; $y \neq 0$ 2. $3y^2 - 2\ln|1 + x^3| = c$; $x \neq -1, y \neq 0$
3. $y^{-1} + \cos x = c$ if $y \neq 0$; also $y = 0$; everywhere
4. $3y + y^2 - x^3 + x = c$; $y \neq -3/2$
5. $2\tan 2y - 2x - \sin 2x = c$ if $\cos 2y \neq 0$; also $y = \pm(2n + 1)\pi/4$ for any integer n;
 everywhere
6. $y = \sin[\ln|x| + c]$ if $x \neq 0$ and $|y| < 1$; also $y = \pm 1$
7. $y^2 - x^2 + 2(e^{y} - e^{-x}) = c$; $y + e^{y} \neq 0$ 8. $3y + y^3 - x^3 = c$; everywhere
9. (a) $y = 1/(x^2 - x - 6)$ (c) $-2 < x < 3$
10. (a) $y = -\sqrt{2x - 2x^2 + 4}$ (c) $-1 < x < 2$
11. (a) $y = [2(1 - x)e^{x} - 1]^{1/2}$ (c) $-1.68 < x < 0.77$ approximately
12. (a) $r = 2/(1 - 2\ln\theta)$ (c) $0 < \theta < \sqrt{e}$
13. (a) $y = -[2\ln(1 + x^2) + 4]^{1/2}$ (c) $-\infty < x < \infty$

14. (a) $y = \left[3 - 2\sqrt{1 + x^2}\right]^{-1/2}$ (c) $|x| < \frac{1}{2}\sqrt{5}$

15. (a) $y = -\frac{1}{2} + \frac{1}{2}\sqrt{4x^2 - 15}$ (c) $x > \frac{1}{2}\sqrt{15}$

16. (a) $y = -\sqrt[3]{(x^2 + 1)/2}$ (c) $-\infty < x < \infty$

17. (a) $y = 5/2 - \sqrt{x^3 - e^x + 13/4}$ (c) $-1.4445 < x < 4.6297$ approximately

18. (a) $y = -\frac{3}{4} + \frac{1}{4}\sqrt{65 - 8e^x - 8e^{-x}}$ (c) $|x| < 2.0794$ approximately

19. (a) $y = [\pi - \arcsin(3\cos^2 x)]/3$ (c) $|x - \pi/2| < 0.6155$

20. (a) $y = \left[\frac{3}{2}(\arcsin x)^2 + 1\right]^{1/3}$ (c) $-1 < x < 1$

21. $y^3 - 3y^2 - x - x^3 + 2 = 0, \quad |x| < 1$

22. $y^3 - 4y - x^3 = -1, \quad |x^3 - 1| < 16/3\sqrt{3}$ or $-1.28 < x < 1.60$

23. $y = -1/(x^2/2 + 2x - 1); \quad x = -2$ 24. $y = -3/2 + \sqrt{2x - e^x + 13/4}; \quad x = \ln 2$

25. $y = -3/2 + \sqrt{\sin 2x + 1/4}; \quad x = \pi/4$ 26. $y = \tan(x^2 + 2x); \quad x = -1$

27. (a) $y \to 4$ if $y_0 > 0; \quad y = 0$ if $y_0 = 0; \quad y \to -\infty$ if $y_0 < 0$

 (b) $T = 3.29527$

28. (a) $y \to 4$ as $t \to \infty$ (b) $T = 2.84367$ (c) $3.6622 < y_0 < 4.4042$

29. $x = \dfrac{c}{a}y + \dfrac{ad - bc}{a^2}\ln|ay + b| + k; \quad a \neq 0, \ ay + b \neq 0$

30. (e) $|y + 2x|^3|y - 2x| = c$ 31. (b) $\arctan(y/x) - \ln|x| = c$

32. (b) $x^2 + y^2 - cx^3 = 0$ 33. (b) $|y - x| = c|y + 3x|^5; \quad$ also $y = -3x$

34. (b) $|y + x|\,|y + 4x|^2 = c$

35. (b) $2x/(x + y) + \ln|x + y| = c; \quad$ also $y = -x$

36. (b) $x/(x + y) + \ln|x| = c; \quad$ also $y = -x$

37. (b) $|x|^3|x^2 - 5y^2| = c$ 38. (b) $c|x|^3 = |y^2 - x^2|$

Section 2.3, page 59

1. $t = 100\ln 100$ min $\cong 460.5$ min 2. $Q(t) = 120\gamma[1 - \exp(-t/60)]; \quad 120\gamma$

3. $Q = 50e^{-0.2}(1 - e^{-0.2})$ lb $\cong 7.42$ lb

4. $Q(t) = 200 + t - [100(200)^2/(200 + t)^2]$ lb, $t < 300; \quad c = 121/125$ lb/gal; $\lim_{t \to \infty} c = 1$ lb/gal

5. (a) $Q(t) = \frac{63{,}150}{2501}e^{-t/50} + 25 - \frac{625}{2501}\cos t + \frac{25}{5002}\sin t$

 (c) level $= 25$; amplitude $= 25\sqrt{2501}/5002 \cong 0.24995$

6. (c) 130.41 sec

7. (a) $(\ln 2)/r$ years (b) 9.90 years (c) 8.66%

8. (a) $k(e^{rt} - 1)/r$ (b) $k \cong \$3930$ (c) 9.77%

9. $k = \$3086.64$/year; \$1259.92

10. (a) \$89,034.79 (b) \$102,965.21

11. (a) $t \cong 135.36$ months (b) \$152,698.56

12. (a) 0.00012097 year^{-1} (b) $Q_0 \exp(-0.00012097t), \quad t$ in years

 (c) 13,305 years

13. $P = 201{,}977.31 - 1977.31e^{(\ln 2)t}, \quad 0 \le t \le t_f \cong 6.6745$ (weeks)

14. (a) $\tau \cong 2.9632$; no (b) $\tau = 10\ln 2 \cong 6.9315$ (c) $\tau = 6.3805$

15. (b) $y_c \cong 0.83$ 16. $t = \ln\frac{13}{8}/\ln\frac{13}{12}$ min $\cong 6.07$ min

17. (a) $u(t) = 2000/(1 + 0.048t)^{1/3}$ (c) $\tau \cong 750.77$ sec

18. (a) $u(t) = ce^{-kt} + T_0 + kT_1(k\cos\omega t + \omega\sin\omega t)/(k^2 + \omega^2)$

 (b) $R \cong 9.11°F; \quad \tau \cong 3.51$ hr

 (c) $R = kT_1/\sqrt{k^2 + \omega^2}; \quad \tau = (1/\omega)\arctan(\omega/k)$

19. (a) $c = k + (P/r) + [c_0 - k - (P/r)]e^{-rt/V}$; $\quad \lim_{t \to \infty} c = k + (P/r)$

 (b) $T = (V \ln 2)/r$; $\quad T = (V \ln 10)/r$

 (c) Superior, $T = 431$ years; Michigan, $T = 71.4$ years; Erie, $T = 6.05$ years;
 Ontario, $T = 17.6$ years

20. (a) 50.408 m (b) 5.248 sec

21. (a) 45.783 m (b) 5.129 sec

22. (a) 48.562 m (b) 5.194 sec

23. (a) 176.7 ft/sec (b) 1074.5 ft (c) 15 ft/sec (d) 256.6 sec

24. (a) $dv/dx = -\mu v$ (b) $\mu = (66/25) \ln 10$ mi$^{-1} \cong 6.0788$ mi^{-1}

 (c) $\tau = 900/(11 \ln 10)$ sec $\cong 35.533$ sec

25. (a) $x_m = -\dfrac{m^2 g}{k^2} \ln\left(1 + \dfrac{kv_0}{mg}\right) + \dfrac{mv_0}{k}$; $\quad t_m = \dfrac{m}{k} \ln\left(1 + \dfrac{kv_0}{mg}\right)$

26. (a) $v = -(mg/k) + [v_0 + (mg/k)]\exp(-kt/m)$

 (b) $v = v_0 - gt$; yes (c) $v = 0$ for $t > 0$

27. (a) $v_L = 2a^2 g(\rho - \rho')/9\mu$ (b) $e = 4\pi a^3 g(\rho - \rho')/3E$

28. (a) 11.58 m/sec (b) 13.45 m (c) $k \geq 0.2394$ kg/sec

29. (a) $v = R\sqrt{2g/(R + x)}$ (b) 50.6 hr

30. (b) $x = ut \cos A$, $y = -gt^2/2 + ut \sin A + h$

 (d) $-16L^2/(u^2 \cos^2 A) + L \tan A + 3 \geq H$

 (e) 0.63 rad $\leq A \leq 0.96$ rad

 (f) $u = 106.89$ ft/sec, $A = 0.7954$ rad

31. (a) $v = (u \cos A)e^{-rt}$, $w = -g/r + (u \sin A + g/r)e^{-rt}$

 (b) $x = u \cos A(1 - e^{-rt})/r$, $y = -gt/r + (u \sin A + g/r)(1 - e^{-rt})/r + h$

 (d) $u = 145.3$ ft/sec, $A = 0.644$ rad

32. (d) $k = 2.193$

Section 2.4, page 75

1. $0 < t < 3$ 2. $0 < t < 4$

3. $\pi/2 < t < 3\pi/2$ 4. $-\infty < t < -2$

5. $-2 < t < 2$ 6. $1 < t < \pi$

7. $2t + 5y > 0$ or $2t + 5y < 0$ 8. $t^2 + y^2 < 1$

9. $1 - t^2 + y^2 > 0$ or $1 - t^2 + y^2 < 0$, $\quad t \neq 0$, $y \neq 0$

10. Everywhere 11. $y \neq 0$, $\quad y \neq 3$

12. $t \neq n\pi$ for $n = 0, \pm 1, \pm 2, \ldots$; 13. $y = \pm\sqrt{y_0^2 - 4t^2}$ if $y_0 \neq 0$;
 $y \neq -1$ $|t| < |y_0|/2$

14. $y = [(1/y_0) - t^2]^{-1}$ if $y_0 \neq 0$; $\quad y = 0$ if $y_0 = 0$; \quad interval is $|t| < 1/\sqrt{y_0}$ if $y_0 > 0$;
 $-\infty < t < \infty$ if $y_0 \leq 0$

15. $y = y_0/\sqrt{2ty_0^2 + 1}$ if $y_0 \neq 0$; $\quad y = 0$ if $y_0 = 0$;
 interval is $-1/2y_0^2 < t < \infty$ if $y_0 \neq 0$; $\quad -\infty < t < \infty$ if $y_0 = 0$

16. $y = \pm\sqrt{\frac{2}{3}\ln(1 + t^3) + y_0^2}$; $\quad -[1 - \exp(-3y_0^2/2)]^{1/3} < t < \infty$

17. $y \to 3$ if $y_0 > 0$; $y = 0$ if $y_0 = 0$; $y \to -\infty$ if $y_0 < 0$

18. $y \to -\infty$ if $y_0 < 0$; $y \to 0$ if $y_0 \geq 0$ 19. $y \to 0$ if $y_0 \leq 9$; $y \to \infty$ if $y_0 > 9$

20. $y \to -\infty$ if $y_0 < y_c \approx -0.019$; otherwise y is asymptotic to $\sqrt{t - 1}$

21. (a) No (b) Yes; set $t_0 = 1/2$ in Eq. (19) in text.

 (c) $|y| \leq (4/3)^{3/2} \cong 1.5396$

22. (a) $y_1(t)$ is a solution for $t \geq 2$; $\quad y_2(t)$ is a solution for all t

 (b) f_y is not continuous at $(2, -1)$

26. (a) $y_1(t) = \dfrac{1}{\mu(t)};\quad y_2(t) = \dfrac{1}{\mu(t)}\displaystyle\int_{t_0}^{t}\mu(s)g(s)\,ds$

28. $y = \pm[5t/(2 + 5ct^5)]^{1/2}$ 29. $y = r/(k + cre^{-rt})$

30. $y = \pm[\epsilon/(\sigma + c\epsilon e^{-2\epsilon t})]^{1/2}$

31. $y = \pm\left\{\mu(t)\Big/\left[2\displaystyle\int_{t_0}^{t}\mu(s)\,ds + c\right]\right\}^{1/2}$, where $\mu(t) = \exp(2\Gamma\sin t + 2Tt)$

32. $y = \frac{1}{2}(1 - e^{-2t})$ for $0 \le t \le 1$; $y = \frac{1}{2}(e^2 - 1)e^{-2t}$ for $t > 1$

33. $y = e^{-2t}$ for $0 \le t \le 1$; $y = e^{-(t+1)}$ for $t > 1$

Section 2.5, page 88

1. $y = 0$ is unstable

2. $y = -a/b$ is asymptotically stable, $y = 0$ is unstable

3. $y = 1$ is asymptotically stable, $y = 0$ and $y = 2$ are unstable

4. $y = 0$ is unstable 5. $y = 0$ is asymptotically stable

6. $y = 0$ is asymptotically stable 7. (c) $y = [y_0 + (1 - y_0)kt]/[1 + (1 - y_0)kt]$

8. $y = 1$ is semistable

9. $y = -1$ is asymptotically stable, $y = 0$ is semistable, $y = 1$ is unstable

10. $y = -1$ and $y = 1$ are asymptotically stable, $y = 0$ is unstable

11. $y = 0$ is asymptotically stable, $y = b^2/a^2$ is unstable

12. $y = 2$ is asymptotically stable, $y = 0$ is semistable, $y = -2$ is unstable

13. $y = 0$ and $y = 1$ are semistable

15. (a) $\tau = (1/r)\ln 4$; 55.452 years
 (b) $T = (1/r)\ln[\beta(1 - \alpha)/(1 - \beta)\alpha]$; 175.78 years

16. (a) $y = 0$ is unstable, $y = K$ is asymptotically stable
 (b) Concave up for $0 < y \le K/e$, concave down for $K/e \le y < K$

17. (a) $y = K\exp\{[\ln(y_0/K)]e^{-rt}\}$ (b) $y(2) \cong 0.7153K \cong 57.6 \times 10^6$ kg
 (c) $\tau \cong 2.215$ years

18. (b) $(h/a)\sqrt{k/\alpha\pi}$; yes (c) $k/\alpha \le \pi a^2$

19. (b) $k^2/2g(\alpha a)^2$

20. (c) $Y = Ey_2 = KE[1 - (E/r)]$ (d) $Y_m = Kr/4$ for $E = r/2$

21. (a) $y_{1,2} = K[1 \mp \sqrt{1 - (4h/rK)}]/2$

22. (a) $y = 0$ is unstable, $y = 1$ is asymptotically stable
 (b) $y = y_0/[y_0 + (1 - y_0)e^{-\alpha t}]$

23. (a) $y = y_0 e^{-\beta t}$ (b) $x = x_0\exp[-\alpha y_0(1 - e^{-\beta t})/\beta]$ (c) $x_0\exp(-\alpha y_0/\beta)$

24. (b) $z = 1/[\nu + (1 - \nu)e^{\beta t}]$ (c) 0.0927

25. (a,b) $a = 0$: $y = 0$ is semistable.
 $a > 0$: $y = \sqrt{a}$ is asymptotically stable and $y = -\sqrt{a}$ is unstable.

26. (a) $a \le 0$: $y = 0$ is asymptotically stable.
 $a > 0$: $y = 0$ is unstable; $y = \sqrt{a}$ and $y = -\sqrt{a}$ are asymptotically stable.

27. (a) $a < 0$: $y = 0$ is asymptotically stable and $y = a$ is unstable.
 $a = 0$: $y = 0$ is semistable.
 $a > 0$: $y = 0$ is unstable and $y = a$ is asymptotically stable.

28. (a) $\displaystyle\lim_{t\to\infty} x(t) = \min(p, q)$; $x(t) = \dfrac{pq[e^{\alpha(q-p)t} - 1]}{qe^{\alpha(q-p)t} - p}$

 (b) $\displaystyle\lim_{t\to\infty} x(t) = p$; $x(t) = \dfrac{p^2\alpha t}{p\alpha t + 1}$

Section 2.6, page 99

1. $x^2 + 3x + y^2 - 2y = c$
2. Not exact
3. $x^3 - x^2 y + 2x + 2y^3 + 3y = c$
4. $x^2 y^2 + 2xy = c$
5. $ax^2 + 2bxy + cy^2 = k$
6. Not exact
7. $e^x \sin y + 2y \cos x = c$; also $y = 0$
8. Not exact
9. $e^{xy} \cos 2x + x^2 - 3y = c$
10. $y \ln x + 3x^2 - 2y = c$
11. Not exact
12. $x^2 + y^2 = c$
13. $y = \left[x + \sqrt{28 - 3x^2} \right] \big/ 2,\quad |x| < \sqrt{28/3}$
14. $y = [x - (24x^3 + x^2 - 8x - 16)^{1/2}]/4,\quad x > 0.9846$
15. $b = 3;\quad x^2 y^2 + 2x^3 y = c$
16. $b = 1;\quad e^{2xy} + x^2 = c$
19. $x^2 + 2 \ln |y| - y^{-2} = c$; also $y = 0$
20. $e^x \sin y + 2y \cos x = c$
21. $xy^2 - (y^2 - 2y + 2)e^y = c$
22. $x^2 e^x \sin y = c$
24. $\mu(t) = \exp \displaystyle\int R(t)\, dt,\quad$ where $t = xy$
25. $\mu(x) = e^{3x};\quad (3x^2 y + y^3)e^{3x} = c$
26. $\mu(x) = e^{-x};\quad y = ce^x + 1 + e^{2x}$
27. $\mu(y) = y;\quad xy + y \cos y - \sin y = c$
28. $\mu(y) = e^{2y}/y;\quad xe^{2y} - \ln |y| = c$; also $y = 0$
29. $\mu(y) = \sin y;\quad e^x \sin y + y^2 = c$
30. $\mu(y) = y^2;\quad x^4 + 3xy + y^4 = c$
31. $\mu(x, y) = xy;\quad x^3 y + 3x^2 + y^3 = c$

Section 2.7, page 107

1. (a) 1.2, 1.39, 1.571, 1.7439
 (b) 1.1975, 1.38549, 1.56491, 1.73658
 (c) 1.19631, 1.38335, 1.56200, 1.73308
 (d) 1.19516, 1.38127, 1.55918, 1.72968

2. (a) 1.1, 1.22, 1.364, 1.5368
 (b) 1.105, 1.23205, 1.38578, 1.57179
 (c) 1.10775, 1.23873, 1.39793, 1.59144
 (d) 1.1107, 1.24591, 1.41106, 1.61277

3. (a) 1.25, 1.54, 1.878, 2.2736
 (b) 1.26, 1.5641, 1.92156, 2.34359
 (c) 1.26551, 1.57746, 1.94586, 2.38287
 (d) 1.2714, 1.59182, 1.97212, 2.42554

4. (a) 0.3, 0.538501, 0.724821, 0.866458
 (b) 0.284813, 0.513339, 0.693451, 0.831571
 (c) 0.277920, 0.501813, 0.678949, 0.815302
 (d) 0.271428, 0.490897, 0.665142, 0.799729

5. Converge for $y \geq 0$; undefined for $y < 0$

6. Converge for $y \geq 0$; diverge for $y < 0$

7. Converge

8. Converge for $|y(0)| < 2.37$ (approximately); diverge otherwise

9. Diverge 10. Diverge

11. (a) 2.30800, 2.49006, 2.60023, 2.66773, 2.70939, 2.73521
 (b) 2.30167, 2.48263, 2.59352, 2.66227, 2.70519, 2.73209
 (c) 2.29864, 2.47903, 2.59024, 2.65958, 2.70310, 2.73053
 (d) 2.29686, 2.47691, 2.58830, 2.65798, 2.70185, 2.72959

12. (a) 1.70308, 3.06605, 2.44030, 1.77204, 1.37348, 1.11925
 (b) 1.79548, 3.06051, 2.43292, 1.77807, 1.37795, 1.12191
 (c) 1.84579, 3.05769, 2.42905, 1.78074, 1.38017, 1.12328
 (d) 1.87734, 3.05607, 2.42672, 1.78224, 1.38150, 1.12411

13. (a) $-1.48849, -0.412339, 1.04687, 1.43176, 1.54438, 1.51971$
 (b) $-1.46909, -0.287883, 1.05351, 1.42003, 1.53000, 1.50549$
 (c) $-1.45865, -0.217545, 1.05715, 1.41486, 1.52334, 1.49879$
 (d) $-1.45212, -0.173376, 1.05941, 1.41197, 1.51949, 1.49490$

14. (a) 0.950517, 0.687550, 0.369188, 0.145990, 0.0421429, 0.00872877
 (b) 0.938298, 0.672145, 0.362640, 0.147659, 0.0454100, 0.0104931
 (c) 0.932253, 0.664778, 0.359567, 0.148416, 0.0469514, 0.0113722
 (d) 0.928649, 0.660463, 0.357783, 0.148848, 0.0478492, 0.0118978

15. (a) $-0.166134, -0.410872, -0.804660, 4.15867$
 (b) $-0.174652, -0.434238, -0.889140, -3.09810$

16. A reasonable estimate for y at $t = 0.8$ is between 5.5 and 6. No reliable estimate is possible at $t = 1$ from the specified data.

17. A reasonable estimate for y at $t = 2.5$ is between 18 and 19. No reliable estimate is possible at $t = 3$ from the specified data.

18. (b) $2.37 < \alpha_0 < 2.38$ 19. (b) $0.67 < \alpha_0 < 0.68$

Section 2.8, page 117

1. $dw/ds = (s + 1)^2 + (w + 2)^2, \quad w(0) = 0$

2. $dw/ds = 1 - (w + 3)^3, \quad w(0) = 0$

3. (a) $\phi_n(t) = \sum_{k=1}^{n} \frac{2^k t^k}{k!}$ (c) $\lim_{n\to\infty} \phi_n(t) = e^{2t} - 1$

4. (a) $\phi_n(t) = \sum_{k=1}^{n} \frac{(-1)^k t^k}{k!}$ (c) $\lim_{n\to\infty} \phi_n(t) = e^{-t} - 1$

5. (a) $\phi_n(t) = \sum_{k=1}^{n} (-1)^{k+1} t^{k+1}/(k+1)! 2^{k-1}$
 (c) $\lim_{n\to\infty} \phi_n(t) = 4e^{-t/2} + 2t - 4$

6. (a) $\phi_n(t) = t - \frac{t^{n+1}}{(n+1)!}$ (c) $\lim_{n\to\infty} \phi_n(t) = t$

7. (a) $\phi_n(t) = \sum_{k=1}^{n} \frac{t^{2k-1}}{1 \cdot 3 \cdot 5 \cdots (2k-1)}$ 8. (a) $\phi_n(t) = -\sum_{k=1}^{n} \frac{t^{3k-1}}{2 \cdot 5 \cdot 8 \cdots (3k-1)}$

9. (a) $\phi_1(t) = \frac{t^3}{3}; \quad \phi_2(t) = \frac{t^3}{3} + \frac{t^7}{7 \cdot 9}; \quad \phi_3(t) = \frac{t^3}{3} + \frac{t^7}{7 \cdot 9} + \frac{2t^{11}}{3 \cdot 7 \cdot 9 \cdot 11} + \frac{t^{15}}{(7 \cdot 9)^2 \cdot 15}$

10. (a) $\phi_1(t) = t; \quad \phi_2(t) = t - \frac{t^4}{4}; \quad \phi_3(t) = t - \frac{t^4}{4} + \frac{3t^7}{4 \cdot 7} - \frac{3t^{10}}{16 \cdot 10} + \frac{t^{13}}{64 \cdot 13}$

11. (a) $\phi_1(t) = t, \quad \phi_2(t) = t - \frac{t^2}{2!} + \frac{t^4}{4!} - \frac{t^6}{6!} + O(t^8)$,

$$\phi_3(t) = t - \frac{t^2}{2!} + \frac{t^3}{3!} + \frac{t^4}{4!} - \frac{7t^5}{5!} + \frac{14t^6}{6!} + O(t^7),$$

$$\phi_4(t) = t - \frac{t^2}{2!} + \frac{t^3}{3!} - \frac{7t^5}{5!} + \frac{31t^6}{6!} + O(t^7)$$

12. (a) $\phi_1(t) = -t - t^2 - \dfrac{t^3}{2}$,

$$\phi_2(t) = -t - \frac{t^2}{2} + \frac{t^3}{6} + \frac{t^4}{4} - \frac{t^5}{5} - \frac{t^6}{24} + O(t^7),$$

$$\phi_3(t) = -t - \frac{t^2}{2} + \frac{t^4}{12} - \frac{3t^5}{20} + \frac{4t^6}{45} + O(t^7),$$

$$\phi_4(t) = -t - \frac{t^2}{2} + \frac{t^4}{8} - \frac{7t^5}{60} + \frac{t^6}{15} + O(t^7)$$

Section 2.9, page 129

1. $y_n = (-1)^n(0.9)^n y_0$; $y_n \to 0$ as $n \to \infty$
2. $y_n = y_0/(n+1)$; $y_n \to 0$ as $n \to \infty$
3. $y_n = y_0\sqrt{(n+2)(n+1)/2}$; $y_n \to \infty$ as $n \to \infty$
4. $y_n = \begin{cases} y_0, & \text{if } n = 4k \text{ or } n = 4k - 1; \\ -y_0, & \text{if } n = 4k - 2 \text{ or } n = 4k - 3; \end{cases}$ y_n has no limit as $n \to \infty$
5. $y_n = (0.5)^n(y_0 - 12) + 12$; $y_n \to 12$ as $n \to \infty$
6. $y_n = (-1)^n(0.5)^n(y_0 - 4) + 4$; $y_n \to 4$ as $n \to \infty$
7. 7.25% 8. \$2283.63 9. \$258.14
10. (a) \$804.62 (b) \$877.57 (c) \$1028.61
11. 30 years: \$804.62/month; \$289,663.20 total
 20 years: \$899.73/month; \$215,935.20 total
12. \$103,624.62 13. 9.73%
16. (b) $u_n \to -\infty$ as $n \to \infty$
19. (a) 4.7263 (b) 1.223% (c) 3.5643 (e) 3.5699

Miscellaneous Problems, page 131

1. $y = (c/x^2) + (x^3/5)$
2. $\arctan(y/x) - \ln\sqrt{x^2 + y^2} = c$
3. $x^2 + xy - 3y - y^3 = 0$
4. $x = ce^y + ye^y$
5. $x^2y + xy^2 + x = c$
6. $y = x^{-1}(1 - e^{1-x})$
7. $(x^2 + y^2 + 1)e^{-y^2} = c$
8. $y = (4 + \cos 2 - \cos x)/x^2$
9. $x^2y + x + y^2 = c$
10. $(y^2/x^3) + (y/x^2) = c$
11. $x^3/3 + xy + e^y = c$
12. $y = ce^{-x} + e^{-x}\ln(1 + e^x)$
13. $2(y/x)^{1/2} - \ln|x| = c$; also $y = 0$
14. $x^2 + 2xy + 2y^2 = 34$
15. $y = c/\cosh^2(x/2)$
16. $(2/\sqrt{3})\arctan[(2y - x)/\sqrt{3}x] - \ln|x| = c$
17. $y = ce^{3x} - e^{2x}$
18. $y = cx^{-2} - x$
19. $3y - 2xy^3 - 10x = 0$
20. $e^x + e^{-y} = c$
21. $e^{-y/x} + \ln|x| = c$
22. $y^3 + 3y - x^3 + 3x = 2$
23. $\dfrac{1}{y} = -x\displaystyle\int \frac{e^{2x}}{x^2}\,dx + cx$; also $y = 0$
24. $\sin^2 x \sin y = c$
25. $x^2/y + \arctan(y/x) = c$
26. $x^2 + 2x^2y - y^2 = c$
27. $\sin x \cos 2y - \frac{1}{2}\sin^2 x = c$
28. $2xy + xy^3 - x^3 = c$
29. $\arcsin(y/x) - \ln|x| = c$; also $y = x$ and $y = -x$
30. $xy^2 - \ln|y| = 0$
31. $x + \ln|x| + x^{-1} + y - 2\ln|y| = c$; also $y = 0$
32. $x^3y^2 + xy^3 = -4$
34. (a) $y = t + (c - t)^{-1}$ (b) $y = t^{-1} + 2t(c - t^2)^{-1}$
 (c) $y = \sin t + (c\cos t - \frac{1}{2}\sin t)^{-1}$

35. (a) $v' - [x(t) + b]v = b$

 (b) $v = \left[b \int \mu(t)\, dt + c \right] \Big/ \mu(t), \quad \mu(t) = \exp[-(at^2/2) - bt]$

36. $y = c_1 t^{-1} + c_2 + \ln t$ 37. $y = c_1 \ln t + c_2 + t$

38. $y = (1/k) \ln |(k - t)/(k + t)| + c_2$ if $c_1 = k^2 > 0$;
 $y = (2/k) \arctan(t/k) + c_2$ if $c_1 = -k^2 < 0$; $y = -2t^{-1} + c_2$ if $c_1 = 0$; also $y = c$

39. $y = \pm\frac{2}{3}(t - 2c_1)\sqrt{t + c_1} + c_2$; also $y = c$ *Hint:* $\mu(v) = v^{-3}$ is an integrating factor.

40. $y = c_1 e^{-t} + c_2 - te^{-t}$

41. $c_1^2 y = c_1 t - \ln|1 + c_1 t| + c_2$ if $c_1 \neq 0$; $y = \frac{1}{2}t^2 + c_2$ if $c_1 = 0$; also $y = c$

42. $y^2 = c_1 t + c_2$ 43. $y = c_1 \sin(t + c_2) = k_1 \sin t + k_2 \cos t$

44. $\frac{1}{3}y^3 - 2c_1 y + c_2 = 2t$; also $y = c$ 45. $t + c_2 = \pm\frac{2}{3}(y - 2c_1)(y + c_1)^{1/2}$

46. $y \ln|y| - y + c_1 y + t = c_2$; also $y = c$ 47. $e^y = (t + c_2)^2 + c_1$

48. $y = \frac{4}{3}(t + 1)^{3/2} - \frac{1}{3}$ 49. $y = 2(1 - t)^{-2}$

50. $y = 3\ln t - \frac{3}{2}\ln(t^2 + 1) - 5\arctan t + 2 + \frac{3}{2}\ln 2 + \frac{5}{4}\pi$

51. $y = \frac{1}{2}t^2 + \frac{3}{2}$

CHAPTER 3 **Section 3.1, page 142**

1. $y = c_1 e^t + c_2 e^{-3t}$ 2. $y = c_1 e^{-t} + c_2 e^{-2t}$

3. $y = c_1 e^{t/2} + c_2 e^{-t/3}$ 4. $y = c_1 e^{t/2} + c_2 e^t$

5. $y = c_1 + c_2 e^{-5t}$ 6. $y = c_1 e^{3t/2} + c_2 e^{-3t/2}$

7. $y = c_1 \exp[(9 + 3\sqrt{5})t/2] + c_2 \exp[(9 - 3\sqrt{5})t/2]$

8. $y = c_1 \exp[(1 + \sqrt{3})t] + c_2 \exp[(1 - \sqrt{3})t]$

9. $y = e^t$; $y \to \infty$ as $t \to \infty$ 10. $y = \frac{5}{2}e^{-t} - \frac{1}{2}e^{-3t}$; $y \to 0$ as $t \to \infty$

11. $y = 12e^{t/3} - 8e^{t/2}$; $y \to -\infty$ as $t \to \infty$

12. $y = -1 - e^{-3t}$; $y \to -1$ as $t \to \infty$

13. $y = \frac{1}{26}(13 + 5\sqrt{13}) \exp[(-5 + \sqrt{13})t/2] + \frac{1}{26}(13 - 5\sqrt{13}) \exp[(-5 - \sqrt{13})t/2]$;
 $y \to 0$ as $t \to \infty$

14. $y = (2/\sqrt{33}) \exp[(-1 + \sqrt{33})t/4] - (2/\sqrt{33}) \exp[(-1 - \sqrt{33})t/4]$; $y \to \infty$ as $t \to \infty$

15. $y = \frac{1}{10}e^{-9(t-1)} + \frac{9}{10}e^{t-1}$; $y \to \infty$ as $t \to \infty$

16. $y = -\frac{1}{2}e^{(t+2)/2} + \frac{3}{2}e^{-(t+2)/2}$; $y \to -\infty$ as $t \to \infty$

17. $y'' + y' - 6y = 0$ 18. $2y'' + 5y' + 2y = 0$

19. $y = \frac{1}{4}e^t + e^{-t}$; minimum is $y = 1$ at $t = \ln 2$

20. $y = -e^t + 3e^{t/2}$; maximum is $y = \frac{9}{4}$ at $t = \ln(9/4)$, $y = 0$ at $t = \ln 9$

21. $\alpha = -2$ 22. $\beta = -1$

23. $y \to 0$ for $\alpha < 0$; y becomes unbounded for $\alpha > 1$

24. $y \to 0$ for $\alpha < 1$; there is no α for which all nonzero solutions become unbounded

25. (a) $y = \frac{1}{5}(1 + 2\beta)e^{-2t} + \frac{1}{5}(4 - 2\beta)e^{t/2}$ (b) $y \cong 0.71548$ when $t = \frac{2}{5}\ln 6 \cong 0.71670$
 (c) $\beta = 2$

26. (a) $y = (6 + \beta)e^{-2t} - (4 + \beta)e^{-3t}$
 (b) $t_m = \ln[(12 + 3\beta)/(12 + 2\beta)]$, $y_m = \frac{4}{27}(6 + \beta)^3/(4 + \beta)^2$
 (c) $\beta = 6(1 + \sqrt{3}) \cong 16.3923$ (d) $t_m \to \ln(3/2)$, $y_m \to \infty$

27. (a) $y = d/c$ (b) $aY'' + bY' + cY = 0$

28. (a) $b > 0$ and $0 < c < b^2/4a$ (b) $c < 0$
 (c) $b < 0$ and $0 < c < b^2/4a$

Section 3.2, page 151

1. $-\frac{7}{2}e^{t/2}$
2. 1
3. e^{-4t}
4. $x^2 e^x$
5. $-e^{2t}$
6. 0
7. $0 < t < \infty$
8. $-\infty < t < 1$
9. $0 < t < 4$
10. $0 < t < \infty$
11. $0 < x < 3$
12. $2 < x < 3\pi/2$
14. The equation is nonlinear.
15. The equation is nonhomogeneous.
16. No
17. $3te^{2t} + ce^{2t}$
18. $te^t + ct$
19. $5W(f,g)$
20. $-4(t\cos t - \sin t)$
21. $y_1(t) = \frac{1}{3}e^{-2t} + \frac{2}{3}e^t$, $\quad y_2(t) = -\frac{1}{3}e^{-2t} + \frac{1}{3}e^t$
22. $y_1(t) = -\frac{1}{2}e^{-3(t-1)} + \frac{3}{2}e^{-(t-1)}$, $\quad y_2(t) = -\frac{1}{2}e^{-3(t-1)} + \frac{1}{2}e^{-(t-1)}$
23. Yes
24. Yes
25. Yes
26. Yes
27. (b) Yes

 (c) $[y_1(t), y_3(t)]$ and $[y_1(t), y_4(t)]$ are fundamental sets of solutions; $[y_2(t), y_3(t)]$ and $[y_4(t), y_5(t)]$ are not

29. Yes, $\quad y = c_1 e^{-x^2/2} \displaystyle\int_{x_0}^x e^{t^2/2}\, dt + c_2 e^{-x^2/2}$

30. No

31. Yes, $\quad y = \dfrac{1}{\mu(x)} \left[c_1 \displaystyle\int_{x_0}^x \dfrac{\mu(t)}{t}\, dt + c_2 \right]$, $\quad \mu(x) = \exp\left[-\displaystyle\int \left(\dfrac{1}{x} + \dfrac{\cos x}{x} \right) dx \right]$

32. Yes, $\quad y = c_1 x^{-1} + c_2 x$
34. $x^2\mu'' + 3x\mu' + (1 + x^2 - \nu^2)\mu = 0$
35. $(1 - x^2)\mu'' - 2x\mu' + \alpha(\alpha + 1)\mu = 0$
36. $\mu'' - x\mu = 0$
38. The Legendre and Airy equations are self-adjoint.

Section 3.3, page 158

1. Independent
2. Dependent
3. Independent
4. Dependent
5. Dependent
6. Independent
7. Independent if origin is interior to interval; otherwise dependent
8. Independent if origin is interior to interval; otherwise dependent
9. Independent; W is not always zero
10. Independent; W is not always zero
11. $W(c_1 y_1, c_2 y_2) = c_1 c_2 W(y_1, y_2) \neq 0$
12. $W(y_3, y_4) = -2W(y_1, y_2)$
13. $a_1 b_2 - a_2 b_1 \neq 0$
15. $ct^2 e^t$
16. $c\cos t$
17. c/x
18. $c/(1 - x^2)$
20. $2/25$
21. $3\sqrt{e} \cong 4.946$
22. $p(t) = 0$ for all t
26. If t_0 is an inflection point, and $y = \phi(t)$ is a solution, then from the differential equation $p(t_0)\phi'(t_0) + q(t_0)\phi(t_0) = 0$.

Section 3.4, page 164

1. $e\cos 2 + ie\sin 2 \cong -1.1312 + 2.4717i$
2. $e^2\cos 3 - ie^2\sin 3 \cong -7.3151 - 1.0427i$
3. -1
4. $e^2\cos(\pi/2) - ie^2\sin(\pi/2) = -e^2 i \cong -7.3891i$
5. $2\cos(\ln 2) - 2i\sin(\ln 2) \cong 1.5385 - 1.2779i$
6. $\pi^{-1}\cos(2\ln\pi) + i\pi^{-1}\sin(2\ln\pi) \cong -0.20957 + 0.23959i$

7. $y = c_1 e^t \cos t + c_2 e^t \sin t$

8. $y = c_1 e^t \cos \sqrt{5}\, t + c_2 e^t \sin \sqrt{5}\, t$

9. $y = c_1 e^{2t} + c_2 e^{-4t}$

10. $y = c_1 e^{-t} \cos t + c_2 e^{-t} \sin t$

11. $y = c_1 e^{-3t} \cos 2t + c_2 e^{-3t} \sin 2t$

12. $y = c_1 \cos(3t/2) + c_2 \sin(3t/2)$

13. $y = c_1 e^{-t} \cos(t/2) + c_2 e^{-t} \sin(t/2)$

14. $y = c_1 e^{t/3} + c_2 e^{-4t/3}$

15. $y = c_1 e^{-t/2} \cos t + c_2 e^{-t/2} \sin t$

16. $y = c_1 e^{-2t} \cos(3t/2) + c_2 e^{-2t} \sin(3t/2)$

17. $y = \frac{1}{2} \sin 2t$; steady oscillation

18. $y = e^{-2t} \cos t + 2e^{-2t} \sin t$; decaying oscillation

19. $y = -e^{t-\pi/2} \sin 2t$; growing oscillation

20. $y = (1 + 2\sqrt{3}) \cos t - (2 - \sqrt{3}) \sin t$; steady oscillation

21. $y = 3e^{-t/2} \cos t + \frac{5}{2} e^{-t/2} \sin t$; decaying oscillation

22. $y = \sqrt{2}\, e^{-(t-\pi/4)} \cos t + \sqrt{2}\, e^{-(t-\pi/4)} \sin t$; decaying oscillation

23. (a) $u = 2e^{t/6} \cos(\sqrt{23}\, t/6) - (2/\sqrt{23})e^{t/6} \sin(\sqrt{23}\, t/6)$
 (b) $t = 10.7598$

24. (a) $u = 2e^{-t/5} \cos(\sqrt{34}\, t/5) + (7/\sqrt{34})\, e^{-t/5} \sin(\sqrt{34}\, t/5)$
 (b) $T = 14.5115$

25. (a) $y = 2e^{-t} \cos \sqrt{5}\, t + [(\alpha + 2)/\sqrt{5}]\, e^{-t} \sin \sqrt{5}\, t$
 (b) $\alpha = 1.50878$
 (c) $t = \{\pi - \arctan[2\sqrt{5}/(2 + \alpha)]\}/\sqrt{5}$
 (d) $\pi/\sqrt{5}$

26. (a) $y = e^{-at} \cos t + ae^{-at} \sin t$
 (b) $T = 1.8763$
 (c) $\alpha = \frac{1}{4},\ T = 7.4284;\quad \alpha = \frac{1}{2},\ T = 4.3003;\quad \alpha = 2,\ T = 1.5116$

35. Yes, $y = c_1 \cos x + c_2 \sin x,\quad x = \displaystyle\int e^{-t^2/2}\, dt$

36. No

37. Yes, $y = c_1 e^{-t^2/4} \cos(\sqrt{3}\, t^2/4) + c_2 e^{-t^2/4} \sin(\sqrt{3}\, t^2/4)$

39. $y = c_1 \cos(\ln t) + c_2 \sin(\ln t)$

40. $y = c_1 t^{-1} + c_2 t^{-2}$

41. $y = c_1 t^{-1} \cos(\frac{1}{2} \ln t) + c_2 t^{-1} \sin(\frac{1}{2} \ln t)$

42. $y = c_1 t^6 + c_2 t^{-1}$

Section 3.5, page 172

1. $y = c_1 e^t + c_2 t e^t$

2. $y = c_1 e^{-t/3} + c_2 t e^{-t/3}$

3. $y = c_1 e^{-t/2} + c_2 e^{3t/2}$

4. $y = c_1 e^{-3t/2} + c_2 t e^{-3t/2}$

5. $y = c_1 e^t \cos 3t + c_2 e^t \sin 3t$

6. $y = c_1 e^{3t} + c_2 t e^{3t}$

7. $y = c_1 e^{-t/4} + c_2 e^{-4t}$

8. $y = c_1 e^{-3t/4} + c_2 t e^{-3t/4}$

9. $y = c_1 e^{2t/5} + c_2 t e^{2t/5}$

10. $y = e^{-t/2} \cos(t/2) + c_2 e^{-t/2} \sin(t/2)$

11. $y = 2e^{2t/3} - \frac{7}{3} t e^{2t/3},\quad y \to -\infty$ as $t \to \infty$

12. $y = 2t e^{3t},\quad y \to \infty$ as $t \to \infty$

13. $y = -e^{-t/3} \cos 3t + \frac{5}{9} e^{-t/3} \sin 3t,\quad y \to 0$ as $t \to \infty$

14. $y = 7e^{-2(t+1)} + 5t e^{-2(t+1)},\quad y \to 0$ as $t \to \infty$

15. (a) $y = e^{-3t/2} - \frac{5}{2} t e^{-3t/2}$ (b) $t = \frac{2}{5}$
 (c) $t_0 = 16/15,\quad y_0 = -\frac{5}{3} e^{-8/5} \cong -0.33649$
 (d) $y = e^{-3t/2} + (b + \frac{3}{2})t e^{-3t/2};\quad b = -\frac{3}{2}$

16. $y = 2e^{t/2} + (b - 1)t e^{t/2};\quad b = 1$

17. (a) $y = e^{-t/2} + \frac{5}{2} t e^{-t/2}$ (b) $t_M = \frac{8}{5},\quad y_M = 5e^{-4/5} \cong 2.24664$
 (c) $y = e^{-t/2} + (b + \frac{1}{2})t e^{-t/2}$
 (d) $t_M = 4b/(1 + 2b) \to 2$ as $b \to \infty$;
 $y_M = (1 + 2b) \exp[-2b/(1 + 2b)] \to \infty$ as $b \to \infty$

18. (a) $y = ae^{-2t/3} + (\frac{2}{3}a - 1)t e^{-2t/3}$ (b) $a = \frac{3}{2}$

23. $y_2(t) = t^3$

24. $y_2(t) = t^{-2}$

25. $y_2(t) = t^{-1} \ln t$

26. $y_2(t) = t e^t$

27. $y_2(x) = \cos x^2$

28. $y_2(x) = x$

29. $y_2(x) = x^{1/4}e^{-2\sqrt{x}}$

30. $y_2(x) = x^{-1/2}\cos x$

32. $y = c_1 e^{-\delta x^2/2} \int_0^x e^{\delta s^2/2}\, ds + c_2 e^{-\delta x^2/2}$

33. $y_2(t) = y_1(t) \int_{t_0}^t y_1^{-2}(s)\exp\left[-\int_{s_0}^s p(r)\, dr\right] ds$

34. $y_2(t) = t^{-1}\ln t$

35. $y_2(t) = \cos t^2$

36. $y_2(x) = x$

37. $y_2(x) = x^{-1/2}\cos x$

39. (b) $y_0 + (a/b)y_0'$

41. $y = c_1 t^2 + c_2 t^2 \ln t$

42. $y = c_1 t^{-1/2} + c_2 t^{-1/2}\ln t$

Section 3.6, page 184

1. $y = c_1 e^{3t} + c_2 e^{-t} - e^{2t}$

2. $y = c_1 e^{-t}\cos 2t + c_2 e^{-t}\sin 2t + \frac{3}{17}\sin 2t - \frac{12}{17}\cos 2t$

3. $y = c_1 e^{3t} + c_2 e^{-t} + \frac{3}{16}te^{-t} + \frac{3}{8}t^2 e^{-t}$

4. $y = c_1 + c_2 e^{-2t} + \frac{3}{2}t - \frac{1}{2}\sin 2t - \frac{1}{2}\cos 2t$

5. $y = c_1\cos 3t + c_2\sin 3t + \frac{1}{162}(9t^2 - 6t + 1)e^{3t} + \frac{2}{3}$

6. $y = c_1 e^{-t} + c_2 te^{-t} + t^2 e^{-t}$

7. $y = c_1 e^{-t} + c_2 e^{-t/2} + t^2 - 6t + 14 - \frac{3}{10}\sin t - \frac{9}{10}\cos t$

8. $y = c_1\cos t + c_2\sin t - \frac{1}{3}t\cos 2t - \frac{5}{9}\sin 2t$

9. $u = c_1\cos\omega_0 t + c_2\sin\omega_0 t + (\omega_0^2 - \omega^2)^{-1}\cos\omega t$

10. $u = c_1\cos\omega_0 t + c_2\sin\omega_0 t + (1/2\omega_0)t\sin\omega_0 t$

11. $y = c_1 e^{-t/2}\cos(\sqrt{15}\,t/2) + c_2 e^{-t/2}\sin(\sqrt{15}\,t/2) + \frac{1}{6}e^t - \frac{1}{4}e^{-t}$

12. $y = c_1 e^{-t} + c_2 e^{2t} + \frac{1}{6}te^{2t} + \frac{1}{8}e^{-2t}$

13. $y = e^t - \frac{1}{2}e^{-2t} - t - \frac{1}{2}$

14. $y = \frac{7}{10}\sin 2t - \frac{19}{40}\cos 2t + \frac{1}{4}t^2 - \frac{1}{8} + \frac{3}{5}e^t$

15. $y = 4te^t - 3e^t + \frac{1}{6}t^3 e^t + 4$

16. $y = e^{3t} + \frac{2}{3}e^{-t} - \frac{2}{3}e^{2t} - te^{2t}$

17. $y = 2\cos 2t - \frac{1}{8}\sin 2t - \frac{3}{4}t\cos 2t$

18. $y = e^{-t}\cos 2t + \frac{1}{2}e^{-t}\sin 2t + te^{-t}\sin 2t$

19. (a) $Y(t) = t(A_0 t^4 + A_1 t^3 + A_2 t^2 + A_3 t + A_4) + t(B_0 t^2 + B_1 t + B_2)e^{-3t} + D\sin 3t + E\cos 3t$

 (b) $A_0 = 2/15$, $A_1 = -2/9$, $A_2 = 8/27$, $A_3 = -8/27$, $A_4 = 16/81$, $B_0 = -1/9$, $B_1 = -1/9$, $B_2 = -2/27$, $D = -1/18$, $E = -1/18$

20. (a) $Y(t) = A_0 t + A_1 + t(B_0 t + B_1)\sin t + t(D_0 t + D_1)\cos t$

 (b) $A_0 = 1$, $A_1 = 0$, $B_0 = 0$, $B_1 = 1/4$, $D_0 = -1/4$, $D_1 = 0$

21. (a) $Y(t) = e^t(A\cos 2t + B\sin 2t) + (D_0 t + D_1)e^{2t}\sin t + (E_0 t + E_1)e^{2t}\cos t$

 (b) $A = -1/20$, $B = -3/20$, $D_0 = -3/2$, $D_1 = -5$, $E_0 = 3/2$, $E_1 = 1/2$

22. (a) $Y(t) = Ae^{-t} + t(B_0 t^2 + B_1 t + B_2)e^{-t}\cos t + t(D_0 t^2 + D_1 t + D_2)e^{-t}\sin t$

 (b) $A = 3$, $B_0 = -2/3$, $B_1 = 0$, $B_2 = 1$, $D_0 = 0$, $D_1 = 1$, $D_2 = 1$

23. (a) $Y(t) = A_0 t^2 + A_1 t + A_2 + t^2(B_0 t + B_1)e^{2t} + (D_0 t + D_1)\sin 2t + (E_0 t + E_1)\cos 2t$

 (b) $A_0 = 1/2$, $A_1 = 1$, $A_2 = 3/4$, $B_0 = 2/3$, $B_1 = 0$, $D_0 = 0$, $D_1 = -1/16$, $E_0 = 1/8$, $E_1 = 1/16$

24. (a) $Y(t) = t(A_0 t^2 + A_1 t + A_2)\sin 2t + t(B_0 t^2 + B_1 t + B_2)\cos 2t$

 (b) $A_0 = 0$, $A_1 = 13/16$, $A_2 = 7/4$, $B_0 = -1/12$, $B_1 = 0$, $B_2 = 13/32$

25. (a) $Y(t) = (A_0 t^2 + A_1 t + A_2)e^t\sin 2t + (B_0 t^2 + B_1 t + B_2)e^t\cos 2t + e^{-t}(D\cos t + E\sin t) + Fe^t$

 (b) $A_0 = 1/52$, $A_1 = 10/169$, $A_2 = -1233/35152$, $B_0 = -5/52$, $B_1 = 73/676$, $B_2 = -4105/35152$, $D = -3/2$, $E = 3/2$, $F = 2/3$

26. (a) $Y(t) = t(A_0 t + A_1)e^{-t}\cos 2t + t(B_0 t + B_1)e^{-t}\sin 2t + (D_0 t + D_1)e^{-2t}\cos t + (E_0 t + E_1)e^{-2t}\sin t$

 (b) $A_0 = 0$, $A_1 = 3/16$, $B_0 = 3/8$, $B_1 = 0$, $D_0 = -2/5$, $D_1 = -7/25$, $E_0 = 1/5$, $E_1 = 1/25$

27. (b) $w = -\frac{2}{5} + c_1 e^{5t}$

28. $y = c_1 \cos \lambda t + c_2 \sin \lambda t + \sum_{m=1}^{N} [a_m/(\lambda^2 - m^2\pi^2)] \sin m\pi t$

29. $y = \begin{cases} t, & 0 \le t \le \pi \\ -(1 + \pi/2)\sin t - (\pi/2)\cos t + (\pi/2)e^{\pi-t}, & t > \pi \end{cases}$

30. $y = \begin{cases} \frac{1}{5} - \frac{1}{10}e^{-t}\sin 2t - \frac{1}{5}e^{-t}\cos 2t, & 0 \le t \le \pi/2 \\ -\frac{1}{5}(1 + e^{\pi/2})e^{-t}\cos 2t - \frac{1}{10}(1 + e^{\pi/2})e^{-t}\sin 2t, & t > \pi/2 \end{cases}$

31. No 34. $y = c_1 e^{4t} + c_2 e^{-t} - \frac{1}{2}e^{2t}$

Section 3.7, page 190

1. $Y(t) = e^t$ 2. $Y(t) = -\frac{2}{3}te^{-t}$
3. $Y(t) = \frac{3}{2}t^2 e^{-t}$ 4. $Y(t) = 2t^2 e^{t/2}$
5. $y = c_1 \cos t + c_2 \sin t - (\cos t)\ln(\tan t + \sec t)$
6. $y = c_1 \cos 3t + c_2 \sin 3t + (\sin 3t)\ln(\tan 3t + \sec 3t) - 1$
7. $y = c_1 e^{-2t} + c_2 t e^{-2t} - e^{-2t}\ln t$
8. $y = c_1 \cos 2t + c_2 \sin 2t + \frac{3}{4}(\sin 2t)\ln \sin 2t - \frac{3}{2}t \cos 2t$
9. $y = c_1 \cos(t/2) + c_2 \sin(t/2) + t \sin(t/2) + 2[\ln \cos(t/2)]\cos(t/2)$
10. $y = c_1 e^t + c_2 t e^t - \frac{1}{2}e^t \ln(1 + t^2) + te^t \arctan t$
11. $y = c_1 e^{2t} + c_2 e^{3t} + \int [e^{3(t-s)} - e^{2(t-s)}]g(s)\,ds$
12. $y = c_1 \cos 2t + c_2 \sin 2t + \frac{1}{2}\int [\sin 2(t-s)]g(s)\,ds$
13. $Y(t) = \frac{1}{2} + t^2 \ln t$ 14. $Y(t) = -2t^2$
15. $Y(t) = \frac{1}{2}(t-1)e^{2t}$ 16. $Y(t) = -\frac{1}{2}(2t-1)e^{-t}$
17. $Y(x) = \frac{1}{6}x^2(\ln x)^3$ 18. $Y(x) = -\frac{3}{2}x^{1/2}\cos x$
19. $Y(x) = \int \frac{xe^t - te^x}{(1-t)^2 e^t}g(t)\,dt$

20. $Y(x) = x^{-1/2}\int t^{-3/2}\sin(x-t)g(t)\,dt$

23. (b) $y = y_0 \cos t + y_0' \sin t + \int_{t_0}^{t}\sin(t-s)g(s)\,ds$

24. $y = (b-a)^{-1}\int_{t_0}^{t}[e^{b(t-s)} - e^{a(t-s)}]g(s)\,ds$

25. $y = \mu^{-1}\int_{t_0}^{t}e^{\lambda(t-s)}\sin \mu(t-s)g(s)\,ds$ 26. $y = \int_{t_0}^{t}(t-s)e^{a(t-s)}g(s)\,ds$

29. $y = c_1 t + c_2 t^2 + 4t^2 \ln t$ 30. $y = c_1 t^{-1} + c_2 t^{-5} + \frac{1}{12}t$
31. $y = c_1(1+t) + c_2 e^t + \frac{1}{2}(t-1)e^{2t}$ 32. $y = c_1 e^t + c_2 t - \frac{1}{2}(2t-1)e^{-t}$

Section 3.8, page 203

1. $u = 5\cos(2t - \delta), \quad \delta = \arctan(4/3) \cong 0.9273$
2. $u = 2\cos(t - 2\pi/3)$
3. $u = 2\sqrt{5}\cos(3t - \delta), \quad \delta = -\arctan(1/2) \cong -0.4636$
4. $u = \sqrt{13}\cos(\pi t - \delta), \quad \delta = \pi + \arctan(3/2) \cong 4.1244$
5. $u = \frac{1}{4}\cos 8t$ ft, t in sec; $\omega = 8$ rad/sec, $T = \pi/4$ sec, $R = 1/4$ ft
6. $u = \frac{5}{7}\sin 14t$ cm, t in sec; $t = \pi/14$ sec

7. $u = (1/4\sqrt{2})\sin(8\sqrt{2}\,t) - \frac{1}{12}\cos(8\sqrt{2}\,t)$ ft, t in sec; $\omega = 8\sqrt{2}$ rad/sec,
$T = \pi/4\sqrt{2}$ sec, $R = \sqrt{11/288} \cong 0.1954$ ft, $\delta = \pi - \arctan(3/\sqrt{2}) \cong 2.0113$

8. $Q = 10^{-6}\cos 2000t$ coulombs, t in sec

9. $u = e^{-10t}[2\cos(4\sqrt{6}\,t) + (5/\sqrt{6})\sin(4\sqrt{6}\,t)]$ cm, t in sec;
$\mu = 4\sqrt{6}$ rad/sec, $T_d = \pi/2\sqrt{6}$ sec, $T_d/T = 7/2\sqrt{6} \cong 1.4289$, $\tau \cong 0.4045$ sec

10. $u = (1/8\sqrt{31})e^{-2t}\sin(2\sqrt{31}\,t)$ ft, t in sec; $t = \pi/2\sqrt{31}$ sec, $\tau \cong 1.5927$ sec

11. $u \cong 0.057198e^{-0.15t}\cos(3.87008\,t - 0.50709)$ m, t in sec; $\mu = 3.87008$ rad/sec,
$\mu/\omega_0 = 3.87008/\sqrt{15} \cong 0.99925$

12. $Q = 10^{-6}(2e^{-500t} - e^{-1000t})$ coulombs; t in sec

13. $\gamma = \sqrt{20/9} \cong 1.4907$

16. $r = \sqrt{A^2 + B^2}$, $r\cos\theta = B$, $r\sin\theta = -A$; $R = r$; $\delta = \theta + (4n+1)\pi/2, n = 0, 1, 2, \ldots$

17. $\gamma = 8$ lb-sec/ft 18. $R = 10^3$ ohms

20. $v_0 < -\gamma u_0/2m$ 22. $2\pi/\sqrt{31}$

23. $\gamma = 5$ lb-sec/ft 24. $k = 6$, $v = \pm 2\sqrt{5}$

25. (a) $\tau \cong 41.715$ (d) $\gamma_0 \cong 1.73$, min $\tau \cong 4.87$
 (e) $\tau = (2/\gamma)\ln(400/\sqrt{4 - \gamma^2})$

26. (a) $u(t) = e^{-\gamma t/2m}\left[u_0\sqrt{4km - \gamma^2}\cos\mu t + (2mv_0 + \gamma u_0)\sin\mu t\right]/\sqrt{4km - \gamma^2}$

 (b) $R^2 = 4m(ku_0^2 + \gamma u_0 v_0 + mv_0^2)/(4km - \gamma^2)$

27. $\rho l u'' + \rho_0 g u = 0$, $T = 2\pi\sqrt{\rho l/\rho_0 g}$

28. (a) $u = \sqrt{2}\sin\sqrt{2}\,t$ (c) clockwise

29. (a) $u = (16/\sqrt{127})e^{-t/8}\sin(\sqrt{127}\,t/8)$ (c) clockwise

30. (b) $u = a\cos(\sqrt{k/m}\,t) + b\sqrt{m/k}\sin(\sqrt{k/m}\,t)$

32. (b) $u = \sin t$, $A = 1$, $T = 2\pi$ (c) $A = 0.98$, $T = 6.07$
 (d) $\epsilon = 0.2$, $A = 0.96$, $T = 5.90$; $\epsilon = 0.3$, $A = 0.94$, $T = 5.74$
 (f) $\epsilon = -0.1$, $A = 1.03$, $T = 6.55$; $\epsilon = -0.2$, $A = 1.06$, $T = 6.90$;
 $\epsilon = -0.3, A = 1.11$, $T = 7.41$

Section 3.9, page 214

1. $-2\sin 8t \sin t$ 2. $2\sin(t/2)\cos(13t/2)$

3. $2\cos(3\pi t/2)\cos(\pi t/2)$ 4. $2\sin(7t/2)\cos(t/2)$

5. $u'' + 256u = 16\cos 3t$, $u(0) = \frac{1}{6}$, $u'(0) = 0$, u in ft, t in sec

6. $u'' + 10u' + 98u = 2\sin(t/2)$, $u(0) = 0$, $u'(0) = 0.03$, u in m, t in sec

7. (a) $u = \frac{151}{1482}\cos 16t + \frac{16}{247}\cos 3t$ (c) $\omega = 16$ rad/sec

8. (a) $u = \frac{1}{153,281}[160e^{-5t}\cos(\sqrt{73}\,t) + \frac{383,443}{7300}e^{-5t}\sin(\sqrt{73}\,t) - 160\cos(t/2) + 3128\sin(t/2)]$
 (b) The first two terms are the transient. (d) $\omega = 4\sqrt{3}$ rad/sec

9. $u = \frac{64}{45}(\cos 7t - \cos 8t) = \frac{128}{45}\sin(t/2)\sin(15t/2)$ ft, t in sec

10. $u = (\cos 8t + \sin 8t - 8t\cos 8t)/4$ ft, t in sec; $1/8$, $\pi/8$, $\pi/4$, $3\pi/8$ sec

11. (a) $\frac{8}{901}(30\cos 2t + \sin 2t)$ ft, t in sec (b) $m = 4$ slugs

12. $u = (\sqrt{2}/6)\cos(3t - 3\pi/4)$ m, t in sec

15. $u = \begin{cases} F_0(t - \sin t), & 0 \leq t \leq \pi \\ F_0[(2\pi - t) - 3\sin t], & \pi < t \leq 2\pi \\ -4F_0\sin t, & 2\pi < t < \infty \end{cases}$

16. $Q(t) = 10^{-6}(e^{-4000t} - 4e^{-1000t} + 3)$ coulombs, t in sec, $Q(0.001) \cong 1.5468 \times 10^{-6}$;
 $Q(0.01) \cong 2.9998 \times 10^{-6}$; $Q(t) \to 3 \times 10^{-6}$ as $t \to \infty$

17. (a) $u = [32(2 - \omega^2)\cos\omega t + 8\omega\sin\omega t]/(64 - 63\omega^2 + 16\omega^4)$
 (b) $A = 8/\sqrt{64 - 63\omega^2 + 16\omega^4}$
 (d) $\omega = 3\sqrt{14}/8 \cong 1.4031$, $A = 64/\sqrt{127} \cong 5.6791$

18. (a) $u = 3(\cos t - \cos \omega t)/(\omega^2 - 1)$
19. (a) $u = [(\omega^2 + 2) \cos t - 3 \cos \omega t]/(\omega^2 - 1) + \sin t$

CHAPTER 4

Section 4.1, page 222

1. $-\infty < t < \infty$ 2. $t > 0$ or $t < 0$
3. $t > 1$, or $0 < t < 1$, or $t < 0$ 4. $t > 0$
5. $\ldots, -3\pi/2 < x < -\pi/2, \quad -\pi/2 < x < 1, \quad 1 < x < \pi/2, \quad \pi/2 < x < 3\pi/2, \ldots$
6. $-\infty < x < -2, \quad -2 < x < 2, \quad 2 < x < \infty$
7. Linearly independent
8. Linearly dependent; $f_1(t) + 3f_2(t) - 2f_3(t) = 0$
9. Linearly dependent; $2f_1(t) + 13f_2(t) - 3f_3(t) - 7f_4(t) = 0$
10. Linearly independent 11. 1
12. 1 13. $-6e^{-2t}$
14. e^{-2t} 15. $6x$
16. $6/x$ 17. $\sin^2 t = \frac{1}{10}(5) - \frac{1}{2}\cos 2t$
19. (a) $a_0[n(n-1)(n-2)\cdots 1] + a_1[n(n-1)\cdots 2]t + \cdots + a_n t^n$
 (b) $(a_0 r^n + a_1 r^{n-1} + \cdots + a_n)e^{rt}$
 (c) $e^t, e^{-t}, e^{2t}, e^{-2t}$; yes, $W(e^t, e^{-t}, e^{2t}, e^{-2t}) \neq 0, \; -\infty < t < \infty$
21. $W = ce^{-2t}$ 22. $W = c$
23. $W = c/t^2$ 24. $W = c/t$
27. $y = c_1 e^t + c_2 t + c_3 t e^t$ 28. $y = c_1 t^2 + c_2 t^3 + c_3(t + 1)$

Section 4.2, page 230

1. $\sqrt{2}\, e^{i[(\pi/4)+2m\pi]}$ 2. $2e^{i[(2\pi/3)+2m\pi]}$
3. $3e^{i(\pi + 2m\pi)}$ 4. $e^{i[(3\pi/2)+2m\pi]}$
5. $2e^{i[(11\pi/6)+2m\pi]}$ 6. $\sqrt{2}e^{i[(5\pi/4)+2m\pi]}$
7. $1, \frac{1}{2}(-1 + i\sqrt{3}), \frac{1}{2}(-1 - i\sqrt{3})$ 8. $2^{1/4}e^{-\pi i/8}, \; 2^{1/4}e^{7\pi i/8}$
9. $1, i, -1, -i$ 10. $(\sqrt{3} + i)/\sqrt{2}, -(\sqrt{3} + i)/\sqrt{2}$
11. $y = c_1 e^t + c_2 t e^t + c_3 e^{-t}$ 12. $y = c_1 e^t + c_2 t e^t + c_3 t^2 e^t$
13. $y = c_1 e^t + c_2 e^{2t} + c_3 e^{-t}$ 14. $y = c_1 + c_2 t + c_3 e^{2t} + c_4 t e^{2t}$
15. $y = c_1 \cos t + c_2 \sin t + e^{\sqrt{3}t/2}(c_3 \cos \frac{1}{2}t + c_4 \sin \frac{1}{2}t) + e^{-\sqrt{3}t/2}(c_5 \cos \frac{1}{2}t + c_6 \sin \frac{1}{2}t)$
16. $y = c_1 e^t + c_2 e^{-t} + c_3 e^{2t} + c_4 e^{-2t}$
17. $y = c_1 e^t + c_2 t e^t + c_3 t^2 e^t + c_4 e^{-t} + c_5 t e^{-t} + c_6 t^2 e^{-t}$
18. $y = c_1 + c_2 t + c_3 e^t + c_4 e^{-t} + c_5 \cos t + c_6 \sin t$
19. $y = c_1 + c_2 e^t + c_3 e^{2t} + c_4 \cos t + c_5 \sin t$
20. $y = c_1 + c_2 e^{2t} + e^{-t}(c_3 \cos \sqrt{3}\, t + c_4 \sin \sqrt{3}\, t)$
21. $y = e^t[(c_1 + c_2 t)\cos t + (c_3 + c_4 t)\sin t] + e^{-t}[(c_5 + c_6 t)\cos t + (c_7 + c_8 t)\sin t]$
22. $y = (c_1 + c_2 t)\cos t + (c_3 + c_4 t)\sin t$
23. $y = c_1 e^t + c_2 e^{(2+\sqrt{5})t} + c_3 e^{(2-\sqrt{5})t}$ 24. $y = c_1 e^{-t} + c_2 e^{(-2+\sqrt{2})t} + c_3 e^{(-2-\sqrt{2})t}$
25. $y = c_1 e^{-t/2} + c_2 e^{-t/3}\cos(t/\sqrt{3}) + c_3 e^{-t/3}\sin(t/\sqrt{3})$
26. $y = c_1 e^{3t} + c_2 e^{-2t} + c_3 e^{(3+\sqrt{3})t} + c_4 e^{(3-\sqrt{3})t}$
27. $y = c_1 e^{-t/3} + c_2 e^{-t/4} + c_3 e^{-t}\cos 2t + c_4 e^{-t}\sin 2t$
28. $y = c_1 e^{-t}\cos t + c_2 e^{-t}\sin t + c_3 e^{-2t}\cos(\sqrt{3}\, t) + c_4 e^{-2t}\sin(\sqrt{3}\, t)$
29. $y = 2 - 2\cos t + \sin t$
30. $y = \frac{1}{2}e^{-t/\sqrt{2}}\sin(t/\sqrt{2}) - \frac{1}{2}e^{t/\sqrt{2}}\sin(t/\sqrt{2})$
31. $y = 2t - 3$ 32. $y = 2\cos t - \sin t$
33. $y = -\frac{2}{3}e^t - \frac{1}{10}e^{2t} - \frac{1}{6}e^{-2t} - \frac{16}{15}e^{-t/2}$ 34. $y = \frac{2}{13}e^{-t} + \frac{24}{13}e^{t/2}\cos t + \frac{3}{13}e^{t/2}\sin t$
35. $y = 8 - 18e^{-t/3} + 8e^{-t/2}$

36. $y = \frac{21}{13}e^{-t}\cos t - \frac{38}{13}e^{-t}\sin t - \frac{8}{13}e^{-2t}\cos(\sqrt{3}\,t) + \dfrac{17\sqrt{3}}{39}e^{-2t}\sin(\sqrt{3}\,t)$

37. $y = \frac{1}{2}(\cosh t - \cos t) + \frac{1}{2}(\sinh t - \sin t)$

38. (a) $W = c$, a constant (b) $W = -8$ (c) $W = 4$

39. (b) $u_1 = c_1\cos t + c_2\sin t + c_3\cos\sqrt{6}\,t + c_4\sin\sqrt{6}\,t$

Section 4.3, page 235

1. $y = c_1 e^t + c_2 t e^t + c_3 e^{-t} + \frac{1}{2}t e^{-t} + 3$
2. $y = c_1 e^t + c_2 e^{-t} + c_3\cos t + c_4\sin t - 3t - \frac{1}{4}t\sin t$
3. $y = c_1 e^{-t} + c_2\cos t + c_3\sin t + \frac{1}{2}t e^{-t} + 4(t-1)$
4. $y = c_1 + c_2 e^t + c_3 e^{-t} + \cos t$
5. $y = c_1 + c_2 t + c_3 e^{-2t} + c_4 e^{2t} - \frac{1}{3}e^t - \frac{1}{48}t^4 - \frac{1}{16}t^2$
6. $y = c_1\cos t + c_2\sin t + c_3 t\cos t + c_4 t\sin t + 3 + \frac{1}{9}\cos 2t$
7. $y = c_1 + c_2 t + c_3 t^2 + c_4 e^{-t} + e^{t/2}[c_5\cos(\sqrt{3}\,t/2) + c_6\sin(\sqrt{3}\,t/2)] + \frac{1}{24}t^4$
8. $y = c_1 + c_2 t + c_3 t^2 + c_4 e^{-t} + \frac{1}{20}\sin 2t + \frac{1}{40}\cos 2t$
9. $y = \frac{3}{16}(1 - \cos 2t) + \frac{1}{8}t^2$
10. $y = (t-4)\cos t - (\frac{3}{2}t + 4)\sin t + 3t + 4$
11. $y = 1 + \frac{1}{4}(t^2 + 3t) - t e^t$
12. $y = -\frac{2}{5}\cos t - \frac{4}{5}\sin t + \frac{1}{20}e^{-t} + \frac{81}{40}e^t + \frac{73}{520}e^{-3t} + \frac{77}{65}\cos 2t - \frac{49}{130}\sin 2t$
13. $Y(t) = t(A_0 t^3 + A_1 t^2 + A_2 t + A_3) + B t^2 e^t$
14. $Y(t) = t(A_0 t + A_1)e^{-t} + B\cos t + C\sin t$
15. $Y(t) = A t^2 e^t + B\cos t + C\sin t$
16. $Y(t) = A t^2 + (B_0 t + B_1)e^t + t(C\cos 2t + D\sin 2t)$
17. $Y(t) = t(A_0 t^2 + A_1 t + A_2) + (B_0 t + B_1)\cos t + (C_0 t + C_1)\sin t$
18. $Y(t) = A e^t + (B_0 t + B_1)e^{-t} + t e^{-t}(C\cos t + D\sin t)$
19. $k_0 = a_0, \quad k_n = a_0\alpha^n + a_1\alpha^{n-1} + \cdots + a_{n-1}\alpha + a_n$

Section 4.4, page 240

1. $y = c_1 + c_2\cos t + c_3\sin t - \ln\cos t - (\sin t)\ln(\sec t + \tan t)$
2. $y = c_1 + c_2 e^t + c_3 e^{-t} - \frac{1}{2}t^2$
3. $y = c_1 e^t + c_2 e^{-t} + c_3 e^{2t} + \frac{1}{30}e^{4t}$
4. $y = c_1 + c_2\cos t + c_3\sin t + \ln(\sec t + \tan t) - t\cos t + (\sin t)\ln\cos t$
5. $y = c_1 e^t + c_2\cos t + c_3\sin t - \frac{1}{5}e^{-t}\cos t$
6. $y = c_1\cos t + c_2\sin t + c_3 t\cos t + c_4 t\sin t - \frac{1}{8}t^2\sin t$
7. $y = c_1 e^t + c_2\cos t + c_3\sin t - \frac{1}{2}(\cos t)\ln\cos t + \frac{1}{2}(\sin t)\ln\cos t - \frac{1}{2}t\cos t - \frac{1}{2}t\sin t$
$\qquad + \frac{1}{2}e^t \displaystyle\int_{t_0}^{t}(e^{-s}/\cos s)\,ds$

8. $y = c_1 + c_2 e^t + c_3 e^{-t} - \ln\sin t + \ln(\cos t + 1) + \frac{1}{2}e^t\displaystyle\int_{t_0}^{t}(e^{-s}/\sin s)\,ds$
$\qquad + \frac{1}{2}e^{-t}\displaystyle\int_{t_0}^{t}(e^{s}/\sin s)\,ds$

9. $c_1 = 0,\ c_2 = 2,\ c_3 = 1$ in answer to Problem 4
10. $c_1 = 2,\ c_2 = \frac{7}{8},\ c_3 = -\frac{7}{8},\ c_4 = \frac{1}{2}$ in answer to Problem 6
11. $c_1 = \frac{3}{2},\ c_2 = \frac{1}{2},\ c_3 = -\frac{5}{2},\ t_0 = 0$ in answer to Problem 7
12. $c_1 = 3,\ c_2 = 0,\ c_3 = -e^{\pi/2},\ t_0 = \pi/2$ in answer to Problem 8
13. $Y(x) = x^4/15$
14. $Y(t) = \frac{1}{2}\displaystyle\int_{t_0}^{t}[e^{t-s} - \sin(t-s) - \cos(t-s)]g(s)\,ds$

15. $Y(t) = \frac{1}{2} \int_{t_0}^{t} [\sinh(t - s) - \sin(t - s)]g(s)\,ds$

16. $Y(t) = \frac{1}{2} \int_{t_0}^{t} e^{(t-s)}(t - s)^2 g(s)\,ds;\quad Y(t) = -te^t \ln|t|$

17. $Y(x) = \frac{1}{2} \int_{x_0}^{x} [(x/t^2) - 2(x^2/t^3) + (x^3/t^4)]g(t)\,dt$

CHAPTER 5

Section 5.1, page 249

1. $\rho = 1$

2. $\rho = 2$

3. $\rho = \infty$

4. $\rho = \frac{1}{2}$

5. $\rho = \frac{1}{2}$

6. $\rho = 1$

7. $\rho = 3$

8. $\rho = e$

9. $\displaystyle\sum_{n=0}^{\infty} \frac{(-1)^n x^{2n+1}}{(2n + 1)!},\quad \rho = \infty$

10. $\displaystyle\sum_{n=0}^{\infty} \frac{x^n}{n!},\quad \rho = \infty$

11. $1 + (x - 1),\quad \rho = \infty$

12. $1 - 2(x + 1) + (x + 1)^2,\quad \rho = \infty$

13. $\displaystyle\sum_{n=1}^{\infty} (-1)^{n+1} \frac{(x - 1)^n}{n},\quad \rho = 1$

14. $\displaystyle\sum_{n=0}^{\infty} (-1)^n x^n,\quad \rho = 1$

15. $\displaystyle\sum_{n=0}^{\infty} x^n,\quad \rho = 1$

16. $\displaystyle\sum_{n=0}^{\infty} (-1)^{n+1}(x - 2)^n,\quad \rho = 1$

17. $y' = 1 + 2^2 x + 3^2 x^2 + 4^2 x^3 + \cdots + (n + 1)^2 x^n + \cdots$
$y'' = 2^2 + 3^2 \cdot 2x + 4^2 \cdot 3x^2 + 5^2 \cdot 4x^3 + \cdots + (n + 2)^2(n + 1)x^n + \cdots$

18. $y' = a_1 + 2a_2 x + 3a_3 x^2 + 4a_4 x^3 + \cdots + (n + 1)a_{n+1}x^n + \cdots$

$\displaystyle = \sum_{n=1}^{\infty} na_n x^{n-1} = \sum_{n=0}^{\infty} (n + 1)a_{n+1}x^n$

$y'' = 2a_2 + 6a_3 x + 12a_4 x^2 + 20a_5 x^3 + \cdots + (n + 2)(n + 1)a_{n+2}x^n + \cdots$

$\displaystyle = \sum_{n=2}^{\infty} n(n - 1)a_n x^{n-2} = \sum_{n=0}^{\infty} (n + 2)(n + 1)a_{n+2}x^n$

21. $\displaystyle\sum_{n=0}^{\infty} (n + 2)(n + 1)a_{n+2}x^n$

22. $\displaystyle\sum_{n=2}^{\infty} a_{n-2}x^n$

23. $\displaystyle\sum_{n=0}^{\infty} (n + 1)a_n x^n$

24. $\displaystyle\sum_{n=0}^{\infty} [(n + 2)(n + 1)a_{n+2} - n(n - 1)a_n]x^n$

25. $\displaystyle\sum_{n=0}^{\infty} [(n + 2)(n + 1)a_{n+2} + na_n]x^n$

26. $a_1 + \displaystyle\sum_{n=1}^{\infty} [(n + 1)a_{n+1} + a_{n-1}]x^n$

27. $\displaystyle\sum_{n=0}^{\infty} [(n + 1)na_{n+1} + a_n]x^n$

28. $a_n = (-2)^n a_0/n!,\quad n = 1, 2, \ldots;\quad a_0 e^{-2x}$

Section 5.2, page 259

1. $a_{n+2} = a_n/(n+2)(n+1)$

$$y_1(x) = 1 + \frac{x^2}{2!} + \frac{x^4}{4!} + \frac{x^6}{6!} + \cdots = \sum_{n=0}^{\infty} \frac{x^{2n}}{(2n)!} = \cosh x$$

$$y_2(x) = x + \frac{x^3}{3!} + \frac{x^5}{5!} + \frac{x^7}{7!} + \cdots = \sum_{n=0}^{\infty} \frac{x^{2n+1}}{(2n+1)!} = \sinh x$$

2. $a_{n+2} = a_n/(n+2)$

$$y_1(x) = 1 + \frac{x^2}{2} + \frac{x^4}{2 \cdot 4} + \frac{x^6}{2 \cdot 4 \cdot 6} + \cdots = \sum_{n=0}^{\infty} \frac{x^{2n}}{2^n n!}$$

$$y_2(x) = x + \frac{x^3}{3} + \frac{x^5}{3 \cdot 5} + \frac{x^7}{3 \cdot 5 \cdot 7} + \cdots = \sum_{n=0}^{\infty} \frac{2^n n! x^{2n+1}}{(2n+1)!}$$

3. $(n+2)a_{n+2} - a_{n+1} - a_n = 0$

$y_1(x) = 1 + \frac{1}{2}(x-1)^2 + \frac{1}{6}(x-1)^3 + \frac{1}{6}(x-1)^4 + \cdots$

$y_2(x) = (x-1) + \frac{1}{2}(x-1)^2 + \frac{1}{2}(x-1)^3 + \frac{1}{4}(x-1)^4 + \cdots$

4. $a_{n+4} = -k^2 a_n/(n+4)(n+3); \quad a_2 = a_3 = 0$

$$y_1(x) = 1 - \frac{k^2 x^4}{3 \cdot 4} + \frac{k^4 x^8}{3 \cdot 4 \cdot 7 \cdot 8} - \frac{k^6 x^{12}}{3 \cdot 4 \cdot 7 \cdot 8 \cdot 11 \cdot 12} + \cdots$$

$$= 1 + \sum_{m=0}^{\infty} \frac{(-1)^{m+1}(k^2 x^4)^{m+1}}{3 \cdot 4 \cdot 7 \cdot 8 \cdots (4m+3)(4m+4)}$$

$$y_2(x) = x - \frac{k^2 x^5}{4 \cdot 5} + \frac{k^4 x^9}{4 \cdot 5 \cdot 8 \cdot 9} - \frac{k^6 x^{13}}{4 \cdot 5 \cdot 8 \cdot 9 \cdot 12 \cdot 13} + \cdots$$

$$= x \left[1 + \sum_{m=0}^{\infty} \frac{(-1)^{m+1}(k^2 x^4)^{m+1}}{4 \cdot 5 \cdot 8 \cdot 9 \cdots (4m+4)(4m+5)} \right]$$

Hint: Let $n = 4m$ in the recurrence relation, $m = 1, 2, 3, \ldots$.

5. $(n+2)(n+1)a_{n+2} - n(n+1)a_{n+1} + a_n = 0, \quad n \geq 1; \quad a_2 = -\frac{1}{2}a_0$

$y_1(x) = 1 - \frac{1}{2}x^2 - \frac{1}{6}x^3 - \frac{1}{24}x^4 + \cdots, \quad y_2(x) = x - \frac{1}{6}x^3 - \frac{1}{12}x^4 - \frac{1}{24}x^5 + \cdots$

6. $a_{n+2} = -(n^2 - 2n + 4)a_n/[2(n+1)(n+2)], \quad n \geq 2; \quad a_2 = -a_0, \quad a_3 = -\frac{1}{4}a_1$

$y_1(x) = 1 - x^2 + \frac{1}{6}x^4 - \frac{1}{30}x^6 + \cdots, \quad y_2(x) = x - \frac{1}{4}x^3 + \frac{7}{160}x^5 - \frac{19}{1920}x^7 + \cdots$

7. $a_{n+2} = -a_n/(n+1), \quad n = 0, 1, 2, \ldots$

$$y_1(x) = 1 - \frac{x^2}{1} + \frac{x^4}{1 \cdot 3} - \frac{x^6}{1 \cdot 3 \cdot 5} + \cdots = 1 + \sum_{n=1}^{\infty} \frac{(-1)^n x^{2n}}{1 \cdot 3 \cdot 5 \cdots (2n-1)}$$

$$y_2(x) = x - \frac{x^3}{2} + \frac{x^5}{2 \cdot 4} - \frac{x^7}{2 \cdot 4 \cdot 6} + \cdots = x + \sum_{n=1}^{\infty} \frac{(-1)^n x^{2n+1}}{2 \cdot 4 \cdot 6 \cdots (2n)}$$

8. $a_{n+2} = -[(n+1)^2 a_{n+1} + a_n + a_{n-1}]/(n+1)(n+2), \quad n = 1, 2, \ldots$

$a_2 = -(a_0 + a_1)/2$

$y_1(x) = 1 - \frac{1}{2}(x-1)^2 + \frac{1}{6}(x-1)^3 - \frac{1}{12}(x-1)^4 + \cdots$

$y_2(x) = (x-1) - \frac{1}{2}(x-1)^2 + \frac{1}{6}(x-1)^3 - \frac{1}{6}(x-1)^4 + \cdots$

9. $(n+2)(n+1)a_{n+2} + (n-2)(n-3)a_n = 0; \quad n = 0, 1, 2, \ldots$

$y_1(x) = 1 - 3x^2, \quad y_2(x) = x - x^3/3$

10. $4(n+2)a_{n+2} - (n-2)a_n = 0;$ $n = 0, 1, 2, \ldots$

$$y_1(x) = 1 - \frac{x^2}{4}, \quad y_2(x) = x - \frac{x^3}{12} - \frac{x^5}{240} - \frac{x^7}{2240} - \cdots - \frac{x^{2n+1}}{4^n(2n-1)(2n+1)} - \cdots$$

11. $3(n+2)a_{n+2} - (n+1)a_n = 0;$ $n = 0, 1, 2, \ldots$

$$y_1(x) = 1 + \frac{x^2}{6} + \frac{x^4}{24} + \frac{5}{432}x^6 + \cdots + \frac{3 \cdot 5 \cdots (2n-1)}{3^n \cdot 2 \cdot 4 \cdots (2n)}x^{2n} + \cdots$$

$$y_2(x) = x + \frac{2}{9}x^3 + \frac{8}{135}x^5 + \frac{16}{945}x^7 + \cdots + \frac{2 \cdot 4 \cdots (2n)}{3^n \cdot 3 \cdot 5 \cdots (2n+1)}x^{2n+1} + \cdots$$

12. $(n+2)(n+1)a_{n+2} - (n+1)na_{n+1} + (n-1)a_n = 0;$ $n = 0, 1, 2, \ldots$

$$y_1(x) = 1 + \frac{x^2}{2} + \frac{x^3}{6} + \frac{x^4}{24} + \cdots + \frac{x^n}{n!} + \cdots, \quad y_2(x) = x$$

13. $2(n+2)(n+1)a_{n+2} + (n+3)a_n = 0;$ $n = 0, 1, 2, \ldots$

$$y_1(x) = 1 - \frac{3}{4}x^2 + \frac{5}{32}x^4 - \frac{7}{384}x^6 + \cdots + (-1)^n \frac{3 \cdot 5 \cdots (2n+1)}{2^n(2n)!}x^{2n} + \cdots$$

$$y_2(x) = x - \frac{x^3}{3} + \frac{x^5}{20} - \frac{x^7}{210} + \cdots + (-1)^n \frac{4 \cdot 6 \cdots (2n+2)}{2^n(2n+1)!}x^{2n+1} + \cdots$$

14. $2(n+2)(n+1)a_{n+2} + 3(n+1)a_{n+1} + (n+3)a_n = 0;$ $n = 0, 1, 2, \ldots$

$$y_1(x) = 1 - \frac{3}{4}(x-2)^2 + \frac{3}{8}(x-2)^3 + \frac{1}{64}(x-2)^4 + \cdots$$

$$y_2(x) = (x-2) - \frac{3}{4}(x-2)^2 + \frac{1}{24}(x-2)^3 + \frac{9}{64}(x-2)^4 + \cdots$$

15. (a) $y = 2 + x + x^2 + \frac{1}{3}x^3 + \frac{1}{4}x^4 + \cdots$ (c) about $|x| < 0.7$

16. (a) $y = -1 + 3x + x^2 - \frac{3}{4}x^3 - \frac{1}{6}x^4 + \cdots$ (c) about $|x| < 0.7$

17. (a) $y = 4 - x - 4x^2 + \frac{1}{2}x^3 + \frac{4}{3}x^4 + \cdots$ (c) about $|x| < 0.5$

18. (a) $y = -3 + 2x - \frac{3}{2}x^2 - \frac{1}{2}x^3 - \frac{1}{8}x^4 + \cdots$ (c) about $|x| < 0.9$

19. $y_1(x) = 1 - \frac{1}{3}(x-1)^3 - \frac{1}{12}(x-1)^4 + \frac{1}{18}(x-1)^6 + \cdots$

$\quad y_2(x) = (x-1) - \frac{1}{4}(x-1)^4 - \frac{1}{20}(x-1)^5 + \frac{1}{28}(x-1)^7 + \cdots$

21. (a) $y_1(x) = 1 - \frac{\lambda}{2!}x^2 + \frac{\lambda(\lambda-4)}{4!}x^4 - \frac{\lambda(\lambda-4)(\lambda-8)}{6!}x^6 + \cdots$

$\quad y_2(x) = x - \frac{\lambda-2}{3!}x^3 + \frac{(\lambda-2)(\lambda-6)}{5!}x^5 - \frac{(\lambda-2)(\lambda-6)(\lambda-10)}{7!}x^7 + \cdots$

(b) $1, x, 1 - 2x^2, x - \frac{2}{3}x^3, 1 - 4x^2 + \frac{4}{3}x^4, x - \frac{4}{3}x^3 + \frac{4}{15}x^5$

(c) $1, 2x, 4x^2 - 2, 8x^3 - 12x, 16x^4 - 48x^2 + 12, 32x^5 - 160x^3 + 120x$

22. (b) $y = x - x^3/6 + \cdots$

Section 5.3, page 265

1. $\phi''(0) = -1,$ $\phi'''(0) = 0,$ $\phi^{(4)}(0) = 3$

2. $\phi''(0) = 0,$ $\phi'''(0) = -2,$ $\phi^{(4)}(0) = 0$

3. $\phi''(1) = 0,$ $\phi'''(1) = -6,$ $\phi^{(4)}(1) = 42$

4. $\phi''(0) = 0,$ $\phi'''(0) = -a_0,$ $\phi^{(4)}(0) = -4a_1$

5. $\rho = \infty,$ $\rho = \infty$ 6. $\rho = 1,$ $\rho = 3,$ $\rho = 1$

7. $\rho = 1,$ $\rho = \sqrt{3}$ 8. $\rho = 1$

9. (a) $\rho = \infty$ (b) $\rho = \infty$ (c) $\rho = \infty$ (d) $\rho = \infty$ (e) $\rho = 1$

 (f) $\rho = \sqrt{2}$ (g) $\rho = \infty$ (h) $\rho = 1$ (i) $\rho = 1$ (j) $\rho = 2$

 (k) $\rho = \sqrt{3}$ (l) $\rho = 1$ (m) $\rho = \infty$ (n) $\rho = \infty$

10. (a) $y_1(x) = 1 - \dfrac{\alpha^2}{2!}x^2 - \dfrac{(2^2 - \alpha^2)\alpha^2}{4!}x^4 - \dfrac{(4^2 - \alpha^2)(2^2 - \alpha^2)\alpha^2}{6!}x^6 - \cdots$

$\qquad - \dfrac{[(2m - 2)^2 - \alpha^2] \cdots (2^2 - \alpha^2)\alpha^2}{(2m)!}x^{2m} - \cdots$

$\qquad y_2(x) = x + \dfrac{1 - \alpha^2}{3!}x^3 + \dfrac{(3^2 - \alpha^2)(1 - \alpha^2)}{5!}x^5 + \cdots$

$\qquad + \dfrac{[(2m - 1)^2 - \alpha^2] \cdots (1 - \alpha^2)}{(2m + 1)!}x^{2m+1} + \cdots$

(b) $y_1(x)$ or $y_2(x)$ terminates with x^n as $\alpha = n$ is even or odd

(c) $n = 0$, $y = 1$; $n = 1$, $y = x$; $n = 2$, $y = 1 - 2x^2$; $n = 3$, $y = x - \frac{4}{3}x^3$

11. $y_1(x) = 1 - \frac{1}{6}x^3 + \frac{1}{120}x^5 + \frac{1}{180}x^6 + \cdots$, $y_2(x) = x - \frac{1}{12}x^4 + \frac{1}{180}x^6 + \frac{1}{504}x^7 + \cdots$, $\rho = \infty$

12. $y_1(x) = 1 - \frac{1}{6}x^3 + \frac{1}{12}x^4 - \frac{1}{40}x^5 + \cdots$, $y_2(x) = x - \frac{1}{12}x^4 + \frac{1}{20}x^5 - \frac{1}{60}x^6 + \cdots$, $\rho = \infty$

13. $y_1(x) = 1 + x^2 + \frac{1}{12}x^4 + \frac{1}{120}x^6 + \cdots$, $y_2(x) = x + \frac{1}{6}x^3 + \frac{1}{60}x^5 + \frac{1}{560}x^7 + \cdots$, $\rho = \pi/2$

14. $y_1(x) = 1 + \frac{1}{6}x^3 + \frac{1}{12}x^4 - \frac{1}{120}x^6 + \cdots$, $y_2(x) = x - \frac{1}{6}x^3 + \frac{1}{24}x^4 + \frac{7}{120}x^5 + \cdots$, $\rho = 1$

15. Cannot specify arbitrary initial conditions at $x = 0$; hence $x = 0$ is a singular point.

16. $y = 1 + x + \dfrac{x^2}{2!} + \cdots + \dfrac{x^n}{n!} + \cdots = e^x$

17. $y = 1 + \dfrac{x^2}{2} + \dfrac{x^4}{2 \cdot 4} + \dfrac{x^6}{2 \cdot 4 \cdot 6} + \cdots + \dfrac{x^{2n}}{2^n \cdot n!} + \cdots$

18. $y = 1 + x + \frac{1}{2}x^2 + \frac{1}{2}x^3 + \cdots$

19. $y = 1 + x + x^2 + \cdots + x^n + \cdots = \dfrac{1}{1 - x}$

20. $y = a_0 \left(1 + x + \dfrac{x^2}{2!} + \cdots + \dfrac{x^n}{n!} + \cdots \right) + 2\left(\dfrac{x^3}{3!} + \dfrac{x^4}{4!} + \cdots + \dfrac{x^n}{n!} + \cdots \right)$

$\qquad = a_0 e^x + 2\left(e^x - 1 - x - \dfrac{x^2}{2} \right) = ce^x - 2 - 2x - x^2$

21. $y = a_0 \left(1 - \dfrac{x^2}{2} + \dfrac{x^4}{2^2 2!} - \dfrac{x^6}{2^3 3!} + \cdots + \dfrac{(-1)^n x^{2n}}{2^n n!} + \cdots \right)$

$\qquad + \left(x + \dfrac{x^2}{2} - \dfrac{x^3}{3} - \dfrac{x^4}{2 \cdot 4} + \dfrac{x^5}{3 \cdot 5} + \cdots \right)$

$\qquad = a_0 e^{-x^2/2} + \left(x + \dfrac{x^2}{2} - \dfrac{x^3}{3} - \dfrac{x^4}{2 \cdot 4} + \dfrac{x^5}{3 \cdot 5} + \cdots \right)$

23. 1, $1 - 3x^2$, $1 - 10x^2 + \frac{35}{3}x^4$; x, $x - \frac{5}{3}x^3$, $x - \frac{14}{3}x^3 + \frac{21}{5}x^5$

24. (a) 1, x, $(3x^2 - 1)/2$, $(5x^3 - 3x)/2$, $(35x^4 - 30x^2 + 3)/8$, $(63x^5 - 70x^3 + 15x)/8$

(c) P_1, 0; P_2, ± 0.57735; P_3, 0, ± 0.77460; P_4, ± 0.33998, ± 0.86114;

$\quad P_5$, 0, ± 0.53847, ± 0.90618

Section 5.4, page 271

1. $x = 0$, regular	2. $x = 0$, regular; $x = 1$, irregular
3. $x = 0$, irregular; $x = 1$, regular	4. $x = 0$, irregular; $x = \pm 1$, regular
5. $x = 1$, regular; $x = -1$, irregular	6. $x = 0$, regular
7. $x = -3$, regular	8. $x = 0$, -1, regular; $x = 1$, irregular
9. $x = 1$, regular; $x = -2$, irregular	10. $x = 0$, 3, regular
11. $x = 1$, -2, regular	12. $x = 0$, regular

CHAPTER 6 **Section 6.1, page 312**

1. Piecewise continuous 2. Neither
3. Continuous 4. Piecewise continuous
5. (a) $1/s^2$, $s > 0$ (b) $2/s^3$, $s > 0$ (c) $n!/s^{n+1}$, $s > 0$
6. $s/(s^2 + a^2)$, $s > 0$ 7. $\dfrac{s}{s^2 - b^2}$, $s > |b|$

8. $\dfrac{b}{s^2 - b^2}$, $s > |b|$ 9. $\dfrac{s - a}{(s - a)^2 - b^2}$, $s - a > |b|$

10. $\dfrac{b}{(s - a)^2 - b^2}$, $s - a > |b|$ 11. $\dfrac{b}{s^2 + b^2}$, $s > 0$

12. $\dfrac{s}{s^2 + b^2}$, $s > 0$ 13. $\dfrac{b}{(s - a)^2 + b^2}$, $s > a$

14. $\dfrac{s - a}{(s - a)^2 + b^2}$, $s > a$ 15. $\dfrac{1}{(s - a)^2}$, $s > a$

16. $\dfrac{2as}{(s^2 + a^2)^2}$, $s > 0$ 17. $\dfrac{s^2 + a^2}{(s - a)^2(s + a)^2}$, $s > |a|$

18. $\dfrac{n!}{(s - a)^{n+1}}$, $s > a$ 19. $\dfrac{2a(3s^2 - a^2)}{(s^2 + a^2)^3}$, $s > 0$

20. $\dfrac{2a(3s^2 + a^2)}{(s^2 - a^2)^3}$, $s > |a|$ 21. Converges

22. Converges 23. Diverges
24. Converges
26. (d) $\Gamma(3/2) = \sqrt{\pi}/2$; $\Gamma(11/2) = 945\sqrt{\pi}/32$

Section 6.2, page 322

1. $\frac{3}{2}\sin 2t$ 2. $2t^2 e^t$
3. $\frac{2}{5}e^t - \frac{2}{5}e^{-4t}$ 4. $\frac{9}{5}e^{3t} + \frac{6}{5}e^{-2t}$
5. $2e^{-t}\cos 2t$ 6. $2\cosh 2t - \frac{3}{2}\sinh 2t$
7. $2e^t\cos t + 3e^t\sin t$ 8. $3 - 2\sin 2t + 5\cos 2t$
9. $-2e^{-2t}\cos t + 5e^{-2t}\sin t$ 10. $2e^{-t}\cos 3t - \frac{5}{3}e^{-t}\sin 3t$
11. $y = \frac{1}{5}(e^{3t} + 4e^{-2t})$ 12. $y = 2e^{-t} - e^{-2t}$
13. $y = e^t\sin t$ 14. $y = e^{2t} - te^{2t}$
15. $y = 2e^t\cos\sqrt{3}\,t - (2/\sqrt{3})e^t\sin\sqrt{3}\,t$ 16. $y = 2e^{-t}\cos 2t + \frac{1}{2}e^{-t}\sin 2t$
17. $y = te^t - t^2 e^t + \frac{2}{3}t^3 e^t$ 18. $y = \cosh t$
19. $y = \cos\sqrt{2}\,t$ 20. $y = (\omega^2 - 4)^{-1}[(\omega^2 - 5)\cos\omega t + \cos 2t]$
21. $y = \frac{1}{5}(\cos t - 2\sin t + 4e^t\cos t - 2e^t\sin t)$
22. $y = \frac{1}{5}(e^{-t} - e^t\cos t + 7e^t\sin t)$ 23. $y = 2e^{-t} + te^{-t} + 2t^2 e^{-t}$

24. $Y(s) = \dfrac{s}{s^2 + 4} + \dfrac{1 - e^{-\pi s}}{s(s^2 + 4)}$ 25. $Y(s) = \dfrac{1}{s^2(s^2 + 1)} - \dfrac{e^{-s}(s + 1)}{s^2(s^2 + 1)}$

26. $Y(s) = (1 - e^{-s})/s^2(s^2 + 4)$ 29. $1/(s - a)^2$
30. $2b(3s^2 - b^2)/(s^2 + b^2)^3$ 31. $n!/s^{n+1}$
32. $n!/(s - a)^{n+1}$ 33. $2b(s - a)/[(s - a)^2 + b^2]^2$
34. $[(s - a)^2 - b^2]/[(s - a)^2 + b^2]^2$
36. (a) $Y' + s^2 Y = s$ (b) $s^2 Y'' + 2sY' - [s^2 + \alpha(\alpha + 1)]Y = -1$

Section 6.3, page 329

7. $F(s) = 2e^{-2s}/s^3$

8. $F(s) = e^{-s}(s^2 + 2)/s^3$

9. $F(s) = \dfrac{e^{-\pi s}}{s^2} - \dfrac{e^{-2\pi s}}{s^2}(1 + \pi s)$

10. $F(s) = \dfrac{1}{s}(e^{-s} + 2e^{-3s} - 6e^{-4s})$

11. $F(s) = s^{-2}[(1 - s)e^{-2s} - (1 + s)e^{-3s}]$

12. $F(s) = (1 - e^{-s})/s^2$

13. $f(t) = t^3 e^{2t}$

14. $f(t) = \frac{1}{3}u_2(t)[e^{t-2} - e^{-2(t-2)}]$

15. $f(t) = 2u_2(t)e^{t-2}\cos(t - 2)$

16. $f(t) = u_2(t)\sinh 2(t - 2)$

17. $f(t) = u_1(t)e^{2(t-1)}\cosh(t - 1)$

18. $f(t) = u_1(t) + u_2(t) - u_3(t) - u_4(t)$

20. $f(t) = 2(2t)^n$

21. $f(t) = \frac{1}{2}e^{-t/2}\cos t$

22. $f(t) = \frac{1}{6}e^{t/3}(e^{2t/3} - 1)$

23. $f(t) = \frac{1}{2}e^{t/2}u_2(t/2)$

24. $F(s) = s^{-1}(1 - e^{-s}), \quad s > 0$

25. $F(s) = s^{-1}(1 - e^{-s} + e^{-2s} - e^{-3s}), \quad s > 0$

26. $F(s) = \dfrac{1}{s}[1 - e^{-s} + \cdots + e^{-2ns} - e^{-(2n+1)s}] = \dfrac{1 - e^{-(2n+2)s}}{s(1 + e^{-s})}, \quad s > 0$

27. $F(s) = \dfrac{1}{s}\displaystyle\sum_{n=0}^{\infty}(-1)^n e^{-ns} = \dfrac{1/s}{1 + e^{-s}}, \quad s > 0$

29. $\mathcal{L}\{f(t)\} = \dfrac{1/s}{1 + e^{-s}}, \quad s > 0$

30. $\mathcal{L}\{f(t)\} = \dfrac{1 - e^{-s}}{s(1 + e^{-s})}, \quad s > 0$

31. $\mathcal{L}\{f(t)\} = \dfrac{1 - (1 + s)e^{-s}}{s^2(1 - e^{-s})}, \quad s > 0$

32. $\mathcal{L}\{f(t)\} = \dfrac{1 + e^{-\pi s}}{(1 + s^2)(1 - e^{-\pi s})}, \quad s > 0$

33. (a) $\mathcal{L}\{f(t)\} = s^{-1}(1 - e^{-s}), \quad s > 0$
 (b) $\mathcal{L}\{g(t)\} = s^{-2}(1 - e^{-s}), \quad s > 0$
 (c) $\mathcal{L}\{h(t)\} = s^{-2}(1 - e^{-s})^2, \quad s > 0$

34. (b) $\mathcal{L}\{p(t)\} = \dfrac{1 - e^{-s}}{s^2(1 + e^{-s})}, \quad s > 0$

Section 6.4, page 337

1. $y = 1 - \cos t + \sin t - u_{\pi/2}(t)(1 - \sin t)$

2. $y = e^{-t}\sin t + \frac{1}{2}u_{\pi}(t)[1 + e^{-(t-\pi)}\cos t + e^{-(t-\pi)}\sin t]$
 $\quad - \frac{1}{2}u_{2\pi}(t)[1 - e^{-(t-2\pi)}\cos t - e^{-(t-2\pi)}\sin t]$

3. $y = \frac{1}{6}[1 - u_{2\pi}(t)](2\sin t - \sin 2t)$

4. $y = \frac{1}{6}(2\sin t - \sin 2t) - \frac{1}{6}u_{\pi}(t)(2\sin t + \sin 2t)$

5. $y = \frac{1}{2} + \frac{1}{2}e^{-2t} - e^{-t} - u_{10}(t)[\frac{1}{2} + \frac{1}{2}e^{-2(t-10)} - e^{-(t-10)}]$

6. $y = e^{-t} - e^{-2t} + u_2(t)[\frac{1}{2} - e^{-(t-2)} + \frac{1}{2}e^{-2(t-2)}]$

7. $y = \cos t + u_{3\pi}(t)[1 - \cos(t - 3\pi)]$

8. $y = h(t) - u_{\pi/2}(t)h(t - \pi/2), \quad h(t) = \frac{4}{25}(-4 + 5t + 4e^{-t/2}\cos t - 3e^{-t/2}\sin t)$

9. $y = \frac{1}{2}\sin t + \frac{1}{2}t - \frac{1}{2}u_6(t)[t - 6 - \sin(t - 6)]$

10. $y = h(t) + u_{\pi}(t)h(t - \pi), \quad h(t) = \frac{4}{17}[-4\cos t + \sin t + 4e^{-t/2}\cos t + e^{-t/2}\sin t]$

11. $y = u_{\pi}(t)[\frac{1}{4} - \frac{1}{4}\cos(2t - 2\pi)] - u_{3\pi}(t)[\frac{1}{4} - \frac{1}{4}\cos(2t - 6\pi)]$

12. $y = u_1(t)h(t - 1) - u_2(t)h(t - 2), \quad h(t) = -1 + (\cos t + \cosh t)/2$

13. $y = h(t) - u_{\pi}(t)h(t - \pi), \quad h(t) = (3 - 4\cos t + \cos 2t)/12$

14. $f(t) = [u_{t_0}(t)(t - t_0) - u_{t_0+k}(t)(t - t_0 - k)](h/k)$

15. $g(t) = [u_{t_0}(t)(t - t_0) - 2u_{t_0+k}(t)(t - t_0 - k) + u_{t_0+2k}(t)(t - t_0 - 2k)](h/k)$

16. (b) $u(t) = ku_{3/2}(t)h(t - \frac{3}{2}) - ku_{5/2}(t)h(t - \frac{5}{2})$,
 $\quad h(t) = 1 - (\sqrt{7}/21)e^{-t/8}\sin(3\sqrt{7}\,t/8) - e^{-t/8}\cos(3\sqrt{7}\,t/8)$
 (d) $k = 2.51$ (e) $\tau = 25.6773$

17. (a) $k = 5$

 (b) $y = [u_5(t)h(t - 5) - u_{5+k}(t)h(t - 5 - k)]/k$, $\quad h(t) = \frac{1}{4}t - \frac{1}{8}\sin 2t$

18. (b) $f_k(t) = [u_{4-k}(t) - u_{4+k}(t)]/2k$;

 $y = [u_{4-k}(t)h(t - 4 + k) - u_{4+k}(t)h(t - 4 - k)]/2k$,

 $h(t) = \frac{1}{4} - \frac{1}{4}e^{-t/6}\cos(\sqrt{143}\,t/6) - (\sqrt{143}/572)\,e^{-t/6}\sin(\sqrt{143}\,t/6)$

19. (b) $y = 1 - \cos t + 2\sum_{k=1}^{n}(-1)^k u_{k\pi}(t)[1 - \cos(t - k\pi)]$

21. (b) $y = 1 - \cos t + \sum_{k=1}^{n}(-1)^k u_{k\pi}(t)[1 - \cos(t - k\pi)]$

23. (a) $y = 1 - \cos t + 2\sum_{k=1}^{n}(-1)^k u_{11k/4}(t)[1 - \cos(t - 11k/4)]$

Section 6.5, page 344

1. $y = e^{-t}\cos t + e^{-t}\sin t + u_\pi(t)e^{-(t-\pi)}\sin(t - \pi)$

2. $y = \frac{1}{2}u_\pi(t)\sin 2(t - \pi) - \frac{1}{2}u_{2\pi}(t)\sin 2(t - 2\pi)$

3. $y = -\frac{1}{2}e^{-2t} + \frac{1}{2}e^{-t} + u_5(t)[-e^{-2(t-5)} + e^{-(t-5)}] + u_{10}(t)[\frac{1}{2} + \frac{1}{2}e^{-2(t-10)} - e^{-(t-10)}]$

4. $y = \cosh(t) - 20u_3(t)\sinh(t - 3)$

5. $y = \frac{1}{4}\sin t - \frac{1}{4}\cos t + \frac{1}{4}e^{-t}\cos\sqrt{2}\,t + (1/\sqrt{2})\,u_{3\pi}(t)e^{-(t-3\pi)}\sin\sqrt{2}(t - 3\pi)$

6. $y = \frac{1}{2}\cos 2t + \frac{1}{2}u_{4\pi}(t)\sin 2(t - 4\pi)$

7. $y = \sin t + u_{2\pi}(t)\sin(t - 2\pi)$ $\qquad\qquad$ 8. $y = u_{\pi/4}(t)\sin 2(t - \pi/4)$

9. $y = u_{\pi/2}(t)[1 - \cos(t - \pi/2)] + 3u_{3\pi/2}(t)\sin(t - 3\pi/2) - u_{2\pi}(t)[1 - \cos(t - 2\pi)]$

10. $y = (1/\sqrt{31})\,u_{\pi/6}(t)\exp[-\frac{1}{4}(t - \pi/6)]\sin(\sqrt{31}/4)(t - \pi/6)$

11. $y = \frac{1}{5}\cos t + \frac{2}{5}\sin t - \frac{1}{5}e^{-t}\cos t - \frac{3}{5}e^{-t}\sin t + u_{\pi/2}(t)e^{-(t-\pi/2)}\sin(t - \pi/2)$

12. $y = u_1(t)[\sinh(t - 1) - \sin(t - 1)]/2$ \qquad 13. (a) $-e^{-T/4}\delta(t - 5 - T)$, $\quad T = 8\pi/\sqrt{15}$

14. (a) $y = (4/\sqrt{15})\,u_1(t)e^{-(t-1)/4}\sin(\sqrt{15}/4)(t - 1)$

 (b) $t_1 \cong 2.3613$, $\quad y_1 \cong 0.71153$

 (c) $y = (8\sqrt{7}/21)\,u_1(t)e^{-(t-1)/8}\sin(3\sqrt{7}/8)\,(t - 1)$; $\quad t_1 \cong 2.4569$, $\quad y_1 \cong 0.83351$

 (d) $t_1 = 1 + \pi/2 \cong 2.5708$, $\quad y_1 = 1$

15. (a) $k_1 \cong 2.8108$ \qquad (b) $k_1 \cong 2.3995$ \qquad (c) $k_1 = 2$

16. (a) $\phi(t, k) = [u_{4-k}(t)h(t - 4 + k) - u_{4+k}(t)h(t - 4 - k)]/2k$, $\quad h(t) = 1 - \cos t$

 (b) $u_4(t)\sin(t - 4)$ $\qquad\qquad\qquad$ (c) Yes

17. (b) $y = \sum_{k=1}^{20}u_{k\pi}(t)\sin(t - k\pi)$ $\qquad\qquad$ 18. (b) $y = \sum_{k=1}^{20}(-1)^{k+1}u_{k\pi}(t)\sin(t - k\pi)$

19. (b) $y = \sum_{k=1}^{20}u_{k\pi/2}(t)\sin(t - k\pi/2)$ \qquad 20. (b) $y = \sum_{k=1}^{20}(-1)^{k+1}u_{k\pi/2}(t)\sin(t - k\pi/2)$

21. (b) $y = \sum_{k=1}^{15}u_{(2k-1)\pi}(t)\sin[t - (2k - 1)\pi]$

22. (b) $y = \sum_{k=1}^{40}(-1)^{k+1}u_{11k/4}(t)\sin(t - 11k/4)$

23. (b) $y = \frac{20}{\sqrt{399}}\sum_{k=1}^{20}(-1)^{k+1}u_{k\pi}(t)e^{-(t-k\pi)/20}\sin[\sqrt{399}(t - k\pi)/20]$

24. (b) $y = \frac{20}{\sqrt{399}}\sum_{k=1}^{15}u_{(2k-1)\pi}(t)e^{-[t-(2k-1)\pi]/20}\sin\{\sqrt{399}[t - (2k - 1)\pi]/20\}$

Section 6.6, page 351

3. $\sin t * \sin t = \frac{1}{2}(\sin t - t \cos t)$ is negative when $t = 2\pi$, for example.

4. $F(s) = 2/s^2(s^2 + 4)$
5. $F(s) = 1/(s + 1)(s^2 + 1)$

6. $F(s) = 1/s^2(s - 1)$
7. $F(s) = s/(s^2 + 1)^2$

8. $f(t) = \frac{1}{6}\int_0^t (t - \tau)^3 \sin \tau \, d\tau$
9. $f(t) = \int_0^t e^{-(t-\tau)} \cos 2\tau \, d\tau$

10. $f(t) = \frac{1}{2}\int_0^t (t - \tau)e^{-(t-\tau)} \sin 2\tau \, d\tau$
11. $f(t) = \int_0^t \sin(t - \tau)g(\tau) \, d\tau$

12. (c) $\int_0^1 u^m(1 - u)^n \, du = \dfrac{\Gamma(m + 1)\,\Gamma(n + 1)}{\Gamma(m + n + 2)}$

13. $y = \dfrac{1}{\omega}\sin \omega t + \dfrac{1}{\omega}\int_0^t \sin \omega(t - \tau)g(\tau) \, d\tau$

14. $y = \int_0^t e^{-(t-\tau)} \sin(t - \tau) \sin \alpha\tau \, d\tau$
15. $y = \frac{1}{8}\int_0^t e^{-(t-\tau)/2} \sin 2(t - \tau)g(\tau) \, d\tau$

16. $y = e^{-t/2}\cos t - \frac{1}{2}e^{-t/2}\sin t + \int_0^t e^{-(t-\tau)/2} \sin(t - \tau)[1 - u_\pi(\tau)] \, d\tau$

17. $y = 2e^{-2t} + te^{-2t} + \int_0^t (t - \tau)e^{-2(t-\tau)}g(\tau) \, d\tau$

18. $y = 2e^{-t} - e^{-2t} + \int_0^t [e^{-(t-\tau)} - e^{-2(t-\tau)}]\cos \alpha\tau \, d\tau$

19. $y = \dfrac{1}{2}\int_0^t [\sinh(t - \tau) - \sin(t - \tau)]g(\tau) \, d\tau$

20. $y = \frac{4}{3}\cos t - \frac{1}{3}\cos 2t + \frac{1}{6}\int_0^t [2\sin(t - \tau) - \sin 2(t - \tau)]g(\tau) \, d\tau$

21. $\Phi(s) = \dfrac{F(s)}{1 + K(s)}$
22. (a) $\phi(t) = \frac{1}{3}(4\sin 2t - 2\sin t)$

23. (a) $\phi(t) = \cos t$
 (b) $\phi''(t) + \phi(t) = 0, \quad \phi(0) = 1, \quad \phi'(0) = 0$

24. (a) $\phi(t) = \cosh t$
 (b) $\phi''(t) - \phi(t) = 0, \quad \phi(0) = 1, \quad \phi'(0) = 0$

25. (a) $\phi(t) = (1 - 2t + t^2)e^{-t}$
 (b) $\phi''(t) + 2\phi'(t) + \phi(t) = 2e^{-t}, \quad \phi(0) = 1, \quad \phi'(0) = -3$

26. (a) $\phi(t) = \frac{1}{3}e^{-t} - \frac{1}{3}e^{t/2}\cos(\sqrt{3}\,t/2) + \frac{1}{\sqrt{3}}e^{t/2}\sin(\sqrt{3}\,t/2)$
 (b) $\phi'''(t) + \phi(t) = 0, \quad \phi(0) = 0, \quad \phi'(0) = 0, \quad \phi''(0) = 1$

27. (a) $\phi(t) = \cos t$
 (b) $\phi^{(4)}(t) - \phi(t) = 0, \quad \phi(0) = 1, \quad \phi'(0) = 0, \quad \phi''(0) = -1, \quad \phi'''(0) = 0$

28. (a) $\phi(t) = 1 - \frac{2}{\sqrt{3}}e^{-t/2}\sin(\sqrt{3}\,t/2)$
 (b) $\phi'''(t) + \phi''(t) + \phi'(t) = 0, \quad \phi(0) = 1, \quad \phi'(0) = -1, \quad \phi''(0) = 1$

CHAPTER 7 Section 7.1, page 360

1. $x_1' = x_2, \quad x_2' = -2x_1 - 0.5x_2$
2. $x_1' = x_2, \quad x_2' = -2x_1 - 0.5x_2 + 3\sin t$

3. $x_1' = x_2, \quad x_2' = -(1 - 0.25t^{-2})x_1 - t^{-1}x_2$
4. $x_1' = x_2, \quad x_2' = x_3, \quad x_3' = x_4, \quad x_4' = x_1$

5. $x_1' = x_2, \quad x_2' = -4x_1 - 0.25x_2 + 2\cos 3t, \quad x_1(0) = 1, \quad x_2(0) = -2$

6. $x_1' = x_2, \quad x_2' = -q(t)x_1 - p(t)x_2 + g(t); \quad x_1(0) = u_0, \quad x_2(0) = u_0'$

7. (a) $x_1 = c_1 e^{-t} + c_2 e^{-3t}$, $x_2 = c_1 e^{-t} - c_2 e^{-3t}$
 (b) $c_1 = 5/2$, $c_2 = -1/2$ in solution in (a)
 (c) Graph approaches origin in the first quadrant tangent to the line $x_1 = x_2$.

8. $x_1 = \frac{11}{3} e^{2t} - \frac{2}{3} e^{-t}$, $x_2 = \frac{11}{6} e^{2t} - \frac{4}{3} e^{-t}$
 Graph is asymptotic to the line $x_1 = 2x_2$ in the first quadrant.

9. $x_1 = -\frac{3}{2} e^{t/2} - \frac{1}{2} e^{2t}$, $x_2 = \frac{3}{2} e^{t/2} - \frac{1}{2} e^{2t}$
 Graph is asymptotic to the line $x_1 = x_2$ in the third quadrant.

10. $x_1 = -7 e^{-t} + 6 e^{-2t}$, $x_2 = -7 e^{-t} + 9 e^{-2t}$
 Graph approaches the origin in the third quadrant tangent to the line $x_1 = x_2$.

11. $x_1 = 3 \cos 2t + 4 \sin 2t$, $x_2 = -3 \sin 2t + 4 \cos 2t$
 Graph is a circle, center at origin, radius 5, traversed clockwise.

12. $x_1 = -2 e^{-t/2} \cos 2t + 2 e^{-t/2} \sin 2t$, $x_2 = 2 e^{-t/2} \cos 2t + 2 e^{-t/2} \sin 2t$
 Graph is a clockwise spiral, approaching the origin.

13. $LRCI'' + LI' + RI = 0$

18. $y_1' = y_3$, $y_2' = y_4$, $m_1 y_3' = -(k_1 + k_2) y_1 + k_2 y_2 + F_1(t)$,
 $m_2 y_4' = k_2 y_1 - (k_2 + k_3) y_2 + F_2(t)$

22. (a) $Q_1' = \frac{3}{2} - \frac{1}{10} Q_1 + \frac{3}{40} Q_2$, $Q_1(0) = 25$
 $Q_2' = 3 + \frac{1}{10} Q_1 - \frac{1}{5} Q_2$, $Q_2(0) = 15$
 (b) $Q_1^E = 42$, $Q_2^E = 36$
 (c) $x_1' = -\frac{1}{10} x_1 + \frac{3}{40} x_2$, $x_1(0) = -17$
 $x_2' = \frac{1}{10} x_1 - \frac{1}{5} x_2$, $x_2(0) = -21$

23. (a) $Q_1' = 3 q_1 - \frac{1}{15} Q_1 + \frac{1}{100} Q_2$, $Q_1(0) = Q_1^0$
 $Q_2' = q_2 + \frac{1}{30} Q_1 - \frac{3}{100} Q_2$, $Q_2(0) = Q_2^0$
 (b) $Q_1^E = 6(9 q_1 + q_2)$, $Q_2^E = 20(3 q_1 + 2 q_2)$
 (c) No (d) $\frac{10}{9} \leq Q_2^E / Q_1^E \leq \frac{20}{3}$

Section 7.2, page 372

1. (a) $\begin{pmatrix} 6 & -6 & 3 \\ 5 & 9 & -2 \\ 2 & 3 & 8 \end{pmatrix}$ (b) $\begin{pmatrix} -15 & 6 & -12 \\ 7 & -18 & -1 \\ -26 & -3 & -5 \end{pmatrix}$

 (c) $\begin{pmatrix} 6 & -12 & 3 \\ 4 & 3 & 7 \\ 9 & 12 & 0 \end{pmatrix}$ (d) $\begin{pmatrix} -8 & -9 & 11 \\ 14 & 12 & -5 \\ 5 & -8 & 5 \end{pmatrix}$

2. (a) $\begin{pmatrix} 1 - i & -7 + 2i \\ -1 + 2i & 2 + 3i \end{pmatrix}$ (b) $\begin{pmatrix} 3 + 4i & 6i \\ 11 + 6i & 6 - 5i \end{pmatrix}$

 (c) $\begin{pmatrix} -3 + 5i & 7 + 5i \\ 2 + i & 7 + 2i \end{pmatrix}$ (d) $\begin{pmatrix} 8 + 7i & 4 - 4i \\ 6 - 4i & -4 \end{pmatrix}$

3. (a) $\begin{pmatrix} -2 & 1 & 2 \\ 1 & 0 & -1 \\ 2 & -3 & 1 \end{pmatrix}$ (b) $\begin{pmatrix} 1 & 3 & -2 \\ 2 & -1 & 1 \\ 3 & -1 & 0 \end{pmatrix}$ (c), (d) $\begin{pmatrix} -1 & 4 & 0 \\ 3 & -1 & 0 \\ 5 & -4 & 1 \end{pmatrix}$

4. (a) $\begin{pmatrix} 3 - 2i & 2 - i \\ 1 + i & -2 + 3i \end{pmatrix}$ (b) $\begin{pmatrix} 3 + 2i & 1 - i \\ 2 + i & -2 - 3i \end{pmatrix}$ (c) $\begin{pmatrix} 3 + 2i & 2 + i \\ 1 - i & -2 - 3i \end{pmatrix}$

5. $\begin{pmatrix} 10 & 6 & -4 \\ 0 & 4 & 10 \\ 4 & 4 & 6 \end{pmatrix}$

6. (a) $\begin{pmatrix} 7 & -11 & -3 \\ 11 & 20 & 17 \\ -4 & 3 & -12 \end{pmatrix}$ (b) $\begin{pmatrix} 5 & 0 & -1 \\ 2 & 7 & 4 \\ -1 & 1 & 4 \end{pmatrix}$ (c) $\begin{pmatrix} 6 & -8 & -11 \\ 9 & 15 & 6 \\ -5 & -1 & 5 \end{pmatrix}$

8. (a) $4i$ (b) $12 - 8i$ (c) $2 + 2i$ (d) 16

10. $\begin{pmatrix} \frac{3}{11} & -\frac{4}{11} \\ \frac{2}{11} & \frac{1}{11} \end{pmatrix}$

11. $\begin{pmatrix} \frac{1}{6} & \frac{1}{12} \\ -\frac{1}{2} & \frac{1}{4} \end{pmatrix}$

12. $\begin{pmatrix} 1 & -3 & 2 \\ -3 & 3 & -1 \\ 2 & -1 & 0 \end{pmatrix}$

13. $\begin{pmatrix} \frac{1}{3} & \frac{1}{3} & 0 \\ \frac{1}{3} & -\frac{1}{3} & \frac{1}{3} \\ -\frac{1}{3} & 0 & \frac{1}{3} \end{pmatrix}$

14. Singular

15. $\begin{pmatrix} \frac{1}{2} & -\frac{1}{4} & \frac{1}{8} \\ 0 & \frac{1}{2} & -\frac{1}{4} \\ 0 & 0 & \frac{1}{2} \end{pmatrix}$

16. $\begin{pmatrix} \frac{1}{10} & \frac{3}{10} & \frac{1}{10} \\ -\frac{2}{10} & \frac{4}{10} & -\frac{2}{10} \\ -\frac{7}{10} & -\frac{1}{10} & \frac{3}{10} \end{pmatrix}$

17. Singular

18. $\begin{pmatrix} 1 & 1 & 0 & 1 \\ 1 & 0 & 1 & 1 \\ 1 & 1 & 1 & 1 \\ 0 & 1 & 0 & 1 \end{pmatrix}$

19. $\begin{pmatrix} 6 & \frac{13}{5} & -\frac{8}{5} & \frac{2}{5} \\ 5 & \frac{11}{5} & -\frac{6}{5} & \frac{4}{5} \\ 0 & -\frac{1}{5} & \frac{1}{5} & \frac{1}{5} \\ -2 & -\frac{4}{5} & \frac{4}{5} & -\frac{1}{5} \end{pmatrix}$

21. (a) $\begin{pmatrix} 7e^t & 5e^{-t} & 10e^{2t} \\ -e^t & 7e^{-t} & 2e^{2t} \\ 8e^t & 0 & -e^{2t} \end{pmatrix}$

(b) $\begin{pmatrix} 2e^{2t} - 2 + 3e^{3t} & 1 + 4e^{-2t} - e^t & 3e^{3t} + 2e^t - e^{4t} \\ 4e^{2t} - 1 - 3e^{3t} & 2 + 2e^{-2t} + e^t & 6e^{3t} + e^t + e^{4t} \\ -2e^{2t} - 3 + 6e^{3t} & -1 + 6e^{-2t} - 2e^t & -3e^{3t} + 3e^t - 2e^{4t} \end{pmatrix}$

(c) $\begin{pmatrix} e^t & -2e^{-t} & 2e^{2t} \\ 2e^t & -e^{-t} & -2e^{2t} \\ -e^t & -3e^{-t} & 4e^{2t} \end{pmatrix}$

(d) $(e-1)\begin{pmatrix} 1 & 2e^{-1} & \frac{1}{2}(e+1) \\ 2 & e^{-1} & -\frac{1}{2}(e+1) \\ -1 & 3e^{-1} & e+1 \end{pmatrix}$

Section 7.3, page 383

1. $x_1 = -\frac{1}{3}$, $x_2 = \frac{7}{3}$, $x_3 = -\frac{1}{3}$ 2. No solution
3. $x_1 = -c$, $x_2 = c + 1$, $x_3 = c$, where c is arbitrary
4. $x_1 = c$, $x_2 = -c$, $x_3 = -c$, where c is arbitrary
5. $x_1 = 0$, $x_2 = 0$, $x_3 = 0$ 6. Linearly independent
7. $\mathbf{x}^{(1)} - 5\mathbf{x}^{(2)} + 2\mathbf{x}^{(3)} = \mathbf{0}$ 8. $2\mathbf{x}^{(1)} - 3\mathbf{x}^{(2)} + 4\mathbf{x}^{(3)} - \mathbf{x}^{(4)} = \mathbf{0}$
9. Linearly independent 10. $\mathbf{x}^{(1)} + \mathbf{x}^{(2)} - \mathbf{x}^{(4)} = \mathbf{0}$
12. $3\mathbf{x}^{(1)}(t) - 6\mathbf{x}^{(2)}(t) + \mathbf{x}^{(3)}(t) = \mathbf{0}$ 13. Linearly independent

15. $\lambda_1 = 2$, $\mathbf{x}^{(1)} = \begin{pmatrix} 1 \\ 3 \end{pmatrix}$; $\lambda_2 = 4$, $\mathbf{x}^{(2)} = \begin{pmatrix} 1 \\ 1 \end{pmatrix}$

16. $\lambda_1 = 1 + 2i$, $\mathbf{x}^{(1)} = \begin{pmatrix} 1 \\ 1 - i \end{pmatrix}$; $\lambda_2 = 1 - 2i$, $\mathbf{x}^{(2)} = \begin{pmatrix} 1 \\ 1 + i \end{pmatrix}$

17. $\lambda_1 = -3$, $\mathbf{x}^{(1)} = \begin{pmatrix} 1 \\ -1 \end{pmatrix}$; $\lambda_2 = -1$, $\mathbf{x}^{(2)} = \begin{pmatrix} 1 \\ 1 \end{pmatrix}$

18. $\lambda_1 = 0$, $\mathbf{x}^{(1)} = \begin{pmatrix} 1 \\ i \end{pmatrix}$; $\lambda_2 = 2$, $\mathbf{x}^{(2)} = \begin{pmatrix} 1 \\ -i \end{pmatrix}$

19. $\lambda_1 = 2$, $\mathbf{x}^{(1)} = \begin{pmatrix} \sqrt{3} \\ 1 \end{pmatrix}$; $\lambda_2 = -2$, $\mathbf{x}^{(2)} = \begin{pmatrix} 1 \\ -\sqrt{3} \end{pmatrix}$

20. $\lambda_1 = -1/2$, $\mathbf{x}^{(1)} = \begin{pmatrix} 3 \\ 10 \end{pmatrix}$; $\lambda_2 = -3/2$, $\mathbf{x}^{(2)} = \begin{pmatrix} 1 \\ 2 \end{pmatrix}$

21. $\lambda_1 = 1$, $\mathbf{x}^{(1)} = \begin{pmatrix} 2 \\ -3 \\ 2 \end{pmatrix}$; $\lambda_2 = 1 + 2i$, $\mathbf{x}^{(2)} = \begin{pmatrix} 0 \\ 1 \\ -i \end{pmatrix}$; $\lambda_3 = 1 - 2i$, $\mathbf{x}^{(3)} = \begin{pmatrix} 0 \\ 1 \\ i \end{pmatrix}$

22. $\lambda_1 = 1$, $\mathbf{x}^{(1)} = \begin{pmatrix} 1 \\ 0 \\ -1 \end{pmatrix}$; $\lambda_2 = 2$, $\mathbf{x}^{(2)} = \begin{pmatrix} -2 \\ 1 \\ 0 \end{pmatrix}$; $\lambda_3 = 3$, $\mathbf{x}^{(3)} = \begin{pmatrix} 0 \\ 1 \\ -1 \end{pmatrix}$

23. $\lambda_1 = 1$, $\mathbf{x}^{(1)} = \begin{pmatrix} 2 \\ -2 \\ -1 \end{pmatrix}$; $\lambda_2 = 2$, $\mathbf{x}^{(2)} = \begin{pmatrix} 2 \\ 1 \\ 2 \end{pmatrix}$; $\lambda_3 = -1$, $\mathbf{x}^{(3)} = \begin{pmatrix} 1 \\ 2 \\ -2 \end{pmatrix}$

24. $\lambda_1 = -1$, $\mathbf{x}^{(1)} = \begin{pmatrix} 1 \\ -4 \\ 1 \end{pmatrix}$; $\lambda_2 = -1$, $\mathbf{x}^{(2)} = \begin{pmatrix} 1 \\ 0 \\ -1 \end{pmatrix}$; $\lambda_3 = 8$, $\mathbf{x}^{(3)} = \begin{pmatrix} 2 \\ 1 \\ 2 \end{pmatrix}$

Section 7.4, page 389

2. (c) $W(t) = c \exp \int [p_{11}(t) + p_{22}(t)]\, dt$

6. (a) $W(t) = t^2$

(b) $\mathbf{x}^{(1)}$ and $\mathbf{x}^{(2)}$ are linearly independent at each point except $t = 0$; they are linearly independent on every interval.

(c) At least one coefficient must be discontinuous at $t = 0$.

(d) $\mathbf{x}' = \begin{pmatrix} 0 & 1 \\ -2t^{-2} & 2t^{-1} \end{pmatrix} \mathbf{x}$

7. (a) $W(t) = t(t - 2)e^t$

(b) $\mathbf{x}^{(1)}$ and $\mathbf{x}^{(2)}$ are linearly independent at each point except $t = 0$ and $t = 2$; they are linearly independent on every interval.

(c) There must be at least one discontinuous coefficient at $t = 0$ and $t = 2$.

(d) $\mathbf{x}' = \begin{pmatrix} 0 & 1 \\ \dfrac{2 - 2t}{t^2 - 2t} & \dfrac{t^2 - 2}{t^2 - 2t} \end{pmatrix} \mathbf{x}$

Section 7.5, page 398

1. $\mathbf{x} = c_1 \begin{pmatrix} 1 \\ 2 \end{pmatrix} e^{-t} + c_2 \begin{pmatrix} 2 \\ 1 \end{pmatrix} e^{2t}$
2. $\mathbf{x} = c_1 \begin{pmatrix} 1 \\ 1 \end{pmatrix} e^{-t} + c_2 \begin{pmatrix} 2 \\ 3 \end{pmatrix} e^{-2t}$

3. $\mathbf{x} = c_1 \begin{pmatrix} 1 \\ 1 \end{pmatrix} e^{t} + c_2 \begin{pmatrix} 1 \\ 3 \end{pmatrix} e^{-t}$
4. $\mathbf{x} = c_1 \begin{pmatrix} 1 \\ -4 \end{pmatrix} e^{-3t} + c_2 \begin{pmatrix} 1 \\ 1 \end{pmatrix} e^{2t}$

5. $\mathbf{x} = c_1 \begin{pmatrix} 1 \\ -1 \end{pmatrix} e^{-3t} + c_2 \begin{pmatrix} 1 \\ 1 \end{pmatrix} e^{-t}$
6. $\mathbf{x} = c_1 \begin{pmatrix} 1 \\ -1 \end{pmatrix} e^{t/2} + c_2 \begin{pmatrix} 1 \\ 1 \end{pmatrix} e^{2t}$

7. $\mathbf{x} = c_1 \begin{pmatrix} 3 \\ 4 \end{pmatrix} + c_2 \begin{pmatrix} 1 \\ 2 \end{pmatrix} e^{-2t}$

8. $\mathbf{x} = c_1 \begin{pmatrix} -2 \\ 1 \end{pmatrix} + c_2 \begin{pmatrix} -3 \\ 1 \end{pmatrix} e^t$

9. $\mathbf{x} = c_1 \begin{pmatrix} 1 \\ i \end{pmatrix} + c_2 \begin{pmatrix} 1 \\ -i \end{pmatrix} e^{2t}$

10. $\mathbf{x} = c_1 \begin{pmatrix} 2+i \\ -1 \end{pmatrix} e^t + c_2 \begin{pmatrix} 1 \\ -1 \end{pmatrix} e^{-it}$

11. $\mathbf{x} = c_1 \begin{pmatrix} 1 \\ 1 \\ 1 \end{pmatrix} e^{4t} + c_2 \begin{pmatrix} 1 \\ -2 \\ 1 \end{pmatrix} e^t + c_3 \begin{pmatrix} 1 \\ 0 \\ -1 \end{pmatrix} e^{-t}$

12. $\mathbf{x} = c_1 \begin{pmatrix} 1 \\ -4 \\ 1 \end{pmatrix} e^{-t} + c_2 \begin{pmatrix} 1 \\ 0 \\ -1 \end{pmatrix} e^{-t} + c_3 \begin{pmatrix} 2 \\ 1 \\ 2 \end{pmatrix} e^{8t}$

13. $\mathbf{x} = c_1 \begin{pmatrix} 4 \\ -5 \\ -7 \end{pmatrix} e^{-2t} + c_2 \begin{pmatrix} 3 \\ -4 \\ -2 \end{pmatrix} e^{-t} + c_3 \begin{pmatrix} 0 \\ 1 \\ -1 \end{pmatrix} e^{2t}$

14. $\mathbf{x} = c_1 \begin{pmatrix} 1 \\ -4 \\ -1 \end{pmatrix} e^t + c_2 \begin{pmatrix} 1 \\ -1 \\ -1 \end{pmatrix} e^{-2t} + c_3 \begin{pmatrix} 1 \\ 2 \\ 1 \end{pmatrix} e^{3t}$

15. $\mathbf{x} = -\dfrac{3}{2} \begin{pmatrix} 1 \\ 3 \end{pmatrix} e^{2t} + \dfrac{7}{2} \begin{pmatrix} 1 \\ 1 \end{pmatrix} e^{4t}$

16. $\mathbf{x} = \dfrac{1}{2} \begin{pmatrix} 1 \\ 1 \end{pmatrix} e^{-t} + \dfrac{1}{2} \begin{pmatrix} 1 \\ 5 \end{pmatrix} e^{3t}$

17. $\mathbf{x} = \begin{pmatrix} 0 \\ -2 \\ 1 \end{pmatrix} e^t + 2 \begin{pmatrix} 1 \\ 1 \\ 0 \end{pmatrix} e^{2t}$

18. $\mathbf{x} = 6 \begin{pmatrix} 1 \\ 2 \\ -1 \end{pmatrix} e^t + 3 \begin{pmatrix} 1 \\ -2 \\ 1 \end{pmatrix} e^{-t} - \begin{pmatrix} 2 \\ 1 \\ -8 \end{pmatrix} e^{4t}$

20. $\mathbf{x} = c_1 \begin{pmatrix} 1 \\ 1 \end{pmatrix} t + c_2 \begin{pmatrix} 1 \\ 3 \end{pmatrix} t^{-1}$

21. $\mathbf{x} = c_1 \begin{pmatrix} 1 \\ 3 \end{pmatrix} t^2 + c_2 \begin{pmatrix} 1 \\ 1 \end{pmatrix} t^4$

22. $\mathbf{x} = c_1 \begin{pmatrix} 3 \\ 4 \end{pmatrix} + c_2 \begin{pmatrix} 1 \\ 2 \end{pmatrix} t^{-2}$

23. $\mathbf{x} = c_1 \begin{pmatrix} 1 \\ 2 \end{pmatrix} t^{-1} + c_2 \begin{pmatrix} 2 \\ 1 \end{pmatrix} t^2$

29. (a) $x_1' = x_2, \quad x_2' = -(c/a)x_1 - (b/a)x_2$

30. (a) $\mathbf{x} = -\dfrac{55}{8} \begin{pmatrix} 3 \\ 2 \end{pmatrix} e^{-t/20} + \dfrac{29}{8} \begin{pmatrix} 1 \\ -2 \end{pmatrix} e^{-t/4}$

(c) $T \cong 74.39$

31. (a) $\mathbf{x} = c_1 \begin{pmatrix} -\sqrt{2} \\ 1 \end{pmatrix} e^{(-2+\sqrt{2})t/2} + c_2 \begin{pmatrix} \sqrt{2} \\ 1 \end{pmatrix} e^{(-2-\sqrt{2})t/2}; \quad r_{1,2} = (-2 \pm \sqrt{2})/2; \quad$ node

(b) $\mathbf{x} = c_1 \begin{pmatrix} -1 \\ \sqrt{2} \end{pmatrix} e^{(-1+\sqrt{2})t} + c_2 \begin{pmatrix} 1 \\ \sqrt{2} \end{pmatrix} e^{(-1-\sqrt{2})t}; \quad r_{1,2} = -1 \pm \sqrt{2}; \quad$ saddle point

(c) $r_{1,2} = -1 \pm \sqrt{\alpha}; \quad \alpha = 1$

32. (a) $\begin{pmatrix} I \\ V \end{pmatrix} = c_1 \begin{pmatrix} 1 \\ 3 \end{pmatrix} e^{-2t} + c_2 \begin{pmatrix} 1 \\ 1 \end{pmatrix} e^{-t}$

33. (a) $\left(\dfrac{1}{CR_2} - \dfrac{R_1}{L} \right)^2 - \dfrac{4}{CL} > 0$

Section 7.6, page 410

1. $\mathbf{x} = c_1 e^t \begin{pmatrix} \cos 2t \\ \cos 2t + \sin 2t \end{pmatrix} + c_2 e^t \begin{pmatrix} \sin 2t \\ -\cos 2t + \sin 2t \end{pmatrix}$

2. $\mathbf{x} = c_1 e^{-t} \begin{pmatrix} 2\cos 2t \\ \sin 2t \end{pmatrix} + c_2 e^{-t} \begin{pmatrix} -2\sin 2t \\ \cos 2t \end{pmatrix}$

3. $\mathbf{x} = c_1 \begin{pmatrix} 5\cos t \\ 2\cos t + \sin t \end{pmatrix} + c_2 \begin{pmatrix} 5\sin t \\ -\cos t + 2\sin t \end{pmatrix}$

4. $\mathbf{x} = c_1 e^{t/2} \begin{pmatrix} 5\cos \frac{3}{2}t \\ 3(\cos \frac{3}{2}t + \sin \frac{3}{2}t) \end{pmatrix} + c_2 e^{t/2} \begin{pmatrix} 5\sin \frac{3}{2}t \\ 3(-\cos \frac{3}{2}t + \sin \frac{3}{2}t) \end{pmatrix}$

5. $\mathbf{x} = c_1 e^{-t} \begin{pmatrix} \cos t \\ 2\cos t + \sin t \end{pmatrix} + c_2 e^{-t} \begin{pmatrix} \sin t \\ -\cos t + 2\sin t \end{pmatrix}$

6. $\mathbf{x} = c_1 \begin{pmatrix} -2\cos 3t \\ \cos 3t + 3\sin 3t \end{pmatrix} + c_2 \begin{pmatrix} -2\sin 3t \\ \sin 3t - 3\cos 3t \end{pmatrix}$

7. $\mathbf{x} = c_1 \begin{pmatrix} 2 \\ -3 \\ 2 \end{pmatrix} e^t + c_2 e^t \begin{pmatrix} 0 \\ \cos 2t \\ \sin 2t \end{pmatrix} + c_3 e^t \begin{pmatrix} 0 \\ \sin 2t \\ -\cos 2t \end{pmatrix}$

8. $\mathbf{x} = c_1 \begin{pmatrix} 2 \\ -2 \\ 1 \end{pmatrix} e^{-2t} + c_2 e^{-t} \begin{pmatrix} -\sqrt{2}\sin\sqrt{2}t \\ \cos\sqrt{2}t \\ -\cos\sqrt{2}t - \sqrt{2}\sin\sqrt{2}t \end{pmatrix} + c_3 e^{-t} \begin{pmatrix} \sqrt{2}\cos\sqrt{2}t \\ \sin\sqrt{2}t \\ \sqrt{2}\cos\sqrt{2}t - \sin\sqrt{2}t \end{pmatrix}$

9. $\mathbf{x} = e^{-t} \begin{pmatrix} \cos t - 3\sin t \\ \cos t - \sin t \end{pmatrix}$

10. $\mathbf{x} = e^{-2t} \begin{pmatrix} \cos t - 5\sin t \\ -2\cos t - 3\sin t \end{pmatrix}$

11. (a) $r = -\frac{1}{4} \pm i$

12. (a) $r = \frac{1}{5} \pm i$
 (b) $\alpha = 0$

13. (a) $r = \alpha \pm i$

14. (a) $r = (\alpha \pm \sqrt{\alpha^2 - 20})/2$
 (b) $\alpha = -\sqrt{20},\ 0,\ \sqrt{20}$

15. (a) $r = \pm\sqrt{4 - 5\alpha}$
 (b) $\alpha = 4/5$

16. (a) $r = \frac{5}{4} \pm \frac{1}{2}\sqrt{3\alpha}$
 (b) $\alpha = 0,\ 25/12$

17. (a) $r = -1 \pm \sqrt{-\alpha}$
 (b) $\alpha = -1,\ 0$

18. (a) $r = -\frac{1}{2} \pm \frac{1}{2}\sqrt{49 - 24\alpha}$
 (b) $\alpha = 2,\ 49/24$

19. (a) $r = \frac{1}{2}\alpha - 2 \pm \sqrt{\alpha^2 + 8\alpha - 24}$
 (b) $\alpha = -4 - 2\sqrt{10},\ -4 + 2\sqrt{10},\ 5/2$

20. (a) $r = -1 \pm \sqrt{25 + 8\alpha}$
 (b) $\alpha = -25/8,\ -3$

21. $\mathbf{x} = c_1 t^{-1} \begin{pmatrix} \cos(\sqrt{2}\ln t) \\ \sqrt{2}\sin(\sqrt{2}\ln t) \end{pmatrix} + c_2 t^{-1} \begin{pmatrix} \sin(\sqrt{2}\ln t) \\ -\sqrt{2}\cos(\sqrt{2}\ln t) \end{pmatrix}$

22. $\mathbf{x} = c_1 \begin{pmatrix} 5\cos(\ln t) \\ 2\cos(\ln t) + \sin(\ln t) \end{pmatrix} + c_2 \begin{pmatrix} 5\sin(\ln t) \\ -\cos(\ln t) + 2\sin(\ln t) \end{pmatrix}$

23. (a) $r = -\frac{1}{4} \pm i,\ -\frac{1}{4}$

24. (a) $r = -\frac{1}{4} \pm i,\ \frac{1}{10}$

25. (b) $\begin{pmatrix} I \\ V \end{pmatrix} = c_1 e^{-t/2} \begin{pmatrix} \cos(t/2) \\ 4\sin(t/2) \end{pmatrix} + c_2 e^{-t/2} \begin{pmatrix} \sin(t/2) \\ -4\cos(t/2) \end{pmatrix}$
 (c) Use $c_1 = 2$, $c_2 = -\frac{3}{4}$ in answer to part (b).
 (d) $\lim_{t \to \infty} I(t) = \lim_{t \to \infty} V(t) = 0$; no

26. (b) $\begin{pmatrix} I \\ V \end{pmatrix} = c_1 e^{-t} \begin{pmatrix} \cos t \\ -\cos t - \sin t \end{pmatrix} + c_2 e^{-t} \begin{pmatrix} \sin t \\ -\sin t + \cos t \end{pmatrix}$
 (c) Use $c_1 = 2$ and $c_2 = 3$ in answer to part (b).
 (d) $\lim_{t \to \infty} I(t) = \lim_{t \to \infty} V(t) = 0$; no

28. (b) $r = \pm i\sqrt{k/m}$
 (d) $|r|$ is the natural frequency.

29. (c) $r_1^2 = -1$, $\boldsymbol{\xi}^{(1)} = \begin{pmatrix} 3 \\ 2 \end{pmatrix}$; $r_2^2 = -4$, $\boldsymbol{\xi}^{(2)} = \begin{pmatrix} 3 \\ -4 \end{pmatrix}$
 (d) $x_1 = 3c_1\cos t + 3c_2\sin t + 3c_3\cos 2t + 3c_4\sin 2t$,
 $x_2 = 2c_1\cos t + 2c_2\sin t - 4c_3\cos 2t - 4c_4\sin 2t$

(e) $x_1' = -3c_1 \sin t + 3c_2 \cos t - 6c_3 \sin 2t + 6c_4 \cos 2t$,
$x_2' = -2c_1 \sin t + 2c_2 \cos t + 8c_3 \sin 2t - 8c_4 \cos 2t$

30. (a) $\mathbf{A} = \begin{pmatrix} 0 & 0 & 1 & 0 \\ 0 & 0 & 0 & 1 \\ -4 & 3 & 0 & 0 \\ 9/4 & -13/4 & 0 & 0 \end{pmatrix}$

(b) $r_1 = i$, $\boldsymbol{\xi}^{(1)} = \begin{pmatrix} 1 \\ 1 \\ i \\ i \end{pmatrix}$; $r_2 = -i$, $\boldsymbol{\xi}^{(2)} = \begin{pmatrix} 1 \\ 1 \\ -i \\ -i \end{pmatrix}$;

$r_3 = \frac{5}{2}i$, $\boldsymbol{\xi}^{(3)} = \begin{pmatrix} 4 \\ -3 \\ 10i \\ -\frac{15}{2}i \end{pmatrix}$; $r_4 = -\frac{5}{2}i$, $\boldsymbol{\xi}^{(4)} = \begin{pmatrix} 4 \\ -3 \\ -10i \\ \frac{15}{2}i \end{pmatrix}$

(c) $\mathbf{y} = c_1 \begin{pmatrix} \cos t \\ \cos t \\ -\sin t \\ -\sin t \end{pmatrix} + c_2 \begin{pmatrix} \sin t \\ \sin t \\ \cos t \\ \cos t \end{pmatrix} + c_3 \begin{pmatrix} 4\cos\frac{5}{2}t \\ -3\cos\frac{5}{2}t \\ -10\sin\frac{5}{2}t \\ \frac{15}{2}\sin\frac{5}{2}t \end{pmatrix} + c_4 \begin{pmatrix} 4\sin\frac{5}{2}t \\ -3\sin\frac{5}{2}t \\ 10\cos\frac{5}{2}t \\ -\frac{15}{2}\cos\frac{5}{2}t \end{pmatrix}$

(e) $c_1 = \frac{10}{7}$, $c_2 = 0$, $c_3 = \frac{1}{7}$, $c_4 = 0$. period $= 4\pi$.

31. (a) $\mathbf{A} = \begin{pmatrix} 0 & 0 & 1 & 0 \\ 0 & 0 & 0 & 1 \\ -2 & 1 & 0 & 0 \\ 1 & -2 & 0 & 0 \end{pmatrix}$

(b) $r_1 = i$, $\boldsymbol{\xi}^{(1)} = \begin{pmatrix} 1 \\ 1 \\ i \\ i \end{pmatrix}$; $r_2 = -i$, $\boldsymbol{\xi}^{(2)} = \begin{pmatrix} 1 \\ 1 \\ -i \\ -i \end{pmatrix}$;

$r_3 = \sqrt{3}i$, $\boldsymbol{\xi}^{(3)} = \begin{pmatrix} 1 \\ -1 \\ \sqrt{3}i \\ -\sqrt{3}i \end{pmatrix}$; $r_4 = -\sqrt{3}i$, $\boldsymbol{\xi}^{(4)} = \begin{pmatrix} 1 \\ -1 \\ -\sqrt{3}i \\ \sqrt{3}i \end{pmatrix}$

(c) $\mathbf{y} = c_1 \begin{pmatrix} \cos t \\ \cos t \\ -\sin t \\ -\sin t \end{pmatrix} + c_2 \begin{pmatrix} \sin t \\ \sin t \\ \cos t \\ \cos t \end{pmatrix} + c_3 \begin{pmatrix} \cos\sqrt{3}t \\ -\cos\sqrt{3}t \\ -\sqrt{3}\sin\sqrt{3}t \\ \sqrt{3}\sin\sqrt{3}t \end{pmatrix} + c_4 \begin{pmatrix} \sin\sqrt{3}t \\ -\sin\sqrt{3}t \\ \sqrt{3}\cos\sqrt{3}t \\ -\sqrt{3}\cos\sqrt{3}t \end{pmatrix}$

(e) $c_1 = 1$, $c_2 = 0$, $c_3 = -2$, $c_4 = 0$.

Section 7.7, page 420

1. $\mathbf{\Phi}(t) = \begin{pmatrix} -\frac{1}{3}e^{-t} + \frac{4}{3}e^{2t} & \frac{2}{3}e^{-t} - \frac{2}{3}e^{2t} \\ -\frac{2}{3}e^{-t} + \frac{2}{3}e^{2t} & \frac{4}{3}e^{-t} - \frac{1}{3}e^{2t} \end{pmatrix}$ 2. $\mathbf{\Phi}(t) = \begin{pmatrix} \frac{1}{2}e^{-t/2} + \frac{1}{2}e^{-t} & e^{-t/2} - e^{-t} \\ \frac{1}{4}e^{-t/2} - \frac{1}{4}e^{-t} & \frac{1}{2}e^{-t/2} + \frac{1}{2}e^{-t} \end{pmatrix}$

3. $\mathbf{\Phi}(t) = \begin{pmatrix} \frac{3}{2}e^{t} - \frac{1}{2}e^{-t} & -\frac{1}{2}e^{t} + \frac{1}{2}e^{-t} \\ \frac{3}{2}e^{t} - \frac{3}{2}e^{-t} & -\frac{1}{2}e^{t} + \frac{3}{2}e^{-t} \end{pmatrix}$ 4. $\mathbf{\Phi}(t) = \begin{pmatrix} \frac{1}{5}e^{-3t} + \frac{4}{5}e^{2t} & -\frac{1}{5}e^{-3t} + \frac{1}{5}e^{2t} \\ -\frac{4}{5}e^{-3t} + \frac{4}{5}e^{2t} & \frac{4}{5}e^{-3t} + \frac{1}{5}e^{2t} \end{pmatrix}$

5. $\mathbf{\Phi}(t) = \begin{pmatrix} \cos t + 2\sin t & -5\sin t \\ \sin t & \cos t - 2\sin t \end{pmatrix}$

6. $\Phi(t) = \begin{pmatrix} e^{-t}\cos 2t & -2e^{-t}\sin 2t \\ \frac{1}{2}e^{-t}\sin 2t & e^{-t}\cos 2t \end{pmatrix}$

7. $\Phi(t) = \begin{pmatrix} -\frac{1}{2}e^{2t} + \frac{3}{2}e^{4t} & \frac{1}{2}e^{2t} - \frac{1}{2}e^{4t} \\ -\frac{3}{2}e^{2t} + \frac{3}{2}e^{4t} & \frac{3}{2}e^{2t} - \frac{1}{2}e^{4t} \end{pmatrix}$

8. $\Phi(t) = \begin{pmatrix} e^{-t}\cos t + 2e^{-t}\sin t & e^{-t}\sin t \\ 5e^{-t}\sin t & e^{-t}\cos t - 2e^{-t}\sin t \end{pmatrix}$

9. $\Phi(t) = \begin{pmatrix} -2e^{-2t} + 3e^{-t} & -e^{-2t} + e^{-t} & -e^{-2t} + e^{-t} \\ \frac{5}{2}e^{-2t} - 4e^{-t} + \frac{3}{2}e^{2t} & \frac{5}{4}e^{-2t} - \frac{4}{3}e^{-t} + \frac{13}{12}e^{2t} & \frac{5}{4}e^{-2t} - \frac{4}{3}e^{-t} + \frac{1}{12}e^{2t} \\ \frac{7}{2}e^{-2t} - 2e^{-t} - \frac{3}{2}e^{2t} & \frac{7}{4}e^{-2t} - \frac{2}{3}e^{-t} - \frac{13}{12}e^{2t} & \frac{7}{4}e^{-2t} - \frac{2}{3}e^{-t} - \frac{1}{12}e^{2t} \end{pmatrix}$

10. $\Phi(t) = \begin{pmatrix} \frac{1}{6}e^{t} + \frac{1}{3}e^{-2t} + \frac{1}{2}e^{3t} & -\frac{1}{3}e^{t} + \frac{1}{3}e^{-2t} & \frac{1}{2}e^{t} - e^{-2t} + \frac{1}{2}e^{3t} \\ -\frac{2}{3}e^{t} - \frac{1}{3}e^{-2t} + e^{3t} & \frac{4}{3}e^{t} - \frac{1}{3}e^{-2t} & -2e^{t} + e^{-2t} + e^{3t} \\ -\frac{1}{6}e^{t} - \frac{1}{3}e^{-2t} + \frac{1}{2}e^{3t} & \frac{1}{3}e^{t} - \frac{1}{3}e^{-2t} & -\frac{1}{2}e^{t} + e^{-2t} + \frac{1}{2}e^{3t} \end{pmatrix}$

11. $\mathbf{x} = \frac{7}{2}\begin{pmatrix} 1 \\ 1 \end{pmatrix}e^{t} - \frac{3}{2}\begin{pmatrix} 1 \\ 3 \end{pmatrix}e^{-t}$

12. $\mathbf{x} = \begin{pmatrix} 3 \\ 1 \end{pmatrix}e^{-t}\cos 2t + \begin{pmatrix} -2 \\ 3/2 \end{pmatrix}e^{-t}\sin 2t$

17. (c) $\mathbf{x} = \begin{pmatrix} u_0 \\ v_0 \end{pmatrix}\cos \omega t + \begin{pmatrix} v_0 \\ -\omega^2 u_0 \end{pmatrix}\frac{\sin \omega t}{\omega}$

Section 7.8, page 428

1. $\mathbf{x} = c_1\begin{pmatrix} 2 \\ 1 \end{pmatrix}e^{t} + c_2\left[\begin{pmatrix} 2 \\ 1 \end{pmatrix}te^{t} + \begin{pmatrix} 1 \\ 0 \end{pmatrix}e^{t}\right]$

2. $\mathbf{x} = c_1\begin{pmatrix} 1 \\ 2 \end{pmatrix} + c_2\left[\begin{pmatrix} 1 \\ 2 \end{pmatrix}t - \begin{pmatrix} 0 \\ \frac{1}{2} \end{pmatrix}\right]$

3. $\mathbf{x} = c_1\begin{pmatrix} 2 \\ 1 \end{pmatrix}e^{-t} + c_2\left[\begin{pmatrix} 2 \\ 1 \end{pmatrix}te^{-t} + \begin{pmatrix} 0 \\ 2 \end{pmatrix}e^{-t}\right]$

4. $\mathbf{x} = c_1\begin{pmatrix} 1 \\ 1 \end{pmatrix}e^{-t/2} + c_2\left[\begin{pmatrix} 1 \\ 1 \end{pmatrix}te^{-t/2} + \begin{pmatrix} 0 \\ \frac{2}{5} \end{pmatrix}e^{-t/2}\right]$

5. $\mathbf{x} = c_1\begin{pmatrix} -3 \\ 4 \\ 2 \end{pmatrix}e^{-t} + c_2\begin{pmatrix} 0 \\ 1 \\ -1 \end{pmatrix}e^{2t} + c_3\left[\begin{pmatrix} 0 \\ 1 \\ -1 \end{pmatrix}te^{2t} + \begin{pmatrix} 1 \\ 0 \\ 1 \end{pmatrix}e^{2t}\right]$

6. $\mathbf{x} = c_1\begin{pmatrix} 1 \\ 1 \\ 1 \end{pmatrix}e^{2t} + c_2\begin{pmatrix} 1 \\ 0 \\ -1 \end{pmatrix}e^{-t} + c_3\begin{pmatrix} 0 \\ 1 \\ -1 \end{pmatrix}e^{-t}$

7. $\mathbf{x} = \begin{pmatrix} 3 + 4t \\ 2 + 4t \end{pmatrix}e^{-3t}$

8. $\mathbf{x} = \begin{pmatrix} 3 \\ -1 \end{pmatrix}e^{-t} - 6\begin{pmatrix} 1 \\ 1 \end{pmatrix}te^{-t}$

9. $\mathbf{x} = \begin{pmatrix} 3 \\ -2 \end{pmatrix}e^{t/2} + \frac{3}{2}\begin{pmatrix} 1 \\ -1 \end{pmatrix}te^{t/2}$

10. $\mathbf{x} = 2\begin{pmatrix} 1 \\ 2 \end{pmatrix} + 14\begin{pmatrix} 3 \\ -1 \end{pmatrix}t$

11. $\mathbf{x} = \begin{pmatrix} -1 \\ 2 \\ -33 \end{pmatrix}e^{t} + 4\begin{pmatrix} 0 \\ 1 \\ -6 \end{pmatrix}te^{t} + 3\begin{pmatrix} 0 \\ 0 \\ 1 \end{pmatrix}e^{2t}$

12. $\mathbf{x} = \frac{4}{3}\begin{pmatrix} 1 \\ 1 \\ 1 \end{pmatrix}e^{-t/2} + \frac{1}{3}\begin{pmatrix} 2 \\ 5 \\ -7 \end{pmatrix}e^{-7t/2}$

13. $\mathbf{x} = c_1\begin{pmatrix} 2 \\ 1 \end{pmatrix}t + c_2\left[\begin{pmatrix} 2 \\ 1 \end{pmatrix}t\ln t + \begin{pmatrix} 1 \\ 0 \end{pmatrix}t\right]$

14. $\mathbf{x} = c_1\begin{pmatrix} 1 \\ 1 \end{pmatrix}t^{-3} + c_2\left[\begin{pmatrix} 1 \\ 1 \end{pmatrix}t^{-3}\ln t - \begin{pmatrix} 0 \\ \frac{1}{4} \end{pmatrix}t^{-3}\right]$

16. (b) $\begin{pmatrix} I \\ V \end{pmatrix} = -\begin{pmatrix} 1 \\ -2 \end{pmatrix}e^{-t/2} + \left[\begin{pmatrix} 1 \\ -2 \end{pmatrix}te^{-t/2} + \begin{pmatrix} 2 \\ 0 \end{pmatrix}e^{-t/2}\right]$

17. (b) $\mathbf{x}^{(1)}(t) = \begin{pmatrix} 0 \\ 1 \\ -1 \end{pmatrix} e^{2t}$ (c) $\mathbf{x}^{(2)}(t) = \begin{pmatrix} 0 \\ 1 \\ -1 \end{pmatrix} te^{2t} + \begin{pmatrix} 1 \\ 1 \\ 0 \end{pmatrix} e^{2t}$

(d) $\mathbf{x}^{(3)}(t) = \begin{pmatrix} 0 \\ 1 \\ -1 \end{pmatrix} (t^2/2)e^{2t} + \begin{pmatrix} 1 \\ 1 \\ 0 \end{pmatrix} te^{2t} + \begin{pmatrix} 2 \\ 0 \\ 3 \end{pmatrix} e^{2t}$

(e) $\mathbf{\Psi}(t) = e^{2t} \begin{pmatrix} 0 & 1 & t+2 \\ 1 & t+1 & (t^2/2)+t \\ -1 & -t & -(t^2/2)+3 \end{pmatrix}$

(f) $\mathbf{T} = \begin{pmatrix} 0 & 1 & 2 \\ 1 & 1 & 0 \\ -1 & 0 & 3 \end{pmatrix}$, $\mathbf{T}^{-1} = \begin{pmatrix} -3 & 3 & 2 \\ 3 & -2 & -2 \\ -1 & 1 & 1 \end{pmatrix}$, $\mathbf{J} = \begin{pmatrix} 2 & 1 & 0 \\ 0 & 2 & 1 \\ 0 & 0 & 2 \end{pmatrix}$

18. (a) $\mathbf{x}^{(1)}(t) = \begin{pmatrix} 1 \\ 0 \\ 2 \end{pmatrix} e^t$, $\mathbf{x}^{(2)}(t) = \begin{pmatrix} 0 \\ 2 \\ -3 \end{pmatrix} e^t$

(d) $\mathbf{x}^{(3)}(t) = \begin{pmatrix} 2 \\ 4 \\ -2 \end{pmatrix} te^t + \begin{pmatrix} 0 \\ 0 \\ -1 \end{pmatrix} e^t$

(e) $\mathbf{\Psi}(t) = e^t \begin{pmatrix} 1 & 0 & 2t \\ 0 & 2 & 4t \\ 2 & -3 & -2t-1 \end{pmatrix}$ or $e^t \begin{pmatrix} 1 & 2 & 2t \\ 0 & 4 & 4t \\ 2 & -2 & -2t-1 \end{pmatrix}$

(f) $\mathbf{T} = \begin{pmatrix} 1 & 2 & 0 \\ 0 & 4 & 0 \\ 2 & -2 & -1 \end{pmatrix}$, $\mathbf{T}^{-1} = \begin{pmatrix} 1 & -1/2 & 0 \\ 0 & 1/4 & 0 \\ 2 & -3/2 & -1 \end{pmatrix}$, $\mathbf{J} = \begin{pmatrix} 1 & 0 & 0 \\ 0 & 1 & 1 \\ 0 & 0 & 1 \end{pmatrix}$

19. (a) $\mathbf{J}^2 = \begin{pmatrix} \lambda^2 & 2\lambda \\ 0 & \lambda^2 \end{pmatrix}$, $\mathbf{J}^3 = \begin{pmatrix} \lambda^3 & 3\lambda^2 \\ 0 & \lambda^3 \end{pmatrix}$, $\mathbf{J}^4 = \begin{pmatrix} \lambda^4 & 4\lambda^3 \\ 0 & \lambda^4 \end{pmatrix}$

(c) $\exp(\mathbf{J}t) = e^{\lambda t} \begin{pmatrix} 1 & t \\ 0 & 1 \end{pmatrix}$ (d) $\mathbf{x} = \exp(\mathbf{J}t)\mathbf{x}^0$

20. (c) $\exp(\mathbf{J}t) = e^{\lambda t} \begin{pmatrix} 1 & 0 & 0 \\ 0 & 1 & t \\ 0 & 0 & 1 \end{pmatrix}$ 21. (c) $\exp(\mathbf{J}t) = e^{\lambda t} \begin{pmatrix} 1 & t & t^2/2 \\ 0 & 1 & t \\ 0 & 0 & 1 \end{pmatrix}$

Section 7.9, page 439

1. $\mathbf{x} = c_1 \begin{pmatrix} 1 \\ 1 \end{pmatrix} e^t + c_2 \begin{pmatrix} 1 \\ 3 \end{pmatrix} e^{-t} + \frac{3}{2} \begin{pmatrix} 1 \\ 1 \end{pmatrix} te^t - \frac{1}{4} \begin{pmatrix} 1 \\ 3 \end{pmatrix} e^t + \begin{pmatrix} 1 \\ 2 \end{pmatrix} t - \begin{pmatrix} 0 \\ 1 \end{pmatrix}$

2. $\mathbf{x} = c_1 \begin{pmatrix} \sqrt{3} \\ 1 \end{pmatrix} e^{2t} + c_2 \begin{pmatrix} 1 \\ -\sqrt{3} \end{pmatrix} e^{-2t} - \begin{pmatrix} 2/3 \\ 1/\sqrt{3} \end{pmatrix} e^t + \begin{pmatrix} -1 \\ 2/\sqrt{3} \end{pmatrix} e^{-t}$

3. $\mathbf{x} = c_1 \begin{pmatrix} 5\cos t \\ 2\cos t + \sin t \end{pmatrix} + c_2 \begin{pmatrix} 5\sin t \\ -\cos t + 2\sin t \end{pmatrix} + \begin{pmatrix} 2 \\ 1 \end{pmatrix} t\cos t - \begin{pmatrix} 1 \\ 0 \end{pmatrix} t\sin t - \begin{pmatrix} 1 \\ 1 \end{pmatrix} \cos t$

4. $\mathbf{x} = c_1 \begin{pmatrix} 1 \\ -4 \end{pmatrix} e^{-3t} + c_2 \begin{pmatrix} 1 \\ 1 \end{pmatrix} e^{2t} - \begin{pmatrix} 0 \\ 1 \end{pmatrix} e^{-2t} + \frac{1}{2} \begin{pmatrix} 1 \\ 0 \end{pmatrix} e^t$

5. $\mathbf{x} = c_1 \begin{pmatrix} 1 \\ 2 \end{pmatrix} + c_2 \left[\begin{pmatrix} 1 \\ 2 \end{pmatrix} t - \frac{1}{2} \begin{pmatrix} 0 \\ 1 \end{pmatrix} \right] - 2 \begin{pmatrix} 1 \\ 2 \end{pmatrix} \ln t + \begin{pmatrix} 2 \\ 5 \end{pmatrix} t^{-1} - \begin{pmatrix} \frac{1}{2} \\ 0 \end{pmatrix} t^{-2}$

6. $\mathbf{x} = c_1 \begin{pmatrix} 1 \\ 2 \end{pmatrix} + c_2 \begin{pmatrix} -2 \\ 1 \end{pmatrix} e^{-5t} + \begin{pmatrix} 1 \\ 2 \end{pmatrix} \ln t + \frac{8}{5} \begin{pmatrix} 1 \\ 2 \end{pmatrix} t + \frac{4}{25} \begin{pmatrix} -2 \\ 1 \end{pmatrix}$

7. $\mathbf{x} = c_1 \begin{pmatrix} 1 \\ 2 \end{pmatrix} e^{3t} + c_2 \begin{pmatrix} 1 \\ -2 \end{pmatrix} e^{-t} + \frac{1}{4} \begin{pmatrix} 1 \\ -8 \end{pmatrix} e^t$

8. $\mathbf{x} = c_1 \begin{pmatrix} 1 \\ 1 \end{pmatrix} e^t + c_2 \begin{pmatrix} 1 \\ 3 \end{pmatrix} e^{-t} + \begin{pmatrix} 1 \\ 0 \end{pmatrix} e^t + 2 \begin{pmatrix} 1 \\ 1 \end{pmatrix} te^t$

9. $\mathbf{x} = c_1 \begin{pmatrix} 1 \\ 1 \end{pmatrix} e^{-t/2} + c_2 \begin{pmatrix} 1 \\ -1 \end{pmatrix} e^{-2t} + \begin{pmatrix} \frac{5}{2} \\ \frac{3}{2} \end{pmatrix} t - \begin{pmatrix} \frac{17}{4} \\ \frac{15}{4} \end{pmatrix} + \begin{pmatrix} \frac{1}{6} \\ \frac{1}{2} \end{pmatrix} e^t$

10. $\mathbf{x} = c_1 \begin{pmatrix} 1 \\ \sqrt{2} \end{pmatrix} e^{-t} + c_2 \begin{pmatrix} -\sqrt{2} \\ 1 \end{pmatrix} e^{-4t} - \frac{1}{3} \begin{pmatrix} \sqrt{2}-1 \\ 2-\sqrt{2} \end{pmatrix} te^{-t} + \frac{1}{9} \begin{pmatrix} 2+\sqrt{2} \\ -1-\sqrt{2} \end{pmatrix} e^{-t}$

11. $\mathbf{x} = c_1 \begin{pmatrix} 5\cos t \\ 2\cos t + \sin t \end{pmatrix} + c_2 \begin{pmatrix} 5\sin t \\ -\cos t + 2\sin t \end{pmatrix} + \begin{pmatrix} 0 \\ 1/2 \end{pmatrix} t\cos t - \begin{pmatrix} 5/2 \\ 1 \end{pmatrix} t\sin t$
$$- \begin{pmatrix} 5/2 \\ 1 \end{pmatrix} \cos t$$

12. $\mathbf{x} = [\frac{1}{5}\ln(\sin t) - \ln(-\cos t) - \frac{2}{5}t + c_1] \begin{pmatrix} 5\cos t \\ 2\cos t + \sin t \end{pmatrix}$
$$+ [\frac{2}{5}\ln(\sin t) - \frac{4}{5}t + c_2] \begin{pmatrix} 5\sin t \\ -\cos t + 2\sin t \end{pmatrix}$$

13. (a) $\Psi(t) = \begin{pmatrix} e^{-t/2}\cos\frac{1}{2}t & e^{-t/2}\sin\frac{1}{2}t \\ 4e^{-t/2}\sin\frac{1}{2}t & -4e^{-t/2}\cos\frac{1}{2}t \end{pmatrix}$　　(b) $\mathbf{x} = e^{-t/2} \begin{pmatrix} \sin\frac{1}{2}t \\ 4 - 4\cos\frac{1}{2}t \end{pmatrix}$

14. $\mathbf{x} = c_1 \begin{pmatrix} 1 \\ 1 \end{pmatrix} t + c_2 \begin{pmatrix} 1 \\ 3 \end{pmatrix} t^{-1} - \begin{pmatrix} 2 \\ 3 \end{pmatrix} + \frac{1}{2} \begin{pmatrix} 1 \\ 3 \end{pmatrix} t - \begin{pmatrix} 1 \\ 1 \end{pmatrix} t\ln t - \frac{1}{3} \begin{pmatrix} 4 \\ 3 \end{pmatrix} t^2$

15. $\mathbf{x} = c_1 \begin{pmatrix} 2 \\ 1 \end{pmatrix} t^2 + c_2 \begin{pmatrix} 1 \\ 2 \end{pmatrix} t^{-1} + \begin{pmatrix} 3 \\ 2 \end{pmatrix} t + \frac{1}{10} \begin{pmatrix} -2 \\ 1 \end{pmatrix} t^4 - \frac{1}{2} \begin{pmatrix} 2 \\ 3 \end{pmatrix}$

CHAPTER 8　Section 8.1, page 449

1. (a) 1.1975, 1.38549, 1.56491, 1.73658
 (b) 1.19631, 1.38335, 1.56200, 1.73308
 (c) 1.19297, 1.37730, 1.55378, 1.72316
 (d) 1.19405, 1.37925, 1.55644, 1.72638

2. (a) 1.59980, 1.29288, 1.07242, 0.930175
 (b) 1.61124, 1.31361, 1.10012, 0.962552
 (c) 1.64337, 1.37164, 1.17763, 1.05334
 (d) 1.63301, 1.35295, 1.15267, 1.02407

3. (a) 1.2025, 1.41603, 1.64289, 1.88590
 (b) 1.20388, 1.41936, 1.64896, 1.89572
 (c) 1.20864, 1.43104, 1.67042, 1.93076
 (d) 1.20693, 1.42683, 1.66265, 1.91802

4. (a) 1.10244, 1.21426, 1.33484, 1.46399
 (b) 1.10365, 1.21656, 1.33817, 1.46832
 (c) 1.10720, 1.22333, 1.34797, 1.48110
 (d) 1.10603, 1.22110, 1.34473, 1.47688

5. (a) 0.509239, 0.522187, 0.539023, 0.559936
 (b) 0.509701, 0.523155, 0.540550, 0.562089
 (c) 0.511127, 0.526155, 0.545306, 0.568822
 (d) 0.510645, 0.525138, 0.543690, 0.566529

6. (a) -0.920498, -0.857538, -0.808030, -0.770038
 (b) -0.922575, -0.860923, -0.812300, -0.774965
 (c) -0.928059, -0.870054, -0.824021, -0.788686
 (d) -0.926341, -0.867163, -0.820279, -0.784275

7. (a) 2.90330, 7.53999, 19.4292, 50.5614
 (b) 2.93506, 7.70957, 20.1081, 52.9779
 (c) 3.03951, 8.28137, 22.4562, 61.5496
 (d) 3.00306, 8.07933, 21.6163, 58.4462

8. (a) 0.891830, 1.25225, 2.37818, 4.07257
 (b) 0.908902, 1.26872, 2.39336, 4.08799
 (c) 0.958565, 1.31786, 2.43924, 4.13474
 (d) 0.942261, 1.30153, 2.42389, 4.11908

9. (a) 3.95713, 5.09853, 6.41548, 7.90174
 (b) 3.95965, 5.10371, 6.42343, 7.91255
 (c) 3.96727, 5.11932, 6.44737, 7.94512
 (d) 3.96473, 5.11411, 6.43937, 7.93424

10. (a) 1.60729, 2.46830, 3.72167, 5.45963
 (b) 1.60996, 2.47460, 3.73356, 5.47774
 (c) 1.61792, 2.49356, 3.76940, 5.53223
 (d) 1.61528, 2.48723, 3.75742, 5.51404

11. (a) -1.45865, -0.217545, 1.05715, 1.41487
 (b) -1.45322, -0.180813, 1.05903, 1.41244
 (c) -1.43600, -0.0681657, 1.06489, 1.40575
 (d) -1.44190, -0.105737, 1.06290, 1.40789

12. (a) 0.587987, 0.791589, 1.14743, 1.70973
 (b) 0.589440, 0.795758, 1.15693, 1.72955
 (c) 0.593901, 0.808716, 1.18687, 1.79291
 (d) 0.592396, 0.804319, 1.17664, 1.77111

15. 1.595, 2.4636

16. $e_{n+1} = [2\phi(\bar{t}_n) - 1]h^2$, $\quad |e_{n+1}| \leq \left[1 + 2\max_{0 \leq t \leq 1} |\phi(t)|\right]h^2$,

 $e_{n+1} = e^{2\bar{t}_n}h^2$, $\quad |e_1| \leq 0.012$, $\quad |e_4| \leq 0.022$

17. $e_{n+1} = [2\phi(\bar{t}_n) - \bar{t}_n]h^2$, $\quad |e_{n+1}| \leq \left[1 + 2\max_{0 \leq t \leq 1} |\phi(t)|\right]h^2$,

 $e_{n+1} = 2e^{2\bar{t}_n}h^2$, $\quad |e_1| \leq 0.024$, $\quad |e_4| \leq 0.045$

18. $e_{n+1} = [\bar{t}_n + \bar{t}_n^2\phi(\bar{t}_n) + \phi^3(\bar{t}_n)]h^2$

19. $e_{n+1} = [19 - 15\bar{t}_n\phi^{-1/2}(\bar{t}_n)]h^2/4$

20. $e_{n+1} = \{1 + [\bar{t}_n + \phi(\bar{t}_n)]^{-1/2}\}h^2/4$

21. $e_{n+1} = \{2 - [\phi(\bar{t}_n) + 2\bar{t}_n^2]\exp[-\bar{t}_n\phi(\bar{t}_n)] - \bar{t}_n\exp[-2\bar{t}_n\phi(\bar{t}_n)]\}h^2/2$

22. (a) $\phi(t) = 1 + (1/5\pi)\sin 5\pi t$ (b) 1.2, 1.0, 1.2
 (c) 1.1, 1.1, 1.0, 1.0 (d) $h < 1/\sqrt{50\pi} \cong 0.08$

24. $e_{n+1} = -\frac{1}{2}\phi''(\bar{t}_n)h^2$

25. (a) 1.55, 2.34, 3.46, 5.07 (b) 1.20, 1.39, 1.57, 1.74
 (c) 1.20, 1.42, 1.65, 1.90

26. (a) 0 (b) 60 (c) -92.16 27. $0.224 \neq 0.225$

Section 8.2, page 456

1. (a) 1.19512, 1.38120, 1.55909, 1.72956
 (b) 1.19515, 1.38125, 1.55916, 1.72965
 (c) 1.19516, 1.38126, 1.55918, 1.72967

2. (a) 1.62283, 1.33460, 1.12820, 0.995445
 (b) 1.62243, 1.33386, 1.12718, 0.994215
 (c) 1.62234, 1.33368, 1.12693, 0.993921

3. (a) 1.20526, 1.42273, 1.65511, 1.90570
 (b) 1.20533, 1.42290, 1.65542, 1.90621
 (c) 1.20534, 1.42294, 1.65550, 1.90634

4. (a) 1.10483, 1.21882, 1.34146, 1.47263
 (b) 1.10484, 1.21884, 1.34147, 1.47262
 (c) 1.10484, 1.21884, 1.34147, 1.47262

5. (a) 0.510164, 0.524126, 0.542083, 0.564251
 (b) 0.510168, 0.524135, 0.542100, 0.564277
 (c) 0.510169, 0.524137, 0.542104, 0.564284

6. (a) −0.924650, −0.864338, −0.816642, −0.780008
 (b) −0.924550, −0.864177, −0.816442, −0.779781
 (c) −0.924525, −0.864138, −0.816393, −0.779725

7. (a) 2.96719, 7.88313, 20.8114, 55.5106
 (b) 2.96800, 7.88755, 20.8294, 55.5758

8. (a) 0.926139, 1.28558, 2.40898, 4.10386
 (b) 0.925815, 1.28525, 2.40869, 4.10359

9. (a) 3.96217, 5.10887, 6.43134, 7.92332
 (b) 3.96218, 5.10889, 6.43138, 7.92337

10. (a) 1.61263, 2.48097, 3.74556, 5.49595
 (b) 1.61263, 2.48092, 3.74550, 5.49589

11. (a) −1.44768, −0.144478, 1.06004, 1.40960
 (b) −1.44765, −0.143690, 1.06072, 1.40999

12. (a) 0.590897, 0.799950, 1.16653, 1.74969
 (b) 0.590906, 0.799988, 1.16663, 1.74992

15. $e_{n+1} = (38h^3/3) \exp(4\bar{t}_n)$, $|e_{n+1}| \leq 37{,}758.8h^3$ on $0 \leq t \leq 2$, $|e_1| \leq 0.00193389$

16. $e_{n+1} = (2h^3/3) \exp(2\bar{t}_n)$, $|e_{n+1}| \leq 4.92604h^3$ on $0 \leq t \leq 1$, $|e_1| \leq 0.000814269$

17. $e_{n+1} = (4h^3/3) \exp(2\bar{t}_n)$, $|e_{n+1}| \leq 9.85207h^3$ on $0 \leq t \leq 1$, $|e_1| \leq 0.00162854$

18. $h \cong 0.071$ 19. $h \cong 0.023$

20. $h \cong 0.081$ 21. $h \cong 0.117$

23. 1.19512, 1.38120, 1.55909, 1.72956 24. 1.62268, 1.33435, 1.12789, 0.995130

25. 1.20526, 1.42273, 1.65511, 1.90570 26. 1.10485, 1.21886, 1.34149, 1.47264

Section 8.3, page 461

1. (a) 1.19516, 1.38127, 1.55918, 1.72968
 (b) 1.19516, 1.38127, 1.55918, 1.72968

2. (a) 1.62231, 1.33362, 1.12686, 0.993839
 (b) 1.62230, 1.33362, 1.12685, 0.993826

3. (a) 1.20535, 1.42295, 1.65553, 1.90638
 (b) 1.20535, 1.42296, 1.65553, 1.90638

4. (a) 1.10484, 1.21884, 1.34147, 1.47262
 (b) 1.10484, 1.21884, 1.34147, 1.47262

5. (a) 0.510170, 0.524138, 0.542105, 0.564286
 (b) 0.520169, 0.524138, 0.542105, 0.564286

6. (a) −0.924517, −0.864125, −0.816377, −0.779706
 (b) −0.924517, −0.864125, −0.816377, −0.779706

7. (a) 2.96825, 7.88889, 20.8349, 55.5957
 (b) 2.96828, 7.88904, 20.8355, 55.5980

8. (a) 0.925725, 1.28516, 2.40860, 4.10350
 (b) 0.925711, 1.28515, 2.40860, 4.10350

9. (a) 3.96219, 5.10890, 6.43139, 7.92338
 (b) 3.96219, 5.10890, 6.43139, 7.92338

10. (a) 1.61262, 2.48091, 3.74548, 5.49587
 (b) 1.61262, 2.48091, 3.74548, 5.49587

11. (a) −1.44764, −0.143543, 1.06089, 1.41008
 (b) −1.44764, −0.143427, 1.06095, 1.41011

12. (a) 0.590909, 0.800000, 1.166667, 1.75000
 (b) 0.590909, 0.800000, 1.166667, 1.75000

Section 8.4, page 467

1. (a) 1.7296801, 1.8934697
 (b) 1.7296802, 1.8934698
 (c) 1.7296805, 1.8934711

2. (a) 0.993852, 0.925764
 (b) 0.993846, 0.925746
 (c) 0.993869, 0.925837

3. (a) 1.906382, 2.179567
 (b) 1.906391, 2.179582
 (c) 1.906395, 2.179611

4. (a) 1.4726173, 1.6126215
 (b) 1.4726189, 1.6126231
 (c) 1.4726199, 1.6126256

5. (a) 0.56428577, 0.59090918
 (b) 0.56428581, 0.59090923
 (c) 0.56428588, 0.59090952

6. (a) −0.779693, −0.753135
 (b) −0.779692, −0.753137
 (c) −0.779680, −0.753089

7. (a) 2.96828, 7.88907, 20.8356, 55.5984
 (b) 2.96829, 7.88909, 20.8357, 55.5986
 (c) 2.96831, 7.88926, 20.8364, 55.6015

8. (a) 0.9257133, 1.285148, 2.408595, 4.103495
 (b) 0.9257124, 1.285148, 2.408595, 4.103495
 (c) 0.9257248, 1.285158, 2.408594, 4.103493

9. (a) 3.962186, 5.108903, 6.431390, 7.923385
 (b) 3.962186, 5.108903, 6.431390, 7.923385
 (c) 3.962186, 5.108903, 6.431390, 7.923385

10. (a) 1.612622, 2.480909, 3.745479, 5.495872
 (b) 1.612622, 2.480909, 3.745479, 5.495873
 (c) 1.612623, 2.480905, 3.745473, 5.495869

11. (a) −1.447639, −0.1436281, 1.060946, 1.410122
 (b) −1.447638, −0.1436762, 1.060913, 1.410103
 (c) −1.447621, −0.1447219, 1.060717, 1.410027

12. (a) 0.5909091, 0.8000000, 1.166667, 1.750000
 (b) 0.5909091, 0.8000000, 1.166667, 1.750000
 (c) 0.5909092, 0.8000002, 1.166667, 1.750001

Section 8.5, page 477

1. (b) $\phi_2(t) - \phi_1(t) = 0.001e^t \to \infty$ as $t \to \infty$
2. (b) $\phi_1(t) = \ln[e^t/(2 - e^t)];$ $\phi_2(t) = \ln[1/(1 - t)]$
3. (a,b) $h = 0.00025$ is sufficient. (c) $h = 0.005$ is sufficient.
4. (a) $y = 4e^{-10t} + (t^2/4)$
 (c) Runge-Kutta is stable for $h = 0.25$ but unstable for $h = 0.3$.
 (d) $h = 5/13 \cong 0.384615$ is small enough.
5. (a) $y = t$ 6. (a) $y = t^2$

Section 8.6, page 480

1. (a) 1.26, 0.76; 1.7714, 1.4824; 2.58991, 2.3703; 3.82374, 3.60413; 5.64246, 5.38885
 (b) 1.32493, 0.758933; 1.93679, 1.57919; 2.93414, 2.66099; 4.48318, 4.22639; 6.84236, 6.56452
 (c) 1.32489, 0.759516; 1.9369, 1.57999; 2.93459, 2.66201; 4.48422, 4.22784; 6.8444, 6.56684

2. (a) 1.451, 1.232; 2.16133, 1.65988; 3.29292, 2.55559; 5.16361, 4.7916; 8.54951, 12.0464
 (b) 1.51844, 1.28089; 2.37684, 1.87711; 3.85039, 3.44859; 6.6956, 9.50309; 15.0987, 64.074
 (c) 1.51855, 1.2809; 2.3773, 1.87729; 3.85247, 3.45126; 6.71282, 9.56846; 15.6384, 70.3792

3. (a) 0.582, 1.18; 0.117969, 1.27344; −0.336912, 1.27382; −0.730007, 1.18572; −1.02134, 1.02371
 (b) 0.568451, 1.15775; 0.109776, 1.22556; −0.32208, 1.20347; −0.681296, 1.10162; −0.937852, 0.937852
 (c) 0.56845, 1.15775; 0.109773, 1.22557; −0.322081, 1.20347; −0.681291, 1.10161; −0.937841, 0.93784

4. (a) −0.198, 0.618; −0.378796, 0.28329; −0.51932, −0.0321025; −0.594324, −0.326801; −0.588278, −0.57545
 (b) −0.196904, 0.630936; −0.372643, 0.298888; −0.501302, −0.0111429; −0.561270, −0.288943; −0.547053, −0.508303
 (c) −0.196935, 0.630939; −0.372687, 0.298866; −0.501345, −0.0112184; −0.561292, −0.28907; −0.547031, −0.508427

5. (a) 2.96225, 1.34538; 2.34119, 1.67121; 1.90236, 1.97158; 1.56602, 2.23895; 1.29768, 2.46732
 (b) 3.06339, 1.34858; 2.44497, 1.68638; 1.9911, 2.00036; 1.63818, 2.27981; 1.3555, 2.5175
 (c) 3.06314, 1.34899; 2.44465, 1.68699; 1.99075, 2.00107; 1.63781, 2.28057; 1.35514, 2.51827

6. (a) 1.42386, 2.18957; 1.82234, 2.36791; 2.21728, 2.53329; 2.61118, 2.68763; 2.9955, 2.83354
 (b) 1.41513, 2.18699; 1.81208, 2.36233; 2.20635, 2.5258; 2.59826, 2.6794; 2.97806, 2.82487
 (c) 1.41513, 2.18699; 1.81209, 2.36233; 2.20635, 2.52581; 2.59826, 2.67941; 2.97806, 2.82488

7. For $h = 0.05$ and 0.025: $x = 10.227, y = -4.9294$; these results agree with the exact solution to five digits.
8. 1.543, 0.0707503; 1.14743, −1.3885
9. 1.99520, −0.662442

CHAPTER 9 **Section 9.1, page 492**

1. (a) $r_1 = -1$, $\xi^{(1)} = (1,2)^T$; $r_2 = 2$, $\xi^{(2)} = (2,1)^T$
 (b) saddle point, unstable
2. (a) $r_1 = 2$, $\xi^{(1)} = (1,3)^T$; $r_2 = 4$, $\xi^{(2)} = (1,1)^T$
 (b) node, unstable
3. (a) $r_1 = -1$, $\xi^{(1)} = (1,3)^T$; $r_2 = 1$, $\xi^{(2)} = (1,1)^T$
 (b) saddle point, unstable
4. (a) $r_1 = r_2 = -3$, $\xi^{(1)} = (1,1)^T$ (b) improper node, asymptotically stable
5. (a) $r_1, r_2 = -1 \pm i$; $\xi^{(1)}, \xi^{(2)} = (2 \pm i, 1)^T$
 (b) spiral point, asymptotically stable
6. (a) $r_1, r_2 = \pm i$; $\xi^{(1)}, \xi^{(2)} = (2 \pm i, 1)^T$ (b) center, stable
7. (a) $r_1, r_2 = 1 \pm 2i$; $\xi^{(1)}, \xi^{(2)} = (1, 1 \mp i)^T$ (b) spiral point, unstable
8. (a) $r_1 = -1$, $\xi^{(1)} = (1,0)^T$; $r_2 = -1/4$, $\xi^{(2)} = (4,-3)^T$
 (b) node, asymptotically stable
9. (a) $r_1 = r_2 = 1$, $\xi^{(1)} = (2,1)^T$ (b) improper node, unstable
10. (a) $r_1, r_2 = \pm 3i$; $\xi^{(1)}, \xi^{(2)} = (2, -1 \pm 3i)^T$
 (b) center, stable
11. (a) $r_1 = r_2 = -1$; $\xi^{(1)} = (1,0)^T$, $\xi^{(2)} = (0,1)^T$
 (b) proper node, asymptotically stable
12. (a) $r_1, r_2 = (1 \pm 3i)/2$; $\xi^{(1)}, \xi^{(2)} = (5, 3 \mp 3i)^T$
 (b) spiral point, unstable
13. $x_0 = 1$, $y_0 = 1$; $r_1 = \sqrt{2}$, $r_2 = -\sqrt{2}$; saddle point, unstable
14. $x_0 = -1$, $y_0 = 0$; $r_1 = -1$, $r_2 = -3$; node, asymptotically stable
15. $x_0 = -2$, $y_0 = 1$; $r_1, r_2 = -1 \pm \sqrt{2}\,i$; spiral point, asymptotically stable
16. $x_0 = \gamma/\delta$, $y_0 = \alpha/\beta$; $r_1, r_2 = \pm\sqrt{\beta\delta}\,i$; center, stable
17. $c^2 > 4km$, node, asymptotically stable; $c^2 = 4km$, improper node, asymptotically stable;
 $c^2 < 4km$, spiral point, asymptotically stable

Section 9.2, page 501

1. $x = 4e^{-t}$, $y = 2e^{-2t}$, $y = x^2/8$
2. $x = 4e^{-t}$, $y = 2e^{2t}$, $y = 32x^{-2}$; $x = 4e^{-t}$, $y = 0$
3. $x = 4\cos t$, $y = 4\sin t$, $x^2 + y^2 = 16$; $x = -4\sin t$, $y = 4\cos t$, $x^2 + y^2 = 16$
4. $x = \sqrt{a}\cos\sqrt{ab}\,t$, $y = -\sqrt{b}\sin\sqrt{ab}\,t$; $(x^2/a) + (y^2/b) = 1$
5. (a, c) $(-\frac{1}{2}, 1)$, saddle point, unstable; $(0,0)$, (proper) node, unstable
6. (a, c) $(-\sqrt{3}/3, -\frac{1}{2})$, saddle point, unstable; $(\sqrt{3}/3, -\frac{1}{2})$, center, stable
7. (a, c) $(0,0)$, node, unstable; $(0,2)$, node, asymptotically stable;
 $(\frac{1}{2}, \frac{1}{2})$, saddle point, unstable; $(1,0)$, node, asymptotically stable
8. (a, c) $(0,0)$, saddle point, unstable; $(0,1)$, spiral point, asymptotically stable;
 $(-2,-2)$, node, asymptotically stable; $(3,-2)$, node, unstable
9. (a, c) $(0,0)$, spiral point, asymptotically stable;
 $(1 - \sqrt{2}, 1 + \sqrt{2})$, saddle point, unstable; $(1 + \sqrt{2}, 1 - \sqrt{2})$, saddle point, unstable
10. (a, c) $(0,0)$, saddle point, unstable; $(2,2)$, spiral point, asymptotically stable;
 $(-1,-1)$, spiral point, asymptotically stable; $(-2,0)$, saddle point, unstable
11. (a, c) $(0,0)$, saddle point, unstable; $(0,1)$, saddle point, unstable;
 $(\frac{1}{2}, \frac{1}{2})$, center, stable; $(-\frac{1}{2}, \frac{1}{2})$, center, stable
12. (a, c) $(0,0)$, saddle point, unstable; $(\sqrt{6},0)$, spiral point, asymptotically stable;
 $(-\sqrt{6},0)$, spiral point, asymptotically stable
13. (a, c) $(0,0)$, saddle point, unstable; $(-2,2)$, node, unstable;
 $(4,4)$, spiral point, asymptotically stable

14. (a, c) $(0,0)$, spiral point, unstable

15. (a) $4x^2 - y^2 = c$

16. (a) $4x^2 + y^2 = c$

17. (a) $(y - 2x)^2(x + y) = c$

18. (a) $\arctan(y/x) - \ln\sqrt{x^2 + y^2} = c$

19. (a) $2x^2y - 2xy + y^2 = c$

20. (a) $x^2y^2 - 3x^2y - 2y^2 = c$

21. (a) $(y^2/2) - \cos x = c$

22. (a) $x^2 + y^2 - (x^4/12) = c$

Section 9.3, page 511

1. linear and nonlinear: saddle point, unstable

2. linear and nonlinear: spiral point, asymptotically stable

3. linear: center, stable; nonlinear: spiral point or center, indeterminate

4. linear: improper node, unstable; nonlinear: node or spiral point, unstable

5. (a, b, c) $(0,0)$; $u' = -2u + 2v$, $v' = 4u + 4v$; $r = 1 \pm \sqrt{17}$;
 saddle point, unstable
 $(-2,2)$; $u' = 4u$, $v' = 6u + 6v$; $r = 4, 6$; node, unstable
 $(4,4)$; $u' = -6u + 6v$, $v' = -8u$; $r = -3 \pm \sqrt{39}\,i$;
 spiral point, asymptotically stable

6. (a, b, c) $(0,0)$; $u' = u$, $v' = 3v$; $r = 1, 3$; node, unstable
 $(1,0)$; $u' = -u - v$, $v' = 2v$; $r = -1, 2$; saddle point, unstable
 $(0, \frac{3}{2})$; $u' = -\frac{1}{2}u$, $v' = -\frac{3}{2}u - 3v$; $r = -\frac{1}{2}, -3$; node, asymptotically stable
 $(-1,2)$; $u' = u + v$, $v' = -2u - 4v$; $r = (-3 \pm \sqrt{17})/2$; saddle point, unstable

7. (a, b, c) $(1,1)$; $u' = -v$, $v' = 2u - 2v$; $r = -1 \pm i$;
 spiral point, asymptotically stable
 $(-1,1)$; $u' = -v$, $v' = -2u - 2v$; $r = -1 \pm \sqrt{3}$; saddle point, unstable

8. (a, b, c) $(0,0)$; $u' = u$, $v' = \frac{1}{2}v$; $r = 1, \frac{1}{2}$; node, unstable
 $(0,2)$; $u' = -u$, $v' = -\frac{3}{2}u - \frac{1}{2}v$; $r = -1, -\frac{1}{2}$; node, asymptotically stable
 $(1,0)$; $u' = -u - v$, $v' = -\frac{1}{4}v$; $r = -1, -\frac{1}{4}$; node, asymptotically stable
 $(\frac{1}{2}, \frac{1}{2})$; $u' = -\frac{1}{2}u - \frac{1}{2}v$, $v' = -\frac{3}{8}u - \frac{1}{8}v$; $r = (-5 \pm \sqrt{57})/16$;
 saddle point, unstable

9. (a, b, c) $(0,0)$; $u' = -u + v$, $v' = 2u$; $r = -2, 1$; saddle point, unstable
 $(0,1)$; $u' = -u - v$, $v' = 3u$; $r = (-1 \pm \sqrt{11}\,i)/2$;
 spiral point, asymptotically stable
 $(-2,-2)$; $u' = -5u + 5v$, $v' = -2v$; $r = -5, -2$;
 node, asymptotically stable
 $(3,-2)$; $u' = 5u + 5v$, $v' = 3v$; $r = 5, 3$; node, unstable

10. (a, b, c) $(0,0)$; $u' = u$, $v' = v$; $r = 1, 1$; node or spiral point, unstable
 $(-1,0)$; $u' = -u$, $v' = 2v$; $r = -1, 2$; saddle point, unstable

11. (a, b, c) $(0,0)$; $u' = 2u + v$, $v' = u - 2v$; $r = \pm\sqrt{5}$; saddle point, unstable
 $(-1.1935, -1.4797)$; $u' = -1.2399u - 6.8393v$, $v' = 2.4797u - 0.80655v$;
 $r = -1.0232 \pm 4.1125i$; spiral point, asymptotically stable

12. (a, b, c) $(0, \pm 2n\pi)$, $n = 0, 1, 2, \ldots$; $u' = v$, $v' = -u$; $r = \pm i$;
 center or spiral point, indeterminate
 $[2, \pm(2n - 1)\pi]$, $n = 1, 2, 3, \ldots$; $u' = -3v$, $v' = -u$; $r = \pm\sqrt{3}$;
 saddle point, unstable

13. (a, b, c) $(0,0)$; $u' = u$, $v' = v$; $r = 1, 1$; node or spiral point, unstable
 $(1,1)$; $u' = u - 2v$, $v' = -2u + v$; $r = 3, -1$; saddle point, unstable

14. (a, b, c) $(1,1)$; $u' = -u - v$, $v' = u - 3v$; $r = -2, -2$;
 node or spiral point, asymptotically stable
 $(-1,-1)$; $u' = u + v$, $v' = u - 3v$; $r = -1 \pm \sqrt{5}$; saddle point, unstable

15. (a, b, c) $(0,0)$; $u' = -2u - v$, $v' = u - v$; $r = (-3 \pm \sqrt{3}\,i)/2$;
 spiral point, asymptotically stable
 $(-0.33076, 1.0924)$ and $(0.33076, -1.0924)$; $u' = -3.5216u - 0.27735v$,
 $v' = 0.27735u + 2.6895v$; $r = -3.5092, 2.6771$; saddle point, unstable
16. (a, b, c) $(0,0)$; $u' = u + v$, $v' = -u + v$; $r = 1 \pm i$; spiral point, unstable
19. (b, c) Refer to Table 9.3.1.
21. (a) $R = A$, $T \cong 3.17$ (b) $R = A$, $T \cong 3.20, \ 3.35, \ 3.63, \ 4.17$
 (c) $T \to \pi$ as $A \to 0$ (d) $A = \pi$
22. (b) $v_c \cong 4.00$ 23. (b) $v_c \cong 4.51$
28. (a) $dx/dt = y$, $dy/dt = -g(x) - c(x)y$
 (b) The linear system is $dx/dt = y$, $dy/dt = -g'(0)x - c(0)y$.
 (c) The eigenvalues satisfy $r^2 + c(0)r + g'(0) = 0$.

Section 9.4, page 525

1. (b, c) $(0,0)$; $u' = \frac{3}{2}u$, $v' = 2v$; $r = \frac{3}{2}, 2$; node, unstable
 $(0,2)$; $u' = \frac{1}{2}u$, $v' = -\frac{3}{2}u - 2v$; $r = \frac{1}{2}, -2$; saddle point, unstable
 $(\frac{3}{2},0)$; $u' = -\frac{3}{2}u - \frac{3}{4}v$, $v' = \frac{7}{8}v$; $r = -\frac{3}{2}, \frac{7}{8}$; saddle point, unstable
 $(\frac{4}{5}, \frac{7}{5})$; $u' = -\frac{4}{5}u - \frac{2}{5}v$, $v' = -\frac{21}{20}u - \frac{7}{5}v$; $r = (-22 \pm \sqrt{204})/20$;
 node, asymptotically stable

2. (b, c) $(0,0)$; $u' = \frac{3}{2}u$, $v' = 2v$; $r = \frac{3}{2}, 2$; node, unstable
 $(0,4)$; $u' = -\frac{1}{2}u$, $v' = -6u - 2v$; $r = -\frac{1}{2}, -2$; node, asymptotically stable
 $(\frac{3}{2},0)$; $u' = -\frac{3}{2}u - \frac{3}{4}v$, $v' = -\frac{1}{4}v$; $r = -\frac{1}{4}, -\frac{3}{2}$; node, asymptotically stable
 $(1,1)$; $u' = -u - \frac{1}{2}v$, $v' = -\frac{3}{2}u - \frac{1}{2}v$; $r = (-3 \pm \sqrt{13})/4$;
 saddle point, unstable

3. (b, c) $(0,0)$; $u' = \frac{3}{2}u$, $v' = 2v$; $r = \frac{3}{2}, 2$; node, unstable
 $(0,2)$; $u' = -\frac{1}{2}u$, $v' = -\frac{9}{4}u - 2v$; $r = -\frac{1}{2}, -2$; node, asymptotically stable
 $(3,0)$; $u' = -\frac{3}{2}u - 3v$, $v' = -\frac{11}{8}v$; $r = -\frac{3}{2}, -\frac{11}{8}$; node, asymptotically stable
 $(\frac{4}{5}, \frac{11}{10})$; $u' = -\frac{2}{5}u - \frac{4}{5}v$, $v' = -\frac{99}{80}u - \frac{11}{10}v$; $r = -1.80475, \ 0.30475$;
 saddle point, unstable

4. (b, c) $(0,0)$; $u' = \frac{3}{2}u$, $v' = \frac{3}{4}v$; $r = \frac{3}{2}, \frac{3}{4}$; node, unstable
 $(0, \frac{3}{4})$; $u' = \frac{3}{4}u$, $v' = -\frac{3}{4}v$; $r = \pm\frac{3}{4}$; saddle point, unstable
 $(3,0)$; $u' = -\frac{3}{2}u - 3v$, $v' = \frac{3}{8}v$; $r = -\frac{3}{2}, \frac{3}{8}$; saddle point, unstable
 $(2, \frac{1}{2})$; $u' = -u - 2v$, $v' = -\frac{1}{16}u - \frac{1}{2}v$; $r = -1.18301, \ -0.31699$;
 node, asymptotically stable

5. (b, c) $(0,0)$; $u' = u$, $v' = \frac{3}{2}v$; $r = 1, \frac{3}{2}$; node, unstable
 $(0, \frac{3}{2})$; $u' = -\frac{1}{2}u$, $v' = -\frac{3}{2}u - \frac{3}{2}v$; $r = -\frac{1}{2}, -\frac{3}{2}$; node, asymptotically stable
 $(1,0)$; $u' = -u - v$, $v' = \frac{1}{2}v$; $r = -1, \frac{1}{2}$; saddle point, unstable

6. (b, c) $(0,0)$; $u' = u$, $v' = \frac{5}{2}v$; $r = 1, \frac{5}{2}$; node, unstable
 $(0, \frac{5}{3})$; $u' = \frac{11}{6}u$, $v' = \frac{5}{12}u - \frac{5}{2}v$; $r = \frac{11}{6}, \ -\frac{5}{2}$; saddle point, unstable
 $(1,0)$; $u' = -u + \frac{1}{2}v$, $v' = \frac{11}{4}v$; $r = -1, \frac{11}{4}$; saddle point, unstable
 $(2,2)$; $u' = -2u + v$, $v' = \frac{1}{2}u - 3v$; $r = (-5 \pm \sqrt{3})/2$;
 node, asymptotically stable

8. (a) Critical points are $x = 0, y = 0$; $x = \epsilon_1/\sigma_1, y = 0$; $x = 0, y = \epsilon_2/\sigma_2$.
 $x \to 0, y \to \epsilon_2/\sigma_2$ as $t \to \infty$; the redear survive.
 (b) Same as part (a) except $x \to \epsilon_1/\sigma_1, y \to 0$ as $t \to \infty$; the bluegill survive.

9. (a) $X = (B - \gamma_1 R)/(1 - \gamma_1\gamma_2)$, $Y = (R - \gamma_2 B)/(1 - \gamma_1\gamma_2)$

 (b) X is reduced, Y is increased; yes, if B becomes less than $\gamma_1 R$, then $x \to 0$ and $y \to R$ as $t \to \infty$.

10. (a) $\sigma_1\epsilon_2 - \alpha_2\epsilon_1 \neq 0$: $(0,0), (0, \epsilon_2/\sigma_2), (\epsilon_1/\sigma_1, 0)$

 $\sigma_1\epsilon_2 - \alpha_2\epsilon_1 = 0$: $(0,0)$, and all points on the line $\sigma_1 x + \alpha_1 y = \epsilon_1$

 (b) $\sigma_1\epsilon_2 - \alpha_2\epsilon_1 > 0$: $(0,0)$ is unstable node; $(\epsilon_1/\sigma_1, 0)$ is saddle point;

 $(0, \epsilon_2/\sigma_2)$ is asymptotically stable node.

 $\sigma_1\epsilon_2 - \alpha_2\epsilon_1 < 0$: $(0,0)$ is unstable node; $(0, \epsilon_2/\sigma_2)$ is saddle point;

 $(\epsilon_1/\sigma_1, 0)$ is asymptotically stable node.

 (c) $(0,0)$ is unstable node; points on the line $\sigma_1 x + \alpha_1 y = \epsilon_1$ are stable, nonisolated critical points.

12. (a) $(0,0)$, saddle point; $(0.15, 0)$, spiral point if $\gamma^2 < 1.11$, node if $\gamma^2 \geq 1.11$; $(2,0)$, saddle point

 (c) $\gamma \cong 1.20$

13. (b) $(2 - \sqrt{4 - \frac{3}{2}\alpha},\ \frac{3}{2}\alpha)$, $(2 + \sqrt{4 - \frac{3}{2}\alpha},\ \frac{3}{2}\alpha)$

 (c) $(1,3)$ is an asymptotically stable node; $(3,3)$ is a saddle point

 (d) $\alpha_0 = 8/3$; critical point is $(2,4)$; $\lambda = 0, -1$

14. (b) $(2 - \sqrt{4 - \frac{3}{2}\alpha},\ \frac{3}{2}\alpha)$, $(2 + \sqrt{4 - \frac{3}{2}\alpha},\ \frac{3}{2}\alpha)$

 (c) $(1,3)$ is a saddle point; $(3,3)$ is an unstable spiral point

 (d) $\alpha_0 = 8/3$; critical point is $(2,4)$; $\lambda = 0, 1$

15. (b) $([3 - \sqrt{9 - 4\alpha}]/2,\ [3 + 2\alpha - \sqrt{9 - 4\alpha}]/2)$,

 $([3 + \sqrt{9 - 4\alpha}]/2,\ [3 + 2\alpha + \sqrt{9 - 4\alpha}]/2)$

 (c) $(1,3)$ is a saddle point; $(2,4)$ is an unstable spiral point

 (d) $\alpha_0 = 9/4$; critical point is $(3/2, 15/4)$; $\lambda = 0, 0$

16. (b) $([3 - \sqrt{9 - 4\alpha}]/2,\ [3 + 2\alpha - \sqrt{9 - 4\alpha}]/2)$,

 $([3 + \sqrt{9 - 4\alpha}]/2,\ [3 + 2\alpha + \sqrt{9 - 4\alpha}]/2)$

 (c) $(1,3)$ is a center of the linear approximation and also of the nonlinear system; $(2,4)$ is a saddle point

 (d) $\alpha_0 = 9/4$; critical point is $(3/2, 15/4)$; $\lambda = 0, 0$

17. (a) $(0,0), (0, 2 + 2\alpha), (4, 0), (2, 2)$

 (b) $\alpha = 0.75$, asymptotically stable node; $\alpha = 1.25$, (unstable) saddle point

 (c) $u' = -2u - 2v$, $v' = -2\alpha u - 2v$

 (d) $r = -2 \pm 2\sqrt{\alpha}$; $\alpha_0 = 1$

Section 9.5, page 534

1. (b, c) $(0,0)$; $u' = \frac{3}{2}u$, $v' = -\frac{1}{2}v$; $r = \frac{3}{2}, -\frac{1}{2}$; saddle point, unstable

 $(\frac{1}{2}, 3)$; $u' = -\frac{1}{4}v$, $v' = 3u$; $r = \pm\sqrt{3}\,i/2$; center or spiral point, indeterminate

2. (b, c) $(0,0)$; $u' = u$, $v' = -\frac{1}{4}v$; $r = 1, -\frac{1}{4}$; saddle point, unstable

 $(\frac{1}{2}, 2)$; $u' = -\frac{1}{4}v$, $v' = u$; $r = \pm\frac{1}{2}i$; center or spiral point, indeterminate

3. (b, c) $(0,0)$; $u' = u$, $v' = -\frac{1}{4}v$; $r = 1, -\frac{1}{4}$; saddle point, unstable

 $(2, 0)$; $u' = -u - v$, $v' = \frac{3}{4}v$; $r = -1, \frac{3}{4}$; saddle point, unstable

 $(\frac{1}{2}, \frac{3}{2})$; $u' = -\frac{1}{4}u - \frac{1}{4}v$, $v' = \frac{3}{4}u$; $r = (-1 \pm \sqrt{11}\,i)/8$;

 spiral point, asymptotically stable

4. (b, c) $(0,0)$; $u' = \frac{9}{8}u$, $v' = -v$; $r = \frac{9}{8}, -1$; saddle point, unstable

 $(\frac{9}{8}, 0)$; $u' = -\frac{9}{8}u - \frac{9}{16}v$, $v' = \frac{1}{8}v$; $r = -\frac{9}{8}, \frac{1}{8}$; saddle point, unstable

 $(1, \frac{1}{4})$; $u' = -u - \frac{1}{2}v$, $v' = \frac{1}{4}u$; $r = (-1 \pm \sqrt{0.5})/2$; node, asymptotically stable

5. (b, c) $(0,0)$; $u' = -u$, $v' = -\frac{3}{2}v$; $r = -1, -\frac{3}{2}$; node, asymptotically stable
 $(\frac{1}{2}, 0)$; $u' = \frac{3}{4}u - \frac{3}{20}v$, $v' = -v$; $r = -1, \frac{3}{4}$; saddle point, unstable
 $(2, 0)$; $u' = -3u - \frac{3}{5}v$, $v' = \frac{1}{2}v$; $r = -3, \frac{1}{2}$; saddle point, unstable
 $(\frac{3}{2}, \frac{5}{3})$; $u' = -\frac{3}{4}u - \frac{9}{20}v$, $v' = \frac{5}{3}u$; $r = (-3 \pm \sqrt{39}\,i)/8$;
 spiral point, asymptotically stable

6. $t = 0, T, 2T, \ldots$:　　　　H is a max.,　　　dP/dt is a max.
 $t = T/4, 5T/4, \ldots$:　　dH/dt is a min.,　　P is a max.
 $t = T/2, 3T/2, \ldots$:　　H is a min.,　　　dP/dt is a min.
 $t = 3T/4, 7T/4, \ldots$:　dH/dt is a max.,　P is a min.

7. (a) $\sqrt{c}\,\alpha/\sqrt{a}\,\gamma$　　(b) $\sqrt{3}$
 (d) The ratio of prey amplitude to predator amplitude increases very slowly as the initial point moves away from the equilibrium point.

8. (a) $4\pi/\sqrt{3} \cong 7.2552$
 (c) The period increases slowly as the initial point moves away from the equilibrium point.

9. (a) $T \cong 6.5$　　　　(b) $T \cong 3.7$, $T \cong 11.5$　　　(c) $T \cong 3.8$, $T \cong 11.1$

11. Trap foxes in half-cycle when $dP/dt > 0$, trap rabbits in half-cycle when $dH/dt > 0$, trap rabbits and foxes in quarter-cycle when $dP/dt > 0$ and $dH/dt > 0$, trap neither in quarter-cycle when $dP/dt < 0$ and $dH/dt < 0$.

12. $dH/dt = aH - \alpha HP - \beta H$, $dP/dt = -cP + \gamma HP - \delta P$, where β and δ are constants of proportionality. New center is $H = (c + \delta)/\gamma > c/\gamma$ and $P = (a - \beta)/\alpha < a/\alpha$, so equilibrium value of prey is increased and equilibrium value of predator is decreased!

13. Let $A = a/\sigma - c/\gamma > 0$. The critical points are $(0,0)$, $(a/\sigma, 0)$, and $(c/\gamma, \sigma A/\alpha)$, where $(0,0)$ is a saddle point, $(a/\sigma, 0)$ is a saddle point, and $(c/\gamma, \sigma A/\alpha)$ is an asymptotically stable node if $(c\sigma/\gamma)^2 - 4c\sigma A \geq 0$ or an asymptotically stable spiral point if $(c\sigma/\gamma)^2 - 4c\sigma A < 0$. $(H, P) \to (c/\gamma, \sigma A/\alpha)$ as $t \to \infty$.

Section 9.7, page 556

1. $r = 1$, $\theta = t + t_0$, asymptotically stable limit cycle

2. $r = 1$, $\theta = -t + t_0$, semistable limit cycle

3. $r = 1$, $\theta = t + t_0$, asymptotically stable limit cycle;
 $r = 3$, $\theta = t + t_0$, unstable periodic solution

4. $r = 1$, $\theta = -t + t_0$, unstable periodic solution;
 $r = 2$, $\theta = -t + t_0$, asymptotically stable limit cycle

5. $r = 2n - 1$, $\theta = t + t_0$, $n = 1, 2, 3, \ldots$, asymptotically stable limit cycle;
 $r = 2n$, $\theta = t + t_0$, $n = 1, 2, 3, \ldots$, unstable periodic solution

6. $r = 2$, $\theta = -t + t_0$, semistable limit cycle;
 $r = 3$, $\theta = -t + t_0$, unstable periodic solution

8. (a) Counterclockwise
 (b) $r = 1$, $\theta = t + t_0$, asymptotically stable limit cycle;
 $r = 2$, $\theta = t + t_0$, semistable limit cycle;
 $r = 3$, $\theta = t + t_0$, unstable periodic solution

9. $r = \sqrt{2}$, $\theta = -t + t_0$, unstable periodic solution

14. (a) $\mu = 0.2$, $T \cong 6.29$;　$\mu = 1$, $T \cong 6.66$;　$\mu = 5$, $T \cong 11.60$

15. (a) $x' = y$, $y' = -x + \mu y - \mu y^3/3$
 (b) $0 < \mu < 2$, unstable spiral point;　$\mu \geq 2$, unstable node
 (c) $A \cong 2.16$, $T \cong 6.65$
 (d) $\mu = 0.2$, $A \cong 1.99$, $T \cong 6.31$;　$\mu = 0.5$, $A \cong 2.03$, $T \cong 6.39$;
 $\mu = 2$, $A \cong 2.60$, $T \cong 7.65$;　$\mu = 5$, $A \cong 4.36$, $T \cong 11.60$

16. (b) $x' = \mu x + y$, $y' = -x + \mu y$; $\lambda = \mu \pm i$; the origin is an asymptotically stable spiral point for $\mu < 0$, a center for $\mu = 0$, and an unstable spiral point for $\mu > 0$.
(c) $r' = r(\mu - r^2)$, $\theta' = -1$

17. (b) $\lambda = \left[-(5/4 - b) \pm \sqrt{(5/4 - b)^2 - 1}\right] \Big/ 2$
(c) $0 < b < 1/4$: asymptotically stable node; $1/4 < b < 5/4$: asymptotically stable spiral point; $5/4 < b < 9/4$: unstable spiral point; $9/4 < b$; unstable node
(d) $b_0 = 5/4$

18. (b) $k = 0$, $(1.1994, -0.62426)$; $k = 0.5$, $(0.80485, -0.13106)$
(c) $k_0 \cong 0.3465$, $(0.95450, -0.31813)$
(d) $k = 0.4$, $T \cong 11.23$; $k = 0.5$, $T \cong 10.37$; $k = 0.6$, $T \cong 9.93$
(e) $k_1 \cong 1.4035$

Section 9.8, page 565

1. (b) $\lambda = \lambda_1$, $\boldsymbol{\xi}^{(1)} = (0, 0, 1)^T$; $\lambda = \lambda_2$, $\boldsymbol{\xi}^{(2)} = (20, 9 - \sqrt{81 + 40r}, 0)^T$;
$\lambda = \lambda_3$, $\boldsymbol{\xi}^{(3)} = (20, 9 + \sqrt{81 + 40r}, 0)^T$
(c) $\lambda_1 \cong -2.6667$, $\boldsymbol{\xi}^{(1)} = (0, 0, 1)^T$; $\lambda_2 \cong -22.8277$, $\boldsymbol{\xi}^{(2)} \cong (20, -25.6554, 0)^T$;
$\lambda_3 \cong 11.8277$, $\boldsymbol{\xi}^{(3)} \cong (20, 43.6554, 0)^T$
2. (c) $\lambda_1 \cong -13.8546$; $\lambda_2, \lambda_3 \cong 0.0939556 \pm 10.1945i$
5. (a) $dV/dt = -2\sigma[rx^2 + y^2 + b(z - r)^2 - br^2]$

CHAPTER 10

Section 10.1, page 575

1. $y = -\sin x$
2. $y = (\cot \sqrt{2}\pi \cos \sqrt{2}x + \sin \sqrt{2}x)/\sqrt{2}$
3. $y = 0$ for all L; $y = c_2 \sin x$ if $\sin L = 0$
4. $y = -\tan L \cos x + \sin x$ if $\cos L \neq 0$; no solution if $\cos L = 0$
5. No solution 6. $y = (-\pi \sin \sqrt{2}x + x \sin \sqrt{2}\pi)/2 \sin \sqrt{2}\pi$
7. No solution 8. $y = c_2 \sin 2x + \frac{1}{3}\sin x$
9. $y = c_1 \cos 2x + \frac{1}{3}\cos x$ 10. $y = \frac{1}{2}\cos x$
11. $y = -\frac{5}{2}x + \frac{3}{2}x^2$ 12. $y = -\frac{1}{9}x^{-1} + \frac{1}{9}(1 - e^3)x^{-1}\ln x + \frac{1}{9}x^2$
13. No solution
14. $\lambda_n = [(2n - 1)/2]^2$, $y_n(x) = \sin[(2n - 1)x/2]$; $n = 1, 2, 3, \ldots$
15. $\lambda_n = [(2n - 1)/2]^2$, $y_n(x) = \cos[(2n - 1)x/2]$; $n = 1, 2, 3, \ldots$
16. $\lambda_0 = 0$, $y_0(x) = 1$; $\lambda_n = n^2$, $y_n(x) = \cos nx$; $n = 1, 2, 3, \ldots$
17. $\lambda_n = [(2n - 1)\pi/2L]^2$, $y_n(x) = \cos[(2n - 1)\pi x/2L]$; $n = 1, 2, 3, \ldots$
18. $\lambda_0 = 0$, $y_0(x) = 1$; $\lambda_n = (n\pi/L)^2$, $y_n(x) = \cos(n\pi x/L)$; $n = 1, 2, 3, \ldots$
19. $\lambda_n = -[(2n - 1)\pi/2L]^2$, $y_n(x) = \sin[(2n - 1)\pi x/2L]$; $n = 1, 2, 3, \ldots$
20. $\lambda_n = 1 + (n\pi/\ln L)^2$, $y_n(x) = x \sin(n\pi \ln x/\ln L)$; $n = 1, 2, 3, \ldots$
21. (a) $w(r) = G(R^2 - r^2)/4\mu$ (c) Q is reduced to 0.3164 of its original value.
22. (a) $y = k(x^4 - 2Lx^3 + L^3x)/24$ (b) $y = k(x^4 - 2Lx^3 + L^2x^2)/24$
(c) $y = k(x^4 - 4Lx^3 + 6L^2x^2)/24$

Section 10.2, page 585

1. $T = 2\pi/5$ 2. $T = 1$
3. Not periodic 4. $T = 2L$
5. $T = 1$ 6. Not periodic
7. $T = 2$ 8. $T = 4$
9. $f(x) = 2L - x$ in $L < x < 2L$; $f(x) = -2L - x$ in $-3L < x < -2L$
10. $f(x) = x - 1$ in $1 < x < 2$; $f(x) = x - 8$ in $8 < x < 9$

11. $f(x) = -L - x$ in $-L < x < 0$

13. (b) $f(x) = \dfrac{2L}{\pi} \displaystyle\sum_{n=1}^{\infty} \dfrac{(-1)^n}{n} \sin \dfrac{n\pi x}{L}$

14. (b) $f(x) = \dfrac{1}{2} - \dfrac{2}{\pi} \displaystyle\sum_{n=1}^{\infty} \dfrac{\sin[(2n-1)\pi x/L]}{2n-1}$

15. (b) $f(x) = -\dfrac{\pi}{4} + \displaystyle\sum_{n=1}^{\infty} \left[\dfrac{2\cos(2n-1)x}{\pi(2n-1)^2} + \dfrac{(-1)^{n+1}\sin nx}{n} \right]$

16. (b) $f(x) = \dfrac{1}{2} + \dfrac{4}{\pi^2} \displaystyle\sum_{n=1}^{\infty} \dfrac{\cos(2n-1)\pi x}{(2n-1)^2}$

17. (b) $f(x) = \dfrac{3L}{4} + \displaystyle\sum_{n=1}^{\infty} \left[\dfrac{2L\cos[(2n-1)\pi x/L]}{(2n-1)^2\pi^2} + \dfrac{(-1)^{n+1}L\sin(n\pi x/L)}{n\pi} \right]$

18. (b) $f(x) = \displaystyle\sum_{n=1}^{\infty} \left[-\dfrac{2}{n\pi} \cos \dfrac{n\pi}{2} + \left(\dfrac{2}{n\pi}\right)^2 \sin \dfrac{n\pi}{2} \right] \sin \dfrac{n\pi x}{2}$

19. (b) $f(x) = \dfrac{4}{\pi} \displaystyle\sum_{n=1}^{\infty} \dfrac{\sin[(2n-1)\pi x/2]}{2n-1}$

20. (b) $f(x) = \dfrac{2}{\pi} \displaystyle\sum_{n=1}^{\infty} \dfrac{(-1)^{n+1}}{n} \sin n\pi x$

21. (b) $f(x) = \dfrac{2}{3} + \dfrac{8}{\pi^2} \displaystyle\sum_{n=1}^{\infty} \dfrac{(-1)^n}{n^2} \cos \dfrac{n\pi x}{2}$

22. (b) $f(x) = \dfrac{1}{2} + \dfrac{12}{\pi^2} \displaystyle\sum_{n=1}^{\infty} \dfrac{\cos[(2n-1)\pi x/2]}{(2n-1)^2} + \dfrac{2}{\pi} \displaystyle\sum_{n=1}^{\infty} \dfrac{(-1)^n}{n} \sin \dfrac{n\pi x}{2}$

23. (b) $f(x) = \dfrac{11}{12} + \dfrac{1}{\pi^2} \displaystyle\sum_{n=1}^{\infty} \dfrac{(-1)^n - 5}{n^2} \cos \dfrac{n\pi x}{2} + \displaystyle\sum_{n=1}^{\infty} \left[\dfrac{4[1-(-1)^n]}{n^3\pi^3} - \dfrac{(-1)^n}{n\pi} \right] \sin \dfrac{n\pi x}{2}$

24. (b) $f(x) = \dfrac{9}{8} + \displaystyle\sum_{n=1}^{\infty} \left[\dfrac{162[(-1)^n - 1]}{n^4\pi^4} - \dfrac{27(-1)^n}{n^2\pi^2} \right] \cos \dfrac{n\pi x}{3} - \displaystyle\sum_{n=1}^{\infty} \dfrac{108(-1)^n + 54}{n^3\pi^3} \sin \dfrac{n\pi x}{3}$

25. $m = 81$

26. $m = 27$

28. $\displaystyle\int_0^x f(t)\,dt$ may not be periodic; for example, let $f(t) = 1 + \cos t$.

Section 10.3, page 592

1. (a) $f(x) = \dfrac{4}{\pi} \displaystyle\sum_{n=1}^{\infty} \dfrac{\sin(2n-1)\pi x}{2n-1}$

2. (a) $f(x) = \dfrac{\pi}{4} - \displaystyle\sum_{n=1}^{\infty} \left[\dfrac{2}{(2n-1)^2\pi} \cos(2n-1)x + \dfrac{(-1)^n}{n} \sin nx \right]$

3. (a) $f(x) = \dfrac{L}{2} + \dfrac{4L}{\pi^2} \displaystyle\sum_{n=1}^{\infty} \dfrac{\cos[(2n-1)\pi x/L]}{(2n-1)^2}$

4. (a) $f(x) = \dfrac{2}{3} + \dfrac{4}{\pi^2} \displaystyle\sum_{n=1}^{\infty} \dfrac{(-1)^{n+1}}{n^2} \cos n\pi x$

5. (a) $f(x) = \dfrac{1}{2} + \dfrac{2}{\pi} \displaystyle\sum_{n=1}^{\infty} \dfrac{(-1)^{n-1}}{2n-1} \cos(2n-1)x$

6. (a) $f(x) = \dfrac{a_0}{2} + \displaystyle\sum_{n=1}^{\infty} (a_n \cos n\pi x + b_n \sin n\pi x);$

 $a_0 = \dfrac{1}{3}, \qquad a_n = \dfrac{2(-1)^n}{n^2 \pi^2}, \qquad b_n = \begin{cases} -1/n\pi, & n \text{ even} \\ 1/n\pi - 4/n^3\pi^3, & n \text{ odd} \end{cases}$

7. (a) $f(x) = -\dfrac{\pi}{4} + \displaystyle\sum_{n=1}^{\infty} \left[\dfrac{1 - \cos n\pi}{\pi n^2} \cos nx - \dfrac{(-1)^n}{n} \sin nx \right]$

 (b) $n = 10;$ $\max|e| = 1.6025$ at $x = \pm\pi$
 $n = 20;$ $\max|e| = 1.5867$ at $x = \pm\pi$
 $n = 40;$ $\max|e| = 1.5788$ at $x = \pm\pi$
 (c) Not possible

8. (a) $f(x) = \dfrac{1}{2} + \dfrac{2}{\pi^2} \displaystyle\sum_{n=1}^{\infty} \dfrac{1 - \cos n\pi}{n^2} \cos n\pi x$

 (b) $n = 10;$ $\max|e| = 0.02020$ at $x = 0, \pm1$
 $n = 20;$ $\max|e| = 0.01012$ at $x = 0, \pm1$
 $n = 40;$ $\max|e| = 0.005065$ at $x = 0, \pm1$
 (c) $n = 21$

9. (a) $f(x) = \dfrac{2}{\pi} \displaystyle\sum_{n=1}^{\infty} \dfrac{(-1)^{n+1}}{n} \sin n\pi x$ (b) $n = 10, 20, 40;$ $\max|e| = 1$ at $x \pm 1$

 (c) Not possible

10. (a) $f(x) = \dfrac{1}{2} + \displaystyle\sum_{n=1}^{\infty} \left[\dfrac{6(1 - \cos n\pi)}{n^2\pi^2} \cos \dfrac{n\pi x}{2} + \dfrac{2\cos n\pi}{n\pi} \sin \dfrac{n\pi x}{2} \right]$

 (b) $n = 10;$ $\text{lub}|e| = 1.0606$ as $x \to 2$
 $n = 20;$ $\text{lub}|e| = 1.0304$ as $x \to 2$
 $n = 40;$ $\text{lub}|e| = 1.0152$ as $x \to 2$
 (c) Not possible

11. (a) $f(x) = \dfrac{1}{6} + \displaystyle\sum_{n=1}^{\infty} \left[\dfrac{2\cos n\pi}{n^2\pi^2} \cos n\pi x - \dfrac{2 - 2\cos n\pi + n^2\pi^2 \cos n\pi}{n^3\pi^3} \sin n\pi x \right]$

 (b) $n = 10;$ $\text{lub}|e| = 0.5193$ as $x \to 1$
 $n = 20;$ $\text{lub}|e| = 0.5099$ as $x \to 1$
 $n = 40;$ $\text{lub}|e| = 0.5050$ as $x \to 1$
 (c) Not possible

12. (a) $f(x) = -\dfrac{12}{\pi^3} \displaystyle\sum_{n=1}^{\infty} \dfrac{(-1)^n}{n^3} \sin n\pi x$

 (b) $n = 10;$ $\max|e| = 0.001345$ at $x = \pm0.9735$
 $n = 20;$ $\max|e| = 0.0003534$ at $x = \pm0.9864$
 $n = 40;$ $\max|e| = 0.00009058$ at $x = \pm0.9931$
 (c) $n = 4$

13. $y = (\omega \sin nt - n \sin \omega t)/\omega(\omega^2 - n^2), \quad \omega^2 \neq n^2$
 $y = (\sin nt - nt \cos nt)/2n^2, \quad \omega^2 = n^2$

14. $y = \sum_{n=1}^{\infty} b_n(\omega \sin nt - n \sin \omega t)/\omega(\omega^2 - n^2), \quad \omega \neq 1, 2, 3, \ldots$

$y = \sum_{\substack{n=1 \\ n \neq m}}^{\infty} b_n(m \sin nt - n \sin mt)/m(m^2 - n^2) + b_m(\sin mt - mt \cos mt)/2m^2, \quad \omega = m$

15. $y = \frac{4}{\pi} \sum_{n=1}^{\infty} \frac{1}{\omega^2 - (2n-1)^2} \left[\frac{1}{2n-1} \sin(2n-1)t - \frac{1}{\omega} \sin \omega t \right]$

16. $y = \cos \omega t + \frac{1}{2\omega^2}(1 - \cos \omega t) + \frac{4}{\pi^2} \sum_{n=1}^{\infty} \frac{\cos(2n-1)\pi t - \cos \omega t}{(2n-1)^2[\omega^2 - (2n-1)^2\pi^2]}$

Section 10.4, page 600

1. Odd
2. Neither
3. Odd
4. Even
5. Even
6. Neither

14. $f(x) = \frac{1}{4} + \frac{4}{\pi^2} \sum_{n=1}^{\infty} \frac{1 - \cos(n\pi/2)}{n^2} \cos \frac{n\pi x}{2}$

$f(x) = \frac{4}{\pi^2} \sum_{n=1}^{\infty} \frac{(n\pi/2) - \sin(n\pi/2)}{n^2} \sin \frac{n\pi x}{2}$

15. $f(x) = \frac{1}{2} + \frac{2}{\pi} \sum_{n=1}^{\infty} \frac{(-1)^{n-1}}{2n-1} \cos \frac{(2n-1)\pi x}{2}$

16. $f(x) = \sum_{n=1}^{\infty} \frac{2}{n\pi} \left(-\cos n\pi + \frac{2}{n\pi} \sin \frac{n\pi}{2} \right) \sin \frac{n\pi x}{2}$

17. $f(x) = 1$

18. $f(x) = \frac{4}{\pi} \sum_{n=1}^{\infty} \frac{\sin(2n-1)x}{2n-1}$

19. $f(x) = \sum_{n=1}^{\infty} \frac{2}{n\pi} \left(\cos \frac{n\pi}{3} + \cos \frac{2n\pi}{3} - 2\cos n\pi \right) \sin \frac{nx}{3}$

20. $f(x) = \frac{1}{2} - \frac{1}{\pi} \sum_{n=1}^{\infty} \frac{\sin 2n\pi x}{n}$

21. $f(x) = \frac{L}{2} + \frac{4L}{\pi^2} \sum_{n=1}^{\infty} \frac{\cos[(2n-1)\pi x/L]}{(2n-1)^2}$

22. $f(x) = \frac{2L}{\pi} \sum_{n=1}^{\infty} \frac{\sin(n\pi x/L)}{n}$

23. (a) $f(x) = \frac{\pi}{4} + \frac{1}{\pi} \sum_{n=1}^{\infty} \left[\frac{2\pi}{n} \sin \frac{n\pi}{2} + \frac{4}{n^2} \left(\cos \frac{n\pi}{2} - 1 \right) \right] \cos \frac{nx}{2}$

24. (a) $f(x) = 2 \sum_{n=1}^{\infty} \frac{(-1)^n}{n} \sin nx$

25. (a) $f(x) = \sum_{n=1}^{\infty} \left[\frac{4n^2\pi^2(1 + \cos n\pi)}{n^3\pi^3} + \frac{16(1 - \cos n\pi)}{n^3\pi^3} \right] \sin \frac{n\pi x}{2}$

26. (a) $f(x) = \frac{4}{3} + \frac{16}{\pi^2} \sum_{n=1}^{\infty} \frac{1 + 3\cos n\pi}{n^2} \cos \frac{n\pi x}{4}$

27. (b) $g(x) = \dfrac{3}{2} + \dfrac{6}{\pi^2} \sum\limits_{n=1}^{\infty} \dfrac{1 - \cos n\pi}{n^2} \cos \dfrac{n\pi x}{3}$

$h(x) = \dfrac{6}{\pi} \sum\limits_{n=1}^{\infty} \dfrac{1}{n} \sin \dfrac{n\pi x}{3}$

28. (b) $g(x) = \dfrac{1}{4} + \sum\limits_{n=1}^{\infty} \dfrac{4\cos(n\pi/2) + 2n\pi \sin(n\pi/2) - 4}{n^2\pi^2} \cos \dfrac{n\pi x}{2}$

$h(x) = \sum\limits_{n=1}^{\infty} \dfrac{4\sin(n\pi/2) - 2n\pi \cos(n\pi/2)}{n^2\pi^2} \sin \dfrac{n\pi x}{2}$

29. (b) $g(x) = -\dfrac{5}{12} + \sum\limits_{n=1}^{\infty} \dfrac{12\cos n\pi + 4}{n^2\pi^2} \cos \dfrac{n\pi x}{2}$

$h(x) = -\dfrac{1}{2} \sum\limits_{n=1}^{\infty} \dfrac{n^2\pi^2(3 + 5\cos n\pi) + 32(1 - \cos n\pi)}{n^3\pi^3} \sin \dfrac{n\pi x}{2}$

30. (b) $g(x) = \dfrac{1}{4} + \sum\limits_{n=1}^{\infty} \dfrac{6n^2\pi^2(2\cos n\pi - 5) + 324(1 - \cos n\pi)}{n^4\pi^4} \cos \dfrac{n\pi x}{3}$

$h(x) = \sum\limits_{n=1}^{\infty} \left[\dfrac{4\cos n\pi + 2}{n\pi} + \dfrac{144\cos n\pi + 180}{n^3\pi^3} \right] \sin \dfrac{n\pi x}{3}$

40. Extend $f(x)$ antisymmetrically into $(L, 2L)$; that is, so that $f(2L - x) = -f(x)$ for $0 \le x < L$. Then extend this function as an even function into $(-2L, 0)$.

Section 10.5, page 610

1. $xX'' - \lambda X = 0, \quad T' + \lambda T = 0$
2. $X'' - \lambda x X = 0, \quad T' + \lambda t T = 0$
3. $X'' - \lambda(X' + X) = 0, \quad T' + \lambda T = 0$
4. $[p(x)X']' + \lambda r(x)X = 0, \quad T'' + \lambda T = 0$
5. Not separable
6. $X'' + (x + \lambda)X = 0, \quad Y'' - \lambda Y = 0$
7. $u(x,t) = e^{-400\pi^2 t} \sin 2\pi x - e^{-2500\pi^2 t} \sin 5\pi x$
8. $u(x,t) = 2e^{-\pi^2 t/16} \sin(\pi x/2) - e^{-\pi^2 t/4} \sin \pi x + 4e^{-\pi^2 t} \sin 2\pi x$
9. $u(x,t) = \dfrac{100}{\pi} \sum\limits_{n=1}^{\infty} \dfrac{1 - \cos n\pi}{n} e^{-n^2\pi^2 t/1600} \sin \dfrac{n\pi x}{40}$
10. $u(x,t) = \dfrac{160}{\pi^2} \sum\limits_{n=1}^{\infty} \dfrac{\sin(n\pi/2)}{n^2} e^{-n^2\pi^2 t/1600} \sin \dfrac{n\pi x}{40}$
11. $u(x,t) = \dfrac{100}{\pi} \sum\limits_{n=1}^{\infty} \dfrac{\cos(n\pi/4) - \cos(3n\pi/4)}{n} e^{-n^2\pi^2 t/1600} \sin \dfrac{n\pi x}{40}$
12. $u(x,t) = \dfrac{80}{\pi} \sum\limits_{n=1}^{\infty} \dfrac{(-1)^{n+1}}{n} e^{-n^2\pi^2 t/1600} \sin \dfrac{n\pi x}{40}$
13. $t = 5, n = 16; \quad t = 20, n = 8; \quad t = 80, n = 4$
14. (d) $t = 673.35$ 15. (d) $t = 451.60$
16. (d) $t = 617.17$
17. (b) $t = 5, x = 33.20; \quad t = 10, x = 31.13; \quad t = 20, x = 28.62; \quad t = 40, x = 25.73;$
 $t = 100, \ x = 21.95; \quad t = 200, x = 20.31$
 (e) $t = 524.81$

18. $u(x,t) = \dfrac{200}{\pi} \displaystyle\sum_{n=1}^{\infty} \dfrac{1 - \cos n\pi}{n} e^{-n^2\pi^2\alpha^2 t/400} \sin \dfrac{n\pi x}{20}$

 (a) 35.91°C (b) 67.23°C (c) 99.96°C

19. (a) 76.73 sec (b) 152.56 sec (c) 1093.36 sec

21. (a) $aw_{xx} - bw_t + (c - b\delta)w = 0$ (b) $\delta = c/b$ if $b \neq 0$

22. $X'' + \mu^2 X = 0, \quad Y'' + (\lambda^2 - \mu^2)Y = 0, \quad T' + \alpha^2\lambda^2 T = 0$

23. $r^2 R'' + rR' + (\lambda^2 r^2 - \mu^2)R = 0, \quad \Theta'' + \mu^2\Theta = 0, \quad T' + \alpha^2\lambda^2 T = 0$

Section 10.6, page 620

1. $u = 10 + \frac{3}{5}x$ 2. $u = 30 - \frac{5}{4}x$

3. $u = 0$ 4. $u = T$

5. $u = 0$ 6. $u = T$

7. $u = T(1 + x)/(1 + L)$ 8. $u = T(1 + L - x)/(1 + L)$

9. (a) $u(x,t) = 3x + \displaystyle\sum_{n=1}^{\infty} \dfrac{70\cos n\pi + 50}{n\pi} e^{-0.86n^2\pi^2 t/400} \sin \dfrac{n\pi x}{20}$

 (d) 160.29 sec

10. (a) $f(x) = 2x, \; 0 \leq x \leq 50; \quad f(x) = 200 - 2x, \; 50 < x \leq 100$

 (b) $u(x,t) = 20 - \dfrac{x}{5} + \displaystyle\sum_{n=1}^{\infty} c_n e^{-1.14n^2\pi^2 t/(100)^2} \sin \dfrac{n\pi x}{100}$,

 $c_n = \dfrac{800}{n^2\pi^2} \sin \dfrac{n\pi}{2} - \dfrac{40}{n\pi}$

 (d) $u(50,t) \to 10$ as $t \to \infty$; 3754 sec

11. (a) $u(x,t) = 30 - x + \displaystyle\sum_{n=1}^{\infty} c_n e^{-n^2\pi^2 t/900} \sin \dfrac{n\pi x}{30}$,

 $c_n = \dfrac{60}{n^3\pi^3}[2(1 - \cos n\pi) - n^2\pi^2(1 + \cos n\pi)]$

12. (a) $u(x,t) = \dfrac{2}{\pi} + \displaystyle\sum_{n=1}^{\infty} c_n e^{-n^2\pi^2\alpha^2 t/L^2} \cos \dfrac{n\pi x}{L}$,

 $c_n = \begin{cases} 0, & n \text{ odd}; \\ -4/(n^2 - 1)\pi, & n \text{ even} \end{cases}$

 (b) $\lim\limits_{t \to \infty} u(x,t) = 2/\pi$

13. (a) $u(x,t) = \dfrac{200}{9} + \displaystyle\sum_{n=1}^{\infty} c_n e^{-n^2\pi^2 t/6400} \cos \dfrac{n\pi x}{40}$,

 $c_n = -\dfrac{160}{3n^2\pi^2}(3 + \cos n\pi)$

 (c) 200/9 (d) 1543 sec

14. (a) $u(x,t) = \dfrac{25}{6} + \displaystyle\sum_{n=1}^{\infty} c_n e^{-n^2\pi^2 t/900} \cos \dfrac{n\pi x}{30}$,

 $c_n = \dfrac{50}{n\pi}\left(\sin \dfrac{n\pi}{3} - \sin \dfrac{n\pi}{6}\right)$

15. (b) $u(x,t) = \sum_{n=1}^{\infty} c_n e^{-(2n-1)^2 \pi^2 \alpha^2 t / 4L^2} \sin \dfrac{(2n-1)\pi x}{2L}$,

$$c_n = \frac{2}{L} \int_0^L f(x) \sin \frac{(2n-1)\pi x}{2L} \, dx$$

16. (a) $u(x,t) = \sum_{n=1}^{\infty} c_n e^{-(2n-1)^2 \pi^2 t / 3600} \sin \dfrac{(2n-1)\pi x}{60}$,

$$c_n = \frac{120}{(2n-1)^2 \pi^2} [2 \cos n\pi + (2n-1)\pi]$$

(c) x_m increases from $x = 0$ and reaches $x = 30$ when $t = 104.4$.

17. (a) $u(x,t) = 40 + \sum_{n=1}^{\infty} c_n e^{-(2n-1)^2 \pi^2 t / 3600} \sin \dfrac{(2n-1)\pi x}{60}$,

$$c_n = \frac{40}{(2n-1)^2 \pi^2} [6 \cos n\pi - (2n-1)\pi]$$

19. $u(x) = \begin{cases} T\dfrac{x}{a} \left[\dfrac{\xi}{\xi + (L/a) - 1} \right], & 0 \le x \le a, \\[3mm] T \left[1 - \dfrac{L-x}{a} \dfrac{1}{\xi + (L/a) - 1} \right], & a \le x \le L, \end{cases}$ where $\xi = \kappa_2 A_2 / \kappa_1 A_1$

20. (e) $u_n(x,t) = e^{-\mu_n^2 \alpha^2 t} \sin \mu_n x$

21. $\alpha^2 v'' + s(x) = 0$; $v(0) = T_1$, $v(L) = T_2$
 $w_t = \alpha^2 w_{xx}$; $w(0,t) = 0$, $w(L,t) = 0$, $w(x,0) = f(x) - v(x)$

22. (a) $v(x) = T_1 + (T_2 - T_1)(x/L) + kLx/2 - kx^2/2$

(b) $w(x,t) = \sum_{n=1}^{\infty} c_n e^{-n^2 \pi^2 t / 400} \sin \dfrac{n\pi x}{20}$, $c_n = \dfrac{160 \, (-1)^n}{n^3 \pi^3}$

23. (a) $v(x) = T_1 + (T_2 - T_1)x/L + kLx/6 - kx^3/6L$

(b) $w(x,t) = \sum_{n=1}^{\infty} c_n e^{-n^2 \pi^2 t / 400} \sin \dfrac{n\pi x}{20}$,

$$c_n = \frac{20}{3} \left[\frac{3m^3 \pi^3 (3 \cos m\pi - 1) + 60 \cos m\pi}{m^4 \pi^4} \right]$$

Section 10.7, page 632

1. (a) $u(x,t) = \dfrac{8}{\pi^2} \sum_{n=1}^{\infty} \dfrac{1}{n^2} \sin \dfrac{n\pi}{2} \sin \dfrac{n\pi x}{L} \cos \dfrac{n\pi at}{L}$

2. (a) $u(x,t) = \dfrac{8}{\pi^2} \sum_{n=1}^{\infty} \dfrac{1}{n^2} \left(\sin \dfrac{n\pi}{4} + \sin \dfrac{3n\pi}{4} \right) \sin \dfrac{n\pi x}{L} \cos \dfrac{n\pi at}{L}$

3. (a) $u(x,t) = \dfrac{32}{\pi^3} \sum_{n=1}^{\infty} \dfrac{2 + \cos n\pi}{n^3} \sin \dfrac{n\pi x}{L} \cos \dfrac{n\pi at}{L}$

4. (a) $u(x,t) = \dfrac{4}{\pi} \sum_{n=1}^{\infty} \dfrac{\sin(n\pi/2) \sin(n\pi/L)}{n} \sin \dfrac{n\pi x}{L} \cos \dfrac{n\pi at}{L}$

5. (a) $u(x,t) = \dfrac{8L}{a\pi^3} \displaystyle\sum_{n=1}^{\infty} \dfrac{1}{n^3} \sin\dfrac{n\pi}{2} \sin\dfrac{n\pi x}{L} \sin\dfrac{n\pi at}{L}$

6. (a) $u(x,t) = \dfrac{8L}{a\pi^3} \displaystyle\sum_{n=1}^{\infty} \dfrac{\sin(n\pi/4) + \sin(3n\pi/4)}{n^3} \sin\dfrac{n\pi x}{L} \sin\dfrac{n\pi at}{L}$

7. (a) $u(x,t) = \dfrac{32L}{a\pi^4} \displaystyle\sum_{n=1}^{\infty} \dfrac{\cos n\pi + 2}{n^4} \sin\dfrac{n\pi x}{L} \sin\dfrac{n\pi at}{L}$

8. (a) $u(x,t) = \dfrac{4L}{a\pi^2} \displaystyle\sum_{n=1}^{\infty} \dfrac{\sin(n\pi/2)\sin(n\pi/L)}{n^2} \sin\dfrac{n\pi x}{L} \sin\dfrac{n\pi at}{L}$

9. $u(x,t) = \displaystyle\sum_{n=1}^{\infty} c_n \sin\dfrac{(2n-1)\pi x}{2L} \cos\dfrac{(2n-1)\pi at}{2L}$,

 $c_n = \dfrac{2}{L} \displaystyle\int_0^L f(x) \sin\dfrac{(2n-1)\pi x}{2L}\, dx$

10. (a) $u(x,t) = \dfrac{8}{\pi} \displaystyle\sum_{n=1}^{\infty} \dfrac{1}{2n-1} \sin\dfrac{(2n-1)\pi}{4} \sin\dfrac{(2n-1)\pi}{2L} \sin\dfrac{(2n-1)\pi x}{2L} \cos\dfrac{(2n-1)\pi at}{2L}$

11. (a) $u(x,t) = \dfrac{512}{\pi^4} \displaystyle\sum_{n=1}^{\infty} \dfrac{(2n-1)\pi + 3\cos n\pi}{(2n-1)^4} \sin\dfrac{(2n-1)\pi x}{2L} \cos\dfrac{(2n-1)\pi at}{2L}$

14. $\phi(x+at)$ represents a wave moving in the negative x direction with speed $a > 0$.

15. (a) 248 ft/sec (b) $49.6\pi n$ rad/sec (c) Frequencies increase; modes are unchanged.

21. $r^2 R'' + rR' + (\lambda^2 r^2 - \mu^2)R = 0$, $\Theta'' + \mu^2 \Theta = 0$, $T'' + \lambda^2 a^2 T = 0$

23. (b) $a_n = a\sqrt{1 + (\gamma^2 L^2/n^2\pi^2)}$ (c) $\gamma = 0$

24. (a) $c_n = \dfrac{20}{n^2\pi^2}\left(2\sin\dfrac{n\pi}{2} - \sin\dfrac{2n\pi}{5} - \sin\dfrac{3n\pi}{5}\right)$

Section 10.8, page 645

1. (a) $u(x,y) = \displaystyle\sum_{n=1}^{\infty} c_n \sin\dfrac{n\pi x}{a} \sinh\dfrac{n\pi y}{a}$, $c_n = \dfrac{2/a}{\sinh(n\pi b/a)} \displaystyle\int_0^a g(x)\sin\dfrac{n\pi x}{a}\, dx$

 (b) $u(x,y) = \dfrac{4a}{\pi^2} \displaystyle\sum_{n=1}^{\infty} \dfrac{1}{n^2} \dfrac{\sin(n\pi/2)}{\sinh(n\pi b/a)} \sin\dfrac{n\pi x}{a} \sinh\dfrac{n\pi y}{a}$

2. $u(x,y) = \displaystyle\sum_{n=1}^{\infty} c_n \sin\dfrac{n\pi x}{a} \sinh\dfrac{n\pi(b-y)}{a}$, $c_n = \dfrac{2/a}{\sinh(n\pi b/a)} \displaystyle\int_0^a h(x)\sin\dfrac{n\pi x}{a}\, dx$

3. (a) $u(x,y) = \displaystyle\sum_{n=1}^{\infty} c_n^{(1)} \sinh\dfrac{n\pi x}{b} \sin\dfrac{n\pi y}{b} + \displaystyle\sum_{n=1}^{\infty} c_n^{(2)} \sin\dfrac{n\pi x}{a} \sinh\dfrac{n\pi(b-y)}{a}$,

 $c_n^{(1)} = \dfrac{2/b}{\sinh(n\pi a/b)} \displaystyle\int_0^b f(y)\sin\dfrac{n\pi y}{b}\, dy$, $c_n^{(2)} = \dfrac{2/a}{\sinh(n\pi b/a)} \displaystyle\int_0^a h(x)\sin\dfrac{n\pi x}{a}\, dx$

 (b) $c_n^{(1)} = \dfrac{2}{n\pi \sinh(n\pi a/b)}$, $c_n^{(2)} = -\dfrac{2}{n^3\pi^3} \dfrac{(n^2\pi^2 - 2)\cos n\pi + 2}{\sinh(n\pi b/a)}$

5. $u(r,\theta) = \dfrac{c_0}{2} + \displaystyle\sum_{n=1}^{\infty} r^{-n}(c_n \cos n\theta + k_n \sin n\theta)$,

 $c_n = \dfrac{a^n}{\pi} \displaystyle\int_0^{2\pi} f(\theta)\cos n\theta\, d\theta$, $k_n = \dfrac{a^n}{\pi} \displaystyle\int_0^{2\pi} f(\theta)\sin n\theta\, d\theta$

6. (a) $u(r,\theta) = \sum_{n=1}^{\infty} c_n r^n \sin n\theta, \qquad c_n = \frac{2}{\pi a^n} \int_0^{\pi} f(\theta) \sin n\theta \, d\theta$

 (b) $c_n = \frac{4}{\pi a^n} \frac{1 - \cos n\pi}{n^3}$

7. $u(r,\theta) = \sum_{n=1}^{\infty} c_n r^{n\pi/\alpha} \sin \frac{n\pi\theta}{\alpha}, \qquad c_n = (2/\alpha) a^{-n\pi/\alpha} \int_0^{\alpha} f(\theta) \sin \frac{n\pi\theta}{\alpha} \, d\theta$

8. (a) $u(x,y) = \sum_{n=1}^{\infty} c_n e^{-n\pi y/a} \sin \frac{n\pi x}{a}, \qquad c_n = \frac{2}{a} \int_0^{a} f(x) \sin \frac{n\pi x}{a} \, dx$

 (b) $c_n = \frac{4a^2}{n^3 \pi^3} (1 - \cos n\pi)$ (c) $y_0 \cong 6.6315$

10. (b) $u(x,y) = c_0 + \sum_{n=1}^{\infty} c_n \cosh \frac{n\pi x}{b} \cos \frac{n\pi y}{b}, \qquad c_n = \frac{2/n\pi}{\sinh(n\pi a/b)} \int_0^{b} f(y) \cos \frac{n\pi y}{b} \, dy$

11. $u(r,\theta) = c_0 + \sum_{n=1}^{\infty} r^n (c_n \cos n\theta + k_n \sin n\theta),$

 $c_n = \frac{1}{n\pi a^{n-1}} \int_0^{2\pi} g(\theta) \cos n\theta \, d\theta, \qquad k_n = \frac{1}{n\pi a^{n-1}} \int_0^{2\pi} g(\theta) \sin n\theta \, d\theta;$

 necessary condition is $\int_0^{2\pi} g(\theta) \, d\theta = 0.$

12. (a) $u(x,y) = \sum_{n=1}^{\infty} c_n \sin \frac{n\pi x}{a} \cosh \frac{n\pi y}{a}, \qquad c_n = \frac{2/a}{\cosh(n\pi b/a)} \int_0^{a} g(x) \sin \frac{n\pi x}{a} \, dx$

 (b) $c_n = \frac{4a \sin(n\pi/2)}{n^2 \pi^2 \cosh(n\pi b/a)}$

13. (a) $u(x,y) = \sum_{n=1}^{\infty} c_n \sinh \frac{(2n-1)\pi x}{2b} \sin \frac{(2n-1)\pi y}{2b},$

 $c_n = \frac{2/b}{\sinh[(2n-1)\pi a/2b]} \int_0^{b} f(y) \sin \frac{(2n-1)\pi y}{2b} \, dy$

 (b) $c_n = \frac{32b^2}{(2n-1)^3 \pi^3 \sinh[(2n-1)\pi a/2b]}$

14. (a) $u(x,y) = \frac{c_0 y}{2} + \sum_{n=1}^{\infty} c_n \cos \frac{n\pi x}{a} \sinh \frac{n\pi y}{a},$

 $c_0 = \frac{2}{ab} \int_0^{a} g(x) \, dx, \qquad c_n = \frac{2/a}{\sinh(n\pi b/a)} \int_0^{a} g(x) \cos \frac{n\pi x}{a} \, dx$

 (b) $c_0 = \frac{2}{b} \left(1 + \frac{a^4}{30} \right), \qquad c_n = -\frac{24a^4(1 + \cos n\pi)}{n^4 \pi^4 \sinh(n\pi b/a)}$

16. (a) $u(x,z) = b + \frac{\alpha a}{2} - \frac{4\alpha a}{\pi^2} \sum_{n=1}^{\infty} \frac{\cos[(2n-1)\pi x/a] \cosh[(2n-1)\pi z/a]}{(2n-1)^2 \cosh[(2n-1)\pi b/a]}$

C H A P T E R 11 **Section 11.1, page 662**

 1. Homogeneous 2. Nonhomogeneous
 3. Nonhomogeneous 4. Homogeneous
 5. Nonhomogeneous 6. Homogeneous

7. $\phi_n(x) = \sin\sqrt{\lambda_n}\,x$, where $\sqrt{\lambda_n}$ satifies $\sqrt{\lambda} = -\tan\sqrt{\lambda}\,\pi$;
 $\lambda_1 \cong 0.6204$, $\lambda_2 \cong 2.7943$, $\lambda_n \cong (2n-1)^2/4$ for large n

8. $\phi_n(x) = \cos\sqrt{\lambda_n}\,x$, where $\sqrt{\lambda_n}$ satisfies $\sqrt{\lambda} = \cot\sqrt{\lambda}$;
 $\lambda_1 \cong 0.7402$, $\lambda_2 \cong 11.7349$, $\lambda_n \cong (n-1)^2\pi^2$ for large n

9. $\phi_n(x) = \sin\sqrt{\lambda_n}\,x + \sqrt{\lambda_n}\cos\sqrt{\lambda_n}\,x$, where $\sqrt{\lambda_n}$ satisfies
 $(\lambda-1)\sin\sqrt{\lambda} - 2\sqrt{\lambda}\cos\sqrt{\lambda} = 0$;
 $\lambda_1 \cong 1.7071$, $\lambda_2 \cong 13.4924$, $\lambda_n \cong (n-1)^2\pi^2$ for large n

10. $\lambda_0 = 0$; $\phi_0(x) = 1 - x$
 For $n = 1, 2, 3, \ldots, \phi_n(x) = \sin\mu_n x - \mu_n\cos\mu_n x$ and $\lambda_n = -\mu_n^2$, where μ_n satisfies
 $\mu = \tan\mu$.
 $\lambda_1 \cong -20.1907$, $\lambda_2 \cong -59.6795$, $\lambda_n \cong -(2n+1)^2\pi^2/4$ for large n

12. $\mu(x) = e^{-x^2}$ 13. $\mu(x) = 1/x$

14. $\mu(x) = e^{-x}$ 15. $\mu(x) = (1 - x^2)^{-1/2}$

16. $X'' + \lambda X = 0$, $T'' + cT' + (k + \lambda a^2)T = 0$

17. (a) $s(x) = e^x$ (b) $\lambda_n = n^2\pi^2$, $\phi_n(x) = e^x\sin n\pi x$; $n = 1, 2, 3, \ldots$

18. Positive eigenvalues $\lambda = \lambda_n$, where $\sqrt{\lambda_n}$ satisfies $\sqrt{\lambda} = \frac{2}{3}\tan 3\sqrt{\lambda}L$; corresponding eigen-
 functions are $\phi_n(x) = e^{-2x}\sin 3\sqrt{\lambda_n}\,x$. If $L = \frac{1}{2}, \lambda_0 = 0$ is eigenvalue, $\phi_0(x) = xe^{-2x}$ is eigen-
 function; if $L \neq \frac{1}{2}, \lambda = 0$ is not eigenvalue. If $L \leq \frac{1}{2}$, there are no negative eigenvalues;
 if $L > \frac{1}{2}$, there is one negative eigenvalue $\lambda = -\mu^2$, where μ is a root of $\mu = \frac{2}{3}\tanh 3\mu L$;
 corresponding eigenfunction is $\phi_{-1}(x) = e^{-2x}\sinh 3\mu x$.

19. No real eigenvalues.

20. Only eigenvalue is $\lambda = 0$; eigenfunction is $\phi(x) = x - 1$.

21. (a) $2\sin\sqrt{\lambda} - \sqrt{\lambda}\cos\sqrt{\lambda} = 0$; $\lambda_1 \cong 18.2738$, $\lambda_2 \cong 57.7075$
 (b) $2\sinh\sqrt{\mu} - \sqrt{\mu}\cosh\sqrt{\mu} = 0$, $\mu = -\lambda$; $\lambda_{-1} \cong -3.6673$

24. (a) $\lambda_n = \mu_n^4$, where μ_n is a root of $\sin\mu L\sinh\mu L = 0$, hence $\lambda_n = (n\pi/L)^4$;
 $\lambda_1 \cong 97.409/L^4$, $\phi_n(x) = \sin(n\pi x/L)$
 (b) $\lambda_n = \mu_n^4$, where μ_n is a root of $\sin\mu L\cosh\mu L - \cos\mu L\sinh\mu L = 0$;
 $$\lambda_1 \cong 237.72/L^4, \quad \phi_n = \frac{\sin\mu_n x\sinh\mu_n L - \sin\mu_n L\sinh\mu_n x}{\sinh\mu_n L}$$
 (c) $\lambda_n = \mu_n^4$, where μ_n is a root of $1 + \cosh\mu L\cos\mu L = 0$; $\lambda_1 \cong 12.362/L^4$,
 $$\phi_n(x) = \frac{(\sin\mu_n x - \sinh\mu_n x)(\cos\mu_n L + \cosh\mu_n L) + (\sin\mu_n L + \sinh\mu_n L)(\cosh\mu_n x - \cos\mu_n x)}{\cos\mu_n L + \cosh\mu_n L}$$

25. (c) $\phi_n(x) = \sin\sqrt{\lambda_n}\,x$, where λ_n satisfies $\cos\sqrt{\lambda_n}\,L - \gamma\sqrt{\lambda_n}\,L\sin\sqrt{\lambda_n}\,L = 0$;
 $\lambda_1 \cong 1.1597/L^2$, $\lambda_2 \cong 13.276/L^2$

Section 11.2, page 675

1. $\phi_n(x) = \sqrt{2}\sin(n - \frac{1}{2})\pi x$; $n = 1, 2, \ldots$

2. $\phi_n(x) = \sqrt{2}\cos(n - \frac{1}{2})\pi x$; $n = 1, 2, \ldots$

3. $\phi_0(x) = 1$, $\phi_n(x) = \sqrt{2}\cos n\pi x$; $n = 1, 2, \ldots$

4. $\phi_n(x) = \dfrac{\sqrt{2}\cos\sqrt{\lambda_n}\,x}{(1 + \sin^2\sqrt{\lambda_n})^{1/2}}$, where λ_n satisfies $\cos\sqrt{\lambda_n} - \sqrt{\lambda_n}\sin\sqrt{\lambda_n} = 0$

5. $\phi_n(x) = \sqrt{2}\,e^x\sin n\pi x$; $n = 1, 2, \ldots$ 6. $a_n = \dfrac{2\sqrt{2}}{(2n-1)\pi}$; $n = 1, 2, \ldots$

7. $a_n = \dfrac{4\sqrt{2}(-1)^{n-1}}{(2n-1)^2\pi^2}$; $n = 1, 2, \ldots$

8. $a_n = \dfrac{2\sqrt{2}}{(2n-1)\pi}\{1 - \cos[(2n-1)\pi/4]\}$; $n = 1, 2, \ldots$

9. $a_n = \dfrac{2\sqrt{2}\sin(n-\frac{1}{2})(\pi/2)}{(n-\frac{1}{2})^2\pi^2}; \quad n = 1,2,\ldots$

In Problems 10 through 13, $\alpha_n = (1 + \sin^2\sqrt{\lambda_n})^{1/2}$ and $\cos\sqrt{\lambda_n} - \sqrt{\lambda_n}\sin\sqrt{\lambda_n} = 0$.

10. $a_n = \dfrac{\sqrt{2}\sin\sqrt{\lambda_n}}{\sqrt{\lambda_n}\alpha_n}; \quad n = 1,2,\ldots$
11. $a_n = \dfrac{\sqrt{2}(2\cos\sqrt{\lambda_n}-1)}{\lambda_n\alpha_n}; \quad n = 1,2,\ldots$

12. $a_n = \dfrac{\sqrt{2}(1-\cos\sqrt{\lambda_n})}{\lambda_n\alpha_n}; \quad n = 1,2,\ldots$
13. $a_n = \dfrac{\sqrt{2}\sin(\sqrt{\lambda_n}/2)}{\sqrt{\lambda_n}\alpha_n}; \quad n = 1,2,\ldots$

14. Not self-adjoint 15. Self-adjoint

16. Not self-adjoint 17. Self-adjoint

18. Self-adjoint

21. (a) If $a_2 = 0$ or $b_2 = 0$, then the corresponding boundary term is missing.

25. (a) $\lambda_1 = \pi^2/L^2; \quad \phi_1(x) = \sin(\pi x/L)$
 (b) $\lambda_1 \cong (4.4934)^2/L^2; \quad \phi_1(x) = \sin\sqrt{\lambda_1}\,x - \sqrt{\lambda_1}\,x\cos\sqrt{\lambda_1}\,L$
 (c) $\lambda_1 = (2\pi)^2/L^2; \quad \phi_1(x) = 1 - \cos(2\pi x/L)$

26. $\lambda_1 = \pi^2/4L^2; \quad \phi_1(x) = 1 - \cos(\pi x/2L)$

27. (a) $X'' - (v/D)X' + \lambda X = 0, \quad X(0) = 0, \quad X'(L) = 0; \qquad T' + \lambda DT = 0$

 (e) $c(x,t) = \displaystyle\sum_{n=1}^{\infty} a_n e^{-\lambda_n Dt} e^{vx/2D}\sin\mu_n x$, where $\lambda_n = \mu_n^2 + (v^2/4D^2)$;

$$a_n = \frac{4D\mu_n^2\displaystyle\int_0^L e^{-vx/2D}f(x)\sin\mu_n x\,dx}{(2LD\mu_n^2 + v\sin^2\mu_n L)}$$

28. (a) $u_t + vu_x = Du_{xx}, \quad u(0,t) = 0, \quad u_x(L,t) = 0, \quad u(x,0) = -c_0$

 (b) $u(x,t) = \displaystyle\sum_{n=1}^{\infty} b_n e^{-\lambda_n Dt}e^{vx/2D}\sin\mu_n x$, where $\lambda_n = \mu_n^2 + (v^2/4D^2)$;

$$b_n = \frac{8c_0 D^2\mu^2(2D\mu_n e^{-vL/2D}\cos\mu_n L + ve^{-vL/2D}\sin\mu_n L - 2D\mu_n)}{(v^2 + 4D^2\mu_n^2)(2LD\mu_n^2 + v\sin^2\mu_n L)}$$

Section 11.3, page 689

1. $y = 2\displaystyle\sum_{n=1}^{\infty}\frac{(-1)^{n+1}\sin n\pi x}{(n^2\pi^2 - 2)n\pi}$

2. $y = 2\displaystyle\sum_{n=1}^{\infty}\frac{(-1)^{n+1}\sin(n-\frac{1}{2})\pi x}{[(n-\frac{1}{2})^2\pi^2 - 2](n-\frac{1}{2})^2\pi^2}$

3. $y = -\dfrac{1}{4} - 4\displaystyle\sum_{n=1}^{\infty}\frac{\cos(2n-1)\pi x}{[(2n-1)^2\pi^2 - 2](2n-1)^2\pi^2}$

4. $y = 2\displaystyle\sum_{n=1}^{\infty}\frac{(2\cos\sqrt{\lambda_n}-1)\cos\sqrt{\lambda_n}\,x}{\lambda_n(\lambda_n - 2)(1 + \sin^2\sqrt{\lambda_n})}$

5. $y = 8\displaystyle\sum_{n=1}^{\infty}\frac{\sin(n\pi/2)\sin n\pi x}{(n^2\pi^2 - 2)n^2\pi^2}$

6–9. For each problem the solution is

$$y = \sum_{n=1}^{\infty}\frac{c_n}{\lambda_n - \mu}\phi_n(x), \qquad c_n = \int_0^1 f(x)\phi_n(x)\,dx, \quad \mu \neq \lambda_n,$$

where $\phi_n(x)$ is given in Problems 1, 2, 3, and 4, respectively, in Section 11.2, and λ_n is the corresponding eigenvalue. In Problem 8 summation starts at $n = 0$.

10. $a = -\dfrac{1}{2}, \quad y = \dfrac{1}{2\pi^2}\cos \pi x + \dfrac{1}{\pi^2}\left(x - \dfrac{1}{2}\right) + c\sin \pi x$

11. No solution 12. a is arbitrary, $y = c\cos \pi x + a/\pi^2$

13. $a = 0, \quad y = c\sin \pi x - (x/2\pi)\sin \pi x$

17. $v(x) = a + (b - a)x$ 18. $v(x) = 1 - \frac{3}{2}x$

19. $u(x,t) = \sqrt{2}\left[-\dfrac{4c_1}{\pi^2} + \left(\dfrac{4c_1}{\pi^2} + \dfrac{1}{\sqrt{2}}\right)e^{-\pi^2 t/4}\right]\sin\dfrac{\pi x}{2}$

$$-\sqrt{2}\sum_{n=2}^{\infty}\dfrac{4c_n}{(2n-1)^2\pi^2}[1 - e^{-(n-1/2)^2\pi^2 t}]\sin(n - \tfrac{1}{2})\pi x,$$

$$c_n = \dfrac{4\sqrt{2}(-1)^{n+1}}{(2n-1)^2\pi^2}, \quad n = 1, 2, \ldots$$

20. $u(x,t) = \sqrt{2}\displaystyle\sum_{n=1}^{\infty}\left[\dfrac{c_n}{\lambda_n - 1}(e^{-t} - e^{-\lambda_n t}) + \alpha_n e^{-\lambda_n t}\right]\dfrac{\cos\sqrt{\lambda_n}\,x}{(1 + \sin^2\sqrt{\lambda_n})^{1/2}},$

$$c_n = \dfrac{\sqrt{2}\sin\sqrt{\lambda_n}}{\sqrt{\lambda_n}(1 + \sin^2\sqrt{\lambda_n})^{1/2}}, \qquad \alpha_n = \dfrac{\sqrt{2}(1 - \cos\sqrt{\lambda_n})}{\lambda_n(1 + \sin^2\sqrt{\lambda_n})^{1/2}},$$

and λ_n satisfies $\cos\sqrt{\lambda_n} - \sqrt{\lambda_n}\sin\sqrt{\lambda_n} = 0$.

21. $u(x,t) = 8\displaystyle\sum_{n=1}^{\infty}\dfrac{\sin(n\pi/2)}{n^4\pi^4}(1 - e^{-n^2\pi^2 t})\sin n\pi x$

22. $u(x,t) = \sqrt{2}\displaystyle\sum_{n=1}^{\infty}\dfrac{c_n(e^{-t} - e^{-(n-1/2)^2\pi^2 t})\sin(n - \tfrac{1}{2})\pi x}{(n - \tfrac{1}{2})^2\pi^2 - 1},$

$$c_n = \dfrac{2\sqrt{2}(2n-1)\pi + 4\sqrt{2}(-1)^n}{(2n-1)^2\pi^2}$$

23. (a) $r(x)w_t = [p(x)w_x]_x - q(x)w, \quad w(0,t) = 0, \quad w(1,t) = 0, \quad w(x,0) = f(x) - v(x)$

24. $u(x,t) = x^2 - 2x + 1 + \dfrac{4}{\pi}\displaystyle\sum_{n=1}^{\infty}\dfrac{e^{-(2n-1)^2\pi^2 t}\sin(2n-1)\pi x}{2n-1}$

25. $u(x,t) = -\cos \pi x + e^{-9\pi^2 t/4}\cos(3\pi x/2)$

31–34. In all cases solution is $y = \displaystyle\int_0^1 G(x,s)f(s)\,ds$, where $G(x,s)$ is given below.

31. $G(x,s) = \begin{cases} 1 - x, & 0 \le s \le x \\ 1 - s, & x \le s \le 1 \end{cases}$

32. $G(x,s) = \begin{cases} s(2 - x)/2, & 0 \le s \le x \\ x(2 - s)/2, & x \le s \le 1 \end{cases}$

33. $G(x,s) = \begin{cases} \cos s\sin(1 - x)/\cos 1, & 0 \le s \le x \\ \sin(1 - s)\cos x/\cos 1, & x \le s \le 1 \end{cases}$

34. $G(x,s) = \begin{cases} s, & 0 \le s \le x \\ x, & x \le s \le 1 \end{cases}$

Section 11.4, page 700

1. $y = \sum_{n=1}^{\infty} \dfrac{c_n}{\lambda_n - \mu} J_0(\sqrt{\lambda_n}\, x)$, $c_n = \displaystyle\int_0^1 f(x)J_0(\sqrt{\lambda_n}\, x)\, dx \Big/ \int_0^1 xJ_0^2(\sqrt{\lambda_n}\, x)\, dx$,

 $\sqrt{\lambda_n}$ satisfies $J_0(\sqrt{\lambda}) = 0$.

2. (c) $y = -\dfrac{c_0}{\mu} + \sum_{n=1}^{\infty} \dfrac{c_n}{\lambda_n - \mu} J_0(\sqrt{\lambda_n}\, x)$,

 $c_0 = 2\displaystyle\int_0^1 f(x)\, dx$, $c_n = \displaystyle\int_0^1 f(x)J_0(\sqrt{\lambda_n}\, x)\, dx \Big/ \int_0^1 xJ_0^2(\sqrt{\lambda_n}\, x)\, dx$; $n = 1, 2, \ldots$,

 $\sqrt{\lambda_n}$ satisfies $J_0'(\sqrt{\lambda}) = 0$.

3. (d) $a_n = \displaystyle\int_0^1 xJ_k(\sqrt{\lambda_n}\, x)f(x)\, dx \Big/ \int_0^1 xJ_k^2(\sqrt{\lambda_n}\, x)\, dx$

 (e) $y = \sum_{n=1}^{\infty} \dfrac{c_n}{\lambda_n - \mu} J_k(\sqrt{\lambda_n}\, x)$, $c_n = \displaystyle\int_0^1 f(x)J_k(\sqrt{\lambda_n}\, x)\, dx \Big/ \int_0^1 xJ_k^2(\sqrt{\lambda_n}\, x)\, dx$

4. (b) $y = \sum_{n=1}^{\infty} \dfrac{c_n}{\lambda_n - \mu} P_{2n-1}(x)$, $c_n = \displaystyle\int_0^1 f(x)P_{2n-1}(x)\, dx \Big/ \int_0^1 P_{2n-1}^2(x)\, dx$

Section 11.5, page 705

1. (b) $u(\xi, 2) = f(\xi + 1)$, $u(\xi, 0) = 0$, $0 \le \xi \le 2$
 $u(0, \eta) = u(2, \eta) = 0$, $0 \le \eta \le 2$

2. $u(r, t) = \sum_{n=1}^{\infty} k_n J_0(\lambda_n r) \sin \lambda_n at$, $k_n = \dfrac{1}{\lambda_n a} \displaystyle\int_0^1 rJ_0(\lambda_n r)g(r)\, dr \Big/ \int_0^1 rJ_0^2(\lambda_n r)\, dr$

3. Superpose the solution of Problem 2 and the solution [Eq. (21)] of the example in the text.

6. $u(r, z) = \sum_{n=1}^{\infty} c_n e^{-\lambda_n z} J_0(\lambda_n r)$, $c_n = \displaystyle\int_0^1 rJ_0(\lambda_n r)f(r)\, dr \Big/ \int_0^1 rJ_0^2(\lambda_n r)\, dr$,

 and λ_n satisfies $J_0(\lambda) = 0$.

7. (b) $v(r, \theta) = \frac{1}{2}c_0 J_0(kr) + \sum_{m=1}^{\infty} J_m(kr)(b_m \sin m\theta + c_m \cos m\theta)$,

 $b_m = \dfrac{1}{\pi J_m(kc)} \displaystyle\int_0^{2\pi} f(\theta) \sin m\theta\, d\theta$; $m = 1, 2, \ldots$

 $c_m = \dfrac{1}{\pi J_m(kc)} \displaystyle\int_0^{2\pi} f(\theta) \cos m\theta\, d\theta$; $m = 0, 1, 2, \ldots$

8. $c_n = \displaystyle\int_0^1 rf(r)J_0(\lambda_n r)\, dr \Big/ \int_0^1 rJ_0^2(\lambda_n r)\, dr$

10. $u(\rho, s) = \sum_{n=0}^{\infty} c_n \rho^n P_n(s)$, where $c_n = \displaystyle\int_{-1}^1 f(\arccos s)P_n(s)\, ds \Big/ \int_{-1}^1 P_n^2(s)\, ds$;

 P_n is the nth Legendre polynomial and $s = \cos\phi$.

Section 11.6, page 715

1. $n = 21$
2. (a) $b_m = (-1)^{m+1} \sqrt{2}/m\pi$ (c) $n = 20$
3. (a) $b_m = 2\sqrt{2}(1 - \cos m\pi)/m^3\pi^3$ (c) $n = 1$
7. (a) $f_0(x) = 1$ (b) $f_1(x) = \sqrt{3}(1 - 2x)$ (c) $f_2(x) = \sqrt{5}(-1 + 6x - 6x^2)$
 (d) $g_0(x) = 1$, $g_1(x) = 2x - 1$, $g_2(x) = 6x^2 - 6x + 1$
8. $P_0(x) = 1$, $P_1(x) = x$, $P_2(x) = (3x^2 - 1)/2$, $P_3(x) = (5x^3 - 3x)/2$

INDEX